Phosphoinositides II: The Diverse Biological Functions

SUBCELLULAR BIOCHEMISTRY

Recent Volumes in this Series

For further volumes:
http://www.springer.com/series/6515

Tamas Balla • Matthias Wymann • John D. York
Editors

Phosphoinositides II: The Diverse Biological Functions

Springer

Editors
Dr. Tamas Balla
National Institutes of Health
NICHD
Bethesda, MD
USA

Dr. Matthias Wymann
University of Basel
Cancer- and Immunobiology
Basel
Switzerland

Dr. John D. York
Duke University Medical Center
Pharmacology and Cancer
Biology
Durham, NC
USA

ISBN 978-94-007-3014-4 e-ISBN 978-94-007-3015-1
DOI 10.1007/978-94-007-3015-1
Springer Dordrecht Heidelberg London New York

Library of Congress Control Number: 2012931684

Printed on acid-free paper

Springer is part of Springer Science+Business Media (www.springer.com)

Preface

When I was approached to shape a book about phosphoinositide signaling, I first felt honored and humbled. On second thought, this appeared to be an impossible task. Phosphoinositides have grown from being just a curious lipid fraction isolated from bovine brain, showing increased radioactive metabolic labeling during intense stimulation protocols, to become the focus of immense interest as key regulatory molecules that penetrate every aspect of eukaryotic biology. The expansion of this field in the last three decades has been enormous: it turned from a basic science exercise of a devoted few to highly translatable science relevant to a large number of human diseases (isn't this the nature of good basic science?). These include cancer, metabolic-, immuno- and neurodegenerative disorders, to name just a few. Reviewing the large number of enzymes that convert phosphoinositides would fill a book—let alone the diverse biological processes in which phosphoinositides play key regulatory roles. Given the interest, a collection of up-to-date reviews compiled in a book is clearly warranted, which was enough to sway me to accept this assignment. As one editor is unable to handle this enormous task, I was delighted when Matthias Wymann and John York were kind enough to join me in this ambitious effort.

When thinking about potential authors, the obvious choice would have been to approach the people whose contributions have been crucial to push and elevate this field to the level it is today. Bob Michell, prophetically placed phosphoinositides in the center of signal transduction in a 1975 Biochem. Biophys. Acta review (Michell 1975), Michael Berridge had a key role in linking phosphoinositides and Ca^{2+} signaling and whose fascinating reviews have inspired many of us (Berridge and Irvine 1984). Robin Irvine, whose group found that InsP_3 was a mixture of two isomers, the active Ins(1,4,5)P_3 and an inactive Ins(1,3,4)P_3, and who described the tetrakisphosphate pathway (Irvine et al. 1986), and who always challenges us with most provocative ideas. Philip Majerus, who has insisted on the importance of inositide phosphatases (Majerus et al. 1999) very early on. The group of Lewis Cantley, with the discovery of PI 3-kinase activities and the mapping of downstream effectors (Whitman et al. 1988; Franke et al. 1997), or the Waterfield lab where the first PI 3-kinase catalytic subunit was isolated and cloned (Otsu et al. 1991; Hiles et al. 1992). Peter Downes, who recognized the translational value of phosphoinositide research. Jeremy Thorner and Scott Emr, whose work in baker's yeast still forms the foundation of our understanding of the role of inositol lipids in trafficking (Strahl and Thorner 2007) or Pietro De Camilli, whose group documented the central role of inositides in brain and synaptic biology (Cremona et al. 1999). There are many others who made valuable or even greater contributions to phosphoinositide research. The above list reflects my bias, as these researchers had the largest impact on my thinking and the directions of my work. Research is, however, a constantly evolving process and we (now Matthias and John being involved) wanted to involve contributions of scientists who represent a second or third wave of researchers infected with the interest in phosphoinositides. We made an effort to recruit authors who have been trainees of these founding laboratories. With this selection our goal was to sample the view of the current and future generation. By selecting their trainees, we feel that we pay tribute to the "Founding Fathers", and show that the research they put in motion is alive and continues with fresh ideas, new ambitions and a translational and therapeutic value.

Phosphoinositide research in the 1980s went hand in hand with research on Ca^{2+}signaling pursued in "non-excitable" cells and was also marked with the discovery of the family of protein kinase C enzymes, regulated by diacylglycerol, one of the products of phosphoinositide-specific phospholipase C enzymes. These areas of research developed and expanded to form their own fields, and could not be discussed here in detail—even though they are linked historically to the development of phosphoinositide signaling. The enormous work of the groups of Yasutomi Nishizuka on protein kinase C, and Katsuhiko Mikoshiba on cloning and characterizing the Ins(1,4,5)P_3 receptors are prime examples of these achievements. Although we could not cover all these areas, we included a chapter on Ca^{2+} signaling via the Ins(1,4,5)P_3 receptor by Colin Taylor, a trainee of the Michael Berridge's lab, where important links between Ca^{2+} release and Ins(1,4,5)P_3 receptor signaling were discovered. We also decided to allocate some space to inositol phosphates, the soluble counterparts of some of the phosphoinositides. These molecules for long had been viewed only as the metabolic products of the second messenger Ins(1,4,5)P_3 but recently gained significant prominence as regulators of important physiological processes. With the discovery of the highly phosphorylated and pyrophosphorylated inositols and the enzymes that produce them, it became clear that this system represents a whole new regulatory paradigm with exciting new developments.

Finally, it was a difficult dilemma whether to include a Chapter on the early history of phosphoinositides. We decided against it for a number of reasons. First, the really interesting history is traced back to studies that preceded the landmark 1975 Bob Michell review and included the work of the Hokins (1987), Bernard Agranoff (2009) and other pioneers of phosphoinositide research. Nobody could tell these early developments better than Bob Michell in his several recollections (Michell 1995) or Robin Irvine who commemorated the 20 years of Ins(1,4,5)P_3 and the period leading to its discovery (Irvine 2003). We encourage the young readers to go back and read these recollections, as they show several examples of how seemingly uninspiring observations formed the beginning of something that became huge as it unfolded. What came after these landmark discoveries is so overwhelming that each one of us has own views and subjective memories and stories to tell on some aspects of it. As Editors we felt that our views should not be elevated above others on these historical aspects, and leave it to the authors of the individual Chapters to elucidate the diversity in this respect. The only exception is a Chapter on the history of PI 3-kinases by Alex Toker that we felt deserves special emphasis as it had the most transforming impact on the field since the late 1980s.

One needs to understand that selection of authors is a subjective process and does not always reflect on who contributed the most in a selected field. However, we are confident that proper credit is given in the individual Chapters to each groups and individuals whose work has moved this field forward. It should also be understood that a field that generates over 10,000 entries in PubMed with each keyword that relates to phosphoinositides cannot be covered without missing some aspects that could be important. However, we trust that this collection will be found useful for both the experts and the novices.

References

Agranoff BW (2009) Turtles all the way: Reflections on myo-Inositol. *J Biol Chem* 284(32):21121–21126

Berridge MJ, Irvine RF (1984) Inositol trisphosphate, a novel second messenger in cellular signal transduction. *Nature* 312:315–321

Cremona O, et al (1999) Essential role of phosphoinositide metabolism in synaptic vesicle recycling. *Cell* 99:179–188

Franke TF, Kaplan DR, Cantley LC, Toker A (1997) Direct regulation of the Akt protooncogene product by PI3,4P2. *Science* 275:665–668

Hiles ID et al (1992) Phosphatidylinositol 3-kinase: structure and expression of the 110 kDa catalytic subunit. *Cell* 70:419–429

Hokin LE (1987) The road to the phosphoinositide-generated second messengers. *Trends Pharmacol Sci* 8:53–56

Irvine RF (2003) 20 years of Ins(1,4,5)P3, and 40 years before. *Nat Rev Mol Cell Biol* 4(7):586–590

Irvine RF, Letcher AJ, Heslop JP, Berridge MJ (1986) The inositol tris/tetrakis phosphate pathway—demonstration of inositol (1,4,5)trisphosphate-3-kinase activity in mammalian tissues. *Nature* 320:631–634

Majerus PW, Kisseleva MV, Norris FA (1999) The role of phosphatases in inositol signaling reactions. *J. Biol Chem* 274:10669–10672

Michell B (1995) Early steps along the road to inositol-lipid-based signalling. *Trends Biochem Sci* 20(8):326–329

Michell RH (1975) Inositol phospholipids and cell surface receptor function. *Biochim Biophys Acta* 415:81–147

Otsu M et al (1991) Characterization of two 85 kDa proteins that associate with receptor tyrosine kinases, middle-T/pp60c-src complexes, and PI3-kinase. *Cell* 65:91–104

Strahl T, Thorner J (2007) Synthesis and function of membrane phosphoinositides in budding yeast, Saccharomyces cerevisiae. *Biochim Biophys Acta* 1771(3):353–404

Whitman M, Downes CP, Keeler M, Keller T, Cantley L (1988) Type-I phosphatidylinositol kinase makes a novel inositol phospholipid, phosphatidylinositol-3-phosphate. *Nature* 332:644–646

Contents

Abbreviations

AD	Alzheimer's disease
AMPK	5′-AMP-activated protein kinase
ALL	acute lymphocytic leukemia
ALS	amyotrophic lateral sclerosis
AML	acute myeloblastic leukemia
ARNO	Arf nucleotide binding site opener
ASK1	apoptosis signal-regulating kinase 1
ATM	ataxia telangiectasia mutated
ATX	arabidopsis trithorax 1
Bad	Bcl-XL/Bcl-2-associated death promoter
BAFF	B cell activation factor of the TNF family
BCR	B cell receptor
Bcr/Abl	break point cluster region/Abelson kinase fusion protein
Btk	Bruton's tyrosine kinase
c-Kit	stem cell growth factor receptor
CAD	caspase activated DNase
CCR(L)	C-C chemokine receptor (ligand) type
CDK	cyclin-dependent kinase
CDKN2A	cyclin-dependent kinase inhibitor 2A
CERT	ceramide transfer protein
CIN85	Cbl-interactin protein of 85kD (also Ruk (regulator of ubiquitous kinase), SETA (SH3 domain-containing gene expressed in tumorigenic astrocytes))
CML	chronic myeloid leukemia
CMT	Charcot-Marie-Tooth
COPI/II	coatomer protein complex I/II
CXCR(L)	C-X-C chemokine receptor (ligand) type
DAAX	death domain-associated protein
DAG	diacylglycerol
DGK	diacylglycerol kinase
DH	dbl-homology
DMSO	dimethyl sulfoxide

DNA-PK_{cs}	DNA-dependent protein kinase, catalytic subunit
DOCK2	dedicator of cytokinesis 2
Dpm1	dolichol phosphate mannosyltransferase
EGF(R)	epidermal growth factor (receptor)
eEF1A	eukaryotic elongation factor 1A
eIF4E	elongation initiation factor 4E
EMT	epithelial-to-mesenchymal transition
EnaC	epithelial sodium channel
ER	estrogen receptor, or endoplasmic reticulum
ErbB1	epidermal growth factor receptor
ERM	ezrin/radexin/moesin
FAK	focal adhesion kinase
FAPP1	phosphoinositol 4-phosphate adaptor protein 1
FAPP2	phosphoinositol 4-phosphate adaptor protein 2
FcεRI	high affinity receptor for Fc fragment of IgE
FOXO	forkhead transcription factor, class O
FYVE	Fab1, YOTB, Vac1, EEA-1 homology
G6P	glucose-6-phosphatase
Gab	Grb2-associated binder
GAP	GTPase-activating protein
GEF	guanine nucleotide exchange factor
GFP	green fluorescent protein
GIST	gastrointestinal stromal tumors
GK	glucokinase
GLUT4	glucose transporter type 4
GM-CSF	granulocyte and macrophage colony stimulating factor
GPCR	G protein-coupled receptors
GRK2	G protein-coupled receptor kinase 2 (also βARK1 (adrenergic receptor kinase 1)
Grp1	general receptor for phosphoinositides
GSK-3	glycogen synthase kinase-3
GST-2xFYVE	glutathione S-transferase-tagged to tandem FYVE domains
HAUSP	herpesvirus-associated ubiquitin-specific protease
Hdac2	histone deacetylase 2
HSCs	hematopoietic stem cells
IκBK	IκB kinase
ING2	inhibitor of growth protein 2
Inpp5e/INPP5E	72 kDa inositol polyphosphate 5-phosphatase
Ins	*myo*-inositol
IGF1(R)	insulin-like growth factor (receptor)
ILK	integrin-linked kinase
Ins(1,4)P_2	inositol 1,4-bisphosphate
Ins(1,4,5)P_3	inositol 1,4,5-trisphosphate; also used $InsP_3$
IPMK	inositol polyphosphate multikinase
IRS	insulin receptor substrate

ITAM	immunoreceptor tyrosine-based activation motif
ITIM	immunoreceptor tyrosine-based inhibitory motif
JAK	Janus-activated kinase
JNK	Jun N-terminal Kinase
Kv1.3	Voltage-gated K^+ channel
LAT	linker for activation of T cells
LOH	loss of heterozygosity
LSCs	leukemic stem cells
LTP	long term potentiation
MAPK	mitogen-activated protein kinase
MAPKAP-2	mitogen-activated protein kinase-activated kinase 2
M-CSF	macrophage colony-stimulating factor
MDM2	murine double minute 2
MDS	myelodysplastic syndrome
MEFs	mouse embryonic fibroblasts
miRNA	microRNA
MPP(+)	1-methyl-4-phenylpyridinium iodide
MSN	medium sized spiny projection neurons
MTM	myotubularin
MTMR	myotubularin related
mTOR	mammalian target of rapamycin, see also TOR
MVB	multivesicular body
MVP	major vault protein
Nedd4	neural-precursor-cell-expressed developmentally down-regulated 4
NFκB	nuclear factor κB
NLS	nuclear localization signal
NMDA(R)	N-methyl-D-aspartate (receptor)
NOS3/eNOS	NO-synthase 3
NTAL	non-T cell activation linker, also named LAB (Linker of activation for B cells) or LAT2
OSBP	oxysterol binding protein
OCRL	oculocerebrorenal syndrome of Lowe
OGD	oxygen–glucose deprivation
PAO	phenyalrsine oxide
PCAF	p300/CBP-associated factor
PDE	phosphodiesterase
PDGF(R)	platelet-derived growth factor (receptor)
PDZ	post synaptic density protein, Drosophila disc large tumor suppressor, zonula occludens-1 protein
PDK1	phosphoinositide-dependent kinase 1
PEPCK	phosphoenolpyruvate carboxy kinase
PEST	proline, glutamic acid, serine, threonine
PH	plecktrin-homology
PHD	plant homeodomain
PH-GRAM	pleckstrin homology glucosyltransferase Rab-like GTPase activator

PHTS	PTEN hamartoma tumor syndrome
PI3K	phosphoinositide 3-kinase; catalytic subunits of class I PI3K are referred to as p110α, p110β, p110γ and p110δ
PI3Kc	PI3K catalytic domain
PI3Kr	PI3K regulatory subunit
PI4K	phosphatidylinositol 4-kinase
PI4KII	type II phosphatidylinositol 4-kinase
PI4KIII	type III phosphatidylinositol 4-kinase
PICS	Pten-loss–induced cellular senescence
PID	phosphoinositide interacting domain
PIKE	PI-3-kinase enhancer
PIKK	phosphoinositide 3-kinase-related kinase
PIP4K	phosphatidylinositol 5-phosphate 4-kinase (also called type II PIP kinase)
PIP5K	phosphatidylinositol 4-phosphate 5-kinase (also called type I PIP kinase)
PIPP	proline-rich inositol polyphosphate 5-phosphatase
PIX	PAK-associated guanine nucleotide exchange factor
PKA	protein kinase A
PKB/Akt	protein kinase B, also called Akt after the transforming kinase encoded by the AKT8 retrovirus
PKC	protein kinase C
PLC	phospholipase C
PLD	phospholipase D
PM	plasma membrane
PML	promyelocytic leukemia protein
PPI	polyphosphoinositide
pRB	retinoblastoma protein
PRD	proline-rich domain
P-Rex	PtdIns(3,4,5)P_3-dependent Rac exchanger
PSD95	post synaptic density protein 95
PtdIns	phosphatidylinositol
PtdIns4*P*	phosphatidylinositol 4-phosphate; short PIP
PtdIns3*P*	phosphatidylinositol 3-phosphate; PIP should not be used here
PtdIns5*P*	phosphatidylinositol 5-phosphate; PIP should not be used here
PtdIns(4,5)P_2	phosphatidylinositol 4,5-bisphosphate; short PIP_2
PtdIns(3,4)P_2	phosphatidylinositol 3,4-bisphosphate; the abbreviation PIP_2 should not be used here.
PtdIns(3,5)P_2	phosphatidylinositol 3,5-bisphosphate; the abbreviation PIP_2 should not be used here.
PtdIns(3,4,5)P_3	phosphatidylinositol 3,4,5-trisphosphate; short PIP_3
PtdOH	phosphatidic acid (also used PA)
PTEN	Phosphatase and Tensin homolog deleted on chromosome Ten, [also MMAC (mutated in multiple advanced cancers), TEP1 (TGF-β-regulated and epithelial cell enriched phosphatase 1)]

PX	Phox-homology
RAN	Ras-related nuclear protein
RID	Rac-induced recruitment domain
RNAi	ribonucleic acid interference
ROS	reactive oxygen species
RSK	ribosomal S6 kinase
R-SMAD	receptor regulated SMAD
RTK	receptor tyrosine kinase
Rb2	retinoblastoma-related gene $p130^{Rb2}$
RYR1	type 1 ryanodine receptor
Sac	suppressor of actin
SCIP	Sac domain-containing inositol phosphatases
SCV	*Salmonella*-containing vacuole
SGK	serum- and glucocorticoid-induced protein kinase
SH2	Src homology 2
SHIP	SH2 domain-containing inositol 5'-phosphatase
SID	Set interacting domain
siRNA	short-interfering RNA
SKICH	SKIP carboxy homology
SKIP	skeletal muscle and kidney enriched inositol phosphatase
SNP	single nucleotide polymorphism
SSC	squamos cell carcinoma
Star-PAP	poly(A) polymerase
Syk	spleen tyrosine kinase, member of the Src tyrosine kinase family
TAC	transverse aortic constriction
TDLU	terminal ductal lobuloalveolar units
TGFβ	transforming growth factor β
Tiam	T-lymphoma invasion and metastasis inducing protein
TNF	tumour necrosis factor
TopoIIα	topoisomerase IIα
TOR	target of rapamycin (also called FRAP or mTOR)
TPIP	TPTE and PTEN homologous inositol lipid phosphatase
TPTE	trans-membrane phosphatase with tensin homology
TRAPs	transmembrane adapter proteins, link immune-receptors to downstream signaling cascades. Examples: LAT, NTAL/LAB
TSC	tuberous sclerosis complex
UTR	untranslated region
Vps34p	vacuolar protein sorting mutant 34 protein
WASP	Wiskott Aldrich Syndrome protein
Wm	wortmannin
WT	wild type

Chapter 1
Ca^{2+} Signalling by IP_3 Receptors

Colin W. Taylor and David L. Prole

Abstract The Ca^{2+} signals evoked by inositol 1,4,5-trisphosphate (IP_3) are built from elementary Ca^{2+} release events involving progressive recruitment of IP_3 receptors (IP_3R), intracellular Ca^{2+} channels that are expressed in almost all animal cells. The smallest events ('blips') result from opening of single IP_3R. Larger events ('puffs') reflect the near-synchronous opening of a small cluster of IP_3R. These puffs become more frequent as the stimulus intensity increases and they eventually trigger regenerative Ca^{2+} waves that propagate across the cell. This hierarchical recruitment of IP_3R is important in allowing Ca^{2+} signals to be delivered locally to specific target proteins or more globally to the entire cell. Co-regulation of IP_3R by Ca^{2+} and IP_3, the ability of a single IP_3R rapidly to mediate a large efflux of Ca^{2+} from the endoplasmic reticulum, and the assembly of IP_3R into clusters are key features that allow IP_3R to propagate Ca^{2+} signals regeneratively. We review these properties of IP_3R and the structural basis of IP_3R behavior.

Keywords Ca^{2+} signalling · Ca^{2+} channel · Endoplasmic reticulum · IP_3 receptor

1.1 Introduction

As the breadth of these volumes confirms, phosphoinositides fulfill many and varied roles in the activities of all eukaryotic cells, whether from unicellular or multicellular organisms (Balla et al. 2011). Pathways for synthesis or uptake of *myo*-inositol, and for synthesis of phosphatidylinositol (PtdIns) appear to have evolved early in a common ancestor of archaea and eukaryotes. Phosphatidylinositol 4,5-bisphosphate (PtdIns(4,5)P_2) and many other polyphosphoinositides, which are ubiquitous in eukaryotes, made their appearance soon after the divergence of eukaryotes and archaea (Michell 2008). The phosphoinositide-specific phospholipases C (PLC),

C. W. Taylor (✉) · D. L. Prole
Department of Pharmacology, Tennis Court Road, CB2 1PD, Cambridge, UK
e-mail: cwt1000@cam.ac.uk

T. Balla et al. (eds.), *Phosphoinositides II: The Diverse Biological Functions*,
Subcellular Biochemistry 59, DOI 10.1007/978-94-007-3015-1_1,

which provide the only route to inositol 1,4,5-trisphosphate (IP_3) in animal cells (Irvine and Schell 2001), are also ancient, although the subtypes known to be regulated by extracellular stimuli appear to be a later invention and are restricted to multicellular animals (Michell 2008). This distribution is broadly similar to that of IP_3 3-kinases (Irvine and Schell 2001), a major route for inactivation of IP_3, and of IP_3 receptors (IP_3R). Genes encoding related IP_3R are probably found in all animals, including all vertebrates, many invertebrates, including arthropods, nematodes, coelenterates and even *Paramecium*, a unicellular ciliate; and in the unicellular green algae, *Chlamydomonas* (Wheeler and Brownlee 2008), but not in fungi or land plants (Krinke et al. 2007). The latter despite evidence that IP_3 can release Ca^{2+} from the intracellular stores of plants, within either the ER or vacuole (Krinke et al. 2007). The identity of the protein(s) through which IP_3 evokes Ca^{2+} signals in plants remains unknown. One possibility is that land plants lost genes encoding IP_3R (and ryanodine receptors) and re-invented intracellular IP_3-gated Ca^{2+} channels after their divergence from animals (Wheeler and Brownlee 2008). We can speculate that the vacuole of plants, with its membrane potential and steep electrochemical gradient for H^+, may have demanded evolution of IP_3R with greater selectivity for Ca^{2+} than is required for IP_3R within the ER of animal cells (Sect. 1.5).

Our focus here is on IP_3 and the receptors (IP_3R) through which it releases Ca^{2+} in animal cells, but there are many additional interactions between phosphoinositides and Ca^{2+} signalling. Examples of the web of interactions that entangle phosphoinositides and Ca^{2+} signalling include regulation of membrane trafficking and cytoskeletal organization by phosphoinositides (Di Paolo and De Camilli 2006); stimulation of ion channels and transporters, including Ca^{2+}-permeable channels and Na^+-Ca^{2+} exchangers, by PtdIns(4,5)P_2 (Suh and Hille 2008); the possibility that PtdIns(4,5)P_2 may be an antagonist of IP_3R (Lupu et al. 1998); and PtdIns(3,4,5)P_3-mediated activation of protein kinase B and thereby phosphorylation of IP_3R (Khan et al. 2006; Szado et al. 2008).

Use of phosphoinositides to control membrane trafficking by defining the identity of different membranes, and signalling via 3-phosphorylated phosphoinositides (Di Paolo and De Camilli 2006) probably evolved before the appearance of IP_3-mediated Ca^{2+} signalling (Michell 2008), but it was the latter that was first experimentally elucidated. A relationship between phospholipid turnover and secretion was first noted in 1953 (Hokin and Hokin 1953), but it was an influential review that first proposed a causal link between PLC-mediated hydrolysis of phosphoinosides and Ca^{2+} signalling (Michell 1975). The proposal initially envisaged a direct link between PLC activity and Ca^{2+} entry across the plasma membrane (Michell 1975), but it was later extended to Ca^{2+} release from intracellular stores. Indeed it is only very recently that the relationship between PLC, intracellular Ca^{2+} stores and Ca^{2+} entry has been resolved with the elucidation of the mechanisms linking empty Ca^{2+} stores to activation of store-operated Ca^{2+} channels in the plasma membrane (Park et al. 2009).

In the years following Michell's provocative review, and amidst some skepticism about the role of PLC in triggering Ca^{2+} signals, evidence progressively accumulated to support his insightful hypothesis (Michell 2009). Key breakthroughs included the demonstration that PLC-mediated catalysis of PtdIns(4,5)P_2 (rather than PtdIns) was the first step in the signalling sequence (Berridge 1983; Michell et al. 1981); evidence from blowfly salivary gland that depletion of phosphoinositides reversibly abrogated 5-HT-evoked Ca^{2+} signals (Fain and Berridge 1979); a demonstration that the ER, rather than mitochondria, was the likely source of the Ca^{2+} released by PLC-linked receptors (Burgess et al. 1983); and finally, in 1983, proof that IP_3 stimulated Ca^{2+} release from a non-mitochondrial store in permeabilized pancreatic acinar cells (Streb et al. 1983). The latter paper—one of the most cited in biology—moved the field into new territory with countless groups rapidly confirming the findings in numerous cell types (Berridge 1987, 1993; Berridge and Irvine 1984). This was followed by purification and functional reconstitution of IP_3R (Ferris et al. 1989), cloning of three vertebrate IP_3R genes (Blondel et al. 1993; Furuichi et al. 1989; Mignery et al. 1989), and structural and functional analyses of IP_3R behavior. We focus on the latter in this review. Several of the key players have published personal perspectives of the historical development of our current understanding of the roles of IP_3R in Ca^{2+} signalling (Berridge 2005; Irvine 2003; Michell 2009; Mikoshiba 2007; Putney 1997), and there are recent reviews on many aspects of IP_3R (Choe and Ehrlich 2006; Foskett et al. 2007; Patterson et al. 2004; Serysheva 2010; Taylor and Dellis 2006; Taylor et al. 1999; Yule et al. 2010).

1.2 Ca^{2+} Is a Versatile and Ubiquitous Intracellular Messenger

At least three features of Ca^{2+} are likely to have contributed to the evolutionary steps leading to it becoming a ubiquitous intracellular messenger. First, it seems likely that as soon as early prokaryotes adopted a phosphate-based energy economy, there was a strong selective pressure to exclude Ca^{2+} from the cytosol to avoid precipitation of calcium phosphate (Kretsinger 1977). The result is that all eukaryotic cells now maintain a low free cytosolic Ca^{2+} concentration, typically about 100 nM. This they achieve by actively extruding Ca^{2+} from the cytosol across both the plasma membrane and the membranes of intracellular organelles. Among the latter, the endoplasmic reticulum (ER) is the most important and, at least in most cells, the major intracellular Ca^{2+} reservoir from which extracellular stimuli can release Ca^{2+}. The important point is that in all eukaryotic cells, the plasma membrane and the membranes of the ER separate the cytosol from much higher concentrations of Ca^{2+}. Both gradients are used to evoke rapid, and usually transient, increases in cytosolic Ca^{2+} concentration. Interactions between these sources of Ca^{2+} are also important. The second key feature of Ca^{2+} is its ability, in the presence of much higher concentrations of other cations, notably Mg^{2+}, to be recognized selectively and with appropriate affinity by simple protein folds—EF-hands are the most common (Celio et al. 1996). These Ca^{2+}-binding proteins provide the means of decoding cytosolic Ca^{2+} signals. Third, Ca^{2+} binds also with relatively low-affinity to countless cellular components—proteins, lipids, small organic anions, etc—causing it to diffuse in cytosol up to 100-times more slowly than expected for a small freely diffusing cation

(Allbritton et al. 1992). The result is that the increases in cytosolic Ca^{2+} concentration that result from opening of Ca^{2+}-permeable channels are, at least initially, restricted to volumes within tens of nm of the channel pore, where the free Ca^{2+} concentration (at least tens of μM) might be some 1000-times higher than that of the bulk cytosol (Shuai et al. 2006). The significant point is that Ca^{2+} signalling can be local, such that Ca^{2+} entering the cytosol via one channel may cause a local increase that is spatially distinct from that evoked by another channel. Different Ca^{2+} channels can thereby direct their Ca^{2+} to different intracellular targets (Berridge et al. 2003; Di Capite et al. 2009; Dick et al. 2008; Neher 1998; Willoughby and Cooper 2007).

That Ca^{2+} can function as either a local messenger, signalling only to proteins in close proximity to specific Ca^{2+} channels, or as a global messenger invading the cell adds enormously to its versatility as an intracellular messenger (Berridge et al. 2000). For intracellular Ca^{2+} channels, notably IP_3R and ryanodine receptors (RyR), the growth of Ca^{2+} signals from local events to larger ones depends on their ability to propagate Ca^{2+} signals regeneratively via Ca^{2+}-induced Ca^{2+} release. This regenerative capacity arises from three key attributes of both major families of intracellular Ca^{2+} channels, namely their permeability to Ca^{2+}, their stimulation by increases in cytosolic Ca^{2+} concentration and their organization into groups of channels spaced to allow effective Ca^{2+}-mediated communication between them. Other Ca^{2+} channels, such as the NAADP-regulated two-pore channels (TPC) in endosomes or lysosomes (Patel et al. 2010; Zhu et al. 2010), or voltage-gated Ca^{2+} channels in the plasma membrane (Berridge et al. 2003) can also provide the Ca^{2+} that recruits the activity of IP_3R or RyR. Here we concentrate on the interactions *between* IP_3R (Fig. 1.1).

Confocal microscopy, and more recently total internal reflection fluorescence microscopy (TIRFM) (Parker and Smith 2010), have shown that in intact cells IP_3-evoked Ca^{2+} signals are built from 'elementary events' with characteristics (durations, amplitudes and spatial spreads) that are relatively unaffected by IP_3 concentration (Rose et al. 2006). The smallest events, 'blips', are the Ca^{2+} signals that arise from opening of single IP_3R; they typically last $\sim$10 ms and spread about $\sim$500 nm from the source. Larger events, 'Ca^{2+} puffs', which are often preceded by a triggering Ca^{2+} blip (Rose et al. 2006), reflect the near-synchronous opening of several IP_3R within a cluster, as Ca^{2+} released by an active IP_3R very rapidly (<20 ms) ignites the activity of its neighbors (Smith and Parker 2009) (Fig. 1.1). Although there is considerable variability, reflecting the stochastic recruitment of IP_3R, a typical puff might last $\sim$100 ms and spread $\sim$1 μm (Marchant et al. 1999; Thomas et al. 1998). The number of IP_3R contributing to a puff is unresolved, but on average it is probably no more than a handful; indeed with high-resolution TIRFM the falling phase of a puff often suggests the stepwise closure of about 5–6 IP_3R (Smith and Parker 2009). The shapes of individual blips and puffs are relatively insensitive to IP_3 concentration, but IP_3 increases the number of IP_3R ready to respond to Ca^{2+} (Sect. 1.4), thereby allowing Ca^{2+} signals to grow with increasing stimulus intensity. Hence, as the IP_3 concentration increases, Ca^{2+} signals grow from blips to puffs and then, as Ca^{2+} diffuses between puff sites, into regenerative Ca^{2+} waves that

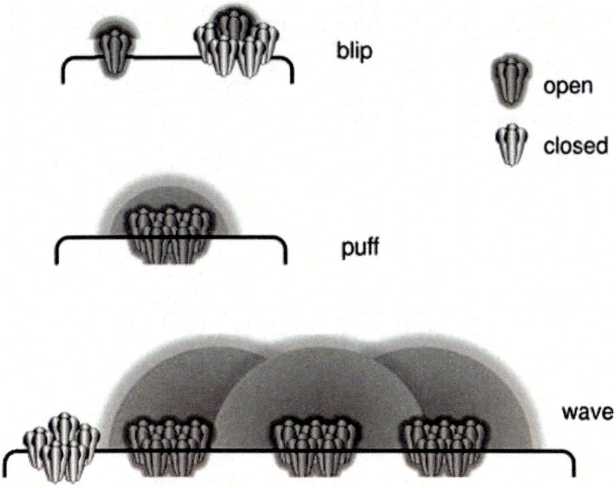

Fig. 1.1 Elementary Ca^{2+} release events mediated by IP_3 receptors. Ca^{2+} blips reflect the opening of single IP_3R, either lone IP_3R (*left*) or a single IP_3R opening within a cluster (*right*). IP_3 causes IP_3R to cluster (Sect. 1.7), and at higher IP_3 concentrations more IP_3R within a cluster are primed to respond to Ca^{2+}, allowing the near simultaneous opening of several IP_3R to give a Ca^{2+} puff. At still higher IP_3 concentrations, Ca^{2+} diffusing from one puff site can ignite the activity of a neighboring cluster of IP_3R to give a regenerative Ca^{2+} wave

can invade the entire cytosol (Bootman and Berridge 1995; Marchant et al. 1999). These Ca^{2+} waves viewed from the perspective of the entire cell appear as repetitive Ca^{2+} spikes, the frequency of which also increases with stimulus intensity (Berridge 1997; Woods et al. 1986).

Many studies, each with their strengths and limitations, suggest that most IP_3R are mobile within ER membranes (Chalmers et al. 2006; Cruttwell et al. 2005; Ferreri-Jacobia et al. 2005; Iwai et al. 2005; Tateishi et al. 2005; Tojyo et al. 2008; Wilson et al. 1998). Our fluorescence recovery after photobleaching (FRAP) analyses of IP_3R1 and IP_3R3, for example, suggest that each IP_3R moves freely within the ER membrane (Pantazaka and Taylor 2011), and our patch-clamp analyses of the same IP_3R in the nuclear envelope likewise suggests they are mobile (Rahman et al. 2009) (Sect. 1.7). But optical measurements of IP_3-evoked Ca^{2+} blips (Smith et al. 2009) and Ca^{2+} puffs (Smith and Parker 2009; Thomas et al. 2000; Tovey et al. 2001) suggest that the events repeatedly initiate at a limited number of fixed sites. How might these apparently conflicting observations be reconciled? A second issue relates to the scarcity of elementary events in cells. A typical mammalian cell, like the SHSY5Y neuroblastoma cells most used for TIRFM analyses of elementary Ca^{2+} release events (Parker and Smith 2010; Smith and Parker 2009; Smith et al. 2009), probably expresses about 10^4–10^5 IP_3R (Tovey et al. 2008). Yet there typically appear to be only 4–5 puff sites in each cell (Smith and Parker 2009), a scarcity that appears not to result from imaging only in the TIRF field (Parker and Smith 2010). Why should less than 1% of a cell's IP_3R contribute to puffs? We return briefly to these issues in Sect. 1.7.

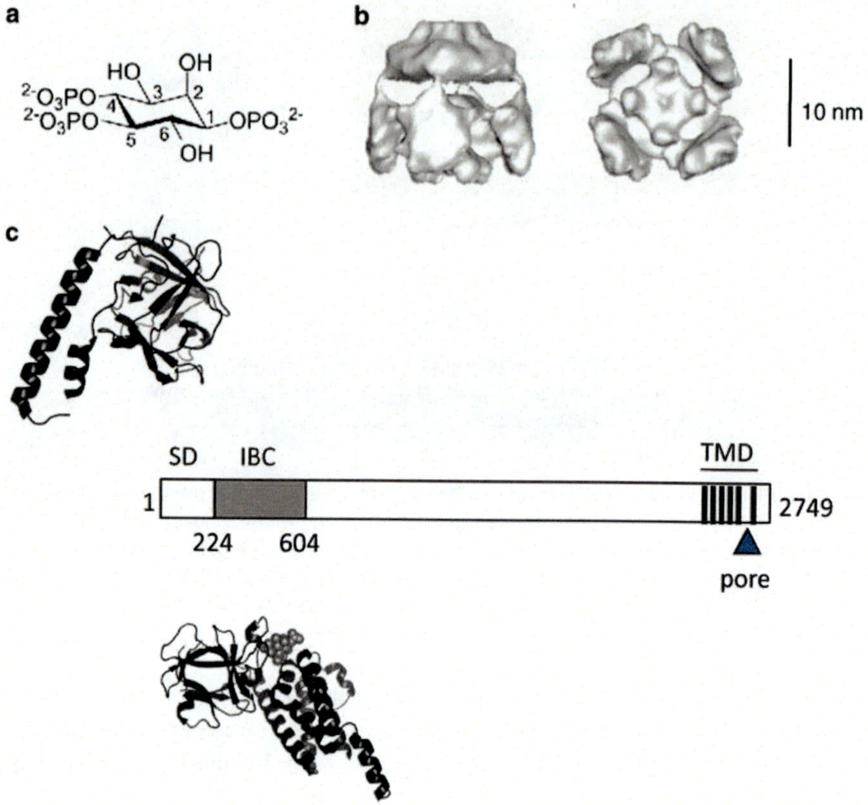

Fig. 1.2 Key structural features of IP_3 receptors. **a** Stucture of IP_3. **b** Low-resolution (~ 30 Å) 3D reconstruction of IP_3R1 derived from single particle reconstruction (da Fonseca et al. 2003). **c** Key regions of a single IP_3R subunit showing the structures of the IBC with IP_3 bound (Protein Data Base, 1N4K) (Bosanac et al. 2002) and of the SD of IP_3R1 (Protein Data Base, 1ZXX) (Bosanac et al. 2005)

1.3 IP_3 Receptors: An Overview

IP_3R are unusually large proteins, each assembled from four homologous subunits arranged around a central pore (Fig. 1.2b) (Taylor et al. 2004). Invertebrate IP_3R are homomeric, but for vertebrates, where there are three genes and multiple splice variants, functional IP_3R may be either homo- or hetero-tetrameric (Taylor et al. 1999). Each subunit comprises about 2700 residues. Although there are differences in the distribution and regulation of IP_3R subtypes, the similarities are generally more striking than the differences. We concentrate on the properties that are either known, or seem very likely, to be shared by all IP_3R.

The IP_3-binding core (IBC), which is entirely responsible for IP_3 recognition, lies towards the N-terminal of the sequence, preceded by the so-called ‘suppressor domain' (SD). Both the IBC and SD are essential for the initial steps in IP_3R activation (Sect. 1.6). High-resolution structures of the IBC with IP_3 bound (Bosanac et al. 2002) and of the SD from IP_3R1 (Bosanac et al. 2005) and IP_3R3 (Chan et al. 2010) have been published (Fig. 1.2c). Towards the C-terminal, there are six predicted transmembrane domains (TMD) in each subunit (Fig. 1.2c). The last pair of TMD (TMD5-6) together with the intervening luminal loop from each of the four subunits assemble to form a central pore (Sect. 1.6). The TMD are also responsible for co-translational targeting of IP_3R to the ER (Joseph 1994; Pantazaka and Taylor 2010; Parker et al. 2004), they play a major role in oligomerization of IP_3R (Galvan et al. 1999), and they retain IP_3R within the ER (Parker et al. 2004). It is, however, noteworthy that although most IP_3R in most cells reside in ER membranes, IP_3 also stimulates Ca^{2+} release from the Golgi apparatus (Pinton et al. 1998), from within the nucleus (Echevarria et al. 2003; Gerasimenko et al. 1995; Marchenko et al. 2005) and perhaps also from secretory vesicles (Gerasimenko et al. 1996). In some cells, small numbers of IP_3R are also selectively targeted to the plasma membrane, where they contribute to the Ca^{2+} entry evoked by physiological stimuli (Dellis et al. 2006; Taylor et al. 2009).

As described in Sect. 1.2, the versatility of Ca^{2+} as an intracellular messenger derives largely from regulating the growth of Ca^{2+} signals from local to global events. For IP_3-evoked Ca^{2+} signals, this depends on the distribution of IP_3R, their Ca^{2+}-permeability and their co-regulation by IP_3 and Ca^{2+}. We therefore concentrate on these features in this short review. It is, however, important to note that there are many other facets to IP_3R behavior. They are, for example, modulated by many additional intracellular signals, like ATP (Betzenhauser et al. 2008), cAMP (Tovey et al. 2008, 2010) and redox state (Higo et al. 2005; Vais et al. 2010); they are substrates for a variety of proteins kinase, which also modulate their activity (deSouza et al. 2007; Fritsch et al. 2004; Szado et al. 2008; Wagner et al. 2008); and many proteins, including calmodulin (Taylor and Laude 2002), Bcl-2 (Li et al. 2007; Rong et al. 2008), presenilins (Cheung et al. 2010), $\beta\gamma$ subunits of G proteins (Zeng et al. 2003) and various luminal proteins (Choe and Ehrlich 2006; Higo et al. 2005) associate with and regulate IP_3R. These modulatory factors, which endow IP_3R with an ability to integrate diverse intracellular signals, effect their influence by tuning the sensitivity of IP_3R to IP_3 or Ca^{2+}, or the interactions between them. It is clear that IP_3R have considerable computational ability: detecting and processing diverse inputs before deciding whether to return them to the cytosol as a Ca^{2+} signal.

Opening of the pore, giving rise to a cytosolic Ca^{2+} signal, is the most important output from IP_3R, but they can also relay information directly to other proteins. IRBIT (IP_3R-binding protein released by IP_3), for example, is a small phosphoprotein that associates with IP_3R1 and appears then to dissociate after IP_3 binding (Ando et al. 2006), freeing IRBIT to then regulate other cellular activities, like Cl^- channels and Na^+-HCO_3^- exchangers in ion-transporting epithelia (Yang et al. 2009). In aortic smooth muscle, IP_3R1 interacts directly with Ca^{2+}-sensitive K^+ channels (BK_{Ca}) and, independent of a Ca^{2+} flux through the IP_3R, promotes their opening and consequent hyperpolarization of the plasma membrane (Zhao et al. 2010). IP_3R

have also been implicated in regulating Ca^{2+} entry via interactions with other Ca^{2+}-permeable channels in the plasma membrane (van Rossum et al. 2004) including some members of the transient receptor potential (trp) channel family (Kiselyov et al. 1999; Tang et al. 2001; Xi et al. 2008; Zhang et al. 2001). These suggestions remain contentious (DeHaven et al. 2009; Taylor et al. 2009).

1.4 Activation of IP_3 Receptors by Ca^{2+} and IP_3

The conformational changes in the IP_3R that lead to opening of its pore are initiated by IP_3 binding to the IP_3-binding core (IBC) (Fig. 1.2a, c). There have been reports of IP_3-independent activation of IP_3R by CaBP1 (Yang et al. 2002), a member of the neuronal Ca^{2+}-sensor family, and by $G\beta\gamma$ subunits of G proteins (Zeng et al. 2003), but the physiological relevance is unclear (Haynes et al. 2004; Nadif Kasri et al. 2004). The current consensus is that binding of IP_3 is essential to initiate activation of all IP_3R, but whether all four IP_3-binding sites of the tetrameric IP_3R must be occupied is unresolved. Positively cooperative responses to IP_3 in some (Dufour et al. 1997; Marchant and Taylor 1997; Tu et al. 2005a; Marchenko et al. 2005), but not all (Finch et al. 1991; Laude et al. 2005; Watras et al. 1991), studies and delays before the first response to IP_3 that decrease with increasing IP_3 concentration (Mak and Foskett 1994; Marchant and Taylor 1997) indicate that channel opening probably requires occupancy of several IP_3-binding sites. However, even heteromeric IP_3R in which at least one subunit has been mutated to prevent IP_3 binding can open in response to IP_3, suggesting that occupancy of fewer than four IP_3-binding sites may be sufficient to cause some gating (Boehning and Joseph 2000).

IP_3R subtypes differ in their affinities for IP_3; the consensus is that IP_3R2 is more sensitive than IP_3R1, and both are more sensitive than IP_3R3 (Iwai et al. 2007; Tu et al. 2005b). Within intact cells, however, differences in expression (Dellis et al. 2006; Tovey et al. 2010), distribution (Petersen et al. 1999), post-transcriptional and post-translational modifications, and association of IP_3R with accessory proteins (Patterson et al. 2004) are likely to be more important determinants of IP_3 sensitivity.

IP_3 is the only known endogenous small ligand of the IBC, but there are many synthetic agonists, all with structures equivalent to the equatorial 6-hydroxyl and the 4- and 5-phosphate groups of IP_3 (Rossi et al. 2010) (Fig. 1.2a). Because neither of the immediate products of IP_3 metabolism, IP_2 and IP_4, binds to the IBC, both metabolic pathways are effective means of inactivating IP_3. The structure of the IBC with IP_3 bound (Bosanac et al. 2002) shows IP_3 held in a clam-like structure with the phosphate groups of IP_3 coordinated by basic residues (Fig. 1.2b). The two sides of the clam, the α- and β-domains, form a network of interactions with the essential groups of IP_3. The 4-phosphate is hydrogen-bonded with residues in the β-domain, the 5-phosphate forms hydrogen bonds with residues that lie predominantly in the α-domain, and the 6-hydroxyl interacts with the backbone of a residue in the α-domain. It is easy to imagine how these interactions might allow IP_3 to pull the α- and β-domains together, causing the clam to close in a manner similar to glutamate binding to ionotropic glutamate receptors (Mayer 2006). In the absence of a structure of the IBC without IP_3 bound, two lines of evidence lend circumstantial support to this proposal. First, the IBC adopts a more constrained structure when it binds IP_3 (Chan et al. 2007). Second, adenophostins, which are high-affinity agonists of IP_3R (Rossi et al. 2010), retain some activity after loss of the 3"-phosphate (analogous to the 5-phosphate of IP_3), probably because their adenine moiety interacts strongly with a residue in the α-domain and thereby partially mimics the clam-closure that would otherwise require binding of the 5-phosphate to the α-domain (Sureshan et al. 2009). We envisage, therefore, that when IP_3 binds to the IBC, the essential vicinal phosphate groups through their contacts with the α- and β-domains cross-bridge the two sides of the clam-like structure causing it to close and thereby initiate the processes that will culminate in opening of the pore (Sect. 1.6).

There are no *specific* antagonists of IP_3R, although with caution some antagonists can be useful (Michelangeli et al. 1995). Heparin is a competitive antagonist of IP_3 (Worley et al. 1987), but it is not membrane-permeant and it has many additional effects (Dasso and Taylor 1991; Ehrlich et al. 1994). 2-aminoethyl diphenylboronate (2-APB) is membrane-permeant and inhibits IP_3-evoked Ca^{2+} release without affecting IP_3 binding (Maruyama et al. 1997); its mechanism of action is unresolved. But 2-APB also inhibits Ca^{2+} uptake and many other Ca^{2+} channels and it is also a modulator of STIM (stromal interaction molecule) and so store-operated Ca^{2+} entry (Goto et al. 2010). Xestospongins are high-affinity membrane-permeant inhibitors of IP_3-evoked Ca^{2+} release that do not affect IP_3 binding (Gafni et al. 1997), but they too have side-effects (Solovyova et al. 2002). High concentrations of caffeine inhibit IP_3-evoked Ca^{2+} release (Parker and Ivorra 1991) without affecting IP_3 binding (Worley et al. 1987), but caffeine also stimulates RyR, inhibits cyclic nucleotide phosphodiesterases and interferes with many Ca^{2+} indicators. Membrane-permeant peptide antagonists of IP_3R may yet provide another source of selective antagonists (Sun and Taylor 2008).

Despite some initially conflicting evidence, it is now accepted that gating of all IP_3R is biphasically regulated by cytosolic Ca^{2+}. Modest increases in Ca^{2+} concentration rapidly potentiate responses to IP_3, while higher concentrations cause a slower inhibition (Finch et al. 1991; Iino 1987, 1990; Marshall and Taylor 1993; Parys et al. 1992; Foskett et al. 2007). Both elements of this Ca^{2+} regulation are widely invoked to explain both the recruitment of elementary Ca^{2+} release events by Ca^{2+}-induced Ca^{2+} release (Sect. 1.2 and 1.7) and to curtail the potentially explosive release of Ca^{2+} resulting from such positive feedback (Smith and Parker 2009).

IP_3 and Ca^{2+} are essential co-agonists of all IP_3R (Adkins and Taylor 1999; Finch et al. 1991; Foskett et al. 2007; Marchant and Taylor 1997; Taylor and Laude 2002), but the interplay between them that leads to channel opening remains incompletely understood. Foskett and colleagues argue from analyses of patch-clamp recordings of nuclear IP_3R that IP_3 decreases the sensitivity of the IP_3R to inhibition by cytosolic Ca^{2+} and that this alone is how IP_3 stimulates channel opening (Ionescu et al. 2006; Mak et al. 1998; Mak et al. 2001). This appealingly simple explanation, where IP_3 serves only to relieve tonic inhibition by resting Ca^{2+} concentrations, is difficult to reconcile with the observation that treatments that abolish Ca^{2+} inhibition do not

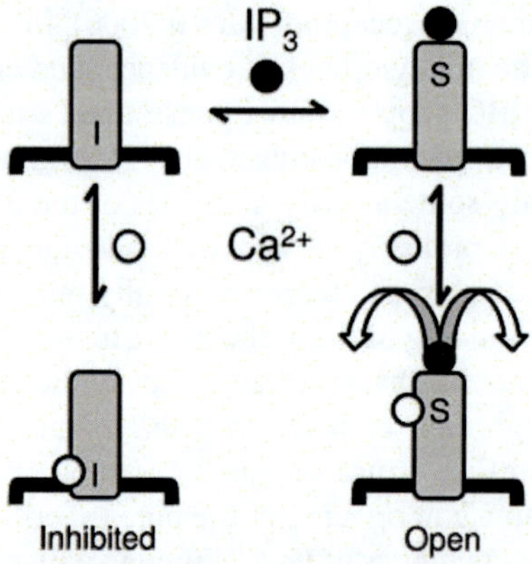

Fig. 1.3 Ca^{2+} regulation of IP_3 receptors. A simplified scheme for the interplay between IP_3 and Ca^{2+} binding is shown. IP_3 binding is proposed to cause both exposure of a stimulatory Ca^{2+}-binding site (*S*) (or massively to increase its affinity), while simultaneously causing an inhibitory Ca^{2+}-binding site (*I*) to be occluded (or massively to reduce its affinity). Opening of the IP_3R requires Ca^{2+} to bind to the stimulatory site

prevent IP_3 from activating IP_3R (Mak et al. 2003). Nor is the scheme compatible with evidence that nuclear IP_3R from cerebellar Purkinje cells show exactly the same biphasic Ca^{2+} concentration-dependence whether activated by low or saturating concentrations of IP_3 (Marchenko et al. 2005). The simple model was later elaborated to include at least three different Ca^{2+} sensors (Mak et al. 2003), but at the core of this revised scheme is a single Ca^{2+}-binding site that switches from being inhibitory in the absence of IP_3 to stimulatory in its presence (Mak et al. 2003). The core of this scheme is consistent with our initial model, derived from rapid superfusion analysis, which suggests that IP_3 both relieves Ca^{2+} inhibition and promotes binding of Ca^{2+} to a stimulatory site (Adkins and Taylor 1999; Marchant and Taylor 1997). The latter is essential for the channel to open. We, however, argue that the stimulatory and inhibitory Ca^{2+}-binding sites are distinct (Marshall and Taylor 1994). We suggest, therefore, that the essential role of IP_3 is to promote Ca^{2+} binding to a stimulatory Ca^{2+}-binding site. IP_3, by priming this site, allows Ca^{2+} to provide instantaneous control over whether the channel opens (Fig. 1.3).

Patch-clamp recordings from excised nuclear patches of insect cells in which the cytosolic surface is accessible to rapid changes in cytosolic IP_3 and Ca^{2+} concentration, have allowed further high-resolution analyses of the interplay between IP_3 and Ca^{2+} (Mak et al. 2007). The results reveal that activation of IP_3R is, as previously suggested (Adkins and Taylor 1999; Marchant and Taylor 1997), much slower than for other ligand-gated ion channels. The reasons for such slow conformational changes in the IP_3R as it passes though different closed states before opening, are unknown. The results also indicate that the interactions of Ca^{2+} with the IP_3R are faster than those with IP_3, consistent with rapid recruitment of IP_3-bound IP_3R by Ca^{2+} within elementary Ca^{2+} release events (Sect. 1.2). The authors claim, from their analyses of the latencies between rapid changes in IP_3 and/or Ca^{2+} concentration and channel opening, that sequential binding of IP_3 and then Ca^{2+} (Marchant and Taylor 1997) is an unlikely explanation for IP_3R activation (Mak et al. 2007). However, the results from these challenging analyses include unexpected findings that are difficult to reconcile with any simple interactions between IP_3 and Ca^{2+} (Mak et al. 2007). We conclude that while IP_3 and Ca^{2+} are clearly both required for activation of IP_3R, no single scheme wherein IP_3 gates IP_3R solely by regulating Ca^{2+} binding to either an inhibitory or stimulatory site is entirely consistent with all published data.

The structural basis for Ca^{2+} regulation of IP_3R is unresolved. It may be direct, via Ca^{2+} binding to a site intrinsic to the IP_3R, or via an accessory Ca^{2+}-binding protein (Taylor et al. 2004). Stimulation of IP_3R by cytosolic Ca^{2+} is universally observed even with purified IP_3R reconstituted into lipid bilayers (Ferris et al. 1989; Hirota et al. 1995; Michikawa et al. 1999), suggesting that this essential Ca^{2+}-binding site probably resides within the primary sequence of the IP_3R. At least seven cytosolic Ca^{2+}-binding sites have been identified within IP_3R1 (Sienaert et al. 1996, 1997), but their physiological relevance is unresolved. Two sites (residues 304–381 and 378–450) are within the IBC, in which there are two surface-exposed clusters of acidic residues that overlap with residues in the second of these Ca^{2+}-binding regions. However, mutation of several of these acidic residues had no effect on Ca^{2+} regulation of IP_3R (Joseph et al. 2005). The remaining Ca^{2+}-binding sites fall within the central region of the IP_3R (Sienaert et al. 1996, 1997), but there is no evidence to link any of them directly to Ca^{2+} regulation of IP_3R. The only tangible link between specific residues and Ca^{2+} regulation comes from mutagenesis of a glutamate residue that is conserved in all IP_3R and RyR. Mutation of this residue (E2100 in IP_3R1) to another acidic residue (D) caused a $\sim$5–10-fold decrease in the Ca^{2+}-sensitivity of the IP_3R to both stimulation and inhibition, abolished oscillatory Ca^{2+} transients in response to agonist stimulation, and reduced the Ca^{2+}-binding affinity of a large fragment of the IP_3R that includes the residue (Miyakawa et al. 2001; Tu et al. 2003). A puzzling aspect of these results is the observation that mutation of a single residue similarly attenuated both stimulation and inhibition by Ca^{2+}, when other evidence suggests that the two effects are mediated by distinct sites (Hajnóczky and Thomas 1997; Marshall and Taylor 1994). This, together with the lack of direct evidence that Ca^{2+} is coordinated by the conserved glutamate, leaves open the possibility that rather than itself contributing to an essential Ca^{2+}-binding site, this residue may allosterically couple to the site.

It remains unclear whether inhibition of IP_3R by Ca^{2+} is mediated by Ca^{2+} binding directly to IP_3R or to an associated protein (Taylor and Laude 2002). The effects of Ca^{2+} on IP_3 binding differ between subtypes (Taylor and Laude 2002), purified IP_3R1 is not inhibited by Ca^{2+} (Benevolensky et al. 1994; Danoff et al. 1988; Lin et al. 2000; Richardson and Taylor 1993) and in some bilayer recordings of reconstituted IP_3R there is no Ca^{2+}-inhibition (Hagar et al. 1998; Michikawa et al. 1999; Ramos-Franco et al. 2000). These observations lend support to the idea that Ca^{2+} inhibition may be mediated by an accessory protein. However, deletion of the suppressor domain (SD, residues 1–223) of IP_3R1, which appears not to include a Ca^{2+}-binding site, abolishes inhibition of IP_3 binding by Ca^{2+} (Sienaert et al. 2002). This suggests that regulation by an accessory protein might require the SD. This is significant because the conformational changes initiated by IP_3 binding to the IBC are proposed to both pass entirely via the SD (Rossi et al. 2009) (Sect. 1.6) and to regulate the behavior of an inhibitory Ca^{2+}-binding site (Adkins and Taylor 1999; Mak et al. 1998) (Fig. 1.3).

Calmodulin (CaM) was considered a candidate for the accessory protein through which Ca^{2+} inhibition is exercised, but that now seems improbable (Nadif Kasri et al. 2002; Taylor and Laude 2002). All IP_3R subtypes are inhibited by Ca^{2+}-CaM (Adkins et al. 2000; Hirota et al. 1999; Michikawa et al. 1999; Missiaen et al. 1999, 2000) and CaM has been shown to restore Ca^{2+} inhibition to purified IP_3R (Hirota et al. 1999; Michikawa et al. 1999; Nosyreva et al. 2002). But it has proven difficult to relate these functional effects of CaM to either its effects on IP_3 binding or to identified CaM-binding sites within IP_3R. CaM inhibits IP_3 binding to IP_3R1 in a Ca^{2+}-independent manner (Cardy and Taylor 1998; Patel et al. 1997), through a site that probably lies within the SD (Adkins et al. 2000; Sienaert et al. 2002), but the properties of this site are inconsistent with the ability of CaM to inhibit IP_3R function only in the presence of Ca^{2+}. There is a high-affinity Ca^{2+}-CaM-binding site within the central region of IP_3R1 and IP_3R2, but it is not present in IP_3R3 (Lin et al. 2000; Yamada et al. 1995). However, mutations that prevent Ca^{2+}-CaM binding to this site have no effect on Ca^{2+}-dependent inhibition of IP_3R (Nosyreva et al. 2002; Zhang and Joseph 2001). This evidence and the absence of the site in IP_3R3 suggest that the central Ca^{2+}-CaM-binding site cannot be responsible for Ca^{2+} inhibition of IP_3R. An additional high-affinity Ca^{2+}-CaM-binding site is created in IP_3R1 after removal of the S2 splice region. This may increase the Ca^{2+}-CaM sensitivity of peripheral $S2^-$ IP_3R1, but it is not a universal candidate for mediating Ca^{2+} inhibition of IP_3R (Islam et al. 1996; Lin et al. 2000).

Recently it was suggested that bound CaM is essential for IP_3R function because a peptide antagonist of CaM inhibited IP_3-evoked Ca^{2+} release (Nadif Kasri et al. 2006). It is now clear that this peptide acts directly on IP_3R with no requirement for CaM (Sun and Taylor 2008). While this eliminates an *essential* role for tethered CaM in activating IP_3R, it raises the intriguing possibility that an endogenous CaM-like structure might be essential for IP_3R activation (Sun and Taylor 2008). In summary, all IP_3R subtypes are inhibited by Ca^{2+}-CaM, but the molecular basis of this inhibition has not been established. The site through which Ca^{2+} inhibits IP_3R remains unresolved, but it is unlikely to be CaM.

Whether Ca^{2+} also regulates IP_3R from the luminal surface is another unresolved issue. Ca^{2+} release by RyR terminates before Ca^{2+} stores are entirely depleted because luminal Ca^{2+} is required to maintain RyR activity (Györke and Györke 1998; Jiang et al. 2008; Launikonis et al. 2006), possibly via its interaction with calsequestrin, a luminal high-capacity Ca^{2+}-binding protein (Launikonis et al. 2006; Terentyev et al. 2006). A similar scheme has been proposed to account for the initiation of Ca^{2+} release after the quiescent interval between repetitive Ca^{2+} spikes (Berridge 2007) and for 'quantal Ca^{2+} release' via IP_3R (Muallem et al. 1989). The latter describes the situation wherein after stimulation with sub-maximally effective concentrations of IP_3, unidirectional Ca^{2+} efflux from intracellular stores terminates before they have fully emptied (Bootman et al. 1992; Brown et al. 1992; Combettes et al. 1992; Ferris et al. 1992; Hirota et al. 1995; Meyer and Stryer 1990; Muallem et al. 1989; Oldershaw et al. 1991; Taylor and Potter 1990). The proposal is that luminal Ca^{2+} sets the gain on the regulation by cytosolic IP_3 and Ca^{2+}, so that as the luminal free Ca^{2+} concentration falls it causes the sensitivity of the IP_3R to IP_3 to fall until, as Ca^{2+} leaks from the ER, the IP_3R closes despite the continued presence of cytosolic IP_3 and residual Ca^{2+} within the ER (Irvine 1990). Conversely, as stores refill between Ca^{2+} spikes in an intact cell, the model predicts that the sensitivity of the IP_3R increases until it exceeds the threshold at which prevailing cytosolic IP_3 and Ca^{2+} concentrations become sufficient to trigger opening. Despite the appeal of the model, evidence that luminal Ca^{2+} directly regulates IP_3R is not yet wholly convincing.

Stores have been shown to become more sensitive to IP_3 as they load with Ca^{2+} in some studies (Combettes et al. 1996; Horne and Meyer 1995; Missiaen et al. 1992, 1994; Nunn and Taylor 1992; Oldershaw and Taylor 1993; Parys et al. 1993; Tanimura and Turner 1996; Yamasaki-Mann et al. 2010), but not in others (Combettes et al. 1992, 1993; Shuttleworth 1992; van de Put et al. 1994). But even the supportive results do not generally eliminate the possibility that the increased sensitivity to IP_3 comes from having Ca^{2+} pass through active IP_3R and increase their sensitivity from the cytosolic surface (Laver 2009; Marchenko et al. 2005). An exception that provides direct evidence for regulation of IP_3R by luminal Ca^{2+}*per se* is provided by work where buffering of luminal Ca^{2+} attenuated IP_3-evoked Ca^{2+} release under conditions where feed-forward regulation via a cytosolic Ca^{2+}-binding site was unlikely (Caroppo et al. 2003). In bilayer recordings of IP_3R1, where essential accessory proteins may be lost, luminal Ca^{2+} either failed to potentiate responses to IP_3 (Bezprozvanny and Ehrlich 1994) or inhibited them (Thrower et al. 2000). Despite the caveats, regulation of IP_3R by luminal Ca^{2+} deserves serious consideration. A high-affinity Ca^{2+}-binding site within the luminal loop linking TMD5 and 6 (Sienaert et al. 1996) contains conserved acidic residues that could mediate luminal Ca^{2+} regulation, although the sub-μM affinity of this site for Ca^{2+} would be poorly suited to detecting likely changes in luminal Ca^{2+} concentration. Luminal accessory proteins, akin to those that regulate RyR, are another possibility, with ERp44 being one candidate. ERp44 belongs to the thioredoxin protein family and regulates IP_3R1 in a pH- and luminal Ca^{2+}-dependent manner (Higo et al. 2005). Binding of ERp44 to the TMD5-6 loop of IP_3R1 inhibits channel activity, and the interaction is disrupted by high concentrations of Ca^{2+}, consistent with the suggestion that luminal Ca^{2+} might enhance IP_3R activity.

To summarize, IP_3 works by tuning the Ca^{2+} sensitivity of the IP_3R, although the details are not resolved. We propose that IP_3 stimulates Ca^{2+} binding to a stimulatory site and inhibits Ca^{2+} binding to an inhibitory site (Fig. 1.3). Binding to the former is the trigger for opening of the pore. Others suggest that IP_3 works solely by preventing Ca^{2+} from binding to an inhibitory site. The identity of neither Ca^{2+}-binding site is known: the stimulatory site probably resides within the IP_3R, but the inhibitory site may require an accessory protein, though this is unlikely to be CaM. Luminal Ca^{2+} may further tune the sensitivity of the IP_3R to regulation by its cytosolic ligands, but this remains unproven.

1.5 Structure and Function of the IP_3 Receptor Pore

Although commonly referred to as 'intracellular Ca^{2+} channels', IP_3R are only modestly selective for bivalent over monovalent cations (P_{Ca}/P_K or $P_{Ba}/P_K \sim 6$–8) (Dellis et al. 2006; Foskett et al. 2007; Mak and Foskett 1994; Marchenko et al. 2005). Within the ER, where Ca^{2+} is probably the only permeant ion with an appreciable electrochemical gradient, there is no need for IP_3R to discriminate between cations. Ion selectivity is delegated to the SR/ER Ca^{2+}-ATPase (SERCA), the Ca^{2+} pump that creates the Ca^{2+} concentration gradient across the ER membrane. Indeed within the ER, the ability to conduct K^+ may allow IP_3R to mediate both Ca^{2+} release into the cytosol and the retrograde movement of K^+ required to maintain electroneutrality and so sustain high rates of Ca^{2+} release (Gillespie and Fill 2008). The weak cation selectivity of IP_3R might therefore be viewed as an adaptation to allow rapid rates of Ca^{2+} release. Within the plasma membrane, where IP_3R are expressed in some cells (Dellis et al. 2006; Kuno and Gardner 1987; Tanimura et al. 2000), the situation is different and opening of such a relatively non-selective cation channel would be expected to cause both Ca^{2+} entry and depolarization.

Although parallel studies of each of the IP_3R subtypes suggest that each has similar ion selectivity and conductance (Tu et al. 2005b), there is considerable, and largely unexplained, variation in published values for the single-channel conductance of IP_3R. One point, however, is clear: all IP_3R have large conductance for monovalent cations (up to $\sim$500 pS) (Boehning et al. 2001a, b; Dellis et al. 2006; Ionescu et al. 2006; Mak and Foskett 1998; Marchenko et al. 2005; Perez et al. 1997; Ramos-Franco et al. 2000) and smaller, though still large, conductance for bivalent cations (up to $\sim$80 pS) (Watras et al. 1991). We need, therefore, to understand how IP_3R allow cations to pass selectively and rapidly through the pore.

IP_3R lacking TMD1-4 assemble to form constitutively active channels that are insensitive to IP_3, but with appropriate permeation and conduction properties (Ramos-Franco et al. 1999). This demonstrates that the single ion-conducting pore, almost certainly lying at the centre of the tetrameric IP_3R, is formed by the TMD5-6 region (Fig. 1.4a). This is consistent with mutations within this region affecting conductance and/or ion selectivity (Boehning et al. 2001b; Dellis et al. 2006, 2008; Schug et al. 2008). Because available 3D reconstructions of the IP_3R are not yet sufficient to resolve details of the pore (Taylor et al. 2004), our present understanding of its structure is inferred from comparisons with K^+ channels and RyR. Each of these tetrameric channels is thought to share a pore structure formed by two TMD from each subunit cradling a selectivity filter, but the K^+ channels for which there are high-resolution structures (eg, KcsA, KirBac1.1 and MthK) have minimal sequence similarity with IP_3R or RyR. Indeed even RyR, the closest relatives of IP_3R, while sharing some sequence similarity with IP_3R within this region, differ from them in both the length and primary sequence of the TMD5-6 region (Fig. 1.4a). High resolution maps ($\sim$ 10 Å) of RyR1 come close to revealing the likely secondary structure of the pore (Ludtke et al. 2005; Samso et al. 2005, 2009). This region appears to have six α-helices contributed by each of the four subunits (Samso et al. 2009), and along the central axis there is a luminal constriction (probably the selectivity filter) and

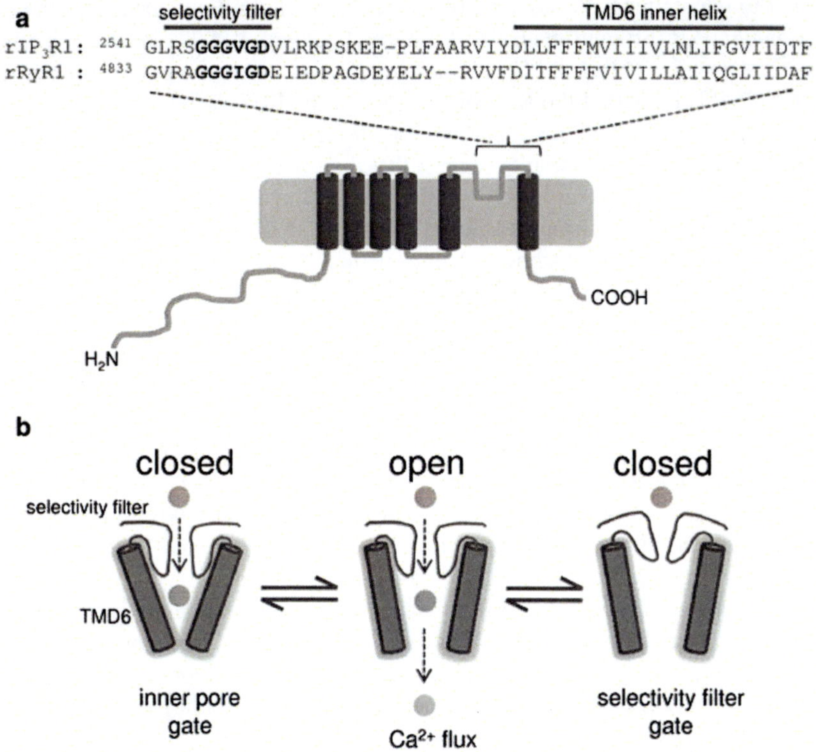

Fig. 1.4 Structure of the pore of the IP_3 receptor. **a** Alignment of the selectivity filter and TMD6 of rat IP_3R1 (ACT21453.1) and RyR1 (XP_001078539). **b** Schematic of the IP_3R pore showing TMD6 from two of the four subunits. Gates located within the inner pore (*left*) or selectivity filter (*right*) my prevent ion flow until the channel opens (*centre*)

a tepee-like assembly of four inner helices (probably TMD6) with the apex pointing into the cytoplasmic structure. In all likelihood the IP_3R forms a similar overall structure.

The only clear sequence similarity between K^+ channels, RyR and IP_3R is that known to form the selectivity filter in K^+ channels (^{2545}GGGXGD^{2550} in IP_3R1, Fig. 1.4a). Mutations within this sequence are also consistent with its role as a selectivity filter in IP_3R. Mutation of D2550 abolishes Ca^{2+} conductance without affecting K^+ conductance (Boehning et al. 2001b; Dellis et al. 2006; van Rossum et al. 2004), although in the mutant, low concentrations of Ca^{2+} block the K^+ conductance (Dellis et al. 2008). Whether this acidic residue binds directly to Ca^{2+}, as occurs for voltage-gated Ca^{2+} channels (Yang et al. 1993), or whether it facilitates coordination of Ca^{2+} by neighboring carbonyl groups, as occurs for cyclic nucleotide-gated (CNG) channels and their bacterial relatives, NaK channels, is unknown (Alam et al. 2007). Mutation of other residues within the putative selectivity filter (G2545A, G2546A, G2547A, G2549C or G2549W of IP_3R1) abolished IP_3-evoked Ca^{2+} release (Schug et al. 2008), and the G2547A mutant substantially reduced the K^+ conductance (Dellis et al. 2006). This suggests that these residues also contribute to either ion-binding sites or to maintaining the structure of the selectivity filter. Equivalent mutations in *Drosophila* IP_3R (Srikanth et al. 2004) and RyR1 (Gao et al. 2000) are also consistent with these results. Mutation of the remaining residue within the putative selectivity filter (V2548I) increased the K^+ conductance of the IP_3R (Boehning et al. 2001b; Dellis et al. 2006), again consistent with the reverse mutation in RyR1 (Gao et al. 2000). Collectively these results suggest that the selectivity filter of the IP_3R controls both cation conductance and Ca^{2+} selectivity (Boehning et al. 2001b). Different residues are likely to contribute to the ability of IP_3R to discriminate between cations and anions because at least some mutations within the selectivity filter (D2550E and V2548I) do not affect this ability (Boehning et al. 2001b). For K^+ channels, the pore helix dipoles largely determine cation-anion selectivity (Doyle et al. 1998; Roux et al. 2000; Roux and MacKinnon 1999), while charged residues in the selectivity filter are important for CNG channels (Qu et al. 2006). It remains to be established which of these mechanisms is most important for cation-anion selectivity in IP_3R.

As they select between ions, IP_3R must also allow rapid transit of cations through the pore. To achieve such a large single-channel cation conductance, a short, wide pore with a large capture radius is required (Williams et al. 2001). For RyR, the estimated diameter at the selectivity filter is ~7 Å (Tinker and Williams 1993), much larger than for KcsA ~ 3.3 Å (Doyle et al. 1998; Zhou et al. 2001). The relative ionic conductance sequence of IP_3R, which corresponds to ion mobility in solution (Bezprozvanny and Ehrlich 1994; Stehno-Bittel et al. 1995; Striggow and Ehrlich 1996), and their Mg^{2+} permeability are each consistent with ions passing through the pore without fully dehydrating and therefore with a pore of large diameter. From analysis of the voltage-dependence of channel blockers, RyR has been proposed to form a short pore (~ 10.4 Å), similar in length to the selectivity filter of KcsA (~12 Å) (Doyle et al. 1998), and with the narrowest part near the luminal surface being extremely short (~ 1 Å long) (Tinker and Williams 1995; Williams et al. 2001). Large vestibules lined with negative charges on either side of the selectivity filter probably provide a large capture radius for both RyR and IP_3R. This is similar to large-conductance Ca^{2+}-activated K^+ channels, which have wide inner vestibules (~ 16–20 Å) enriched with acidic residues (Brelidze et al. 2003). Ca^{2+} channels, inward-rectifier K^+ channels and MthK also have wide inner pores (Lu et al. 1999; Zhen et al. 2005) (~20 Å in the latter) (Jiang et al. 2002b). Structures of RyR reconstructed from electron microscopy images also reveal a wide inner pore (Ludtke et al. 2005; Samso et al. 2009), consistent with the suggestion that IP_3R also have such a structure. In summary, the pore of the IP_3R has only a modest selectivity for Ca^{2+} but a large conductance. These properties are determined by a selectivity filter within the TMD5-6 loop and a short, wide pore with a large capture radius (Fig. 1.4).

1.6 How Does IP_3 Binding Cause the Pore to Open?

Several relatively low resolution (~ 30 Å) 3D structures of the entire IP_3R1 have been published, each derived from single particle analysis of images from electron microscopy (da Fonseca et al. 2003; Hamada and Mikoshiba 2002; Hamada et al. 2003;

Jiang et al. 2002a; Sato et al. 2004; Serysheva et al. 2003; Wolfram et al. 2010). These structures confirm the tetrameric state of IP_3R (Fig. 1.2b), but they do little to reveal the structural basis of IP_3R activation (Taylor et al. 2004). The situation is better for RyR, where higher resolution structures ($\sim$ 10 Å) suggest that large movements of cytoplasmic domains occur around hinges linking them to relatively immobile domains (Jones et al. 2008; Orlova et al. 1996; Samso et al. 2009; Serysheva et al. 2005; Wang et al. 2007). Our limited, and often speculative, understanding of the structural basis of IP_3R activation derives largely from comparisons with other channels (supported by analysis of mutants) and from analyses of key fragments of IP_3R. The key question is to understand how IP_3 binding to the IBC leads to binding of Ca^{2+} to the IP_3R and thereby opening of the pore.

The suppressor domain (SD, residues 1–223 of IP_3R1, Fig. 1.2c), which is connected to the IBC by a flexible linkage (Chan et al. 2007), plays an essential role in IP_3R activation. The clearest evidence is that IP_3 binds to IP_3R without an SD, but it fails to open the pore (Szlufcik et al. 2006; Uchida et al. 2003). The SD derives its name from the observation that although it makes no direct contact with IP_3 its presence decreases the affinity of IP_3R for IP_3 (Uchida et al. 2003). We interpret this to reflect the use of energy from binding of IP_3 to the IBC to cause conformational changes in the SD (Rossi et al. 2009). Our argument is that energy provided by agonist binding drives both the conformational changes that lead to receptor activation and tighter binding of the agonist to its receptor (Burgen 1981). This interpretation gains support from our analysis of the interactions of synthetic partial agonists with native and mutated IP_3R. Partial agonists, because they less effectively activate the receptor, divert more binding energy into stabilizing their binding; while full agonists evoke more substantial conformational changes and less binding energy therefore remains to stabilize binding. Our results show that although full and partial agonists bind with similar affinities to the IBC, the SD causes the affinity of full agonists to decrease more than for partial agonists (Rossi et al. 2009). We have provided evidence that addition of charged or bulky groups to the 2-position of IP_3 cause a reduction in efficacy because the substituent interrupts essential communication between the IBC and SD. IP_3 binding is now less effectively communicated to the pore and the channel opens less frequently. The structural relationship between the IBC and the SD is unknown because the only available high-resolution structures are of the two isolated regions (Bosanac et al. 2002, 2005; Chan et al. 2010) (Fig. 1.2c). By modeling possible relationships between the SD and IBC, we have suggested that mutated residues within the SD that cause the IP_3R to gate less effectively when activated by IP_3 (Rossi et al. 2009; Yamasaki-Mann et al. 2010) may occupy a similar position within the N-terminal structure as the 2-substituents of the partial agonists. Our conclusion is that bulky attachments, whether provided by the ligand or the SD, block effective communication between the IBC and the pore and thereby reduce the effectiveness with which agonists gate the channel (Rossi et al. 2009). Quantitative analyses of these results lead to the conclusion that the most energetically costly conformational change in the IP_3R evoked by IP_3 occurs within its N-terminal (residues 1–604) and that these conformational changes pass entirely via the SD to the pore region (Rossi et al. 2009). We suggest, therefore, that the SD is the essential link between IP_3 binding to the IBC and the subsequent conformational changes that lead

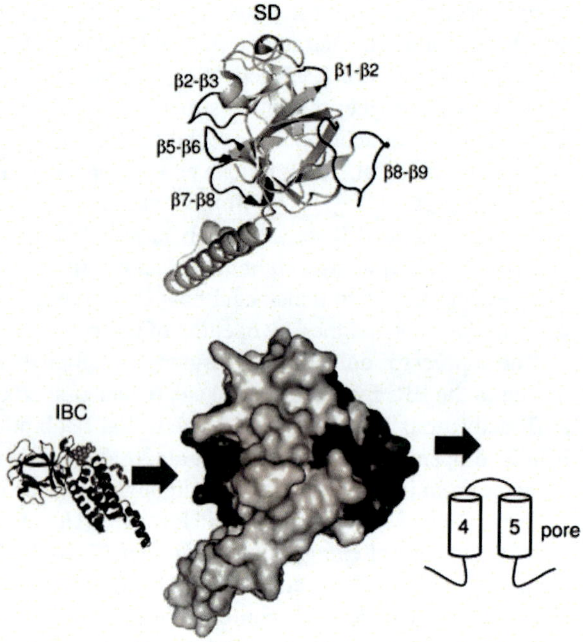

Fig. 1.5 Signalling from IP_3 via the IBC and SD to the pore. The structure of SD from IP_3R1 (Protein Data Base, 1ZXX) (Bosanac et al. 2005) is shown highlighting key loops proposed to mediate communication between the IBC and pore region. The lower representation of the structure highlights (*black*) the regions through which the IBC activated by IP_3 is proposed to communicate with the SD, and the regions of the SD thought to interact with the TMD4-5 loop close to the pore

to opening of the pore. At present, we can only speculate on the physical relationship between the IBC and SD, but our results (Rossi et al. 2009) and those from others (Chan et al. 2010; Yamazaki et al. 2010) are consistent with three exposed loops of the SD (β2-β3, β5-β6 and β7-β8, Fig. 1.5) being likely sites of interaction with the IBC.

Despite their low sequence identities (∼30%), crystal structures of the SD from IP_3R1 (Bosanac et al. 2005) and of the analogous N-terminal regions from RyR1 and RyR2 (Amador et al. 2009; Lobo and Van Petegem 2009) are extremely similar. Several mutations associated with malignant hyperthermia and central core disease (RyR1) or catecholaminergic polymorphic ventricular tachycardia (RyR2), all of which impair the normal regulation of gating, are clustered in an exposed loop (β8-β9) of the N-terminal of RyR (Amador et al. 2009). Furthermore, and consistent with the N-terminal of the RyR mediating essential inter-domain interactions, a peptide derived from this region causes RyR2 to open spontaneously, apparently by uncoupling an interaction between the endogenous loop and a central region of the RyR (Oda et al. 2005; Tateishi et al. 2009). In light of the conservation of structure between IP_3R and RyR, it is tempting to speculate that the same loop in the SD of the IP_3R (β8-β9, Fig. 1.5) may mediate transfer of conformational changes onwards towards the pore. Co-immunoprecipitation studies have suggested an interaction between the N-terminal of IP_3R1 (most likely the SD) and the pore region of an adjacent subunit (Boehning and Joseph 2000), perhaps mediated by the cytosolic loop linking TMD4 to TMD5 (Schug and Joseph 2006). An attractive possibility, therefore is that the SD (perhaps its β8-β9 loop) interacts directly with the short cytosolic helix linking TMD4 and TMD5 and thereby gates the pore (Rossi et al. 2009; Schug and Joseph 2006). Recent evidence lends direct support to this hypothesis because Y167 of IP_3R1 within the β8-β9 loop of its SD is both required for IP_3R activation and appears also to stabilize association of the SD with the TMD4-5 loop (Yamazaki et al. 2010; Chan et al. 2010). E20 within the β1-β2 loop and S217 within the β12 strand appear also to contribute to stabilizing the interaction between the SD and pore region (Chan et al. 2010). Such interactions would require that the SD comes very close to the pore in the native IP_3R, but the location of the SD within the 3D structure of the IP_3R is unknown.

The conformational changes within the channel region that cause the pore to switch from a closed to an open state are unknown. Other ion channels have been proposed to have gates located within the selectivity filter and/or the inner pore helix (TMD6 in IP_3R) (Cordero-Morales et al. 2006; Blunck et al. 2006; Obejero-Paz et al. 2004; Yellen 2002). Contrasting structures of RyR show either a narrow inner pore which is dilated in the open state, consistent with this region acting as a gate (Samso et al. 2009), or a wide inner pore in a putative closed state, consistent with other regions, such as the selectivity filter, acting as the gate (Ludtke et al. 2005) (Fig. 1.4b). Mutation of a conserved hydrophobic residue in TMD6 of IP_3R1 that may be near the narrowest part of the inner pore (F2592) abolished IP_3-evoked Ca^{2+} release (Schug et al. 2008). This residue may contribute to a hydrophobic girdle that moves during gating to allow cations to pass (Fig. 1.4b). The alternative—gating at the selectivity filter—is also consistent with the effects of mutations in this region. For example, mutation of G2546A in IP_3R1 (and the equivalent residue in RyR1) (Gao et al. 2000), reduced channel open probability (Schug et al. 2008) and mutation of residues in RyR1 equivalent to those within the selectivity filter of IP_3R1 (G2546, V2548 and D2550) affected gating (Gao et al. 2000). The location of the gate in IP_3R is not therefore resolved, with the limited available evidence raising the possibility that separate gating processes occur at both the selectivity filter and TMD6 (Fig. 1.4b). For other ion channels, a gating hinge, often formed by a glycine or proline residue within the pore-lining helices has been proposed to allow their movement during gating (del Camino et al. 2000; Doyle 2004; Jiang et al. 2002c; Yellen 2002; Zhao et al. 2004). In both IP_3R and RyR, TMD6 has a conserved glycine at its cytosolic end (G2586 of IP_3R1). Mutation of this residue in RyR1 (G4864A) had no functional effect (Wang et al. 2003), while in IP_3R1 mutation either attenuated (G2586A) or abolished (G2586P) IP_3-evoked Ca^{2+} release, although the latter might have come from constitutive activity draining the Ca^{2+} stores (Schug et al. 2008). The role of a gating hinge in TMD6 of IP_3R1, therefore, remains to be firmly resolved.

We suggest that IP_3R activation is initiated when IP_3 binds to the IBC and triggers closure of its clam-like structure. That conformational change, which must also initiate the events that allow Ca^{2+} to bind to a stimulatory site, is passed to the rest of the IP_3R via the SD. The location of that Ca^{2+}-binding site and so the structural links between it and the SD are unresolved. One face of the SD appears to interact directly with the IBC and the opposite face with a cytoplasmic loop linking TMD4 and TMD5, through which conformational changes pass to the pore. Structural changes of TMD6 and/or the selectivity filter open the pore and so allow passage of Ca^{2+} to the cytosol and of electrically compensating K^+ in the opposite direction.

1.7 Clustered IP_3 Receptors and Elementary Events

The elegant schemes proposed to allow IP_3-evoked Ca^{2+} signals to grow from tiny local Ca^{2+} signals reflecting the activity of single IP_3R to larger local events, and eventually to global Ca^{2+} waves (Sect. 1.2, Fig. 1.1) have hitherto supposed that IP_3R are pre-assembled into small clusters and that the only role of IP_3 is to regulate gating of IP_3R (Sect. 1.4 and 1.6) (Bootman et al. 1997; Horne and Meyer 1997; Lipp and Niggli 1996; Parker and Yao 1996; Sun et al. 1998). Our recent analyses using the patch-clamp technique to record the activity of IP_3R in the nuclear envelope, which is continuous with the ER, suggest that IP_3 contributes also to the assembly of IP_3R clusters and to re-tuning the regulation of IP_3R as they cluster (Rahman and Taylor 2010). These studies use DT40 cells in which the genes for all endogenous IP_3R have been disrupted, thereby allowing the properties of heterologously expressed rat IP_3R to be characterized free of pollution from native IP_3R (Taylor et al. 2009) (Fig. 1.6a, b).

The number of active IP_3R detected within a patch was low (mean ~ 1 IP_3R/patch), but the distribution was random and each IP_3R within the patch gated independently (Rahman et al. 2009). A surprising observation was that in those patches that fortuitously captured more than one IP_3R, and despite each IP_3R gating independently, the open probability (P_o) and mean channel open time (τ_o) of these IP_3R was only half that of patches with only a single IP_3R. We concluded, from additional evidence showing that IP_3 causes IP_3R rapidly and reversibly to assemble into clusters of about five IP_3R, that IP_3 dynamically regulates the assembly of IP_3R into small clusters and that clustered IP_3R are both less sensitive to IP_3 and less active when they do bind IP_3 (Rahman and Taylor 2009; Rahman et al. 2009) (Fig. 1.6c). We propose, therefore, that there are two fundamental units of IP_3R signalling: lone IP_3R and small clusters of about five IP_3R, within which the IP_3R are probably in physical contact. Lone IP_3R are insulated from the Ca^{2+}-releasing activities of their distant neighbors by high concentrations of cytosolic Ca^{2+} buffers (Sect. 1.2), whereas clusters are likely to be exposed immediately to Ca^{2+} released from a neighbor (Fig. 1.6c). IP_3 dynamically regulates the switch between these signalling units.

For lone IP_3R, increasing the free cytosolic Ca^{2+} concentration from 200 nM (typical of a resting cell) to 1 μM (to mimic that within a cluster containing an active IP_3R) almost doubled P_o of IP_3R activated by IP_3. But clustered IP_3R behaved differently: now the increase in Ca^{2+} concentration caused P_o to increase by about fourfold as it both stimulated gating (as it does for lone IP_3R), but also reversed the inhibition imposed by clustering at resting Ca^{2+}. The gating of clustered IP_3R exposed to increased Ca^{2+} is also coupled: IP_3R are more likely to open and close together. This coordinated gating is not caused by local increases in cytosolic Ca^{2+} concentration (because only K^+ passes through open IP_3R in these experiments), but must instead result from a physical coupling of IP_3R. Under physiological conditions, clustered IP_3R are more likely to experience increased Ca^{2+} concentration because their neighbors may release it, but they are also tuned to respond better to it. We suggest that clustering, by first suppressing IP_3R activity, increases the impact of a subsequent increase in the local Ca^{2+} concentration. These interactions exaggerate the effect of Ca^{2+} within a cluster and may thereby allow the rapid recruitment of IP_3R that gives rise to a Ca^{2+} puff (Figs. 1.1 and 1.6).

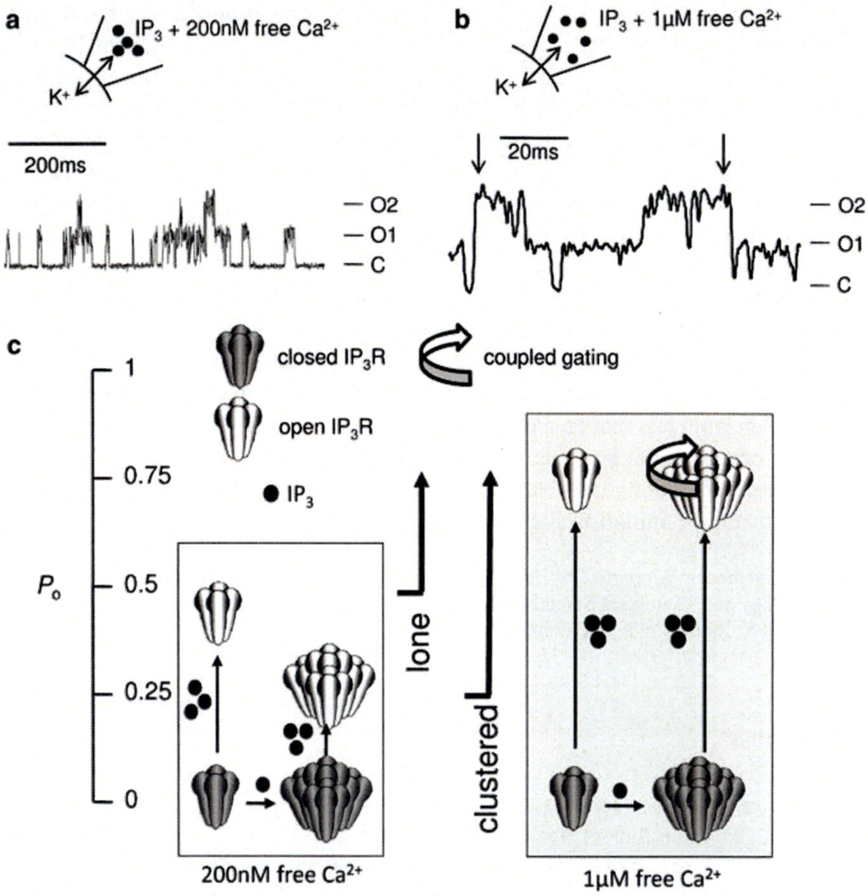

Fig. 1.6 IP_3-evoked clustering of IP_3 receptors tunes their regulation. **a, b** Typical recordings from excised nuclear patches of DT40 cells expressing IP_3R3, each with two active IP_3R. Closed (C) and open (O1 and O2) states are shown for IP_3R stimulated with IP_3 and either 200 nM (**a**) or 1 μM (**b**) free Ca^{2+}. At low Ca^{2+}, IP_3R gate independently, but at elevated Ca^{2+}, there is evidence (*arrows* in **b**) of coupled gating. **c** At resting cytosolic Ca^{2+} concentrations, low concentrations of IP_3 cause IP_3R rapidly to assemble into small clusters within which their responses to IP_3 are attenuated (*left panel*). Within such clusters, IP_3R are more likely to be exposed to the Ca^{2+} released by another IP_3R, but they are also tuned to respond more effectively to a local increase in cytosolic Ca^{2+} concentration. This is shown in the *right panel*, where the increase in Ca^{2+} is shown to both reverse the inhibition caused by clustering and to promote coupled gating (*curved arrow*). The result is that the effects of Ca^{2+} within a cluster are much greater than for a lone IP_3R (*central arrows*). IP_3-evoked clustering both positions IP_3R where they will be exposed to Ca^{2+} and it primes them to respond to it

How might these proposals for the genesis of IP_3-evoked elementary Ca^{2+} release events (Rahman et al. 2009), which require dynamic assembly of mobile IP_3R (Sect. 1.2), be reconciled with suggestions from high-resolution optical analyses of intact cells suggesting that Ca^{2+} blips (Smith et al. 2009) and Ca^{2+} puffs (Smith and Parker 2009; Thomas et al. 2000; Tovey et al. 2001) may consistently originate from relatively fixed positions? We need also to understand why only a tiny fraction of a cell's complement of IP_3R are selectively activated by concentrations of IP_3 that evoke elementary events (Sect. 1.2). One possibility is that there are 'hotspots' within a cell where the local environment, whether the membrane composition, luminal or cytosolic composition, favors IP_3R activation. The relatively fixed location of elementary Ca^{2+} signals need not then reflect the behavior of fixed IP_3R. Instead, we can envisage that IP_3R, whether lone or clustered, that transiently associate with a 'hot spot' are more sensitive to IP_3 and so more likely to initiate Ca^{2+} release events. Individual IP_3R, however, might both dynamically exchange with the 'hot spot' and dynamically assemble into clusters in response to IP_3. However, in seeking to explain the tiny numbers of IP_3R that contribute to apparently immobile Ca^{2+} initiation sites, we face a problem. We need either to explain why a very tiny fraction of IP_3R are both immobile and uniquely sensitive (for fixed IP_3R initiating the events) or we must explain what it is that creates a 'hot spot', wherein those IP_3R that encounter it become more sensitive. The latter, of these still vaguely defined possibilities, has the merit of accommodating the abundant evidence that IP_3R appear to be mobile with evidence that Ca^{2+} initiation sites seem rather stable.

Acknowledgements Supported by the Wellcome Trust, Medical Research Council UK and Biotechnology and Biological Sciences Research Council UK. DLP is a Meres Senior Research Associate of St John's College, Cambridge.

References

Adkins CE, Taylor CW (1999) Lateral inhibition of inositol 1,4,5-trisphosphate receptors by cytosolic Ca^{2+}. Curr Biol 9:1115–1118

Adkins CE, Morris SA, De Smedt H, Török K, Taylor CW (2000) Ca^{2+}-calmodulin inhibits Ca^{2+} release mediated by type-1, -2 and -3 inositol trisphosphate receptors. Biochem J 345:357–363

Alam A, Shi N, Jiang Y (2007) Structural insight into Ca^{2+} specificity in tetrameric cation channels. Proc Natl Acad Sci U S A 104:15334–15339

Allbritton NL, Meyer T, Stryer L (1992) Range of messenger action of calcium ion and inositol 1,4,5-trisphosphate. Science 258:1812–1815

Amador FJ, Liu S, Ishiyama N, Plevin MJ, Wilson A, Maclennan DH, Ikura M (2009) Crystal structure of type I ryanodine receptor amino-terminal b-trefoil domain reveals a disease-associated mutation "hot spot" loop. Proc Natl Acad Sci U S A 106:11040–11044

Ando H, Mizutani A, Kiefer H, Tsuzurugi D, Michikawa T, Mikoshiba K (2006) IRBIT suppresses IP_3 receptor activity by competing with IP_3 for the common binding site on the IP_3 receptor. Mol Cell 22:795–806

Balla T, Wymann M, York J (2011) Phosphoinositides. Springer, Heidelberg

Benevolensky D, Moraru II, Watras J (1994) Micromolar calcium decreases the affinity of inositol trisphosphate receptor in vascular smooth muscle. Biochem J 299:631–636

Berridge MJ (1983) Rapid accumulation of inositol trisphosphate reveals that agonists hydrolyse polyphosphoinositides instead of phosphatidylinositol. Biochem J 212:849–858

Berridge MJ (1987) Inositol trisphosphate and diacylglycerol: two interacting second messengers. Annu Rev Biochem 56:159–193

Berridge MJ (1993) Inositol trisphosphate and calcium signalling. Nature 361:315–325

Berridge MJ (1997) Elementary and global aspects of calcium signalling. J Physiol 499: 291–306

Berridge MJ (2005) Unlocking the secrets of cell signaling. Annu Rev Physiol 67:1–21
Berridge MJ (2007) Inositol trisphosphate and calcium oscillations. Biochem Soc Symp 74: 1–7
Berridge MJ, Irvine RF (1984) Inositol trisphosphate a novel second messenger in cellular signal transduction. Nature 312:315–321
Berridge MJ, Lipp P, Bootman MD (2000) The versatility and universality of calcium signalling. Nat Rev Mol Cell Biol 1:11–21
Berridge MJ, Bootman MD, Roderick HL (2003) Calcium signalling: dynamics, homeostasis and remodelling. Nat Rev Mol Cell Biol 4:517–529
Betzenhauser MJ, Wagner LE 2nd, Iwai M, Michikawa T, Mikoshiba K, Yule DI (2008) ATP modulation of Ca^{2+} release by type-2 and type-3 $InsP_3R$: differing ATP sensitivities and molecular determinants of action. J Biol Chem 283:21579–21587
Bezprozvanny I, Ehrlich BE (1994) Inositol (1,4,5)-trisphosphate ($InsP_3$)-gated Ca channels from cerebellum: conduction properties for divalent cations and regulation by intraluminal calcium. J Gen Physiol 104:821–856
Blondel O, Takeda J, Janssen H, Seino S, Bell GI (1993) Sequence and functional characterization of a third inositol trisphosphate receptor subtype, IP_3R-3, expressed in pancreatic islets, kidney, gastrointestinal tract, and other tissues. J Biol Chem 268:11356–11363
Blunck R, Cordero-Morales JF, Cuello LG, Perozo E, Bezanilla F (2006) Detection of the opening of the bundle crossing in KcsA with fluorescence lifetime spectroscopy reveals the existence of two gates for ion conduction. J Gen Physiol 128:569–581
Boehning D, Joseph SK (2000) Direct association of ligand-binding and pore domains in homo- and heterotetrameric inositol 1,4,5-trisphosphate receptors. EMBO J 19:5450–5459
Boehning D, Joseph SK, Mak DO, Foskett JK (2001a) Single-channel recordings of recombinant inositol trisphosphate receptors in mammalian nuclear envelope. Biophys J 81:117–124
Boehning D, Mak D-O, D, Foskett JK, Joseph SK (2001b) Molecular determinants of ion permeation and selectivity in inositol 1,4,5-trisphosphate receptor Ca^{2+} channels. J Biol Chem 276:13509–13512
Bootman MD, Berridge MJ (1995) The elemental principles of calcium signaling. Cell 83: 675–678
Bootman MD, Berridge MJ, Taylor CW (1992) All-or-nothing Ca^{2+} mobilization from the intracellular stores of single histamine-stimulated HeLa cells. J Physiol 450:163–178
Bootman MD, Niggli E, Berridge MJ, Lipp P (1997) Imaging the hierarchical Ca^{2+} signalling system in HeLa cells. J Physiol 499:307–314
Bosanac I, Alattia J-R, Mal TK, Chan J, Talarico S, Tong FK, Tong KI, Yoshikawa F, Furuichi T, Iwai M, Michikawa T, Mikoshiba K, Ikura M (2002) Structure of the inositol 1,4,5-trisphosphate receptor binding core in complex with its ligand. Nature 420:696–700
Bosanac I, Yamazaki H, Matsu-ura T, Michikawa M, Mikoshiba K, Ikura M (2005) Crystal structure of the ligand binding suppressor domain of type 1 inositol 1,4,5-trisphosphate receptor. Mol Cell 17:193–203
Brelidze TI, Niu X, Magleby KL (2003) A ring of eight conserved negatively charged amino acids doubles the conductance of BK channels and prevents inward rectification. Proc Natl Acad Sci U S A 100:9017–9022
Brown GR, Sayers LG, Kirk CJ, Michell RH, Michelangeli F (1992) The opening of the inositol 1,4,5-trisphosphate-sensitive Ca^{2+} channel in rat cerebellum is inhibited by caffeine. Biochem J 282:309–312
Burgen ASV (1981) Conformational changes and drug action. Fed Proc 40:2723–2728
Burgess GM, McKinney JS, Fabiato A, Leslie BA, Putney JW Jr (1983) Calcium pools in saponin-permeabilized guinea pig hepatocytes. J Biol Chem 258:15336–15345
Cardy TJA, Taylor CW (1998) A novel role for calmodulin: Ca^{2+}-independent inhibition of type-1 inositol trisphosphate receptors. Biochem J 334:447–455
Caroppo R, Colella M, Colasuonno A, DeLuisi A, Debellis L, Curci S, Hofer AM (2003) A reassessment of the effects of luminal [Ca^{2+}] on inositol 1,4,5-trisphosphate-induced Ca^{2+} release from internal stores. J Biol Chem 278:39503–39508

Celio MR, Pauls TL, Schwaller B (1996) Guidebook to the calcium-binding proteins. Oxford University Press, Oxford

Chalmers M, Schell MJ, Thorn P (2006) Agonist-evoked inositol trisphosphate receptor (IP_3R) clustering is not dependent on changes in the structure of the endoplasmic reticulum. Biochem J 394:57–66

Chan J, Whitten AE, Jeffries CM, Bosanac I, Mal TK, Ito J, Porumb H, Michikawa T, Mikoshiba K, Trewhella J, Ikura M (2007) Ligand-induced conformational changes via flexible linkers in the amino-terminal region of the inositol 1,4,5-trisphosphate receptor. J Mol Biol 373:1269–1280

Chan J, Yamazaki H, Ishiyama N, Seo MD, Mal TK, Michikawa T, Mikoshiba K, Ikura M (2010) Structural studies of inositol 1,4,5-trisphosphate receptor: coupling ligand binding to channel gating. J Biol Chem 285:36092–36099

Cheung KH, Mei L, Mak DO, Hayashi I, Iwatsubo T, Kang DE, Foskett JK (2010) Gain-of-function enhancement of IP_3 receptor modal gating by familial Alzheimer's disease-linked presenilin mutants in human cells and mouse neurons. Sci Signal 3:ra22

Choe C, Ehrlich BE (2006) The inositol 1,4,5-trisphosphate receptor (IP_3R) and its regulators: sometimes good and sometimes bad teamwork. Sci STKE re15

Combettes L, Claret M, Champeil P (1992) Do submaximal $InsP_3$ concentrations only induce partial release discharge of permeabilized hepatocyte calcium pools because of the concomitant reduction of intraluminal Ca^{2+} concentration? FEBS Lett 301:287–290

Combettes L, Claret M, Champeil P (1993) Calcium control of $InsP_3$-induced discharge of calcium from permeabilised hepatocyte pools. Cell Calcium 14:279–292

Combettes L, Cheek TR, Taylor CW (1996) Regulation of inositol trisphosphate receptors by luminal Ca^{2+} contributes to quantal Ca^{2+} mobilization. EMBO J 15:2086–2093

Cordero-Morales JF, Cuello LG, Zhao Y, Jogini V, Cortes DM, Roux B, Perozo E (2006) Molecular determinants of gating at the potassium-channel selectivity filter. Nat Struct Mol Biol 13:311–318

Cruttwell C, Bernard J, Hilly M, Nicolas V, Tunwell RE, Mauger JP (2005) Dynamics of the Ins(1,4,5)P_3 receptor during polarization of MDCK cells. Biol Cell 97:699–707

da Fonseca PCA, Morris SA, Nerou EP, Taylor CW, Morris EP (2003) Domain organisation of the type 1 inositol 1,4,5-trisphosphate receptor as revealed by single-particle analysis. Proc Natl Acad Sci U S A 100:3936–3941

Danoff SK, Supattapone S, Snyder SH (1988) Characterization of a membrane protein from brain mediating the inhibition of inositol 1,4,5-trisphosphate receptor binding by calcium. Biochem J 254:701–705

Dasso LLT, Taylor CW (1991) Heparin and other polyanions uncouple α1-adrenoceptors from G-proteins. Biochem J 280:791–795

DeHaven WI, Jones BF, Petranka JG, Smyth JT, Tomita T, Bird GS, Putney JW Jr (2009) TRPC channels function independently of STIM1 and Orai1. J Physiol 587:2275–2298

del Camino D, Holmgren M, Liu Y, Yellen G (2000) Blocker protection in the pore of a voltage-gated K^+ channel and its structural implications. Nature 403:321–325

Dellis O, Dedos S, Tovey SC, Rahman T-U, Dubel SJ, Taylor CW (2006) Ca^{2+} entry through plasma membrane IP_3 receptors. Science 313:229–233

Dellis O, Rossi AM, Dedos SG, Taylor CW (2008) Counting functional IP_3 receptors into the plasma membrane. J Biol Chem 283:751–755

deSouza N, Cui J, Dura M, McDonald TV, Marks AR (2007) A function for tyrosine phosphorylation of type 1 inositol 1,4,5-trisphosphate receptor in lymphocyte activation. J Cell Biol 179:923–934

Di Capite J, Ng SW, Parekh AB (2009) Decoding of cytoplasmic Ca^{2+} oscillations through the spatial signature drives gene expression. Curr Biol 19:853–858

Di Paolo G, De Camilli P (2006) Phosphoinositides in cell regulation and membrane dynamics. Nature 443:651–657

Dick IE, Tadross MR, Liang H, Tay LH, Yang W, Yue DT (2008) A modular switch for spatial Ca^{2+} selectivity in the calmodulin regulation of Cav channels. Nature 451:830–834

Doyle DA (2004) Structural changes during ion channel gating. Trends Neurosci 27:298–302

Doyle DA, Morais Cabral J, Pfuetzner RA, Kuo A, Gulbis JM, Cohen SL, Chait BT, MacKinnon R (1998) The structure of the potassium channel: molecular basis of K^+ conduction and selectivity. Science 280:69–77

Dufour J-F, Arias IM, Turner TJ (1997) Inositol 1,4,5-trisphosphate and calcium regulate the calcium channel function of the hepatic inositol 1,4,5-trisphosphate receptor. J Biol Chem 272:2675–2681

Echevarria W, Leite MF, Guerra MT, Zipfel WR, Nathanson MH (2003) Regulation of calcium signals in the nucleus by a nucleoplasmic reticulum. Nat Cell Biol 5:440–446

Ehrlich BE, Kaftan E, Bezprozvannaya S, Bezprozvanny I (1994) The pharmacology of intracellular Ca^{2+}-release channels. Trends Pharmacol Sci 15:145–149

Fain JN, Berridge MJ (1979) Relationship between phosphatidylinositol synthesis and recovery of 5-hydroxtryphtamine-responsive Ca^{2+} flux in blowfly salivary glands. Biochem J 180:655–661

Ferreri-Jacobia M, Mak D-O, D, Foskett JK (2005) Translational mobility of the type 3 inositol 1,4,5-trisphosphate receptor Ca^{2+} release channel in endoplasmic reticulum membrane. J Biol Chem 280:3824–3831

Ferris CD, Huganir RL, Supattapone S, Snyder SH (1989) Purified inositol 1,4,5-trisphosphate receptor mediates calcium flux in reconstituted lipid vesicles. Nature 342:87–89

Ferris CD, Cameron AM, Huganir RL, Snyder SH (1992) Quantal calcium release by purified reconstituted inositol 1,4,5-trisphosphate receptors. Nature 356:350–352

Finch EA, Turner TJ, Goldin SM (1991) Calcium as a coagonist of inositol 1,4,5-trisphosphate-induced calcium release. Science 252:443–446

Foskett JK, White C, Cheung KH, Mak DO (2007) Inositol trisphosphate receptor Ca^{2+} release channels. Physiol Rev 87:593–658

Fritsch RM, Saur D, Kurjak M, Oesterle D, Schlossmann J, Geiselhoringer A, Hofmann F, Allescher HD (2004) $InsP_3R$-associated cGMP kinase substrate (IRAG) is essential for nitric oxide-induced inhibition of calcium signaling in human colonic smooth muscle. J Biol Chem 279:12551–12559

Furuichi T, Yoshikawa S, Miyawaki A, Wada K, Maeda M, Mikoshiba K (1989) Primary structure and functional expression of the inositol 1,4,5-trisphosphate-binding protein P_{400}. Nature 342:32–38

Gafni J, Munsch JA, Lam TH, Catlin MC, Costa LG, Molinski TF, Pessah IN (1997) Xestospongins: potent membrane permeable blockers of the inositol 1,4,5-trisphosphate receptor. Neuron 19:723–733

Galvan DL, Borrego-Diaz E, Perez PJ, Mignery GA (1999) Subunit oligomerization, and topology of the inositol 1,4,5-trisphosphate receptor. J Biol Chem 274:29483–29492

Gao L, Balshaw D, Xu L, Tripathy A, Xin C, Meissner G (2000) Evidence for a role of the lumenal M3-M4 loop in skeletal muscle Ca^{2+} release channel (ryanodine receptor) activity and conductance. Biophys J 79:828–840

Gerasimenko OV, Gerasimenko JV, Tepikin AV, Petersen OH (1995) ATP-dependent accumulation and inositol trisphosphate- or cyclic ADP-ribose-mediated release of Ca^{2+} from the nuclear envelope. Cell 80:439–444

Gerasimenko OV, Gerasimenko JV, Belan PV, Petersen OH (1996) Inositol trisphosphate and cyclic ADP-ribose-mediated release of Ca^{2+} from single isolated pancreatic zymogen granules. Cell 84:473–480

Gillespie D, Fill M (2008) Intracellular calcium release channels mediate their own countercurrent: the ryanodine receptor case study. Biophys J 95:3706–3714

Goto J, Suzuki AZ, Ozaki S, Matsumoto N, Nakamura T, Ebisui E, Fleig A, Penner R, Mikoshiba K (2010) Two novel 2-aminoethyl diphenylborinate (2-APB) analogues differentially activate and inhibit store-operated Ca^{2+} entry via STIM proteins. Cell Calcium 47:1–10

Györke I, Györke S (1998) Regulation of the cardiac ryanodine receptor channel by luminal Ca^{2+} involves luminal Ca^{2+} sensing sites. Biophys J 75:2801–2810

Hagar RE, Burgstahler AD, Nathanson MH, Ehrlich BE (1998) Type III $InsP_3$ receptor channel stays open in the presence of increased calcium. Nature 296:81–84

Hajnóczky G, Thomas AP (1997) Minimal requirements for calcium oscillations driven by the IP_3 receptor. EMBO J 16:3533–3543
Hamada K, Mikoshiba K (2002) Two-state conformational changes in inositol 1,4,5-trisphosphate receptor regulated by calcium. J Biol Chem 277:21115–21118
Hamada K, Terauchi A, Mikoshiba K (2003) Three-dimensional rearrangements with inositol 1,4,5-trisphosphate receptor by calcium. J Biol Chem 278:52881–52889
Haynes LP, Tepikin AV, Burgoyne RD (2004) Calcium-binding protein 1 is an inhibitor of agonist-evoked, inositol 1,4,5-trisphosphate-mediated calcium signaling. J Biol Chem 279:547–555
Higo T, Hattori M, Nakamura T, Natsume T, Michikawa T, Mikoshiba K (2005) Subtype-specific and ER lumenal environment-dependent regulation of inositol 1,4,5-trisphosphate receptor type 1 by ERp44. Cell 120:85–98
Hirota J, Michikawa T, Miyawaki A, Furuichi T, Okura I, Mikoshiba K (1995) Kinetics of calcium release by immunoaffinity-purified inositol 1,4,5-trisphosphate receptor in reconstituted lipid vesicles. J Biol Chem 270:19046–19051
Hirota J, Michikawa T, Natsume T, Furuichi T, Mikoshiba K (1999) Calmodulin inhibits inositol 1,4,5-trisphosphate-induced calcium release through the purified and reconstituted inositol 1,4,5-trisphosphate receptor type 1. FEBS Lett 456:322–326
Hokin MR, Hokin LE (1953) Enzyme secretion and the incorporation of P^{32} into phospholipides of pancreas slices. J Biol Chem 203:967–977
Horne JH, Meyer T (1995) Luminal calcium regulates the inositol trisphosphate receptor of rat basophilic leukemia cells at the cytosolic side. Biochemistry 34:12738–12746
Horne JH, Meyer T (1997) Elementary calcium-release units induced by inositol trisphosphate. Science 276:1690–1694
Iino M (1987) Calcium dependent inositol trisphosphate-induced calcium release in the guinea-pig taenia caeci. Biochem Biophys Res Commun 142:47–52
Iino M (1990) Biphasic Ca^{2+} dependence of inositol 1,4,5-trisphosphate-induced Ca^{2+} release in smooth muscle cells of the guinea pig taenia caeci. J Gen Physiol 95:1103–1122
Ionescu L, Cheung KH, Vais H, Mak DO, White C, Foskett JK (2006) Graded recruitment and inactivation of single $InsP_3$ receptor Ca^{2+}-release channels: implications for quantal Ca^{2+} release. J Physiol 573:645–662
Irvine RF (1990) "Quantal" Ca^{2+} release and the control of Ca^{2+} entry by inositol phosphates—a possible mechanism. FEBS Lett 262:5–9
Irvine RF (2003) 20 yrs of Ins(1,4,5)P_3, and 40 yrs before. Nat Rev Mol Cell Biol 4:586–590
Irvine RF, Schell MJ (2001) Back in the water: the return of the inositol phosphates. Nat Rev Mol Cell Biol 2:327–338
Islam MO, Yoshida Y, Koga T, Kojima M, Kanagawa K, Imai S (1996) Isolation and characterization of vascular smooth muscle inositol 1,4,5-trisphosphate receptor. Biochem J 316:295–302
Iwai M, Tateishi Y, Hattori M, Mizutani A, Nakamura T, Futatsugi A, Inoue T, Furuichi T, Michikawa T, Mikoshiba K (2005) Molecular cloning of mouse type 2 and type 3 inositol 1,4,5-trisphosphate receptors and identification of a novel type 2 receptor splice variant. J Biol Chem 280:10305–10317
Iwai M, Michikawa T, Bosanac I, Ikura M, Mikoshiba K (2007) Molecular basis of the isoform-specific ligand-binding affinity of inositol 1,4,5-trisphosphate receptors. J Biol Chem 282:12755–12764
Jiang Q-X, Thrower EC, Chester DW, Ehrlich BE, Sigworth FJ (2002a) Three-dimensional structure of the type 1 inositol 1,4,5-trisphosphate receptor at 24 Å resolution. EMBO J 21:3575–3581
Jiang Y, Lee A, Chen J, Cadene M, Chait BT, MacKinnon R (2002b) Crystal structure and mechanism of a calcium-gated potassium channel. Nature 417:515–522
Jiang Y, Lee A, Chen J, Cadene M, Chait BT, MacKinnon R (2002c) The open pore conformation of potassium channels. Nature 417:523–526
Jiang D, Chen W, Xiao J, Wang R, Kong H, Jones PP, Zhang L, Fruen B, Chen SR (2008) Reduced threshold for luminal Ca^{2+} activation of RyR1 underlies a causal mechanism of porcine malignant hyperthermia. J Biol Chem 283:20813–20820

Jones PP, Meng X, Xiao B, Cai S, Bolstad J, Wagenknecht T, Liu Z, Chen SR (2008) Localization of PKA phosphorylation site, Ser2030, in the three-dimensional structure of cardiac ryanodine receptor. Biochem J 410:261–270

Joseph SK (1994) Biosythesis of the inositol trisphosphate receptor in WB rat liver epithelial cells. J Biol Chem 269:5673–5679

Joseph SK, Brownell S, Khan MT (2005) Calcium regulation of inositol 1,4,5-trisphosphate receptors. Cell Calcium 38:539–546

Khan MT, Wagner L 2nd, Yule DI, Bhanumathy C, Joseph SK (2006) Akt kinase phosphorylation of inositol 1,4,5-trisphosphate receptors. J Biol Chem 281:3731–3737

Kiselyov K, Mignery GA, Zhu MX, Muallem S (1999) The N-terminal domain of the IP_3 receptor gates store-operated hTrp3 channels. Mol Cell 4:423–429

Kretsinger RH (1977) Evolution of the informational role of calcium in eukaryotes. In: Wasserman R, Corradino R, Carafoli E, Kretsinger RH, MacLennan D, Siegel F (eds) Calcium binding proteins and calcium function. Elsevier/North Holland, Amsterdam, pp 63–72

Krinke O, Novotna Z, Valentova O, Martinec J (2007) Inositol trisphosphate receptor in higher plants: is it real? J Exp Bot 58:361–376

Kuno M, Gardner P (1987) Ion channels activated by inositol 1,4,5-trisphosphate in plasma membrane of human T-lymphocytes. Nature 326:301–304

Laude AJ, Tovey SC, Dedos S, Potter BVL, Lummis SCR, Taylor CW (2005) Rapid functional assays of recombinant IP_3 receptors. Cell Calcium 38:45–51

Launikonis BS, Zhou J, Royer L, Shannon TR, Brum G, Rios E (2006) Depletion “skraps” and dynamic buffering inside the cellular calcium store. Proc Natl Acad Sci U S A 103:2982–2987

Laver DR (2009) Luminal Ca^{2+} activation of cardiac ryanodine receptors by luminal and cytoplasmic domains. Eur J Biophys 39:19–26

Li C, Wang X, Vais H, Thompson CB, Foskett JK, White C (2007) Apoptosis regulation by Bcl-xL modulation of mammalian inositol 1,4,5-trisphosphate receptor channel isoform gating. Proc Natl Acad Sci U S A 104:12565–12570

Lin C, Widjaja J, Joseph SK (2000) The interaction of calmodulin with alternatively spliced isoforms of the type-I inositol trisphosphate receptor. J Biol Chem 275:2305–2311

Lipp P, Niggli E (1996) A hierarchical concept of cellular and subcellular Ca^{2+}-signalling. Prog Biophys Mol Biol 65:265–296

Lobo PA, Van Petegem F (2009) Crystal structures of the N-terminal domains of cardiac and skeletal muscle ryanodine receptors: insights into disease mutations. Structure 17:1505–1514

Lu T, Nguyen B, Zhang X, Yang J (1999) Architecture of a K^+ channel inner pore revealed by stoichiometric covalent modification. Neuron 22:571–580

Ludtke SJ, Serysheva II, Hamilton SL, Chiu W (2005) The pore structure of the closed RyR1 channel. Structure 13:1203–1211

Lupu VD, Kaznacheyeva E, Krishna UM, Falck JR, Bezprozvanny I (1998) Functional coupling of phosphatidylinositol 4,5-bisphosphate to inositol 1,4,5-trisphosphate receptors. J Biol Chem 273:14067–14070

Mak D-O, D, Foskett JK (1994) Single channel inositol 1,4,5-trisphosphate receptor currents revealed by patch clamp of isolated Xenopus oocyte nuclei. J Biol Chem 269:29375–29378

Mak DO, Foskett JK (1998) Effects of divalent cations on single-channel conduction properties of Xenopus IP_3 receptor. Am J Physiol 275: C179–188

Mak D-O, D, McBride S, Foskett JK (1998) Inositol 1,4,5-tris-phosphate activation of inositol trisphosphate receptor Ca^{2+} channel by ligand tuning of Ca^{2+} inhibition. Proc Natl Acad Sci U S A 95:15821–15825

Mak D-O, McBride S, Foskett JK (2001) Regulation by Ca^{2+} and inositol 1,4,5-trisphosphate ($InsP_3$) of single recombinant type 3 $InsP_3$ receptor channels: Ca^{2+} activation uniquely distinguishes types 1 and 3 $InsP_3$ receptors. J Gen Physiol 117:435–446

Mak D-O, McBride SMJ, Petrenko NB, Foskett JK (2003) Novel regulation of calcium inhibition of the inositol 1,4,5-trisphosphate receptor calcium-release channel. J Gen Physiol 122:569–581

Mak DO, Pearson JE, Loong KP, Datta S, Fernandez-Mongil M, Foskett JK (2007) Rapid ligand-regulated gating kinetics of single inositol 1,4,5-trisphosphate receptor Ca^{2+} release channels. EMBO Rep 8:1044–1051
Marchant JS, Taylor CW (1997) Cooperative activation of IP_3 receptors by sequential binding of IP_3 and Ca^{2+} safeguards against spontaneous activity. Curr Biol 7:510–518
Marchant J, Callamaras N, Parker I (1999) Initiation of IP_3-mediated Ca^{2+} waves in Xenopus oocytes. EMBO J 18:5285–5299
Marchenko SM, Yarotskyy VV, Kovalenko TN, Kostyuk PG, Thomas RC (2005) Spontaneously active and $InsP_3$-activated ion channels in cell nuclei from rat cerebellar Purkinje and granule neurones. J Physiol 565:897–910
Marshall ICB, Taylor CW (1993) Biphasic effects of cytosolic calcium on Ins(1,4,5)P_3-stimulated Ca^{2+} mobilization in hepatocytes. J Biol Chem 268:13214–13220
Marshall ICB, Taylor CW (1994) Two calcium-binding sites mediate the interconversion of liver inositol 1,4,5-trisphosphate receptors between three conformational states. Biochem J 301:591–598
Maruyama T, Kanaji T, Nakade S, Kanno T, Mikoshiba K (1997) 2APB, 2-aminoethoxydiphenyl borate, a membrane-penetrable modulator of Ins(1,4,5)P_3-induced Ca^{2+} release. J Biochem 122:498–505
Mayer ML (2006) Glutamate receptors at atomic resolution. Nature 440:456–462
Meyer T, Stryer L (1990) Transient calcium release induced by successive increments of inositol 1,4,5-trisphosphate. Proc Natl Acad Sci U S A 87:3841–3845
Michelangeli F, Mezna M, Tovey S, Sayers LG (1995) Pharmacological modulators of the inositol 1,4,5-trisphosphate receptor. Neuropharmacology 34:1111–1122
Michell RH (1975) Inositol phospholipids and cell surface receptor function. Biochim Biophys Acta 415:81–147
Michell RH (2008) Inositol derivatives: evolution and functions. Nat Rev Mol Cell Biol 9:151–161
Michell RH (2009) First came the link between phosphoinositides and Ca^{2+} signalling, and then a deluge of other phosphoinositide functions. Cell Calcium 45:521–526
Michell RH, Kirk CJ, Jones LM, Downes CP, Creba JA (1981) The stimulation of inositol lipid metabolism that accompanies calcium mobilization in stimulated cells: defined characteristics and unanswered questions. Philos Trans R Soc Lond B 296:123–137
Michikawa T, Hirota J, Kawano S, Hiraoka M, Yamada M, Furuichi T, Mikoshiba K (1999) Calmodulin mediates calcium-dependent inactivation of the cerebellar type 1 inositol 1,4,5-trisphosphate receptor. Neuron 23:799–808
Mignery GA, Südhof TC, Takei K, De Camilli P (1989) Putative receptor for inositol 1,4,5-trisphosphate similar to ryanodine receptor. Nature 342:192–195
Mikoshiba K (2007) IP_3 receptor/Ca^{2+} channel: from discovery to new signaling concepts. J Neurochem 102:1426–1446
Missiaen L, Taylor CW, Berridge MJ (1992) Luminal Ca^{2+} promoting spontaneous Ca^{2+} release from inositol trisphosphate-sensitive stores of rat hepatocytes. J Physiol 455:623–640
Missiaen L, De Smedt H, Parys JB, Casteels R (1994) Co-activation of inositol trisphosphate-induced Ca^{2+} release by cytosolic Ca^{2+} is loading-dependent. J Biol Chem 269:7238–7242
Missiaen L, Parys JB, Weidema AF, Sipma H, Vanlingen S, De Smet P, Callewaert G, De Smedt H (1999) The bell-shaped Ca^{2+}-dependence of the inositol 1,4,5-trisphosphate induced Ca^{2+} release is modulated by Ca^{2+}/calmodulin. J Biol Chem 274:13748–13751
Missiaen L, De Smedt H, Bultynck G, Vanlingen S, De Smet P, Callewaert G, Parys J (2000) Calmodulin increases the sensitivity of type 3 inositol 1,4,5-trisphosphate receptors to Ca^{2+} inhibition in human bronchial mucosal cells. Mol Pharmacol 57:564–567
Miyakawa T, Mizushima A, Hirose K, Yamazawa T, Bezprozvanny I, Kurosaki T, Iino M (2001) Ca^{2+}-sensor region of IP_3 receptor controls intracellular Ca^{2+} signaling. EMBO J 20:1674–1680
Muallem S, Pandol SJ, Beeker TG (1989) Hormone-evoked calcium release from intracellular stores is a quantal process. J Biol Chem 264:205–212

Nadif Kasri N, Bultynck G, Sienaert I, Callewaert G, Erneux C, Missiaen L, Parys JB, De Smedt H (2002) The role of calmodulin for inositol 1,4,5-trisphosphate receptor function. Biochim Biophys Acta 1600:19–31

Nadif Kasri N, Holmes AM, Bultynck G, Parys JB, Bootman MD, Rietdorf K, Missiaen L, McDonald F, De Smedt H, Conway SJ, Holmes AB, Berridge MJ, Roderick HL (2004) Regulation of $InsP_3$ receptor activity by neuronal Ca^{2+}-binding proteins. EMBO J 23:312–321

Nadif Kasri N, Torok K, Galione A, Garnham C, Callewaert G, Missiaen LJ, BP, De Smedt H (2006) Endogenously bound calmodulin is essential for the function of the inositol 1,4,5-trisphosphate receptor. J Biol Chem 281:8332–8338

Neher E (1998) Vesicle pools and Ca^{2+} microdomains: new tools for understanding their roles in neurotransmitter release. Neuron 20:389–399

Nosyreva E, Miyakawa T, Wang Z, Glouchankova L, Iino M, Bezprozvanny I (2002) The high-affinity calcium-calmodulin-binding site does not play a role in the modulation of type 1 inositol 1,4,5-trisphosphate receptor function by calcium and calmodulin. Biochem J 365:659–667

Nunn DL, Taylor CW (1992) Luminal Ca^{2+} increases the sensitivity of Ca^{2+} stores to inositol 1,4,5-trisphosphate. Mol Pharmacol 41:115–119

Obejero-Paz CA, Gray IP, Jones SW (2004) Y^{3+} block demonstrates an intracellular activation gate for the alpha1G T-type Ca^{2+} channel. J Gen Physiol 124:631–640

Oda T, Yano M, Yamamoto T, Tokuhisa T, Okuda S, Doi M, Ohkusa T, Ikeda Y, Kobayashi S, Ikemoto N, Matsuzaki M (2005) Defective regulation of interdomain interactions within the ryanodine receptor plays a key role in the pathogenesis of heart failure. Circulation 111:3400–3410

Oldershaw KA, Taylor CW (1993) Luminal Ca^{2+} increases the affinity of inositol 1,4,5-trisphosphate for its receptor. Biochem J 292:631–633

Oldershaw KA, Nunn DL, Taylor CW (1991) Quantal Ca^{2+} mobilization stimulated by inositol 1,4,5-trisphosphate in permeabilized hepatocytes. Biochem J 278:705–708

Orlova EV, Serysheva II, Heel M van, Hamilton SL, Chiu W (1996) Two structural configurations of the skeletal muscle calcium release channel. Nat Struct Biol 3:547–552

Pantazaka E, Taylor CW (2010) Targeting of inositol 1,4,5-trisphosphate receptor to the endoplasmic reticulum by its first transmembrane domain. Biochem J 425:61–69

Pantazaka E, Taylor CW (2011) Differential distribution, clustering and lateral diffusion of subtypes of inositol 1,4,5-trisphosphate receptor. J Biol Chem 286:23378–23387

Park CY, Hoover PJ, Mullins FM, Bachhawat P, Covington ED, Raunser S, Walz T, Garcia KC, Dolmetsch RE, Lewis RS (2009) STIM1 clusters and activates CRAC channels via direct binding of a cytosolic domain to Orai1. Cell 136:876–890

Parker I, Ivorra I (1991) Caffeine inhibits inositol trisphosphate-mediated liberation of intracellular calcium in Xenopus oocytes. J Physiol 433:229–240

Parker I, Smith IF (2010) Recording single-channel activity of inositol trisphosphate receptors in intact cells with a microscope, not a patch clamp. J Gen Physiol 136:119–127

Parker I, Yao Y (1996) Ca^{2+} transients associated with openings of inositol trisphosphate-gated channels in Xenopus oocytes. J Physiol 491 3:663–668

Parker AKT, Gergely FV, Taylor CW (2004) Targeting of inositol 1,4,5-trisphosphate receptors to the endoplasmic reticulum by multiple signals within their transmembrane domains. J Biol Chem 279:23797–23805

Parys JB, Sernett SW, DeLisle S, Snyder PM, Welsh MJ, Campbell KP (1992) Isolation, characterization, and localization of the inositol 1,4,5-trisphosphate receptor protein in Xenopus laevis oocytes. J Biol Chem 267:18776–18782

Parys JB, Missiaen L, De Smedt H, Casteels R (1993) Loading dependence of inositol 1,4,5-trisphosphate-induced Ca^{2+} release in the clonal cell line A7r5. J Biol Chem 268:25206–25212

Patel S, Morris SA, Adkins CE, O'Beirne G, Taylor CW (1997) Ca^{2+}-independent inhibition of inositol trisphosphate receptors by calmodulin: redistribution of calmodulin as a possible means of regulating Ca^{2+} mobilization. Proc Natl Acad Sci U S A 94:11627–11632

Patel S, Marchant JS, Brailoiu E (2010) Two-pore channels: regulation by NAADP and customized roles in triggering calcium signals. Cell Calcium 47:480–490

Patterson RL, Boehning D, Snyder SH (2004) Inositol 1,4,5-trisphosphate receptors as signal integrators. Annu Rev Biochem 73:437–465

Perez PJ, Ramos-Franco J, Fill M, Mignery GA (1997) Identification and functional reconstitution of the type 2 inositol 1,4,5-trisphosphate receptor from ventricular cardiac myocytes. J Biol Chem 272:23961–23969

Petersen OH, Burdakov D, Tepikin AV (1999) Polarity in intracellular calcium signalling. Bioessays 21:851–860

Pinton P, Pozzan T, Rizzuto R (1998) The Golgi apparatus is an inositol 1,4,5-trisphosphate-sensitive Ca^{2+} store, with functional properties distinct from those of the endoplasmic reticulum. EMBO J 17:5298–5308

Putney JW Jr (1997) Capacitative calcium entry. R.G. Landes Company, Austin

Qu W, Moorhouse AJ, Chandra M, Pierce KD, Lewis TM, Barry PH (2006) A single P-loop glutamate point mutation to either lysine or arginine switches the cation-anion selectivity of the CNGA2 channel. J Gen Physiol 127:375–389

Rahman T, Taylor CW (2009) Dynamic regulation of IP_3 receptor clustering and activity by IP_3. Channels 3:336–332

Rahman T, Taylor CW (2010) Nuclear patch-clamp recording from inositol 1,4,5-trisphosphate receptors. In: Whittaker M (ed) Calcium in living cells. Elsevier, Amsterdam, pp 199–224.

Rahman TU, Skupin A, Falcke M, Taylor CW (2009) Clustering of IP_3 receptors by IP_3 retunes their regulation by IP_3 and Ca^{2+}. Nature 458:655–659

Ramos-Franco J, Galvan D, Mignery GA, Fill M (1999) Location of the permeation pathway in the recombinant type-1 inositol 1,4,5-trisphosphate receptor. J Gen Physiol 114:243–250

Ramos-Franco J, Bare D, Caenepeel S, Nani A, Fill M, Mignery G (2000) Single-channel function of recombinant type 2 inositol 1,4,5-trisphosphate receptor. Biophys J 79:1388–1399

Richardson A, Taylor CW (1993) Effects of Ca^{2+} chelators on purified inositol 1,4,5-trisphosphate ($InsP_3$) receptors and $InsP_3$-stimulated Ca^{2+} mobilization. J Biol Chem 268:11528–11533

Rong YP, Aromolaran AS, Bultynck G, Zhong F, Li X, McColl K, Matsuyama S, Herlitze S, Roderick HL, Bootman MD, Mignery GA, Parys JB, De Smedt H, Distelhorst CW (2008) Targeting Bcl-2-IP_3 receptor interaction to reverse Bcl-2's inhibition of apoptotic calcium signals. Mol Cell 31:255–265

Rose HJ, Dargan S, Shuai J, Parker I (2006) 'Trigger' events precede calcium puffs in Xenopus oocytes. Biophys J 91:4024–4032

Rossi AM, Riley AM, Tovey SC, Rahman T, Dellis O, Taylor EJA, Veresov VG, Potter BVL, Taylor CW (2009) Synthetic partial agonists reveal key steps in IP_3 receptor activation. Nat Chem Biol 5:631–639

Rossi AM, Riley AM, Potter BVL, Taylor CW (2010) Adenophostins: high-affinity agonists of IP_3 receptors. Curr Top Membr 66:209–233

Roux B, MacKinnon R (1999) The cavity and pore helices in the KcsA K^+ channel: electrostatic stabilization of monovalent cations. Science 285:100–102

Roux B, Berneche S, Im W (2000) Ion channels, permeation, and electrostatics: insight into the function of KcsA. Biochemistry 39:13295–13306

Samso M, Wagenknecht T, Allen PD (2005) Internal structure and visualization of transmembrane domains of the RyR1calcium release channel by cryo-EM. Nat Struct Mol Biol 6:539–544

Samso M, Feng W, Pessah IN, Allen PD (2009) Coordinated movement of cytoplasmic and transmembrane domains of RyR1 upon gating. PLoS Biol 7:e85

Sato C, Hamada K, Ogura T, Miyazawa A, Iwasaki K, Hiroaki Y, Tani K, Terauchi A, Fujiyoshi Y, Mikoshiba K (2004) Inositol 1,4,5-trisphosphate receptor contains multiple cavities and L-shaped ligand-binding domains. J Mol Biol 336:155–164

Schug ZT, Joseph SK (2006) The role of the S4-S5 linker and C-terminal tail in inositol 1,4,5-trisphosphate receptor function. J Biol Chem 281:24431–24440

Schug ZT, da Fonseca PC, Bhanumathy CD, Wagner L 2nd, Zhang X, Bailey B, Morris EP, Yule DI, Joseph SK (2008) Molecular characterization of the inositol 1,4,5-trisphosphate receptor pore-forming segment. J Biol Chem 283:2939–2948

Serysheva I (ed) (2010) Structure and function of calcium release channels. Academic Press, San Diego

Serysheva II, Bare DJ, Ludtke SJ, Kettlun CS, Chiu W, Mignery GA (2003) Structure of the type 1 inositol 1,4,5-trisphosphate receptor revealed by cryomicroscopy. J Biol Chem 278:21319–21322

Serysheva II, Hamilton SL, Chiu W, Ludtke SJ (2005) Structure of a Ca^{2+} release channel at 14 Å resolution. J Struct Biol 345:427–431

Shuai J, Rose HJ, Parker I (2006) The number and spatial distribution of IP_3 receptors underlying calcium puffs in Xenopus oocytes. Biophys J 91:4033–4044

Shuttleworth TJ (1992) Ca^{2+} release from inositol trisphosphate-sensitive stores is not modulated by intraluminal [Ca^{2+}]. J Biol Chem 267:3573–3576

Sienaert I, De Smedt H, Parys JB, Missiaen L, Vanlingen S, Sipma H, Casteels R (1996) Characterization of a cytosolic and a luminal Ca^{2+} binding site in the type I inositol 1,4,5-trisphosphate receptor. J Biol Chem 271:27005–27012

Sienaert I, Missiaen L, De Smedt H, Parys JB, Sipma H, Casteels R (1997) Molecular and functional evidence for multiple Ca^{2+}-binding domains on the type 1 inositol 1,4,5-trisphosphate receptor J Biol Chem 272:25899–25906

Sienaert I, Kasri NN, Vanlingen S, Parys J, Callewaert G, Missiaen L, De Smedt H (2002) Localization and function of a calmodulin/apocalmodulin binding domain in the N-terminal part of the type 1 inositol 1,4,5-trisphosphate receptor. Biochem J 365:269–277

Smith IF, Parker I (2009) Imaging the quantal substructure of single IP_3R channel activity during Ca^{2+} puffs in intact mammalian cells. Proc Natl Acad Sci U S A 106:6404–6409

Smith IF, Wiltgen SM, Shuai J, Parker I (2009) Ca^{2+} puffs originate from preestablished stable clusters of inositol trisphosphate receptors. Sci Signal 2:ra77

Solovyova N, Fernyhough P, Glazner G, Verkhratsky A (2002) Xestospongin C empties the ER calcium store but does not inhibit $InsP_3$-induced Ca^{2+} release in cultured dorsal root ganglia neurones. Cell Calcium 32:49–52

Srikanth S, Wang Z, Hasan G, Bezprozvanny I (2004) Functional properties of a pore mutant in the Drosophila melanogaster inositol 1,4,5-trisphosphate receptor. FEBS Lett 575:95–98

Stehno-Bittel L, Luckhoff A, Clapham DE (1995) Calcium release from the nucleus by $InsP_3$ receptor channels. Neuron 14:163–167

Streb H, Irvine RF, Berridge MJ, Schulz I (1983) Release of Ca^{2+} from a nonmitochondrial store of pancreatic acinar cells by inositol-1,4,5-trisphosphate. Nature 306:67–69

Striggow F, Ehrlich BE (1996) The inositol 1,4,5-trisphosphate receptor of cerebellum. Mn^{2+} permeability and regulation by cytosolic Mn^{2+}. J Gen Physiol 108:115–124

Suh BC, Hille B (2008) PIP_2 is a necessary cofactor for ion channel function: how and why? Annu Rev Biophys 37:175–195

Sun Y, Taylor CW (2008) A calmodulin antagonist reveals a calmodulin-independent interdomain interaction essential for activation of inositol 1,4,5-trisphosphate receptors. Biochem J 416:243–253

Sun X-P, Callamaras N, Marchant JS, Parker I (1998) A continuum of $InsP_3$-mediated elementary Ca^{2+} signalling events in Xenopus oocytes. J Physiol 509:67–80

Sureshan KM, Riley AM, Rossi AM, Tovey SC, Dedos SG, Taylor CW, Potter BVL (2009) Activation of IP_3 receptors by synthetic bisphosphate ligands. Chem Commun 14:1204–1206

Szado T, Vanderheyden V, Parys JB, De Smedt H, Rietdorf K, Kotelevets L, Chastre E, Khan F, Landegren U, Soderberg O, Bootman MD, Roderick HL (2008) Phosphorylation of inositol 1,4,5-trisphosphate receptors by protein kinase B/Akt inhibits Ca^{2+} release and apoptosis. Proc Natl Acad Sci U S A 105:2427–2432

Szlufcik K, Bultynck G, Callewaert G, Missiaen L, Parys JB, De Smedt H (2006) The suppressor domain of inositol 1,4,5-trisphosphate receptor plays an essential role in the protection against apoptosis. Cell Calcium 39:325–336

Tang J, Lin Y, Zhang Z, Tikunova S, Birnbaumer L, Zhu MX (2001) Identification of common binding sites for calmodulin and inositol 1,4,5-trisphosphate receptors on the carboxyl termini of trp channels. J Biol Chem 276:21303–21310

Tanimura A, Turner RJ (1996) Calcium release in HSY cells conforms to a steady-state mechanism involving regulation of the inositol 1,4,5-trisphosphate receptor Ca^{2+} channel by luminal $[Ca^{2+}]$. J Cell Biol 132:607–616

Tanimura A, Tojyo Y, Turner RJ (2000) Evidence that type I, II and III inositol-trisphosphate receptors can occur as integral membrane proteins. J Biol Chem 275:27488–27498

Tateishi Y, Hattori M, Nakayama T, Iwai M, Bannai H, Nakamura T, Michikawa T, Inoue T, Mikoshiba K (2005) Cluster formation of inositol 1,4,5-trisphosphate receptor requires its transition to open state. J Biol Chem 280:6816–6822

Tateishi H, Yano M, Mochizuki M, Suetomi T, Ono M, Xu X, Uchinoumi H, Okuda S, Oda T, Kobayashi S, Yamamoto T, Ikeda Y, Ohkusa T, Ikemoto N, Matsuzaki M (2009) Defective domain-domain interactions within the ryanodine receptor as a critical cause of diastolic Ca^{2+} leak in failing hearts. Cardiovasc Res 81:536–545

Taylor CW, Dellis O (2006) Plasma membrane IP_3 receptors. Biochem Soc Trans 34:910–912

Taylor CW, Laude AJ (2002) IP_3 receptors and their regulation by calmodulin and cytosolic Ca^{2+}. Cell Calcium 32:321–334

Taylor CW, Potter BVL (1990) The size of inositol 1,4,5-trisphosphate-sensitive Ca^{2+} stores depends on inositol 1,4,5-trisphosphate concentration. Biochem J 266:189–194

Taylor CW, Genazzani AA, Morris SA (1999) Expression of inositol trisphosphate receptors. Cell Calcium 26:237–251

Taylor CW, da Fonseca PCA, Morris EP (2004) IP_3 receptors: the search for structure. Trends Biochem Sci 29:210–219

Taylor CW, Rahman T, Tovey SC, Dedos SG, Taylor EJA, Velamakanni S (2009) IP_3 receptors: some lessons from DT40 cells. Immunol Rev 231:23–44

Terentyev D, Nori A, Santoro M, Viatchenko-Karpinski S, Kubalova Z, Gyorke I, Terentyeva R, Vedamoorthyrao S, Blom NA, Valle G, Napolitano C, Williams SC, Volpe P, Priori SG, Gyorke S (2006) Abnormal interactions of calsequestrin with the ryanodine receptor calcium release channel complex linked to exercise-induced sudden cardiac death. Circ Res 98:1151–1158

Thomas D, Lipp P, Berridge MJ, Bootman MD (1998) Hormone-evoked elementary Ca^{2+} signals are not stereotypic, but reflect activation of different size channel clusters and variable recruitment of channels within a cluster. J Biol Chem 273:27130–27136

Thomas D, Lipp P, Tovey SC, Berridge MJ, Li W, Tsien RY, Bootman MD (2000) Microscopic properties of elementary Ca^{2+} release sites in non-excitable cells. Curr Biol 10:8–15

Thrower EC, Mobasheri H, Dargen S, Marius P, Lea EJA, Dawson AP (2000) Interaction of luminal calcium and cytosolic ATP in the control of type 1 inositol (1,4,5)-trisphosphate receptor channels. J Biol Chem 275:36049–36055

Tinker A, Williams AJ (1993) Probing the structure of the conduction pathway of the sheep cardiac sarcoplasmic reticulum calcium-release channel with permeant and impermeant organic cations. J Gen Physiol 102:1107–1129

Tinker A, Williams AJ (1995) Measuring the length of the pore of the sheep cardiac sarcoplasmic reticulum calcium-release channel using related trimethylammonium ions as molecular calipers. Biophys J 68:111–120

Tojyo Y, Morita T, Nezu A, Tanimura A (2008) The clustering of inositol 1,4,5-trisphosphate (IP_3) receptors is triggered by IP_3 binding and facilitated by depletion of the Ca^{2+} store. J Pharm Sci 107:138–150

Tovey SC, De Smet P, Lipp P, Thomas D, Young KW, Missiaen L, De Smedt H, Parys J, Berridge MJ, Thuring J, Holmes A, Bootman MD (2001) Calcium puffs are generic $InsP_3$-activated elementary calcium signals and are downregulated by prolonged hormonal stimulation to inhibit cellular calcium responses. J Cell Sci 114:3979–3989

Tovey S, Dedos SG, Taylor EJA, Church JE, Taylor CW (2008) Selective coupling of type 6 adenylyl cyclase with type 2 IP_3 receptors mediates a direct sensitizaton of IP_3 receptors by cAMP. J Cell Biol 183:297–311

Tovey SC, Dedos SG, Rahman T, Taylor EJA, Pantazaka E, Taylor CW (2010) Regulation of inositol 1,4,5-trisphosphate receptors by cAMP independent of cAMP-dependent protein kinase. J Biol Chem 285:12979–12989

Tu H, Nosyreva E, Miyakawa T, Wang Z, Mizushima A, Iino M, Bezprozvanny I (2003) Functional and biochemical analysis of the type 1 inositol (1,4,5)-trisphosphate receptor calcium sensor. Biophys J 85:290–299

Tu H, Wang Z, Bezprozvanny I (2005a) Modulation of mammalian inositol 1,4,5-trisphosphate receptor isoforms by calcium: a role of calcium sensor region. Biophys J 88:1056–1069

Tu H, Wang Z, Nosyreva E, De Smedt H, Bezprozvanny I (2005b) Functional characterization of mammalian inositol 1,4,5-trisphosphate receptor isoforms. Biophys J 88:1046–1055

Uchida K, Miyauchi H, Furuichi T, Michikawa T, Mikoshiba K (2003) Critical regions for activation gating of the inositol 1,4,5-trisphosphate receptor. J Biol Chem 278:16551–16560

Vais H, Siebert AP, Ma Z, Fernandez-Mongil M, Foskett JK, Mak DO (2010) Redox-regulated heterogeneous thresholds for ligand recruitment among $InsP_3R$ Ca^{2+}-release channels. Biophys J 99:407–416

Put FHMM van de, De Pont JJHHM, Willems PHGM (1994) Heterogeneity between intracellular Ca^{2+} stores as the underlying principle of quantal Ca^{2+} release by inositol 1,4,5-trisphosphate in permeabilized pancreatic acinar cells. J Biol Chem 269:12438–12443

Rossum DB van, Patterson RL, Kiselyov K, Boehning D, Barrow RK, Gill DL, Snyder SH (2004) Agonist-induced Ca^{2+} entry determined by inositol 1,4,5-trisphosphate recognition. Proc Natl Acad Sci U S A 101:2323–2327

Wagner LE 2nd, Joseph SK, Yule DI (2008) Regulation of single inositol 1,4,5-trisphosphate receptor channel activity by protein kinase A phosphorylation. J Physiol 586:3577–3596

Wang R, Zhang L, Bolstad J, Diao N, Brown C, Ruest L, Welch W, Williams AJ, Chen SR (2003) Residue Gln4863 within a predicted transmembrane sequence of the Ca^{2+} release channel (ryanodine receptor) is critical for ryanodine interaction. J Biol Chem 278:51557–51565

Wang R, Chen W, Cai S, Zhang J, Bolstad J, Wagenknecht T, Liu Z, Chen SR (2007) Localization of an NH_2-terminal disease-causing mutation hot spot to the "clamp" region in the three-dimensional structure of the cardiac ryanodine receptor. J Biol Chem 282:17785–17793

Watras J, Bezprozvanny I, Ehrlich BE (1991) Inositol 1,4,5-trisphosphate-gated channels in cerebellum: presence of multiple conductance states. J Neurosci 11:3239–3245

Wheeler GL, Brownlee C (2008) Ca^{2+} signalling in plants and green algae—changing channels. Trends Plant Sci 13:506–514

Williams AJ, West DJ, Sitsapesan R (2001) Light at the end of the Ca^{2+}-release channel tunnel: structures and mechanisms involved in ion translocation in ryanodine receptor channels. Q Rev Biophys 34:61–104

Willoughby D, Cooper DM (2007) Organization and Ca^{2+} regulation of adenylyl cyclases in cAMP microdomains. Physiol Rev 87:965–1010

Wilson BS, Pfeiffer JR, Smith AJ, Oliver JM, Oberdorf JA, Wojcikiewicz RJH (1998) Calcium-dependent clustering of inositol 1,4,5-trisphosphate receptors. Mol Biol Cell 9:1465–1478

Wolfram F, Morris E, Taylor CW (2010) Three-dimensional structure of recombinant type 1 inositol 1,4,5-trisphosphate receptor. Biochem J 428:483–489

Woods NM, Cuthbertson KSR, Cobbold PH (1986) Repetitive transient rises in cytoplasmic free calcium in hormone-stimulated hepatocytes. Nature 319:600–602

Worley PF, Baraban JM, Supattapone S, Wilson VS, Snyder SH (1987) Characterization of inositol trisphosphate receptor binding in brain. Regulation by pH and calcium. J Biol Chem 262:12132–12136

Xi Q, Adebiyi A, Zhao G, Chapman KE, Waters CM, Hassid A, Jaggar JH (2008) IP_3 constricts cerebral arteries via IP_3 receptor-mediated TRPC3 channel activation and independently of sarcoplasmic reticulum Ca^{2+} release. Circ Res 102:1118–11126

Yamada M, Miyawaki A, Saito K, Yamamoto-Hino M, Ryo Y, Furuichi T, Mikoshiba K (1995) The calmodulin-binding domain in the mouse type 1 inositol 1,4,5-trisphosphate receptor. Biochem J 308:83–88

Yamasaki-Mann M, Demuro A, Parker I (2010) Modulation of ER Ca^{2+} store filling by cADPR promotes IP_3-evoked Ca^{2+} signals. J Biol Chem 285:25053–25061
Yamazaki H, Chan J, Ikura M, Michikawa T, Mikoshiba K (2010) Tyr-167/Trp-168 in type1/3 inositol 1,4,5-trisphosphate receptor mediates functional coupling between ligand binding and channel opening. J Biol Chem 285:36081–36091
Yang J, Ellinor PT, Sather WA, Zhang J-F, Tsien RW (1993) Molecular determinants of Ca^{2+} selectivity and ion permeation in L-type Ca^{2+} channels. Nature 366:158–161
Yang J, McBride S, Mak D-O, D, Vardi N, Palczewski K, Haeseleer F, Foskett JK (2002) Identification of a family of calcium sensors as protein ligands of the inositol trisphosphate receptor Ca^{2+} release channels. Proc Natl Acad Sci U S A 99:7711–7716
Yang D, Shcheynikov N, Zeng W, Ohana E, So I, Ando H, Mizutani A, Mikoshiba K, Muallem S (2009) IRBIT coordinates epithelial fluid and HCO_3- secretion by stimulating the transporters pNBC1 and CFTR in the murine pancreatic duct. J Clin Invest 119:193–202
Yellen G (2002) The voltage-gated potassium channels and their relatives. Nature 419:35–42
Yule DI, Betzenhauser MJ, Joseph SK (2010) Linking structure to function: recent lessons from inositol 1,4,5-trisphosphate receptor mutagenesis. Cell Calcium 47:469–479
Zeng W, Mak DD, Li Q, Shin DM, Foskett JK, Muallem S (2003) A new mode of Ca^{2+} signaling by G protein-coupled receptors: gating of IP_3 receptor Ca^{2+} release channels by $G\beta\gamma$. Curr Biol 13:872–876
Zhang X, Joseph SK (2001) Effect of mutation of a calmodulin-binding sites on Ca^{2+} regulation of inositol trisphosphate receptors. Biochem J 360:395–400
Zhang Z, Tang J, Tikunova S, Johnson JD, Chen Z, Qin N, Dietrich A, Stefani E, Birnbaumer L, Zhu MX (2001) Activation of Trp3 by inositol 1,4,5-trisphosphate receptors through displacement of inhibitory calmodulin from a common binding domain. Proc Natl Acad Sci U S A 98:3168–3173
Zhao Y, Yarov-Yarovoy V, Scheuer T, Catterall WA (2004) A gating hinge in Na^+ channels; a molecular switch for electrical signaling. Neuron 41:859–865
Zhao G, Neeb ZP, Leo MD, Pachuau J, Adebiyi A, Ouyang K, Chen J, Jaggar JH (2010) Type 1 IP_3 receptors activate BK_{Ca} channels via local molecular coupling in arterial smooth muscle cells. J Gen Physiol 136:283–291
Zhen XG, Xie C, Fitzmaurice A, Schoonover CE, Orenstein ET, Yang J (2005) Functional architecture of the inner pore of a voltage-gated Ca^{2+} channel. J Gen Physiol 126:193–204
Zhou Y, Morais-Cabral JH, Kaufman A, MacKinnon R (2001) Chemistry of ion coordination and hydration revealed by a K^+ channel-Fab complex at 2.0 Å resolution. Nature 414:43–48
Zhu MX, Ma J, Parrington J, Calcraft PJ, Galione A, Evans AM (2010) Calcium signaling via two-pore channels: local or global, that is the question. Am J Physiol 298:C430–C441

Chapter 2
Phosphoinositide Signaling During Membrane Transport in *Saccharomyces Cerevisiae*

Amber L. Schuh and Anjon Audhya

Abstract Phosphatidylinositol (PI) is distinct from other phospholipids, possessing a head group that can be modified by phosphorylation at multiple positions to generate unique signaling molecules collectively known as phosphoinositides. The set of kinases and phosphatases that regulate PI metabolism are conserved throughout eukaryotic evolution, and numerous studies have demonstrated that phosphoinositides regulate a diverse spectrum of cellular processes, including vesicle transport, cell proliferation, and cytoskeleton organization. Over the past two decades, nearly all PI derivatives have been shown to interact directly with cellular proteins to affect their localization and/or activity. Additionally, there is growing evidence, which suggests that phosphoinositides may also affect local membrane topology. Here, we focus on the role of phosphoinositides in membrane trafficking and underscore the significant role that yeast has played in the field.

Keywords Lysosomal/vacuolar trafficking · Autophagy · Endocytosis · Protein secretion · Actin cytoskeleton

2.1 Phosphoinositide Metabolism in the Yeast *Saccharomyces Cerevisiae*

In wild-type *Saccharomyces cerevisiae*, four major phosphorylated derivatives of PI have been identified: PI3P, PI4P, PI3,5P_2 and PI4,5P_2 (Fig. 2.1). A single soluble Class III PI 3-kinase, encoded by *VPS34*, generates the total cellular pool of PI3P, which accumulates mostly in the limiting membranes of endosomes and lysosome-like vacuoles (Schu et al. 1993; Stack and Emr 1994; Stenmark and Gillooly 2001).

A. Audhya (✉) · A. L. Schuh
Department of Biomolecular Chemistry, University of Wisconsin-Madison Medical School, 1300 University Avenue, WI, 53706 Madison, USA
e-mail: audhya@wisc.edu

T. Balla et al. (eds.), *Phosphoinositides II: The Diverse Biological Functions*, Subcellular Biochemistry 59, DOI 10.1007/978-94-007-3015-1_2,

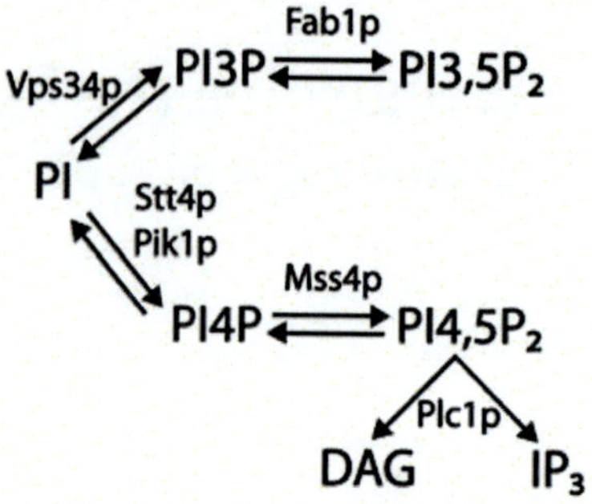

Fig. 2.1 Phosphoinositide metabolism in yeast. The synthesis pathways for the 4 phosphoinositides known to exist in yeast are shown. Major PI kinases are highlighted, as is the phospholipase C isoform Plc1p, which hydrolyzes $PI4,5P_2$. Not pictured are phosphoinositide phosphatases, which mediate the dephosphorylation of various PI derivatives

Vps34p functions as a subunit of at least two distinct protein complexes, both of which also contain the serine/threonine-protein kinase Vps15p. Myristoylation anchors Vps15p to membranes and assists in the recruitment of cytosolic Vps34p to appropriate intracellular membranes (Herman et al. 1991; Stack et al. 1993). Vps15p is considered to be a regulatory subunit within PI 3-kinase complexes and contains HEAT domains responsible for additional protein-protein interactions (Vanhaesebroeck et al. 2010). Despite the essential nature of the Vps15p kinase domain to Vps34p recruitment and activation, direct substrates of Vps15p remain undefined.

The stability of PI3P is regulated by both catabolic and anabolic pathways. Two major routes of PI3P degradation have been characterized. First, PI3P contained in the limiting membrane of endosomes is a substrate for proteolytic enzymes in the vacuole lumen. Inhibition of membrane transport to the vacuole thereby causes an increase in cellular PI3P levels (Wurmser and Emr 1998). Alternatively, PI3P is subject to dephosphorylation by a group of lipid phosphatases, which include the myotubularin-related enzyme Ymr1p and two synaptojanin-like proteins, Sjl2p and Sjl3p (Stolz et al. 1998; Parrish et al. 2004). Although PI3P synthesis is not essential to yeast cell viability, depletion of all three PI3P phosphatases results in lethality (Parrish et al. 2005). It remains unclear whether elevated PI3P levels are solely responsible for this effect, but the result highlights the importance of regulated phosphoinositide turnover for normal cell proliferation.

PI3P is also a substrate for the Fab1p lipid kinase, which specifically phosphorylates the D5 position in the inositol ring to generate $PI3,5P_2$. The majority of Fab1p localizes to the limiting membrane of the yeast vacuole, where it functions together with Vac7p, a transmembrane regulatory factor, to generate the total cellular pool of $PI3,5P_2$ (Bonangelino et al. 1997; Gary et al. 2002; Botelho et al. 2008). In the absence of Vac7p, levels of $PI3,5P_2$ become undetectable. However, deletion of Fig4p, a $PI3,5P_2$ 5-phosphatase, can suppress phenotypes exhibited by loss of Vac7p, indicating that Fab1p remains functional in the absence of its regulator (Gary et al. 2002). Consistent with this finding, mutant isoforms of Fab1p have been characterized, which bypass the requirement for Vac7p in $PI3,5P_2$ synthesis.

In contrast to the unique functions of Vps34p to generate 3′ phosphoinositides and Fab1p to generate $PI3,5P_2$, three PI 4-kinases have been characterized in yeast and include the type IIIa PI 4-kinase *STT4*, the type IIIb PI 4-kinase *PIK1*, and the type II PI 4-kinase *LSB6*. Both Stt4p and Pik1p are essential genes in yeast, suggesting they possess non-overlapping functions and generate unique pools of PI4P, which cannot substitute for one another (Audhya et al. 2000). Consistent with this idea, Stt4p localizes primarily to the plasma membrane, while the majority of Pik1p accumulates on Golgi membranes (Walch-Solimena and Novick 1999; Audhya and Emr 2002). At the cell surface, Stt4p appear to coalesce into discrete patches that are enriched in the mother cell. Purification of Stt4p has revealed a number of interacting proteins that regulate Stt4p function. These include Sfk1p and Efr3p, two putative transmembrane proteins and Ypp1p, a soluble protein containing two tetratricopeptide repeat domains (Audhya and Emr 2002; Baird et al. 2008; Zhai et al. 2008). Inhibition of each regulator has been shown to decrease PI4P levels at the cell surface and perturb the localization of Stt4p. Unlike Sfk1p, both Efr3p and Ypp1p are essential for yeast cell viability, suggesting they may regulate unique biochemical pathways downstream of Stt4p, but further studies are required to address this issue.

Distinct from plasma membrane synthesis of PI4P, Pik1p generates an essential pool of PI4P on Golgi membranes. Proper targeting of Pik1p to the Golgi requires an interaction with Frq1p, a myristoylated regulator of Pik1p-mediated PI4P synthesis (Hendricks et al. 1999). Loss of Frq1p is lethal, but overproduction of Pik1p can bypass the requirement of Frq1p, suggesting that additional mechanisms exist to target Pik1p to Golgi membranes. In addition to its role at the Golgi, Pik1p undergoes nucleo-cytoplasmic shuttling and has been postulated to generate a nuclear pool of PI4P (Garcia-Bustos et al. 1994; Strahl et al. 2005). Frq1p is not required for Pik1p function in the nucleus, and a specific role for nuclear PI4P has yet to be defined.

Unlike Stt4p and Pik1p, Lsb6p activity is not essential for normal growth of *S. cerevisiae*, and its loss does not impact the total cellular levels of PI4P under standard conditions (Han et al. 2002; Shelton et al. 2003). However, overproduction of Lsb6p weakly suppresses the loss of Stt4p, indicating that Lsb6p can function as a PI 4-kinase *in vivo*. Localization studies have placed Lsb6p at the plasma membrane, consistent with its ability to suppress deletion of *STT4*, and the limiting membrane of the vacuole. Additionally, Lsb6p has been shown to regulate endosome motility (Chang et al. 2005). However, this function of Lsb6p is independent of its lipid kinase activity.

Similar to PI3P, metabolism of PI4P is mediated by a set of lipid phosphatases and a single lipid kinase, the PI4P 5-kinase Mss4p. The plasma membrane pool of PI4P generated by Stt4p is largely regulated by the Sac1p lipid phosphatase (Foti et al. 2001). Biochemical and localization studies indicate that Sac1p is a type II membrane protein that localizes mainly to the ER under normal growth conditions (Foti et al. 2001; Faulhammer et al. 2005). Recent findings indicate that Sac1p hydrolyzes PI4P at sites of ER-plasma membrane contact, acting from the ER in trans on its plasma membrane substrate (Stefan et al. 2011). The retention of Sac1p in the ER is dependent on the dolichol phosphate mannose synthase Dpm1p, and loss of Dpm1p results in the accumulation of Sac1p in the Golgi (Faulhammer et al.

2005). In the absence of Sac1p, PI4P levels on the cell surface dramatically increase in an Stt4p-dependent fashion (Roy and Levine 2004).

PI4P generated by Stt4p can also be converted into $PI4,5P_2$ by the action of Mss4p. Similar to Stt4p, Mss4p is largely restricted to discrete foci on the plasma membrane (Homma et al. 1998; Audhya and Emr 2002). However, patches of Stt4p and Mss4p do not overlap, suggesting that Mss4p can utilize alternative sources of PI4P at the cell surface (Audhya and Emr 2002). Consistent with this idea, elimination of Stt4p activity only diminishes $PI4,5P_2$ levels by ~50% (Audhya et al. 2000). The remaining pool of $PI4,5P_2$ is synthesized from PI4P initially generated at the Golgi by Pik1p, which is likely transported to the plasma membrane within the membranes of secretory vesicles. Golgi PI4P is further regulated by multiple phosphoinositide phosphatases, including members of the synaptojanin-like family of lipid phosphatases, Sjl2p and Sjl3p, as well as Sac1p (Guo et al. 1999a; Foti et al. 2001; Faulhammer et al. 2005).

Although the majority of Mss4p localizes to the plasma membrane, the PI4P 5-kinase also undergoes nucleo-cytoplasmic shuttling similar to Pik1p (Audhya and Emr 2003). The factors that mediate nuclear import and export of Mss4p (Kap123p and Bcp1p) are distinct from those that regulate Pik1p nucleo-cytoplasmic transport (Kap95p and Msn5p) (Strahl et al. 2005). Also unlike Pik1p, inhibition of Mss4p nuclear entry fails to affect normal cellular growth, suggesting that nuclear $PI4,5P_2$ is not essential in yeast (Audhya and Emr 2003). Instead, targeting of Mss4p to the nucleus may function to regulate its activity on the cell surface and thereby control cytoplasmic $PI4,5P_2$ synthesis. Although factors required specifically for cell surface Mss4p patch formation have yet to be defined, plasma membrane targeting of Mss4p is partially dependent on its phosphorylation by yeast casein kinase I activity (Audhya and Emr 2003). Additionally, the small calcium-binding protein calmodulin appears to regulate Mss4p lipid kinase activity, although the mechanistic basis for this affect remains undefined (Desrivieres et al. 2002). Levels of $PI4,5P_2$ are further regulated by lipid phosphatases, including the synaptojanin-like proteins Sjl1p and Sjl2p, the ER-localized inositol 5-phosphatase Inp54p, and the phospholipase C isoform Plc1p, which specifically hydrolyzes $PI4,5P_2$ to generate diacylglycerol (DAG) and IP_3 (Flick and Thorner 1993; Wiradjaja et al. 2001; Stefan et al. 2002).

Through these highly regulated biosynthetic and degradation pathways, individual phosphoinositides are spatially restricted within the cell. PI3P and $PI3,5P_2$ are enriched in endosomal and vacuolar membranes, while PI4P and $PI4,5P_2$ are concentrated in the Golgi and plasma membrane (Fig. 2.2). In the following sections, we will highlight mechanisms by which phosphoinositides function to regulate virtually all of the membrane trafficking pathways that have been described in yeast.

2.2 Binding Domains and Effector Proteins of PI3P

PI3P carries out its cellular functions by recruiting and/or activating a subset of proteins to specific internal membrane compartments where PI3P is synthesized. These proteins function in signal transduction, vesicle trafficking, and cytoskeletal rearrangements. Despite these differences in function, the majority of proteins known to

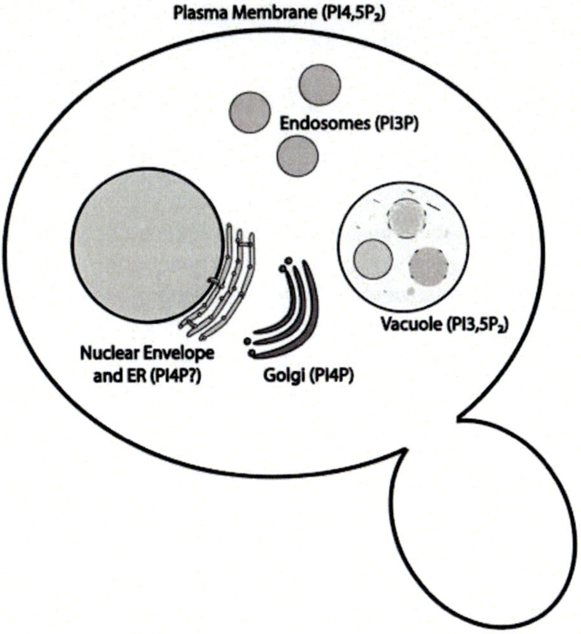

Fig. 2.2 Distribution of phosphoinositides in yeast. A cartoon depicting a yeast cell, with various organelles that harbor phosphoinositides highlighted. The plasma membrane is enriched with $PI4,5P_2$, the vacuole with $PI3,5P_2$, endosomes with PI3P, and the Golgi with PI4P. The pool of PI4P synthesized on the plasma membrane is not shown for simplicity

bind to PI3P share a small subset of domains that interact directly with the phosphoinositide (Stenmark and Gillooly 2001, Seet and Hong 2006). We will first highlight roles for PI3P generated at the endosome by PI3K Complex II, which is composed of Vps15p, Vps34p, Vps30p and Vps38p (Kihara et al. 2001). In particular, effectors of PI3P in both anterograde transport from the endosome to the vacuole and retrograde transport from the endosome to the Golgi will be discussed. Additionally, we will describe alternative roles for PI3P in constitutive and starvation-induced autophagy pathways, which utilize PI3K Complex I, containing Vps15p, Vps34p, Vps30p, and Atg14p (Kametaka et al. 1998; Kihara et al. 2001).

2.2.1 The FYVE Domain

The FYVE domain is composed of ~70 conserved residues that specifically bind PI3P. This domain is cysteine rich and coordinates two zinc ions, which are essential for structural integrity and PI3P binding (Gaullier et al. 1998; Burd and Emr 1998; Kutateladze et al. 1999; Misra and Hurley 1999). Three conserved sequences have been identified in all FYVE domains: the amino-terminal WxxD motif, the central (R/K)(R/K) HHCR motif, and the carboxyl-terminal RVC motif. These three regions are crucial in forming a concave binding pocket for PI3P association. In addition, the FYVE domain contains a hydrophobic protrusion, commonly referred to as the membrane insertion loop (MIL) domain, adjacent to the (R/K)(R/K) HHCR motif (Kutateladze et al. 1999; Misra and Hurley 1999).

The FYVE domain uses a multivalent mechanism to anchor itself to a target membrane, which includes specific interactions with PI3P as well as other lipids. Based upon the calculated electrostatic properties of several FYVE domains, a strong positive potential exists around the MIL domain, which supports interactions with acidic phospholipids (Diraviyam et al. 2003). Consistent with this finding, studies have shown that the acidic phospholipid phosphatidylserine (PS) specifically enhances the affinity of FYVE domains for lipid bilayers (Stahelin et al. 2002). In addition, these nonspecific electrostatic interactions likely help to align the FYVE domain in an optimal position for membrane penetration of the MIL domain (Stahelin et al. 2002; Kutateladze et al. 2004). Insertion of the hydrophobic MIL domain is dependent upon the presence of PI3P, and this process has been shown to significantly increase the affinity between lipid bilayers and FYVE domains (Kutateladze et al. 2004). Mutagenesis of select hydrophobic residues within the MIL region drastically reduces the affinity for membrane bound PI3P and disrupts normal function of FYVE domain containing proteins (Kutateladze et al. 1999; Stahelin et al. 2002).

The association of FYVE domains with membranes also appears to be pH dependent. In a neutral buffer (pH 7.0), FYVE domain containing proteins in yeast localize to endosomal and vacuolar membranes. However, when cells are shifted into a mildly basic buffer (pH 8.0), FYVE domain containing proteins largely redistribute to the cytoplasm (Lee et al. 2005; He et al. 2009). These data support a model in which the two histidine residues within the conserved (R/K)(R/K) HHCR motif must be protonated to form hydrogen bonds with the 3′ phosphate of PI3P (Dumas et al. 2001; Kutateladze 2006).

In total, five FYVE domain containing proteins have been identified in yeast: Vac1p, Fab1p, Vps27p, Pib1p, and Pib2p. Vac1p is required for the transport of vesicles from the Golgi to prevacuolar endosomes (Webb et al. 1997). Similar to other FYVE domains, the Vac1p FYVE domain binds directly to PI3P and localizes to endosomes when expressed as a GFP fusion. However, in cells lacking PI3P, Vac1p continues to accumulate on membranes, suggesting that the interaction between Vac1p and PI3P does not solely influence it localization (Tall et al. 1999). Instead, PI3P may regulate Vac1p activity, potentially by influencing its interactions with other proteins. Consistent with this idea, both the Rab-type GTPase Vps21p and the Sec1-like protein Vps45p, interact with Vac1p in a FYVE domain-dependent manner (Peterson et al. 1999; Tall et al. 1999).

Fab1p, a PI3P 5-kinase, and Vps27p, a component of the ESCRT (endosomal sorting complex required for transport) machinery, have both been implicated in the formation of multivesicular endosomes (MVEs), which are specialized organelles essential for the degradation of many transmembrane proteins (Odorizzi et al. 1998; Katzmann et al. 2003; Bilodeau et al. 2003; Raiborg and Stenmark 2009). In this pathway, ubiquitin-modified cargoes are initially sequestered within vesicles that bud into the endosome lumen. Upon MVE fusion with the vacuole, cargo-laden vesicles are transferred directly into the hydrolytic environment of the vacuole lumen, resulting in protein and vesicle degradation. In the absence of PI3P, neither Vps27p nor Fab1p localize properly, and MVE biogenesis is dramatically inhibited. Unlike the FYVE domain of Vps27p, which localizes to endosomes when expressed as a GFP fusion, the Fab1p FYVE domain associates specifically with the vacuolar limiting membrane (Botelho et al. 2008). Although it has been shown to bind PI3P directly, these data suggest that the Fab1p FYVE domain may also bind another factor, which directs its accumulation onto the vacuole membrane. The identity of such a factor remains unknown.

In addition to the interaction between PI3P and the FYVE domains of Vps27p and Fab1p, the ESCRT-II subunit Vps36p, which also functions in MVE biogenesis, binds to PI3P through a noncanonical binding pocket within its amino-terminal GLUE (GRAM-like, ubiquitin binding in EAP45) domain (Teo et al. 2006). The GLUE domain exhibits a split PH (pleckstrin homology) domain fold with two curved beta sheets and a single long alpha helix. A highly basic pocket is delineated by three variable loops within the GLUE domain, a configuration that is distinct from most PH domains that interact with phosphoinositides. Mutations within the basic region inhibit protein sorting into MVEs, indicating association of ESCRT-II with PI3P is critical for function (Teo et al. 2006; Im and Hurley 2008).

Pib1p may also function in the MVE pathway as an E3 RING-type ubiquitin ligase. Studies indicate that Pib1p localizes to both endosomal and vacuolar membranes, and its distribution is dependent solely on its interaction with PI3P (Shin et al. 2001). Although specific Pib1p-dependent cargoes have yet to be identified, it is likely that Pib1p functions at the initial stages of cargo selection/modification in the MVE pathway. The role of Pib2p remains unknown, although localization studies indicate that it accumulates on the vacuolar membrane under steady state conditions (Huh et al. 2003).

2.2.2 The Phox Homology (PX) Domain

The PX domain is composed of approximately 130 amino acids and is found in proteins that function in vesicle trafficking, protein sorting, and lipid modification (Seet and Hong 2006). Typically, the PX domain folds into a compact structure composed of three beta strands followed by three alpha helices. A conserved RR(F/Y)S(D/E)F motif and three additional basic residues located nearby are proposed to form a binding pocket for PI3P (Cheever et al. 2001; Bravo et al. 2001; Sato et al. 2001). In addition, many of the PX domains contain a polyproline motif (PxxP), which is predicted to interact with SH3 domains (Xu et al. 2001). PX domains also interact with membranes by using a multivalent mechanism, which includes non-specific electrostatic interactions, hydrophobic insertion, and oligomerization. In particular, oligomerization is crucial for increasing the affinity of PX domains that otherwise would be unable to localize to membranes due to low affinities in their monomeric state (Kutateladze 2007).

In yeast, 15 proteins that harbor a PX domain have been identified, and all have an affinity for PI3P. Using surface plasmon resonance (SPR), four yeast PX domains have been classified as having a high affinity for PI3P, ranging from 0.15 to 0.5 μM, while the rest exhibit affinities greater than 100 μM and are classified as low affinity PX domains (Yu and Lemmon 2001). These affinity measurements suggest that only the four high affinity PX domains are capable of membrane association independently of other binding partners, while the remaining PX domains must require oligomerization or interaction with additional proteins to mediate their localization.

The four yeast proteins with high affinity for PI3P are Mdm1p, Snx3p, Vam7p and Ypt35p. Mdm1p is required for mitochondrial and nuclear inheritance, and although its PX domain has been shown to bind to PI3P with a high affinity, the function of this interaction remains unknown (Yu and Lemmon 2001). Snx3p is a member of the sorting nexin family and functions in the retrograde transport of a subset of cargoes from endosomes to the Golgi (Strochlic et al. 2007). Importantly, the retrieval of certain transmembrane receptors from late endosomes is required to maintain the proper sorting of hydrolases to the vacuole. One of the best studied receptors is the type I membrane protein Vps10p, which is essential for the normal trafficking of the soluble vacuolar hydrolase carboxypeptidase Y (CPY). At the endosome, CPY dissociates from Vps10p for ultimate delivery to the vacuole, while Vps10p undergoes recycling to the trans Golgi network (TGN), becoming available for another round of CPY transport (Marcusson et al. 1994; Stack et al. 1995). The high affinity of Snx3p for PI3P targets it to tubular endosomes where it can associate with Vps10p and mediate its retrieval. However, mutations in the Snx3p PX domain that inhibit PI3P binding result in the missorting of receptors to the vacuole, inhibiting further transport of CPY and other cargoes (Strochlic et al. 2007). These data implicate PI3P in retrograde trafficking from endosomes to the Golgi. Snx3p functions together with the retromer complex, a set of proteins also required for retrograde trafficking from endosomes to the Golgi. Two components of the retromer complex, Vps5p and Vps17p, also contain PX domains, each with a low affinity for PI3P. However, when co-assembled, the multiple low affinity interactions with PI3P maintain an endosomal distribution for the retromer complex, and further highlight a role for PI3P in orchestrating receptor recycling from the endosomal system (Burda et al. 2002).

The third high affinity PI3P interacting protein Vam7p is a target SNARE (Soluble NSF Attachment Protein Receptor) that functions during the docking and fusion steps of membrane transport to the vacuole (Cheever et al. 2001; Song et al. 2001). An intact PX domain is required for this function, implicating PI3P in the late stages of membrane transport to the vacuole. Finally, Ypt35p also binds to PI3P with high affinity and has been found to localize to endosomal membranes. However, the specific function of Ypt35p remains unknown (Yu and Lemmon 2001).

Among the other low affinity PI3P-interacting PX domains, Mvp1p, Spo14p, Snx4p, Snx41p, and Atg20p each possess characterized roles in membrane trafficking. Mvp1p is an endosomal protein required for protein sorting to the vacuole. Although its precise role has not been clearly identified, genetic studies indicate that Mvp1p functions with the dynamin-like protein Vps1p in retrograde protein transport from endosomes to the Golgi (Ekena and Stevens 1995). Spo14p is a phospholipase D isoform that has been implicated in protein secretion (Sreenivas et al. 1998). During vegetative growth, Spo14p localizes to endosomes, but its specific activity there is poorly characterized, and it would be premature to suggest that PI3P functions directly in the secretory pathway (Sciorra et al. 2002). Snx4p (also known as Atg24p), Snx41p, and Atg20p (also known as Snx42p) all are sorting nexins that function in a common complex to sort receptors from early endosomes to the Golgi (Hettema et al. 2003). Even though the binding affinities for individual PX domains within the individual sorting nexins are low, their assembly into a complex leads to PX domain multimerization and an increased affinity for PI3P on endosomes.

In addition, both Snx4p/Atg24p and Atg20p are required for selective autophagy, a constitutive cytoplasm to vacuole targeting (CVT) pathway in which a perivacuolar phagophore assembly site (PAS) initiates the biogenesis of an autophagosome, a double-membrane vesicle (Nice et al. 2002; He and Klionsky 2009). This pathway depends on a large group of proteins that assist in the elongation of a double-membrane structure known as an isolation membrane that ultimately circularizes to form an autophagosome. The autophagosome then fuses with the vacuole, leading to the degradation of its contents (Huang and Klionsky 2002). When yeast cells are under starvation, an alternative bulk autophagy pathway is initiated to break down proteins into amino acids for energy (Burman and Ktistakis 2010). The CVT and bulk autophagy pathways share many of the same proteins, including the PI3K Complex I (Vps15p, Vps34p, Vps30p, and Atg14p). The Atg14p subunit of this complex provides specificity in directing Vps34p to PASs (Farre et al. 2009). Interactions between PI3P and the PX domains of Snx4p/Atg24p and Atg20p are required for normal CVT pathway function, directly implicating PI3P in the constitutive formation of autophagosomes (Nice et al. 2002).

Further highlighting a role for phosphoinositides in the CVT pathway are two additional PI3P effectors called Atg21p and Atg27p. Unlike other PI3P binding proteins, neither Atg21p nor Atg27p harbor a PX or FYVE domain. Instead, PI3P binding is mediated by short basic stretches of amino acids, FRRG in Atg21p and KKPAKK in Atg27p (Wurmser and Emr 2002; Stromhaug et al. 2004; Krick et al. 2006; Nair et al. 2010). Mutations in these motifs, which inhibit PI3P binding, block CVT pathway function. In addition, deletion of *VPS34* or *VPS15* has also been shown to block starvation-induced bulk autophagy (Wurmser and Emr 2002). One phosphoinositide effector potentially responsible for this effect is Atg18p, which also contains an FRRG motif capable of binding to PI3P (Barth et al. 2001; Dove et al. 2004). Although the degree to which mutations in the Atg18p PI3P-binding motif affects bulk autophagy remains controversial (Krick et al. 2006), the participation of PI3P in autophagosome biogenesis is incontrovertible.

In summary, studies in yeast have clearly illustrated the various roles PI3P plays in membrane trafficking, both in directing localization of effector molecules and regulating their activities. Although it is likely that several new PI3P effectors still await characterization, most will likely function at endosomes, vacuoles, or autophagosomes, where PI3P is highly enriched. We predict that many of these new molecules will exhibit a low affinity for PI3P, but together with other protein-protein interactions, specificity for endosomal/autophagosomal signaling pathways can be achieved.

2.3 Effectors of $PI3,5P_2$ and Their Roles in Membrane Trafficking

In yeast, a single PI3P 5-kinase called Fab1p has been identified and is responsible for the generation of $PI3,5P_2$ (Cooke et al. 1998; Gary et al. 1998). The Fab1p lipid kinase contains three conserved protein domains: an amino-terminal FYVE domain, a Cpn60/TCP-1 chaperonin family (CCT) domain, and a carboxyl-terminal lipid kinase domain (Gary et al. 1998; Efe et al. 2005). $PI3,5P_2$ is largely generated on the vacuolar limiting membrane, where it has a role in retrograde trafficking from the vacuole and vacuolar homeostasis (Dove and Johnson 2007). Deletion of *FAB1* leads to several phenotypes including enlarged vacuoles, defects in MVE biogenesis, a lack of vacuolar acidification and slow temperature sensitive growth (Yamamoto et al. 1995; Cooke et al. 1998; Gary et al. 1998; Odorizzi et al. 1998). Here, we will discuss the effectors of $PI3,5P_2$ that have been discovered and how they relate to the effects seen upon loss of Fab1p activity.

Two classes of $PI3,5P_2$ binding domains have been identified. The first is found in a family of seven bladed β-propeller proteins that bind phosphoinositides (PROPPINs), and the second is contained within members of the epsin family. In yeast, three PROPPIN proteins have been identified, including Atg18p, Atg21p and Hsv2p (Michell et al. 2006). As discussed earlier, both Atg18p and Atg21p harbor a FRRG motif capable of interacting with PI3P. However, SPR studies indicate they can also bind $PI3,5P_2$ with high affinity (~500 nM) (Dove et al. 2004). Moreover, loss of Atg18p causes a dramatic increase in vacuole size, similar to the phenotype observed following *FAB1* deletion, suggesting a role downstream of $PI3,5P_2$ signaling (Dove et al. 2004; Cooke et al. 2004). Localization studies indicate that all three proteins associate with both endosomal membranes and the limiting membrane of the vacuole. It is possible that PROPPIN proteins associate with both PI3P and $PI3,5P_2$ and exhibit distinct activities depending on the lipid to which they are bound. For example, interactions between PI3P and Atg18p and Atg21p are likely important for constitutive autophagy, which does not depend on Fab1p-mediated $PI3,5P_2$ production (Gary et al. 1998; Wurmser and Emr 2002). In contrast, interaction between $PI3,5P_2$ and Atg18p is probably required for normal vacuole homeostasis and retrograde trafficking from the vacuole to the Golgi via an endosomal intermediate (Dove et al. 2004). Notably, in cells lacking Fab1p, the PROPPIN proteins continue to localize to the endosome and vacuolar membranes (Efe et al. 2007). Nevertheless, $PI3,5P_2$ binding to these factors may regulate their function as opposed to their localization. Hsv2p is the least well characterized PROPPIN in yeast, but studies indicate that the protein participates in a unique autophagic pathway responsible for the turnover of nuclear membranes (Krick et al. 2008). It remains unclear whether $PI3,5P_2$ may function to regulate this activity.

As discussed earlier, Fab1p also regulates the biogenesis of lumenal vesicles within endosomes (Odorizzi et al. 1998). While none of the PROPPIN proteins exhibit a function at the MVE, two other effectors of $PI3,5P_2$ have been implicated in MVE-mediated protein sorting. Both Ent3p, which contains a phosphoinositide binding motif called the ENTH (epsin N-terminal homology) domain, and Ent5p, which harbors a related ANTH (AP180 N-terminal homology) domain, bind to $PI3,5P_2$

in vitro and require Fab1p activity for localization *in vivo* (Friant et al. 2003; Eugster et al. 2004). Additionally, the simultaneous loss of Ent3p and Ent5p interferes with the trafficking of integral membrane proteins to the vacuole lumen, similar to the phenotype exhibited by *fab1* mutant cells (Eugster et al. 2004). Moreover, both Ent3p and Ent5p have been implicated in AP-1 (adaptor protein 1)-dependent sorting of chitin synthase, a process previously shown to require Fab1p activity, further suggesting a role for the epsin-like proteins downstream of $PI3,5P_2$ signaling (Costaguta et al. 2006; Phelan et al. 2006; Copic et al. 2007).

The phenotypes exhibited by mutant yeast cells lacking Fab1p cannot be explained by the few effector molecules that have been discovered to date. Therefore, it is highly probable that additional $PI3,5P_2$-binding proteins await characterization. In particular, the proteins that mediate Fab1p-dependent vacuolar acidification remain unknown. Although speculative, the vacuolar ATPase, which is known to regulate the acidification of vacuoles, is a likely candidate for this function. Presumably, there are also other $PI3,5P_2$ effectors, in addition to Atg18p, which participate in the retrograde transport of proteins from the vacuole to endosomes, since deletion of *ATG18* does not phenocopy the effects of *FAB1* deletion in this pathway (Efe et al. 2007). The limited production of $PI3,5P_2$ in yeast cells has hindered progress to identify key effectors of this lipid, but further genetic and biochemical studies will almost certainly uncover new proteins that harbor binding domains specific for $PI3,5P_2$.

2.4 Roles for PI4P in Membrane Transport

At least two non-redundant pools of PI4P are synthesized in yeast cells, one at the Golgi apparatus and a second at the plasma membrane. Each functions in multiple membrane trafficking pathways, involving both protein and lipid transport. The type IIIb PI 4-kinase Pik1p regulates PI4P production at the Golgi, and plays critical roles in maintaining secretory protein export to the cell surface, trafficking of cargoes to the vacuole, and endocytic protein transport (Hama et al. 1999; Walch-Solimena and Novick 1999; Audhya et al. 2000). The type IIIa PI 4-kinase controls PI4P synthesis at the plasma membrane and has also been implicated in endocytic trafficking, likely through its regulation of actin cytoskeleton organization (Audhya and Emr 2002; Tahirovic et al. 2005). Additionally, roles for PI4P extend to the endoplasmic reticulum, both in protein secretion from this compartment and aminophospholipid transport (Trotter et al. 1998; Lorente-Rodriguez and Barlowe 2011). Here, we will discuss the various roles of PI4P at different intracellular compartments, specifying effectors that have been identified thus far, which function in membrane transport.

2.4.1 Functions of PI4P Synthesized by Pik1p in the Golgi

A function for PI4P in secretion from the Golgi in yeast was initially suggested by studies that focused on the PI transfer protein Sec14p, which is essential for the biogenesis of Golgi transport vesicles (Kearns et al. 1997). In a screen for bypass suppressors of a *sec14* temperature sensitive allele, a mutant isoform of the Sac1p phosphoinositide phosphatase was identified. Loss of Sac1p activity led to an increase in PI4P that was necessary for *sec14* suppression, suggesting a role for this lipid in Golgi secretion. Consistent with this finding, overexpression of the PI 4-kinase *PIK1* partially restored the growth of *sec14* mutant cells at elevated temperature (Hama et al. 1999). Moreover, loss of Sec14p activity led to diminished production of PI4P, suggesting a role for Sec14p in regulating Pik1p activity. Since subsequent studies have demonstrated that *sac1* mutant cells accumulate PI4P mostly at the cell surface in an Stt4p-dependent manner, the precise nature of Sac1p-mediated Sec14p bypass remains unclear (Foti et al. 2001; Roy and Levine 2004). However, additional studies further confirmed an essential function for PI4P synthesis at the Golgi. Specifically, multiple *pik1* mutant alleles have been isolated, and each confers a defect in protein transport from the Golgi (Hama et al. 1999; Walch-Solimena and Novick 1999). In some cases, Golgi to plasma membrane secretion is specifically inhibited by loss of Pik1p activity, but other *pik1* mutant strains exhibit more severe defects in Golgi function that affect both secretory protein transport and the trafficking of biosynthetic cargoes from the Golgi to endosomes and the vacuole (Audhya et al. 2000; Sciorra et al. 2005; Lorente-Rodriguez and Barlowe 2011). Additionally, protein glycosylation in the Golgi is adversely affected by the absence of Pik1p activity, suggesting that multiple cisternae require PI4P synthesis for normal assembly and/or function (Audhya et al. 2000).

Several effectors of PI4P that function at the Golgi have been described in yeast. The first were members of the oxysterol binding protein (OSBP) family, which likely function in the transfer of lipids between biological membranes (Li et al. 2002). There are a total of 7 OSBPs in yeast, Osh1p-Osh7p, which share a common essential function, and each binds promiscuously to phosphoinositides (Beh et al. 2001; Schulz and Prinz 2007). Osh1p, Osh2p and Osh3p each harbor an amino-terminal PH domain, a known phosphoinositide interacting motif. In general, PH domains are composed of approximately 120 amino acids that share a common structure consisting of two perpendicular anti-parallel beta sheets, followed by a carboxyl-terminal amphipathic helix (Lemmon 2008). Although diverse in amino acid composition, a single tryptophan located within the helix serves to nucleate the core of the domain. A survey of more than 33 PH domains in yeast revealed that most bind to phosphoinositides with little specificity and low affinity, and proteins that harbor PH domains typically require additional interactions for proper intracellular targeting (Yu et al. 2004). By analyzing GFP fusions to the PH domain-containing OSBPs, Osh1p was shown to localize in part to the Golgi and is potentially involved in sterol transfer with other organelles (Levine and Munro 2001). In contrast, neither Osh2p nor Osh3p discernibly accumulated on Golgi membranes, although the PH domain of Osh2p alone binds to PI4P on the Golgi in a Pik1p-dependent fashion (Levine and Munro 2002). These data highlight the role of additional protein-protein interactions in specifying the distribution of PH domain containing proteins in yeast.

In the case of Osh4p, phosphoinositide binding is mediated by the conserved sterol binding domain, which is found in all OSBPs (Li et al. 2002). Studies indicate that Osh4p localizes to the Golgi in a PI4P-dependent fashion and may mediate the movement of sterols to and from this organelle, thereby regulating lipid homeostasis and vesicle biogenesis. Additionally, recent findings indicate that Osh4p negatively regulates Pik1p activity (Fairn et al. 2007). These data suggest that Osh4p may "sense" PI4P levels on Golgi membranes. For example, if a high concentration of PI4P accumulated on the Golgi, Osh4p recruitment would increase to inhibit Pik1p and slow PI4P synthesis. This type of feedback inhibition may be especially important to regulate secretion in response to changing growth conditions and environmental stress (Faulhammer et al. 2007).

In addition to members of the OSBP family, the gamma-ear-containing, ADP-ribosylation factor binding protein Gga2p has also been shown to interact with PI4P at the Golgi (Demmel et al. 2008). Gga2p functions as an adaptor for clathrin recruitment and participates in transport between the Golgi and endosomes (Black and Pelham 2000; Costaguta et al. 2001). Loss of Gga2p function phenocopied several of the morphological and secretory defects observed in *pik1* mutant cells, including the accumulation of abnormal, cup-shaped membranous structures in the cytoplasm termed "Berkeley bodies." These data suggest that Gga2p may be the most relevant Pik1p-dependent PI4P effector in Golgi to endosome trafficking (Demmel et al. 2008). PI4P binding is mediated by the VHS (Vps27p/Hrs/STAM) domain of Gga2p, which shares significant similarity to the structure of phosphoinositide-binding ANTH/ENTH domains (Demmel et al. 2008). In general, VHS domains contain approximately 150 amino acids and consist of 8 helices arranged in a super-helix (Mao et al. 2000). In Gga2p, the loop preceding helix 8 exhibits a pattern of charged and aromatic residues, similar to those found in the ANTH domain of CALM, a clathrin adaptor that interacts with $PI4,5P_2$ at the plasma membrane (Stahelin et al. 2003; Demmel et al. 2008). Mutations within this basic region of Gga2p disrupted its association with PI4P and diminished Gga2p association with the Golgi. However, a GFP fusion to the VHS domain of Gga2p showed that it alone was insufficient to localize to Golgi membranes. Instead, the neighboring Arf1p GTPase-interacting GAT domain was also required. Therefore, similar to most PH domain containing proteins in yeast, both lipid-protein and protein-protein interactions are required for proper Gga2p localization (Zhdankina et al. 2001; Demmel et al. 2008).

As noted earlier, protein glycosylation in the secretory pathway is sensitive to impaired Pik1p function, suggesting that PI4P may be required for the proper function or localization of Golgi glycosyltransferases. In many cases, retention of these enzymes in the Golgi requires the function of Vps74p, an oligomeric protein that binds to the cytosolic domains of glycosyltransferases and restricts their transport to other organelles (Schmitz et al. 2008; Tu et al. 2008). Based on crystallographic data, Vps74p harbors four alpha-helices that form a central core, which is surrounded by several additional amphipathic alpha-helices and four beta-strands (Schmitz et al. 2008). The recruitment of Vps74p to the Golgi is mediated by PI4P generated by Pik1p (Wood et al. 2009). In *pik1* mutant cells, Vps74p becomes cytosolic, and glycosyltransferases are no longer retained in the Golgi, thus leading to defects in secretory cargo glycosylation. The PI4P-binding motif within Vps74p is composed of a conserved basic region near the amino-terminus of helix 6, and mutations in this domain inhibit Vps74p-mediated retention of Golgi glycosyltransferases. Although the loss of other PI4P effectors may further contribute to the defects in protein glycosylation observed in *pik1* mutant cells, these data confirm a role for PI4P in this process and further demonstrate that Pik1p-mediated PI4P production is required for retrograde transport of cargoes in the Golgi (Wood et al. 2009).

Effectors of PI4P in yeast that function during secretory membrane transport have been arguably the most challenging to identify. Genetic studies strongly suggest that Pik1p functions together with multiple components of the secretory pathway, including at least three Rab-type GTPases (Ypt31p, Ypt32p, and Sec4p), two tethering complexes (TRAPPII and exocyst), components of the actin-mysoin network, and the phospholipid flippase Drs2p (Walch-Solimena and Novick 1999; Sciorra et al. 2005). In particular, Rab GTPases are known regulators of membrane trafficking, which have been shown to interface with phosphoinositide signaling during endosomal sorting (Zerial and McBride 2001). One attractive model for Golgi secretion involves a cascade of Rab activation, in which the Golgi-localized Ypt32p GTPase recruits the guanine nucleotide exchange factor (GEF) for the subsequent Rab GTPase that acts in the pathway (Mizuno-Yamasaki et al. 2010). Indeed, the active GTP-bound form of Ypt32p, generated by the GEF activity of the TRAPPII tethering complex, binds directly to Sec2p, which catalyzes GTP exchange on Sec4p that is present on secretory vesicles (Ortiz et al. 2002). Active Sec4p can then recruit components of the exocyst complex, which are necessary for vesicle fusion with the plasma membrane (Guo et al. 1999b). The Rab cascade is further regulated by Sec15p, a component of the exocyst complex, which competes with Ypt32p for Sec2p binding (Mizuno-Yamasaki et al. 2010). The precise mechanism by which Sec15p replaces Ypt32p remains unknown, but recent evidence implicates a role for PI4P in this switch. Specifically, Sec2p was found to interact directly with PI4P generated by Pik1p on Golgi membranes. Three basic patches within the Sec2p GEF were found to be important for PI4P binding *in vitro*, and a mutant isoform of Sec2p containing mutations in these regions failed to localize properly (Mizuno-Yamasaki et al. 2010). Both Ypt31p and Ypt32p were also found to be important for Sec2p localization, suggesting that a combination of protein-protein and lipid-protein interactions were necessary for proper Sec2p targeting. Consistent with this idea, Sec2p was shown to form a ternary complex with both PI4P and Ypt32p *in vitro*. In contrast, PI4P inhibited the association of Sec2p with Sec15p, suggesting a role for PI4P in stabilizing the association between Sec2p and Ypt32p at the Golgi to drive vesicle formation (Mizuno-Yamasaki et al. 2010). Subsequent to vesicle budding, Sec2p may catalyze the formation of active, GTP-bound Sec4p, initiating the recruitment of exocyst subunits including Sec15p, which ultimately displaces Ypt32p. Although this idea is speculative, the data clearly define a novel function for PI4P in secretion from the Golgi and help to explain several of the genetic interactions defined previously using *pik1* mutant cells.

In addition to the role of Rab-type GTPases, vesicle biogenesis from the Golgi also requires the action of Drs2p, a type IV P-type ATPase, which catalyzes the translocation of aminophospholipids from one leaflet of the lipid bilayer to the other. Similar to Pik1p, Drs2p has been implicated in the formation of vesicles destined for the plasma membrane as well as the endosomal/vacuolar system (Chen et al. 1999; Hua et al. 2002; Gall et al. 2002). Recently, the flippase activity of Drs2p was shown to be dependent on PI4P generated by Pik1p (Natarajan et al. 2009). A phosphoinositide binding site within Drs2p was identified and found to exhibit similarity to the PI3P-interacting split PH domain of Vps36p, a component of the ESCRT-II complex. A basic motif within this region (RMKKQR) was critical for PI4P binding *in vitro*, and mutations in this region prohibited complementation in *drs2* mutant cells. Unlike other roles for PI4P in targeting effectors to the Golgi, these data highlight a unique function for this lipid in regulating an enzymatic activity necessary for Golgi vesicle formation.

Since secretory vesicles move from the Golgi to the plasma membrane along actin cables in yeast, it is not surprising that mutations affecting actin cytoskeleton organization are lethal to *pik1* mutant cells (Pruyne et al. 2004; Walch-Solimena and Novick 1999). Further study into this connection has demonstrated that PI4P present in secretory vesicles is critical for vesicle movement mediated by the type V myosin Myo2p. Under normal conditions, directed vesicle movement also requires interactions between Myo2p and the Rab-type GTPases Ypt31p, Ypt32p, and Sec4p. However, by enhancing the association between Myo2p and PI4P, binding to the Rab-type GTPases becomes dispensable (Santiago-Tirado et al. 2011). Although Myo2p has not been demonstrated to interact directly with PI4P, these findings strongly suggest that coincidence detection of PI4P and Rab GTPases is important for myosin-dependent transport of secretory vesicles. In the future, it will be critical to understand the mechanism by which Myo2p recognizes vesicles containing PI4P.

While the majority of studies have focused on a role for Pik1p-mediated PI4P production at the trans Golgi network, a cell free assay used to study ER to Golgi transport uncovered a critical role for PI4P in this pathway. Specifically, the presence of PI4P in the cis-Golgi was found to be necessary for COPII vesicle fusion (Lorente-Rodriguez and Barlowe 2011). Although vesicle tethering did not require the presence of PI4P, inhibitors of PI4P reduced SNARE complex assembly. Analysis of *pik1* mutant cells further suggested a role for PI4P in anterograde transport between the ER and Golgi. However, a specific effector of PI4P at this early step of the secretory pathway remains unidentified. In a similar fashion, several studies have suggested a potential role for PI4P at the endosome, but confirmation of such a function awaits the characterization of a PI4P-binding protein, which regulates endosomal trafficking (Walch-Solimena and Novick 1999; Audhya et al. 2000).

2.4.2 Functions of PI4P Synthesized by Stt4p at the Plasma Membrane

Although a significant portion of PI4P generated by Stt4p is rapidly metabolized to $PI4,5P_2$ (discussed in the next section), the unique phenotypes of *stt4* mutant cells suggest that the plasma membrane pool of PI4P has specific effectors. Consistent with this finding, the Cla4p protein kinase, which is involved in actin cytoskeleton organization, polarized cell growth and cell division, harbors a PH domain that binds to PI4P generated by Stt4p (Eby et al. 1998; Wild et al. 2004). In cells lacking Stt4p kinase activity, Cla4p is mislocalized from the cell surface, suggesting that its interaction with PI4P is critical for its normal targeting. Like other yeast proteins that harbor PH domains, interaction with PI4P alone is insufficient to direct localization of Cla4p. However, together with another interacting protein, the Cdc42p Rho-type GTPase, Cla4p is able to maintain a polarized distribution at the plasma membrane (Wild et al. 2004). These data again highlight coincident roles for lipid and protein interactions to maintain the cellular distribution of phosphoinositide interacting molecules. Furthermore, these findings illustrate that PI4P directly regulates actin polarity by controlling Cla4p recruitment to the plasma membrane. Notably, in the absence of Stt4p function, directed membrane transport to the bud is disrupted by perturbations in actin organization, resulting in isotropic cell growth (Audhya et al. 2000). Similar phenotypes have been observed in cells lacking Cla4p and a related protein kinase, Ste20p (Holly and Blumer 1999). Although Ste20p has not been shown to interact with phosphoinositides, these are the first data demonstrating that PI4P generated by Stt4p exhibits functions in actin organization beyond its role as a precursor to Mss4p-mediated $PI4,5P_2$ synthesis.

In addition to its role at the plasma membrane, Stt4p has also been implicated in the transport of the phospholipid PS from the ER to the Golgi. Specifically, inhibition of Stt4p kinase activity leads to an accumulation of PS in the ER, which under normal conditions is metabolized to form phosphatidylethanolamine (PE) in the Golgi (Trotter et al. 1998). These data suggest that PI4P generated by Stt4p regulates the movement of phospholipids in the early secretory pathway, although an effector of PI4P in this process remains unknown. Additionally, these findings raise the possibility that PI4P generated by Stt4p at the plasma membrane can be transferred to the ER, potentially at sites of ER-plasma membrane contact. Such a process may be mediated by members of the OSBP family, which all exhibit the capacity to bind phosphoinositides. In particular, Osh3p appears to be specifically enriched at sites of ER-plasma membrane contact, and this localization is dependent on an interaction between its PH domain and PI4P (Stefan et al. 2011). Although a model in which OSBPs directly transfer phosphoinositides between organelles is highly speculative, it is supported by studies suggesting that PI4P generated by Stt4p can accumulate in the ER following inactivation of the Sac1p lipid phosphatase (Li et al. 2002; Wood et al. 2009). Further studies are clearly required to confirm or reject this possibility.

Beyond the putative role of OSBPs in lipid transfer, this family of phosphoinositide-binding proteins also regulates PI4P turnover. In mutant cells deficient in OSBP function, PI4P levels rise substantially, similar to the phenotype exhibited by cells lacking Sac1p phosphatase activity (Stefan et al. 2011). Furthermore, OSBPs have been shown to activate the phosphatase activity of Sac1p *in vitro*. Thus, PI4P autoregulates its stability by recruiting effectors that increase PI4P hydrolysis. This type of feedback inhibition may be necessary to restrict PI4P signaling on the plasma membrane and prevent the mistargeting of PI4P binding proteins required for function downstream of Pik1p at the Golgi.

As mentioned earlier, Osh3p is an effector of PI4P generated on the plasma membrane by Stt4p and localizes to sites of ER-plasma membrane contact (Stefan et al. 2011). Since PI4P accumulates homogenously on the cell surface, these data suggest that Osh3p distribution is also regulated by additional interactions. Consistent with this idea, two ER membrane proteins that regulate phospholipid biosynthesis, Scs2p and Scs22p, also bind to Osh3p and control its accumulation at sites of ER-plasma membrane contact (Stefan et al. 2011). Notably, Scs2p has been shown to bind PI4P *in vitro*, and mutations that disrupt its ability to associate with phosphoinositides diminish its function in regulating the metabolism of ER phospholipids (Kagiwada and Hashimoto 2007). Together, these data suggest that PI4P generated by Stt4p plays an important role in generating and/or stabilizing contact sites between the plasma membrane and ER through the recruitment of multiple effector proteins. These contact sites potentially allow for the direct transfer of lipids between these distinct organelles, allowing for rapid changes in membrane content independently of vesicular transport. Further studies are necessary to determine whether inactivation of Stt4p may lead to the disruption of ER-plasma membrane contact and what affect this has on cell growth and viability.

Characterization of *stt4* mutant cells also revealed a role for Stt4p-mediated production of PI4P in vacuolar membrane homeostasis. Following loss of Stt4p activity, vacuoles undergo a dramatic reduction in volume, while their overall surface area does not appear to change significantly (Audhya et al. 2000). One possibility is that PI4P generated at the plasma membrane is required for controlling cellular osmolarity, and defects in PI4P production lead to changes in osmotic pressure within the cell that cause vacuoles to lose volume. Importantly, such an effect is not seen following loss of Mss4p-mediated PI4,5P$_2$ production, indicating that PI4P does not simply act as a precursor lipid in this pathway. In the future, it will be important to define specific effectors of PI4P, which control cellular osmolarity and/or vacuole size and shape.

2.5 Roles for PI4,5P$_2$ in Membrane Transport

Although the single yeast PI4P 5-kinase Mss4p undergoes nucleo-cytoplasmic shuttling, studies indicate that the essential cellular pool of PI4,5P$_2$ is synthesized on the plasma membrane (Audhya and Emr 2003). At this location, PI4,5P$_2$ regulates a number of distinct processes, which include endocytosis, exocytosis, cytokinesis, maintenance of cell polarity, and actin cytoskeleton organization. In several cases, PI4,5P$_2$ functions as a localization determinant. However, many effectors rely on PI4,5P$_2$ binding to regulate their activities directly. We will discuss a variety of mechanisms by which PI4,5P$_2$ modulates cellular function, with an emphasis on its diverse roles in membrane trafficking.

During endocytosis, a number of different factors must be recruited to the cell surface in a coordinated fashion to drive membrane invagination and scission. In the case of clathrin-mediated endocytosis, adaptor proteins arrive soon after clathrin marks an endocytic site (Liu et al. 2010). Several of these factors require the presence of $PI4,5P_2$ to associate with the plasma membrane. For example, the alpha subunit of the mammalian AP-2 adaptor protein complex has been shown to bind $PI4,5P_2$ through a conserved amino-terminal basic region, and this interaction is required for its localization (Collins et al. 2002). Although the yeast AP-2 alpha subunit Apl3p exhibits only 30% sequence identity with its human homolog, the basic residues within the amino-terminus are almost entirely conserved, suggesting that Apl3p also binds $PI4,5P_2$. However, in contrast to human AP-2, the role of the yeast complex is less clear as mutant cells lacking all AP-2 adaptor subunits fail to exhibit significant defects in clathrin-mediated endocytosis (Huang et al. 1999; Yeung et al. 1999). These data suggest that alternative proteins likely function in a redundant manner to the AP-2 complex in yeast.

The epsins, Ent1p and Ent2p, are adaptor proteins, which may fulfill such a function. Unlike AP-2, epsin function is essential for yeast cell viability, and specifically plays an important role during clathrin-mediated endocytosis (Wendland et al. 1999). Both Ent1p and Ent2p harbor ENTH (epsin N-terminal homology) domains, which bind to $PI4,5P_2$ (Aguliar et al. 2003). In general, ENTH domains are approximately 150 amino acids in length and are composed of 9 alpha helices connected by loops of varying sizes. Based on the structure of the ENTH domain from mammalian epsin, $PI4,5P_2$ interacts with basic amino acids in loop 1 and helices 3 and 4 (Itoh et al. 2001). The conservation of this region in Ent1p and Ent2p strongly suggests a common mechanism for phosphoinositide binding. Additionally, genetic studies indicate that the ENTH domain from either Ent1p or Ent2p is sufficient to complement deletions of both genes (Aguliar et al. 2006). These data suggest additional functions beyond phosphoinositide binding for this region. Consistent with this idea, the ENTH domain has been found to be a coincident detector of both $PI4,5P_2$ and GEFs for the Rho-type GTPase Cdc42p, which play an important role in polarized cell growth (Aguliar et al. 2006). Thus, a combination of protein-protein and protein-lipid interactions coordinates spatial and temporal regulation of endocytosis with maintenance of cell polarity.

Yeast cells also express two epsin-related proteins, Yap1801p and Yap1802p, which harbor ANTH (AP180 N-terminal homology) domains. The ANTH domain exhibits a similar overall structure to the ENTH domain, containing 9 helices connected by loops (Legendre-Guillemin et al. 2004). Additionally, a $PI4,5P_2$ binding site within the ANTH domain has been defined (Stahelin et al. 2003), and contains the conserved sequence $Kx_9Kx(K/R)(H/Y)$. Although initial studies failed to identify endocytic defects in cells lacking Yap1801p and Yap1802p, subsequent findings indicated that the adaptor proteins possess cargo specific roles in endocytosis (Wendland and Emr 1998; Huang et al. 1999; Burston et al. 2009). These data further implicate $PI4,5P_2$ in regulating endocytic trafficking in yeast.

In addition to Yap1801p and Yap1802p, a third ANTH domain containing protein Sla2p has been implicated in endocytosis. Furthermore, Sla2p also functions to regulate actin organization, connecting membrane transport from the cell surface to the underlying cytoskeleton (Wesp et al. 1997). The conserved basic region within the Sla2p ANTH domain was shown to be critical for $PI4,5P_2$ interaction, and mutations in this region perturbed endocytic function in cells otherwise lacking Sla2p (Sun et al. 2005). However, loss of the basic region failed to impact localization of Sla2p, indicating that PI4,5P_2 is not essential for targeting of Sla2p to the plasma membrane. Instead, PI4,5P_2 likely regulates the activity of Sla2p during endocytosis. Since Sla2p binds to components of both the clathrin coat and the actin cytoskeleton, modulations in Sla2p function by PI4,5P_2 may be critical to appropriately couple vesicle formation with actin-dependent vesicle internalization. However, further studies are necessary to pinpoint the precise consequence of PI4,5P_2 interaction with the Sla2p ANTH domain.

The intimate connection between endocytosis and cytoskeletal organization implicates several additional PI4,5P_2 binding proteins, which modulate actin assembly, in membrane trafficking. For example in mammalian cells, the actin severing protein gelsolin is rapidly inactivated in the presence of PI4,5P_2, while PI4,5P_2 binding to the basic motif in WASP family members stimulates actin polymerization mediated by the Arp2/3 complex (Janmey and Stossel 1987; Rohatgi et al. 2000; Papayannopoulos et al. 2005). Thus, using several mechanisms, PI4,5P_2 accelerates the formation of actin networks, which play a key role in endocytic vesicle budding. In total, more than 30 actin filament binding proteins have been reported to associate directly with phosphoinositides (Yin and Janmey 2003).

Similar to gelsolin, PI4,5P_2 also exhibits an inhibitory effect on the actin depolymerizing factor (ADF)/cofilin known as Cof1p in yeast. In the absence of PI4,5P_2, Cof1p promotes rapid actin dynamics by driving the severing of actin filaments, but its ability to associate with actin and its depolymerizing activity is strongly diminished by PI4,5P_2 (Carlier et al. 1997; Gorbatyuk et al. 2006). Based on a series of *cof1* mutant alleles, a highly conserved basic helix in Cof1p, as well as other positively charged residues found throughout the molecule, appear to be important for PI4,5P_2 binding (Ojala et al. 2001). These findings suggest that Cof1p may bind multiple PI4,5P_2 molecules simultaneously. However, work using ADF/cofilin from another organism indicated that PI4,5P_2 binding was mediated by the carboxyl-terminus of the protein (Kusano et al. 1999). Additional work is necessary to define the precise mechanism by which PI4,5P_2 binds cofilin and inhibits its activity.

Actin polymerization can also be terminated by the function of capping protein, an evolutionarily conserved, heterodimeric complex consisting of alpha and beta subunits, encoded by *CAP1* and *CAP2* in yeast. Through its association with the ends of actin filaments, capping protein inhibits further actin dynamics (Saarikangas et al. 2010). Both subunits contain highly basic residues that are critical for PI4,5P_2 binding, and association with PI4,5P_2 interferes with the interaction between capping protein and actin, promoting actin dynamics necessary for endocytosis (Kim et al. 2007). The importance of the interaction between capping protein and PI4,5P_2 is further highlighted by work indicating that a partial loss of function *mss4* allele exhibits synthetic lethality with deletion of either *CAP1* or *CAP2* (Audhya et al. 2004).

Profilin, encoded by *PFY1* in yeast, is another PI4,5P_2 binding protein that interacts specifically with monomeric actin. Under certain cellular conditions, profilin can act as an actin monomer sequestering molecule, inhibiting actin filament assembly (Witke 2004). Interaction with PI4,5P_2 disrupts the association of profilin with actin, thus inhibiting the effect of profilin on actin polymerization. Several regions of profilin have been implicated in $PI4,5P_2$ binding, and the interaction appears to be electrostatic in nature (Lassing and Lindberg 1985; Richer et al. 2008). In a similar fashion, $PI4,5P_2$ also inhibits the actin monomer sequestering activity of the yeast twinfilin Twf1p (Palmgren et al. 2001). Thus, by maintaining free pools of monomeric actin, the presence of $PI4,5P_2$ generates an environment conducive to actin polymerization and endocytosis.

In several cases, $PI4,5P_2$ regulates cytoskeletal organization through effectors that do not bind actin directly. For example, the Rho-type GTPase Rho1p is required for polarity of the actin cytoskeleton in yeast, as well as polarized secretion (Levin 2005). The Rho1p GEF Rom2p harbors a PH domain that has been shown to interact with $PI4,5P_2$ (Audhya and Emr 2002). Inhibition of Stt4p or Mss4p, which diminish $PI4,5P_2$ synthesis on the plasma membrane, results in a defect in Rom2p localization and leads to depolarization of the actin cytoskeleton, suggesting that phosphoinositide signaling regulates Rho1p activity. Furthermore, a polybasic sequence within Rho1p, which plays a role in targeting the GTPase to sites of polarized growth independent of its GEFs, has also been shown to interact with $PI4,5P_2$ (Yoshida et al. 2009). Together, these findings demonstrate that phosphoinositide and Rho-type GTPase signaling are tightly coupled.

Mss4p-dependent $PI4,5P_2$ synthesis is also required for the normal localization of Slm1p and Slm2p, two additional regulators of actin organization in yeast. Both Slm1p and Slm2p contain PH domains capable of interacting with $PI4,5P_2$, and mutations in the Slm1p PH domain that inhibit phosphoinositide interaction result in the mislocalization of Slm1p to the cytoplasm (Audhya et al. 2004; Fadri et al. 2005). Although the precise mechanism underlying Slm1p and Slm2p regulation of actin cytoskeleton organization remains unknown, these data further highlight the diversity of $PI4,5P_2$ effectors that control actin polarity. Furthermore, Slm1p and Slm2p also play roles in sphingolipid biosynthesis, which function in several membrane trafficking pathways including endocytosis (Tabuchi et al. 2006). These data suggest that phosphoinositide and sphingolipid signaling likely intersect, potentially to coordinate vesicle transport with membrane homeostasis.

In addition to its role in endocytosis, the actin cytoskeleton also performs a key function in directed protein secretion and exocytosis. As discussed earlier, PI4P generated on the Golgi by Pik1p plays several important roles in generating secretory vesicles, which move along actin cables to sites of polarized growth. Fusion of vesicles with the plasma membrane requires the activity of a tethering complex known as the exocyst. Of the 8 exocyst subunits, two have been shown to interact with $PI4,5P_2$ and direct their localization to the plasma membrane, Exo70p and Sec3p (He et al. 2007; Zhang et al. 2008). In the case of Exo70p, $PI4,5P_2$ binding is mediated by a basic patch of amino acids in the carboxyl-terminus. Similarly, a basic region in the amino-terminus of Sec3p, which has been shown to adopt a PH domain fold, also binds to $PI4,5P_2$. Consistent with a role for $PI4,5P_2$ in regulating exocyst localization on the cell surface, inhibition of Mss4p activity led to the accumulation of the exocyst complex in the cytoplasm. Notably, unlike the uniform distribution of $PI4,5P_2$ on the plasma membrane, the exocyst complex is highly polarized, suggesting additional interactions are necessary for proper exocyst distribution. Indeed, several Rho-type GTPases, which localize in an asymmetric fashion, bind to components of the exocyst, and these interactions also play an important role in recruiting the complex to the plasma membrane (Zhang et al. 2008). Taken together, these data again highlight the importance of dual targeting mechanisms, which combine protein and lipid signals, in determining the localization of phosphoinositide effectors in cells.

Finally, $PI4,5P_2$ has also been implicated in the final stages of cell division in yeast by regulating several components of the septin complex, which orchestrate membrane remodeling events necessary for cytokinesis. The septins Cdc3p, Cdc10p, Cdc11p, and Cdc12p have been demonstrated to form filaments that can function as a diffusion barrier as cells divide (Weirich et al. 2008). Studies indicate that each septin harbors a basic motif capable of binding to $PI4,5P_2$. Mutations in basic residues found in Cdc3p, which mediate $PI4,5P_2$ binding, cause defects in cell division (Votin et al. 2009). *In vitro*, $PI4,5P_2$ promotes septin filament assembly and organization, suggesting that protein-lipid interactions in this case function beyond membrane recruitment (Bertin et al. 2008, 2010). Further work is necessary to understand how phosphoinositide binding may affect septin dynamics during the process of cell division.

2.6 Perspectives and Conclusions

The recognized roles of phosphoinositides in membrane transport pathways have grown significantly over the last two decades. Although many more effectors likely await characterization, it has become clear that phosphoinositide signaling plays key roles in the movement of cargoes between virtually all cellular compartments. Two major themes have emerged from the study of PI metabolism. First, the localization of phosphoinositides is highly regulated, and their distributions rarely overlap. Thus, phosphoinositides function as organelle specific labels, enabling membrane compartments to be distinguished from one another and preventing crosstalk between distinct biochemical processes within a common cytoplasm. Second, phosphoinositides infrequently function alone, and instead regulate cellular pathways in concert with other factors. Such a mechanism provides the high specificity necessary for directed membrane transport between organelles. In the future, the diversity of phosphoinositide effectors will likely continue to grow, hopefully shedding additional light on the complex mechanisms by which membrane trafficking is appropriately coordinated.

References

Aguliar RC, Watson HA, Wendland B (2003) The yeast Ent1 is recruited to membranes through multiple independent interactions. J Biol Chem 278:10737–10743

Aguliar RC, Longhi SA, Shaw JD, Yeh LY, Kim S, Schon A, Freire E, Hsu A, McCormick WK, Watson HA, Wendland B (2006) Epsin N-terminal homology domains perform an essential function regulating Cdc42 through binding Cdc42 GTPase-activating proteins. Proc Natl Acad Sci U S A 103:4116–4121

Audhya A, Emr SD (2002) Stt4 PI 4-kinase localizes to the plasma membrane and functions in the Pkc1-mediated MAP kinase cascade. Dev Cell 2:593–605

Audhya A, Emr SD (2003) Regulation of PI4,5P2 synthesis by nuclear-cytoplasmic shuttling of the Mss4 lipid kinase. EMBO J 22:4223–4236

Audhya A, Foti M, Emr SD (2000) Distinct roles for the yeast phosphatidylinositol 4-kinases, Stt4p and Pik1p, in secretion, cell growth, and organelle membrane dynamics. Mol Biol Cell 11:2673–2689

Audhya A, Loewith R, Parsons AB, Gao L, Tabuchi M, Zhou J, Boone C, Hall MN, Emr SD (2004) Genome-wide lethality screen identifies new PI4,5P2 effectors that regulate the actin cytoskeleton. EMBO J 23:3747–3757

Baird D, Stefan C, Audhya A, Weys S, Emr SD (2008) Assembly of the PtdIns 4-kinase Stt4 complex at the plasma membrane require Ypp1 and Efr3. J Cell Biol 183:1061–1074

Barth H, Meiling-Wesse K, Epple UD, Thumm M (2001) Autophagy and the cytoplasm to vacuole targeting pathway both require Aut10p. FEBS Lett 508:23–28

Beh CT, Cool L, Phillips J, Rine J (2001) Overlapping functions of the yeast oxysterol-binding protein homologues. Genetics 157:1117–1140

Bertin A, McMurray MA, Grob P, Park SS, Garcia G, Patanwala I, Ng HL, Alber T, Thorner J, Nogales E (2008) Saccharomyces cerevisiae septins: supramolecular organization of heterooligomers and the mechanism of filament assembly. Proc Natl Acad Sci U S A 105:8274–8279

Bertin A, McMurray MA, Thai L, Garcia G, Votin V, Grob P, Allyn T, Thorner J, Nogales E (2010) Phosphatidylinositol-4,5-bisphosphate promotes budding yeast septin filament assembly and organization. J Mol Biol 404:711–731

Bilodeau PS, Winistorfer SC, Kearney WR, Robertson AD, Piper RC (2003) Vps27-Hse1 and ESCRT-I complexes cooperate to increase efficiency of sorting ubiquitinated proteins at the endosome. J Cell Biol 163:237–243

Black MW, Pelham HR (2000) A selective transport route from Golgi to late endosomes that requires the yeast GGA proteins. J Cell Biol 151:587–600

Bonangelino CJ, Catlett NL, Weisman LS (1997) Vac7p, a novel vacuolar protein, is required for normal vacuole inheritance and morphology. Mol Cell Biol 17:6847–6858

Botelho RJ, Efe JA, Teis D, Emr SD (2008) Assembly of a Fab1 phosphoinositide kinase signaling complex requires the Fig4 phosphoinositide phosphatase. Mol Biol Cell 19:4273–4286

Bravo J, Karathanassis D, Pacold CM, Pacold ME, Ellson CD, Anderson KE, Butler PJ, Lavenir I, Perisic O, Hawkins PT, Stephens L, Williams RL (2001) The crystal structure of the PX domain from p40(phox) bound to phosphatidylinositol 3-phosphate. Mol Cell 8:829–839

Burd CG, Emr SD (1998) Phosphatidylinositol(3)-phosphate signaling mediated by specific binding to RING FYVE domains. Mol Cell 2:157–162

Burda P, Padilla SM, Sarkar S, Emr SD (2002) Retromer function in endosome-to-Golgi retrograde transport is regulated by the yeast Vps34 PtdIns 3-kinase. J Cell Sci 115:3889–3900

Burman C, Ktistakis NT (2010) Regulation of autophagy by phosphatidylinositol 3-phosphate. FEBS Lett 584:1302–1312

Burston HE, Maldonado-Baez L, Davey M, Montpetit B, Schluter C, Wendland B, Conibear E (2009) Regulators of yeast endocytosis identified by systematic quantitative analysis. J Cell Biol 185:1097–1110

Carlier MF, Laurent V, Santolini J, Melki R, Didry D, Xia GX, Hong Y, Chua NH, Pantaloni D (1997) Actin depolymerizing factor (ADF/cofilin) enhances the rate of filament turnover: implication in actin-based motility. J Cell Biol 136:1307–1322

Chang FS, Han GS, Carman GM, Blumer KJ (2005) A WASP-binding type II phosphatidylinositol 4-kinase required for actin polymerization-driven endosome motility. J Cell Biol 171:133–142

Cheever ML, Sato TK, Beer T de, Kutateladze TG, Emr SD, Overduin M (2001) Phox domain interaction with PtdIns(3)P targets the Vam7 tSNARE to vacuole membranes. Nat Cell Biol 3:613–618

Chen CY, Ingram MF, Rosal PH, Graham TR (1999) Role for Drs2p, a P-type ATPase and potential aminophospholipid translocase, in yeast late Golgi function. J Cell Biol 147:1223–1236

Collins BM, McCoy AJ, Kent HM, Evans PR, Owen DJ (2002) Molecular architecture and functional model of the endocytic AP2 complex. Cell 109:523–535

Cooke FT, Dove SK, McEwen RK, Painter G, Holmes AB, Hall MN, Michell RH, Parker PJ (1998) The stress-activated phosphatidylinositol 3-phosphate 5-kinase Fab1p is essential for vacuole function in S. cerevisiae. Curr Biol 8:1219–1222

Cooke FT, Michell RH, Parker PJ, Lemmon MA (2004) Svp1p defines a family of phosphatidylinositol 3,5-bisphosphate effectors. EMBO J 23:1922–1933

Copic A, Starr TL, Schekman R (2007) Ent3p and Ent5p exhibit cargo-specific functions in trafficking proteins between the trans-Golgi network and the endosomes in yeast. Mol Biol Cell 18:1803–1815

Costaguta G, Stefan CJ, Bensen ES, Emr SD, Payne GS (2001) Yeast Gga coat proteins function with clathrin in Golgi to endosome transport. Mol Biol Cell 12:1885–1896

Costaguta G, Duncan MC, Fernandez GE, Huang GH, Payne GS (2006) Distinct roles for TGN/endosome epsin-like adaptors Ent3p and Ent5p. Mol Biol Cell 17:3907–3920

Demmel L, Gravert M, Ercan E, Habermann B, Muller-Reichert T, Kukhtina V, Baust T, Sohrmann M, Kalaidzidis Y, Klose C, Beck M, Peter M, Walch-Solimena C (2008) The clathrin adaptor Gga2p is a phosphatidylinositol 4-phosphate effector at the Golgi exit. Mol Biol Cell 19:1991–2002

Desrivieres S, Cooke FT, Morales-Johansson H, Parker PJ, Hall MN (2002) Calmodulin controls organization of the actin cytoskeleton via regulation of phosphatidylinositol (4,5)-bisphosphate synthesis in Saccharomyces cerevisiae. Biochem J 15:945–951

Diraviyam K, Stahelin RV, Cho W, Murray D (2003) Computer modeling of the membrane interaction of FYVE domains. J Mol Biol 328:721–736

Dove SK, Johnson ZE (2007) Our FABulous VACation: a decade of phosphatidylinositol 3,5-bisphosphate. Biochem Soc Symp 74:129–139

Dove SK, Piper RC, McEwen RK, Yu JW, King MC, Hughes DC, Thuring J, Holmes AB, Bilodeau PS, Windistorfer SC, Kearney WR, Robertson AD, Piper RC (2003) Vps27-Hse1 and ESCRT-I complexes cooperate to increase efficiency of sorting ubiquitinated proteins at the endosome. J Cell Biol 163:237–243

Dumas JJ, Merithew E, Sudharshan E, Rajamani D, Hayes S, Lawe D, Corvera S, Lambright DG (2001) Multivalent endosome targeting by homodimeric EEA1. Mol Cell 8:947–958

Eby JJ, Holly SP, Drogen F van, Grishin AV, Peter M, Drubin DG, Blumer KJ (1998) Actin cytoskeleton organization regulated by the PAK family of protein kinases. Curr Biol 8:967–970

Efe JA, Botelho RJ, Emr SD (2005) The Fab1 phosphatidylinositol kinase pathway in the regulation of vacuole morphology. Curr Opin Cell Biol 17:402–408

Efe JA, Botelho RJ, Emr SD (2007) Atg18 regulates organelle morphology and Fab1 kinase activity independent of its membrane recruitment by phosphatidylinositol 3,5-bisphosphate. Mol Biol Cell 18:4232–4244

Eugster A, Pecheur E, Michel F, Winsor B, Letourneur F, Friant S (2004) Ent5p is required with Ent3p and Vps27p for ubiquitin-dependent protein sorting into the multivesicular body. Mol Biol Cell 15:3031–3041

Ekena K, Stevens TH (1995) The Saccharomyces cerevisiae MVP1 gene interacts with VPS1 and is required for vacuolar protein sorting. Mol Cell Biol 15:1671–1678

Fadri M, Daquinag A, Wang S, Xue T, Kunz J (2005) The pleckstrin homology domain proteins Slm1 and Slm2 are required for actin cytoskeleton organization in yeast and bind phosphatidylinositol-4,5-bisphosphate and TORC2. Mol Biol Cell 16:1883–1900

Fairn GD, Curwin AJ, Stefan CJ, McMaster CR (2007) The oxysterol binding protein Kes1p regulates Golgi apparatus phosphatidylinositol-4-phosphate function. Proc Natl Acad Sci U S A 104:15352–15357

Farre JC, Krick R, Subramani S, Thumm M (2009) Turnover of organelles by autophagy in yeast. Curr Opin Cell Biol 21:522–530

Faulhammer F, Konrad G, BranKatschk B, Tahirovic S, Knodler A, Mayinger P (2005) Cell growth-dependent coordination of lipid signaling and glycosylation is mediated by interactions between Sac1p and Dpm1p. J Cell Biol 168:185–191

Faulhammer F, Kanjilal-Kolar S, Knodler A, Lo J, Lee Y, Konrad G, Mayinger P (2007) Growth control of Golgi phosphoinositides by reciprocal localization of sac1 lipid phosphatase and pik1 4-kinase. Traffic 8:1554–1567

Flick JS, Thorner J (1993) Genetic and biochemical characterization of a phosphatidylinositol-specific phospholipase C in Saccharomyces cerevisiae. Mol Cell Biol 13:5861–5876

Foti M, Audhya A, Emr SD (2001) Sac1 lipid phosphatase and Stt4 phosphatidylinositol 4-kinase regulate a pool of phosphatidylinositol 4-phosphate that functions in the control of the actin cytoskeleton and vacuole morphology. Mol Biol Cell 12:2396–2411

Friant S, Pecheur E, Eugster A, Michel F, Lefkir Y, Nourrisson D, Letourneur F (2003) Ent3p is a PtdIns(3,5)P2 effector required for protein sorting to the multivesicular body. Dev Cell 5:499–511

Gall WE, Geething NC, Hua Z, Ingram MF, Liu K, Chen SI, Graham TR (2002) Drs2p-dependent formation of exocytic clathrin-coated vesicles in vivo. Curr Biol 12:1623–1627

Garcia-Bustos JF, Marini F, Stevenson I, Frei C, Hall MN (1994) PIK1, an essential phosphatidylinositol 4-kinase associated with the yeast nucleus. EMBO J 13:2352–2361

Gary JD, Wurmser AE, Bonangelino CJ, Weisman LS, Emr SD (1998) Fab1p is essential for PtdIns(3)P 5-kinase activity and the maintenance of vacuolar size and membrane homeostasis. J Cell Biol 143:65–79

Gary JD, Sato TK, Stefan CJ, Bonangelino CJ, Weisman LS, Emr SD (2002) Regulation of Fab1 phosphatidylinositol 3-phosphate 5-kinase pathway by Vac7 protein and Fig4, a polyphosphoinositide phosphatase family member. Mol Biol Cell 13:1238–1251

Gaullier JM, Simonsen A, D'Arrigo A, Bremnes B, Stenmark H, Aasland R (1998) FYVE fingers bind PtdIns(3)P. Nature 394:432–433

Gorbatyuk VY, Nosworthy NJ, Robson SA, Bains NP, Maciejewski MW, Dos Remedious CG, King GF (2006) Mapping the phosphoinositide-binding site on chick cofilin explains how PIP2 regulates the cofilin-actin interaction. Mol Cell 24:511–522

Guo S, Stolz LE, Lemrow SM, York JD (1999a) SAC1-like domains of yeast SAC1, INP52, and INP53 and of human synaptojanin encode polyphosphoinositide phosphatases. J Biol Chem 274:12990–12995

Guo W, Roth D, Walch-Solimena C, Novick P (1999b) The exocyst is an effector of Sec4p, targeting secretory vesicles to sites of exocytosis. EMBO J 18:1071–1080

Hama H, Schnieders EA, Thorner J, Takemoto JY, DeWald DB (1999) Direct involvement of phosphatidylinositol 4-phosphate in secretion in the yeast Saccharomyces cerevisiae. J Biol Chem 274:34294–34300

Han GS, Audhya A, Markley DJ, Emr SD, Carman GM (2002) The Saccharomyces cerevisiae LSB6 gene encodes phosphatidylinositol 4-kinase activity. J Biol Chem 277:47709–47718

He C, Klionsky DJ (2009) Regulation mechanisms and signaling pathways of autophagy. Annu Rev Genet 43:67–93

He B, Xi F, Zhang X, Zhang J, Guo W (2007) Exo70 interacts with phospholipids and mediates the targeting of the exocyst to the plasma membrane. EMBO J 26:4053–4065

He J, Vora M, Haney RM, Filonov GS, Musselman CA, Burd CG, Kutateladze AG, Verkhusha VV, Stahelin RV, Kutateladze TG (2009) Membrane insertion of the FYVE domain is modulated by pH. Proteins 76:852–860

Hendricks KB, Wang BQ, Schnieders EA, Thorner J (1999) Yeast homologue of neuronal frequenin is a regulator of phosphatidylinositol-4-OH kinase. Nat Cell Biol 1:234–241

Herman PK, Stack JH, DeModena JA, Emr SD (1991) A novel protein kinase homolog essential for protein sorting to the yeast lysosome-like vacuole. Cell 64:425–437

Hettema EH, Lewis MJ, Black MW, Pelham HR (2003) Retromer and the sorting nexins Snx4/41/42 mediate distinct retrieval pathways from yeast endosomes. EMBO J 22:548–557

Holly SP, Blumer KJ (1999) PAK-family kinases regulate cell and actin polarization throughout the cell cycle of Saccharomyces cerevisiae. J Cell Biol 147:845–856

Homma K, Terui S, Minemura M, Qadota H, Anraku Y, Kanaho Y, Ohya Y (1998) Phosphatidylinositol-4-phosphate 5-kinase localized on the plasma membrane is essential for yeast cell morphogenesis. J Biol Chem 273:15770–15786

Hua Z, Fatheddin P, Graham TR (2002) An essential subfamily of Drs2p-related P-type ATPases is required for protein trafficking between Golgi complex and endosomal/vacuolar system. Mol Biol Cell 13:3162–3177

Huang WP, Klionsky DJ (2002) Autophagy in yeast: a review of the molecular machinery. Cell Struct Funct 27:409–420

Huang KM, D'Hondt K, Riezman H, Lemmon SK (1999) Clathrin functions in the absence of heterotetrameric adaptors and AP180-related proteins in yeast. EMBO J 18:3897–3908

Huh WK, Falvo JV, Gerke LC, Carroll AS, Howson RW, Weissman JS, O'Shea EK (2003) Global analysis of protein localization in budding yeast. Nature 425:686–691

Im YJ, Hurley JH (2008) Integrated structural model and membrane targeting mechanism of the human ESCRT-II complex. Dev Cell 14:902–913

Itoh T, Koshiba S, Kigawa T, Kikuchi A, Yokoyama S, Takenawa T (2001) Role of the ENTH domain in phosphatidylinositol-4,5-bisphosphate binding and endocytosis. Science 291:1047–1051

Janmey PA, Stossel TP (1987) Modulation of gelsolin function by phosphatidylinositol 4,5-bisphosphate. Nature 325:362–364

Kagiwada S, Hashimoto M (2007) The yeast VAP homolog Scs2p has a phosphoinositide-binding ability that is correlated with its activity. Biochem Biophys Res Commun 364:870–876

Kametaka S, Okano T, Ohsumi M, Ohsumi Y (1998) Apg14p and Apg6p/Vps30p form a protein complex essential for autophagy in the yeast, Saccharomyces cerevisiae. J Biol Chem 273:22284–22291

Katzmann DJ, Stefan CJ, Babst M, Emr SD (2003) Vps27 recruits ESCRT machinery to endosomes during MVB sorting. J Cell Biol 162:413–423

Kearns BG, McGee TP, Mayinger P, Gedvilaite A, Phillips SE, Kagiwada S, Bankaitis VA (1997) Essential role for diacylglycerol in protein transport from the yeast Golgi complex. Nature 387:101–105

Kihara A, Noda T, Ishihara N, Ohsumi Y (2001) Two distinct Vps34 phosphatidylinositol 3-kinase complexes function in autophagy and carboxypeptidase Y sorting in Saccharomyces cerevisiae. J Cell Biol 152:519–530

Kim K, McCully ME, Bhattacharya N, Butler B, Sept D, Cooper JA (2007) Structure/function analysis of the interaction of phosphatidylinositol 4,5-bisphosphate with actin-capping protein: implications for how capping protein binds the actin filament. J Biol Chem 282:5871–5879

Krick R, Tolstrup J, Appelles A, Henke S, Thumm M (2006) The relevance of the phosphatidylinositol phosphate-binding motif FRRGT of Atg18 and Atg21 for the CVT pathway and autophagy. FEBS Lett 580:4632–4638

Krick R, Henke S, Tolstrup J, Thumm M (2008) Dissecting the localization and function of Atg18, Atg21, and Ygr223c. Autophagy 4:896–910

Kusano K, Abe H, Obinata T (1999) Detection of a sequence involved in actin-binding and phosphoinositide-binding in the N-terminal side of cofilin. Mol Cell Biochem 190:133–141

Kutateladze TG (2006) Phosphatidylinositol 3-phosphate recognition and membrane docking by the FYVE domain. Biochim Biophys Acta 1761:868–877

Kutateladze TG (2007) Mechanistic similarities in docking of the FYVE and PX domains to phosphatidylinositol 3-phosphate containing membranes. Prog Lipid Res 46:315–327

Kutateladze TG, Ogburn KD, Watson WT, Beer T de, Emr SD, Burd CG, Overduin M (1999) Phosphatidylinositol 3-phosphate recognition by the FYVE domain. Mol Cell 3:805–811

Kutateladze TG, Capelluto DG, Ferguson CG, Cheever ML, Kutateladze AG, Prestwich GD, Overduin M (2004) Multivalent mechanism of membrane insertion by the FYVE domain. J Biol Chem 279:3050–3057

Lassing I, Lindberg U (1985) Specific interaction between phosphatidylinositol 4,5-bisphosphate and profilactin. Nature 314:472–474
Lee SA, Eyeson R, Cheever ML, Geng J, Verkhusha VV, Burd C, Overduin M, Kutateladze TG (2005) Targeting of the FYVE domain to endosomal membranes is regulated by a histidine switch. Proc Natl Acad Sci U S A 102:13052–13057
Legendre-Guillemin V, Wasiak S, Hussain NK, Angers A, McPherson PS (2004) ENTH/ANTH proteins and clathrin-mediated membrane budding. J Cell Sci 117:9–18
Lemmon MA (2008) Membrane recognition by phospholipid-binding domains. Nat Rev Mol Cell Biol 9:99–111
Levin DE (2005) Cell wall integrity signaling in Saccharomyces cerevisiae. Microbiol Mol Biol Rev 69:262–291
Levine TP, Munro S (2001) Dual targeting of Osh1p, a yeast homologue of oxysterol-binding protein, to both the Golgi and the nucleus-vacuole junction. Mol Biol Cell 12:1633–1644
Levine TP, Munro S (2002) Targeting of Golgi-specific pleckstrin homology domains involves both PtdIns 4-kinase-dependent and –independent components. Curr Biol 12:695–704
Li X, Rivas MP, Fang M, Marchena J, Mehrotra B, Chaudhary A, Feng L, Prestwich GD, Bankaitis VA (2002) Analysis of oxysterol binding protein homologue Kes1p function in regulation of Sec14p-dependent protein transport from the yeast Golgi complex. J Cell Biol 157:63–77
Liu J, Sun Y, Oster GF, Drubin DG (2010) Mechanochemical crosstalk during endocytic vesicle formation. Curr Opin Cell Biol 22:36–43
Lorente-Rodriguez A, Barlowe C (2011) Requirement for Golgi-localized PI(4)P in fusion of COPII vesicles with Golgi compartments. Mol Biol Cell 22:216–229
Mao Y, Nickitenko A, Duan X, Lloyd TE, Wu MN, Bellen H, Quiocho FA (2000) Crystal structure of the VHS and FYVE tandem domains of Hrs, a protein involved in membrane trafficking and signal transduction. Cell 100:447–456
Marcusson EG, Horazdovsky BF, Cereghino JL, Gharakhanian E, Emr SD (1994) The sorting receptor for yeast vacuole carboxypeptidase Y is encoded by the VPS10 gene. Cell 77:579–586
Michell RH, Heath VL, Lemmon MA, Dove SK (2006) Phosphatidylinositol 3,5-bisphosphate: metabolism and cellular functions. Trends Biochem Sci 31:52–63
Misra S, Hurley JH (1999) Crystal structure of a phosphatidylinositol 3-phosphate-specific membrane-targeting motif, the FYVE domain of Vps27p. Cell 97:657–666
Mizuno-Yamasaki E, Medkova M, Coleman J, Novick P (2010) Phosphatidylinositol 4-phosphate controls both membrane recruitment and a regulatory switch of the Rab GEF Sec2p. Dev Cell 18:828–840
Nair U, Cao T, Xie Z, Klionsky DJ (2010) Roles of the lipid-binding motifs of Atg18 and Atg21 in the cytoplasm to vacuole targeting pathway and autophagy. J Biol Chem 285:11476–11488
Natarajan P, Liu K, Patil DV, Sciorra VA, Jackson CL, Graham TR (2009) Regulation of a Golgi flippase by phosphoinositides and an ArfGEF. Nat Cell Biol 11:1421–1426
Nice DC, Sato TK, Stomhaug PE, Emr SD, Klionsky DJ (2002) Cooperative binding of the cytoplasm to vacuole targeting pathway proteins, Cvt13 and Cvt20, to phosphatidylinositol 3-phosphate at the pre-autophagosomal structure is required for selective autophagy. J Biol Chem 277:30198–30207
Odorizzi G, Babst M, Emr SD (1998) Fab1p PtdIns(3)P 5-kinase function essential for protein sorting in the multivesicular body. Cell 95:847–858
Ojala PJ, Paavilainen V, Lappalainen P (2001) Identification of yeast cofilin residues specific for actin monomer and PIP2 binding. Biochemistry 40:15562–15569
Ortiz D, Medkova M, Walch-Solimena C, Novick P (2002) Ypt32 recruits the Sec4p guanine nucleotide exchange factor, Sec2p, to secretory vesicles; evidence for a Rab cascade in yeast. J Cell Biol 157:1005–1015
Palmgren S, Ojala PJ, Wear MA, Cooper JA, Lappalainen P (2001) Interactions with PIP2, ADP-actin monomers, and capping protein regulate the activity and localization of yeast twinfilin. J Cell Biol 155:251–260
Papayannopoulos V, Co C, Prehoda KE, Snapper S, Taunton J, Lim WA (2005) A polybasic motif allows N-WASP to act as a sensor of PIP(2) density. Mol Cell 17:181–191

Parrish WR, Stefan CJ, Emr SD (2004) Essential role for the myotubularin-related phosphatase Ymr1p and the synaptojanin-like phosphatases Sjl2p and Sjl3p in regulation of phosphatidylinositol 3-phosphate in yeast. Mol Biol Cell 15:3567–3579

Parrish WR, Stefan CJ, Emr SD (2005) PtdIns(3)P accumulation in triple lipid-phosphatase-deletion mutants triggers lethal hyperactivation of the Rho1p/Pkc1p cell-integrity MAP kinase pathway. J Cell Sci 118:5589–5601

Peterson MR, Burd CG, Emr SD (1999) Vac1p coordinates Rab and phosphatidylinositol 3-kinase signaling in Vps45p-dependent vesicle docking/fusion at the endosome. Curr Biol 9:159–162

Phelan JP, Millson SH, Parker PJ, Piper PW, Cooke FT (2006) Fab1p and AP-1 are required for trafficking of endogenously ubiquitylated cargoes to the vacuole lumen in S. cerevisiae. J Cell Sci 119:4225–4234

Pruyne D, Legesse-Miller A, Gao L, Dong Y, Bretscher A (2004) Mechanisms of polarized growth and organelle segregation in yeast. Annu Rev Cell Dev Biol 20:559–591

Raiborg C, Stenmark H (2009) The ESCRT machinery in endosomal sorting of ubiquitylated membrane proteins. Nature 458:444–452

Richer SM, Stewart NK, Tomaszewski JW, Stone MJ, Oakley MG (2008) NMR investigation of the binding between human profiling I and inositol 1,4,5-triphosphate, the soluble headgroup of phosphatidylinositol 4,5-bisphosphate. Biochemistry 47:13455–13462

Rohatgi R, Ho HY, Kirschner MW (2000) Mechanism of N-WASP activation by CDC42 and phosphatidylinositol 4,5-bisphosphate. J Cell Biol 150:1299–1310

Roy A, Levine TP (2004) Multiple pools of phosphatidylinositol 4-phosphate detected using the pleckstrin homology domain of Osh2p. J Biol Chem 279:44683–44689

Saarikangas J, Zhao H, Lappalainen P (2010) Regulation of the actin cytoskeleton-plasma membrane interplay by phosphoinositides. Physiol Rev 90:259–289

Santiago-Tirado FH, Legesse-Miller A, Schott D, Bretscher A (2011) PI4P and Rab inputs collaborate in myosin-V-dependent transport of secretory compartments in yeast. Dev Cell 20:47–59

Sato TK, Overduin M, Emr SD (2001) Location, location, location: membrane targeting directed by PX domains. Science 294:1881–1885

Schu PV, Takegawa K, Fry MJ, Stack JH, Waterfield MD, Emr SD (1993) Phosphatidylinositol 3-kinase encoded by yeast VPS34 gene essential for protein sorting. Science 260:88–91

Schulz TA, Prinz WA (2007) Sterol transport in yeast and the oxysterol binding protein homologue (OSH) family. Biochim Biophys Acta 1771:769–780

Sciorra VA, Rudge SA, Wang J, McLaughlin S, Engebrecht J, Morris AJ (2002) Dual role for phosphoinositides in regulation of yeast and mammalian phospholipase D enzymes. J Cell Biol 159:1039–1049

Sciorra VA, Audhya A, Parsons AB, Negev N, Boone C, Emr SD (2005) Synthetic genetic array analysis of the PtdIns 4-kinase Pik1p identifies components in a Golgi-specific Ypt31/rab-GTPase signaling pathway. Mol Biol Cell 16:776–793

Schmitz KR, Liu J, Li S, Setty TG, Wood CS, Burd CG, Ferguson KM (2008) Golgi localization of glycosyltransferases requires a Vps74p oligomer. Dev Cell 14:523–534

Shin ME, Ogburn KD, Varban OA, Gilbert PM, Burd CG (2001) FYVE domain targets Pib1p ubiquitin ligase to endosome and vacuolar membranes. J Biol Chem 276:41388–41393

Seet LF, Hong W (2006) The Phox (PX) domain proteins and membrane traffic. Biochim Biophys Acta 1761:878–896

Shelton SN, Barylko B, Binns DD, Horazdovsky BF, Albanesi JP, Goodman JM (2003) Saccharomyces cerevisiae contains a Type II phosphoinositide 4-kinase. Biochem J 371:533–540

Song X, Xu W, Zhang A, Huang G, Liang X, Virbasius JV, Czech MP, Zhou GW (2001) Phox homology domains specifically bind phosphatidylinositol phosphates. Biochemistry 40:8940–8944

Sreenivas A, Patton-Vogt JL, Bruno V, Griac P, Henry SA (1998) A role for phospholipase D (Pld1p) in growth, secretion, and regulation of membrane lipid synthesis in yeast. J Biol Chem 273:16635–16638

Stack JH, Emr SD (1994) Vps34p required for yeast vacuolar protein sorting is a multiple specificity kinase that exhibits both protein kinase and phosphatidylinositol-specific PI 3-kinase activities. J Biol Chem 269:31552–31562

Stack JH, Herman PK, Schu PV, Emr SD (1993) A membrane-associated complex containing the Vps15 protein kinase and the Vps34 PI 3-kinase is essential for protein sorting to the yeast lysosome-like vacuole. EMBO J 12:2195–2204

Stack JH, Horazdovsky B, Emr SD (1995) Receptor-mediated protein sorting to the vacuole in yeast: roles for a protein kinase, a lipid kinase and GTP-binding proteins. Annu Rev Cell Dev Biol 11:1–33

Stahelin RV, Long F, Diraviyam K, Bruzik KS, Murray D, Cho W (2002) Phosphatidylinositol 3-phosphate induces the membrane penetration of the FYVE domains of Vps27p and Hrs. J Biol Chem 277:26379–26388

Stahelin RV, Long F, Peter BJ, Murray D, DeCamilli P, McMahon HT, Cho W (2003) Contrasting membrane interaction mechanisms of AP180 N-terminal homology (ANTH) and epsin N-terminal homology (ENTH) domains. J Biol Chem 278:28993–28999

Stefan CJ, Audhya A, Emr SD (2002) The yeast synaptojanin-like proteins control the cellular distribution of phosphatidylinositol (4,5)-bisphosphate. Mol Biol Cell 13:542–557

Stefan CJ, Manford AG, Baird D, Yamada-Hanff J, Mao Y, Emr SD (2011) Osh proteins regulate phosphoinositide metabolism at ER-plasma membrane contact sites. Cell 144:1–13

Stenmark H, Gillooly DJ (2001) Intracellular trafficking and turnover of phosphatidylinositol 3-phosphate. Semin Cell Dev Biol 12:193–199

Stolz LE, Huynh CV, Thorner J, York JD (1998) Identification and characterization of an essential family of inositol polyphosphate 5-phosphatases (INP51, INP52, and INP53 gene products) in the yeast Saccharomyces cerevisiae. Genetics 148:1715–1729

Strahl T, Hama H, DeWald DB, Thorner J (2005) Yeast phosphatidylinositol 4-kinase, Pik1, has essential roles at the Golgi and in the nucleus. J Cell Biol 171:967–979

Strochlic TI, Setty TG, Sitaram A, Burd CG (2007) Grd19/Snx3p functions as a cargo-specific adapter for retromer-dependent endocytic recycling. J Cell Biol 177:115–125

Stromhaug PE, Reggiori F, Guan J, Wang CW, Klionsky DJ (2004) Atg21 is a phosphoinositide binding protein required for efficient lipidation and localization of Atg8 during uptake of aminopeptidase I by selective autophagy. Mol Biol Cell 15:3553–3566

Sun Y, Kaksonen M, Madden DR, Schekman R, Drubin DG (2005) Interaction of Sla2p's ANTH domain with PtdIns(4,5)P2 is important for actin-dependent endocytic internalization. Mol Biol Cell 16:717–730

Tabuchi M, Audhya A, Parsons AB, Boone C, Emr SD (2006) The phosphatidylinositol 4,5-bisphosphate and TORC2 binding proteins Slm1 and Slm2 function in sphingolipid regulation. Mol Cell Biol 26:5861–5875

Tahirovic S, Schorr M, Mayinger P (2005) Regulation of intracellular phosphatidylinositol-4-phosphate by the Sac1 lipid phosphatase. Traffic 6:116–130

Tall CG, Hama H, DeWald DB, Horazdovsky BF (1999) The phosphatidylinositol 3-phosphate binding protein Vac1p interacts with a Rab GTPase and a Sec1p homologue to facilitate vacuolar protein sorting. Mol Biol Cell 10:1873–1889

Teo H, Gill DJ, Sun J, Perisic O, Veprintsev DB, Vallis Y, Emr SD, Williams RL (2006) ESCRT-I core and ESCRT-II GLUE domain structures reveal role for GLLE in linking to ESCRT-I and membranes. Cell 125:99–111

Trotter PJ, Wu WI, Pedretti J, Yates R, Voelker DR (1998) A genetic screen for aminophospholipid transport mutants identifies the phosphatidylinositol 4-kinase, STT4p, as an essential component in phosphatidylserine metabolism. J Biol Chem 273:13189–13196

Tu L, Tai WC, Chen L, Banfield DK (2008) Signal-mediated dynamic retention of glycosyltransferases in the Golgi. Science 321:404–407

Vanhaesebroeck B, Guillermet-Guibert J, Graupera M, Bilanges B (2010) The emerging mechanisms of isoform-specific PI3K signaling. Nat Rev Mol Cell Biol 11:329–341

Votin V, Allyn T, Thorner J (2009) Binding of PI4,5P2 by septin complexes is required for their essential function in cytokinesis in budding yeast. FASEB J 23:697.5

Walch-Solimena C, Novick P (1999) The yeast phosphatidylinositol-4-OH kinase pik1 regulates secretion at the Golgi. Nat Cell Biol 1:523–525

Webb GC, Zhang J, Garlow SJ, Wesp A, Riezman H, Jones EW (1997) Pep7p provides a novel protein that functions in vesicle-mediated transport between the yeast Golgi and endosome. Mol Biol Cell 8:871–895

Weirich CS, Erzberger JP, Barral Y (2008) The septin family of GTPases: architecture and dynamics. Nat Rev Mol Cell Biol 9:478–489

Wendland B, Emr SD (1998) Pan1p, yeast eps15, functions as a multivalent adaptor that coordinates protein-protein interactions essential for endocytosis. J Cell Biol 141:71–84

Wendland B, Steece KE, Emr SD (1999) Yeast epsins contain an essential N-terminal ENTH domain, bind clathrin and are required for endocytosis. EMBO J 18:4383–4393

Wesp A, Hicke L, Palecek J, Lombardi R, Aust T, Munn AL, Riezman H (1997) End4p/Sla2p interacts with actin-associated proteins for endocytosis in Saccharomyces cerevisiae. Mol Biol Cell 8:2291–2306

Wild AC, Yu JW, Lemmon MA, Blumer KJ (2004) The p21-activated protein kinase-related kinase Cla4 is a coincidence detector of signaling by Cdc42 and phosphatidylinositol 4-phosphate. J Biol Chem 279:17101–17110

Wiradjaja F, Ooms LM, Whisstock JC, McColl B, Heifenbaum L, Sanbrook JF, Gething MJ, Mitchell CA (2001) The yeast inositol polyphosphate 5-phosphatase Inp54p localizes to the endoplasmic reticulum via a C-terminal hydrophobic anchoring tail: regulation of secretion from the endoplasmic reticulum. J Biol Chem 276:7643–7653

Witke W (2004) The role of profiling complexes in cell motility and other cellular processes. Trends Cell Biol 14:461–469

Wood CS, Schmitz KR, Bessman NJ, Setty TG, Ferguson KM, Burd CG (2009) PtdIns4P recognition by Vps74/GOLPH3 links PtdIns 4-kinase signaling to retrograde Golgi trafficking. J Cell Biol 187:967–975

Wurmser AE, Emr SD (1998) Phosphoinositide signaling and turnover: PtdIns(3)P, a regulator of membrane traffic, is transported to the vacuole and degraded by a process that requires lumenal vacuolar hydrolase activities. EMBO J 17:4930–4942

Wurmser AE, Emr SD (2002) Novel PtdIns(3)P-binding protein Etf1 functions as an effector of the Vps34 PtdIns 3-kinase in autophagy. J Cell Biol 158:761–772

Xu Y, Seet LF, Hanson B, Hong W (2001) The Phox homology (PX) domain, a new player in phosphoinositide signaling. Biochem J 360:513–530

Yamamoto A, DeWald DB, Boronenkov IV, Anderson RA, Emr SD, Koshland D (1995) Novel PI(4)P 5-kinase homologue, Fab1p, essential for normal vacuole function and morphology in yeast. Mol Biol Cell 6:525–539

Yeung BG, Phan HL, Payne GS (1999) Adaptor complex-independent clathrin function in yeast. Mol Biol Cell 10:3643–3659

Yin HL, Janmey PA (2003) Phosphoinositide regulation of the actin cytoskeleton. Annu Rev Physiol 65:761–789

Yoshida S, Bartolini S, Pellman D (2009) Mechanisms for concentrating Rho1 during cytokinesis. Genes Dev 23:810–823

Yu JW, Lemmon MA (2001) All phox homology (PX) domains from Saccharomyces cerevisiae specifically recognize phosphatidylinositol 3-phosphate. J Biol Chem 276:44179–44184

Yu JW, Mendrola JM, Audhya A, Singh S, Keleti D, DeWald DB, Murray D, Emr SD, Lemmon MA (2004) Genome-wide analysis of membrane targeting by S. cerevisiae pleckstrin homology domains. Mol Cell 13:677–688

Zerial M, McBride H (2001) Rab proteins as membrane organizers. Nat Rev Mol Cell Biol 2:107–117

Zhai C, Li K, Markaki V, Phelan JP, Bowers K, Cooke FT, Panaretou B (2008) Ypp1/YGR198w plays an essential role in phosphoinositide signaling at the plasma membrane. Biochem J 415:455–466

Zhang X, Orlando K, He B, Xi F, Zhang J, Zajac A, Guo W (2008) Membrane association and functional regulation of Sec3 by phospholipids and Cdc42. J Cell Biol 180:145–158

Zhdankina O, Strand NL, Redmond JM, Boman AL (2001) Yeast GGA proteins interact with GTP-bound Arf and facilitate transport through the Golgi. Yeast 18:1–18

Chapter 3
Phosphoinositides in the Mammalian Endo-lysosomal Network

Peter J. Cullen and Jeremy G. Carlton

Abstract The endo-lysosomal system is an interconnected tubulo-vesicular network that acts as a sorting station to process and distribute internalised cargo. This network accepts cargoes from both the plasma membrane and the biosynthetic pathway, and directs these cargos either towards the lysosome for degradation, the peri-nuclear recycling endosome for return to the cell surface, or to the *trans*-Golgi network. These intracellular membranes are variously enriched in different phosphoinositides that help to shape compartmental identity. These lipids act to localise a number of phosphoinositide-binding proteins that function as sorting machineries to regulate endosomal cargo sorting. Herein we discuss regulation of these machineries by phosphoinositides and explore how phosphoinositide-switching contributes toward sorting decisions made at this platform.

Keywords Endosome · Membrane traffic · Phosphoinositide · Endosomal sorting

3.1 Introduction—The Endo-lysosomal Network

The endo-lysosomal network begins at the plasma membrane after transmembrane cargo becomes internalized through the process of endocytosis (Mayor and Pagano 2007). After internalization, cargo proteins enter the early endosome, an endomembrane compartment characterised by interconnected morphologically distinct vacuolar and tubular elements. Here cargo sorting occurs (Cullen 2008). Cargoes destined for degradation, such as the epidermal growth factor receptor (EGFR), are

P. J. Cullen (✉)
Henry Wellcome Integrated Signaling Laboratories, School of Biochemistry, Medical Sciences Building, University of Bristol, BS8 1TD, Bristol, United Kingdom
e-mail: Pete.Cullen@bris.ac.uk

J.G. Carlton
Department of Infectious Diseases, King's College London, SE1 9RT and Division of Cancer Studies, King's College London, SE1 1UL, United Kingdom
e-mail: jeremy.carlton@kcl.ac.uk

T. Balla et al. (eds.), *Phosphoinositides II: The Diverse Biological Functions*, Subcellular Biochemistry 59, DOI 10.1007/978-94-007-3015-1_3,

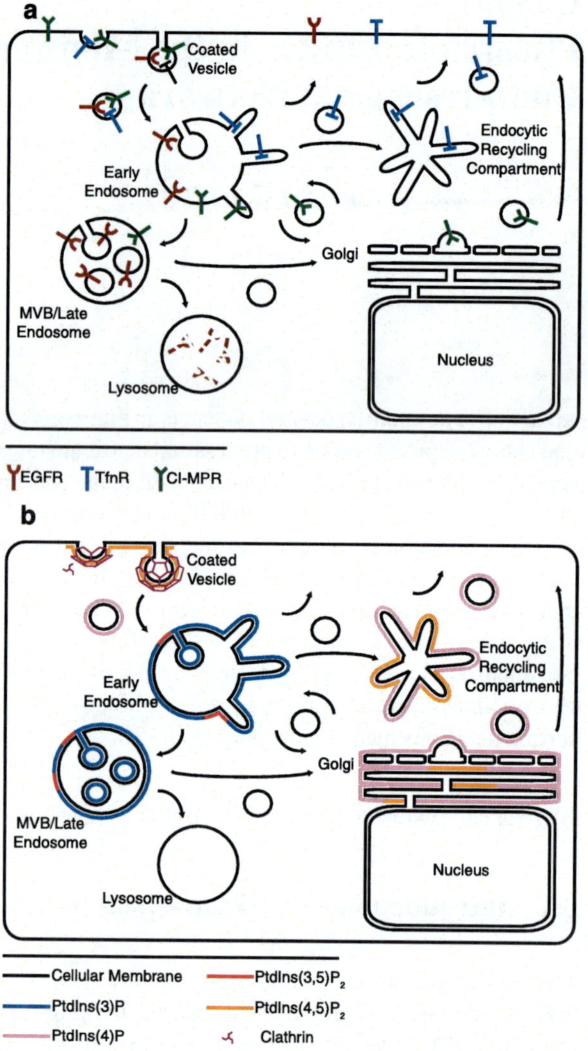

Fig. 3.1 The endo-lysosomal network. **a** The interconnected membranous network that constitutes the endo-lysosomal network. **b** Steady state distribution of phosphoinositide isomers within the network: the 'phosphoinositide map'

sorted onto intraluminal vesicles (ILVs) of endosomes, which, through a process of compartment maturation give rise to the late endosome/multivesicular body (MVB) (Piper and Katzmann 2007; Williams and Urbe 2007; Hurley and Hanson 2010). In parallel, other cargoes are retrieved from this pathway. Transferrin receptors (TfnRs), integrins and a variety of other cargo undergo sorting into tubular-branched structures from where they are recycled back to the plasma membrane via a direct fast recycling route or, more slowly, through the juxtanuclear endocytic recycling compartment (ERC) (Grant and Donaldson 2009; Hsu and Prekeris 2010). Yet other cargoes, such as cation-independent mannose 6-phosphate receptors (CI-MPRs) are sorted into distinct tubular sub-domains for retrieval to the *trans*-Golgi network

(TGN) (Bonifacino and Rojas 2006; Johannes and Popoff 2008). When all recycled material has been removed, late endosomes/MVBs fuse with lysosomes thereby degrading cargoes associated with the intraluminal vesicles (Saftig and Klumperman 2009) (Fig. 3.1a).

Interest in this membranous network stemmed from academic curiosity into how the complexities of the network are established and maintained. Subsequent medical importance of endo-lysosomal sorting has emerged from an ever-growing body of evidence implicating adverse sorting as an underlying mechanism for a number of diseases. Moreover, adventitious viruses, pathogens and toxins exploit this pathway in order to enter target cells and gain access to the cytosol (Gruenberg and van der Goot 2006; Johannes and Romer 2010; Mercer et al. 2010). Research into this network is therefore moving increasingly towards defining the molecular mechanisms of endosomal related diseases, and nowhere is this more evident than in the study of endosomal phosphoinositides (Nicot and Laporte 2008; McCrea and De Camilli 2009; Liu and Bankaitis 2010).

3.2 Subcellular Distribution of Phosphoinositides

Phosphoinositides are not uniformly distributed throughout cellular membranes rather they adopt a compartmentalised distribution which is dependent on the spatial and temporal regulation of specific kinases and phosphatases. While the exact phosphoinositide composition of individual membrane compartments is not comprehensively understood, imaging based studies employing fluorescently-tagged isomer-specific phosphoinositide biosensors have described a general map of phosphoinositide distributions (Balla and Várnai 2009; Cooke 2009). For the endo-lysosomal network this is defined by the enrichment of PtdIns(4,5)P_2 at sites of clathrin-mediated endocytosis (Haucke 2005), while PtdIns(3)P is localized to the cytosolic face of the limiting membrane of early endosomes and those intraluminal vesicles contained within MVBs (Gillooly et al. 2000a). PtdIns(3,5)P_2 on the other hand has been proposed to be primarily localised to the limiting membrane of late endosomes although the evidence for this remains incomplete (Michell et al. 2006). Finally, membranes of the ERC, Golgi apparatus and cytoplasmic vesicles are thought to contain PtdIns(4)P and PtdIns(4,5)P_2 (De Matteis et al. 2005; Hammond et al. 2009; Jovic et al. 2009) (Fig. 3.1b).

Such a phosphoinositide map should always be taken as a generality. It is clear that under certain physiological conditions specific phosphoinositide can be generated on membranes other than those described within the map. Take for example the early endosomal pool of PtdIns(3)P. It can be generated on the inner leaflet of the plasma membrane during insulin stimulation (Maffucci et al. 2003), appears to be associated with the outer plasma membrane leaflet (Kale et al. 2010), and is present in the endoplasmic reticulum during the formation of autophagosomes (Axe et al. 2008).

3.3 Endosomal Phosphoinositide Kinases and Phosphatases

A vital element in the biology of phosphoinositide is their rapid inter-conversion through the actions of specific kinases and phosphatases (see (Balla and Balla 2006; Robinson and Dixon 2006; Ooms et al. 2009; van den Bout and Divecha 2009; Liu and Bankaitis 2010; Vanhaesebroeck et al. 2010) for detailed reviews). Briefly, mammalian cells express families of specific kinases and phosphatases which add or remove phosphates from the D3, D4 and D5 position of the inositol head-group of specific phosphoinositide isomers. Thus, Class I PI 3-kinase generates PtdIns(3,4,5)P_3 by catalysing the addition of a phosphate to the 3-position of PtdIns(4,5)P_2, while PTEN is a 3-phosphatase that dephosphorylates PtdIns(3,4,5)P_3 re-generating PtdIns(4,5)P_2 (Chalhoub and Baker 2009). Broadly speaking, the molecular architecture of phosphoinositide kinases and phosphatases are organised so as to target the catalytic region to different subcellular membranes through mechanisms that, for the majority of enzymes, remain to be completely defined. This has the effect of spatially restricting the activity of individual kinases and phosphatases to the membrane:cytosol interface of specified compartments. If one adds to this the fact that these enzymes are also designed to respond to changes in the cellular state, for example receptor activation and changes to the cell cycle, then a picture emerges in which the metabolic turnover of specific phosphoinositides can be spatially and temporal restricted to unique membrane compartments (i.e. the phosphoinositide map, Fig. 3.1b).

To further illustrate and expand this, consider the endo-lysosomal network. The central axis of this pathway classically begins with the formation of PtdIns(4,5)P_2-enriched clathrin-coated endocytic pits, proceeds through PtdIns(3)P-containing early endosomes before ending with PtdIns(3,5)P_2-containing late endosomes. Importantly, the transition from clathrin-coated pits to late endosomes occurs through compartment maturation—clathrin-coated pits mature and then fuse with early endosomes, which in turn mature into late endosomes (Rink et al. 2005; Poteryaev et al. 2010). Along this maturation axis therefore, there appears to be three phosphoinositide switches—PtdIns(4,5)P_2-to-PtdIns(4)P, PtdIns-to-PtdIns(3)P and PtdIns(3)P-to-PtdIns(3,5)P_2 (Joly et al. 1995; Dove et al. 1997; Shin et al. 2005; Zoncu et al. 2009).

3.3.1 *PtdIns(4,5)P_2-to-PtdIns(4)P Switch*

The formation of plasma membrane PtdIns(4,5)P_2 through the actions of PtdIns(4)P 5-kinases is an essential localisation cue required for clathrin-mediated endocytosis (Jost et al. 1998; Wenk et al. 2001). Early genetic studies in mice revealed the importance of Type I PtdIns(4)P 5-kinase-γ for this process during synaptic transmission (Paolo et al. 2004). This class of phosphoinositide kinases are able to directly associate with the clathrin-binding cargo adaptor AP-2 (Krauss et al. 2006; Nakano-Kobayashi et al. 2007; Thieman et al. 2009), an intriguing observation given that initial endocytic assembly zones are nucleated at the plasma membrane through low affinity interactions between AP-2 and discrete, localised patches of

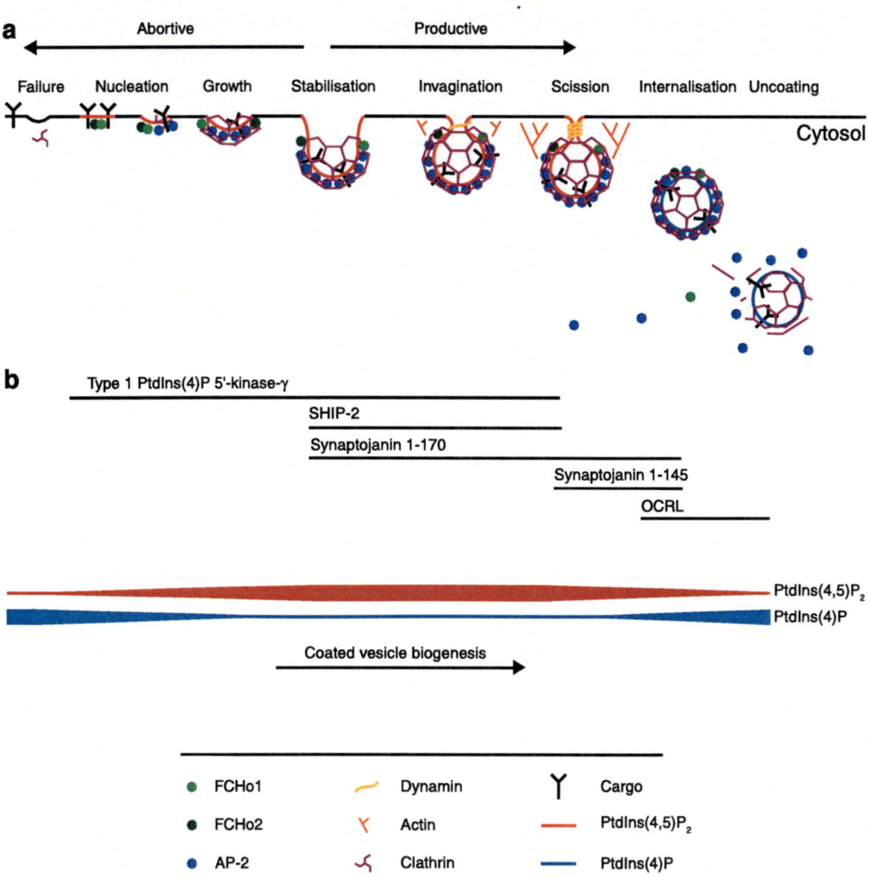

Fig. 3.2 Clathrin-mediated endocytosis. **a** Schematic depicting stages of clathrin-coated vesicle (CCV) formation. **b** Phosphoinositide kinases and phosphatases involved in shaping the phosphoinositide profile of the nascent CCV and predicted PtdIns(4)P—PtdIns(4,5)P_2 transition during this process

plasma membrane PtdIns(4,5)P_2 (Traub 2009b) (Fig. 3.2a, b) (see Sect. 3.5.1 for more detail).

As for inositol 5-phosphatases, a number, including two isoforms of synaptojanin 1 (synaptojanin 1–170 and synaptojanin 1–145), synaptojanin 2, SHIP2, INPP5B and OCRL (McPherson et al. 1996; Shin et al. 2005; Hyvola et al. 2006; Perera et al. 2006; Erdmann et al. 2007; Nakatsu et al. 2010), are recruited to forming clathrin-coated endocytic structures, further establishing the importance of PtdIns(4,5)P_2 turnover in the dynamics of this process. While there is no evidence to suggest a direct association between Type I PtdIns(4)P 5-kinases and PtdIns(4,5)P_2 5-phosphatases, synaptojanin 1–170 does associate with the AP-2 adaptor (Haffner et al. 2000; Jha et al. 2004), arguing for a close spatial relationship between these two catalytic activities. Interestingly, while synaptojanin 1–170 remains associated with clathrin-coated pits from early nucleation through to clathrin uncoating after fission (Perera et al. 2006), the association of other 5-phosphatases is dependent on the maturation state of the clathrin-coated pit (Nakatsu et al. 2010). SHIP2 associates at early stages during nucleation and growth but dissociates prior to scission (Perera et al. 2006; Nakatsu et al. 2010), and synaptojanin 1–145 and OCRL both arrive at late stages of pit formation. After clathrin uncoating, an event triggered by PtdIns(4,5)P_2-to-PtdIns(4)P switching, only OCRL remains associated with the newly formed endosomal vesicle (Perera et al. 2006; Erdmann et al. 2007) (see Sect. 3.5 for a discussion of clathrin-mediated endocytosis).

Overall, the requirement for Type I PtdIns(4)P 5-kinases and the sequential association of multiple 5-phosphatases is consistent with the continuous turnover of PtdIns(4,5)P_2 and PtdIns(4)P during maturation of clathrin-coated pits and vesicles, which helps create a plastic system allowing for regulation of the maturation process (Fig. 3.2a, b). One interpretation of these data is that inclusion of individual enzymes within discrete interaction networks within the global clathrin coat interactome may, at specified times during the maturation pathway, modify on a local scale the level of PtdIns(4,5)P_2.

3.3.2 *PtdIns-to-PtdIns(3)P Switch*

Clathrin derived endocytic vesicles (and also those from clathrin-independent endocytosis) on entering the endosomal network contain Rab5 and APPL, but lack detectable levels of PtdIns(3)P (Zoncu et al. 2009): presumably these vesicle also contain PtdIns and residual PtdIns(4)P from dephosphorylation of PtdIns(4,5)P_2. Conversion of these intermediate endosomes to PtdIns(3)P-positive early endosomes correlates with displacement of APPL and the acquisition of PtdIns(3)P-binding proteins such as EEA1 (Zoncu et al. 2009). How this conversion is controlled is unclear although competition with Rab5 may be one possibility (Zoncu et al. 2009): APPL directly binds Rab5 (Miaczynska et al. 2004) and also the 5-phosphatase OCRL which, as described above, remains associated with the newly formed endosomal vesicle after clathrin-mediated endocytosis (Hyvola et al. 2006; Perera et al. 2006; Erdmann et al. 2007).

Active GTP-bound Rab5 is a key regulatory of the principal PtdIns 3-kinase in mammalian cells, VPS34 (Christoforidis et al. 1999; Backer 2008). VPS34 is associated with a HEAT domain-containing regulatory subunit VPS15, and it is the interaction with Rab5 that targets this dimer to endosomes, catalysing the formation of PtdIns(3)P (Christoforidis et al. 1999). The antagonistic PtdIns(3)P 3-phosphatases are members of the myotubularin family: a large group of lipid phosphatases, mutations in which are linked to various genetic diseases affecting skeletal muscle and the neuronal system (Robinson and Dixon 2006; Liu and Bankaitis 2010). Myotubularin 1 (MTM1) (defective in X-linked myotubular myopathy) and myotubularin-related 2 (MTMR2) (defective in Charcot-Marie-Tooth disease type 4B (CMT4B1)) are both able to associate with the VPS15:VPS34 complex through an interaction with VPS15 (Cao et al. 2007, 2008). In these complexes both 3-kinase and 3-phosphatase activities are inactivated, effectively shutting off the turnover of PtdIns(3)P (Cao et al. 2007, 2008) and stabilising domains of this lipid upon endosomes. Rab5 and Rab7 however compete with MTM1 and MTMR2 for binding to the VPS15, releasing the inhibition and allowing PtdIns(3)P turnover (Cao et al. 2007, 2008). Imaging studies are consistent with the VPS15:VPS34:MTM1 complex residing at the Rab5 early endosome and the VPS15:VPS34:MTMR2 complex localising to Rab7 late endosomes (Cao et al. 2007, 2008). Perhaps competition of Rab5 with the VPS15:VPS34:MTM1 complex on APPL-labelled intermediate endosomes may be one component in switching to a PtdIns(3)P positive early endosome. Understanding how the activation statuses of Rab5 (and Rab7) modulate the assembly of these PtdIns 3-kinase:PtdIns(3)P 3-phosphatase complexes will surely reveal interesting insight into the co-ordination of PtdIns(3)P turnover and early-to-late endosomal maturation.

3.3.3 *PtdIns(3)P-to-PtdIns(3,5)P_2 Switch*

In mammalian cells the PtdIns(3)P-to-PtdIns(3,5)P_2 switch is catalysed by the PtdIns(3)P 5-kinase PIKfyve (Sbrissa et al. 1999; Shisheva et al. 1999), an enzyme localised to early and late endosomes (Cabezas et al. 2006; Ikonomov et al. 2006; Rutherford et al. 2006) through its PtdIns(3)P-binding FYVE domain. The reverse reaction, converting PtdIns(3,5)P_2-to-PtdIns(3)P, is driven by the 5-phosphatase mFig4/Sac3 (Sbrissa et al. 2007). These two enzymes associate in a complex alongside the associated regulator of PIKfyve (ArPIKfyve/hVac14) (Sbrissa et al. 2004; Jin et al. 2008), a HEAT domain-containing scaffold protein which regulates the catalytic activities of both PIKfyve and mFig4 (this complex is evolutionarily conserved being also present in yeast (Bonangelino et al. 2002; Dove et al. 2002; Gary et al. 2002; Rudge et al. 2004; Duex et al. 2006; Botelho et al. 2008)). Like many other examples where antagonistic or complementary enzymatic activities are associated within a single complex, the PIKfyve/mFig4/hVac14 complex appears designed to allow co-ordination and fine regulation of the relative 5-kinase versus 5-phosphatase activities during the PtdIns(3)P-to-PtdIns(3,5)P_2 switch (Botelho 2009). Again, how this regulation is co-ordinated with endosomal sorting and maturation will be an interesting area for further research.

That this complex is of central importance for normal endo-lysosomal function is however clear (see Sect. 3.7.3 for more detailed discussion). Perturbing PtdIns(3,5)P_2 leads to a variety of phenotypes one of which is the striking formation of numerous swollen endosomal vacuoles that arise from a defect in the biology of late endosomes and lysosomes (Michell et al. 2006). In mice the loss of Fig4 or Vac14 leads to vacuolation of neurons and neurodegeneration (Chow et al. 2007; Zhang et al. 2007). Moreover, mutations in Fig4 are observed in patients with recessive peripheral-nerve disorder CMT4J, amyotrophic lateral sclerosis (ALS) and primary lateral sclerosis (PLS) (Chow et al. 2007, 2009; Zhang et al. 2008), in which at least for CMT4J patients, isolated fibroblasts have swollen vacuoles and impaired trafficking (Zhang et al. 2008).

3.3.4 Other Phosphoinositide Switches

Besides these switches one needs to consider that PtdIns(3,4,5)P_3 also plays an important role during certain endocytic events including, abeit perhaps in a more minor capacity than PtdIns(4,5)P_2, clathrin-mediated endocytosis (Shin et al. 2005; Nakatsu et al. 2010). Here a Rab5-GTP regulated phosphatase cascade acts to sequentially convert PtdIns(3,4,5)P_3 to PtdIns(3,4)P_2 before dephosphorylation of PtdIns(3,4)P_2 generates PtdIns(3)P—employing type II inositol 5-phosphatase and type 1α PI 4-phosphatase activities respectively (Shin et al. 2005). Rab5 therefore controls the generation of PtdIns(3)P through two distinct routes—direct activation of the VPS34 Class III PI 3-kinase and indirect switching of PtdIns(3,4,5)P_3-to-PtdIns(3)P.

Away from the central axis of the endo-lysosomal degradative pathway, other endosomal trafficking routes require phosphoinositide switches for their maturation. Switching of PtdIns-PtdIns(4)P-PtdIns(4,5)P_2 is necessary for recycling through the slow juxtanuclear endocytic recycling compartment. Here, the activity of a type II PtdIns 4-kinase generates PtdIns(4)P, and the ARF6-mediated activation of PtdIns(4)P 5-kinase leads to the production of PtdIns(4,5)P_2 in order to regulate exit from the compartment back to the plasma membrane (Brown et al. 2001; D'Angelo et al. 2008; Grant and Donaldson 2009; Jovic et al. 2009). Finally, one can speculate that retrograde retrieval pathways will also require phosphoinositide switching as membrane containing carriers are transported from the donor endosomes to the target membranes. For example, retrograde carriers en route to the *trans*-Golgi network will, depending on whether they originated from early or late endosomal donor compartments, be predicted to contain respectively PtdIns(3)P or PtdIns(3,5)P_2. As the *trans*-Golgi network is not enriched in these phosphoinositides, switching must occur to maintain phosphoinositide compartmental identity.

In summary: to maintain compartmental identity the phosphoinositide signature of the proceeding compartment must be erased while the new phosphoinositide identity is developed (Botelho 2009). These switches appear tightly regulated. Indeed, as a general concept one can argue that an inability to efficiently maintain phosphoinositide switching, and the resultant effects on compartment identity may well underlie many of the diseases that are linked to endosomal phosphoinositides.

3.4 Sensing the Phosphoinositide Identity Code

With the distinction between phosphoinositides residing mostly in the organization of phosphate groups on the inositol ring, the ability of proteins to bind these lipids is mediated primarily through electrostatic interactions with basic amino acids. The ways in which these amino acids are presented to the phosphoinositide are however distinct, ranging from structurally well organized binding pockets (*e.g.* PH (pleckstrin homology) domains, FYVE (Fab1, YOTB, Vac1, EEA1 homology) domains, PX (phox homology) domains, ENTH (epsin N-terminal homology) and ANTH (AP180 N-terminal homology) domains) through to unstructured surfaces defined by amino acid clusters (e.g. MARCKS proteins (McLaughlin et al. 2002)) (for structural details of these and other phosphoinositide-binding domains, plus broader discussion of their roles in signaling and membrane trafficking see (Cullen et al. 2001; Balla 2005; Lemmon 2008)).

Phosphoinositide binding sites and surfaces are associated with integral proteins such as ion channels or, more commonly, peripheral cytosolic proteins. For the later, binding is often characterized by relatively low (often micromolar) affinities (Carlton and Cullen 2005). This favors a rapid, reversible mode of binding where the phosphoinositide-binding protein is constantly sampling the membrane surface for the present of the corresponding phosphoinositide, enriching at sites where the lipid resides (Carlton and Cullen 2005). By itself such binding of individual molecules to membranous phosphoinositides may not be of sufficient strength to lead to stable membrane association. Here the affinity for phosphoinositide recognition needs to be enhanced by combination with another effect to drive productive membrane localization. This additional localization signal can simply be interaction with a prelocalised binding partner (Carlton and Cullen 2005), or can be the avidity-mediated enhancement in affinity that occurs through generation of multivalent interactions with membranes. To refer back to a previously described analogue (Carlton and Cullen 2005)—this avidity-mediated binding can be likened to the ability of Velcro™ to generate a high-strength bond even though the interaction of each individual hook and loop is relatively weak. Thus, although phosphoinositides are commonly regarded as membrane association cues that target peripheral proteins to membrane surfaces, it is increasingly clear that other interactions besides recognition of phosphoinositides function to restrict the association of phosphoinositide-binding proteins to cellular membranes. Such avidity-dependent, phosphoinositide-mediated membrane association can take multiple forms. For example, the PH domain of dynamin has a low millimolar affinity for membranes containing PtdIns(4,5)P_2 (Klein et al. 1998; Lemmon 2008). However, upon self-assembly into oligomers the apparent affinity shifts towards the micro/nanomolar range leading to stabilisation of dynamin oligomers on PtdIns(4,5)P_2 enriched regions of the plasma membrane during the process of endocytic scission (Fig. 3.3a) (Klein et al. 1998; Lemmon 2008). Other modes of avidity-based membrane association include, where a given phosphoinositide-binding protein can associates with another membrane-bound protein such as integral cargo protein or a lipid-modified small GTPase (e.g., PTB domains from Dab1/2, numb and ARH all bind both NPXY-containing cargo and PtdIns(4,5)P_2 (Traub 2003)), with an additional membrane lipid (*e.g.* phosphatidic acid or phosphatidyl serine in the case of the PX domain of p47phox) (Fig. 3.3b) (Karathanassis et al. 2002; Stahelin et al. 2003), or can sense the geometry of the membrane surface (*e.g.* the combination of membrane curvature sensing BAR domain and PtdIns(3)P-binding PX domain of the sorting nexins SNX1 and 2 (Fig. 3.3c) ((Carlton et al. 2004, 2005) see Sect. 3.7.1.1)). An intriguing example of how these associations can be regulated is observed for HIV-1 Gag; here a polybasic region drives association with PtdIns(4,5)P_2 (Ono et al. 2004; Chukkapalli et al. 2008) for viral assembly at the plasma membrane. Association with PtdIns(4,5)P_2 induces exposure of a sequestered myristate moiety (Saad et al. 2006) and is thought to stabilize Gag associated at the

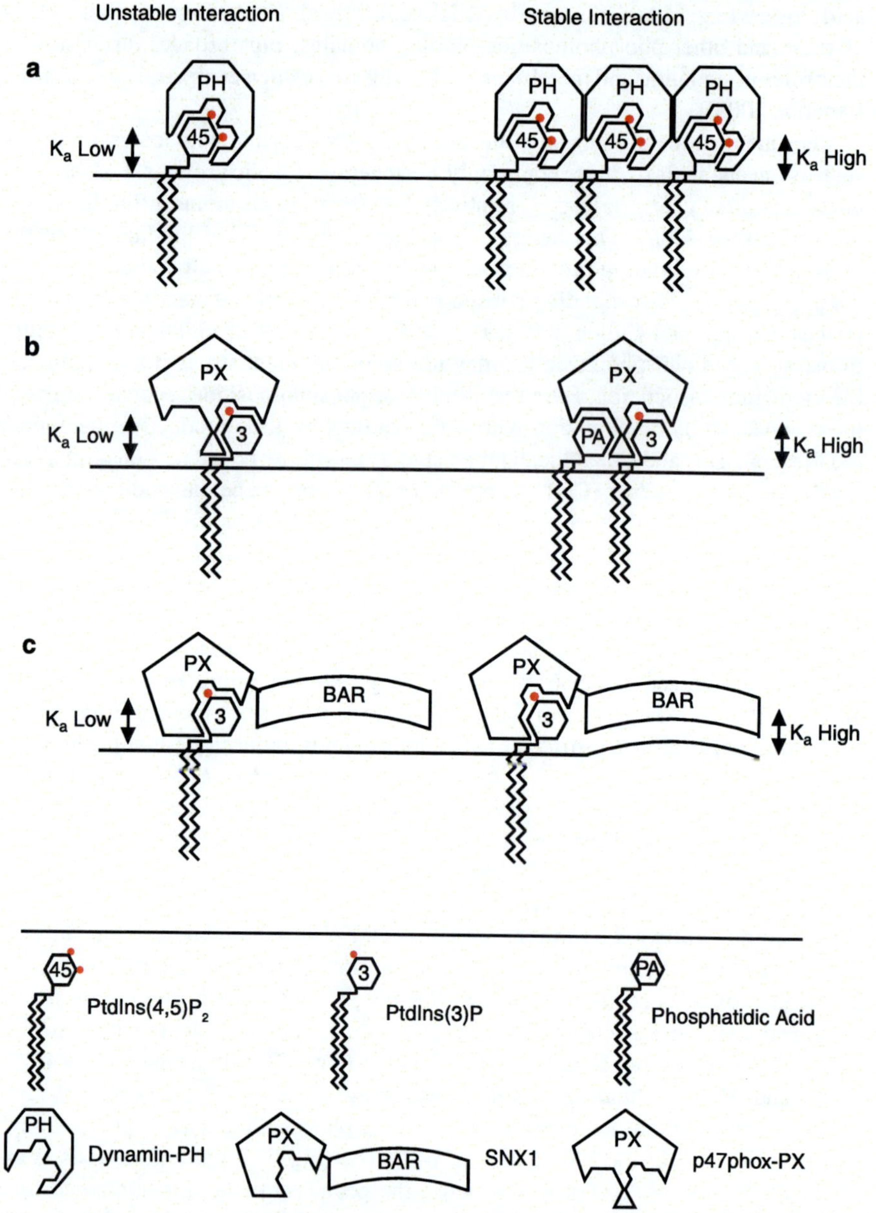

Fig. 3.3 Mechanisms of enhancing the strength of interaction between phosphoinositides and membranes. **a** Oligomerisation of dynamin's PH domain strengthens its interaction with membranes through avidity-mediated enhancement of affinity. **b** The PX-domain of p47phox binds to both PtdIns(3)P and acidic phospholipids such as PtdSer/PtdOH (PA). Ligation of both classes of lipid enhances membrane association. **c** Sorting nexins contain tandem membrane-binding PX and BAR domains. Membrane association requires co-incident recognition of both PtdIns(3)P and curved membranes

plasma membrane. Similarly, in the case of p47phox's PX domain, phospholipids binding sites are masked by self-interaction and phospho-mimic mutations that relieve this self-interaction greatly enhance membrane association, suggesting that this interaction is regulatable (Stahelin et al. 2003).

Why incorporate phosphoinositide-binding into a membrane association system that is based on avidity? Low affinity gives rise to a highly plastic system, defined by dynamic instability, that allows rapid re-modelling in response to, for example, fluctuations in the level phosphoinositides. Moreover, the fact that membrane association depends upon multivalent interactions, builds into the system a property referred to as co-incidence detection (Carlton and Cullen 2005). Here, two or more signals, one of which is phosphoinositide recognition, are required to be present at the same time within the same membrane in order to achieve membrane association of the phosphoinositide-binding protein. In turn, co-incidence detection affords a greater element of regulation as manipulation of signals, other than the level of phosphoinositides, can have pronounced effects on the membrane association and hence the function of phosphoinositide-binding proteins. Avidity-based membrane association thereby gives rise to a greater spatial and temporal regulation of phosphoinositide-mediated cellular regulation. In the following we expand on the importance of avidity in phosphoinositide biology by discussing the four major steps within the endo-lysosomal network: endocytosis, the biogenesis of MVBs, and cargo sorting into the retrograde endosome-to-Golgi and endosomal recycling pathways.

3.5 PtdIns(4,5)P_2 in Endocytosis

The function of PtdIns(4,5)P_2 in clathrin-mediated endocytosis can be viewed as a paradigm for the role of phosphoinositide in membrane trafficking. All membrane trafficking pathways follow a similar series of co-ordinated events. At the donor membrane, coat complexes package with high fidelity selected transmembrane spanning cargo into membrane regions that they deform into vesicular or tubular shaped buds. Bud fission generates coated cargo-enriched carriers, which undergo uncoating, either prior to or after, short or long range movement towards their target membrane. The correct target membrane is detected through interaction with tethering complexes before SNARE-mediated membrane fusion allows lateral movement of the cargo into the target compartment.

Mechanistically, clathrin-mediated endocytosis can be considered to comprise three overlapping stages: (i) co-ordination of cargo selection and sorting with the assembly of the coat complex; (ii) generation and stabilization of membrane curvature; and (iii) fission to liberate the cargo-enriched clathrin-coated vesicle.

3.5.1 Initiation and Assembly of Clathrin-coat Pits

As discussed previously, formation of endocytic assembly zones appears to occur through low affinity interactions between AP-2 and discrete, localised patches of plasma membrane PtdIns(4,5)P_2 (Fig. 3.2a, b) (Traub 2009b). Actual nucleation of these assembly zones may however be dependent on the F-BAR domain-containing protein family FCHo1/2 (Fer/Cip4 homology domain-only proteins 1 and 2) (Henne et al. 2010). FCHo proteins, which associate with the plasma membrane through binding to PtdIns(4,5)P_2 (Henne et al. 2010), contain an F-BAR domain that is distinct to other BAR domains in being able to recognise less extreme membrane curvature (Henne et al. 2007; Shimada et al. 2007). Importantly, overexpression of FCHo1/2 increases the rate of clathrin-coated pit nucleation (Henne et al. 2010), and FCho2 associates with eps15 and intersectin, proteins that directly interact with AP-2 but not clathrin (Henne et al. 2010). FCHo1/2 proteins therefore drive the initial membrane sculpturing required for nucleating pit formation prior to the recruitment of AP-2 (Henne et al. 2010). Once recruited to these nucleation 'hot-spots', AP-2 leads to clathrin recruitment and self-assembly and mediates cargo capture (Traub 2009b).

A key feature in the formation of clathrin-coated pits is the reliance on multiple low affinity interactions in order to generate the necessary avidity to drive stable coat assembly (Schmid and McMahon 2007). A consequence of avidity is the generation of dynamic instability, leading to the formation of pathway checkpoints (Loerke et al. 2009). For example, there is a dependence on a threshold of PtdIns(4,5)P_2 being breached to ensure sufficient AP-2 is membrane associated to engage enough cargo. Indeed, a significant number of initial assemblies, rather than proceeding to productive clathrin-coated pits, undergo early or late abortive events (Loerke et al. 2009; Saffarian et al. 2009). Furthermore, manipulating the level of cargo can alter the lifetime and shift the relative levels of abortive to productive assemblies (Mettlen et al. 2010). A beautiful study suggests that abortive events may correspond to incomplete opening of the AP-2 complex, a process necessary for multiple simultaneous PtdIns(4,5)P_2-binding events and cargo recognition through linear sorting motifs present on the cytosolic tail of cargo proteins (Jackson et al. 2010).

3.5.2 *Maturation of Clathrin-coated Pits*

Co-ordinated with the initial assembly of the clathrin-coated pit, the association of a large number of accessory proteins (many of which associate with PtdIns(4,5)P_2) act to constrain clathrin self-assembly and further sculpture membrane bending (Fig. 3.2a). Interacting with the core clathrin coat components are proteins such as dynamin, amphiphysin, epsin and sorting nexin-9 (SNX9), which through the presence of BAR and ENTH (epsin N-terminal homology) domains, sense and/or generate membrane curvature (Schmid and McMahon 2007). These drive and/or stabilize further membrane re-modelling, allowing maturation towards highly invaginated clathrin-coated pits. Again, recruitments are based on multivalent interactions, which generate dynamic instability, allowing for pathway progression (Schmid and McMahon 2007). Within these accessory proteins, the GTPase dynamin appears to play a key role in monitoring the maturation state of the clathrin-coated pit (Loerke et al. 2009). Dynamin itself is PtdIns(4,5)P_2-regulated, and its association with PtdIns(4,5)P_2 containing membranes is driven by avidity-mediated oligomerisation (Fig. 3.3a) (Klein et al. 1998). Manipulating dynamin levels has the effect of either enhancing or decreasing the turnover rate of abortive clathrin-coated pit assemblies (Loerke et al. 2009).

3.5.3 Formation of Clathrin-coated Vesicles

Fission of the mature invaginated clathrin coated pit requires GTP hydrolysis and the self-assembly of a dynamin spiral at the collar of the pit (Hinshaw and Schmid 1995; Takei et al. 1995) (see (Mettlen et al. 2009; Pucadyil and Schmid 2009; Traub 2009a, b) for extensive reviews on dynamin regulation and function). Dynamin is recruited at late stages of clathrin-coated vesicles formation through an interaction of its PH domain with PtdIns(4,5)P_2 (Salim et al. 1996; Merrifield et al. 2002; Rappoport et al. 2008). In solution, dynamin can spontaneously assemble into helices and rings, and in the presence of protein-free, PtdIns(4,5)P_2-containing liposomes can drive the formation of membrane tubules (Hinshaw and Schmid 1995; Sweitzer and Hinshaw 1998). *In vivo* however, it is predicted that with the level of membrane tensions observed in cellular membranes, dynamin may not deform membranes into tubules (Roux et al. 2010b). Rather, dynamin is recruited to the neck of clathrin-coated pits only when the curvature of the neck is sufficiently high (Roux et al. 2010b). Subsequent polymerisation of dynamin constricts the neck further down to a radius of approximately 10 nm (Roux et al. 2010b). Assembly of the dynamin collar also triggers cooperative GTP hydrolysis. This is argued to lead to fission either via mechanochemical constriction, extension and twisting of the collar (Stowell et al. 1999; Roux et al. 2006), or through a regulated cycle of dynamin assembly and disassembly which, through imposing curvature stress, allows spontaneous stochastic fission (Bashkirov et al. 2008).

Also aiding the sculpturing and fission of mature clathrin-coated pits is the regulated polymerisation of actin (Orth and McNiven 2003; Schafer 2004). Indeed, a number of dynamin interacting proteins are actin-regulatory proteins (Orth and McNiven 2003; Schafer 2004). An example is the BAR domain-containing protein sorting nexin-9 (SNX9). This PtdIns(4,5)P_2-binding protein directly binds dynamin, aiding its recruitment to the highly invaginated necks of mature clathrin-coated pits (Lundmark and Carlsson 2003, 2004, 2009). SNX9 also associates with and activates, in a membrane-dependent manner, N-WASP (Shin et al. 2007, 2008; Yarar et al. 2007, 2008), thereby driving actin polymerisation via the Arp2/3 actin nucleating complex. The localised, co-ordinated burst of actin polymerisation may assist in the efficiency of dynamin-mediated fission through the generation of longitudinal force (Fig. 3.2a) (Ferguson et al. 2009; Lundmark and Carlsson 2009).

Following liberation from the plasma membrane, the cargo-enriched clathrin-coated vesicles undergo uncoating prior to delivery to early elements of the endosomal network. Here, one needs to overcome the increased stability arising from the avidity-based interactions that have developed during maturation of the clathrin-coated vesicle, in order to allow recycling of the molecular components for further rounds of endocytosis (Schmid and McMahon 2007). For clathrin-coated vesicles, uncoating is triggered by recruitment of the auxilin and Hsc70 ATPase (Fotin et al. 2004) along with the switching of PtdIns(4,5)P_2 to PtdIns(4)P (Fig. 3.2a, b) (see Sect. 3.3.1).

3.6 PtdIns(3)P and MVB Biogenesis

Cargos targeted for lysosomal degradation are tagged with ubiquitin after receptor activation at the plasma membrane, suggesting cargo fate is determined early in its endocytic itinerary. Ubiquitinated cargos are retained within specialized subdomains of the endosome and, through the actions of endosomal sorting machinery, become incorporated onto intraluminal endosomal vesicles (ILV) to generate an organelle called a multivesicular body (MVB) (Fig. 3.1a) (Gruenberg and Stenmark 2004; Piper and Katzmann 2007). Subsequent fusion of MVBs with the yeast vacuole or mammalian lysosome results in the delivery of these cargoes to the hydrolytic and proteolytic interior. A key event in degradative sorting is the separation of cargoes destined for degradation onto ILVs, preventing them from being recycled. ILV biogenesis requires deformation of the limiting endosomal membrane into the endosomal lumen with transfer of sequestered ubiquitinated cargoes onto the developing bud (Gruenberg and Stenmark 2004; Piper and Katzmann 2007). Extension of these buds leads to the development of intraluminal membrane invaginations and scission of the necks of these invaginations leads to the release of nascent ILV into the MVB interior. Importantly, these activities are in topological apposition to other cellular budding events (Fig. 3.4a) and must be effected through the actions of a cytoplasmic machinery.

Isolation of the yeast Vacuolar Protein Sorting (VPS) mutants provided initial insights into mechanisms of degradative sorting (Bankaitis et al. 1986; Robinson et al. 1988; Raymond et al. 1992). The Class E subset displayed clear defects in ILV biogenesis and accompanying failure to properly sort cargo to the vacuole. Proteins encoding these genes were subsequently found to form three macromolecular complexes, termed ESCRT-I, -II and -III (Katzmann et al. 2001; Babst et al. 2002a, b) (Table 3.1). Genetic evidence suggested that these complexes acted sequentially in the trafficking of ubiquitinated cargo onto the yeast vacuole and homologues of these proteins were identified in mammalian cells and an evolutionarily conserved activity of degradative protein sorting was ascribed to these complexes (Williams and Urbe 2007; Carlton 2010). In addition to a role for ESCRT proteins in regulating degradative sorting, ESCRT-proteins were shown to be required for the release of retroviruses such as HIV-1 (Morita and Sundquist 2004; Carlton and Martin-Serrano 2009). Here, they are recruited to sites of particle assembly at the plasma membrane through direct interaction with viral structural proteins and provide an activity allowing scission of membranous stalks connecting the viral particles to the cell and subsequent release of the viruses (Morita and Sundquist 2004; Carlton and Martin-Serrano 2009; Carlton 2010). Further, the ESCRT-machinery has been shown to be important for the terminal phase of cell division where it plays a topologically

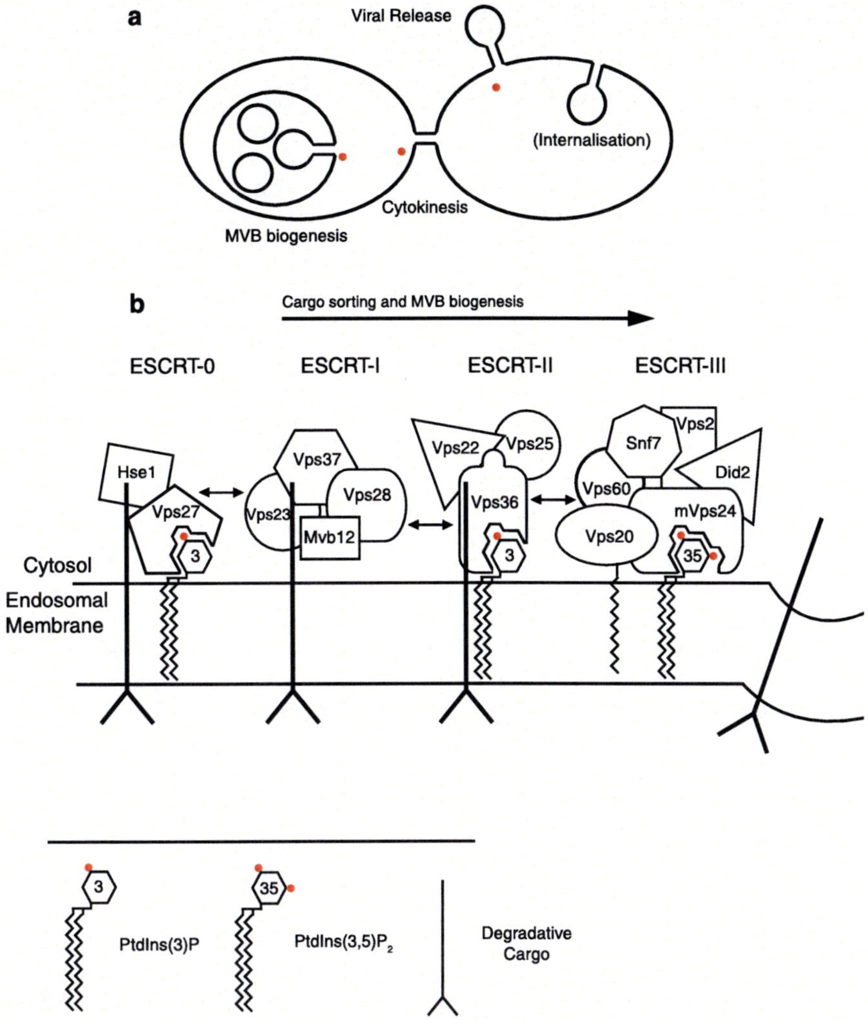

Fig. 3.4 Phosphoinositide-dependent regulation of ESCRT activity. **a** Topological equivalence of sites of ESCRT-activity. The ESCRT-machinery regulates membrane fission events leading to MVB biogenesis, cytokinesis and viral release, indicated by the *red asterisks*. Note topological equivalence of these events and contrast with topology of internalisation. **b** Known phosphoinositide-binding activities within discrete ESCRT-complexes. Yeast nomenclature used for simplicity (barring mVps24). It is unknown whether yeast Vps24 shares PtdIns(3,5)P_2 binding specificity

equivalent role in severing the membranous stalk connecting daughter cells at the end of cytokinesis (Carlton and Martin-Serrano 2007; Morita et al. 2007; Carlton et al. 2008). The topological equivalence of the membrane scission event required for ILV biogenesis, viral particle release and cytokinesis suggests a common mechanism is involved (Fig. 3.4a). That all three processes require the ESCRT-machinery to effect

Table 3.1 ESCRT-subunits in mammals and yeast

Complex	Yeast	Mammals
ESCRT-0	Vps27p	HRS
	Hse1p	STAM1, STAM2
ESCRT-I	Vps23p	TSG101
	Vps28p	VPS28
	Vps37p	VPS37a, VPS37b, VPS37c, VPS37d
ESCRT-II	Vps22p	EAP30
	Vps25p	EAP20
	Vps36p	EAP45
ESCRT-III	Did2p/Vps46p	CHMP1a, CHMP1b
	Vps2p	CHMP2a, CHMP2b
	Vps24p	CHMP3
	Snf7p	CHMP4a, CHMP4b, CHMP4c
	Vps60p	CHMP5
	Vps20p	CHMP6
	?	CHMP7
ESCRT-associated	Vps4p	VPS4a, VPS4b
	Vta1p	LIP5
	Ist1p	IST-1
	Bro1p	ALIX, HD-PTP

this scission suggests further that this machinery provides this membrane fission activity. Thus, a picture emerges of the ESCRT-machinery as a transplantable membrane deformation and scission apparatus, able to impose its topologically unique activity upon cellular membranes.

In the context of endosomal sorting, the *vps34* and *vps15* mutants in yeast display clear defects in the trafficking hydrolases to the yeast vacuole (Schu et al. 1993; Stack et al. 1993). Vps34p is the sole PI 3-kinase in yeast, and Vps15p its regulatory subunit, suggesting that 3-phosphoinositides are important in regulating degradative sorting. In mammalian cells, treatment with PI 3-Kinase inhibitors such as wortmannin or microinjection of antisera targeting the Class-III PI 3-kinase produce a striking defect in MVB biogenesis with endosomes lacking ILVs (Fernandez-Borja et al. 1999; Futter et al. 2001). These data suggested that products of this Class 3 PI 3-kinase controlled key events allowing MVB biogenesis and subsequent degradative sorting, findings confirmed by an analysis of sorting in cells in which endosomal PtdIns(3)P was sequestered (Petiot et al. 2003). This endosomal PI 3-kinase activity is important for generating a localization cue for a variety of endosomal regulators. As described previously, Rab5-GTP can associate with the endosomal PI 3-kinase complex, coupling Rab5 activity to PtdIns(3)P generation (Christoforidis et al. 1999). Further, PtdIns(3)P-binding Rab5 effectors such as EEA1 and Rabenosyn5 can be recruited to these microdomains, and their associated proteins such as the Rab5 GEF Rabex-5 can further stimulate Rab5 activation and thus PI 3-kinase activity, suggesting how microdomains of PtdIns(3)P may be generated upon the endosomal membrane (Zerial and McBride 2001). These microdomains have been visualized in mammalian cells (Gillooly et al. 2000b, 2003) and act as localization signals for PtdIns(3)P-binding endosomal regulators. Generation of these domains appears dependent upon the localised activation of PI 3-kinases and 3′-phosphatases, which, as discussed earlier are subject to feed-forward control through the actions of Rab5. These findings support the generation of highly plastic endomembrane domains to support nucleation of the ESCRT-machinery. Here we will examine 3-phosphoinositide binding domains present within distinct ESCRT-complexes and will consider how these endosomal lipids may regulate ESCRT-dependent cargo sorting to the MVB.

3.6.1 ESCRT-0

ESCRT-0 is a complex of HRS, STAM-1 and -2 in mammalian cells (Vps27p and Hse1p in yeast) (Williams and Urbe 2007). A FYVE domain within HRS confers high affinity PtdIns(3)P-binding upon this complex (Burd and Emr 1998; Kutateladze et al. 1999; Raiborg et al. 2001; Sankaran et al. 2001), directing ESCRT-assembly to PtdIns(3)P-enriched microdomains (Fig. 3.4b). Multiple VHS (Vps27, Hrs, Stam) and UIM domains within ESCRT-0 components bind strongly and co-operatively to ubiquitinated cargo (Ren and Hurley), ensuring that these cargoes and downstream ESCRT-complexes are retained within the ESCRT-0 containing PtdIns(3)P microdomain. Subsequent interactions of ESCRT-0 with deubiquitinating enzymes such as AMSH or UBPY (McCullough et al. 2004, 2006; Mizuno et al. 2006), explain how ubiquitin removal and recycling is coupled to cargo sequestration. ESCRT-0 thus appears a PtdIns(3)P-localised endosomal adaptor for subsequent ESCRT-recruitment.

3.6.2 ESCRT-I and -II

ESCRT-I and -II play important roles in linking ESCRT-III with upstream adaptors: in the case of endosomal sorting—the ESCRT-0 complex, and in the case of viral release or cytokinesis, viral Gags or the cytokinetic apparatus respectively (Carlton and Martin-Serrano 2009). ESCRT-I functions as a 1:1:1:1 heterotetramer (Vps23p, Vps28p, Vps37p, Mvb12p) in yeast; paralogues in mammalian cells) that comprises a ubiquitin-binding UEV-domain from TSG101/Vps23p (Sundquist et al. 2004; Teo et al. 2004) linked to an elongated stalk made from all 4 ESCRT-I subunits (Kostelansky et al. 2006, 2007; Teo et al. 2006; Gill et al. 2007). This stalk terminates in a headpiece that connects to ESCRT-II via interactions between Vps28p and the GLUE domain of Vps36p/EAP45 within ESCRT-II (Teo et al. 2006; Gill et al. 2007; Im and Hurley 2008). The presence of ubiquitin binding domains within ESCRT-I and -II (the UEV domain of TSG101 (Sundquist et al. 2004; Teo et al. 2004), non-canonical Ub-binding domain within Mvb12p (Shields et al. 2009), the GLUE domain of ESCRT-II (Slagsvold et al. 2005; Alam et al. 2006; Hirano et al. 2006) suggest how ubiquitinated cargo can be transported through these complexes). No defined phosphoinositide binding domains have been observed within ESCRT-I and consistent with this, the recombinant ESCRT-I complex binds only to modestly to acidic membranes with moderate enhancement given by the presence of PtdIns(3)P (Kostelansky et al. 2007). Whilst unable to bind PtdIns(3)P when in isolation, mixture of ESCRT-I and -II produced a PtdIns(3)P-binding complex, suggesting that interaction with ESCRT-II is sufficient to drive ESCRT-I onto PtdIns(3)P containing membranes (Fig. 3.4b) (Teo et al. 2006).

The ESCRT-II core binds to acidic membranes in a manner requiring the GLUE domain of Vps36p. This domain was found to bind PtdIns(3)P (Table 3.1), suggesting how ESCRT-II could localize independently to PtdIns(3)P enriched endosomes (Fig. 3.4b) (Teo et al. 2006). Indeed, in mammalian cells, ESCRT-II localizes to endosomal membranes and could bind 3-phosphoinositides (Slagsvold et al. 2005) (although in this case, interaction with PtdIns(3,4,5)P_3 was observed via fat-blot). The GLUE domain has a PH-domain like fold (Slagsvold et al. 2005; Teo et al. 2006). In fungi, two NZF fingers make a large 150-residue insertion into this fold, suggesting that it may be a split PH domain. The 1st NZF-finger binds Vps28 in ESCRT-I, the second NZF-finger binds ubiquitinated cargo (Teo et al. 2006). Split phosphoinositide binding domains comprise a relatively new class of PI binding module, formed by intra- or inter-molecular interactions of half-domains that preclude their identification by sequence-based searching (van Rossum et al. 2005). Crystalisation of the yeast Vps36p GLUE domain lacking this insertion reveals a PH domain with a non-canonical PtdIns(3)P binding pocket, which is responsible for endosomal localization (Teo et al. 2006). The spatial relationship between the NZF-fingers and the PtdIns(3)P binding surfaces is unclear in the context of holo-Vps36, although interactions between lipid, ESCRT-I and ubiquitinated cargo can be simultaneously observed, suggesting that the ESCRT-II GLUE domain is a master regulator of ESCRT-dependent traffic and that these co-incident localization cues may act as foci for ESCRT-assembly (Fig. 3.4b). Indeed, expression of a Vps36p mutant rendered unable to bind PtdIns(3)P induced a severe defect in degradative sorting, emphasizing the critical role 3-phosphoinositides play in this process (Teo et al. 2006). Importantly, mammalian EAP45 lacks these NZF fingers, yet retains an ability to bind ESCRT-I and ubiquitinated cargo, suggesting a functional conservation of ESCRT-II GLUE domain activity in the process of lysosomal sorting (Slagsvold et al. 2005). In addition to the 3′-phosphoinositide binding GLUE domain, the mammalian ESCRT-II subunit VPS22/EAP30 contains an acidic lipid-binding N-terminal helix. This helix is conserved from yeast to humans and cooperates with the 3-phosphoinositide binding GLUE domain to allow full binding of ESCRT-II to cellular membranes (Im and Hurley 2008).

3.6.3 ESCRT-III

ESCRT-III has long been considered the machinery providing membrane fission activity. ESCRT-III subunits are all highly related Charged Multivesicular Body Proteins (CHMP) with a highly basic N-terminus and a highly acidic C-terminus

(Williams and Urbe 2007). They are thought to fold back into a 'closed' or autoinhibited state, rendering them inactive. Like other ESCRT-subunits, ESCRT-III proteins are mostly peripheral membrane proteins. The exception to this Vps20p/CHMP6; this protein is myristolated and thought to be pre-localised to endosomal membranes. Subsequent interaction of CHMPs with membranes or other CHMP proteins has been postulated to drive them into an active conformation, allowing them to drive membrane fission (Williams and Urbe 2007; Teis et al. 2008; Saksena et al. 2009). ESCRT-III proteins are able to bind to membranes and deform them into inward invaginations, recapitulating key features of ESCRT-activity (Hanson et al. 2008; Saksena et al. 2009). The ESCRT-III subunit CHMP3 was isolated by virtue of its ability to bind 3-phosphoinositides including PtdIns(3,5)P_2 (Fig. 3.4b, Table 3.1) (Whitley et al. 2003). Again, no detectable phosphoinositide-binding domains were observed, suggesting that this interaction occurred through a non-canonical PI-binding motif. Indeed, other ESCRT-III proteins that are thought to form part of the core membrane fission machinery (CHMP2a, CHMP4b) could be isolated from cytosolic extracts upon PtdIns(3,5)P_2 liposomes (Catimel et al. 2008), suggesting that PtdIns(3,5)P_2 recognition may be a conserved, if unexplained, feature of this complex that tallies with the predicted location of this lipid (Fig. 3.1b). Crystallization of the central region of CHMP3 demonstrated this protein folded into a multimeric helical arrangement with a flat membrane targeting face (Muziol et al. 2006). Both basic membrane targeting and multimerisation activities were required for CHMP3 function, and sequence analysis of other ESCRT-III subunits revealed similar electrostatic potentials, suggesting that these proteins function as a membrane bound lattice to effect ESCRT-activity (Muziol et al. 2006). Thus, whilst CHMP3 and possibly other CHMPs can bind to 3-phosphoinositides, it is possible that the ESCRT-III complex acts simply on cellular membranes to effect membrane scission with upstream ESCRT-complexes (ESCRT-0 and ESCRT-II) directing its assembly in the endosomal context to regions of endosomal PtdIns(3)P. Consistent with this possibility is the finding that ESCRT-III can function effectively at the plasma membrane—usually devoid of PtdIns(3)P—to effect viral particle release. However, it should be noted that PtdIns(3)P has been detected at the midbody of dividing cells and localizes a PtdIns(3)P-binding protein called FYVE-CENT to this structure (Sagona et al. 2010). FYVE-CENT is required for cytokinesis and can interact with ESCRT-III via TTC19 (Sagona et al. 2010). These data suggest that under some conditions, PtdIns(3)P may be generated upon non-endosomal membranes where it may assist ESCRT-dependent events.

3.6.4 In Vitro Reconstitution of ESCRT Function

Recent elegant biophysical studies have reconstituted ESCRT-activity upon artificial giant unilamellar vesicles (GUVs) *in vitro* (Wollert et al. 2009; Wollert and Hurley 2010). Here, ordered addition of ESCRT-0, -I, -II and -III acted to sequester ubiquitinated cargo, generate invaginations and effect ILV release from the limiting membrane of GUVs containing small molar percentages of PtdIns(3)P. These data recapitulated numerous observed functions of the ESCRT-machinery, with ESCRT-0 required for cargo sequestration, ESCRT-I and -II acting to produce inward invaginations and ESCRT-III acting to release the ILV. An isolated study of ESCRT-III-dependent ILV release revealed that PtdIns(3)P or PtdIns(3,5)P_2 provided only a twofold enhancement of ILV release, suggesting that 3-phosphoinoitides are not absolutely required for ESCRT-III activity (Wollert et al. 2009) and may instead play regulatory roles in this activity and/or specifying endosomal localisation of this machinery through coordination of upstream regulators.

3.6.5 ESCRT-independent, Phosphoinositide-mediated Sorting into MVBs

Traffic to the lysosome appears PtdIns(3)P-dependent (Petiot et al. 2003) and the ESCRT proteins are localized to endosomes in a 3-phosphoinositide-dependent manner. However, ESCRT-dependent routes are not the only way of generating MVBs; Gruenberg and colleagues have recently demonstrated that the PtdIns(3)P-binding protein SNX3 is required for MVB biogenesis, suggesting either that SNX3 may coordinate the formation of a distinct class of MVB from those formed downstream of Hrs, or that SNX3 acts synergistically with the ESCRTs to produce ILVs. Yet other data have shown that MVBs are still present in cells undergoing long term depletion of numerous ESCRT-proteins (Stuffers et al. 2009), that some proteins can access the MVB lumen independently of ESCRTs (Theos et al. 2006) and that enveloped viruses such as influenza (Bruce et al. 2009; Rossman et al. 2010) can exit cells in an ESCRT-independent manner. All of these data point to the existence of ESCRT-independent mechanism for ILV biogenesis and membrane fission. Whether SNX3 co-ordinates such an activity remains an exciting area of future study.

3.7 Phosphoinositides and Retrograde Transport

Retrograde trafficking is the process by which proteins and lipids are transported between endosomes and the biosynthetic/secretory compartments including the *trans*-Golgi network (TGN) and Golgi apparatus (Fig. 3.5a) (Bonifacino and Rojas 2006; Johannes and Popoff 2008). It defines not a single pathway, but a number of mechanistically distinct transport routes that may function in parallel or sequentially. Retrograde transport has a variety of roles in normal physiology, which include the retrieval of receptors and cargoes from the endo-lysosomal degradative pathway, cellular adaptation to fluctuations in the nutrient environment and the establishment of morphogenic gradients during tissue patterning(Johannes and Popoff 2008).

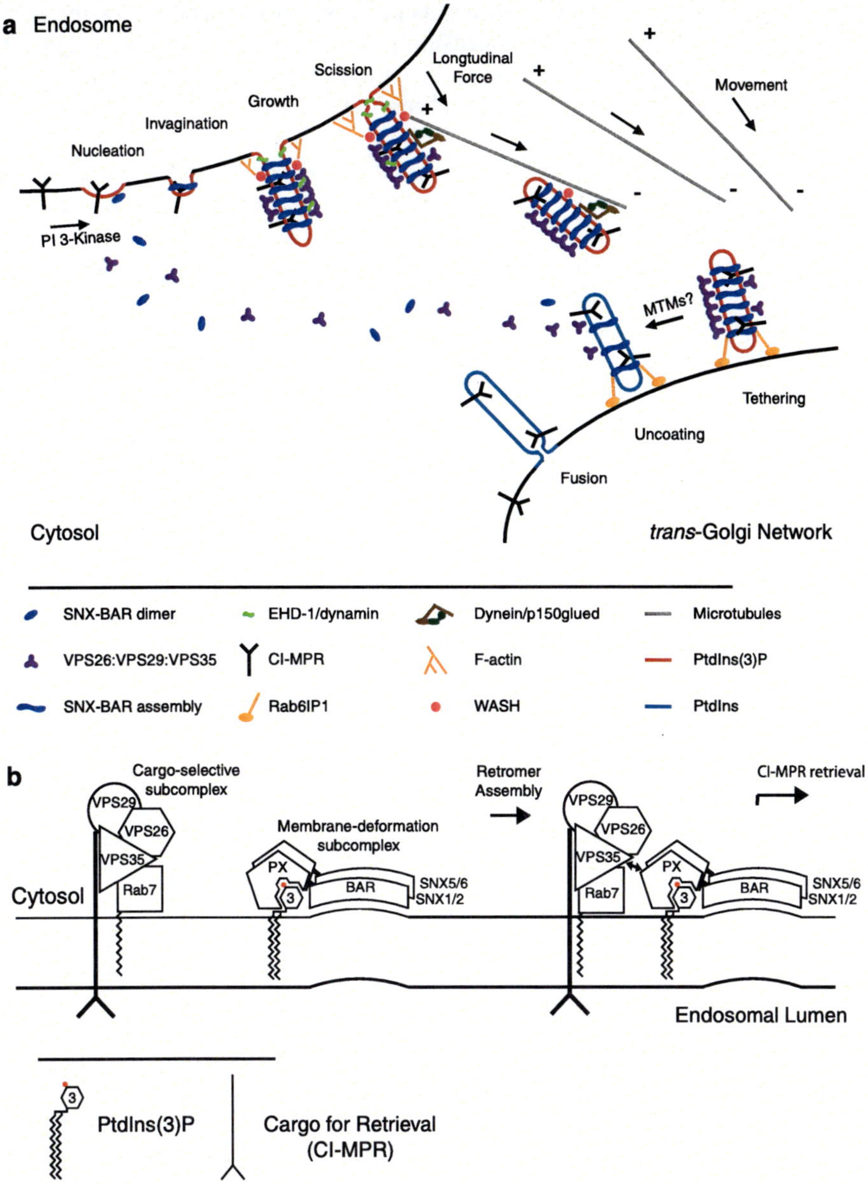

Fig. 3.5 Retromer-mediated retrograde endosome-to-TGN transport. **a** TGN resident cargoes (CI-MPR) are retrieved from endosomes, in part through the actions of the retromer complex. WASH and EHD-1 assist this process. F-actin polymerisation may drive carrier biogenesis, Directed movement to the TGN is accomplished through dynein-mediated transport to the perinuclear region where Rab6IP1 acts to tether incident cargo and dephosphorylation of PtdIns(3)P is thought to allow removal and recycling of retromer components. **b** Schematic depiction of retromer assembly. Retromer comprises 2 subcomplexes—a membrane deforming SNX-BAR protein containing subcomplex and a cargo-binding VPS26:VPS29:VPS35 subcomplex. Rab7-GTP is thought to help localise the cargo-binding subcomplex. 3′-phosphoinositides are thought to localise the SNX-BAR subcomplex

Moreover, endosome-to-Golgi transport is perturbed in various diseases and is required for the cellular entry of pathogens and pathogenic toxins (Johannes and Popoff 2008). Defining the mechanistic basis of retrograde transport is therefore important in understanding human disease and constitutes a novel avenue for therapeutic development.

A number of molecular entities play important roles in retrograde transport (see Johannes and Popoff (2008) for an extensive list). Included are clathrin (Lauvrak et al. 2004; Saint-Pol et al. 2004; Esk et al. 2010), clathrin-binding proteins such as epsinR, AP-1, AP-3, and OCRL, the clathrin uncoating ATPase Hsc70 (Folsch et al. 2001; Medigeshi and Schu 2003; Saint-Pol et al. 2004; Choudhury et al. 2005; Popoff et al. 2009; Shi et al. 2009), and various small GTPases including ARF1 and Rab9 along with their regulatory proteins (Lombardi et al. 1993; Diaz et al. 1997; Shiba et al. 2010). In addition, the VPS34 PtdIns 3-kinase (Burda et al. 2002), the PtdIns(3)P 5-kinase PIKfyve (Rutherford et al. 2006), the PtdIns(4,5)P_2 5-phosphatase OCRL (Choudhury et al. 2005) have all been implicated in retrograde transport as have the PtdIns(4)P-binding proteins AP-1 (Wang et al. 2003) and epsinR (Mills et al. 2003) and a number of PtdIns(3)P and PtdIns(3,5)P_2-binding proteins that make up components of the multimeric retromer complex (Cozier et al. 2002). That a number of these phosphoinositide regulatory and binding proteins also associate with clathrin and small GTPases hints at a complex level of communication between multiple retrograde transport pathways.

3.7.1 The Retromer Complex

A major advance in our understanding of the role played by phosphoinositides in the regulation of retrograde transport came with the identification of the yeast retromer complex (Seaman et al. 1998). Here retromer is composed of two sub-complexes: a stable cargo-binding Vps26p-Vps29p-Vps35p trimer and a membrane deforming coat comprising the PtdIns(3)P-binding proteins Vps5p and Vps17p (Fig. 3.5b) (Burda et al. 2002). Retromer is evolutionarily conserved, being present in the most primitive of eukaryotic ancestors (Dacks et al. 2009), and functions to regulate endosome-to-Golgi retrograde transport of cargo proteins including the yeast carboxypeptidase Y receptor Vps10p (reviewed in (Bonifacino and Hurley 2008; Verges 2008; Attar and Cullen 2010)). The mammalian retromer is more complex and comprises a VPS26-VPS29-VPS35 trimer (including two isoforms of VPS26) (Edgar and Polak 2000; Haft et al. 2000; Kerr et al. 2005) and gene duplication appears to have resulted in two Vps5p's, sorting nexin-1 (SNX1) and SNX2, and two, possibly three Vps17p's, SNX5 and SNX6 (and possibly SNX32) (Teasdale et al. 2001; Wassmer et al. 2007, 2009). Like its yeast counterpart, mammalian retromer regulates retrograde endosome-to-Golgi transport of an increasingly diverse group of cargo proteins which include the cation-independent mannose 6-phosphate receptor (CI-MPR) (Arighi et al. 2004; Carlton et al. 2004; Seaman 2004), sortilin (Seaman

2007; Mari et al. 2008), the Wntless receptor/chaperone for Wnt developmental morphogens (Belenkaya et al. 2008; Franch-Marro et al. 2008; Pan et al. 2008; Port et al. 2008), polymeric immunoglobulin receptor (Verges et al. 2004), the iron transporter DMT1-II (Tabuchi et al. 2010), amyloid precursor protein (He et al. 2005; Small et al. 2005) and Shiga toxin (Bujny et al. 2007; Popoff et al. 2007). Retromer has also been linked with trafficking between mitochondria and peroxisomes (Braschi et al. 2010), and in the clearance of apoptotic bodies in *C. elegans* (Chen et al. 2010).

3.7.1.1 Membrane Deformation Within the Retromer Pathway

The retromer sorting nexins all share a common domain architecture defined by an amino-terminal phosphoinositide-binding PX domain (which in most cases binds the early endosomal phosphoinositide PtdIns(3)P but can also associated with PtdIns(3,5)P_2 (Cozier et al. 2002)) and a carboxy-terminal BAR domain (Cullen 2008): a domain that through dimerisation generates a rigid 'banana-shaped' structure, the concave face of which preferentially associates with membranes of high positive curvature (Frost et al. 2009). The combined ability to bind PtdIns(3)P and the curvature sensing mode of the BAR domain generates a co-incidence signal that effectively targets retromer sorting nexins to high curvature sub-domains of the PtdIns(3)P-enriched early endosome (Fig. 3.3c) (Carlton et al. 2004).

Like other BAR domain-containing proteins, retromer sorting nexins have an inherent potential to drive the formation of membrane tubules (Carlton et al. 2004; Gokool et al. 2007). Here, as co-incidence detection increases their effective membrane concentration, retromer sorting nexins undergo spontaneous assembly into a helical coat (Frost et al. 2008). In forming this coat, the rigid nature of the BAR domain imposes membrane curvature which is accommodated by the formation of membrane tubules. In other words, on increasing their effective membrane concentration retromer sorting nexins may switch from a curvature sensing to curvature inducing mode (Roux et al. 2010a). That said, although these proteins can induce tubulation in a protein-free reconstituted membrane system (Carlton et al. 2004), nucleation *in vivo* may require additional factors.

One factor that lies upstream of retromer and therefore may play a role in retromer nucleation *in vivo* is clathrin (Johannes and Popoff 2008). Evidence from endosome-to-TGN transport of Shiga toxin and CI-MPR, suggests that clathrin and retromer act sequentially in sorting from endosomes (Popoff et al. 2007, 2009). Indeed, a two-step model for retrograde sorting has been presented in which retromer drives the formation of a membrane tubule from the edge of a flat, cargo-enriched clathrin containing lattice or, alongside other clathrin adaptors such as AP-1 and epsinR, clathrin generates a dynamic flat, cargo-enriched lattice that nucleates the formation of high membrane curvature which is subsequently stabilized and processed into cargo-enriched carriers by retromer (Fig. 3.5a) (Popoff et al. 2007; Johannes and Popoff 2008). Irrespective of the subtilties of these models, evidence that retromer sorting nexins can form complexes with the clathrin-binding protein HRS (Chin et al. 2001; Popoff et al. 2009), the clathrin uncoating ATPase Hsc70 (Popoff et al. 2009;

Shi et al. 2009), and clathrin heavy chain 22 (Towler et al. 2004; Esk et al. 2010), all point to a dynamic relationship between clathrin and retromer. Clarifying the precise nature of these relationships will be an important area for future research into retrograde transport.

3.7.1.2 Cargo Selection—The Mammalian VPS26-VPS29-VPS35 Sub-complex

One route for endosomal association of the trimeric cargo-selective retromer sub-complex is through binding to the GTP-bound form of Rab7 (Fig. 3.5b) (Rojas et al. 2008; Seaman et al. 2009): this is entirely consistent with retromer-decorated tubules occurring at a relative late stage of early endosome maturation close to the transition with Rab7-labeled late endosome (Mari et al. 2008). The VPS35 subunit has been implicated in direct recognition of sorting motifs within the cytosolic region of cargo proteins: for CI-MPR and sortilin this comprises a linear tripeptide motif of [*FM*]*L*[*MV*] with similar motifs being present in the cytosolic tails of sorLA, Wntless and DMT1-II (Seaman 2007; Tabuchi et al. 2010). Regions lying upstream and/or downstream of these motifs also appear important for efficient cargo binding and, at least for CI-MPR and sortilin, palmitoylation by DHHC-15 is required for the interaction with VPS35 (McCormick et al. 2008). Structurally, VPS35 is arranged as an α-solenoid composed of a predicted series of 17 HEAT-like α-helical repeats (Hierro et al. 2007). Onto this core component the VPS29 and VPS26 subunits bind independently at respectively the carboxy and amino distal ends to form an extended trimeric complex (Hierro et al. 2007; Norwood et al. 2011). VPS29 itself displays a phosphatase-like fold, although this appears to be non-functional (Collins et al. 2005; Wang et al. 2005; Damen et al. 2006; Hierro et al. 2007), and VPS26 has an arrestin-like fold, leading to the suggestion that it may also play a role in cargo recognition (Shi et al. 2006; Collins et al. 2008). Currently there is no data to support this conclusion.

Away from the process of cargo selection, the VPS26-VPS29-VPS35 heterotrimer also functions as a hub for the scaffolding of additional complexes that play important regulatory roles in endosomal sorting (Harbour et al. 2010). Interacting proteins include TBC1D5, a Rab GTPase-activating protein that probably functions on Rab7 (Harbour et al. 2010), FKBP15 (Harbour et al. 2010), and, most notably, the WASH complex. In humans this is composed of WASH and a core unit, recently termed the 'WASH regulatory complex' (Jia et al. 2010), comprising FAM21, SWIP, Strumpellin and CCDC53 (coiled-coil domain containing protein 53) (Derivery et al. 2009; Gomez and Billadeau 2009; Harbour et al. 2010; Jia et al. 2010). VPS26-VPS29-VPS35 associates through direct binding of VPS35 to FAM21 (Harbour et al. 2010): additional binding of VPS35 and SNX1/SNX2 to WASH and FAM21 may stabilize the interaction (Gomez and Billadeau 2009; Harbour et al. 2010). Importantly, while FAM21 does bind to a variety of phosphoinositides including PtdIns(3)P and PtdIns(3,5)P_2 (Jia et al. 2010) (a note of caution here—data derived using fat blots), the presence of VPS26-VPS29-VPS35 is necessary for the association of the WASH complex to endosomes (Harbour et al. 2010).

Functionally, suppression of the WASH complex results in the formation of extended and exaggerated retromer-labelled tubules leading to a defect in retrograde sorting of the CI-MPR (Derivery et al. 2009; Gomez and Billadeau 2009; Harbour et al. 2010). WASH itself is a member of the Wiskott-Aldrich syndrome protein (WASP) family member, and shares with these proteins an ability to regulate the actin nucleating ability of the Arp2/3 complex (Derivery et al. 2009; Gomez and Billadeau 2009; Jia et al. 2010) (there are obvious parallels here with the role of actin polymerization in clathrin-mediated endocytosis). Importantly, while isolated WASH is constitutively active in driving Arp2/3-dependent actin nucleation, the addition of the regulatory complex is inhibitory (Jia et al. 2010). Moreover, the WASH regulatory complex also interacts, although more weakly, with CAPZ (Harbour et al. 2010; Jia et al. 2010). This binds to the barbed ends of actin filaments, preventing further elongation and promoting increased branch density—characteristics required for force generation. A model emerges therefore where the VPS26-VPS29-VPS35 sub-complex regulates the association of the WASH complex to sites of endosomal tubulation (Harbour et al. 2010). From here, Arp2/3-dependent nucleation leads to actin polymerization suitable for generating longitudinal force, thereby aiding the efficiency of tubule scission (Fig. 3.5a) (Derivery et al. 2009; Gomez and Billadeau 2009). Overall, given that the VPS26-VPS29-VPS35 sub-complex is also required for the endosomal association of TBC1D5 and FKBP15 (Harbour et al. 2010), one should consider that, besides a role in cargo selection, this sub-complex acts to scaffold the recruitment of accessory proteins that regulate the dynamics of tubular profiles during retromer-mediated endosomal transport.

Modes of cargo selection that are independent of the VPS26-VPS29-VPS35 are also evident in the retromer pathway. In yeast the PtdIns(3)P-binding sorting nexin, Grd19/Snx3p, acts as a cargo-specific adaptor, which, by ‘piggy-backing’ on the retromer sorts cargo to the Golgi (Strochlic et al. 2007). Whether the equivalent proteins in mammalian cells, sorting nexin-3 (SNX3) and sorting nexin-12 (SNX12), function in a similar manner remains to be determined. However, if this does prove to be the case, and as SNX3 has been implicated in the biogenesis of MVBs ((Pons et al. 2008), see Sect. 3.6.5), this sorting nexin may form a nexus between PtdIns(3)P-mediated endo-lysosomal sorting and retrograde transport. Finally, retromer sorting nexins also directly associate with cargoes including EGFR, TGFβR and various G-protein-coupled receptors (Kurten et al. 1996; Otsuki et al. 1999; Parks et al. 2001; Pons et al. 2003; Heydorn et al. 2004; Gullapalli et al. 2006; Yoon et al. 2006; Nisar et al. 2010). Together, these VPS26-VPS29-VPS35-independent routes for cargo selection allow for a far greater repertoire of retromer-dependent cargoes.

3.7.1.3 Processing of Retromer Tubules

As the retromer sorting nexins, SNX1 and SNX2, can association with VPS35 and VPS29 (Haft et al. 2000; Collins et al. 2005), the classic model describing retromer function argues that it acts by co-ordinating cargo selection with an ability to drive membrane tubulation, thus enriching cargo into tubular carriers destined for retrograde transport. Once cargo has been enriched into the tubular profiles membrane scission occurs via a mechanism that has yet to be described, but may require the interaction of retromer with dynamin and/or the dynamin-like protein EHD1 (Fig. 3.5a) (Daumke et al. 2007; Gokool et al. 2007). However, what is emerging is that the generation of longitudinal tension is an important element for efficient processing of retromer tubules. As discussed above, retromer associates with the WASH complex thereby driving the polymerization of actin (Derivery et al. 2009; Gomez and Billadeau 2009), and SNX5 and SNX6 are able to bind the p150^{glued} component of the dynein-dynactin minus-end directed microtubule motor complex (Hong et al. 2009; Wassmer et al. 2009). In both cases, uncoupling of retromer from actin polymerization or motor function leads to the formation of extend tubular profiles that appear unable to undergo efficient processing into tubular carriers (Gomez and Billadeau 2009; Wassmer et al. 2009).

In the final steps of retromer-mediated retrograde transport, coupling to the dynein-dynactin motor allows for long-range movement of cargo-enriched carriers from the cell periphery towards the *trans*-Golgi network (Fig. 3.5a) (Wassmer et al. 2009). The carriers themselves, referred to as ETC's (endosome-to-TGN transport carriers), are devoid of clathrin and appear as non-branched tubules on average 170–230 nm in length and 20–50 nm in diameter(Mari et al. 2008). Recognition of ETCs at the TGN is mediated, at least in part, through the binding of SNX1 and SNX2 by Rab6IP1 (Fig. 3.5a) (Wassmer et al. 2009): a TGN resident protein proposed to act as a tethering factor for carriers from the endosomal network (Miserey-Lenkei et al. 2007). That components of the retromer coat are required during the tethering process suggests that uncoating of ETCs occurs after arrival at the TGN: a scenario similar to the recently described interaction of the COPII coat with the TRAPPI tethering complex (Cai et al. 2007). The nature of the signal(s) that leads to uncoating remains to be defined, although based on analogue with clathrin uncoating during endocytosis, one can speculate on the involvement of phosphoinositide phosphatases and the need for energy input to destabilize the assembled sorting nexin coat complex.

3.7.2 *Phosphoinositide Regulation of Retromer-independent Retrograde Transport*

Other phosphoinositide-binding sorting nexins have been implicated in retrograde transport. RNAi suppression of early endosomal associated sorting nexin-8 (SNX8) leads to an inhibition of retrograde ricin transport whilst elevating endosome-to-TGN transport of Shiga toxin (Dyve et al. 2009). In addition, suppression of sorting nexin-4 (SNX4) leads to an inhibition of retrograde ricin transport, which is enhanced by retromer co-suppression (Skanland et al. 2007) (see Sect. 3.8.2 for more detailed discussion of SNX4). Finally, a paralog of SNX9, sorting nexin-18 (SNX18), has been localised to early endosomes positive for the AP-1 adaptor and PACS1 (Haberg et al. 2008). This sorting nexin displays an *in vitro* binding preference for PtdIns(4,5)P_2, has the ability to drive membrane tubulation (it contains a BAR domain) and forms complexes with dynamin-2 and AP-1 (Haberg et al. 2008). SNX18 may therefore scaffold a membrane-fission unit that mediates carrier formation during AP-1-positive endosomal trafficking (Haberg et al. 2008). That said, others have questioned this interpretation, preferring a model in which SNX18 plays a redundant function with SNX9 in modulating endocytosis at the plasma membrane (Park et al. 2010).

3.7.3 PtdIns(3,5)P_2 in Retrograde Transport

To completely define the role of phosphoinositides in endosomal sorting, one needs to complement the growing repertoire of molecular complexes regulated by PtdIns(3)P, with those regulated by PtdIns(3,5)P_2. As eluded to previously, this phosphoinositide clearly plays a key role in endosomal homeostasis: perturbing PtdIns(3,5)P_2 levels in a variety of organisms results in multiple endosomal phenotypes including endosomal swelling, and defects in acidification, MVB sorting, endosome-to-Golgi transport and the late endosome-to-lysosome axis(Michell et al. 2006; Jefferies et al. 2008; Shisheva 2008).

A major hurdle in our appreciation of the role played by this phosphoinositide is the lack of well defined, *bona-fida* PtdIns(3,5)P_2 effectors. While CHMP3, Ent3p and Ent5p have been suggested as PtdIns(3,5)P_2-binding proteins their true phosphoinositide specificity remains controversial(Michell et al. 2006). The best characterized specific PtdIns(3,5)P_2-binding protein is yeast Svp1p/ATG18 which regulates traffic from the vacuole (the yeast equivalent of the mammalian lysosome) back to the endosome (Dove et al. 2004). A mammalian ATG18 is WIPI49 which binds PtdIns(3)P and PtdIns(3,5)P_2, is localized to TGN/endosomes and regulates endosome-to-TGN transport of the CI-MPR(Jeffries et al. 2004). Since the retromer sorting nexins, SNX1 and SNX2 also bind with similar affinities PtdIns(3)P and PtdIns(3,5)P_2 (Cozier et al. 2002), these proteins have been independently isolated in a PtdIns(3,5)P_2 interactome screen (Catimel et al. 2008), the role of PtdIns(3,5)P_2 in mammalian retromer biology remains controversial, as does retromer's relationship with WIPI49. However, it is clear that defective retrograde endosome-to-TGN transport, through retromer-dependent and -independent pathways is observed upon manipulation of PtdIns(3,5)P_2 (Rutherford et al. 2006; Zhang et al. 2007; de Lartigue et al. 2009). Some of the phenotypes observed upon perturbing the level of this phosphoinositide may arise indirectly from the miss-sorting of important components in multiple pathways.

Moving away from retrograde transport, recent studies have linked PtdIns(3,5)P_2 with promoting the targeting of CaV1.2 channels to lysosomes, thereby promoting their degradation and protecting neuronal cells from excitotoxicity (Tsuruta et al. 2009), regulating the open probability of single RyR2 channels and hence modulating cardiac contractility (Touchberry et al. 2010), and perhaps most intriguingly, activating the endo-lysosome-localized mucolipin TRP (TRPML) channel (Dong et al. 2010). Interestingly, in cells lacking TRPLM1, late endocytic trafficking defects are observed along with the generation of swollen endo-lysosomes (Dong et al. 2010). The PtdIns(3,5)P_2-mediated regulation of the TRPLM1 channel is evolutionarily conserved being observed in yeast upon hyper-osmotic shock (Dong et al. 2010). As TRPLM1 channels regulate Ca^{2+} release from the endo-lysosomal network, a enticing aspect of PtdIns(3,5)P_2 biology may be the regulation of localized Ca^{2+} levels required for fusion and fission reactions during membrane trafficking (Dong et al. 2010).

3.8 Phosphoinositides in Endosomal Recycling

Cargos that are to be returned to the cell surface are rescued from the degradative route via endosomal recycling pathways. Recycling cargoes are returned to the plasma membrane from the early endosome either via a rapid recycling route or transit more slowly through a tubular juxtanuclear endocytic recycling compartment (the ERC) (Fig. 3.1a) (Maxfield and McGraw 2004). By virtue of their high surface area to volume ratio, recycling tubules are thought to effectively partition transmembrane receptors whilst capturing little luminal content (Maxfield and McGraw 2004). Although recycling has long been assumed to constitute a default sorting mechanism occurring by virtue of iterative geometric sorting, recent data have identified new protein and lipid regulators of this pathway, suggesting that machineries exist to recognize, retain and sort cargoes into these recycling routes. Endosomal recycling may therefore not be as simple as was once thought.

3.8.1 EHD Proteins and Recycling Tubules

Perhaps the clearest example of regulators of endosomal recycling is found in *C. elegans*. Here, Receptor-Mediated Endocytosis-1 (Rme-1) was identified as a regulator of endosomal recycling in the worm (Grant et al. 2001; Lin et al. 2001). Mammalian cells express four orthologues of Rme-1– the Eps15 homology domain (EHD) containing proteins EHD1–4 and of these, EHD-1 is the best characterized, being required for endosomal recycling of cargoes including the TfnR receptor, MHC-I molecule and numerous other recycling cargoes (Lin et al. 2001; Caplan et al. 2002; Grant and Caplan 2008). EHD-1 is a peripheral membrane protein that localizes to membrane tubules exiting the recycling endosome (Caplan et al. 2002). EHD-1 localisation to these tubules is abolished either upon expression of a PIP5K1γ, a PtdIns(4)P 5-kinase, or upon expression of SacI-1, a PtdIns(4)P 4′ phosphatase, suggesting that PtdIns(4)P controls localisation of EHD-1 to these tubules. The EHD proteins have a broad phosphoinositide-binding specificity in both fat-blots and liposome-based assays (Blume et al. 2007; Daumke et al. 2007; Naslavsky et al. 2007), binding promiscuously to many phosphoinositides, with some preference for the 4′-position. NMR-based analysis of the EHD-1 EH domain has demonstrated it can indeed bind PtdIns(4)P and a targeted mutation within this domain abolished EHD-1's ability to localize to endosomal recycling tubules and, importantly, prevented it from rescuing transferrin and β1 integrin recycling in EHD-1 depleted cells (Jovic et al. 2009). Interestingly, this mutant EHD-1 still localized to punctate endosomal structures, suggesting that EH-domain-independent mechanisms of endomembrane binding exist, but that 4′-phosphoinositide binding is important for recruitment to recycling tubules. In worm and mammals, association of EHD1/Rme1 with BAR-domain containing proteins (AMPH-1 in *C- elegans*, BIN-1 in mammals) and Rab8 effectors such as MICAL-L1 (Sharma et al. 2009) also assisted the tubular localization of these proteins and their function in endocytic recycling (Pant et al. 2009). Indeed, the localization of this complex to phosphoinositide-enriched tubules courtesy of coincident phosphoinositide:EH domain interactions and BAR domain:membrane interactions has parallels with the localization of endosomal retrieval complexes such as retromer ((Carlton and Cullen 2005), see Sect. 3.7.1.1) and suggest that cells may employ common mechanisms of localizing sorting complexes to endosomal exit sites via coincident localization signals.

Other EHD proteins also regulate various aspects of early endosome trafficking; EHD3 assists cargo retrieval to the Golgi (Naslavsky et al. 2009), (a pathway that EHD1 participates in through interactions with retromer (Gokool et al. 2007)), EHD1 and 3 interact with the Rab11-FIP2 proteins to regulate export from early endosomes to the endocytic recycling compartment (Naslavsky et al. 2006) and EHD4 is involved in the trafficking of proteins at the early endosome (Sharma et al. 2008). Given the role of the EH-domain in binding phosphoinositides, these data define the family of EHD proteins as important regulators of phosphoinositide-mediated endosomal trafficking.

3.8.2 *Sorting Nexins—Another Route for the Generation of Recycling Tubules*

PtdIns(3)P is another phosphoinositide important for traffic through recycling endosomes: for example, PtdIns(3)P consumption through overexpression of the 3′-phosphatase MTMR4 inhibits Tfn recycling (Naughtin et al. 2010). Here, an important component of the PtdIns(3)P-dependent recycling pathway is the sorting nexin, sorting nexin-4 (SNX4). Localised to both the early endosome and to membranes of the juxtanuclear ERC, SNX4 acts to regulate the formation of membrane tubules required for the geometric based recycling of TfnR away from the degradative route, allowing their return to the cell surface (Traer et al. 2007). Like the previously described role of retromer sorting nexins (see Sect. 3.7.1.1), SNX4 contains a PtdIns(3)P-binding PX domain and a curvature-sensing BAR domain, which can switch to a curvature inducing mode, driving the formation of recycling membrane tubules (Traer et al. 2007). Furthermore, nucleation of SNX4 tubules may again be intimately linked with clathrin dynamics, given that this sorting nexin interacts with clathrin (Skanland et al. 2009). Again, as with the retromer SNX4 also associates with the minus-end directed microtubule motor protein dynein to aid the efficiency of tubule fission (Traer et al. 2007; Skanland et al. 2009), although whether it also interacts with elements that regulate actin polymerization remains to be established.

In yeast, Snx4p is part of a complex with two additional sorting nexins, Snx41p and Snx42p, and acts to retrieve Snc1p from a post-Golgi endosome (Hettema et al. 2003), although in this case, the complex functions in parallel with retromer, returning cargoes to the yeast Golgi. As discussed previously (see Sect. 3.7.2), mammalian SNX4 is also thought to play a role in endosome-to-Golgi trafficking of toxins such as ricin (Skanland et al. 2007), suggesting that it may bisect both recycling and retrieval pathways. Recent data suggesting that the endocytic recycling compartment lies on a route between early endosomes and the Golgi may reconcile these itineraries (Ang et al. 2004). Whether mammalian cells also contain a similar SNX4 complex, the most closely related proteins to Snx41p and Snx42p would appear to be SNX7 and SNX30, remains to be explored. Given that SNX7 and SNX30 each contain predicted phosphoinositide-binding PX domains, studying their function alongside SNX4 may reveal exciting new insight into phosphoinositide-mediated endosomal recycling.

3.8.3 Sorting Nexins and Cargo Recognition

Besides the role of SNX4 in driving the formation of endosomal recycling tubules (Traer et al. 2007), other members of the sorting nexin family have emerged as cargo adaptors in endosomal recycling. Sorting nexin-17 (SNX17), which contains a PtdIns(3)P-binding PX domain and a truncated FERM (4.1, ezrin, radixin, and moesin) domain (it lacks a BAR domain), is associated with recycling tubules emanating from the early endosome (Florian et al. 2001; van Kerkhof et al. 2005). As an adaptor, SNX17 associates directly with a number of recycling proteins including low-density lipoprotein (LDL) receptor-related protein (LRP) (here binding is via an NPxY motif in the cytosolic domain of the receptor) (Stockinger et al. 2002; Burden et al. 2004; van Kerkhof et al. 2005), amyloid precursor protein (APP) (recognition via YXNPXY motif) (Lee et al. 2008), the scavenger receptor FEEL-1/stabilin-1 (possible involvement of an NPxF motif) (Adachi and Tsujimoto 2010), and P-selectin (recognition motif FxNaa(F/Y)) (Florian et al. 2001; Williams et al. 2004; Knauth et al. 2005). Where trafficking itineries have been explored, data is consistent with SNX17 sorting receptors into an endosomal recycling pathway (*e.g.* see (van Kerkhof et al. 2005)). Whether this is achieved by 'piggy-backing' onto pre-existing SNX4 derived tubules, in a manner that would be analogous to the role of SNX3 in Fet3p-Ftr1p sorting onto the retromer pathway (Strochlic et al. 2007), is an interesting possibility that certainly merits exploration.

Another PtdIns(3)P-binding sorting nexin with a role in endosomal recycling is sorting nexin-27 (SNX27) (Lunn et al. 2007; Rincon et al. 2007). In addition to the early endosomal targeting PX domain, SNX27 contains a PDZ domain and, possibly, an association domain for the small GTPase Ras (Fujiyama et al. 2003;

Kajii et al. 2003). In the case of the β2-adrenoreceptor (β2-AR), the PDZ domain of SNX27 associates with the β2-AR tail sequence and is required for the sorting of the receptor into the endosome-to-plasma membrane recycling pathway (this is also dependent on the ability of SNX27 to bind PtdIns(3)P) (Lauffer et al. 2010). SNX27 also associates, again via its PDZ domain, with other cargoes, including G protein-gated potassium (Kir3) channels (Lunn et al. 2007), and the α-variant of the 5-hydroxytryptamine type 4 receptor (5-HT4R) (Joubert et al. 2004). With regard to these cargoes, the interaction with SNX27 is argued to regulate endocytosis or endo-lysosomal sorting rather than recycling (Joubert et al. 2004; Lunn et al. 2007). Clarifying these somewhat distinct conclusions will require further studies. In addition, as with the role of SNX17, it will be interesting to determine whether the adaptor function of SNX27 in cargo recognition is co-ordinated to transport through other phosphoinositide-mediated endosomal recycling pathways.

3.8.4 Carrier Movement and Fusion with the Plasma Membrane

From the growing body of molecular evidence, it is clear that PtdIns(3)P can regulate recycling pathways that exit the early endosome, and it is possible that transition of PtdIns(3)P to PtdIns(4)P may allow for co-ordination of separable machineries along the endocytic recycling pathway(s). Alternatively, these phosphoinositides may sequentially regulate an individual protein. In this regard, the ACCH (Amot coiled-coil homology) domain-containing Amot family of proteins may be of relevance (Heller et al. 2010). Amot associates with both PtdIns(3)P and PtdIns(4)P, is localized to the juxtanuclear ERC and, through its ACCH domain, has the ability to sense membrane curvature and drive membrane tabulation (Heller et al. 2010). Moreover, Amot can re-model the juxtanuclear endocytic recycling compartment (Heller et al. 2010). These data are entirely consistent with an important role for Amot family members in phosphoinositide-mediated endosomal recycling.

Whilst phosphoinositides function in endosomal recycling pathways and the formation of carriers carrying recycling cargos, it is currently unknown what signals are required to allow fusion of these endosomal recycling carriers with the plasma membrane. Identification of targeting and fusion machineries that regulate this step will be instrumental in our understanding of mechanisms underlying endosomal recycling. The PX domain-containing kinesin, KIF16B regulates the microtubule plus-end-directed movement of endosomal recycling carriers towards the plasma membrane through binding to PtdIns(3)P (Hoepfner et al. 2005). As for fusion, some insight has come from depleting cellular levels of PtdIns(4,5)P_2 through overexpression of the 5′-phosphatase domain of synaptojanin, depletion of PtdIns(4)P 5′-kinases or induced PtdIns(4,5)P_2 hydrolysis via recruitment of a rapamycin-sensitive PtdIns(4,5)$P_2$5′-phosphatase to the plasma membrane. In all cases, recycling of Tfn was slowed (Abe et al. 2008; Kim et al. 2006). Interestingly, in the latter, Tfn accumulated beneath the plasma membrane (Abe et al. 2008) suggesting that PtdIns(4,5)P_2 may regulate fusion of these recycling carriers with the plasma membrane.

3.9 Future Perspectives

The embryonic lethality that occurs upon manipulation of enzymes that regulate endosomal phosphoinositide, and the increasing number of genetic diseases linked to mutations in these enzymes, clearly highlights the importance of phosphoinositides for the function of the endo-lysosomal network. Several of these genetic disorders are linked to defects in the sculpturing of membranes that occur during all aspects of endosomal sorting. Further analysis of these, and related proteins, will certainly extend our appreciation of the basic mechanisms through which membrane re-modelling is co-ordinated with the process of cargo capture and sorting. Moreover, characterizing additional phosphoinositide-binding proteins, an in particular those for PtdIns(3,5)P_2, will inevitably generate greater insight to human disease and may identify new targets for therapeutic intervention.

Acknowledgements Phosphoinositide research in the Cullen lab is supported by the Wellcome Trust, the Biotechnology and Biological Sciences Research Council, the Royal Society and the Swiss Science Federation. J.G. Carlton is a Wellcome Trust VIP and RCDF fellow. We thank all of our colleagues, past and present, for their intellectual input.

References

Abe N, Inoue T, Galvez T, Klein L, Meyer T (2008) Dissecting the role of PtdIns(4,5)P2 in endocytosis and recycling of the transferrin receptor. J Cell Sci 121:1488–1494

Adachi H, Tsujimoto M (2010) Adaptor protein sorting nexin 17 interacts with the scavenger receptor FEEL-1/stabilin-1 and modulates its expression on the cell surface. Biochim Biophys Acta 1803:553–563

Alam SL, Langelier C, Whitby FG, Koirala S, Robinson H, Hill CP, Sundquist WI (2006) Structural basis for ubiquitin recognition by the human ESCRT-II EAP45 GLUE domain. Nat Struct Mol Biol 13:1029–1030

Ang AL, Taguchi T, Francis S, Folsch H, Murrells LJ, Pypaert M, Warren G, Mellman I (2004) Recycling endosomes can serve as intermediates during transport from the Golgi to the plasma membrane of MDCK cells. J Cell Biol 167:531–543

Arighi CN, Hartnell LM, Aguilar RC, Haft CR, Bonifacino JS (2004) Role of the mammalian retromer in sorting of the cation-independent mannose 6-phosphate receptor. J Cell Biol 165:123–133

Attar N, Cullen PJ (2010) The retromer complex. Adv Enzyme Regul 50:216–236

Axe EL, Walker SA, Manifava M, Chandra P, Roderick HL, Habermann A, Griffiths G, Ktistakis NT (2008) Autophagosome formation from membrane compartments enriched in phosphatidylinositol 3-phosphate and dynamically connected to the endoplasmic reticulum. J Cell Biol 182:685–701

Babst M, Katzmann DJ, Estepa-Sabal EJ, Meerloo T, Emr SD (2002a) Escrt-III: an endosome-associated heterooligomeric protein complex required for mvb sorting. Dev Cell 3:271–282

Babst M, Katzmann DJ, Snyder WB, Wendland B, Emr SD (2002b) Endosome-associated complex, ESCRT-II, recruits transport machinery for protein sorting at the multivesicular body. Dev Cell 3:283–289

Backer JM (2008) The regulation and function of Class III PI3Ks: novel roles for Vps34. Biochem J 410:1–17

Balla T (2005) Inositol-lipid binding motifs: signal integrators through protein-lipid and protein-protein interactions. J Cell Sci 118:2093–2104

Balla A, Balla T (2006) Phosphatidylinositol 4-kinases: old enzymes with emerging functions. Trends Cell Biol 16:351–361

Balla T, Várnai P (2009) Visualization of cellular phosphoinositide pools with GFP-fused protein-domains. Wiley, New York

Bankaitis VA, Johnson LM, Emr SD (1986) Isolation of yeast mutants defective in protein targeting to the vacuole. Proc Natl Acad Sci U S A 83:9075–9079

Bashkirov PV, Akimov SA, Evseev AI, Schmid SL, Zimmerberg J, Frolov VA (2008) GTPase cycle of dynamin is coupled to membrane squeeze and release, leading to spontaneous fission. Cell 135:1276–1286

Belenkaya TY, Wu Y, Tang X, Zhou B, Cheng L, Sharma YV, Yan D, Selva EM, Lin X (2008) The retromer complex influences Wnt secretion by recycling wntless from endosomes to the trans-Golgi network. Dev Cell 14:120–131

Blume JJ, Halbach A, Behrendt D, Paulsson M, Plomann M (2007) EHD proteins are associated with tubular and vesicular compartments and interact with specific phospholipids. Exp Cell Res 313:219–231

Bonangelino CJ, Nau JJ, Duex JE, Brinkman M, Wurmser AE, Gary JD, Emr SD, Weisman LS (2002) Osmotic stress-induced increase of phosphatidylinositol 3,5-bisphosphate requires Vac14p, an activator of the lipid kinase Fab1p. J Cell Biol 156:1015–1028

Bonifacino JS, Hurley JH (2008) Retromer. Curr Opin Cell Biol 20:427–436

Bonifacino JS, Rojas R (2006) Retrograde transport from endosomes to the trans-Golgi network. Nat Rev Mol Cell Biol 7:568–579

Botelho RJ (2009) Changing phosphoinositides "on the fly": how trafficking vesicles avoid an identity crisis. BioEssays 31:1127–1136

Botelho RJ, Efe JA, Teis D, Emr SD (2008) Assembly of a Fab1 phosphoinositide kinase signaling complex requires the Fig4 phosphoinositide phosphatase. Mol Biol Cell 19:4273–4286

Braschi E, Goyon V, Zunino R, Mohanty A, Xu L, McBride HM (2010) Vps35 mediates vesicle transport between the mitochondria and peroxisomes. Curr Biol 20:1310–1315

Brown FD, Rozelle AL, Yin HL, Balla T, Donaldson JG (2001) Phosphatidylinositol 4,5-bisphosphate and Arf6-regulated membrane traffic. J Cell Biol 154:1007–1017

Bruce EA, Medcalf L, Crump CM, Noton SL, Stuart AD, Wise HM, Elton D, Bowers K, Digard P (2009) Budding of filamentous and non-filamentous influenza A virus occurs via a VPS4 and VPS28-independent pathway. Virology 390:268–278

Bujny MV, Popoff V, Johannes L, Cullen PJ (2007) The retromer component sorting nexin-1 is required for efficient retrograde transport of Shiga toxin from early endosome to the trans Golgi network. J Cell Sci 120:2010–2021

Burd CG, Emr SD (1998) Phosphatidylinositol(3)-phosphate signaling mediated by specific binding to RING FYVE domains. Mol Cell 2:157–162

Burda P, Padilla SM, Sarkar S, Emr SD (2002) Retromer function in endosome-to-Golgi retrograde transport is regulated by the yeast Vps34 PtdIns 3-kinase. J Cell Sci 115:3889–3900

Burden JJ, Sun XM, Garcia AB, Soutar AK (2004) Sorting motifs in the intracellular domain of the low density lipoprotein receptor interact with a novel domain of sorting nexin-17. J Biol Chem 279:16237–16245

Cabezas A, Pattni K, Stenmark H (2006) Cloning and subcellular localization of a human phosphatidylinositol 3-phosphate 5-kinase, PIKfyve/Fab1. Gene 371:34–41

Cai H, Yu S, Menon S, Cai Y, Lazarova D, Fu C, Reinisch K, Hay JC, Ferro-Novick S (2007) TRAPPI tethers COPII vesicles by binding the coat subunit Sec23. Nature 445:941–944

Cao C, Laporte J, Backer JM, Wandinger-Ness A, Stein MP (2007) Myotubularin lipid phosphatase binds the hVPS15/hVPS34 lipid kinase complex on endosomes. Traffic 8:1052–10567

Cao C, Backer JM, Laporte J, Bedrick EJ, Wandinger-Ness A (2008) Sequential actions of myotubularin lipid phosphatases regulate endosomal PI(3)P and growth factor receptor trafficking. Mol Biol Cell 19:3334–3346

Caplan S, Naslavsky N, Hartnell LM, Lodge R, Polishchuk RS, Donaldson JG, Bonifacino JS (2002) A tubular EHD1-containing compartment involved in the recycling of major histocompatibility complex class I molecules to the plasma membrane. EMBO J 21:2557–2567

Carlton JG (2010) The ESCRT machinery—a cellular complex for sorting and scission. Biochem Soc Trans 38:1397–1412

Carlton JG, Cullen PJ (2005) Coincidence detection in phosphoinositide signaling. Trends Cell Biol 15:540–547

Carlton JG, Martin-Serrano J (2007) Parallels between cytokinesis and retroviral budding: a role for the ESCRT machinery. Science 316:1908–1912

Carlton JG, Martin-Serrano J (2009) The ESCRT machinery: new functions in viral and cellular biology. Biochem Soc Trans 37:195–199

Carlton J, Bujny M, Peter BJ, Oorschot VM, Rutherford A, Mellor H, Klumperman J, McMahon HT, Cullen PJ (2004) Sorting nexin-1 mediates tubular endosome-to-TGN transport through coincidence sensing of high-curvature membranes and 3-phosphoinositides. Curr Biol 14:1791–1800

Carlton JG, Bujny MV, Peter BJ, Oorschot VM, Rutherford A, Arkell RS, Klumperman J, McMahon HT, Cullen PJ (2005) Sorting nexin-2 is associated with tubular elements of the early endosome, but is not essential for retromer-mediated endosome-to-TGN transport. J Cell Sci 118:4527–4539

Carlton JG, Agromayor M, Martin-Serrano J (2008) Differential requirements for Alix and ESCRT-III in cytokinesis and HIV-1 release. Proc Natl Acad Sci U S A 105:10541–10546

Catimel B, Schieber C, Condron M, Patsiouras H, Connolly L, Catimel J, Nice EC, Burgess AW, Holmes AB (2008) The PI(3,5)P2 and PI(4,5)P2 interactomes. J Proteome Res 7:5295–5313

Chalhoub N, Baker SJ (2009) PTEN and the PI3-kinase pathway in cancer. Ann Rev Pathol Mech Dis 4:127–150

Chen D, Xiao H, Zhang K, Wang B, Gao Z, Jian Y, Qi X, Sun J, MIAO L, Yang C (2010) Retromer is required for apoptotic cell clearance by phagocytic receptor recycling. Science 327:1261–1264

Chin LS, Raynor MC, Wei X, Chen HQ, Li L (2001) Hrs interacts with sorting nexin 1 and regulates degradation of epidermal growth factor receptor. J Biol Chem 276:7069–7078

Choudhury R, Diao A, Zhang F, Eisenberg E, Saint-Pol A, Williams C, Konstantakopoulos A, Lucocq J, Johannes L, Rabouille C, Greene LE, Lowe M (2005) Lowe syndrome protein OCRL1 interacts with clathrin and regulates protein trafficking between endosomes and the trans-Golgi network. Mol Biol Cell 16:3467–3479

Chow CY, Zhang Y, Dowling JJ, Jin N, Adamska M, Shiga K, Szigeti K, Shy ME, Li J, Zhang X, Lupski JR, Weisman LS, Meisler MH (2007) Mutation of FIG4 causes neurodegeneration in the pale tremor mouse and patients with CMT4J. Nature 448:68–72

Chow CY, Landers JE, Bergren SK, Sapp PC, Grant AE, Jones JM, Everett L, Lenk GM, McKenna-Yasek DM, Weisman LS, Figlewicz D, Brown RH, Meisler MH (2009) Deleterious variants of FIG4, a phosphoinositide phosphatase, in patients with ALS. Am J Hum Genet 84:85–88

Christoforidis S, Miaczynska M, Ashman K, Wilm M, Zhao L, Yip SC, Waterfield MD, Backer JM, Zerial M (1999) Phosphatidylinositol-3-OH kinases are Rab5 effectors. Nat Cell Biol 1:249–252

Chukkapalli V, Hogue IB, Boyko V, Hu WS, Ono A (2008) Interaction between the human immunodeficiency virus type 1 Gag matrix domain and phosphatidylinositol-(4,5)-bisphosphate is essential for efficient gag membrane binding. J Virol 82:2405–2417

Collins BM, Skinner CF, Watson PJ, Seaman MN, Owen DJ (2005) Vps29 has a phosphoesterase fold that acts as a protein interaction scaffold for retromer assembly. Nat Struct Mol Biol 12:594–602

Collins BM, Norwood SJ, Kerr MC, Mahony D, Seaman MN, Teasdale RD, Owen DJ (2008) Structure of Vps26B and mapping of its interaction with the retromer protein complex. Traffic 9:366–379

Cooke FT (2009) Measurement of polyphosphoinositides in cultured mammalian cells. Methods Mol Biol 462:43–58

Cozier GE, Carlton J, McGregor AH, Gleeson PA, Teasdale RD, Mellor H, Cullen PJ (2002) The phox homology (PX) domain-dependent, 3-phosphoinositide-mediated association of sorting nexin-1 with an early sorting endosomal compartment is required for its ability to regulate epidermal growth factor receptor degradation. J Biol Chem 277:48730–48736

Cullen PJ (2008) Endosomal sorting and signalling: an emerging role for sorting nexins. Nat Rev Mol Cell Biol 9:574–582

Cullen PJ, Cozier GE, Banting G, Mellor H (2001) Modular phosphoinositide-binding domains—their role in signalling and membrane trafficking. Curr Biol 11:R882–R893

D'angelo G, Vicinanza M, Di Campli A, De Matteis MA (2008) The multiple roles of PtdIns(4)P—not just the precursor of PtdIns(4,5)P2. J Cell Sci 121:1955–1963

Dacks JB, Peden AA, Field MC (2009) Evolution of specificity in the eukaryotic endomembrane system. Int J Biochem Cell Biol 41:330–340

Damen E, Krieger E, Nielsen JE, Eygensteyn J, Van Leeuwen JE (2006) The human Vps29 retromer component is a metallo-phosphoesterase for a cation-independent mannose 6-phosphate receptor substrate peptide. Biochem J 398:399–409

Daumke O, Lundmark R, Vallis Y, Martens S, Butler PJ, McMahon H T (2007) Architectural and mechanistic insights into an EHD ATPase involved in membrane remodelling. Nature 449:923–927

Lartigue J de, Polson H, Feldman M, Shokat K, Tooze SA, Urbe S, Clague MJ (2009) PIKfyve regulation of endosome-linked pathways. Traffic 10:883–893

De Matteis MA, Di Campli A, Godi A (2005) The role of the phosphoinositides at the Golgi complex. Biochim Biophys Acta 1744:396–405

Derivery E, Sousa C, Gautier JJ, Lombard B, Loew D, Gautreau A (2009) The Arp2/3 activator WASH controls the fission of endosomes through a large multiprotein complex. Dev Cell 17:712–723

Diaz E, Schimmoller F, Pfeffer SR (1997) A novel Rab9 effector required for endosome-to-TGN transport. J Cell Biol 138:283–290

Dong XP, Shen D, Wang X, Dawson T, Li X, Zhang Q, Cheng X, Zhang Y, Weisman LS, Delling M, Xu H (2010) PI(3,5)P2 controls membrane trafficking by direct activation of mucolipin Ca^{2+} release channels in the endolysosome. Nat Commun 1:38. doi:10.1038/ncomms1037

Dove SK, Cooke FT, Douglas MR, Sayers LG, Parker PJ, Michell RH (1997) Osmotic stress activates phosphatidylinositol-3,5-bisphosphate synthesis. Nature 390:187–192

Dove SK, McEwen RK, Mayes A, Hughes DC, Beggs JD, Michell RH (2002) Vac14 controls PtdIns(3,5)P2 synthesis and Fab1-dependent protein trafficking to the multivesicular body. Curr Biol 12:885–893

Dove SK, Piper RC, McEwen RK, Yu JW, King MC, Hughes DC, Thuring J, Holmes AB, Cooke FT, Michell RH, Parker PJ, Lemmon MA (2004) Svp1p defines a family of phosphatidylinositol 3,5-bisphosphate effectors. EMBO J 23:1922–1933

Duex JE, Nau JJ, Kauffman EJ, Weisman LS (2006) Phosphoinositide 5-phosphatase Fig4p is required for both acute rise and subsequent fall in stress-induced phosphatidylinositol 3,5-bisphosphate levels. Eukaryot Cell 5:723–731

Dyve AB, Bergan J, Utskarpen A, Sandvig K (2009) Sorting nexin 8 regulates endosome-to-Golgi transport. Biochem Biophys Res Commun 390:109–114

Edgar AJ, Polak JM (2000) Human homologues of yeast vacuolar protein sorting 29 and 35. Biochem Biophys Res Commun 277:622–630

Erdmann KS, Mao Y, McErea HJ, Zoncu R, Lee S, Paradise S, Modregger J, Biemesderfer D, Toomre D, De Camilli P (2007) A role of the Lowe syndrome protein OCRL in early steps of the endocytic pathway. Dev Cell 13:377–390

Esk C, Chen CY, Johannes L, Brodsky FM (2010) The clathrin heavy chain isoform CHC22 functions in a novel endosomal sorting step. J Cell Biol 188:131–144

Ferguson SM, Raimondi A, Paradise S, Shen H, Mesaki K, Ferguson A, Destaing O, Ko G, Takasaki J, Cremona O, E O T, De Camilli P (2009) Coordinated actions of actin and BAR proteins upstream of dynamin at endocytic clathrin-coated pits. Dev Cell 17:811–822

Fernandez-Borja M, Wubbolts R, Calafat J, Janssen H, Divecha N, Dusseljee S, Neefjes J (1999) Multivesicular body morphogenesis requires phosphatidyl-inositol 3-kinase activity. Curr Biol 9:55–58
Florian V, Schluter T, Bohnensack R (2001) A new member of the sorting nexin family interacts with the C-terminus of P-selectin. Biochem Biophys Res Commun 281:1045–1050
Folsch H, Pypaert M, Schu P, Mellman I (2001) Distribution and function of AP-1 clathrin adaptor complexes in polarized epithelial cells. J Cell Biol 152:595–606
Fotin A, Cheng Y, Grigorieff N, Walz T, Harrison SC, Kirchhausen T (2004) Structure of an auxilin-bound clathrin coat and its implications for the mechanism of uncoating. Nature 432:649–653
Franch-Marro X, Wendler F, Guidato S, Griffith J, Baena-Lopez A, Itasaki N, Maurice MM, Vincent JP (2008) Wingless secretion requires endosome-to-Golgi retrieval of Wntless/Evi/Sprinter by the retromer complex. Nat Cell Biol 10:170–177
Frost A, Perera R, Roux A, Spasov K, Destaing O, Egelman EH, De Camilli P, Unger VM (2008) Structural basis of membrane invagination by F-BAR domains. Cell 132:807–817
Frost A, Unger VM, De Camilli P (2009) The BAR domain superfamily: membrane-molding macromolecules. Cell 137:191–196
Fujiyama K, Kajii Y, Hiraoka S, Nishikawa T (2003) Differential regulation by stimulants of neocortical expression of mrt1, arc, and homer1a mRNA in the rats treated with repeated methamphetamine. Synapse 49:143–149
Futter CE, Collinson LM, Backer JM, Hopkins CR (2001) Human VPS34 is required for internal vesicle formation within multivesicular endosomes. J Cell Biol 155:1251–1264
Gary JD, Sato TK, Stefan CJ, Bonangelino CJ, Weisman LS, Emr SD (2002) Regulation of Fab1 phosphatidylinositol 3-phosphate 5-kinase pathway by Vac7 protein and Fig4, a polyphosphoinositide phosphatase family member. Mol Biol Cell 13:1238–1251
Gill DJ, Teo H, Sun J, Perisic O, Veprintsev DB, Emr SD, Williams RL (2007) Structural insight into the ESCRT-I/-II link and its role in MVB trafficking. EMBO J 26:600–612
Gillooly DJ, Morrow IC, Lindsay M, Gould R, Bryant NJ, Gaullier J-M, Parton RG, Stenmark H (2000a) Localization of phosphatidylinositol 3-phosphate in yeast and mammalian cells. EMBO J 19:4577–4588
Gillooly DJ, Morrow IC, Lindsay M, Gould R, Bryant NJ, Gaullier JM, Parton RG, Stenmark H (2000b) Localization of phosphatidylinositol 3-phosphate in yeast and mammalian cells. EMBO J 19:4577–4588
Gillooly DJ, Raiborg C, Stenmark H (2003) Phosphatidylinositol 3-phosphate is found in microdomains of early endosomes. Histochem Cell Biol 120:445–453
Gokool S, Tattersall D, Seaman MN (2007) EHD1 interacts with retromer to stabilize SNX1 tubules and facilitate endosome-to-Golgi retrieval. Traffic 8:1873–1886
Gomez TS, Billadeau DD (2009) A FAM21-containing WASH complex regulates retromer-dependent sorting. Dev Cell 17:699–711
Grant BD, Caplan S (2008) Mechanisms of EHD/RME-1 protein function in endocytic transport. Traffic 9:2043–2052
Grant BD, Donaldson JG (2009) Pathways and mechanisms of endocytic recycling. Nat Rev Mol Cell Biol 10:597–608
Grant B, Zhang Y, Paupard MC, Lin SX, Hall DH, Hirsh D (2001) Evidence that RME-1, a conserved C. elegans EH-domain protein, functions in endocytic recycling. Nat Cell Biol 3:573–579
Gruenberg J, Stenmark H (2004) The biogenesis of multivesicular endosomes. Nat Rev Mol Cell Biol 5:317–323
Gruenberg J, Goot FG Van der (2006) Mechanisms of pathogen entry through the endosomal compartments. Nat Rev Mol Cell Biol 7:495–504
Gullapalli A, Wolfe BL, Griffin CT, Magnuson T, Trejo J (2006) An essential role for SNX1 in lysosomal sorting of protease-activated receptor-1: evidence for retromer-, Hrs-, and Tsg101-independent functions of sorting nexins. Mol Biol Cell 17:1228–1238
Haberg K, Lundmark R, Carlsson SR (2008) SNX18 is an SNX9 paralog that acts as a membrane tubulator in AP-1-positive endosomal trafficking. J Cell Sci 121:1495–1505

Haffner C, Di Paolo G, Rosenthal JA, De Camilli P (2000) Direct interaction of the 170 kDa isoform of synaptojanin 1 with clathrin and with the clathrin adaptor AP-2. Curr Biol 10:471–474
Haft CR, De la Luz Sierra M, Bafford R, Lesniak MA, Barr VA, Taylor SI (2000) Human orthologs of yeast vacuolar protein sorting proteins Vps26, 29, and 35: assembly into multimeric complexes. Mol Biol Cell 11:4105–4116
Hammond GR, Schiavo G, Irvine RF (2009) Immunocytochemical techniques reveal multiple, distinct cellular pools of PtdIns4P and PtdIns(4,5)P(2). Biochem J 422:23–35
Hanson PI, Roth R, Lin Y, Heuser JE (2008) Plasma membrane deformation by circular arrays of ESCRT-III protein filaments. J Cell Biol 180:389–402
Harbour ME, Breusegem SY, Antrobus R, Freeman C, Reid E, Seaman MN (2010) The cargo-selective retromer complex is a recruiting hub for protein complexes that regulate endosomal tubule dynamics. J Cell Sci 123:3703–3717
Haucke V (2005) Phosphoinositide regulation of clathrin-mediated endocytosis. Biochem Soc Trans 33:1285–1289
He X, Li F, Chang WP, Tang J (2005) GGA proteins mediate the recycling pathway of memapsin 2 (BACE). J Biol Chem 280:11696–11703
Heller B, Adu-Gyamfi E, Smith-Kinnaman W, Babbey C, Vora M, Xue Y, Bittman R, Stahelin RV, Wells CD (2010) Amot recognizes a juxtanuclear endocytic recycling compartment via a novel lipid binding domain. J Biol Chem 285:12308–12320
Henne WM, Kent HM, Ford MG, Hegde BG, Daumke O, Butler PJ, Mittal R, Langen R, Evans PR, McMahon HT (2007) Structure and analysis of FCHo2 F-BAR domain: a dimerizing and membrane recruitment module that effects membrane curvature. Structure 15:839–852
Henne WM, Boucrot E, Meinecke M, Evergren E, Vallis Y, Mittal R, McMahon HT (2010) FCHo proteins are nucleators of clathrin-mediated endocytosis. Science 328:1281–1284
Hettema EH, Lewis MJ, Black MW, Pelham HR (2003) Retromer and the sorting nexins Snx4/41/42 mediate distinct retrieval pathways from yeast endosomes. EMBO J 22:548–557
Heydorn A, Sondergaard BP, Ersboll B, Holst B, Nielsen FC, Haft CR, Whistler J, Schwartz TW (2004) A library of 7TM receptor C-terminal tails. Interactions with the proposed post-endocytic sorting proteins ERM-binding phosphoprotein 50 (EBP50), N-ethylmaleimide-sensitive factor (NSF), sorting nexin 1 (SNX1), and G protein-coupled receptor-associated sorting protein (GASP). J Biol Chem 279:54291–54303
Hierro A, Rojas AL, Rojas R, Murthy N, Effantin G, Kajava AV, Steven AC, Bonifacino JS, Hurley JH (2007) Functional architecture of the retromer cargo-recognition complex. Nature 449:1063–1067
Hinshaw JE, Schmid SL (1995) Dynamin self-assembles into rings suggesting a mechanism for coated vesicle budding. Nature 374:190–192
Hirano S, Suzuki N, Slagsvold T, Kawasaki M, Trambaiolo D, Kato R, Stenmark H, Wakatsuki S (2006) Structural basis of ubiquitin recognition by mammalian Eap45 GLUE domain. Nat Struct Mol Biol 13:1031–1032
Hoepfner S, Severin F, Cabezas A, Habermann B, Runge A, Gillooly D, Stenmark H, Zerial M (2005) Modulation of receptor recycling and degradation by the endosomal kinesin KIF16B. Cell 121:437–450
Hong Z, Yang Y, Zhang C, Niu Y, Li K, Zhao X, Liu JJ (2009) The retromer component SNX6 interacts with dynactin p150(Glued) and mediates endosome-to-TGN transport. Cell Res 19:1334–1349
Hsu VW, Prekeris R (2010) Transport at the recycling endosome. Curr Opin Cell Biol 22:528–534
Hurley JH, Hanson PI (2010) Membrane budding and scission by the ESCRT machinery: it's all in the neck. Nat Rev Mol Cell Biol 11:556–566
Hyvola N, Diao A, McKenzie E, Skippen A, Cockcroft S, Lowe M (2006) Membrane targeting and activation of the Lowe syndrome protein OCRL1 by rab GTPases. EMBO J 25:3750–3761
Ikonomov OC, Sbrissa D, Shisheva A (2006) Localized PtdIns 3,5-P2 synthesis to regulate early endosome dynamics and fusion. Am J Physiol Cell Physiol 291:C393–C404

Im YJ, Hurley JH (2008) Integrated structural model and membrane targeting mechanism of the human ESCRT-II complex. Dev Cell 14:902–913

Jackson LP, Kelly BT, McCoy AJ, Gaffry T, James LC, Collins BM, Honing S, Evans PR, Owen DJ (2010) A large-scale conformational change couples membrane recruitment to cargo binding in the AP2 clathrin adaptor complex. Cell 141:1220–1229

Jefferies HB, Cooke FT, Jat P, Boucheron C, Koizumi T, Hayakawa M, Kaizawa H, Ohishi T, Workman P, Waterfield MD, Parker PJ (2008) A selective PIKfyve inhibitor blocks PtdIns(3,5)P(2) production and disrupts endomembrane transport and retroviral budding. EMBO Rep 9:164–170

Jeffries TR, Dove SK, Michell RH, Parker PJ (2004) PtdIns-specific MPR pathway association of a novel WD40 repeat protein, WIPI49. Mol Biol Cell 15:2652–2663

Jha A, Agostinelli NR, Mishra SK, Keyel PA, Hawryluk MJ, Traub LM (2004) A Novel AP-2 adaptor interaction motif initially identified in the long-splice isoform of synaptojanin 1, SJ170. J Biol Chem 279:2281–2290

Jia D, Gomez TS, Metlagel Z, Umetani J, Otwinowski Z, Rosen MK, Billadeau DD (2010) WASH and WAVE actin regulators of the Wiskott-Aldrich syndrome protein (WASP) family are controlled by analogous structurally related complexes. Proc Natl Acad Sci U S A 107:10442–10447

Jin N, Chow CY, Liu L, Zolov SN, Bronson R, Davisson M, Petersen JL, Zhang Y, Park S, Duex JE, Goldowitz D, Meisler MH, Weisman LS (2008) VAC14 nucleates a protein complex essential for the acute interconversion of PI3P and PI(3,5)P2 in yeast and mouse. EMBO J 27:3221–3234

Johannes L, Popoff, V (2008) Tracing the retrograde route in protein trafficking. Cell 135:1175–1187

Johannes L, Romer W (2010) Shiga toxins [mdash] from cell biology to biomedical applications. Nat Rev Micro 8:105–116

Joly M, Kazlauskas A, Corvera S (1995) Phosphatidylinositol 3-kinase activity is required at a postendocytic step in platelet-derived growth factor receptor trafficking. J Biol Chem 270:13225–13230

Jost M, Simpson F, Kavran JM, Lemmon MA, Schmid SL (1998) Phosphatidylinositol-4,5-bisphosphate is required for endocytic coated vesicle formation. Curr Biol 8:1399–1404

Joubert L, Hanson B, Barthet G, Sebben M, Claeysen S, Hong W, Marin P, Dumuis A, Bockaert J (2004) New sorting nexin (SNX27) and NHERF specifically interact with the 5-HT4a receptor splice variant: roles in receptor targeting. J Cell Sci 117:5367–5379

Jovic M, Kieken F, Naslavsky N, Sorgen PL, Caplan S (2009) Eps15 homology domain 1-associated tubules contain phosphatidylinositol-4-phosphate and phosphatidylinositol-(4,5)-bisphosphate and are required for efficient recycling. Mol Biol Cell 20:2731–2743

Kajii Y, Muraoka S, Hiraoka S, Fujiyama K, Umino A, Nishikawa T (2003) A developmentally regulated and psychostimulant-inducible novel rat gene mrt1 encoding PDZ-PX proteins isolated in the neocortex. Mol Psychiatry 8:434–444

Kale SD, Gu B, Capelluto DG, Dou D, Feldman E, Rumore A, Arredondo FD, Hanlon R, Fudal I, Rouxel T, Lawrence CB, Shan W, Tyler BM (2010) External lipid PI3P mediates entry of eukaryotic pathogen effectors into plant and animal host cells. Cell 142:284–295

Karathanassis D, Stahelin RV, Bravo J, Perisic O, Pacold CM, Cho W, Williams RL (2002) Binding of the PX domain of p47(phox) to phosphatidylinositol 3,4-bisphosphate and phosphatidic acid is masked by an intramolecular interaction. EMBO J 21:5057–5068

Katzmann DJ, Babst M, Emr SD (2001) Ubiquitin-dependent sorting into the multivesicular body pathway requires the function of a conserved endosomal protein sorting complex, ESCRT-I. Cell 106:145–155

Kerr MC, Bennetts JS, Simpson F, Thomas EC, Flegg C, Gleeson PA, Wicking C, Teasdale RD (2005) A novel mammalian retromer component, Vps26B. Traffic 6:991–1001

Kim S, Kim H, Chang B, Ahn N, Hwang S, Di PAOLO G, Chang S (2006) Regulation of transferrin recycling kinetics by PtdIns[4,5]P2 availability. FASEB J 20:2399–2401

Klein DE, Lee A, Frank DW, Marks MS, Lemmon MA (1998) The pleckstrin homology domains of dynamin isoforms require oligomerization for high affinity phosphoinositide binding. J Biol Chem 273:27725–27733

Knauth P, Schluter T, Czubayko M, Kirsch C, Florian V, Schreckenberger S, Hahn H, Bohnensack R (2005) Functions of sorting nexin 17 domains and recognition motif for P-selectin trafficking. J Mol Biol 347:813–825

Kostelansky MS, Sun J, Lee S, Kim J, Ghirlando R, Hierro A, Emr SD, Hurley JH (2006) Structural and functional organization of the ESCRT-I trafficking complex. Cell 125:113–126

Kostelansky MS, Schluter C, Tam YY, Lee S, Ghirlando R, Beach B, Conibear E, Hurley JH (2007) Molecular architecture and functional model of the complete yeast ESCRT-I heterotetramer. Cell 129:485–498

Krauss M, Kukhtina V, Pechstein A, Haucke V (2006) Stimulation of phosphatidylinositol kinase type I-mediated phosphatidylinositol (4,5)-bisphosphate synthesis by AP-2μ-cargo complexes. Proc Natl Acad Sci 103:11934–11939

Kurten RC, Cadena DL, Gill GN (1996) Enhanced degradation of egf receptors by a sorting nexin, SNX1. Science 272:1008–1010

Kutateladze TG, Ogburn KD, Watson WT, De Beer T, Emr SD, Burd CG, Overduin M (1999) Phosphatidylinositol 3-phosphate recognition by the FYVE domain. Mol Cell 3:805–811

Lauffer BE, Melero C, Temkin P, Lei C, Hong W, Kortemme T, Von Zastrow M (2010) SNX27 mediates PDZ-directed sorting from endosomes to the plasma membrane. J Cell Biol 190:565–574

Lauvrak SU, Torgersen ML, Sandvig K (2004) Efficient endosome-to-Golgi transport of Shiga toxin is dependent on dynamin and clathrin. J Cell Sci 117:2321–2331

Lee J, Retamal C, Cuitino L, Caruano-Yzermans A, Shin JE, Van Kerkhof P, Marzolo MP, Bu G (2008) Adaptor protein sorting nexin 17 regulates amyloid precursor protein trafficking and processing in the early endosomes. J Biol Chem 283:11501–11508

Lemmon MA (2008) Membrane recognition by phospholipid-binding domains. Nat Rev Mol Cell Biol 9:99–111

Lin SX, Grant B, Hirsh D, Maxfield FR (2001) Rme-1 regulates the distribution and function of the endocytic recycling compartment in mammalian cells. Nat Cell Biol 3:567–572

Liu Y, Bankaitis VA (2010) Phosphoinositide phosphatases in cell biology and disease. Prog Lipid Res 49:201–217

Loerke D, Mettlen M, Yarar D, Jaqaman K, Jaqaman H, Danuser G, Schmid SL (2009) Cargo and dynamin regulate clathrin-coated pit maturation. PLoS Biol 7:e1000057

Lombardi D, Soldati T, Riederer MA, Goda Y, Zerial M, Pfeffer SR (1993) Rab9 functions in transport between late endosomes and the trans Golgi network. EMBO J 12:677–682

Lundmark R, Carlsson SR (2003) Sorting nexin 9 participates in clathrin-mediated endocytosis through interactions with the core components. J Biol Chem 278:46772–46781

Lundmark R, Carlsson SR (2004) Regulated membrane recruitment of dynamin-2 mediated by sorting nexin 9. J Biol Chem 279:42694–42702

Lundmark R, Carlsson SR (2009) SNX9—a prelude to vesicle release. J Cell Sci 122:5–11

Lunn ML, Nassirpour R, Arrabit C, Tan J, McLeod I, Arias CM, Sawchenko PE, Yates JR 3RD, Slesinger PA (2007) A unique sorting nexin regulates trafficking of potassium channels via a PDZ domain interaction. Nat Neurosci 10:1249–1259

Maffucci T, Brancaccio A, Piccolo E, Stein RC, Falasca M (2003) Insulin induces phosphatidylinositol-3-phosphate formation through TC10 activation. EMBO J 22:4178–4189

Mari M, Bujny MV, Zeuschner D, Geerts WJ, Griffith J, Petersen CM, Cullen PJ, Klumperman J, Geuze HJ (2008) SNX1 defines an early endosomal recycling exit for sortilin and mannose 6-phosphate receptors. Traffic 9:380–393

Maxfield FR, McGraw TE (2004) Endocytic recycling. Nat Rev Mol Cell Biol 5:121–132

Mayor S, Pagano RE (2007) Pathways of clathrin-independent endocytosis. Nat Rev Mol Cell Biol 8:603–612

McCormick PJ, Dumaresq Doiron K, Pluviose AS, Pichette V, Tosato G, Lefrancois S (2008) Palmitoylation controls recycling in lysosomal sorting and trafficking. Traffic 9:1984–1997

McCrea HJ, De Camilli P (2009) Mutations in phosphoinositide metabolizing enzymes and human disease. Physiology 24:8–16

McCullough J, Clague MJ, Urbe S (2004) AMSH is an endosome-associated ubiquitin isopeptidase. J Cell Biol 166:487–492

McCullough J, Row PE, Lorenzo O, Doherty M, Beynon R, Clague MJ, Urbe S (2006) Activation of the endosome-associated ubiquitin isopeptidase AMSH by STAM, a component of the multivesicular body-sorting machinery. Curr Biol 16:160–165

McLaughlin S, Wang J, Gambhir A, Murray D (2002) PIP(2) and proteins: interactions, organization, and information flow. Annu Rev Biophys Biomol Struct 31:151–175

McPherson PS, Garcia EP, Slepnev VI, David C, Zhang X, Grabs D, Sossin WS, Bauerfeind R, Nemoto Y, De Camilli P (1996) A presynaptic inositol-5-phosphatase. Nature 379:353–357

Medigeshi GR, Schu P (2003) Characterization of the in vitro retrograde transport of MPR46. Traffic 4:802–811

Mercer J, Schelhaas M, Helenius A (2010) Virus entry by endocytosis. Ann Rev Biochem 79:803–833

Merrifield CJ, Feldman ME, Wan L, Almers W (2002) Imaging actin and dynamin recruitment during invagination of single clathrin-coated pits. Nat Cell Biol 4:691–698

Mettlen M, Pucadyil T, Ramachandran R, Schmid SL (2009) Dissecting dynamin's role in clathrin-mediated endocytosis. Biochem Soc Trans 37:1022–1026

Mettlen M, Loerke D, Yarar D, Danuser G, Schmid SL (2010) Cargo- and adaptor-specific mechanisms regulate clathrin-mediated endocytosis. J Cell Biol 188:919–933

Miaczynska M, Christoforidis S, Giner A, Shevchenko A, Uttenweiler-Joseph S, Habermann B, Wilm M, Parton RG, Zerial M (2004) APPL proteins link Rab5 to nuclear signal transduction via an endosomal compartment. Cell 116:445–456

Michell RH, Heath VL, Lemmon MA, Dove SK (2006) Phosphatidylinositol 3,5-bisphosphate: metabolism and cellular functions. Trends Biochem Sci 31:52–63

Mills IG, Praefcke GJ, Vallis Y, Peter BJ, Olesen LE, Gallop JL, Butler PJ, Evans PR, McMahon HT (2003) EpsinR: an AP1/clathrin interacting protein involved in vesicle trafficking. J Cell Biol 160:213–222

Miserey-Lenkei S, Waharte F, Boulet A, Cuif MH, Tenza D, El Marjou A, Raposo G, Salamero J, Heliot L, Goud B, Monier S (2007) Rab6-interacting protein 1 links Rab6 and Rab11 function. Traffic 8:1385–1403

Mizuno E, Kobayashi K, Yamamoto A, Kitamura N, Komada M (2006) A deubiquitinating enzyme UBPY regulates the level of protein ubiquitination on endosomes. Traffic 7:1017–1031

Morita E, Sundquist WI (2004) Retrovirus budding. Annu Rev Cell Dev Biol 20:395–425

Morita E, Sandrin V, Chung HY, Morham SG, Gygi SP, Rodesch CK, Sundquist WI (2007) Human ESCRT and ALIX proteins interact with proteins of the midbody and function in cytokinesis. EMBO J 26:4215–4227

Muziol T, Pineda-Molina E, Ravelli RB, Zamborlini A, Usami Y, Gottlinger H, Weissenhorn W (2006) Structural basis for budding by the ESCRT-III factor CHMP3. Dev Cell 10:821–830

Nakano-Kobayashi A, Yamazaki M, Unoki T, Hongu T, Murata C, Taguchi R, Katada T, Frohman MA, Yokozeki T, Kanaho Y (2007) Role of activation of PIP5K[gamma]661 by AP-2 complex in synaptic vesicle endocytosis. EMBO J 26:1105–1116

Nakatsu F, Perera RM, Lucast L, Zoncu R, Domin J, Gertler FB, Toomre D, De Camilli P (2010) The inositol 5-phosphatase SHIP2 regulates endocytic clathrin-coated pit dynamics. J Cell Biol 190:307–315

Naslavsky N, Rahajeng J, Sharma M, Jovic M, Caplan S (2006) Interactions between EHD proteins and Rab11-FIP2: a role for EHD3 in early endosomal transport. Mol Biol Cell 17:163–177

Naslavsky N, Rahajeng J, Chenavas S, Sorgen PL, Caplan S (2007) EHD1 and Eps15 interact with phosphatidylinositols via their Eps15 homology domains. J Biol Chem 282:16612–16622

Naslavsky N, McKenzie J, Altan-Bonnet N, Sheff D, Caplan S (2009) EHD3 regulates early-endosome-to-Golgi transport and preserves Golgi morphology. J Cell Sci 122:389–400

Naughtin MJ, Sheffield DA, Rahman P, Hughes WE, Gurung R, Stow JL, Nandurkar HH, Dyson JM, Mitchell CA (2010) The myotubularin phosphatase MTMR4 regulates sorting from early endosomes. J Cell Sci 123:3071–3083

Nicot AS, Laporte J (2008) Endosomal phosphoinositides and human diseases. Traffic 9:1240–1249
Nisar S, Kelly E, Cullen PJ, Mundell SJ (2010) Regulation of P2Y1 receptor traffic by sorting Nexin 1 is retromer independent. Traffic 11:508–519
Norwood SJ, Shaw DJ, Cowieson NP, Owen DJ, Teasdale RD, Collins BM (2011) Assembly and solution structure of the core retromer protein complex. Traffic 12:56–71
Ono A, Ablan SD, Lockett SJ, Nagashima K, Freed EO (2004) Phosphatidylinositol (4,5) bis-phosphate regulates HIV-1 Gag targeting to the plasma membrane. Proc Natl Acad Sci U S A 101:14889–14894
Ooms LM, Horan KA, Rahman P, Seaton G, Gurung R, Kethesparan DS, Mitchell CA (2009) The role of the inositol polyphosphate 5-phosphatases in cellular function and human disease. Biochem J 419:29–49.
Orth JD, McNiven MA (2003) Dynamin at the actin-membrane interface. Curr Opin Cell Biol 15:31–39
Otsuki T, Kajigaya S, Ozawa K, Liu JM (1999) SNX5, a new member of the sorting nexin family, binds to the Fanconi anemia complementation group A protein. Biochem Biophys Res Commun 265:630–635
Pan CL, Baum PD, Gu M, Jorgensen EM, Clark SG, Garriga G (2008) C. elegans AP-2 and retromer control Wnt signaling by regulating mig-14/Wntless. Dev Cell 14:132–139
Pant S, Sharma M, Patel K, Caplan S, Carr CM, Grant BD (2009) AMPH-1/Amphiphysin/Bin1 functions with RME-1/Ehd1 in endocytic recycling. Nat Cell Biol 11:1399–1410
Paolo GD, Moskowitz HS, Gipson K, Wenk MR, Voronov S, Obayashi M, Flavell R, Fitzsimonds RM, Ryan TA, Camilli PD (2004) Impaired PtdIns(4,5)P2 synthesis in nerve terminals produces defects in synaptic vesicle trafficking. Nature 431:415–422
Park J, Kim Y, Lee S, Park JJ, Park ZY, Sun W, Kim H, Chang S (2010) SNX18 shares a redundant role with SNX9 and modulates endocytic trafficking at the plasma membrane. J Cell Sci 123:1742–1750
Parks WT, Frank DB, Huff C, Renfrew HAFT C, Martin J, Meng X, De Caestecker MP, McNally JG, Reddi A, Taylor SI, Roberts AB, Wang T, Lechleider RJ (2001) Sorting nexin 6, a novel SNX, interacts with the transforming growth factor-beta family of receptor serine-threonine kinases. J Biol Chem 276:19332–19339
Perera RM, Zoncu R, Lucast L, De Camilli P, Toomre D (2006) Two synaptojanin 1 isoforms are recruited to clathrin-coated pits at different stages. Proc Natl Acad Sci 103:19332–19337
Petiot A, Faure J, Stenmark H, Gruenberg J (2003) PI3P signaling regulates receptor sorting but not transport in the endosomal pathway. J Cell Biol 162:971–979
Piper RC, Katzmann DJ (2007) Biogenesis and function of multivesicular bodies. Annu Rev Cell Dev Biol 23:519–547
Pons V, Hullin-Matsuda F, Nauze M, Barbaras R, Peres C, Collet X, Perret B, Chap H, Gassama,-DIAGNE A (2003) Enterophilin-1, a new partner of sorting nexin 1, decreases cell surface epidermal growth factor receptor. J Biol Chem 278:21155–21161
Pons V, Luyet PP, Morel E, Abrami L, Van der Goot FG, Parton RG, Gruenberg J (2008) Hrs and SNX3 functions in sorting and membrane invagination within multivesicular bodies. PLoS Biol 6:e214
Popoff V, Mardones GA, Tenza D, Rojas R, Lamaze C, Bonifacino JS, Raposo G, Johannes L (2007) The retromer complex and clathrin define an early endosomal retrograde exit site. J Cell Sci 120:2022–2031
Popoff V, Mardones GA, Bai SK, Chambon V, Tenza D, Burgos PV, Shi A, Benaroch P, Urbe S, Lamaze C, Grant BD, Raposo G, Johannes L (2009) Analysis of articulation between clathrin and retromer in retrograde sorting on early endosomes. Traffic 10:1868–1880
Port F, Kuster M, Herr P, Furger E, Banziger C, Hausmann G, Basler K (2008) Wingless secretion promotes and requires retromer-dependent cycling of Wntless. Nat Cell Biol 10:178–185
Poteryaev D, Datta S, Ackema K, Zerial M, Spang A (2010) Identification of the switch in early-to-late endosome transition. Cell 141:497–508

Pucadyil TJ, Schmid SL (2009) Conserved functions of membrane active GTPases in coated vesicle formation. Science 325:1217–1220

Raiborg C, Bremnes B, Mehlum A, Gillooly DJ, D'ARRIGO A, Stang E, Stenmark H (2001) FYVE and coiled-coil domains determine the specific localisation of Hrs to early endosomes. J Cell Sci 114:2255–2263

Rappoport JZ, Heyman KP, Kemal S, Simon SM (2008) Dynamics of dynamin during clathrin mediated endocytosis in PC12 cells. PLoS One 3:e2416

Raymond CK, Howald-Stevenson I, Vater CA, Stevens TH (1992) Morphological classification of the yeast vacuolar protein sorting mutants: evidence for a prevacuolar compartment in class E vps mutants. Mol Biol Cell 3:1389–1402

Ren X, Hurley JH (2010) VHS domains of ESCRT-0 cooperate in high-avidity binding to polyubiquitinated cargo. EMBO J 29:1045–1054

Rincon E, Santos T, Avila-Flores A, Albar JP, Lalioti V, Lei C, Hong W, Merida I (2007) Proteomics identification of sorting nexin 27 as a diacylglycerol kinase zeta-associated protein: new diacylglycerol kinase roles in endocytic recycling. Mol Cell Proteomics 6:1073–1087

Rink J, Ghigo E, Kalaidzidis Y, Zerial M (2005) Rab conversion as a mechanism of progression from early to late endosomes. Cell 122:735–749

Robinson FL, Dixon JE (2006) Myotubularin phosphatases: policing 3-phosphoinositides. Trends Cell Biol 16:403–412

Robinson JS, Klionsky DJ, Banta LM, Emr SD (1988) Protein sorting in Saccharomyces cerevisiae: isolation of mutants defective in the delivery and processing of multiple vacuolar hydrolases. Mol Cell Biol 8:4936–4948

Rojas R, Van Vlijmen T, Mardones GA, Prabhu Y, Rojas AL, Mohammed S, Heck AJ, Raposo G, Van der Sluijs P, Bonifacino JS (2008) Regulation of retromer recruitment to endosomes by sequential action of Rab5 and Rab7. J Cell Biol 183:513–526

Rossman JS, Jing X, Leser GP, Lamb RA (2010) Influenza virus M2 protein mediates ESCRT-independent membrane scission. Cell 142:902–913

Roux A, Uyhazi K, Frost A, De Camilli P (2006) GTP-dependent twisting of dynamin implicates constriction and tension in membrane fission. Nature 441:528–531

Roux A, Koster G, Lenz M, Sorre B, Manneville JB, Nassoy P, Bassereau P (2010a) Membrane curvature controls dynamin polymerization. Proc Natl Acad Sci U S A 107:4141–4146

Roux AL, Koster G, Lenz M, Sorre BT, Manneville J-B, Nassoy P, Bassereau P (2010b) Membrane curvature controls dynamin polymerization. Proc Natl Acad Sci 107:4141–4146

Rudge SA, Anderson DM, Emr SD (2004) Vacuole size control: regulation of PtdIns(3,5)P2 levels by the vacuole-associated Vac14-Fig4 complex, a PtdIns(3,5)P2-specific phosphatase. Mol Biol Cell 15:24–36

Rutherford AC, Traer C, Wassmer T, Pattni K, Bujny MV, Carlton JG, Stenmark H, Cullen PJ (2006) The mammalian phosphatidylinositol 3-phosphate 5-kinase (PIKfyve) regulates endosome-to-TGN retrograde transport. J Cell Sci 119:3944–3957

Saad JS, Miller J, Tai J, Kim A, Ghanam RH, Summers MF (2006) Structural basis for targeting HIV-1 Gag proteins to the plasma membrane for virus assembly. Proc Natl Acad Sci U S A 103:11364–11369

Saffarian S, Cocucci E, Kirchhausen T (2009) Distinct dynamics of endocytic clathrin-coated pits and coated plaques. PLoS Biol 7:e1000191

Saftig P, Klumperman J (2009) Lysosome biogenesis and lysosomal membrane proteins: trafficking meets function. Nat Rev Mol Cell Biol 10:623–635

Sagona AP, Nezis IP, Pedersen NM, Liestol K, Poulton J, Rusten TE, Skotheim RI, Raiborg C, Stenmark H (2010) PtdIns(3)P controls cytokinesis through KIF13A-mediated recruitment of FYVE-CENT to the midbody. Nat Cell Biol 12:362–371

Saint-Pol A, Yelamos B, Amessou M, Mills IG, Dugast M, Tenza D, Schu P, Antony C, McMahon HT, Lamaze C, Johannes L (2004) Clathrin adaptor epsinR is required for retrograde sorting on early endosomal membranes. Dev Cell 6:525–538

Saksena S, Wahlman J, Teis D, Johnson AE, Emr SD (2009) Functional reconstitution of ESCRT-III assembly and disassembly. Cell 136:97–109
Salim K, Bottomley MJ, Querfurth E, Zvelebil MJ, Gout I, Scaife R, Margolis RL, Gigg R, Smith CI, Driscoll PC, Waterfield MD, Panayotou G (1996) Distinct specificity in the recognition of phosphoinositides by the pleckstrin homology domains of dynamin and Bruton's tyrosine kinase. EMBO J 15:6241–6250
Sankaran VG, Klein DE, Sachdeva MM, Lemmon MA (2001) High-affinity binding of a FYVE domain to phosphatidylinositol 3-phosphate requires intact phospholipid but not FYVE domain oligomerization. Biochemistry 40:8581–8587
Sbrissa D, Ikonomov OC, Shisheva A (1999) PIKfyve, a mammalian ortholog of yeast Fab1p lipid kinase, synthesizes 5-phosphoinositides. Effect of insulin. J Biol Chem 274:21589–21597
Sbrissa D, Ikonomov OC, Strakova J, Dondapati R, Mlak K, Deeb R, Silver R, Shisheva A (2004) A mammalian ortholog of Saccharomyces cerevisiae Vac14 that associates with and up-regulates PIKfyve phosphoinositide 5-kinase activity. Mol Cell Biol 24:10437–10447
Sbrissa D, Ikonomov OC, Fu Z, Ijuin T, Gruenberg J, Takenawa T, Shisheva A (2007) Core protein machinery for mammalian phosphatidylinositol 3,5-bisphosphate synthesis and turnover that regulates the progression of endosomal transport. J Biol Chem 282:23878–23891
Schafer DA (2004) Regulating actin dynamics at membranes: a focus on dynamin. Traffic 5:463–469
Schmid EM, McMahon HT (2007) Integrating molecular and network biology to decode endocytosis. Nature 448:883–888
Schu PV, Takegawa K, Fry MJ, Stack JH, Waterfield MD, Emr SD (1993) Phosphatidylinositol 3-kinase encoded by yeast VPS34 gene essential for protein sorting. Science 260:88–91
Seaman MN (2004) Cargo-selective endosomal sorting for retrieval to the Golgi requires retromer. J Cell Biol 165:111–122
Seaman MN (2007) Identification of a novel conserved sorting motif required for retromer-mediated endosome-to-TGN retrieval. J Cell Sci 120:2378–2389
Seaman MN, McCaffery JM, Emr SD (1998) A membrane coat complex essential for endosome-to-Golgi retrograde transport in yeast. J Cell Biol 142:665–681
Seaman MN, Harbour ME, Tattersall D, Read E, Bright N (2009) Membrane recruitment of the cargo-selective retromer subcomplex is catalysed by the small GTPase Rab7 and inhibited by the Rab-GAP TBC1D5. J Cell Sci 122:2371–2382
Sharma M, Naslavsky N, Caplan S (2008) A role for EHD4 in the regulation of early endosomal transport. Traffic 9:995–1018
Sharma M, Giridharan SS, Rahajeng J, Naslavsky N, Caplan S (2009) MICAL-L1 links EHD1 to tubular recycling endosomes and regulates receptor recycling. Mol Biol Cell 20:5181–5194
Shi H, Rojas R, Bonifacino JS, Hurley JH (2006) The retromer subunit Vps26 has an arrestin fold and binds Vps35 through its C-terminal domain. Nat Struct Mol Biol 13:540–548
Shi A, Sun L, Banerjee R, Tobin M, Zhang Y, Grant BD (2009) Regulation of endosomal clathrin and retromer-mediated endosome to Golgi retrograde transport by the J-domain protein RME-8. EMBO J 28:3290–3302
Shiba Y, Romer W, Mardones GA, Burgos PV, Lamaze C, Johannes L (2010) AGAP2 regulates retrograde transport between early endosomes and the TGN. J Cell Sci 123:2381–2390
Shields SB, Oestreich AJ, Winistorfer S, Nguyen D, Payne JA, Katzmann DJ, Piper R (2009) ESCRT ubiquitin-binding domains function cooperatively during MVB cargo sorting. J Cell Biol 185:213–224
Shimada A, Niwa H, Tsujita K, Suetsugu S, Nitta K, Hanawa-Suetsugu K, Akasaka R, Nishino Y, Toyama M, Chen L, Liu ZJ, Wang BC, Yamamoto M, Terada T, Miyazawa A, Tanaka A, Sugano S, Shirouzu M, Nagayama K, Takenawa T, Yokoyama S (2007) Curved EFC/F-BAR-domain dimers are joined end to end into a filament for membrane invagination in endocytosis. Cell 129:761–772
Shin HW, Hayashi M, Christoforidis S, Lacas-Gervais S, Hoepfner S, Wenk MR, Modregger J, Uttenweiler-Joseph S, Wilm M, Nystuen A, Frankel WN, Solimena M, De Camilli P, Zerial

M (2005) An enzymatic cascade of Rab5 effectors regulates phosphoinositide turnover in the endocytic pathway. J Cell Biol 170:607–618.

Shin N, Lee S, Ahn N, Kim SA, Ahn SG, Yongpark Z, Chang S (2007) Sorting nexin 9 interacts with dynamin 1 and N-WASP and coordinates synaptic vesicle endocytosis. J Biol Chem 282:28939–28950

Shin N, Ahn N, Chang-Ileto B, Park J, Takei K, Ahn SG, Kim SA, Di Paolo G, Chang S (2008) SNX9 regulates tubular invagination of the plasma membrane through interaction with actin cytoskeleton and dynamin 2. J Cell Sci 121:1252–1263

Shisheva A (2008) PIKfyve: Partners, significance, debates and paradoxes. Cell Biol Int 32:591–604

Shisheva A, Sbrissa D, Ikonomov O (1999) Cloning, characterization, and expression of a novel Zn2 + -binding FYVE finger-containing phosphoinositide kinase in insulin-sensitive cells. Mol Cell Biol 19:623–634

Skanland SS, Walchli S, Utskarpen A, Wandinger-Ness A, Sandvig K (2007) Phosphoinositide-regulated retrograde transport of ricin: crosstalk between hVps34 and sorting nexins. Traffic 8:297–309

Skanland SS, Walchli S, Brech A, Sandvig K (2009) SNX4 in complex with clathrin and dynein: implications for endosome movement. PLoS One 4:e5935

Slagsvold T, Aasland R, Hirano S, Bache KG, Raiborg C, Trambaiolo D, Wakatsuki S, Stenmark H (2005) Eap45 in mammalian ESCRT-II binds ubiquitin via a phosphoinositide-interacting GLUE domain. J Biol Chem 280:19600–19606

Small SA, Kent K, Pierce A, Leung C, Kang MS, Okada H, Honig L, Vonsattel JP, Kim TW (2005) Model-guided microarray implicates the retromer complex in Alzheimer's disease. Ann Neurol 58:909–919

Stack JH, Herman PK, Schu PV, Emr SD (1993) A membrane-associated complex containing the Vps15 protein kinase and the Vps34 PI 3-kinase is essential for protein sorting to the yeast lysosome-like vacuole. EMBO J 12:2195–2204

Stahelin RV, Burian A, Bruzik KS, Murray D, Cho W (2003) Membrane binding mechanisms of the PX domains of NADPH oxidase p40phox and p47phox. J Biol Chem 278:14469–14479

Stockinger W, Sailler B, Strasser V, Recheis B, Fasching D, Kahr L, Schneider WJ, Nimpf J (2002) The PX-domain protein SNX17 interacts with members of the LDL receptor family and modulates endocytosis of the LDL receptor. EMBO J 21:4259–4267

Stowell MH, Marks B, Wigge P, McMahon HT (1999) Nucleotide-dependent conformational changes in dynamin: evidence for a mechanochemical molecular spring. Nat Cell Biol 1:27–32

Strochlic TI, Setty TG, Sitaram A, Burd CG (2007) Grd19/Snx3p functions as a cargo-specific adapter for retromer-dependent endocytic recycling. J Cell Biol 177:115–125

Stuffers S, Sem WEGNER C, Stenmark H, Brech A (2009) Multivesicular endosome biogenesis in the absence of ESCRTs. Traffic 10:925–937

Sundquist WI, Schubert HL, Kelly BN, Hill GC, Holton JM, Hill CP (2004) Ubiquitin recognition by the human TSG101 protein. Mol Cell 13:783–789

Sweitzer SM, Hinshaw JE (1998) Dynamin undergoes a GTP-dependent conformational change causing vesiculation. Cell 93:1021–1029

Tabuchi M, Yanatori I, Kawai Y, Kishi F (2010) Retromer-mediated direct sorting is required for proper endosomal recycling of the mammalian iron transporter DMT1. J Cell Sci 123:756–766

Takei K, McPherson PS, Schmid SL, De Camilli P (1995) Tubular membrane invaginations coated by dynamin rings are induced by GTP-gamma S in nerve terminals. Nature 374:186–190

Teasdale RD, Loci D, Houghton F, Karlsson L, Gleeson PA (2001) A large family of endosome-localized proteins related to sorting nexin 1. Biochem J 358:7–16

Teis D, Saksena S, Emr SD (2008) Ordered assembly of the ESCRT-III complex on endosomes is required to sequester cargo during MVB formation. Dev Cell 15:578–589

Teo H, Veprintsev DB, Williams RL (2004) Structural insights into endosomal sorting complex required for transport (ESCRT-I) recognition of ubiquitinated proteins. J Biol Chem 279:28689–28696

Teo H, Gill DJ, Sun J, Perisic O, Veprintsev DB, Vallis Y, Emr SD, Williams RL (2006) ESCRT-I core and ESCRT-II GLUE domain structures reveal role for GLUE in linking to ESCRT-I and membranes. Cell 125:99–111

Theos AC, Truschel ST, Tenza D, Hurbain I, Harper DC, Berson JF, Thomas PC, Raposo G, Marks MS (2006) A lumenal domain-dependent pathway for sorting to intralumenal vesicles of multivesicular endosomes involved in organelle morphogenesis. Dev Cell 10:343–354

Thieman JR, Mishra SK, Ling K, Doray B, Anderson RA, Traub LM (2009) Clathrin regulates the association of PIPKIγ661 with the AP-2 adaptor β2 appendage. J Biol Chem 284:13924–13939

Touchberry CD, Bales IK, Stone JK, Rohrberg TJ, Parelkar NK, Nguyen T, Fuentes O, Liu X, Qu CK, Andresen JJ, Valdivia HH, Brotto M, Wacker MJ (2010) Phosphatidylinositol 3,5-bisphosphate (PI(3,5)P2) potentiates cardiac contractility via activation of the ryanodine receptor. J Biol Chem 285:40312–40321

Towler MC, Gleeson PA, Hoshino S, Rahkila P, Manalo V, Ohkoshi N, Ordahl C, Parton RG, Brodsky FM (2004) Clathrin isoform CHC22, a component of neuromuscular and myotendinous junctions, binds sorting nexin 5 and has increased expression during myogenesis and muscle regeneration. Mol Biol Cell 15:3181–3195

Traer CJ, Rutherford AC, Palmer KJ, Wassmer T, Oakley J, Attar N, Carlton JG, Kremerskothen J, Stephens DJ, Cullen PJ (2007) SNX4 coordinates endosomal sorting of TfnR with dynein-mediated transport into the endocytic recycling compartment. Nat Cell Biol 9:1370–1380

Traub LM (2003) Sorting it out: AP-2 and alternate clathrin adaptors in endocytic cargo selection. J Cell Biol 163:203–208

Traub LM (2009a) Clathrin couture: fashioning distinctive membrane coats at the cell surface. PLoS Biol 7:e1000192

Traub LM (2009b) Tickets to ride: selecting cargo for clathrin-regulated internalization. Nat Rev Mol Cell Biol 10:583–596

Tsuruta F, Green EM, Rousset M, Dolmetsch RE (2009) PIKfyve regulates CaV1.2 degradation and prevents excitotoxic cell death. J Cell Biol 187:279–294

van Den Bout I, Divecha N (2009) PIP5K-driven PtdIns(4,5)P2 synthesis: regulation and cellular functions. J Cell Sci 122:3837–3850

van Kerkhof P, Lee J, McCormick L, Tetrault E, Lu W, Schoenfish M, Oorschot V, Strous GJ, Klumperman J, Bu G (2005) Sorting nexin 17 facilitates LRP recycling in the early endosome. EMBO J 24:2851–2861

van Rossum DB, Patterson RL, Sharma S, Barrow RK, Kornberg M, Gill DL, Snyder SH (2005) Phospholipase Cgamma1 controls surface expression of TRPC3 through an intermolecular PH domain. Nature 434:99–104

Vanhaesebroeck B, Guillermet-Guibert J, Graupera M, Bilanges B (2010) The emerging mechanisms of isoform-specific PI3K signalling. Nat Rev Mol Cell Biol 11:329–341

Verges M (2008) Retromer: multipurpose sorting and specialization in polarized transport. Int Rev Cell Mol Biol 271:153–198

Verges M, Luton F, Gruber C, Tiemann F, Reinders LG, Huang L, Burlingame AL, Haft CR, Mostov KE (2004) The mammalian retromer regulates transcytosis of the polymeric immunoglobulin receptor. Nat Cell Biol 6:763–769

Wang YJ, Wang J, Sun HQ, Martinez M, Sun YX, Macia E, Kirchhausen T, Albanesi JP, Roth MG, Yin HL (2003) Phosphatidylinositol 4 phosphate regulates targeting of clathrin adaptor AP-1 complexes to the Golgi. Cell 114:299–310

Wang D, Guo M, Liang Z, Fan J, Zhu Z, Zang J, Li X, Teng M, Niu L, Dong Y, Liu P (2005) Crystal structure of human vacuolar protein sorting protein 29 reveals a phosphodiesterase/nuclease-like fold and two protein-protein interaction sites. J Biol Chem 280:22962–22967

Wassmer T, Attar N, Bujny MV, Oakley J, Traer CJ, Cullen PJ (2007) A loss-of-function screen reveals SNX5 and SNX6 as potential components of the mammalian retromer. J Cell Sci 120:45–54

Wassmer T, Attar N, Harterink M, Van Weering JR, Traer CJ, Oakley J, Goud B, Stephens DJ, Verkade P, Korswagen HC, Cullen PJ (2009) The retromer coat complex coordinates endosomal sorting and dynein-mediated transport, with carrier recognition by the trans-Golgi network. Dev Cell 17:110–122

Wenk MR, Pellegrini L, Klenchin VA, Di Paolo G, Chang S, Daniell L, Arioka M, Martin TF, De Camilli P (2001) PIP kinase Igamma is the major PI(4,5)P(2) synthesizing enzyme at the synapse. Neuron 32:79–88

Whitley P, Reaves BJ, Hashimoto M, Riley AM, Potter BV, Holman GD (2003) Identification of mammalian Vps24p as an effector of phosphatidylinositol 3,5-bisphosphate-dependent endosome compartmentalization. J Biol Chem 278:38786–38795

Williams RL, Urbe S (2007) The emerging shape of the ESCRT machinery. Nat Rev Mol Cell Biol 8:355–368

Williams R, Schluter T, Roberts MS, Knauth P, Bohnensack R, Cutler DF (2004) Sorting nexin 17 accelerates internalization yet retards degradation of P-selectin. Mol Biol Cell 15:3095–3105

Wollert T, Hurley JH (2010) Molecular mechanism of multivesicular body biogenesis by ESCRT complexes. Nature 464:864–869

Wollert T, Wunder C, Lippincott-Schwartz J, Hurley JH (2009) Membrane scission by the ESCRT-III complex. Nature 458:172–177

Yarar D, Waterman-Storer CM, Schmid SL (2007) SNX9 couples actin assembly to phosphoinositide signals and is required for membrane remodeling during endocytosis. Dev Cell 13:43–56

Yarar D, Surka MC, Leonard MC, Schmid SL (2008) SNX9 activities are regulated by multiple phosphoinositides through both PX and BAR domains. Traffic 9:133–146

Yoon T, Kim M, Lee K (2006) Inhibition of Na, K-ATPase-suppressive activity of translationally controlled tumor protein by sorting nexin 6. FEBS Lett 580:3558–3564

Zerial M, McBride H (2001) Rab proteins as membrane organizers. Nat Rev Mol Cell Biol 2:107–117

Zhang Y, Zolov SN, Chow CY, Slutsky SG, Richardson SC, Piper RC, Yang B, Nau JJ, Westrick RJ, Morrison SJ, Meisler MH, Weisman LS (2007) Loss of Vac14, a regulator of the signaling lipid phosphatidylinositol 3,5-bisphosphate, results in neurodegeneration in mice. Proc Natl Acad Sci 104:17518–17523

Zhang X, Chow CY, Sahenk Z, Shy ME, Meisler MH, Li J (2008) Mutation of FIG4 causes a rapidly progressive, asymmetric neuronal degeneration. Brain 131:1990–2001

Zoncu R, Perera RM, Balkin DM, Pirruccello M, Toomre D, De Camilli P (2009) A phosphoinositide switch controls the maturation and signaling properties of APPL endosomes. Cell 136:1110–1121

Chapter 4
Role of $PI(4,5)P_2$ in Vesicle Exocytosis and Membrane Fusion

Thomas F. J. Martin

Abstract A role for phosphatidylinositol 4,5-bisphosphate ($PI(4,5)P_2$) in membrane fusion was originally identified for regulated dense-core vesicle exocytosis in neuroendocrine cells. Subsequent studies demonstrated essential roles for $PI(4,5)P_2$ in regulated synaptic vesicle and constitutive vesicle exocytosis. For regulated dense-core vesicle exocytosis, $PI(4,5)P_2$ appears to be primarily required for priming, a stage in vesicle exocytosis that follows vesicle docking and precedes Ca^{2+}-triggered fusion. The priming step involves the organization of SNARE protein complexes for fusion. A central issue concerns the mechanisms by which $PI(4,5)P_2$ exerts an essential role in membrane fusion events at the plasma membrane. The observed microdomains of $PI(4,5)P_2$ in the plasma membrane of neuroendocrine cells at fusion sites has suggested possible direct effects of the phosphoinositide on membrane curvature and tension. More likely, $PI(4,5)P_2$ functions in vesicle exocytosis as in other cellular processes to recruit and activate $PI(4,5)P_2$-binding proteins. CAPS and Munc13 proteins, which bind $PI(4,5)P_2$ and function in vesicle priming to organize SNARE proteins, are key candidates as effectors for the role of $PI(4,5)P_2$ in vesicle priming. Consistent with roles prior to fusion that affect SNARE function, subunits of the exocyst tethering complex involved in constitutive vesicle exocytosis also bind $PI(4,5)P_2$. Additional roles for $PI(4,5)P_2$ in fusion pore dilation have been described, which may involve other $PI(4,5)P_2$-binding proteins such as synaptotagmin. Lastly, the SNARE proteins that mediate exocytic vesicle fusion contain highly basic membrane-proximal domains that interact with acidic phospholipids that likely affect their function.

Keywords CAPS · Munc13 · $PI(4,5)P_2$ microdomains · SNARE proteins · Vesicle exocytosis

T. F. J. Martin (✉)
Department of Biochemistry, University of Wisconsin-Madison, 433 Babcock Drive, 53706, Madison, WI, U.S.A
e-mail: tfmartin@wisc.edu

T. Balla et al. (eds.), *Phosphoinositides II: The Diverse Biological Functions*, Subcellular Biochemistry 59, DOI 10.1007/978-94-007-3015-1_4,

4.1 Introduction

The phosphoinositide $PI(4,5)P_2$ serves many roles in cellular function. As the substrate for receptor-regulated phospholipase C (PLC)-mediated hydrolysis, its cleavage generates the signaling molecules $Ins(1,4,5)P_3$ and DAG. The metabolism of $PI(4,5)P_2$ also gives rise to PI(4)P or $PI(3,4,5)P_3$ as signaling lipids. But possibly the most extensive role that $PI(4,5)P_2$ plays is as an intact phospholipid that is characteristic of the plasma membrane. Whereas the total membrane composition of a cell consists of 1 mol% $PI(4,5)P_2$, this lipid can achieve high local concentrations (~5 mol%) where its unique properties of high charge density and large hydrated headgroup can exert direct physical effects. Of likely greater significance for its signaling role, $PI(4,5)P_2$ serves to recruit to or activate proteins or protein complexes in the plasma membrane. A large number of proteins have structured domains such as a PH domain or a C2 domain that interact stereoselectively with $PI(4,5)P_2$ (Lemmon 2003, 2008). An even larger number of proteins contain Arg/Lys-rich+ hydrophobic regions that interact electrostatically with $PI(4,5)P_2$ (McLaughlin et al. 2002). $PI(4,5)P_2$ involvement in plasma membrane function extends to actin cytoskeletal regulation (Yin and Janmey 2003), channel and transporter regulation (Balla 2009; Suh and Hille 2008), virus budding (Saad et al. 2006), exocytosis (Martin 2001), phagocytosis (Grinstein 2010) and endocytosis (Martin 2001; Di Paolo and De Camilli 2006).

$PI(4,5)P_2$ regulates vectorial membrane trafficking to and from the plasma membrane. In the anterograde direction, both constitutive and regulated vesicle exocytosis require $PI(4,5)P_2$. Following an initial discussion of these exocytic pathways and the early discoveries that $PI(4,5)P_2$ plays a role in membrane fusion, we will discuss mechanisms by which $PI(4,5)P_2$ participates directly as a membrane constituent or as a cofactor for protein function in vesicle exocytosis.

4.2 Background on Membrane Fusion in Vesicle Exocytosis

$PI(4,5)P_2$ at the plasma membrane functions in the vectorial process of exocytic vesicle fusion. All cells have an essential constitutive secretory pathway in which cargo in vesicles leaves the Golgi and transits directly or indirectly via endosomal intermediates to the plasma membrane (De Matteis and Luini 2008). In these pathways, the exocytic fusion of vesicles with the plasma membrane does not require cellular Ca^{2+} elevations. A second set of post-Golgi pathways found in neural, endocrine, exocrine and hematopoietic secretory cells constitute the regulated secretory pathway in which dense-core vesicles (DCVs) fuse with the plasma membrane only upon Ca^{2+} elevation. Additional regulated secretory pathways utilize endosome-derived vesicles such as the synaptic vesicles (SVs) in neurons that undergo Ca^{2+}-dependent exocytosis. In regulated secretory pathways, vesicles are commonly staged at the plasma membrane prior to exocytosis in a docked configuration (Verhage and Sorensen 2008). Several lines of evidence indicate that vesicles undergo an obligatory priming step

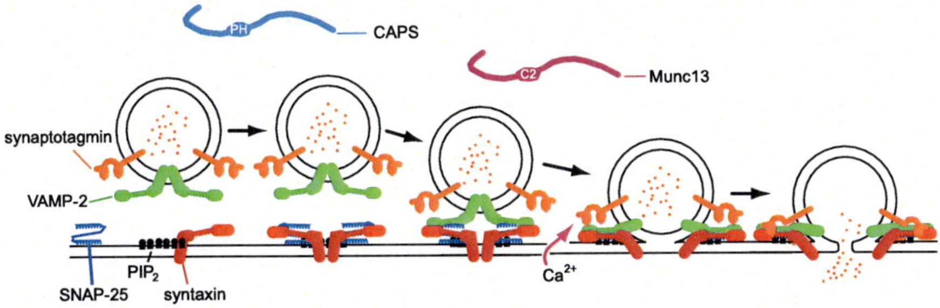

Fig. 4.1 Sequential priming and fusion steps of Ca^{2+}-triggered vesicle exocytosis are depicted with proteins that bind PI(4,5)P$_2$. A hypothetical sequence of SNARE complex formation involving plasma membrane SNAP-25 and syntaxin and vesicle VAMP-2 is shown from left to right. Syntaxin is shown interacting with PI(4,5)P$_2$ in PI(4,5)P$_2$-rich microdomains with progressive segregation of PI(4,5)P$_2$ from fusion sites. CAPS (via its PH domain) and Munc13 (via its Ca^{2+}-dependent C2 domain) exhibit PI(4,5)P$_2$ binding, which may mediate recruitment of these proteins to fusion sites for promoting SNARE complex formation. Synaptotagmin is shown to trigger fusion by interacting with SNAREs and membrane upon Ca^{2+} binding

that renders them capable of engaging in Ca^{2+}-triggered fusion (Rettig and Neher 2002). Priming is a regulated step between vesicle docking and fusion for which a number of distinct molecular constituents have been identified.

During the preceding two decades, many of the molecular constituents for vesicle exocytosis and its regulation have been identified (Jahn and Scheller 2006) (see Fig. 4.1). The core exocytic machinery consists of SNARE proteins present on vesicles and plasma membrane. The SNAREs constitute a minimal sufficient set of proteins to catalyze membrane fusion as demonstrated in liposome fusion assays (Weber et al. 1998). In cells, many other factors regulate and modulate SNARE protein function (Jahn and Scheller 2006). The pathway for membrane bilayer fusion consists of the initial merger of contacting leaflets to form a hemi-fused stalk intermediate. This is followed by the merger of non-contacting leaflets to form a fusion pore (Cohen and Melikyan 2004). The route to stalk formation and its resolution into a fusion pore involves considerable membrane bending (Chernomordik and Kozlov 2008). While membrane fusion is driven by proteins, studies over the last two decades identified phospholipids that play active roles in the membrane fusion process (van Meer and Sprong 2004; Salaun et al. 2004). This chapter will focus on PI(4,5)P$_2$ and the role(s) it plays directly or indirectly (via proteins) in membrane fusion.

4.3 Discovery of a Role for PI(4,5)P$_2$ in Trafficking to the Plasma Membrane

Evidence for the involvement of PI(4,5)P$_2$ in membrane fusion first emerged from studies of regulated DCV exocytosis in permeable neuroendocrine cells. Eberhard et al. (1990) found that that treatment of digitonin-permeabilized chromaffin cells with a bacterial PI-specific phospholipase C (PLC) decreased PI levels and inhibited Ca^{2+}-triggered catecholamine secretion. Because inhibition was preferential for an ATP-dependent stage of DCV exocytosis, it was suggested that polyphosphoinositides may be required. Hay and Martin (1992) reported that sequential ATP-dependent and Ca^{2+}-triggered reactions for DCV exocytosis in mechanically-permeabilized PC12 cells required distinct cytosolic protein factors. The cytosolic protein factors required for the ATP-dependent reactions were identified as phosphatidylinositol transfer protein (PITP) (Hay and Martin 1993) and phosphatidylinositol 4-monophosphate 5-kinase type Iγ (Hay et al. 1995). The identification of these factors, which mediate the ATP-dependent restoration of PI(4,5)P_2 in permeable cell incubations, provided direct evidence that PI(4,5)P_2 was essential for regulated DCV exocytosis. Hay et al. (1995) found that the addition of PI(4,5)P_2-specific antibodies or PLCδ1 strongly inhibited regulated exocytosis. These results indicated that the intact lipid PI(4,5)P_2 played a central role in a late step in the DCV exocytic pathway. Consistent with this, products derived from PI(4,5)P_2 by hydrolysis (DAG, inositol phosphates, fatty acids) did not affect regulated DCV exocytosis in the Ca^{2+}-buffered permeable PC12 cell system (Hay et al. 1995). Subsequent studies in permeable chromaffin cells by Wiedemann et al. (1996) suggested that PI 4-kinase activity on the secretory granules in chromaffin cells was also essential for regulated DCV exocytosis.

At the time of these initial discoveries, a role for highly phosphorylated inositides in a membrane fusion process was surprising although emerging studies in yeast were revealing a requirement for a PI 3-kinase (Vps34) in protein sorting to the vacuole (Schu et al. 1993). Studies on PI(4,5)P_2 from this point forward were directed at assessing its importance for regulated vesicle exocytosis in living cells and evaluating the precise steps in the regulated secretory pathway at which it functioned. Along with the advancing understanding of the role of PI(4,5)P_2 in cellular function, later studies probed the mechanism(s) by which PI(4,5)P_2 participates in membrane fusion mechanisms. These developments are reviewed below.

4.4 A Role for PI(4,5)P_2 in the Priming Reactions of Regulated DCV Exocytosis

Several studies in intact cells confirmed an essential role for PI(4,5)P_2 in regulated DCV exocytosis. PI(4,5)P_2 is mainly present at the plasma membrane in cells and Holz et al. (2000) showed that the PH domain of PLCδ1 localized to the plasma membrane of chromaffin cells where it inhibited regulated DCV exocytosis. PI(4,5)P_2 was not detected on DCVs although PI(4)P is present because of their origin in the Golgi. In more recent studies, 3-phosphorylated inositides have also been localized to DCVs (Wen et al. 2011). This might imply that there is trafficking between DCVs and endosomes. PI(4)P 5-kinase 1γ is one of the major 5-kinases responsible for PI(4,5)P_2 synthesis at the plasma membrane (Wenk et al. 2001). Increasing PI(4)P 5-kinase activity (Aikawa and Martin 2003; Aoyagi et al. 2005) or decreasing it

(Lawrence and Birnbaum 2003; Gong et al. 2005; Waselle et al. 2005) in cells with corresponding changes in cellular $PI(4,5)P_2$ levels was associated with increased or decreased rates of DCV exocytosis, respectively. These studies extended to living cells the conclusions that PI(4)P 5-kinase and $PI(4,5)P_2$ regulate vesicle exocytosis.

Studies in permeable PC12 cells had indicated that $PI(4,5)P_2$ was needed for a priming step in DCV exocytosis (Hay et al. 1995; Grishanin et al. 2004). Priming of DCV exocytosis in neuroendocrine cells was observed to be ATP-dependent (Rettig and Neher 2002). To determine the site in the sequential DCV exocytic pathway at which $PI(4,5)P_2$ was required, high resolution capacitance studies were conducted in neuroendocrine cells in which $PI(4,5)P_2$ levels were altered. Evoked capacitance changes in response to cellular Ca^{2+} rises are biphasic exhibiting burst and sustained phases. The burst phase represents the exocytosis of primed DCVs from a ready release pool (RRP) whereas the sustained phase is interpreted as priming reactions that refill the RRP. Olsen et al. (2003) recorded the immediate (<2s) capacitance increase in patch-clamped pancreatic β cells in response to depolarization-elicited Ca^{2+} entry as a measure of the RRP. Maintenance of and refilling of the RRP required ATP and was inhibited by phenylarsine oxide, a non-specific inhibitor of PI 4-kinase (Wiedemann et al. 1996), or by antibodies to PI(4)P or $PI(4,5)P_2$. Remarkably the ATP requirement for priming DCV exocytosis was by-passed by direct microinjection of PI(4)P or $PI(4,5)P_2$. This evidence indicated that the priming of DCV exocytosis in pancreatic β cells involved the synthesis of PI(4)P and $PI(4,5)P_2$. CAPS antibody was found to block priming in response to $PI(4,5)P_2$ injection, which suggested that CAPS was an important effector for the role of $PI(4,5)P_2$ in DCV exocytosis (see below).

Similar studies were conducted in chromaffin cells following the perturbation of $PI(4,5)P_2$ levels (Milosevic et al. 2005). $PI(4,5)P_2$ levels were measured in plasma membrane sheets prepared from cells and reacted with a GFP-PH-PLCδ1 fusion protein. Overexpression of PI(4)P 5-kinase 1γ or direct microinjection of $PI(4,5)P_2$ was used to increase cellular $PI(4,5)P_2$ levels. Increases in $PI(4,5)P_2$ levels correlated with the increased size of the RRP and with increased rates of DCV (re)priming. Conversely, expression of a phosphatase domain of syntaptojanin-1 was utilized to decrease plasma membrane $PI(4,5)P_2$ levels, which strongly reduced the RRP and inhibited DCV (re)priming rates. More recent studies of chromaffin cells from a PI(4)P 5-kinase 1γ knockout mouse reached similar conclusions that a reduction in plasma membrane $PI(4,5)P_2$ levels mainly reduced the RRP and DCV (re)priming while slightly elevating the number of docked DCVs (Gong et al. 2005). Fusion pore expansion was also somewhat delayed in the PI(4)P 5-kinase 1γ knockout chromaffin cells. Overall these studies confirmed the importance of $PI(4,5)P_2$ for priming DCV exocytosis although the basis for the critical role for this lipid remained to be elucidated.

4.5 A Role for $PI(4,5)P_2$ in Other Forms of Vesicle Exocytosis

Regulated synaptic vesicle (SV) exocytosis utilizes an assembly of proteins very similar to that employed for DCV exocytosis. Whereas DCVs are directly Golgi-derived, SVs are derived from recycling endosomes. SVs were reported to possess a type II PI 4-kinase (Guo et al. 2003) similar to that reported for DCVs (Barylko et al. 2001). Previous work (Wenk et al. 2001) had established that PI(4)P 5-kinase 1γ is a major PI(4,5)P_2-synthesizing enzyme in synapses that could potentially utilize the PI(4)P although the actual source of PI(4)P for PI(4,5)P_2 synthesis at the presynaptic plasma membrane is unclear. Early studies on whether PI(4,5)P_2 was required for Ca^{2+}-triggered SV exocytosis in synaptosome preparations produced conflicting results (Khvotchev and Sudhof 1998; Zheng et al. 2004). Because endocytosis is strongly dependent upon PI(4,5)P_2 and SVs rapidly recycle, a requirement for PI(4,5)P_2 in SV exocytosis has been difficult to demonstrate. Di Paolo et al. (2004) reported that evoked synaptic transmission in cortical neurons from the PI(4)P 5-kinase 1γ knockout mouse was normal but there was a reduced RRP for SVs, and the RRP underwent accelerated depletion at high frequency stimulation. A delay in the recycling/repriming time for SVs and a slowing of endocytosis in the synapses from knockout mice was also observed. These results suggested that PI(4,5)P_2 may be required for the evoked exocytosis of SVs.

Vesicle exocytosis in the constitutive secretory pathway is also dependent upon plasma membrane PI(4,5)P_2 although this has yet to be thoroughly examined in mammalian cells. By contrast, extensive genetic evidence in yeast indicates an essential role of PI(4,5)P_2 in post-Golgi vesicle exocytosis and for cell polarity mechanisms involving the actin cytoskeleton (Yakir-Tamang and Gerst 2009b; He and Guo 2009). MSS4 corresponds to the single PI(4)P 5-kinase in yeast. At the non-permissive temperature, *Mss4* cells with a temperature-sensitive PI(4)P 5-kinase exhibit defects in actin localization and in secretion (Yakir-Tamang and Gerst 2009a). Conversely, *MSS4* overexpression was capable of rescuing growth defects and secretion in a number of late *sec* gene mutants including those that encode exocyst subunits and a plasma membrane SNARE protein Sec9p (Yakir-Tamang and Gerst 2009a; Routt et al. 2005) (see below). Reminiscent of the original findings in neuroendocrine cells, overexpression of *SFH5*, a phosphatidylinositol-specific PITP, was found to suppress growth defects in late *sec* gene mutants (Routt et al. 2005; Yakir-Tamang and Gerst 2009a). The evidence indicates that *SFH5* functions in a pathway involving the *Stt4* PI 4-kinase and *Mss4* PI(4)P 5-kinase to synthesize plasma membrane PI(4,5)P_2 and this is required for the function of the exocyst complex and SNAREs in the constitutive secretory pathway (see below). The results support a key role for PI(4,5)P_2 in the constitutive exocytosis of post-Golgi vesicles.

4.6 Is PI(4,5)P_2 Spatially Segregated to Sites of Exocytosis?

Several studies in neuroendocrine cells have found that plasma membrane PI(4,5)P_2 is spatially inhomogeneous and distributed in microdomains (Laux et al. 2000; Caroni 2001; Milosevic et al. 2005; Aoyagi et al. 2005; James et al. 2008). This was in part demonstrated in plasma membrane lawns using a GFP-PH fusion protein from PLCδ1, which binds PI(4,5)P_2 without clustering it (James et al. 2008; Milosevic et al. 2005; Aoyagi et al. 2005). In studies with PC12 cell membrane lawns, the fluorescent probe was calibrated with $PI(4,5)P_2$-containing supported bilayers to infer a microdomain concentration for $PI(4,5)P_2$ corresponding to ~6 mol% (James et al. 2008). Although it had been argued that apparent sites of $PI(4,5)P_2$ enrichment may represent membrane infoldings (van Rheenen et al. 2005), the studies in PC12 cell membranes showed that non-specific lipid staining was not increased at sites of $PI(4,5)P_2$ enrichment (James et al. 2008; Milosevic et al. 2005). Moreover, the inferred concentrations of $PI(4,5)P_2$ detected were proportional to ATP-dependent synthesis (James et al. 2008). In this study, many of the $PI(4,5)P_2$-enriched microdomains corresponded to sites of DCV docking (~35%). About 50% of CAPS, which is a $PI(4,5)P_2$-binding protein required for DCV priming (see below), co-localized at microdomains of $PI(4,5)P_2$ that contained docked DCVs.

Earlier studies by Aoyagi et al. (2005) had found that ~13% of the docked DCVs in PC12 cells resided at membrane sites that were enriched for both syntaxin-1 and $PI(4,5)P_2$. Brief depolarization to elicit DCV exocytosis reduced this co-localization to 3%. The extent of co-localization of DCVs with syntaxin-1/$PI(4,5)P_2$ clusters increased with cellular overexpression of PI(4)P 5-kinase, which also increased Ca^{2+}-triggered DCV exocytosis (Aoyagi et al. 2005). Overall these studies (Aoyagi et al. 2005; James et al. 2008) suggested that plasma membrane sites for DCV docking, priming and fusion may be enriched for $PI(4,5)P_2$. This work on isolated plasma membrane lawns has not yet been extended to living cells. Bodipy TMR-$PI(4,5)P_2$ microinjected into cells was shown to exhibit ~3-fold reduced diffusion compared to the diffusion of other lipids leading the authors (Golebiewska et al. 2008) to conclude that ~2/3 of the $PI(4,5)P_2$ was reversibly bound. However, it will be important to directly image $PI(4,5)P_2$ in cells at sites of exocytosis to determine if membrane fusion occurs in $PI(4,5)P_2$-rich membrane microdomains. The tools available currently to detect $PI(4,5)P_2$ in living cells (e.g., PH-GFP) simultaneously inhibit Ca^{2+}-triggered DCV exocytosis (Holz et al. 2000) so additional methods to detect and quantify $PI(4,5)P_2$ in living cells will be needed.

While there is considerable evidence for independent pools of $PI(4,5)P_2$ in the plasma membrane (Janmey and Lindberg 2004), the basis for $PI(4,5)P_2$ microdomains in the plasma membrane is unknown. Even at concentrated sites of synthesis, diffusion is expected to rapidly dissipate concentration gradients of the lipid. $PI(4,5)P_2$ would need to be "captured" at such sites. This might be achieved by interactions with proteins that have specific $PI(4,5)P_2$-binding domains such as dynamin with its PH domain that in turn could oligomerize and cluster $PI(4,5)P_2$ (Bethoney et al. 2009). Alternatively, the electrostatic clustering of $PI(4,5)P_2$ by proteins that contain basic/hydrophobic regions could alter the diffusion of $PI(4,5)P_2$ away from localized sites of synthesis (McLaughlin and Murray 2005). Proteins such as GAP-43, MARCKS, CAP-23, and NAP-22 contain "basic effector domains" capable of electrostatically sequestering $PI(4,5)P_2$. The 13 basic residues in the MARCKS effector domain sequesters three $PI(4,5)P_2$ molecules (McLaughlin and Murray 2005). Indeed overexpression of MARCKS in PC12 cells was found to increase $PI(4,5)P_2$ clusters in the plasma membrane whereas overexpression of a dominant interfering mutant was found to decrease $PI(4,5)P_2$ clusters (Laux et al.

2000). Many transmembrane proteins have Lys/Arg-rich segments on their cytoplasmic membrane-proximal domains, which would enable formation of a diversity of distinct $PI(4,5)P_2$ microdomains containing different protein clusters. SNARE proteins such as syntaxin-1 that undergo cholesterol-dependent clustering at sites of DCV exocytosis have basic juxtamembrane regions that might sequester $PI(4,5)P_2$ into associated microdomains (see below).

$PI(4,5)P_2$ microdomains on the cytoplasmic leaflet have been suggested to align with extracellular leaflet liquid-ordered lipid rafts enriched in sphingolipids and cholesterol. This was based on biochemical methods isolating detergent-resistant membranes (Hope and Pike 1996). The unsaturated sn-2 acyl chain of $PI(4,5)P_2$ renders this unlikely given the tight packing of saturated acyl chains in the classical lipid raft. However, recent work has indicated that proteins with highly basic domains that sequester $PI(4,5)P_2$ may also partition into raft domains because of their myristoylation or palmitoylation. Studies of the HIV Gag protein suggested that the binding of $PI(4,5)P_2$ by the Gag protein displaces a myristate buried in a hydrophobic pocket of the protein that inserts into a raft domain (Saad et al. 2006). *In vitro* studies of a palmitoylated GAP-43 peptide showed that it partitioned $PI(4,5)P_2$ into liquid-ordered domains on giant unilamellar liposomes (Tong et al. 2008). Additional studies will be needed to determine the relationship, if any, between cytoplasmic leaflet $PI(4,5)P_2$ microdomains and the lipid raft domains in the extracellular leaflet.

4.7 Mechanisms for $PI(4,5)P_2$ Function in Membrane Fusion

A central question concerns the mechanism(s) by which $PI(4,5)P_2$ affects membrane fusion. $PI(4,5)P_2$ plays a strong positive role in regulated DCV exocytosis (Hay et al. 1995) where it regulates a priming step. As discussed below, there may be additional roles for $PI(4,5)P_2$ at later steps in DCV exocytosis. Below we consider a number of suggested mechanisms for both positive and negative effects of $PI(4,5)P_2$ on membrane fusion. Firstly, if $PI(4,5)P_2$ is localized at membrane fusion sites at the high concentrations (~6 mol%) detected (James et al. 2008), it would contribute bulk properties to the local membrane environment including curvature and charge density. High local concentrations and domain segregation may affect membrane tension in fusion mechanisms. Secondly, $PI(4,5)P_2$ is a substrate for enzymatic conversion as well as an activator of enzymes that generate lipid products (DAG, PA) that affect membrane curvature, fluidity and fusion. Thirdly, and the most generally established mechanism for $PI(4,5)P_2$ in cellular processes, is that the lipid recruits cytosolic proteins to specific locations on a membrane surface (Martin 1998; Lemmon 2003; 2008; Kutateladze 2010). Regulation of integral membrane protein function is also well-characterized (Balla 2009; Suh and Hille 2008). The functional diversity of $PI(4,5)P_2$-binding proteins is enormous and could contribute to membrane fusion by a variety of mechanisms. We discuss mechanisms that operate at vesicle priming as well as later steps in vesicle exocytosis.

4.8 Direct Effects of Membrane PI(4,5)P_2

Membrane preparations from PC12 cells exhibit spatially-restricted microdomains of PI(4,5)P_2 near docked DCVs (James et al. 2008; Aoyagi et al. 2005; Milosevic et al. 2005). PI(4,5)P_2 concentrations in microdomains may exceed 5 mol% in contrast to interdomain regions at ~2 mol% (James et al. 2008). PI(4,5)P_2 is considered to be an inverted cone-shaped lipid that would exert positive curvature in a localized region (Chernomordik and Kozlov 2008). PI(4,5)P_2 at 5 mol% in either v-SNARE donor or t-SNARE acceptor liposomes was found to inhibit SNARE-dependent liposome fusion. Vicogne and co-workers also found that PI(4,5)P_2 was inhibitory when included in t-SNARE liposomes (Vicogne et al. 2006). Inhibition by PI(4,5)P_2 was comparable to that by another inverted cone-shaped lipid, lysophosphatidylcholine, at 5 mol% and was attributed to the positive curvature-promoting properties of PI(4,5)P_2 that would counter formation of a stalk intermediate (James et al. 2008). This inhibitory mechanism observed in liposomes was partially counteracted by the sequestration of PI(4,5)P_2 by a basic charge-rich linker domain in syntaxin-1. There may be other mechanisms in cells for which PI(4,5)P_2 exerts stimulatory effects on fusion.

Classical (Chandler and Heuser 1980) and more recent studies (Anantharam et al. 2010) indicate that the plasma membrane invaginates toward DCVs during membrane fusion. The induction of local curvature in the plasma membrane by PI(4,5)P_2 at fusion sites could play a role in promoting bilayer apposition as well as creating tension in the plasma membrane to facilitate fusion (Kozlov et al. 2010). PI(4,5)P_2 microdomains in the plasma membrane may exhibit positive curvature but in addition many PI(4,5)P_2-binding proteins undergo hydrophobic insertion, which would further amplify positive curvature. Many types of PI(4,5)P_2-binding proteins exhibit bilayer insertion including PH domain-containing proteins such as CAPS and dynamin (Ramachandran et al. 2009), tandem C2 domain-containing proteins such as synaptotagmin (Martens et al. 2007), and ENTH domain containing proteins such as epsin (Ford et al. 2002). There is evidence that Ca^{2+}-triggered membrane insertion of synaptotagmin into the plasma membrane during fusion increases membrane curvature and tension to promote fusion pore dilation (Martens et al. 2007; Lynch et al. 2008; Hui et al. 2009). Thus, the overall local membrane curvature imparted by PI(4,5)P_2 within plasma membrane microdomains and enhanced by protein insertion could play a significant positive role in promoting membrane transitions during fusion.

4.9 Role of PI(4,5)P_2-Derived or Activated Metabolites

Under mild Ca^{2+} stimulation conditions, DCV exocytosis requires PI(4,5)P_2 as the intact phospholipid (Eberhard et al. 1990; Hay et al. 1995). However, under strong stimulation conditions, PI(4,5)P_2 can be metabolized by phospholipase C (PLC) (Micheva et al. 2001). One of the metabolites of PI(4,5)P_2, DAG, has been strongly linked to activation mechanisms for regulated vesicle exocytosis. Protein kinase C and brain isoforms of Munc13 have DAG-binding C1 domains that mediate activation of these proteins (Brose et al. 2004). It has also been suggested that the transformation of PI(4,5)P_2 to DAG could exert dramatic effects on the shape of membranes to trigger fusion (Janmey and Kinnunen 2006) but an essential role for PLCs in fusion *per se* remains to be demonstrated. Whether DAG is generated at exocytic fusion sites and whether DAG, as a cone-shaped lipid, has additional positive roles in affecting membrane curvature remain to be explored.

Phospholipase D (PLD), which is a PH domain-containing, PI(4,5)P_2-activated enzyme that hydrolyzes PC to PA, has been strongly implicated both in regulated DCV exocytosis and in constitutive vesicle exocytosis (Bader and Vitale 2009). PA is a cone-shaped phospholipid so its presence in the cytoplasmic leaflet could enhance the transition of merged membranes into a stalk intermediate to promote fusion. In PC12 cells in which Ca^{2+} entry was stimulated by depolarization, an accumulation of PA at the plasma membrane was detected using a PA-binding protein-GFP fusion protein (Zeniou-Meyer et al. 2007) although this was delayed compared to evoked DCV exocytosis. Nonetheless, the down regulation of PLD1 by siRNA was found to block PA accumulation as well as evoked DCV exocytosis. Capacitance recordings in chromaffin cells indicated that PLD1 siRNA reduced the RRP size as well as DCV priming. Application of lysophosphatidylcholine, an inverted cone-shaped lipid, to the extracellular leaflet reversed the inhibitory effect of PA depletion on DCV exocytosis in PC12 cells. The authors (Zeniou-Meyer et al. 2007) suggested that PLD1 activation resulted in membrane bending through the generation of PA. This might be expected to function in DCV fusion rather than in DCV priming. These studies suggested that PLD1 is an important effector for the role of PI(4,5)P_2 in DCV exocytosis. Other studies have suggested that SCAMP2, a membrane tetraspanin protein that binds PI(4,5)P_2 and PLD1, may regulate a late step in DCV exocytosis involving fusion pore formation (Liao et al. 2007).

4.10 Protein Recruitment and Activation by PI(4,5)P_2

At present, the best established mechanisms for the function of PI(4,5)P_2 in actin polymerization (Janmey and Lindberg 2004) and endocytosis (Di Paolo and De Camilli 2006) involve protein recruitment. In each of these cases, proteins interact with PI(4,5)P_2 either through specific binding domains such as PH domains or through electrostatic interactions with domains that are rich in basic and hydrophobic residues. PI(4,5)P_2-binding proteins with PH domains, C2 domains, or Lys/Arg-rich regions play a major role in various steps of vesicle exocytosis including priming.

4.11 SNARE Protein Interactions with Acidic Phospholipids

SNARE proteins, the core constituents of the fusion machinery, are directly regulated by the acidic phospholipids in the cytoplasmic leaflet of membranes. Syntaxin-1/SNAP-25 t-SNARE heterodimers were reported to exhibit reduced mobility in supported bilayers that contained $PI(4,5)P_2$ (Wagner and Tamm 2001). $PI(4,5)P_2$ itself exhibits reduced mobility in supported bilayers (Baumann et al. 2010), which suggests that direct interactions with $PI(4,5)P_2$ may reduce the mobility of t-SNAREs to organize them at sites in the membrane. $PI(4,5)P_2$ may also activate syntaxin-1 for assembly with SNAP-25 as recent studies (Murray and Tamm 2009) indicated that the cholesterol-dependent self-clustering of syntaxin-1 in liposomes was decreased by the inclusion of $PI(4,5)P_2$ at 1–5 mol%. Direct binding of the cytoplasmic domain of syntaxin-1 to acidic phospholipids has been demonstrated (Lam et al. 2008).

A conserved binding site for $PI(4,5)P_2$ (or PA) among exocytic syntaxins consists of $K^{252}KAVKYQSKARRKK^{265}$ (for syntaxin-1) in the membrane-proximal linker domain that is C-terminal to the SNARE motif. Mutations of K residues in this juxtamembrane segment results in a loss of evoked DCV exocytosis in cells and in decreased SNARE-dependent fusion on $PI(4,5)P_2$-containing liposomes *in vitro* (Lam et al. 2008; James et al. 2008). Both of these results indicate that syntaxin interactions with $PI(4,5)P_2$ (James et al. 2008) or PA (Lam et al. 2008) play a positive role in membrane fusion. As noted previously, interactions with syntaxin were proposed to segregate $PI(4,5)P_2$ in the membrane to prevent the steric inhibition of fusion (James et al. 2008) (see Fig. 4.1). Alternatively, for the cellular studies, it was suggested that syntaxin interacted with PA to concentrate this negative curvature-preferring lipid at the periphery of contacting leaflets to reduce the energy requirement for stalk formation (Lam et al. 2008).

Interactions of the juxtamembrane segment with acidic phospholipids could also drive conformational changes in syntaxin. Soluble versions of syntaxin adopt a closed configuration that blocks the interaction of syntaxin with other SNARE proteins (Chen et al. 2008). The conformation of syntaxin in the membrane could be affected by juxtamembrane segment interactions with acidic phospholipids. Alternatively, $PI(4,5)P_2$ interactions with syntaxin could play a role in localizing the protein on the membrane or in promoting SNAP-25 interactions (Aoyagi et al. 2005; Murray and Tamm 2009). Either of these effects might explain a positive role for $PI(4,5)P_2$ in priming DCV exocytosis. While these studies indicate an important role for the highly basic linker domain of syntaxin in interactions with acidic phospholipids, many roles for this interaction seem possible and need further evaluation. Based on the effects of $PI(4.5)P_2$ in SNARE-dependent liposome fusion (James et al. 2008), the mechanisms discussed here are unlikely to provide a complete explanation for the strong role for $PI(4,5)P_2$ in priming DCV exocytosis (see below).

The vesicle SNARE VAMP-2 also interacts with acidic phospholipids through membrane-proximal linker segments containing $K^{83}LKRKYWWKNLK^{94}$ (for VAMP-2) (Williams et al. 2009; Kweon et al. 2003; De Haro et al. 2003). Seagar and co-workers (De Haro et al. 2003; Quetglas et al. 2002) reported that a region of VAMP-2 overlapping this one binds Ca^{2+}/calmodulin and acidic phospholipids in a mutually exclusive manner. They provided evidence that Ca^{2+}/calmodulin binding to VAMP-2 switched its *cis* interactions with vesicle membrane lipids to *trans* interactions with the plasma membrane. These interactions might be expected to promote fusion but recent liposome fusion assay studies showed that Ca^{2+}/calmodulin inhibited SNARE-dependent fusion (Di Giovanni et al. 2010). Williams et al. (2009)

reported that the overexpression of a VAMP-2 K85E/R86D mutant inhibited evoked DCV exocytosis and they suggested that the basic juxtamembrane region of wild-type VAMP-2 acts in *trans* to counteract charge repulsion between the bilayers at approaches of < 1 nm. The principle electrostatic interaction for VAMP-2 in *trans* would be with $PI(4,5)P_2$ in the plasma membrane. It was also proposed (Williams et al. 2009) that the basic juxtamembrane regions on both VAMP-2 and syntaxin-1 may function symmetrically through nonspecific electrostatic interactions in *trans* to promote close membrane apposition and *trans* SNARE complex assembly. These studies indicate an important role for basic charge-containing residues in the membrane-proximal region of VAMP-2 but the role these play remain uncertain.

Interactions with acidic phospholipids for exocytic SNARE proteins are quite general. For example, the yeast syntaxin Sso1p binds acidic phospholipids via membrane-proximal basic residues. About half of the stimulation of SNARE-dependent liposome fusion by PA was attributed to this interaction (Liu et al. 2007b). Additional studies are needed to determine whether there are common mechanisms at work in SNARE protein-lipid interactions and what function they play.

4.12 CAPS and Munc13 as Lipid-Binding Proteins for Priming Vesicle Exocytosis

As indicated previously, the major role for $PI(4,5)P_2$ in the regulated secretory pathway relates to a function in priming DCV exocytosis. Two of the major priming proteins for regulated vesicle exocytosis, CAPS and Munc13, are regulated by $PI(4,5)P_2$ and by $PI(4,5)P_2$ or DAG, respectively. CAPS was discovered as a protein in rat brain cytosol that reconstitutes Ca^{2+}-triggered DCV exocytosis in mechanically-permeabilized PC12 cells (Walent et al. 1992). The activity of CAPS in permeable cells is only evident after ATP-dependent reactions involving PITP and PI(4)P 5-kinase that restore $PI(4,5)P_2$ have gone to completion (Grishanin et al. 2004). CAPS binds $PI(4,5)P_2$ in part through its central PH domain, which is required for CAPS activity in evoked DCV exocytosis (Grishanin et al. 2002, 2004; Loyet et al. 1998). Recent studies reconstituted part of the function of CAPS in a SNARE protein-dependent liposome fusion assay (James et al. 2008, 2009). CAPS activity in the liposome fusion assay requires that $PI(4,5)P_2$ is present in the acceptor liposomes that contain the plasma membrane t-SNAREs syntaxin-1 and SNAP-25. By contrast, $PI(4,5)P_2$ in the donor VAMP-2-containing liposomes failed to support CAPS function. As anticipated for essential interactions with $PI(4,5)P_2$, the PH domain of CAPS was required for its activity in liposome fusion (James et al. 2008). CAPS functions in vesicle priming where it likely promotes the assembly of SNARE protein complexes in advance of triggered fusion (James et al. 2009). On liposomes, heterotrimeric SNARE complex formation is accelerated by CAPS but only when $PI(4,5)P_2$ is present on the t-SNARE liposomes. This contrasts with the lack of a requirement for $PI(4,5)P_2$ for CAPS binding to SNARE proteins (Daily et al. 2010). Thus, the current evidence indicates that CAPS may function in vesicle priming through dual interactions with PI(4,5)P_2 via its PH domain and with SNARE proteins via a C-terminal domain in CAPS. Anchorage in the membrane through PI(4,5)P_2 interactions may allow CAPS to exert force on the SNARE proteins to mediate rearrangements. This model can account for a positive role for PI(4,5)P_2 in priming Ca^{2+}-triggered DCV exocytosis (see Fig. 4.1).

Studies in PC12 cells, chromaffin cells and neurons indicate that C2 domain-containing proteins cofunction with CAPS in vesicle priming reactions (Liu et al. 2010; Jockusch et al. 2007). Munc13 proteins exhibit sequence homology to CAPS in C-terminal regions that mediate SNARE interactions (Koch et al. 2000). Genetic disruption of Munc13 isoforms in mice strongly inhibits neurotransmitter release at the stage of priming SVs (Varoqueaux et al. 2002). Brain-specific isoforms of Munc13 lack a PH domain but contain three C2 domains and a C1 domain. The second C2 domain of Munc13-1 binds Ca^{2+} and exhibits Ca^{2+}-dependent PI(4,5)P_2 binding (Shin et al. 2010). A gain of function C2B domain mutant of Munc13 exhibited increased neurotransmitter release evoked by single action potentials whereas a C2B mutant abrogated for Ca^{2+} binding showed decreased release with trains of action potentials (Shin et al. 2010). As noted previously, brain Munc13 isoforms also contain a C1 domain that binds DAG. Munc13 with a C1 domain mutation is dysfunctional in potentiating SV or DCV exocytosis (Bauer et al. 2007; Rhee et al. 2002; Rosenmund et al. 2002). Munc13 as a priming factor may be recruited to sites of exocytosis, either to PI(4,5)P_2 during Ca^{2+} rises, or to DAG arising from Ca^{2+} activation of PLC (Rosenmund et al. 2002; Rhee et al. 2002) (see Fig. 4.1). Thus, for some forms of regulated vesicle exocytosis, PI(4,5)P_2 hydrolysis may be required for function to generate DAG (Hammond et al. 2006). Overall, both major priming proteins that function in regulated vesicle exocytosis in neural and endocrine cells, CAPS and Munc13, utilize PI(4,5)P_2 or its metabolite DAG for activation. In future studies, it will be important to determine the plasma membrane sites for PI(4,5)P_2 and DAG synthesis relative to vesicle exocytosis and establish whether CAPS and Munc13 proteins are recruited to these sites.

4.13 Roles for Other PI(4,5)P_2-Binding Proteins in Regulated Vesicle Exocytosis

Additional steps in vesicle exocytosis beyond priming may require PI(4,5)P_2 and PI(4,5)P_2-binding proteins. In capacitance recordings of DCV exocytosis in chromaffin cells (Milosevic et al. 2005), modulation of PI(4,5)P_2 levels affected the RRP and rates of [re]priming. However, rates of evoked exocytosis were not affected, which implies that proteins required for fusion *per se* or its Ca^{2+} triggering were not strongly dependent upon PI(4,5)P_2. Similar findings emerged in the capacitance studies of chromaffin cells from PI(4)P 5-kinase 1γ knockout mice (Di Paolo et al. 2004). Synaptotagmins, the major Ca^{2+} sensors for regulated vesicle exocytosis, exhibit

Ca^{2+}-dependent binding to PI(4,5)P_2 *in vitro* (Bai et al. 2004). Loss of synaptotagmin function is associated with decreased rates of triggered exocytosis (Voets et al. 2001). Possibly residual levels of PI(4,5)P_2 in cells from the PI(4)P 5-kinase 1γ knockout mice are sufficient to maintain synaptotagmin function. Additional studies in the knockout mice indicated that the number of DCVs docked at the plasma membrane were unaltered, which implies that proteins involved in DCV docking are not greatly affected by PI(4,5)P_2. However, in amperometric measurements of catecholamine secretion, subtle differences were observed in the amperometric spikes from chromaffin cells of control and knockout mice (Di Paolo et al. 2004). In the latter, a longer duration pre-spike foot was observed, which may indicate altered fusion pore dynamics in cells with decreased PI(4,5)P_2. Proteins that bind PI(4,5)P_2 and regulate fusion pore dynamics might include dynamin, a PH domain-containing protein (Tsuboi et al. 2004), synaptotagmin, a tandem C2 domain-containing protein (Wang et al. 2001; Lynch et al. 2008), and SCAMP2, a PI(4,5)P_2-binding tetraspanin protein (Liao et al. 2007). In addition, a decrease of PI(4,5)P_2 would result in decreased F actin polymerization, which would alter fusion pore dilation (Berberian et al. 2009).

4.14 Tethering Complexes Bind PI(4,5)P_2 in Constitutive Vesicle Exocytosis

Considerable genetic evidence indicates that plasma membrane PI(4,5)P_2 is an important component for establishing the polarity of the actin cytoskeleton and the selection of exocytic fusion sites for post-Golgi vesicles (reviewed in (Yakir-Tamang and Gerst 2009b; He and Guo 2009)). Studies of the constitutive exocytic pathway reinforce the view that PI(4,5)P_2 plays an important role in recruiting proteins to the target membrane. In the budding yeast, post-Golgi vesicles are delivered on a polarized F actin cytoskeleton to bud sites on the plasma membrane. As noted previously, mutants in MSS4, the sole PI(4)P 5-kinase in yeast, exhibit defects in actin localization and defects in secretion (Yakir-Tamang and Gerst 2009a). The former results in part from failure to recruit the PH domain-containing Rho GEF Rom2 to the plasma membrane. The latter results in part due to mislocalization of a vesicle tethering complex called the exocyst complex. The exocyst complex consists of 8 subunits encoded by late Sec genes (Sec3, Sec5, Sec6, Sec8, Sec10, Sec15, Exo70, Exo84) (He and Guo 2009). Sec3 and Exo70 localize to the plasma membrane largely independent of actin whereas other exocyst subunits are vesicle-associated. Actin-dependent delivery of vesicles to the plasma membrane completes formation of the exocyst complex, which activates SNARE complexes for fusion. Sec3 was found to bind PI(4,5)P_2 via N-terminal polybasic sequences (Zhang et al. 2008) and Exo70 to bind PI(4,5)P_2 via a C-terminal domain (He et al. 2007). These subunits also interact with GTPases required for their localization and function (He and Guo 2009; Yakir-Tamang and Gerst 2009b). Recent studies demonstrated that the mammalian exocyst Exo70 also interacts with PI(4,5)P_2, which was essential for the docking and fusion of post-Golgi secretory vesicles in the constitutive secretory pathway (Liu et al. 2007a). These studies indicate that plasma membrane PI(4,5)P_2 plays an important role in recruiting subunits that enable the assembly of an essential tethering complex that activates SNARE-dependent vesicle fusion.

At each stage of vesicle trafficking in the secretory pathway, a diverse set of tethering factors or tethering complexes mediate contact between an incoming vesicle and a target membrane (Sztul and Lupashin 2006). It was recently suggested that CAPS and Munc13 exhibit significant homology to other tethering factor subunits such as exocyst Sec6 suggesting a common ancestral origin (Pei et al. 2009). Many other tethering factors bind to the phosphoinositides that are characteristic of the target membrane. For example, EEA1 in endosome tethering binds PI(3)P (Gaullier et al. 1999). The HOPS complex in vacuole tethering binds PI(3)P and other phosphoinositides (Stroupe et al. 2006). A general prediction for vesicle exocytosis is that proteins involved in vesicle tethering and priming at the plasma membrane will bind PI(4,5)P_2.

4.15 Conclusions

As a specific constituent characteristic of the plasma membrane in resting cells, PI(4,5)P_2 likely participates in all vectorial processes involving the plasma membrane. As an abundant highly-charged constituent in the cytoplasmic leaflet, PI(4,5)P_2 affects many plasma membrane processes through electrostatic interactions with commonly-occurring Arg/Lys/hydrophobic sequences in proteins or through specific PH or C2 domains. The major role for PI(4,5)P_2 in vesicle exocytosis involves protein recruitment and activation. The possibility that this abundant lipid may be concentrated in enriched microdomains where it could exert direct effects on membrane curvature and tension needs to be further assessed. PI(4,5)P_2-binding proteins such as CAPS and Munc13 play a major role in vesicle priming reactions where the principal role of PI(4,5)P_2 is exerted. Other proteins such as PLD1, synaptotagmin and SNAREs may mediate the regulation by PI(4,5)P_2 at other stages of vesicle exocytosis.

References

Aikawa Y, Martin TF (2003) ARF6 regulates a plasma membrane pool of phosphatidylinositol(4,5)bisphosphate required for regulated exocytosis. J Cell Biol 162:647–659

Anantharam A, Axelrod D, Holz RW (2010) Polarized TIRFM reveals changes in plasma membrane topology before and during granule fusion. Cell Mol Neurobiol 30:1343–1349

Aoyagi K, Sugaya T, Umeda M, Yamamoto S, Terakawa S, Takahashi M (2005) The activation of exocytotic sites by the formation of phosphatidylinositol 4,5-bisphosphate microdomains at syntaxin clusters. J Biol Chem 280:17346–17352

Bader MF, Vitale N (2009) Phospholipase D in calcium-regulated exocytosis: lessons from chromaffin cells. Biochim Biophys Acta 1791:936–941

Bai J, Tucker WC, Chapman ER (2004) PIP2 increases the speed of response of synaptotagmin and steers its membrane-penetration activity toward the plasma membrane. Nat Struct Mol Biol 11:36–44

Balla T (2009) Regulation of Ca^{2+} entry by inositol lipids in mammalian cells by multiple mechanisms. Cell Calcium 45:527–534

Barylko B, Gerber SH, Binns DD, Grichine N, Khvotchev M, Sudhof TC, Albanesi JP (2001) A novel family of phosphatidylinositol 4-kinases conserved from yeast to humans. J Biol Chem 276:7705–7708

Bauer CS, Woolley RJ, Teschemacher AG, Seward EP (2007) Potentiation of exocytosis by phospholipase C-coupled G-protein-coupled receptors requires the priming protein Munc13-1. J Neurosci 27:212–219

Baumann MK, Amstad E, Mashaghi A, Textor M, Reimhult E (2010) Characterization of supported lipid bilayers incorporating the phosphoinositides phosphatidylinositol 4,5-biphosphate and phosphoinositol-3,4,5-triphosphate by complementary techniques. Biointerphases 5:114–119

Berberian K, Torres AJ, Fang Q, Kisler K, Lindau M (2009) F-actin and myosin II accelerate catecholamine release from chromaffin granules. J Neurosci 29:863–870

Bethoney KA, King MC, Hinshaw JE, Ostap EM, Lemmon MA (2009) A possible effector role for the pleckstrin homology (PH) domain of dynamin. Proc Natl Acad Sci U S A 106:13359–13364

Brose N, Betz A, Wegmeyer H (2004) Divergent and convergent signaling by the diacylglycerol second messenger pathway in mammals. Curr Opin Neurobiol 14:328–340

Caroni P (2001) New EMBO members' review: actin cytoskeleton regulation through modulation of PI(4,5)P(2) rafts. EMBO J 20:4332–4336

Chandler DE, Heuser JE (1980) Arrest of membrane fusion events in mast cells by quick-freezing. J Cell Biol 86:666–674

Chen X, Lu J, Dulubova I, Rizo J (2008) NMR analysis of the closed conformation of syntaxin-1. J Biomol NMR 41:43–54

Chernomordik LV, Kozlov MM (2008) Mechanics of membrane fusion. Nat Struct Mol Biol 15:675–683

Cohen FS, Melikyan GB (2004) The energetics of membrane fusion from binding, through hemifusion, pore formation, and pore enlargement. J Membr Biol 199:1–14

Daily NJ, Boswell KL, James DJ, Martin TF (2010) Novel interactions of CAPS (Ca^{2+}-dependent activator protein for secretion) with the three neuronal SNARE proteins required for vesicle fusion. J Biol Chem 285:35320–35329

De Haro L, Quetglas S, Iborra C, Leveque C, Seagar M (2003) Calmodulin-dependent regulation of a lipid binding domain in the v-SNARE synaptobrevin and its role in vesicular fusion. Biol Cell 95:459–464

De Matteis MA, Luini A (2008) Exiting the Golgi complex. Nat Rev Mol Cell Biol 9:273–284

Di Giovanni J, Iborra C, Maulet Y, Leveque C, El Far O, Seagar M (2010) Calcium-dependent regulation of SNARE-mediated membrane fusion by calmodulin. J Biol Chem 285:23665–23675

Di Paolo G, De Camilli P (2006) Phosphoinositides in cell regulation and membrane dynamics. Nature 443:651–657

Di Paolo G, Moskowitz HS, Gipson K, Wenk MR, Voronov S, Obayashi M, Flavell R, Fitzsimonds RM, Ryan TA, De Camilli P (2004) Impaired PtdIns(4,5)P2 synthesis in nerve terminals produces defects in synaptic vesicle trafficking. Nature 431:415–422

Eberhard DA, Cooper CL, Low MG, Holz RW (1990) Evidence that the inositol phospholipids are necessary for exocytosis. Loss of inositol phospholipids and inhibition of secretion in permeabilized cells caused by a bacterial phospholipase C and removal of ATP. Biochem J 268:15–25

Ford MG, Mills IG, Peter BJ, Vallis Y, Praefcke GJ, Evans PR, McMahon HT (2002) Curvature of clathrin-coated pits driven by epsin. Nature 419:361–366

Gaullier JM, Simonsen A, D'arrigo A, Bremnes B, Stenmark H (1999) FYVE finger proteins as effectors of phosphatidylinositol 3-phosphate. Chem Phys Lipids 98:87–94

Golebiewska U, Nyako M, Woturski W, Zaitseva I, McLaughlin S (2008) Diffusion coefficient of fluorescent phosphatidylinositol 4,5-bisphosphate in the plasma membrane of cells. Mol Biol Cell 19:1663–1669
Gong LW, Di Paolo G, Diaz E, Cestra G, Diaz ME, Lindau M, De Camilli P, Toomre D (2005) Phosphatidylinositol phosphate kinase type I gamma regulates dynamics of large dense-core vesicle fusion. Proc Natl Acad Sci U S A 102:5204–5209
Grinstein S (2010) Imaging signal transduction during phagocytosis: phospholipids, surface charge, and electrostatic interactions. Am J Physiol Cell Physiol 299:C876–C881
Grishanin RN, Klenchin VA, Loyet KM, Kowalchyk JA, Ann K, Martin TF (2002) Membrane association domains in Ca^{2+}-dependent activator protein for secretion mediate plasma membrane and dense-core vesicle binding required for Ca^{2+}-dependent exocytosis. J Biol Chem 277:22025–22034
Grishanin RN, Kowalchyk JA, Klenchin VA, Ann K, Earles CA, Chapman ER, Gerona RR, Martin TF (2004) CAPS acts at a prefusion step in dense-core vesicle exocytosis as a PIP2 binding protein. Neuron 43:551–562
Guo J, Wenk MR, Pellegrini L, Onofri F, Benfenati F, De Camilli P (2003) Phosphatidylinositol 4-kinase type IIalpha is responsible for the phosphatidylinositol 4-kinase activity associated with synaptic vesicles. Proc Natl Acad Sci U S A 100:3995–4000
Hammond GR, Dove SK, Nicol A, Pinxteren JA, Zicha D, Schiavo G (2006) Elimination of plasma membrane phosphatidylinositol (4,5)-bisphosphate is required for exocytosis from mast cells. J Cell Sci 119:2084–2094
Hay JC, Martin TF (1992) Resolution of regulated secretion into sequential MgATP-dependent and calcium-dependent stages mediated by distinct cytosolic proteins. J Cell Biol 119:139–151
Hay JC, Martin TF (1993) Phosphatidylinositol transfer protein required for ATP-dependent priming of Ca(2+)-activated secretion. Nature 366:572–575
Hay JC, Fisette PL, Jenkins GH, Fukami K, Takenawa T, Anderson RA, Martin TF (1995) ATP-dependent inositide phosphorylation required for Ca(2+)-activated secretion. Nature 374:173–177
He B, Guo W (2009) The exocyst complex in polarized exocytosis. Curr Opin Cell Biol 21:537–542
He B, Xi F, Zhang X, Zhang J, Guo W (2007) Exo70 interacts with phospholipids and mediates the targeting of the exocyst to the plasma membrane. EMBO J 26:4053–4065
Holz RW, Hlubek MD, Sorensen SD, Fisher SK, Balla T, Ozaki S, Prestwich GD, Stuenkel EL, Bittner MA (2000) A pleckstrin homology domain specific for phosphatidylinositol 4, 5-bisphosphate (PtdIns-4,5-P2) and fused to green fluorescent protein identifies plasma membrane PtdIns-4,5-P2 as being important in exocytosis. J Biol Chem 275:17878–17885
Hope HR, Pike LJ (1996) Phosphoinositides and phosphoinositide-utilizing enzymes in detergent-insoluble lipid domains. Mol Biol Cell 7:843–851
Hui E, Johnson CP, Yao J, Dunning FM, Chapman ER (2009) Synaptotagmin-mediated bending of the target membrane is a critical step in Ca(2+)-regulated fusion. Cell 138:709–721
Jahn R, Scheller RH (2006) SNAREs—engines for membrane fusion. Nat Rev Mol Cell Biol 7:631–643
James DJ, Khodthong C, Kowalchyk JA, Martin TF (2008) Phosphatidylinositol 4,5-bisphosphate regulates SNARE-dependent membrane fusion. J Cell Biol 182:355–366
James DJ, Kowalchyk J, Daily N, Petrie M, Martin TF (2009) CAPS drives trans-SNARE complex formation and membrane fusion through syntaxin interactions. Proc Natl Acad Sci U S A 106:17308–17313
Janmey PA, Kinnunen PK (2006) Biophysical properties of lipids and dynamic membranes. Trends Cell Biol 16:538–546
Janmey PA, Lindberg U (2004) Cytoskeletal regulation: rich in lipids. Nat Rev Mol Cell Biol 5:658–666
Jockusch WJ, Speidel D, Sigler A, Sorensen JB, Varoqueaux F, Rhee JS, Brose N (2007) CAPS-1 and CAPS-2 are essential synaptic vesicle priming proteins. Cell 131:796–808

Khvotchev M, Sudhof TC (1998) Newly synthesized phosphatidylinositol phosphates are required for synaptic norepinephrine but not glutamate or gamma-aminobutyric acid (GABA) release. J Biol Chem 273:21451–21454

Koch H, Hofmann K, Brose N (2000) Definition of Munc13-homology-domains and characterization of a novel ubiquitously expressed Munc13 isoform. Biochem J 349:247–253

Kozlov MM, McMahon HT, Chernomordik LV (2010) Protein-driven membrane stresses in fusion and fission. Trends Biochem Sci 35:699–706

Kutateladze TG (2010) Translation of the phosphoinositide code by PI effectors. Nat Chem Biol 6:507–513

Kweon DH, Kim CS, Shin YK (2003) Regulation of neuronal SNARE assembly by the membrane. Nat Struct Biol 10:440–447

Lam AD, Tryoen-Toth P, Tsai B, Vitale N, Stuenkel EL (2008) SNARE-catalyzed fusion events are regulated by Syntaxin 1A-lipid interactions. Mol Biol Cell 19:485–497

Laux T, Fukami K, Thelen M, Golub T, Frey D, Caroni P (2000) GAP43, MARCKS, and CAP23 modulate PI(4,5)P(2) at plasmalemmal rafts, and regulate cell cortex actin dynamics through a common mechanism. J Cell Biol 149:1455–1472

Lawrence JT, Birnbaum MJ (2003) ADP-ribosylation factor 6 regulates insulin secretion through plasma membrane phosphatidylinositol 4,5-bisphosphate. Proc Natl Acad Sci U S A 100:13320–13325

Lemmon MA (2003) Phosphoinositide recognition domains. Traffic 4:201–213

Lemmon MA (2008) Membrane recognition by phospholipid-binding domains. Nat Rev Mol Cell Biol 9:99–111

Liao H, Ellena J, Liu L, Szabo G, Cafiso D, Castle D (2007) Secretory carrier membrane protein SCAMP2 and phosphatidylinositol 4,5-bisphosphate interactions in the regulation of dense core vesicle exocytosis. Biochemistry 46:10909–10920

Liu J, Zuo X, Yue P, Guo W (2007a) Phosphatidylinositol 4,5-bisphosphate mediates the targeting of the exocyst to the plasma membrane for exocytosis in mammalian cells. Mol Biol Cell 18:4483–4492

Liu S, Wilson KA, Rice-Stitt T, Neiman AM, McNew JA (2007b) In vitro fusion catalyzed by the sporulation-specific t-SNARE light-chain Spo20p is stimulated by phosphatidic acid. Traffic 8:1630–1643

Liu Y, Schirra C, Edelmann L, Matti U, Rhee J, Hof D, Bruns D, Brose N, Rieger H, Stevens DR, Rettig J (2010) Two distinct secretory vesicle-priming steps in adrenal chromaffin cells. J Cell Biol 190:1067–1077

Loyet KM, Kowalchyk JA, Chaudhary A, Chen J, Prestwich GD, Martin TF (1998) Specific binding of phosphatidylinositol 4,5-bisphosphate to calcium-dependent activator protein for secretion (CAPS), a potential phosphoinositide effector protein for regulated exocytosis. J Biol Chem 273:8337–8343

Lynch KL, Gerona RR, Kielar DM, Martens S, McMahon HT, Martin TF (2008) Synaptotagmin-1 utilizes membrane bending and SNARE binding to drive fusion pore expansion. Mol Biol Cell 19:5093–5103

Martens S, Kozlov MM, McMahon HT (2007) How synaptotagmin promotes membrane fusion. Science 316:1205–1208

Martin TF (1998) Phosphoinositide lipids as signaling molecules: common themes for signal transduction, cytoskeletal regulation, and membrane trafficking. Annu Rev Cell Dev Biol 14:231–264

Martin TF (2001) PI(4,5)P(2) regulation of surface membrane traffic. Curr Opin Cell Biol 13:493–499

McLaughlin S, Murray D (2005) Plasma membrane phosphoinositide organization by protein electrostatics. Nature 438:605–611

McLaughlin S, Wang J, Gambhir A, Murray D (2002) PIP(2) and proteins: interactions, organization, and information flow. Annu Rev Biophys Biomol Struct 31:151–175

Micheva KD, Holz RW, Smith SJ (2001) Regulation of presynaptic phosphatidylinositol 4,5-biphosphate by neuronal activity. J Cell Biol 154:355–368

Milosevic I, Sorensen JB, Lang T, Krauss M, Nagy G, Haucke V, Jahn R, Neher E (2005) Plasmalemmal phosphatidylinositol-4,5-bisphosphate level regulates the releasable vesicle pool size in chromaffin cells. J Neurosci 25:2557–2565

Murray DH, Tamm LK (2009) Clustering of syntaxin-1A in model membranes is modulated by phosphatidylinositol 4,5-bisphosphate and cholesterol. Biochemistry 48:4617–4625

Olsen HL, Hoy M, Zhang W, Bertorello AM, Bokvist K, Capito K, Efanov AM, Meister B, Thams P, Yang SN, Rorsman P, Berggren PO, Gromada J (2003) Phosphatidylinositol 4-kinase serves as a metabolic sensor and regulates priming of secretory granules in pancreatic beta cells. Proc Natl Acad Sci U S A 100:5187–5192

Pei J, Ma C, Rizo J, Grishin NV (2009) Remote homology between Munc13 MUN domain and vesicle tethering complexes. J Mol Biol 391:509–517

Quetglas S, Iborra C, Sasakawa N, De Haro L, Kumakura K, Sato K, Leveque C, Seagar M (2002) Calmodulin and lipid binding to synaptobrevin regulates calcium-dependent exocytosis. EMBO J 21:3970–3979

Ramachandran R, Pucadyil TJ, Liu YW, Acharya S, Leonard M, Lukiyanchuk V, Schmid SL (2009) Membrane insertion of the pleckstrin homology domain variable loop 1 is critical for dynamin-catalyzed vesicle scission. Mol Biol Cell 20:4630–4639

Rettig J, Neher E (2002) Emerging roles of presynaptic proteins in Ca^{++}-triggered exocytosis. Science 298:781–785

Rhee JS, Betz A, Pyott S, Reim K, Varoqueaux F, Augustin I, Hesse D, Sudhof TC, Takahashi M, Rosenmund C, Brose N (2002) Beta phorbol ester- and diacylglycerol-induced augmentation of transmitter release is mediated by Munc13s and not by PKCs. Cell 108:121–133

Rosenmund C, Sigler A, Augustin I, Reim K, Brose N, Rhee JS (2002) Differential control of vesicle priming and short-term plasticity by Munc13 isoforms. Neuron 33:411–424

Routt SM, Ryan MM, Tyeryar K, Rizzieri KE, Mousley C, Roumanie O, Brennwald PJ, Bankaitis VA (2005) Nonclassical PITPs activate PLD via the Stt4p PtdIns-4-kinase and modulate function of late stages of exocytosis in vegetative yeast. Traffic 6:1157–1172

Saad JS, Miller J, Tai J, Kim A, Ghanam RH, Summers MF (2006) Structural basis for targeting HIV-1 Gag proteins to the plasma membrane for virus assembly. Proc Natl Acad Sci U S A 103:11364–11369

Salaun C, James DJ, Chamberlain LH (2004) Lipid rafts and the regulation of exocytosis. Traffic 5:255–264

Schu PV, Takegawa K, Fry MJ, Stack JH, Waterfield MD, Emr SD (1993) Phosphatidylinositol 3-kinase encoded by yeast VPS34 gene essential for protein sorting. Science 260:88–91

Shin OH, Lu J, Rhee JS, Tomchick DR, Pang ZP, Wojcik SM, Camacho-Perez M, Brose N, Machius M, Rizo J, Rosenmund C, Sudhof TC (2010) Munc13 C2B domain is an activity-dependent Ca^{2+} regulator of synaptic exocytosis. Nat Struct Mol Biol 17:280–288

Stroupe C, Collins KM, Fratti RA, Wickner W (2006) Purification of active HOPS complex reveals its affinities for phosphoinositides and the SNARE Vam7p. EMBO J 25:1579–1589

Suh BC, Hille B (2008) PIP2 is a necessary cofactor for ion channel function: how and why? Annu Rev Biophys 37:175–195

Sztul E, Lupashin V (2006) Role of tethering factors in secretory membrane traffic. Am J Physiol Cell Physiol 290:C11–C26

Tong J, Nguyen L, Vidal A, Simon SA, Skene JH, McIntosh TJ (2008) Role of GAP-43 in sequestering phosphatidylinositol 4,5-bisphosphate to Raft bilayers. Biophys J 94:125–133

Tsuboi T, McMahon HT, Rutter GA (2004) Mechanisms of dense core vesicle recapture following "kiss and run" ("cavicapture") exocytosis in insulin-secreting cells. J Biol Chem 279:47115–47124

Van Meer G, Sprong H (2004) Membrane lipids and vesicular traffic. Curr Opin Cell Biol 16:373–378

Van Rheenen J, Achame EM, Janssen H, Calafat J, Jalink K (2005) PIP2 signaling in lipid domains: a critical re-evaluation. EMBO J 24:1664–1673

Varoqueaux F, Sigler A, Rhee JS, Brose N, Enk C, Reim K, Rosenmund C (2002) Total arrest of spontaneous and evoked synaptic transmission but normal synaptogenesis in the absence of Munc13-mediated vesicle priming. Proc Natl Acad Sci U S A 99:9037–9042

Verhage M, Sorensen JB (2008) Vesicle docking in regulated exocytosis. Traffic 9:1414–1424

Vicogne J, Vollenweider D, Smith JR, Huang P, Frohman MA, Pessin JE (2006) Asymmetric phospholipid distribution drives in vitro reconstituted SNARE-dependent membrane fusion. Proc Natl Acad Sci U S A 103:14761–14766

Voets T, Moser T, Lund PE, Chow RH, Geppert M, Sudhof TC, Neher E (2001) Intracellular calcium dependence of large dense-core vesicle exocytosis in the absence of synaptotagmin I. Proc Natl Acad Sci U S A 98:11680–11685

Wagner ML, Tamm LK (2001) Reconstituted syntaxin1a/SNAP25 interacts with negatively charged lipids as measured by lateral diffusion in planar supported bilayers. Biophys J 81:266–275

Walent JH, Porter BW, Martin TF (1992) A novel 145 kd brain cytosolic protein reconstitutes Ca(2+)-regulated secretion in permeable neuroendocrine cells. Cell 70:765–775

Wang CT, Grishanin R, Earles CA, Chang PY, Martin TF, Chapman ER, Jackson MB (2001) Synaptotagmin modulation of fusion pore kinetics in regulated exocytosis of dense-core vesicles. Science 294:1111–1115

Waselle L, Gerona RR, Vitale N, Martin TF, Bader MF, Regazzi R (2005) Role of phosphoinositide signaling in the control of insulin exocytosis. Mol Endocrinol 19:3097–3106

Weber T, Zemelman BV, McNew JA, Westermann B, Gmachl M, Parlati F, Sollner TH, Rothman JE (1998) SNAREpins: minimal machinery for membrane fusion. Cell 92:759–772

Wen PJ, Osborne SL, Meunier FA (2011) Dynamic control of neuroexocytosis by phosphoinositides in health and disease. Prog Lipid Res 50:52–61

Wenk MR, Pellegrini L, Klenchin VA, Di Paolo G, Chang S, Daniell L, Arioka M, Martin TF, De Camilli P (2001) PIP kinase Igamma is the major PI(4,5)P(2) synthesizing enzyme at the synapse. Neuron 32:79–88

Wiedemann C, Schafer T, Burger MM (1996) Chromaffin granule-associated phosphatidylinositol 4-kinase activity is required for stimulated secretion. EMBO J 15:2094–2101

Williams D, Vicogne J, Zaitseva I, McLaughlin S, Pessin JE (2009) Evidence that electrostatic interactions between vesicle-associated membrane protein 2 and acidic phospholipids may modulate the fusion of transport vesicles with the plasma membrane. Mol Biol Cell 20:4910–4919

Yakir-Tamang L, Gerst JE (2009a) A phosphatidylinositol-transfer protein and phosphatidylinositol-4-phosphate 5-kinase control Cdc42 to regulate the actin cytoskeleton and secretory pathway in yeast. Mol Biol Cell 20:3583–3597

Yakir-Tamang L, Gerst JE (2009b) Phosphoinositides, exocytosis and polarity in yeast: all about actin? Trends Cell Biol 19:677–684

Yin HL, Janmey PA (2003) Phosphoinositide regulation of the actin cytoskeleton. Annu Rev Physiol 65:761–789

Zeniou-Meyer M, Zabari N, Ashery U, Chasserot-Golaz S, Haeberle AM, Demais V, Bailly Y, Gottfried I, Nakanishi H, Neiman AM, Du G, Frohman MA, Bader MF, Vitale N (2007) Phospholipase D1 production of phosphatidic acid at the plasma membrane promotes exocytosis of large dense-core granules at a late stage. J Biol Chem 282:21746–21757

Zhang X, Orlando K, He B, Xi F, Zhang J, Zajac A, Guo W (2008) Membrane association and functional regulation of Sec3 by phospholipids and Cdc42. J Cell Biol 180:145–158

Zheng Q, McFadden SC, Bobich JA (2004) Phosphatidylinositol 4,5-bisphosphate promotes both [3H]-noradrenaline and [14C]-glutamate exocytosis from nerve endings. Neurochem Int 44:243–250

Chapter 5
Role of Phosphoinositides at the Neuronal Synapse

Samuel G. Frere, Belle Chang-Ileto and Gilbert Di Paolo

Abstract Synaptic transmission is amongst the most sophisticated and tightly controlled biological phenomena in higher eukaryotes. In the past few decades, tremendous progress has been made in our understanding of the molecular mechanisms underlying multiple facets of neurotransmission, both pre- and postsynaptically. Brought under the spotlight by pioneer studies in the areas of secretion and signal transduction, phosphoinositides and their metabolizing enzymes have been increasingly recognized as key protagonists in fundamental aspects of neurotransmission. Not surprisingly, dysregulation of phosphoinositide metabolism has also been implicated in synaptic malfunction associated with a variety of brain disorders. In the present chapter, we summarize current knowledge on the role of phosphoinositides at the neuronal synapse and highlight some of the outstanding questions in this research field.

Keywords Synaptojanin · Synaptic vesicle · Exocytosis · Endocytosis · Glutamate receptors · Synaptic plasticity · Phospholipid · Membrane trafficking

5.1 Introduction

Chemical synapses are intercellular junctions through which neurons efficiently transfer electrical signals to target cells. Synapses consist of two juxtaposed structures, the pre- and postsynaptic compartments, which are separated by the synaptic cleft. The presynaptic compartment is specialized for the fast release of

G. Di Paolo (✉) · S. G. Frere · B. Chang-Ileto
Department of Pathology and Cell Biology,
Taub Institute for Research on Alzheimer's Disease and the Aging Brain,
Columbia University Medical Center,
630 West 168th Street, P&S 12-420C, 10032, New York, USA
e-mail: gil.dipaolo@columbia.edu

T. Balla et al. (eds.), *Phosphoinositides II: The Diverse Biological Functions*,
Subcellular Biochemistry 59, DOI 10.1007/978-94-007-3015-1_5,

neurotransmitters in response to action potentials propagating along axons towards nerve terminals and the opening of voltage-dependent Ca^{2+} channels. Ca^{2+}-triggered release of neurotransmitters occurs through fast exocytosis of synaptic vesicles (SVs) at specialized sites called the active zones and is generally followed by a slower retrieval of SV membrane by endocytosis (Fig. 5.1). The released neurotransmitters, which are mainly glutamate and gamma-aminobutyric acid (GABA) in the central nervous system, bind to ionotropic and metabotropic postsynaptic receptors. This translates the chemical signal in the form of neurotransmitters into inhibitory and excitatory electrical events as well as into intracellular signaling cascades, thus transmitting action potentials in the target cell. The strength of synaptic transmission can vary across a broad range in a phenomenon called synaptic plasticity, which is typically associated with drastic morphological changes at the postsynapse of excitatory neurons (*i.e.*, dendritic spines) and believed to underlie key neurobehavioral responses, such as learning and memory. The past 60 years has witnessed a worldwide and massive effort to investigate the molecular and cellular bases of synaptic transmission leading to substantial progress in our understanding of how this phenomenon works. However, despite the fact that many aspects of synaptic transmission critically depend on signaling across lipid bilayers and membrane trafficking, less is known about the role of lipids in this process. In the past two decades, lipids such as phosphorylated derivatives of phosphatidylinositol, also called phosphoinositides, have progressively come to center stage due to their growing implication in fundamental aspects of neurotransmission.

Phosphoinositides are quantitatively a minor lipid class in cellular membranes; nevertheless, they play a crucial role in many aspects of cellular physiology, including synaptic function. There are seven known phosphoinositides, one of which, phosphatidylinositol-4,5-bisphosphate [PtdIns(4,5)P_2], has been extensively studied due to its abundance and its historical implication in signal transduction as a precursor for second messengers as well as in secretion. This book chapter largely focuses on the role of PtdIns(4,5)P_2 metabolism at the neuronal synapse, with particular emphasis on its function in the traffic of SVs at the presynapse. The role of other critical phosphoinositides, such as phosphatidylinositol-3,4,5-trisphosphate [PtdIns(3,4,5)P_3], will also be discussed, based on its implication in synaptic plasticity. Phosphoinositides are also important regulators of many ion channels and transporters, such as voltage-gated Ca^{2+} channels or potassium channels, and this regulation is central to neuronal excitability and synaptic transmission. However, space limitation prevents us from elaborating on this topic, which is covered by a series of recent review articles to which we refer the reader (Hilgemann 2007; Suh and Hille 2008; Logothetis et al. 2010). In the present chapter, we describe the main lipid enzymes controlling the metabolism of the two main signaling phosphoinositides at the synapse, PtdIns(4,5)P_2 and PtdIns(3,4,5)P_3. We then highlight the mechanisms by which these lipids control fundamental aspects of synaptic transmission at the pre- and post-synapse. Finally, we discuss the implications of dysregulation of synaptic phosphoinositide metabolism in disease-related processes, such as in Down syndrome and Alzheimer's disease.

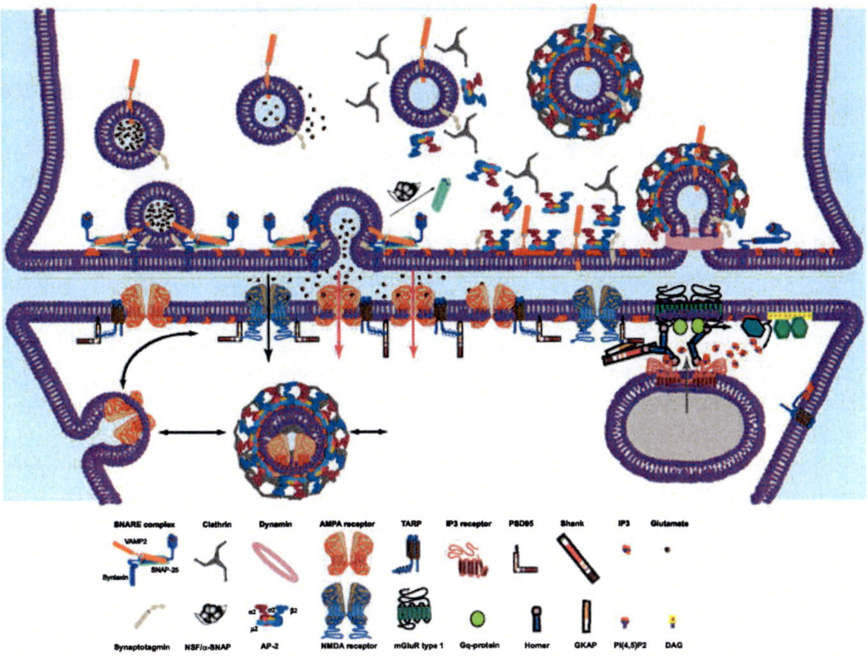

Fig. 5.1 Neurotransmission at glutamatergic synapses: Presynaptically, synaptic vesicles (*SVs*) filled with glutamate are localized in proximity to the plasma membrane (*PM*) and are docked at the active zone after the assembling of the SNARE complex (and other proteins such as Munc18, not depicted here) composed of the PM-associated syntaxin-1, the cytosolic SNAP-25 and the SV membrane-associated synaptobrevin. Upon the arrival of a calcium (Ca^{2+}) influx, SVs fuse with the PM and the content of the SV is released in the synaptic cleft. Following SV collapse, the SNARE complex is disassembled by the ATPase N-ethylmaleimide-sensitive factor (NSF) and its adaptor α-SNAP. The fused vesicle is retrieved by clathrin-mediated endocytosis, which starts by the recruitment of the clathrin adaptors, AP-2 to PM enriched in PtdIns(4,5)P_2 and by its interaction with Syt-1. Clathrin molecules assemble into a lattice structure at the endocytic site and, along with a variety of tubulating factors, permit PM invagination and the formation of a CCP. Dynamin oligomers formed a helix that encircles the bud of the pit and triggers the fission and the individuation of a clathrin-coated vesicles. Quickly, the clathrin coat is removed and the AP-2-clathrin complex is disassembled. The SV is refilled with neurotransmitters and can undergo another cycle of exocytosis and endocytosis.

Postsynaptically, released glutamate binds to the ionotropic AMPA and NMDA receptors and to the Gq-coupled metabotropic glutamate receptor (mGluR) type 1 (the other types of mGluR are not depicted). Gating of the AMPA receptor by glutamate generates a net current influx that depolarizes the postsynaptic PM allowing the gating of glutamate-bound NMDA receptor and a slower influx of sodium and calcium. mGluR type 1 activation releases the Gq proteins and the hydrolysis of PtdIns(4,5)P_2 by phospholipase Cγ into diacylglycerol (DAG) and inositol (1,4,5)-trisphosphate (Ins(1,4,5)P_3). DAG recruits to the PM and activates the protein kinase C, while Ins(1,4,5)P_3 binds to and opens the Ins(1,4,5)P_3 receptor (Ins(1,4,5)P_3R), which liberates the calcium stored in the endoplasmic reticulum (ER). In the figure, interactions of the receptors with the structural proteins that organize the postsynaptic density (PSD) is emphasized. NMDA binds directly to PSD-95 with a PDZ-binding motif localized at the C-terminal extremity of the receptor. AMPA receptor and mGluR type 1 are indirectly anchored to the PSD-95 complex via the transmembrane AMPA receptor regulatory proteins (TARP) and via Homer/Shank/GKAP, respectively. Homer also interacts with the IP_3R and allows the anchoring of the receptor in proximity to the mGluR type 1. Finally, AMPA receptors are mobile and traffic between the synaptic and extrasynaptic zone, where they can be internalized by CME and be recycled via the endosome

5.2 Phosphoinositide Metabolism at the Synapse

While PtdIns(4,5)P_2 and PtdIns4P are the most abundant phosphoinositides in cells, the former has been extensively characterized in the past few decades and is now known to play fundamental roles at the plasma membrane. Indeed, PtdIns(4,5)P_2 controls both exocytosis and endocytosis, actin dynamics, signal transduction as well as the function of many ion channels and transporters (Di Paolo and De Camilli 2006; Saarikangas et al. 2008; Suh and Hille 2008). Although the synaptic plasma membrane contains highly specialized structures, such as the active zone and the post-synaptic density, and a distinct protein and lipid composition, PtdIns(4,5)P_2 is known to regulate these very same biological processes at these membranes. Steady-state PtdIns(4,5)P_2 levels at the synapse are controlled by a specific set of lipid enzymes, which, at mammalian synapses, include PtdInsP kinases type 1γ (Wenk et al. 2001; Di Paolo et al. 2004) and polyphosphoinositide phosphatase Synj1 (Cremona et al. 1999; Voronov et al. 2008). Additionally, a variety of enzymes, such as phosphoinositide 3-kinases (PI3K), phospholipase C (PLC) and A2, play both critical and pleiotropic roles in phosphoinositide-dependent signaling. Besides PtdIns(4,5)P_2, the low abundance phosphoinositide PtdIns(3,4,5)P_3 is known to regulate many cellular processes, some of which are critical for synaptic physiology and plasticity. The following section will primarily focus on the enzymes controlling PtdIns(4,5)P_2 and PtdIns(4,5)P_2 metabolism at the synapse. The precise role of these two phosphoinositides at synapses will be discussed in Sect. 5.3 and 5.4.

5.2.1 Regulation of PtdIns(4,5)P_2 Metabolism at the Synapse

PtdIns(4,5)P_2 can be synthesized by type 1 PtdInsP kinases (PtdInsPK1s), which phosphorylate PtdIns4P on the 5' position, and by type 2 PtdInsP kinases, which phosphorylate PtdIns5P on the 4' position (Doughman et al. 2003; Anderson et al. 1999). The activation loop located at the COOH-terminus of the kinase core domain controls the substrate specificity of either class (Kunz et al. 2000). Each type of PtdInsPKs is comprised of three catalytically-active isoforms, α, β, γ (Ishihara et al. 1996; Anderson et al. 1999). A kinase-dead homolog, PtdInsPKH, has also been reported (Chang et al. 2004). Biochemical measurements using immunodepleted or knockout (KO) brain extracts have demonstrated that the main pathway for the synthesis of PtdIns(4,5)P_2 in the brain and at the synapse involves the PtdInsPK1γ isoform (Wenk et al. 2001; Di Paolo et al. 2004; Volpicelli-Daley et al. 2010). Its substrate, PtdIns4P, is produced by a PtdIns 4-kinase (PI4K), which was identified as a type IIα enzyme (PI4KIIα), at least in the brain and at neuronal synapses (Guo et al. 2003). In mouse brain, PtdInsPK1γ occurs as 3 splice variants, PtdInsPK1γ635 (or PtdInsPK1γ87), PtdInsPK1γ661 (or PtdInsPK1γ90) and PtdInsPK1γ687 (Ishihara et al. 1998; Giudici et al. 2004). At synapses, PtdInsPK1γ661 is the most abundant isoform and the main kinase for the production of PtdIns(4,5)P_2 (Wenk et al. 2001; Di Paolo et al. 2004; Volpicelli-Daley et al. 2010). In mice, the extended COOH-terminal tail of

PtdInsPK1γ661 binds to the Four-point-one, Ezrin, Radixin, Moesin (FERM) domain of the head region of talin (Di Paolo et al. 2002; Ling et al. 2002) and with the clathrin adaptor complex protein-2 (AP-2) (Bairstow et al. 2006; Nakano-Kobayashi et al. 2007; Krauss et al. 2006; Kahlfeldt et al. 2010). Three other types of interactors have been identified for PtdInsPK1γ: (i) Rho GTPases (Chatah and Abrams 2001; Weernink et al. 2004); (ii) Arf GTPases (Krauss and Haucke 2005; Krauss et al. 2003); and (iii) phospholipase D (Divecha et al. 2000; Jarquin-Pardo et al. 2007; Moritz et al. 1992). Similar to a group of proteins involved in SV endocytosis called the "dephosphins", PtdInsPK1γ undergoes depolarization-dependent dephosphorylation by calcineurin, a phenomenon known to promote the assembly of endocytic factors (Cousin and Robinson 2001; Slepnev et al. 1998; Wenk et al. 2001). The phosphorylation of PtdInsPK1γ is mediated at least in part by cyclin-dependent kinase-5 (Cdk-5), a proline-directed serine/threonine kinase that is critically important for SV trafficking and the control of the recycling pool size (Lee et al. 2005; Kim and Ryan 2010). Phosphorylation is a primary mechanism by which the activity of PtdInsPK1γ is locally controlled. For instance, phosphorylation of residue S645 in the COOH tail of PtdInsPK1γ661 decreases the binding of the lipid kinase with talin and AP-2 (Lee et al. 2005; Nakano-Kobayashi et al. 2007). Additionally, phosphorylation of the S645 is reduced by phosphorylation of the neighbor residue, Y649, by non-receptor tyrosine kinase Src (Lee et al. 2005).

The main PtdIns(4,5)P_2 phosphatase at the mammalian synapse is synaptojanin 1 (Synj1) (McPherson et al. 1996; Cremona et al. 1999). Two different Synj-encoding genes, *Synj1* and *Synj2*, have been reported in mammals, although Synj1 is known to be the predominant activity in the brain (Cremona et al. 1999; Voronov et al. 2008). Additionally, two splice variants for Synj1 have been described, Synj1-145 and Synj1-170 (Ramjaun and McPherson 1996; Haffner et al. 1997, 2000; McPherson et al. 1996). However, only the shorter variant, Synj1-145, is highly expressed in mature brain and synapses, while Synj1-170 is ubiquitously expressed but at lower levels. Synj1 contains three protein regions (McPherson et al. 1996; Blero et al. 2007): (i) a central inositol 5-phosphatase domain that can hydrolyze PtdIns(4,5)P_2 to release phosphate from the 5' position of the inositol ring; (ii) an NH_2-terminal Sac1 region that can also function as a phosphoinositide phosphatase towards a variety of substrates, including monophosphorylated phosphoinositides and PtdIns(3,5)P_2 (Guo et al. 1999); and (iii) a COOH-terminal proline-rich domain (PRD) involved in the binding of Src-homology type 3 (SH3) domain-containing proteins, such as Growth factor Receptor-Bound protein 2 (Grb2), amphiphysin, endophilin and intersectin (Simpson et al. 1999; Yamabhai et al. 1998; Ringstad et al. 1997; de Heuvel et al. 1997; Cestra et al. 1999; Pechstein et al. 2010). It is commonly accepted that the 5-phosphatase activity mediates most of Synj1's actions, although a recent functional analysis of a Sac1 activity-deficient mutant of Synj1 suggests some roles for this domain in SV recycling (Mani et al. 2007).

As a member of the dephosphin protein group and similar to PtdInsPK1γ, Synj1 is constitutively phosphorylated in nerve terminals by the Cdk-5 and undergoes stimulation-dependent dephosphorylation by calcineurin (Marks and McMahon 1998; Lee et al. 2004b). Cdk-5 phosphorylates the COOH-terminal tail of Synj1 on a serine residue, which disrupts its interaction with endophilin and amphiphysin (Lee et al. 2004b). Additionally, Synj1 is a substrate of Dyrk1a, a homolog of the fly *minibrain* kinase (Adayev et al. 2006), and phosphorylation of Synj1 by this kinase regulates its interactions with the SH3 domain of amphiphysin and intersectin, but does not affect its enzymatic activity (Adayev et al. 2006). More relevant to the postsynaptic actions of Synj1, three tyrosine residues in Synj1's PRD domain can be phosphorylated by the activated EphB receptor, which reduces the interaction between Synj1 and endophilin and affects glutamate receptor internalization (Irie et al. 2005). Finally, in addition to phosphorylation/dephosphorylation mechanisms, our recent studies have shown that membrane curvature and endophilin control both the recruitment of Synj1 to membranes and its PtdIns(4,5)P_2 phosphatase activity (Chang-Ileto et al. 2011).

5.2.2 *Regulation of PtdIns(3,4,5)P_3 Metabolism of the Synapse*

The metabolism of PtdIns(3,4,5)P_3 is intimately linked to that of PtdIns(4,5)P_2, because class I PtdIns 3-kinases (PI3K), which use the latter as a substrate, are believed to be the main enzymes for the synthesis of PtdIns(3,4,5)P_3. The phosphatase PTEN (*p*hosphatase and *ten*sin homolog located on chromosome 10) catalyzes the converse reaction by dephosphorylating PtdIns(3,4,5)P_3 on the 3' position. While PtdIns(4,5)P_2 has been predominantly studied in its role at the presynapse, PtdIns(3,4,5)P_3 has been mostly characterized at the postsynapse.

Class I PI3Ks are heterodimers composed of a regulatory and a catalytic subunit. Four genes encoding the catalytic subunits have been reported (p110 or PI3Kα, β, δ and γ). PI3Kα, β, δ compose the class Ia PI3Ks and are mainly activated by receptor tyrosine kinases, while PI3Kγ is the sole member of class Ib (Hawkins et al. 2006; Marone et al. 2008). For the class Ia PI3Ks, several genes have been found to encode for the regulatory subunits, collectively referred to as p85 family members, despite the size diversity. These include *pik3r* (p85α, p55α, p50α), *pik3r2* (p85β) and *pik3r3* (p55γ). Two different genes, *pik3r5* and *pik3r6*, code for the p101 and p84 subunits, which form a complex with PI3Kγ. The role of PI3K at the synapse has been mostly assessed using broad specificity pharmacological agents, such as wortmannin (i.e., an inhibitor of class I and III PI3Ks) and LY294002 (i.e., an inhibitor which is fairly specific for class Ia PI3Ks). Generally, little information is known regarding the subtype of PI3K complexes expressed at synapses. Nevertheless, in the brain, the p110β and the p85α seem to be the predominant heterodimer of the class Ia (Geering et al. 2007). At synapses, PI3K activity can also be activated by the complex formed by the GTPase PI3K enhancer-Long (PIKE-L) with Homer 1c and an activated mGluR of class I, which promotes neuronal survival (Rong et al. 2003).

PTEN is a 403 amino-acid protein composed of two large domains, the phosphatase domain and a C2 domain. The phosphatase domain is flanked at the NH_2-terminus by a short PtdIns(4,5)P_2-binding domain composed of several basic amino acids and the C2 domain is flanked at the COOH-terminus by a PDZ-binding

Thr-Lys-Val motif (Bonifant et al. 2007). PTEN is the only 3-phosphatase identified that can mediate the PtdIns(3,4,5)P_3-to-PtdIns(4,5)P_2 conversion, thereby arresting PI3K-dependent signaling. PTEN is widely expressed in mouse brain, and preferentially in neurons, particularly Purkinje cells, olfactory mitral and large pyramidal neurons, where it is present in dendrites and spines (Perandones et al. 2004; Chang et al. 2007).

5.3 Presynaptic Roles of Phosphoinositides

Phosphoinositides are known to play a multitude of roles in cell physiology, including multiple aspects of neuronal function. Because changes in the metabolism of these lipids were originally studied in the context of secretion, the first clues that these lipids may play a critical role in secretion and particularly, secretory granule exocytosis, were obtained from work carried out in chromaffin and PC12 cells (Martin 2003; Di Paolo and De Camilli 2006). However, the identification of Synj as a major regulator of SV recycling contributed to the remarkable expansion of our knowledge on the role of Synj's main substrate, PtdIns(4,5)P_2, in endocytosis. There is now robust molecular, physiological and genetic evidence indicating that PtdIns(4,5)P_2, controls both the exocytic and endocytic limbs of SV trafficking, and that perturbation of its metabolism alters presynaptic function in neurons. The following section highlights the functional consequences of phosphoinositide perturbations on the SV cycle as well as the molecular basis underlying this process.

5.3.1 Synaptic Vesicle Exocytosis

Release of neurotransmitters is triggered by the fusion of SVs with the presynaptic plasma membranes in response to action potentials and Ca^{2+} entry (Fig. 5.1). In the past two decades, the field has achieved a remarkable understanding of this phenomenon through the identification and characterization of proteins mediating the fusion process using molecular, biochemical, structural and genetic approaches (Sudhof and Malenka 2008). The fusion machinery consists of three or four SNARE proteins and one sec1-Munc18-like protein (or SM protein) that are controlled by a variety of accessory factors, including members of the synaptotagmin (Syt) family. These proteins serve as the Ca^{2+}-sensors that mediate the bulk of neurotransmitter release under normal stimulation conditions in all synapses. There is now increasing evidence that phosphoinositides play an important modulatory role in several aspects of the fusion of SVs, as hinted by seminal studies on secretory granule exocytosis 20 years ago.

5.3.1.1 Exocytic Defects in PtdIns(4,5)P_2-deficient Synapses and Neurosecretory Cells

Studies from the groups of Holz and Martin in the early 1990s showed a critical role for inositol lipids and PtdIns(4,5)P_2 in the exocytosis of secretory granules in broken (or permeabilized) chromaffin or PC12 cells (Eberhard et al. 1990; Hay et al. 1995; Martin 1997). Importantly, a phosphatidylinositol transfer protein (Hay and Martin 1993) and a PtdInsP kinase (Hay et al. 1995) were identified as factors required for the ATP- and Ca^{2+}-dependent priming step in the release of large dense core vesicles (LDCVs) in PC12 cells, while a PtdIns 4-kinase activity was shown to be essential for secretion in chromaffin cells (Wiedemann et al. 1996). Remarkably, these early studies suggested that PtdIns(4,5)P_2 plays a role in LDCV exocytosis as an intact molecule rather than as a cleavage product of PLC, a biochemical reaction previously implicated in secretion (Osborne et al. 2006). Subsequently, a series of exocytic factors were found to bind PtdIns(4,5)P_2, demonstrating the importance of this lipid as an intact molecule in the fusion process. These include Syt (Schiavo et al. 1996; Bai et al. 2004), rabphilin (Chung et al. 1998), Ca^{2+}-dependent activator protein for secretion (CAPS) (Loyet et al. 1998), SNARE proteins (Lam et al. 2008), and secretory carrier membrane protein 2 (SCAMP2) (Liao et al. 2007).

While a role for PtdIns(4,5)P_2 in the fusion of LDCV had been well established by the Holz and Martin studies, a reliance of SV exocytosis on this lipid has only been suggested by a couple of studies. In the first study, reducing the synthesis of PtdIns4P, the main precursor of PtdIns(4,5)P_2, by phenylarside oxide was shown to decrease the release of glutamate from depolarized synaptosomes (Wiedemann et al. 1998). The second study involved the functional characterization of mice lacking PtdInsPK1γ, the main PtdIns(4,5)P_2-synthesizing enzyme at the synapse (Di Paolo et al. 2004). Indeed, deletion of PtdInsPK1γ was shown to produce defects in neurotransmitter release in cultured cortical neurons, based on the reduced frequency of miniature EPSPs and inhibitory postsynaptic potentials and the smaller readily-releasable pool (RRP) observed upon brief applications of hypertonic solutions (Di Paolo et al. 2004). Consistent with a smaller RRP, genetic ablation of PtdInsPK1γ also enhanced rapid depression during prolonged high-frequency stimulation. It was hypothesized that these exocytic defects may be accounted for, in part, by a delay in the replacement of docked and primed vesicles that have undergone exocytosis with new fusion-competent vesicles (Di Paolo et al. 2004). Defects in SV recycling also occur in mutant synapses and are discussed below.

A follow-up study on primary chromaffin cells lacking PtdInsPK1γ confirmed that reduced levels of PtdIns(4,5)P_2 decrease the size of the RRP of LDCVs and its refilling rate. Together with electron microscopy data showing an increase in morphologically-docked vesicles, data from this study indicated that PtdInsPK1γ controls the priming of LDCVs (Gong et al. 2005). Finally, amperometry measurements in mutant chromaffin cells revealed a delay in the expansion of the fusion pore, thus implicating PtdIns(4,5)P_2 in the regulation of LDCV fusion pore dynamics (Gong et al. 2005). This study confirms the notion that PtdIns(4,5)P_2 controls the priming step in the exocytosis of LDCVs (Eberhard et al. 1990; Hay et al. 1995) and is in partial agreement with an independent study in PC12 cells in which acute manipulations of PtdIns(4,5)P_2 levels were shown to affect the rate of fusion of LDCVs (Milosevic et al. 2005).

5.3.1.2 Molecular Basis for the Actions of PtdIns(4,5)P_2 in Exocytosis

Since the discovery that PtdIns(4,5)P_2 promotes exocytosis independently of its metabolic conversion to diacylglycerol (DAG) and inositol-1,4,5-trisphosphate [Ins(1,4,5)P_3] by PLC (Eberhard 1990; Hay 1995), the idea that PtdIns(4,5)P_2 effectors may mediate the fusion process gained in popularity, particularly in light of the identification of peptides or protein modules binding to this lipid with high selectivity and affinity. Two such PtdIns(4,5)P_2 effectors were extensively characterized in this context, CAPS and Syt, with significant implications for the priming and Ca^{2+}-sensing stages of the exocytic process.

CAPS and the Priming Step

CAPS was originally discovered as a 145 KDa cytosolic factor promoting Ca^{2+}-dependent exocytosis of LDCVs (Walent et al. 1992). Following the identification of CAPS as a homolog of Unc-31 (Ann et al. 1997), studies on Unc-31 and on mammalian CAPS in semi-broken synaptosomes confirmed the involvement of this gene family in LDCV exocytosis, but not in the fusion of SVs (Speese et al. 2007; Tandon et al. 1998; Gracheva et al. 2007). Although synaptic transmission at the fly and worm neuromuscular junction (NMJ) was found to be altered in CAPS mutants, these phenotypes were interpreted as secondary to defects in LDCV exocytosis (Speese et al. 2007; Renden et al. 2001; Gracheva et al. 2007). In the mouse, ablation of CAPS-1 was shown to perturb the uptake of catecholamines into LDCVs and the exocytosis of LDCVs (Speidel et al. 2005), while deletion of CAPS-2 interferes with the development of the cerebellum and short-term plasticity (Sadakata et al. 2007). A role for CAPS in SV exocytosis was unambiguously shown in synapses derived from mice lacking CAPS-1 and 2 (Jockusch et al. 2007). The double KO mice exhibited dramatic reduction in the number of fusion-competent SVs concomitantly with a normal number of vesicles docked at the plasma membrane, thus implying an essential role of CAPS in the priming step in excitatory neurons. Further supporting this view, deficits in the mutant mice were rescued, at least transiently, by hyperactivation of Munc13, a well-established priming factor (see below) (Jockusch et al. 2007; Augustin et al. 1999; Varoqueaux et al. 2002; Siksou et al. 2009) but see also (Siksou et al. 2009) for a role in Munc13 in SV docking. CAPS also binds to membranes in a Ca^{2+}-dependent manner via its C2 domain and to phosphoinositides, such as PtdIns4P, PtdIns(4,5)P_2 and PtdIns(3,4)P_2, via its PH domain (Loyet et al. 1998; James et al. 2008). Binding of CAPS to PtdIns(4,5)P_2 appears to be critical for the fusion-promoting action of CAPS because it decreases the intrinsic ability of the inverted cone shape of PtdIns(4,5)P_2 to block SNARE-mediated fusion and may directly affect the fusogenic properties of syntaxin/Soluble NSF Attachment Protein-25 (SNAP-25) (Grishanin et al. 2004; James et al. 2008). Syntaxin itself can directly bind to phospholipids, particularly "cone shape" lipid phosphatidic acid (PA), which is fusogenic, and PtdIns(4,5)P_2, via a polybasic juxtamembrane region. Indeed, in PC12 cells, PtdIns(4,5)P_2 co-clusters with syntaxin at sites of exocytosis (Aoyagi et al. 2005). Since mutations of the juxtamembrane domain decrease the fusion pore size in PC12 cells and this effect is mediated by PA, it was suggested that this syntaxin-1 domain clusters PA at the site of fusion (Lam et al. 2008).

The function of priming factor Munc13 heavily relies on the ability of this family of proteins to bind to DAG via their C1 domain, suggesting the interesting possibility that PtdIns(4,5)P_2 hydrolysis by PLC also regulates the priming step of SV exocytosis (Rhee et al. 2002; Lou et al. 2008). Additionally, recent work has shown that Munc13 also binds phosphoinositides in a Ca^{2+}-dependent manner via its central C2B domain and that this property allows Munc13 to potentiate SV exocytosis during repeated action potentials, thereby minimizing synaptic depression induced by SV depletion (Shin et al. 2010).

Role of Synaptotagmin in Ca^{2+}-sensing and Synaptic Vesicle Fusion

Synaptotagmin is considered the main Ca^{2+} sensor for Ca^{2+}-triggered (fast) synchronous release of neurotransmitter at central synapses, but not for the (slow) asynchronous release (Pang and Sudhof 2010; Perin et al. 1991; Brose et al. 1992; Geppert et al. 1994; Chapman 2008). Originally identified as a 65 KDa protein enriched in SVs and LDCV (Matthew et al. 1981; Chapman 2008), synaptotagmin-1 (Syt-1) is the founding member of a large family that includes seventeen other proteins (Chapman 2008). Syt-1 is the most extensively studied Syt member and is believed to be the main isoform controlling SV exocytosis. Work carried out in the past 15 years, including genetic studies in various species and recent cell-free fusion assays, suggest a complex involvement of Syt-1 in the fusion process, which can be summarized as follows: Syt-1 is believed to act as a pre-fusion "clamp" in the absence of Ca^{2+} (Martens et al. 2007; Chicka et al. 2008) and to accelerate SNARE-mediated fusion in the presence of the divalent cation, in part by regulating fusion pore dynamics and the final steps of fusion (Stein et al. 2007; Chapman 2008).

Since the primary sequence of Syt-1 revealed, in addition to the single NH_2-terminal transmembrane domain anchored to the SV membrane, a large cytodomain consisting of tandem C2 domains, C2A and C2B, separated by a 9-residue linker, the idea that Syt may bind and "sense" Ca^{2+} was put forward (Brose et al. 1992; Geppert et al. 1994). Indeed, while the C2A domain can bind to three Ca^{2+} ions, the C2B domain binds to two ions, in agreement with the previous notion that the Ca^{2+} sensor for synchronous release exhibits an apparent cooperativity of five Ca^{2+} ions (Bollmann et al. 2000; Schneggenburger and Neher 2000). In addition to its role in Ca^{2+} sensing, SV-bound Syt-1 has been shown to play a crucial role in docking the vesicles to the plasma membrane via its interaction with SNAP-25, presumably as part of an assembled Munc18/syntaxin-1/SNAP-25 acceptor complex (de Wit et al. 2009).

C2 domains often bind phospholipids, thus suggesting a relationship between Ca^{2+} binding and membrane interactions. Indeed, the C2B domain of Syt-1 was shown to bind to phosphoinositides and the specificity of interaction of this domain for PtdIns(4,5)P_2 or PtdIns(3,4,5)P_3 depends on intracellular levels of Ca^{2+} (Schiavo et al. 1996). The relationship between the C2 domains and phospholipids has been extensively dissected by several groups using biochemical, structural and functional assays in combination with site-directed mutagenesis of the C2 domains. Indeed, in the presence of Ca^{2+}, the C2A domain of Syt-1 interacts with anionic phospholipids, such as phosphatidylserine (PS), which triggers the insertion of the Ca^{2+}-binding loops of C2A into lipid bilayers. It was shown that the C2B domain alone also binds weakly to PS, but binds preferentially to PtdIns(4,5)P_2. C2B does not penetrate into the bilayer as a result of this interaction, it was thus suggested that the C2B domain may simply serve to tether Syt-1 to the plasma membrane (Bai et al. 2004). However, dimers of the C2A and C2B domains primarily mediate the Ca^{2+}-dependent binding of Syt-1 to PtdIns(4,5)P_2- or PtdIns4P-containing membranes and the Ca^{2+}-dependent evoked release (Hui et al. 2009). After its insertion into the plasma membrane, bending of the membrane by Syt-1 is dependent on the C2B domain. The positive membrane curvature induced by Ca^{2+}-bound Syt-1 facilitates the fusion of SVs to the plasma membrane by reducing the distance between the membranes and by reducing the energy barrier for their fusion (Shin et al. 2010).

5.3.2 *Synaptic Vesicle Recycling*

Sustained release of neurotransmitter is crucial for neuronal communication. To prevent SV depletion and a rundown of neurotransmission, SVs must be efficiently recycled in nerve terminals, particularly during prolonged stimulations. Several recycling pathways have been described, including those involving 'kiss-and-run' mechanisms and clathrin-mediated endocytosis (CME), which has been extensively characterized at the molecular level. This latter pathway occurs at the periphery of the active zone (also called periactive zone) and is typically triggered upon full collapse of SVs with the plasma membrane during the release process. An important implication of full fusion of SVs with the plasma membrane is that SV-associated components undergo rapid diffusion at the synaptic membrane. Thus, precise and efficient molecular mechanisms must be in place to ensure the appropriate capture and sorting of SV-associated components into endocytic vesicles in order to preserve the molecular composition and functionality of SVs.

Work carried out in the past decade has provided overwhelming evidence showing that phosphoinositides and specifically PtdIns(4,5)P_2 regulate multiple aspects of SV recycling. In particular, studies on Synj1 and PtdInsP kinase type 1γ, have established that both an excess and a deficiency of PtdIns(4,5)P_2 interfere with the SV cycle. At the molecular level, dysregulation of PtdIns(4,5)P_2 metabolism at the synapse alters the function of a large variety of proteins that control multiple aspects of SV trafficking, including the recycling of SVs via the pathway of CME. The following section highlights how a loss of the main PtdIns(4,5)P_2-metabolizing enzymes affects this process and the physiology of nerve terminals. This followed by a discussion of the mechanistic aspects underlying the pleiotropic roles of PtdIns(4,5)P_2 in the process of SV recycling.

5.3.2.1 Synaptic Vesicle Recycling Defects upon PtdIns(4,5)P_2 Imbalance

Consequences of Synj Loss on Nerve Terminal Function

The identification of Synj1 by De Camilli et al. as a partner for the SH3 domain of Grb2 (and amphiphysin), along with the endocytic fission factor dynamin, indirectly suggested an involvement of this enzyme in SV recycling (McPherson et al. 1994, 1996). Conclusive evidence for a direct role of Synj in this process emanated from a genetic study in the mouse, where a null mutant was shown to produce defects in synaptic vesicle recycling (Cremona et al. 1999). At the *Synj1*$^{-/-}$ synapse, increased PtdIns(4,5)P_2 levels correlated with an excess of clathrin-coated vesicles (CCVs), suggesting a delay in the uncoating reaction (Cremona et al. 1999). In agreement with defects of recycling, a deeper depression of neurotransmission was observed in hippocampal slices as well as in primary cortical cultures from null mice during a prolonged stimulation (Cremona et al. 1999; Luthi et al. 2001). Fluorescent dye (FM1-43) uptake and release assays in cultured neurons confirmed that newly endocytosed vesicles recycle with slower kinetics in nerve terminals, likely reflecting the accumulation of CCVs (Cremona et al. 1999; Kim et al. 2002; Mani et al. 2007). Strikingly, endocytic defects resulting from inactivation of Synj function were also observed in other species, such as the budding yeast (Singer-Kruger et al. 1998; Stefan et al. 2002), the worm (Harris et al. 2000) and the fly (Dickman et al. 2005; Verstreken et al. 2003). Importantly, nerve terminals from lower organisms lacking the only Synj ortholog showed pleiotropic defects in SV recycling, including an accumulation of coated and, in some cases, even non-coated pits at various stages of invagination (Verstreken et al. 2003; Harris et al. 2000). Consistent with the abovementioned studies, an experimental manipulation of the lamprey reticulospinal (giant) synapse showed an accumulation of clathrin-coated pits (CCPs) and free CCVs upon blockade of the interaction of Synj with the SH3 domain of its main interactor, endophilin, utilizing a proline-rich peptide derived from the COOH-terminal tail of mammalian Synj1 (Gad et al. 2000). Additionally, in the zebrafish nrc mutant, defects in SV trafficking were observed in the ribbon synapses of fish photoreceptors as a result of a stop mutation the *Synj1* gene, although in this case, an accumulation of coated intermediates was not reported (Van Epps et al. 2004). Instead, nrc cone photoreceptor pedicle exhibited unanchored ribbons as well as a reduction in SV number and an abnormal distribution of these organelles. A more recent electron tomography study from De Camilli et al. not only showed striking evidence for an accumulation of CCVs in *Synj1*$^{-/-}$ synapse, but also highlighted a stronger requirement for Synj1 in GABAergic neurons, at least based on a morphological assessment (Hayashi et al. 2008). In summary, the endocytic function of

Synj appears to be largely conserved across evolution, although the precise actions of the phosphoinositide phosphatase in this process depend on the cell or synapse type, perhaps reflecting the pleiotropic role of phosphoinositides in cell physiology.

Based on studies of Synj function in various organisms as well as of the role of PtdIns(4,5)P_2 in multiple experimental systems, a main function of Synj appears to be the elimination of PtdIns(4,5)P_2 from membranes during the endocytic process (Stefan et al. 2002; Cremona and De Camilli 2001; Cremona et al. 1999; Di Paolo and De Camilli 2006). However, Synj family members also contain an NH_2-terminal Sac1 domain, which is also present in other proteins and has been shown to dephosphorylate phosphoinositides other than PtdIns(4,5)P_2. These include PtdIns3P, PtdIns4P, PtdIns5P and PtdIns(3,5)P_2 (Guo et al. 1999; Hughes et al. 2000). Although the original study by Cremona et al. only shows an increase in PtdIns(4,5)P_2 in *Synj1*$^{-/-}$ primary cortical neurons, only two phosphoinositides, namely the most abundant, were analyzed: PtdIns(4,5)P_2 and PtdIns4P. Consequently, changes in the levels of other Synj1 substrates cannot be ruled out and PtdIns(4,5)P_2-independent phenotypes in synapses lacking Synj cannot be excluded either. Supporting this idea, the Sac1 domain of yeast Synj-like protein 2 (Sjl2) and 3 (Sjl3) was shown to hydrolyze PtdIns3P in a physiological context, although concomitant deletion of myotubularin ortholog Ymr1p (*i.e.*, an inositol 3-phosphatase) was required to unmask this function (Parrish et al. 2004). Importantly, a more recent study on *Synj1*$^{-/-}$ cultured neurons expressing various Synj1 mutant constructs confirmed the essential nature of the inositol 5-phosphatase domain of this enzyme, suggesting that PtdIns(4,5)P_2 dephosphorylation is central to its function [although PtdIns(3,4,5)P_3 dephosphorylation may also play a role]. This study also unmasked a role for the Sac1 domain of Synj1 in SV internalization, but only for brief stimuli (Mani et al. 2007). However, the physiological substrate(s) of Synj1's Sac1 domain at synapses is (are) still undetermined. Finally, the role of the other Synj isoform, Synj2, has not been addressed at synapses.

Consequences of PtdInsPK1γ Loss on Nerve Terminal Function

Studies on Synj rapidly became a driving force to further explore the role of PtdIns(4,5)P_2 at the synapse and specifically, to identify the main enzymatic source of PtdIns(4,5)P_2 in this compartment. Of the three type 1 PtdInsP kinase isoforms known, PtdInsPK1γ was shown to represent the main activity in the brain and was thus further characterized (Wenk et al. 2001). Evidence for a role of this enzyme in synaptic function originated from a mouse genetic study, where ablation of PtdInsPK1γ was shown to cause a deficiency of CCVs upon stimulation and an increase in surface area of horseradish peroxidase-laden endosome-like structures (likely corresponding to bulk invaginations of the plasma membrane) (Di Paolo et al. 2004). Functional correlates of these morphological phenotypes were a decrease in the rate of SV endocytosis, as measured by the synaptopHluorin technology (Sankaranarayanan and Ryan 2000; Miesenbock et al. 1998), as well as a reduced rate of SV recycling, based on FM1-43 dye uptake and release assays (Di Paolo et al. 2004). Similarly to $Synj1^{-/-}$ synapses, basal synaptic transmission and short-term plasticity were preserved after ablation of PtdInsPK1γ. However, faster depression of inhibitory postsynaptic currents was observed during the first hundred action potentials of a train of stimuli (Di Paolo et al. 2004). Other electrophysiological changes consistent with exocytic defects were described in Sect. 5.3.1.1.

Further suggesting an important role of PtdInsPK1γ in SV recycling was a study in the lamprey giant synapse showing that blocking the interaction between the COOH-terminal tail of this lipid kinase and the FERM domain of talin with a PtdInsPK1γ peptide (Di Paolo et al. 2002), caused the appearance of aberrant CCPs and a decrease in synaptic F-actin upon prolonged stimulation (Morgan et al. 2004). In contrast to the case of Synj, there are no studies on lower organisms indicating a role for PtdInsPK in SV trafficking. There is however a fly mutant (*tweek*) which is associated with lower synaptic levels and aberrant distribution of PtdIns(4,5)P_2 as well as with defects in SV trafficking (Verstreken et al. 2009). Because the underlying gene does not encode a PtdInsP kinase, the link with PtdIns(4,5)P_2 metabolism has remained elusive.

5.3.2.2 Molecular Basis for the Actions of PtdIns(4,5)P_2 in Synaptic Vesicle Recycling

A major function of PtdIns(4,5)P_2 metabolism is to control the process of CME. While this phenomenon is far from being specific to nerve terminals, most of our understanding of the role of PtdIns(4,5)P_2 originates from studies of clathrin-mediated recycling of SVs at the synapse. Another key function of this lipid is to regulate actin dynamics. We summarize the current knowledge on how PtdIns(4,5)P_2 regulates these two processes at the presynapse with a focus on molecular details.

Clathrin Coat Recruitment

Several studies in the late 1990s indicated that clathrin coat components (e.g., AP-2) as well as other endocytic proteins, such as dynamin, bind to PtdIns(4,5)P_2 (Gaidarov and Keen 1999; Jost et al. 1998). However, the physiological significance of these interactions was best highlighted by a mouse genetic study on $Synj1^{-/-}$ synapse, which showed that increased PtdIns(4,5)P_2 levels facilitate clathrin coat assembly *in vitro* and conversely, delay coat shedding *in vivo* (Cremona et al. 1999). Furthermore, ablation of the main PtdIns(4,5)P_2-synthesizing enzyme at the synapse, PtdInsPK1γ, leads to a decreased association of the clathrin coat proteins with membranes in cell-free assays (Wenk et al. 2001) and a reduced number of CCVs in stimulated cultured neurons (Di Paolo et al. 2004). As a result of these studies, a model emerged in which PtdIns(4,5)P_2 is a key parameter controlling the affinity of coat proteins for synaptic membranes and thus the efficacy and kinetics of SV recycling.

The clathrin coat includes both the heavy and light chains of clathrin as well as adaptor proteins, which are essential for the recruitment of clathrin to membranes (in addition to mediating the sorting of cargo proteins into CCPs). The clathrin coat is organized into an assembly of triskelia, which consist of three heavy chains (CHC) and three light chains (CLC), at a stoichiometry of 1:1. Triskelia self-assemble into a basket-like polyhedral protein lattice of pentagons and hexagons that coats the endocytic vesicle (Kirchhausen 2000). The initiation of CME at the plasma membrane occurs by the membrane association of a variety of adaptor proteins, all of which bind PtdIns(4,5)P_2 and clathrin. The main clathrin adaptors at the synapse are discussed below, with emphasis on the best-characterized adaptor at the molecular/structural level, namely AP-2.

AP-2 is a heterotetramer consisting of α, $\beta 2$, $\mu 2$, and $\sigma 2$ subunits (also called adaptins) and plays a central role in clathrin-mediated internalization of multiple cargoes (Robinson 2004). Indeed, AP-2 is considered a "hub" in the endocytic network due to its multiple interactions with clathrin, accessory factors and transmembrane protein cargoes. Although several studies have indicated a critical role for AP-2 in clathrin mediated-receptor internalization, its implication in SV endocytosis was first demonstrated in a fly mutant in the α-adaptin gene (Gonzalez-Gaitan and Jackle 1997). However, more recent studies in the worm (Gu et al. 2008) and dissociated hippocampal neurons showed that ablation/silencing of the $\mu 2$ subunit of AP-2 fails to eliminate cargo retrieval and SV recycling, although slower endocytic kinetics are observed in AP-2 depleted hippocampal synapses (Dittman and Ryan 2009; Kim and Ryan 2010).

In the current model, PtdIns(4,5)P_2-containing membranes play a pivotal role in switching AP-2 from a closed (or locked) conformation to an open, ligand-bound conformation (Jackson et al. 2010b; Kelly et al. 2008; Collins et al. 2002). The first step in AP-2 activation involves its recruitment to the plasma membrane through an interaction of basic residues from its α and $\beta 2$ subunits with PtdIns(4,5)P_2. These electrostatic interactions facilitate the binding of $\mu 2$'s COOH terminus with the membrane, which subsequently allows AP-2 to adopt an open conformation and interact with sorting signals from transmembrane cargo proteins (Jackson et al. 2010b). A key effect of this activation is to dislodge the $\beta 2$ subunit, which acts as a "latch" and can no longer block the two peptide ligand-binding sites in the active conformation. These in turn can freely bind to YxxF or [ED]xxxL[LI] motifs that belong to cargo proteins, resulting in high affinity (*i.e.*, low nM) interactions of the AP-2 complex with the plasma membrane (Jackson et al. 2010b). Further underscoring the importance of PtdIns(4,5)P_2 in the activation of AP-2, this clathrin adaptor has been shown to directly bind to PtdInsPK1γ, suggesting the occurrence of a positive feedback loop controlling PtdIns(4,5)P_2 production in proximity to sites of AP-2 recruitment. Specifically, PtdInsPK1γ was shown to bind to the $\mu 2$ subunit of AP-2, and this binding stimulates its kinase activity (Krauss et al. 2006; Bairstow et al. 2006). Additionally, PtdInsPK1γ661 also interacts with the appendage domain of the $\beta 2$ subunit of AP-2, promoting its catalytic activity only if its COOH-terminal extension is in a dephosphorylated state (Nakano-Kobayashi et al. 2007), This same binding platform of $\beta 2$ also interacts with the heavy chain of clathrin in a mutually exclusive manner (Thieman et al. 2009). It is of note that the other two PtdInsPK1 isoforms, α and β, also interact with AP-2, suggesting that the ability to interact with the clathrin coat is a general feature of PtdInsPK1s (Krauss et al. 2006).

Recruitment of clathrin also occurs via monomeric adaptors, such as AP180 and Clathrin Assembly Lymphoid Myeloid (CALM), which both harbor an *A*P180 *N-t*erminal *h*omology (ANTH) domain. This NH_2-terminal module contains a series of surface basic residues that interact with phosphoinositides, including PtdIns(4,5)P_2, while the central and COOH-terminal portions of AP180 bind to CHC and AP-2, via a variety of peptide motifs. Functional evidence for a role of AP180/CALM in SV recycling has so far been only obtained in the worm and the fly, where mutations in the only ortholog result in an alteration of the size of CCVs and SVs as well as missorting and higher cell surface levels of the SV protein, synaptobrevin (Harel et al. 2008; Dittman and Kaplan 2006; Zhang et al. 1998; Nonet et al. 1999). Thus, AP180/CALM likely regulates both cargo recruitment and the size of the membrane area destined to be internalized at synapses. Other clathrin adaptors, such as epsins, are endowed with membrane deforming properties and will be discussed in Sect. 5.5.2.

Finally, the F-BAR domain-containing Fer/Cip4 homology domain-only proteins 1 and 2 (FCHo1/2) may play an important role in the regulation of the number of endocytic sites. These proteins possess an NH_2-terminal F-BAR domain (see Sect. 5.5.2) and a COOH-terminal μ2-homology domain. Remarkably, FCHo1/2 are recruited to CCPs concomitantly with Eps15, intersectin and AP-2, but prior to clathrin. Their F-BAR domain is extended with a region that is enriched in basic residues and preferentially interacts with PtdIns(4,5)P_2, targeting these proteins to the plasma membrane (Henne et al. 2010).

Membrane Curvature Generation and Stabilization

Generation of membrane curvature during the endocytic process is believed to be driven by clathrin assembly *per se* and by the action of various families of proteins endowed with membrane deforming capacity. This latter property was originally shown for the fission factor dynamin (Takei et al. 1995) and its SH3 domain-containing interactor amphiphysin, a *B*in/*A*mphiphysin/*R*vs (BAR) protein (Takei et al. 1999). However, the past decade has witnessed an expansion in the catalog of endocytic proteins with similar properties and share the ability to bind PtdIns(4,5)P_2, but with differential selectivity and affinity.

The discovery of the membrane deforming property of BAR proteins, such as amphiphysin (Takei et al. 1999) and subsequently endophilin (Farsad et al. 2001), preceded the structure determination of the BAR domain (Peter et al. 2004). However, the latter not only provided a structural basis for the observed effects of this domain on lipid bilayers, but it also introduced the concept of curvature sensing/stabilization for BAR proteins. Indeed, the BAR domain of amphiphysin was originally identified as a crescent-shaped dimer with positively charged residues lining its concave surface. This banana shape of the BAR domain was shown to prefer highly curved substrate liposomes and primarily uses electrostatic forces to bind negatively charged lipids, such as PS or PtdIns(4,5)P_2 (Peter et al. 2004). The BAR family of proteins has now been expanded and can be subdivided into several classes, based on differing structural features: (i) N-BAR proteins (e.g., amphiphysin, endophilin); (ii) F-BAR proteins (e.g., syndapin); (iii) I-BAR proteins; and (iv) PX-BAR proteins represented by several isoforms of the Sorting Nexin (SNX) family (SNX 1,2,5,6 and 9) (Shimada et al. 2007; Frost et al. 2008; Weissenhorn 2005; Gallop et al. 2006; Peter et al. 2004; Frost et al. 2009). Members in each of these classes have been implicated in membrane trafficking and/or endocytic processes. The N-BAR proteins, in particular, play an important role in phosphoinositide metabolism not only because their N-BAR domain binds these lipids, but also because well-characterized members of this subfamily, such as endophilin and amphiphysin, possess a COOH-terminal SH3 domain that physically interacts with Synj (Chang-Ileto et al. 2011; Gad et al. 2000; Ringstad et al. 1997). Importantly, several studies have established the functional significance of the Synj1-endophilin interaction in various species, including the worm (Jorgensen et al. 1995), the fly (Verstreken et al. 2003); the lamprey (Gad et al. 2000) and the mouse (Mani et al. 2007).

Although endophilin's N-BAR domain binds to PtdIns(4,5)P_2, it interacts *in vitro* with various acidic phospholipids via the concave face of this module, which is lined with basic residues (Chang-Ileto et al. 2011; Weissenhorn 2005; Mattila et al. 2007; Tsujita et al. 2006; Itoh et al. 2005; Peter et al. 2004; Gallop et al. 2006). Since the interaction with membranes is largely electrostatic in nature and that the plasma membrane has an overall negative surface charge [in part because of the greater content in PtdIns(4,5)P_2], endophilin primarily tubulates the plasma membrane (Chang-Ileto et al. 2011). While the relatively late recruitment of endophilin to CCPs (*i.e.,* at a time immediately preceding the fission process) argue against a primary role of the tubulating activity of its N-BAR domain in the budding process, it may be involved in stabilizing membrane curvature at sites of endocytosis, potentially in concert with other BAR proteins and/or the fission factor, dynamin (Perera et al. 2006). A key function of endophilin, however, appears to be the recruitment of Synj to sites of endocytosis, thus controlling PtdIns(4,5)P_2 elimination during the endocytic process. This idea is supported by biochemical, morphological and physiological data from various species and research groups (Micheva et al. 1997; Gad et al. 2000; Verstreken et al. 2003; Schuske et al. 2003; Dickman et al. 2005; Chang-Ileto et al. 2011; Perera et al. 2006). Ablation of endophilin recaptitulates many of the features observed upon ablation/inactivation of Synj, including the accumulation of CCPs at various stages, including free CCVs. The interaction between endophilin 1 and Synj1 is regulated by phosphorylation/dephosphorylation of Synj1 by the protein kinase Cdk-5 and the phosphatase calcineurin. Additionally, endophilin stimulates the PtdIns(4,5)P_2 5-phosphatase activity of Synj1 (Chang-Ileto et al. 2011; Lee et al. 2004b) and this phenomenon is enhanced by small liposomes (i.e., 50 nm in diameter), suggesting that robust stimulation of Synj1 by endophilin 1 may occur at sites of high curvature, namely the neck (or perhaps the bud) of the endocytic pit (Chang-Ileto et al. 2011) (Fig. 5.2). Endophilin may also have additional functions at nerve terminals which are unrelated to endocytosis, as suggested by a recent study (Bai et al. 2010). Other BAR proteins, such as PX-BAR member SNX9, also bind to phosphoinositides,

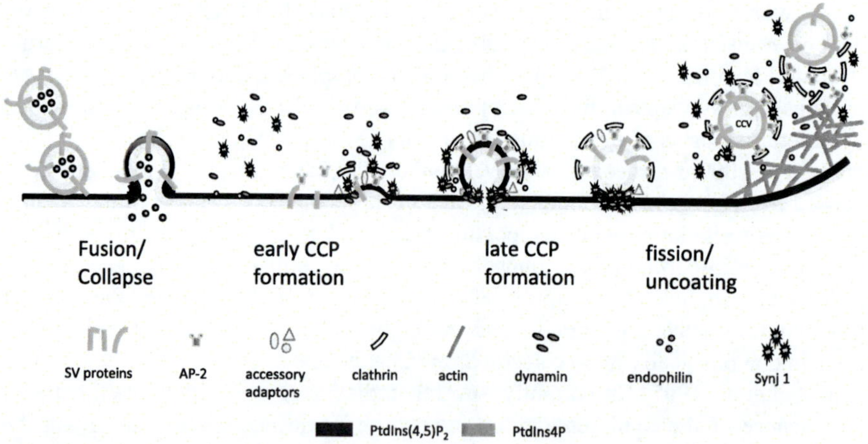

Fig. 5.2 PtdIns(4,5)P_2 role in endocytosis. The PtdIns4P-enriched membrane of the SV fuse and merge with the PtdIns(4,5)P_2-enriched PM liberating the neurotransmitter extracellularly. AP-2 is recruited by PtdIns(4,5)P_2 with other accessory clathrin adaptors to the surface of the PM. The PM buds and invaginates to form a Ω-shaped pit covered by clathrin lattice and that comprised the SV proteins. Fission factor, dynamin gets recruited to the bud with endophilin and synaptojanin. Dynamin mediates the scission of its neck to release a free clathrin-coated vesicle (CCV), possibly facilitated by the hydrolysis of PtdIns(4,5)P_2 by synaptojanin1 (Synj1). The PtdIns(4,5)P_2 present on the CCV is dephosphorylated to PtdIns4P by Synj1, thereby promoting the shading of the adaptors from the membrane and the uncoating reaction

although both PX and BAR domains are involved in these interactions with lipids and are required for the proper localization of SNX9 to CCPs (Yarar et al. 2008).

In addition to BAR proteins, epsin (*Eps*15 *in*teracting protein) family members may also contribute to curvature generation/stabilization at sites of endocytosis, particularly epsin 1. Epsins interact with the CHC, the appendage domain of α-adaptin and the EH domain of Eps15. Epsins also harbor ubiquitin-interacting motifs, which may help to internalize ubiquitinated cargo proteins (Shih et al. 2002). Importantly, the NH_2-terminal ENTH domain of epsins, which is distinct from the ANTH domain of AP180/CALM, contains an unstructured NH_2-terminal sequence that folds into an α-helix upon interaction with PtdIns(4,5)P_2. This α-helix inserts itself into lipid bilayers, thus providing ENTH domains with the ability to deform membranes *in vitro*, a function that may be important for the budding of CCPs in physiological contexts (Ford et al. 2002; Itoh et al. 2001). Consistent with this idea, inactivation of epsin by antibody microinjection into the lamprey giant synapse produces a stimulation-dependent accumulation of large CCPs (Jakobsson et al. 2008). However, loss of epsin in the fly mutant "Liquid Facets" does not seem to alter SV recycling (Bao et al. 2008) and it is unclear whether epsin null mice have SV endocytic defects (Chen et al. 2009).

In conclusion, membrane curvature needed for CCV formation is likely to be generated through clathrin assembly as well as via the coordinated action of several proteins containing ENTH, F-BAR and N-BAR modules. PtdIns(4,5)P_2 not only plays an important role in the targeting of these proteins to the cell surface, but also, in the case of epsins, in altering the conformation of these proteins/modules, which in turn facilitate curvature generation.

Endocytic Fission

The fission of CCPs during the process of SV endocytosis is believed to be mediated by the large GTPase dynamin. This protein was first identified as critical in neuronal function in temperature-sensitive mutants in the fly ("Shibire") (Chen et al. 1991; van der Bliek and Meyerowitz 1991; Grigliatti et al. 1973; Suzuki et al. 1971) and purified through its association with microtubules *in vitro* (Obar et al. 1990; Shpetner and Vallee 1989). Dynamin contains an NH_2-terminal GTPase domain, a PH domain that binds to PtdIns(4,5)P_2, a middle GTPase effector domain (GED) that is important for oligomerization, and a COOH-terminal proline-rich domain, which interacts with the SH3 domain of several endocytic proteins, including endophilin, amphiphysin, intersectin, syndapin and SNX9 (Slepnev and De Camilli 2000). Dynamin has been established as an endocytic factor critical for membrane fission based on the expression of dominant-negative mutants (Yamashita et al. 2005; Newton and Messing 2006; Koenig and Ikeda 1999). However, deletion of dynamin1, the major dynamin isoform in the brain only partially impairs the retrieval of SVs during strong stimulation (Ferguson et al. 2007). SV endocytosis occurs after its cessation or during mild stimulation, most likely due to the low expression of dynamin 3, which is relocalized at the synapse in absence of dynamin 1 (Ferguson et al. 2007).

The relationships between dynamin and phophoinositide metabolism are threefold. First, as mentioned above, the PH domain of dynamin binds to PtdIns(4,5)P_2 and this interaction is critical for dynamin's function, at least in part because it allows dynamin to interact with the plasma membrane (Roux et al. 2006; Lee et al. 1999). Second, PtdIns(4,5)P_2 stimulates the GTPase activity of dynamin (Zheng et al. 1996). Finally, our recent study has shown that acutely-induced PtdIns(4,5)P_2 dephosphorylation of endophilin-coated tubules by the 5-phosphatase domain of Synj1 triggers their fragmentation in a dynamin-dependent fashion, suggesting that PtdIns(4,5)P_2 hydrolysis facilitates membrane fission (Chang-Ileto et al. 2011). While the precise molecular basis for this phenomenon is unclear, we have hypothesized that an acute loss of PtdIns(4,5)P_2 may help to disassemble dynamin scaffolds from membranes. This disassembly of dynamin, which is dependent on dynamin's own GTPase activity, appears to permit membrane fission by destabilizing the underlying constricted bilayer membranes (Pucadyil and Schmid 2008; Bashkirov et al. 2008). In addition, rapid depletion of PtdIns(4,5)P_2 by Synj1 selectively at endocytic sites [presumably at the bud neck (Sundborger et al. 2010)] may produce a transient gradient of this lipid between the globular part of the bud and its neck, which in turn may cause lipid phase separation between these compartments (Liu et al. 2006). The predicted outcome of this rapidly-induced heterogeneity in lipid composition is the generation of interfacial forces at the CCP-bud neck interface, resulting in the squeezing of the lipid domain boundary and facilitating membrane fission (Liu et al. 2010).

Clathrin Coat Shedding

Once a CCV has been released by the fission process, the free vesicle undergoes uncoating in order to recycle clathrin coat components (as well as other endocytic factors) and to produce a new SV. The uncoating of the clathrin coat is triggered by the concerted action of the molecular chaperone Hsc70 and the DNA-J factor auxilin, which mediates the recruitment of the ATPase to the CCV (Eisenberg and Greene 2007; Yim et al. 2010). A major breakthrough in the field of SV endocytosis was the original characterization of nerve terminals from *Synj1*$^{-/-}$ mouse neurons showing an increased number of CCVs. Drawing upon studies showing that clathrin adaptor AP-2 and dynamin bind to PtdIns(4,5)P_2 (Jost et al. 1998; Gaidarov and Keen 1999), the hypothesis that PtdIns(4,5)P_2 elimination by Synj1 may facilitate the shedding of clathrin adaptors was proposed (Cremona et al. 1999). Subsequent studies have confirmed this phenotype, although loss of Synj1 (or its orthologs in other genetic models) is now known to produce pleiotropic defects in SV recycling, as testified by the report of aberrant numbers of clathrin-coated structures at multiple stages of invagination in affected synapses. This is consistent with the complexity and multitude of PtdIns(4,5)P_2 actions in the recycling process.

While the hydrolysis of PtdIns(4,5)P_2 may suffice to destabilize the clathrin coat, it may not be sufficient to trigger its complete disassembly, partly because clathrin spontaneously forms triskelia together with its adaptors. As mentioned above, Hsc70 facilitates the uncoating of clathrin alongside auxilin. Interestingly, auxilin contains an inactive PTEN-homology domain that is necessary for its targeting to clathrin-coated membranes and binds to phosphoinositides, most notably to PtdIns4P and, to a lower extent, PtdIns(4,5)P_2 (Guan et al. 2010; Massol et al. 2006). A burst of recruitment of auxilin to the CCVs occurs after the peak of recruitment of dynamin at the CCVs via the binding of the PTEN-homology domain to membranes (Massol et al. 2006). Because Synj1 is recruited to CCPs concomitantly with dynamin (Perera et al. 2006), it is tempting to speculate that the PtdIns(4,5)P_2-to-PtdIns4P conversion mediated by Synj (following its recruitment by endophilin) may not only facilitate membrane fission (Chang-Ileto et al. 2011), but also represent a signal for the recruitment and activation of auxilin, and thus, the uncoating process (Guan et al. 2010). This mechanism would also allow for a tight coupling between the fission and uncoating steps of CCVs.

Actin Dynamics

It has long been known that phosphoinositides are primary regulators of actin dynamics, largely because a significant number of actin regulatory proteins bind to these lipids, PtdInds(4,5)P_2 in particular (Saarikangas et al. 2008; Yin and Janmey 2003). A commonly accepted view is that PtdInds(4,5)P_2 predisposes actin for polymerization by the nucleation of F-actin by the actin-related protein 2 and 3 (Arp2/3) complex, the removal of capping proteins and the dissociation of actin monomer-profilin complexes (Saarikangas et al. 2008; Yin and Janmey 2003). Actin plays fundamental roles in multiple aspects of neuronal function, including neurite outgrowth and pathfinding, organelle trafficking, and dendritic spine morphogenesis. It also plays an important role presynaptically, where it functions in the structural organization of the active zone. At the subcellular level, F-actin localizes predominantly around the pool of releasable SVs and in the endocytic (periactive) zones, at least in some large synapses (Gaffield et al. 2006; Richards et al. 2004; Schafer 2002; Morgan et al. 2004; Bloom et al. 2003). Accordingly, several key endocytic proteins, such as amphiphysin (Yamada et al. 2009), either interact with actin regulators or are actin regulators themselves (review by (Schafer 2002)). Remarkably, acute manipulations of actin dynamics in mammalian hippocampal synapses with actin drugs have relatively minor consequences on the SV cycle (Dittman and Ryan 2009; Sankaranarayanan et al. 2003). Indeed, the only phenotype caused by a treatment with the actin depolymerizing drug latrunculin-A in this model system is a slight acceleration of the rate of SV exocytosis, with no effects on the rate of SV internalization (Sankaranarayanan et al. 2003). However, a clear role for actin in SV recycling was demonstrated at other types of synapses, such as the lamprey giant synapse, where injections of phalloidin (i.e., an F-actin stabilizing drug) and the Clostridium botulinum C2 toxin (i.e., an actin ADP-ribosylation factor that prevents the polymerization of actin) were shown to cause the accumulation of CCPs with wide neck and the expansion of the plasma membrane (Shupliakov et al. 2002). It is thus not surprising that in such synapses with a strong requirement for actin a role for PtdIns(4,5)P_2 was also demonstrated. Indeed, injection of a peptide that competes for the binding of PtdInsPK1γ to the FERM domain of talin, an adaptor between integrin and the actin cytoskeleton (Di Paolo et al. 2002), results in the disorganization of the actin network in the periactive zone and produces both an increase in the number of unconstricted CCPs and a depletion of SVs upon stimulation (Morgan et al. 2004). Furthermore, in recent work at the drosophila neuromuscular junction, PtdIns(4,5)P_2 was shown to restrict the size of nerve terminals by controlling the localization of WASP, a stimulator of actin nucleation by the Arp2/3 complex (Khuong et al. 2010). Finally, consistent with the ability of this lipid to also promote actin polymerization (Di Paolo and De Camilli 2006; Yin and Janmey 2003), abnormally high levels of actin-like cytomatrix were found in nerve terminals from *Synj1*$^{-/-}$ mice (Cremona et al. 1999) and in the giant synapse from the lamprey following a microinjection of inhibitory antibodies against Synj (Gad et al. 2000), although these studies did not directly address the consequences of such cytoskeletal anomalies on the traffic of SVs.

5.4 Postsynaptic Roles of Phosphoinositides at Excitatory Synapses

Although PtdIns(4,5)P_2 is enriched at the dendritic spines (Horne and Dell'Acqua 2007), little is known about its metabolism and the role of this lipid at the postsynapse. PtdIns(4,5)P_2 has been mostly characterized in light of its role in signal transduction as a PLC substrate downstream of stimulated metabotropic receptors (Rebecchi and

Pentyala 2000). However, as at the presynapse, PtdIns(4,5)P_2 metabolism controls actin dynamics, which, at the postsynapse, mediates changes in dendritic spine morphology, as well the traffic of alpha-amino-3-hydroxy-5-methyl-4-isoxazole-propionic acid (AMPA) receptors. Critical for some paradigms of synaptic plasticity is the phosphorylation of PtdIns(4,5)P_2 to PtdIns(3,4,5)P_3 by PI3Ks and the converse reaction by PTEN. These aspects are discussed in this section.

5.4.1 Regulation of Spine Morphology and Actin Assembly

As mentioned above, PtdIns(4,5)P_2 is a critical regulator of actin dynamics as most actin regulatory proteins interact with and are controlled by this lipid. Dendritic spines are sites of intense actin dynamics, which, together with intracellular and synaptic membrane remodeling, contribute to changes in spine morphology. A major regulator of spine dynamics appears to be Myristoylated-Alanine-Rich C Kinase Substrate (MARCKS), which is an F-actin and Ca^{2+}/Calmodulin (CaM)-binding protein that interacts with the plasma membrane via the dual actions of a hydrophobic, myristoylated NH2 terminus and a polybasic stretch mediating electrostatic interactions with anionic phospholipids (McLaughlin et al. 2002). Through these interactions, MARCKS can laterally sequester PtdIns(4,5)P_2 molecules within the membrane. Several serine residues of the effector domain of MARCKS can be phosphorylated by the protein kinase C (PKC), which decreases the electrostatic interaction with phosphoinositides and triggers the translocation of MARCKS into the cytosol (Arbuzova et al. 2002). Interfering with MARCKS expression and function through RNAi or expression of dominant-negative mutants results in destabilization of the spines and disruption of the actin cytoskeleton (Calabrese and Halpain 2005). A model arising is that synaptic activity-induced activation of PKC leads to the phosphorylation and relocation of MARCKS into the cytosol. This in turns causes the release of PtdIns(4,5)P_2 and the reorganization of the actin skeleton, possibly via a pathway involving the N-WASP-Cdc42-Arp2/3 complex and forming new F-actin branches on mother filaments. Decrease of Arp2/3 by silencing the p34 subunit of this complex leads to a reduction in the number of spines and the elongation of the remaining ones. A similar phenotype is observed by silencing N-WASP or Cdc42 (Wegner et al. 2008). Other postsynaptic PtdIns(4,5)P_2-binding proteins, such as AKAP79/150, may operate through mechanisms similar to those controlling MARCKS function (Gomez et al. 2002; Dell'Acqua et al. 1998; Horne and Dell'Acqua 2007).

Among many actin- and phosphoinositide-binding proteins, *Co*sediment with *Fil*amentous Act*in* (cofilin, also known as actin depolymerizing factor) has been extensively studied in the postsynapse. Cofilin family members bind to ADP-actin (in the G- or F forms) and promote the depolymerization at the pointed ends of actin filament (Ono 2007). They also sever actin filament in a Ca^{2+}-independent manner, which increases the number of filament ends (dos Remedios et al. 2003). The actin-binding domains of cofilin also interacts with PtdIns(4,5)P_2 (Yonezawa et al. 1991; Kusano et al. 1999) and thus the actin-binding ability of cofilin is inhibited by this lipid (Yonezawa et al. 1991). LIM kinase and the phosphatase Slingshot phosphorylate/dephosphorylate cofilin, which leads to its inactivation or its activation, respectively (Niwa et al. 2002; Yuen et al. 2010). During long-term potentiation (LTP), cofilin is inactivated by phosphorylation, which promotes actin polymerization (Fukazawa et al. 2003). In contrast, during long-term depression (LTD), dephosphorylation of cofilin is required for spine shrinkage (Zhou et al. 2004) and for the depression of N-Methyl-D-aspartate (NMDA) receptor-dependent current (Morishita et al. 2005). During chemically-induced synaptic potentiation, cofilin is first dephosphosylated, which coincides with insertion of AMPA receptors to the synaptic membrane and an increase in the number of actin barbed ends. When cofilin is re-phosphorylated, enlargement of the spines is observed (Gu et al. 2010). Forebrain-specific ablation of cofilin in the mouse has recently been show to cause an impairment of associative learning, demonstrating genetically the role of actin dynamics during memory processes (Rust et al. 2010).

5.4.2 *PtdIns(4,5)P_2 in Signaling Mechanisms Downstream of Metabotropic Glutamate Receptors*

Neurotransmitters regulate post-synaptic elements with different kinetics depending on whether they activate ionotropic receptors or metabotropic receptors. Generally, the latter types are receptors coupled to heteromeric G-proteins and characterized by seven α-helical transmembrane domains. Those receptors mediate their effects primarily by the activation of different small G-proteins and a large variety of downstream effectors. Class I metabotropic glutamate receptors (mGluR), which include mGluR1 and mGluR5, are key modulators of synaptic transmission and plasticity (Ferraguti et al. 2008). They localize to the periphery of the post-synaptic density (PSD), in part through an interaction of their proline-rich COOH-terminus with scaffolding proteins Homer, which in turn interacts with Shank and PSD-95 (Brakeman et al. 1997; Tu et al. 1999) (Fig. 5.1). At excitatory synapses, type 1 mGluRs mainly exert their actions through stimulation of PLCβ, which follows the release and activation of Gq proteins (Ferraguti et al. 2008). PLCβ, similar to other phosphoinositide-specific PLC isoforms, hydrolyzes PtdIns(4,5)P_2 to generate DAG and Ins(1,4,5)P_3. These two second messengers initiate distinct signal transduction pathways through activation of PKC (as well as other C1 domain-containing DAG effectors) and of intracellular Ca^{2+} release via activation of the Ins(1,4,5)P_3 receptor, respectively. Other metabotropic receptors, such as the muscarinic acetylcholine receptors, also operate via the Gαq/PLC pathway and are important for the modulation of synaptic transmission (Seol et al. 2007; Giessel and Sabatini 2010). At some synapses, LTD is not induced when PLC is blocked pharmacologically or upon genetic ablation of PLCβ1 (Choi et al. 2005; Reyes-Harde and Stanton 1998). Importantly, activation of type 1 mGluRs and the NMDA receptor can independently stimulate PLCs (with strong stimulation of NMDA receptor activating the Ca^{2+}-sensitive isoform PLCδ), thus leading to the stimulation of PKC in dendrites

(Codazzi et al. 2006). PLC is also important for NMDA-induced spine shrinkage and synaptic depression via the depletion of PtdIns(4,5)P_2 and the decrease of F-actin (Horne and Dell'Acqua 2007).

A fundamental aspect of G-protein coupled receptor (GPCR) regulation involves their desensitization following their stimulation by ligands (DeWire et al. 2007). The mechanism of ligand-dependent silencing or desensitization of the GPCR signaling is highly conserved and occurs via the phosphorylation of COOH-terminal serine/threonine residues via G-protein receptor kinases, followed by the recruitment of phosphoinositide binding protein β-arrestin. Binding of β-arrestin occludes the sites of the activated receptor that interact with G proteins and therefore limits the responsiveness of the GPCRs to repeated stimulations. Importantly, the binding of β-arrestin to the GPCRs exposes the β-arrestin C-terminus, which can then bind to both clathrin and the β-subunit of AP-2 and trigger the internalization of the phospho-GPCRs via CME. Similar to other endocytic adaptors for clathrin, β-arrestin interacts with phosphoinositides and PtdIns(4,5)P_2 and PtdIns(3,4,5)P_3 in particular. Mutation of the phosphoinositide-binding pocket of β-arrestin does not affect its interaction with clathrin or GPCRs, but it prevents the internalization of the activated β-adrenergic receptors (Gaidarov et al. 1999). Moreover, β-arrestins do not only interact with phosphoinositides but they also directly recruit PtdInsPK1α to sites of GPCR endocytosis (Nelson et al. 2008). While much less studied compared to the β-adrenergic receptor, the agonist-induced internalization of mGluR1 likely occurs via the β-arrestin and dynamin pathway in neurons, as shown in heterologous systems (Dale et al. 2001; Mundell et al. 2001). In addition, β-arrestin might be responsible for the termination of DAG signal by recruiting a diacylglycerol kinase in a mechanism similar to that observed after activation of the Gq protein-coupled M1 receptor (Nelson et al. 2007).

5.4.3 Role of PtdIns(4,5)P_2 in the Trafficking of AMPA Receptors

At excitatory synapses, fast response to glutamate is mainly mediated by the AMPA receptor. The second ionotropic glutamate receptor, NMDA receptor, is activated secondarily to AMPA-dependent depolarization and permits Ca^{2+} influx that has an important role in modulating synaptic function. There are four AMPA receptor genes (GluA1–4), whose products form a heterotetrameric cationic channel consisting generally of two different subunits (e.g., GluA1/GluA2 and GluA2/GluA3) are the most abundant in mature synapses (Wenthold et al. 1992; Lu et al. 2009). The COOH-terminal domains of the four subunits are divergent in length and sequence and are the main regions by which the AMPA receptor binds to a variety of auxiliary and regulatory proteins (Fukata et al. 2005). These proteins control key aspects of the AMPA receptor function, including its trafficking and signaling properties.

Role of Phosphoinositides in the Control of Cell Surface Levels of the AMPA Receptor

Because the AMPA receptor exclusively operates at the postsynaptic plasma membrane, a key aspect of its regulation involves the control of its cell surface levels via exo-endocytosis (Newpher and Ehlers 2008). The internalization of the AMPA receptor largely occurs through clathrin- and dynamin-mediated endocytosis (Lee et al. 2004a; Morishita et al. 2005; Kastning et al. 2007) and thus the molecular machinery mediating this process shares many features in common with that controlling SV endocytosis. Just like endocytic zones are segregated from the exocytic (or active) zones at the presynapse, the postsynaptic sites of endocytosis appear to be extra-synaptic as CCPs are mostly observed outside of the PSD (Blanpied et al. 2002). Constitutive endocytosis of the AMPA receptor in proximity to the PSD is required to recapture the receptors that have "escaped" from the PSD through lateral diffusion (Lu et al. 2007). The COOH-terminal domain of the short-tailed subunit, GluA2, directly binds to the μ2 subunit of the AP-2 complex and this interaction is required for AMPA receptor internalization (Kastning et al. 2007; Lee et al. 2004a). Consistent with the principles governing the clathrin-mediated retrieval of SVs, a role for PtdIns(4,5)P_2 was demonstrated by studies on Synj1. While the first study showed a role for Synj1 downstream of ephrinB-EphB signaling in the internalization of AMPA receptor (Irie et al. 2005), the second, more recent study utilized *Synj1* KO neurons to show enhanced amplitudes of AMPA responses as well as a decrease in NMDA-induced internalization of AMPA receptor in mutant synapses (Gong and De Camilli 2008).

The COOH-terminal tail of GluA2 and GluA3 also interacts with PDZ domain-containing proteins, such as PICK1 (*P*rotein *I*nteracting with *CK*inase 1) and GRIP1/2 (*G*lutamate *R*eceptor *I*nteracting *P*rotein 1 and 2). The former was shown to interact with phosphoinositides (Jin et al. 2006). PICK1 and GRIP1/2 antagonistically control the association of AMPA receptors with the PSD and the cell surface with fundamental implications for synaptic plasticity. While GRIP1/2 stabilizes the AMPA receptor at the synaptic membrane (Osten et al. 2000) or facilitates the reinsertion at the surface of internalized AMPA receptor via the exocyst complex (Mao et al. 2010), PICK1 facilitates the removal of the receptor from the cell surface and maintains it in an intracellular compartment (Lin and Huganir 2007). A switch from GRIP1/2 to PICK1 and a resulting untethering of GluA2 from the PSD occurs when the serine 880 of the receptor is phosphorylated by protein kinases, such as PKCα (Chung et al. 2000; Seidenman et al. 2003). Conversely, PICK1 blocks the reinsertion of internalized AMPA receptor at the PSD (Citri et al. 2010), a process that is important for NMDA-dependent LTD (Terashima et al. 2008). The interaction of GluA2 with GRIP or PICK is also important for the replacement of GluA2-lacking receptor newly inserted in the perisynaptic region with GluA2-containing receptor (Yang et al. 2010) and with the maintenance of Ca^{2+}-permeable AMPA receptor at sensory synapses (Clem et al. 2010). PICK1 contains an NH_2-terminal PDZ domain and a central BAR domain, both of which bind to phosphoinositides. The BAR domain preferentially interacts with monophosphorylated phosphoinositides, such as PtdIns3P, via a series of acidic residues on the concave side of the crescent-like domain (Jin et al. 2006). The BAR domain is crucial for the localization of PICK1 (Jin et al. 2006) and for the induction of LTD (Jin et al. 2006; Steinberg et al. 2006). Similarly, the PDZ domain of PICK1 binds to PtdIns(4,5)P_2 and PtdIns(3,4,5)P_3 via three basic residues and a neighboring hydrophobic Cys-Pro-Cys motif. Mutation of the two cysteine residues into glycines impairs the formation of PICK1 clusters in non-neuronal cells and at excitatory synapses (Pan et al. 2007).

Phosphoinositide-dependent Extrasynaptic Clustering of TARP-AMPA Receptor Complexes

TARPs (*T*ransmembrane *A*MPA Receptor *R*egulatory *P*roteins) are auxiliary subunits that regulate the trafficking and the kinetic properties of AMPA receptors (Milstein and Nicoll 2008). Genetic deletion of TARPs, such as γ2 (stargazin) and γ8, results in the reduction of glutamatergic neurotransmission (Hashimoto et al. 1999; Rouach et al. 2005). Phospholipids, such as phosphoinositides, play a central role in this regulation. TARPs bind to the transmembrane region and the extracellular portions of all the AMPA receptor subunits and these interactions are required for the proper trafficking of the AMPA receptor from the endoplasmic reticulum to the synapse. The last four amino-acids of the cytosolic COOH-terminus (TTPV) of TARPs bind to the PDZ domain of PSD-95, a master regulator of the PSD, and this interaction is necessary (but not sufficient) for efficient targeting of the AMPA receptors to the PSD (Cuadra et al. 2004). The COOH-terminus of stargazin binds to phosphoinositides and PA via conserved arginine residues and this interaction is competed by PSD-95 (Sumioka et al. 2010) or disrupted by Calcium/Calmodulin-dependent Kinase II (CamKII)- or PKC-mediated phosphorylation of adjacent serine residues. Phosphorylation of this tail reduces the electrostatic interaction of stargazin with the anionic phospholipids within the membranes, thus permitting the diffusion and the binding of TARP with the postsynaptic scaffolding proteins, such as PSD-95. As a result, AMPA receptors are both immobilized and activated at the PSD (Opazo and Choquet 2010). In summary, the regulation of the interaction between AMPA receptor and TARPs by phosphoinositides and phosphorylation-based mechanisms appears to be a key method to control the mobility of AMPA receptors in or out of the synapses during synaptic plasticity.

5.4.4 *Role of PtdIns(3,4,5)P_3 in Synaptic Plasticity*

Class 1 PI3Ks and their downstream effectors are present at the postsynapses (Raymond et al. 2002) and are part of the complex organized around the AMPA receptor (Man et al. 2003; Peineau et al. 2007). In fact, recent work has shown that a constant production of PtdIns(3,4,5)P_3 is essential for the maintenance of

AMPA receptor at the synapse. Both in hippocampal slices and in cultured hippocampal neurons, manipulations of the metabolism of PtdIns(3,4,5)P_3 alter AMPA receptor-dependent responses. Expression of constitutively active (i.e., membrane-bound) PI3K increases AMPA currents, while inhibition of PI3K or sequestering PtdIns(3,4,5)P_3 with the PH domain of *G*eneral *R*eceptor for *P*hosphoinositides 1 (GRP1) reduces the amplitude of evoked currents (Arendt et al. 2010). Mechanistically, basal PtdIns(3,4,5)P_3 levels appear to control the stability of PSD-95 in the spine and the mobility of AMPA receptor at the PSD. Depleting PtdIns(3,4,5)P_3 promotes the relocation and the accumulation of the AMPA receptor in the extrasynaptic zones of the membrane (Arendt et al. 2010).

Similar to CamKII, PI3K is also an essential regulator of synaptic plasticity. Pharmacological inhibition of PI3K prevents the formation of LTP at various synapses, such as at the CA3-CA1 synapse (Raymond et al. 2002; Sanna et al. 2002), at cortical synapses in the lateral amygdala (Lin et al. 2001) and at parallel fibers-Purkinje cell synapses (Jackson et al. 2010a). LTP is also impaired after sequestering PtdIns(3,4,5)P_3 with overexpression of PH-GRP1 (Arendt et al. 2010). These results are consistent with the impairment in learning and memory observed in mice treated with PI3K inhibitors (Lin et al. 2001; Chen et al. 2005). During LTP, PI3K activity is increased by the small GTPase Ras (Qin et al. 2005), a pathway that is also impaired in the Fragile X syndrome, which is associated with mental retardation (Hu et al. 2008). During glycine-induced LTP, the activity of PI3K associated with the AMPA-receptor is increased, which regulates the insertion of AMPA receptor at the synaptic membrane (Man et al. 2003). PI3Ks are also involved in some forms of LTD through their association with the mGluR5-Homer complex, which mediates synaptic plasticity in the nucleus accumbens during alcohol consumption (Cozzoli et al. 2009) or is important to limit LTD at the synapse receiving the conditioning stimulus (Daw et al. 2002).

The downstream effectors of the PI3K pathway involved in synaptic plasticity have also been investigated. It is well accepted that PtdIns(3,4,5)P_3 generated by PI3Ks recruits to membranes phosphoinositide-dependent kinase 1 (PDK1) and Akt (or PKB) via their PH domain (Pearce et al. 2010). Downstream of phosphorylated Akt, *G*lycogen *s*ynthase *k*inase-*3*β (GSK3β) has a central role in the balance between NMDA-dependent LTD and LTP. Indeed, the LTP induced by activation of PI3K/Akt promotes the phosphorylation of Ser9 of GSK3β, thereby inhibiting its activity and the induction LTD for a period of an hour. During LTD, activation of phosphatase PP1 and/or inhibition of Akt result in the dephosphorylation of Ser9 and the activation of GSK3β, which is necessary for LTD induction (Peineau et al. 2007).

Downstream of PI3K also lies the mTOR/p70^{S6K} pathway, which plays an important role at the postsynapse. mTOR activation during LTP requires the coincident activation of the PI3K pathway with the Erk pathway (Tsokas et al. 2007). mTOR regulates the transcription of mRNAs encoding proteins involved in protein translation, primarily cap-dependent translation initiation. Maintenance of LTP for more than 2–3 hours requires protein synthesis (Fonseca et al. 2006; Huang et al. 1996), a phase that is inhibited by rapamycin, an inhibitor of mTOR, when applied during the induction of LTP (Tang et al. 2002; Cammalleri et al. 2003). Gene deletion of FKBP12, an inhibitor of mTOR, increases mTOR activity, which promotes LTP and enhances contextual fear memory but alters the capacity of the transgenic mice to remodel memory (Hoeffer et al. 2008). Similarly, mGluR-dependent LTD required *de novo* synthesis of protein, the initiation of which seems to be dependent on PI3K and mTOR (Hou and Klann 2004).

PTEN is the enzyme that counterbalances PI3K by removing the 3-phosphate from PtdIns(3,4,5)P_3 (Maehama and Dixon 1999; Sulis and Parsons 2003). Conditional ablation of PTEN in a discrete number of mature neurons in the cortex results in a tremendous impairment of social interaction among other behavioral abnormalities in mice (Kwon et al. 2006). Morphologically, ablation or silencing of PTEN induces an increase in axonal and dendritic growth, thus phenocopying the consequences of PI3K overexpression and hyperactivation of mTOR (Jaworski et al. 2005; Chow et al. 2009). Abnormal synapses with a higher number of SVs and a higher density of spines have also been reported. In the mutant mice, all the downstream targets of Akt, including GSK3β were found to be hyperphosphorylated (Chow et al. 2009). A similar phenotype is observed in a *PTEN* haploinsufficient mice and this phenotype is exacerbated by the haploinsufficiency of the serotonin transporter *Slca4* gene (Page et al. 2009). PTEN can be recruited to the spine by binding to the PDZ domain of PSD-95 after NMDA receptor activation and regulates the expression of LTD by affecting AMPA receptor localization (Jurado et al. 2010). PTEN may also directly bind to extrasynaptic NMDA receptors and positively regulates signaling downstream of this glutamate receptor (Ning et al. 2004).

5.5 Phosphoinositide Imbalance in Synaptic Dysfunction and Brain Disorders

Due to the pleiotropic roles of phosphoinositides, their metabolism is subject to tight regulation. Not surprisingly, several phosphoinositide-metabolizing enzymes have been implicated in a number of human diseases, including genetic disorders (Di Paolo and De Camilli 2006; Pendaries et al. 2003). Recently, our group has linked PtdIns(4,5)P_2 metabolism and in particular Synj1 to Down syndrome (DS) and Alzheimer's disease (AD).

5.5.1 Synaptojanin1 Overexpression in Down Syndrome

Phosphoinositides have been shown to be directly or indirectly linked to several conditions associated with mental retardation (MR), including the Fragile-X and DS. As described by J. Lejeune in the 1950s, DS is the result of the trisomy of the chromosome 21 and occurs in 1 birth out of 800. One of the stringent phenotypes of DS is MR, which reflects the dosage imbalance of genes expressed in the brain. The gene encoding Synj1 is located on human chromosome 21 (mapped on 21q22.2) and the presence of 3 copies in individuals with DS results in a higher expression of Synj1 in DS brain (Arai et al. 2002). A similar result was also observed in the partial trisomy mouse Ts65Dn (Voronov et al. 2008), which has a partial triplication of mouse chromosome 16, which is highly syntenic with human chromosome 21 (Reeves et al. 1995). We have used the Ts65Dn mice, which recapitulates many features of individuals with DS, and a BAC transgenic mouse that overexpresses Synj1 [Tg(Synj1)] uniquely due to 3 gene copies (Voronov et al. 2008). In the brain of both mouse models, the PtdIns(4,5)P_2 5-phosphatase activity was increased and the levels of PtdIns(4,5)P_2 were reduced. Removal of one copy of Synj1 from the Ts65Dn background rescued the defects in PtdIns(4,5)P_2 metabolism (Voronov et al. 2008). While the Ts65Dn mice have important cognitive deficits (Seregaza et al. 2006), the Tg(Synj1) mice were shown to perform poorly in the Morris-water maze task, which assesses spatial learning. However, as expected, cognitive deficits resulting from *Synj1* overexpression are milder than those observed in Ts65Dn, further suggesting that multiple genes contribute to MR in DS.

5.5.2 *PtdIns(4,5)P_2 Dysregulation in Alzheimer's Disease*

AD is the most common form of late-onset dementia. It is characterized by an accumulation of amyloid-beta (Aβ) and neurofibrillary tangles and results in age-dependent cognitive decline. Several studies on animals models of AD described defects in excitatory neurotransmission (Hsieh et al. 2006), and synaptic plasticity (Walsh et al. 2002; Venkitaramani et al. 2007), in synaptic loss (Hsieh et al. 2006; Lue et al. 1999) and a neuron network imbalance with an increased synchronic activity (Palop et al. 2007; Busche et al. 2008; Palop and Mucke 2010). Recent studies have suggested that phosphoinositide imbalance plays an important role in the pathophysiological processes involving Aβ. Presenilin-1 is the catalytic subunit of the proteolytic complex γ-secretase that cleaves the COOH-terminal part of the transmembrane domain of the amyloid precursor protein (APP) producing peptides of different sizes but particularly Aβ40 and in lower amount Aβ42, which is more aggregate prone and synaptotoxic (Haass 2004). Mutations of presenilin-1 observed in familial forms of AD cause an imbalance in PtdIns(4,5)P_2 metabolism as observed with the reduction of the calcium-permeable transient receptor potential melastatin 7-associated magnesium-inhibited cation current, that is positively regulated by PtdIns(4,5)P_2. PtdIns(4,5)P_2 level variation induced by altering PLC activity correlates negatively with the levels of Aβ42 (Landman et al. 2006). Furthermore, incubation of cortical cultures with submicromolar concentration of oligomeric Aβ42 was shown to reduce the levels of PtdIns(4,5)P_2 in a NMDA- and PLC-dependent manner. Compensating for this deficiency of PtdIns(4,5)P_2 by deleting a copy of Synj1 rescued Aβ-induced LTP defects in hippocampal slices (Berman et al. 2008). Our findings suggest a hypothesis whereby PtdIns(4,5)P_2 dyshomeostasis underlies early neurotoxic effects of Aβ at synapses and that Synj1 haploinsufficiency protects against the actions of this peptide. More generally, we hypothesize that

Aβ-induced synaptic dysfunction and cognitive deficits can be ameliorated by treatments that prevent PtdIns(4,5)P_2 downregulation in AD. Finally, as mentioned earlier, a key feature of DS is MR, whose onset typically starts in early childhood (Roizen and Patterson 2003). Superimposed to these deficits is the progressive development of AD pathology, which leads to further cognitive decline in middle-aged individuals with DS (Lott and Head 2001). A key causative agent for AD in DS is the overexpression of chromosome 21-linked APP, the precursor of Aβ. Our studies point to the occurrence in DS of a "dual hit" on PtdIns(4,5)P_2, accounted for by Synj1 trisomy and by Aβ elevation, which could potentiate AD-linked cognitive decline in middle-aged adults with DS.

5.6 Conclusions and Perspectives

As highlighted in this book chapter, phosphoinositides mediate a multitude of biological functions at the neuronal synapse, both pre- and postsynaptically. Emphasis was put on PtdIns(4,5)P_2 and PtdIns(3,4,5)P_3 because these two lipids have been extensively studied and shown to regulate important synaptic functions. Their enrichment at the plasma membrane, along with a rich set of effectors and metabolizing enzymes operating at this compartment (or in close proximity thereof), predisposes them for specialized functions occurring at the cell surface, including the fusion and internalization of SVs, the control of cell surface levels of ion channels, signaling of ligand-bound receptors and regulation of actin dynamics. A fundamental question in the field (and one that is pertinent to the broad field of phosphoinositide research) is how one or two lipids is/are able to control so many critical functions within a single membrane compartment. A partial answer proposed by several investigators is that phosphoinositide metabolism may be controlled locally, in micro- or perhaps even nanodomains, by tightly regulated recruitment and/or activation of lipid enzymes, by lipid sequestration/release mechanisms or by changes in membrane properties (e.g., curvature). The fast diffusion rates reported for membrane lipids, including PtdIns(4,5)P_2, are difficult to reconcile with the concept of local control of phosphoinositide metabolism, unless mechanisms that significantly slow down diffusion rates or *bona fide* "fences" exist. In this respect, the ability to visualize lipids with genetically-encoded fluorescent probes, such as PH domains, in combination with super resolution fluorescence microscopy will be informative.

While PtdIns(4,5)P_2 and PtdIns(3,4,5)P_3 have been extensively characterized, little is known about the role of the other five phosphoinositides at synapses, and particularly, the low abundance ones, such as PtdIns(3,4)P_2, PtdIns(3,5)P_2, and PtdIns5P. Answers may come from the characterization of various KO or mutant animals that are defective for enzymes involved in their metabolism, although in some cases, the development of better probes for low abundance phosphoinositides and identification of synapse-specific effectors may be required to have a better understanding of what these lipids do at the neuronal synapse and elsewhere.

Acknowledgment Work on phosphoinositides in the Di Paolo lab is funded by NIH grants R01 NS056049, R01 R01HD05547 and R03 AG033212 and by the McKnight Endowment Fund. Belle Chang-Ileto is funded by the NIH (F31 NS058096).

References

Adayev T, Chen-Hwang MC, Murakami N, Wang R, Hwang YW (2006) MNB/DYRK1A phosphorylation regulates the interactions of synaptojanin 1 with endocytic accessory proteins. Biochem Biophys Res Commun 351:1060–1065

Anderson RA, Boronenkov IV, Doughman SD, Kunz J, Loijens JC (1999) Phosphatidylinositol phosphate kinases, a multifaceted family of signaling enzymes. J Biol Chem 274:9907–9910

Ann K, Kowalchyk JA, Loyet KM, Martin TF (1997) Novel Ca^{2+}-binding protein (CAPS) related to UNC-31 required for Ca^{2+}-activated exocytosis. J Biol Chem 272:19637–19640

Aoyagi K, Sugaya T, Umeda M, Yamamoto S, Terakawa S, Takahashi M (2005) The activation of exocytotic sites by the formation of phosphatidylinositol 4,5-bisphosphate microdomains at syntaxin clusters. J Biol Chem 280:17346–17352

Arai Y, Ijuin T, Takenawa T, Becker LE, Takashima S (2002) Excessive expression of synaptojanin in brains with down syndrome. Brain Dev 24:67–72

Arbuzova A, Schmitz AA, Vergeres G (2002) Cross-talk unfolded: MARCKS proteins. Biochem J 362:1–12

Arendt KL, Royo M, Fernandez-Monreal M, Knafo S, Petrok CN, Martens JR, Esteban JA (2010) PIP3 controls synaptic function by maintaining AMPA receptor clustering at the postsynaptic membrane. Nat Neurosci 13:36–44

Augustin I, Rosenmund C, Sudhof TC, Brose N (1999) Munc13-1 is essential for fusion competence of glutamatergic synaptic vesicles. Nature 400:457–461

Bai J, Tucker WC, Chapman ER (2004) PIP2 increases the speed of response of synaptotagmin and steers its membrane-penetration activity toward the plasma membrane. Nat Struct Mol Biol 11:36–44

Bai J, Hu Z, Dittman JS, Pym EC, Kaplan JM (2010) Endophilin functions as a membrane-bending molecule and is delivered to endocytic zones by exocytosis. Cell 143:430–441

Bairstow SF, Ling K, Su X, Firestone AJ, Carbonara C, Anderson RA (2006) Type Igamma661 phosphatidylinositol phosphate kinase directly interacts with AP2 and regulates endocytosis. J Biol Chem 281:20632–20642

Bao H, Reist NE, Zhang B (2008) The Drosophila epsin 1 is required for ubiquitin-dependent synaptic growth and function but not for synaptic vesicle recycling. Traffic 9:2190–2205

Bashkirov PV, Akimov SA, Evseev AI, Schmid SL, Zimmerberg J, Frolov VA (2008) GTPase cycle of dynamin is coupled to membrane squeeze and release, leading to spontaneous fission. Cell 135:1276–1286

Berman DE, Dall'Armi C, Voronov SV, McIntire LB, Zhang H, Moore AZ, Staniszewski A, Arancio O, Kim TW, Di Paolo G (2008) Oligomeric amyloid-beta peptide disrupts phosphatidylinositol-4,5-bisphosphate metabolism. Nat Neurosci 11:521–617

Blanpied TA, Scott DB, Ehlers MD (2002) Dynamics and regulation of clathrin coats at specialized endocytic zones of dendrites and spines. Neuron 36:435–449

Blero D, Payrastre B, Schurmans S, Erneux C (2007) Phosphoinositide phosphatases in a network of signalling reactions. Pflugers Arch 455:31–44

Bloom O, Evergren E, Tomilin N, Kjaerulff O, Low P, Brodin L, Pieribone VA, Greengard P, Shupliakov O (2003) Colocalization of synapsin and actin during synaptic vesicle recycling. J Cell Biol 161:737–747

Bollmann JH, Sakmann B, Borst JG (2000) Calcium sensitivity of glutamate release in a calyx-type terminal. Science 289:953–957

Bonifant CL, Kim JS, Waldman T (2007) NHERFs, NEP, MAGUKs, and more: interactions that regulate PTEN. J Cell Biochem 102(4):878–885

Brakeman PR, Lanahan AA, O'Brien R, Roche K, Barnes CA, Huganir RL, Worley PF (1997) Homer: a protein that selectively binds metabotropic glutamate receptors. Nature 386:284–288
Brose N, Petrenko AG, Sudhof TC, Jahn R (1992) Synaptotagmin: a calcium sensor on the synaptic vesicle surface. Science 256:1021–1025
Busche MA, Eichhoff G, Adelsberger H, Abramowski D, Wiederhold KH, Haass C, Staufenbiel M, Konnerth A, Garaschuk O (2008) Clusters of hyperactive neurons near amyloid plaques in a mouse model of Alzheimer's disease. Science 321:1686–1689
Calabrese B, Halpain S (2005) Essential role for the PKC target MARCKS in maintaining dendritic spine morphology. Neuron 48:77–90
Cammalleri M, Lutjens R, Berton F, King AR, Simpson C, Francesconi W, Sanna PP (2003) Time-restricted role for dendritic activation of the mTOR-p70S6K pathway in the induction of late-phase long-term potentiation in the CA1. Proc Natl Acad Sci U S A 100:14368–14373
Cestra G, Castagnoli L, Dente L, Minenkova O, Petrelli A, Migone N, Hoffmuller U, Schneider-Mergener J, Cesareni G (1999) The SH3 domains of endophilin and amphiphysin bind to the proline-rich region of synaptojanin 1at distinct sites that display an unconventional binding specificity. J Biol Chem 274:32001–32007
Chang JD, Field SJ, Rameh LE, Carpenter CL, Cantley LC (2004) Identification and characterization of a phosphoinositide phosphate kinase homolog. J Biol Chem 279:11672–11679
Chang N, El-Hayek YH, Gomez E, Wan Q (2007) Phosphatase PTEN in neuronal injury and brain disorders. Trends Neurosci 30:581–586
Chang-Ileto B, Frere SG, Chan RB, Voronov SV, Roux A, Di Paolo G (2011) Synaptojanin 1-mediated PI(4,5)P(2) hydrolysis is modulated by membrane curvature and facilitates membrane fission. Dev Cell 20:206–218
Chapman ER (2008) How does synaptotagmin trigger neurotransmitter release? Annu Rev Biochem 77:615–641
Chatah NE, Abrams CS (2001) G-protein-coupled receptor activation induces the membrane translocation and activation of phosphatidylinositol-4-phosphate 5-kinase I alpha by a Rac- and Rho-dependent pathway. J Biol Chem 276:34059–34065
Chen MS, Obar RA, Schroeder CC, Austin TW, Poodry CA, Wadsworth SC, Vallee RB (1991) Multiple forms of dynamin are encoded by shibire, a Drosophila gene involved in endocytosis. Nature 351:583–586
Chen X, Garelick MG, Wang H, Lil V, Athos J, Storm DR (2005) PI3 kinase signaling is required for retrieval and extinction of contextual memory. Nat Neurosci 8:925–931
Chen H, Ko G, Zatti A, Di Giacomo G, Liu L, Raiteri E, Perucco E, Collesi C, Min W, Zeiss C, De Camilli P, Cremona O (2009) Embryonic arrest at midgestation and disruption of Notch signaling produced by the absence of both epsin 1 and epsin 2 in mice. Proc Natl Acad Sci U S A 106:13838–13843
Chicka MC, Hui E, Liu H, Chapman ER (2008) Synaptotagmin arrests the SNARE complex before triggering fast, efficient membrane fusion in response to Ca^{2+}. Nat Struct Mol Biol 15:827–835
Choi SY, Chang J, Jiang B, Seol GH, Min SS, Han JS, Shin HS, Gallagher M, Kirkwood A (2005) Multiple receptors coupled to phospholipase C gate long-term depression in visual cortex. J Neurosci 25:11433–11443
Chow DK, Groszer M, Pribadi M, Machniki M, Carmichael ST, Liu X, Trachtenberg JT (2009) Laminar and compartmental regulation of dendritic growth in mature cortex. Nat Neurosci 12:116–118
Chung SH, Song WJ, Kim K, Bednarski JJ, Chen J, Prestwich GD, Holz RW (1998) The C2 domains of Rabphilin3A specifically bind phosphatidylinositol 4,5-bisphosphate containing vesicles in a Ca^{2+}-dependent manner. In vitro characteristics and possible significance. J Biol Chem 273:10240–10248
Chung HJ, Xia J, Scannevin RH, Zhang X, Huganir RL (2000) Phosphorylation of the AMPA receptor subunit GluR2 differentially regulates its interaction with PDZ domain-containing proteins. J Neurosci 20:7258–7267
Citri A, Bhattacharyya S, Ma C, Morishita W, Fang S, Rizo J, Malenka RC (2010) Calcium binding to PICK1 is essential for the intracellular retention of AMPA receptors underlying long-term depression. J Neurosci 30:16437–16452

Clem RL, Anggono V, Huganir RL (2010) PICK1 regulates incorporation of calcium-permeable AMPA receptors during cortical synaptic strengthening. J Neurosci 30:6360–6366
Codazzi F, Di Cesare A, Chiulli N, Albanese A, Meyer T, Zacchetti D, Grohovaz F (2006) Synergistic control of protein kinase Cgamma activity by ionotropic and metabotropic glutamate receptor inputs in hippocampal neurons. J Neurosci 26:3404–3411
Collins BM, McCoy AJ, Kent HM, Evans PR, Owen DJ (2002) Molecular architecture and functional model of the endocytic AP2 complex. Cell 109:523–535
Cousin MA, Robinson PJ (2001) The dephosphins: dephosphorylation by calcineurin triggers synaptic vesicle endocytosis. Trends Neurosci 24:659–665
Cozzoli DK, Goulding SP, Zhang PW, Xiao B, Hu JH, Ary AW, Obara I, Rahn A, Abou-Ziab H, Tyrrel B, Marini C, Yoneyama N, Metten P, Snelling C, Dehoff MH, Crabbe JC, Finn DA, Klugmann M, Worley PF, Szumlinski KK (2009) Binge drinking upregulates accumbens mGluR5-Homer2-PI3K signaling: functional implications for alcoholism. J Neurosci 29:8655–8668
Cremona O, De Camilli P (2001) Phosphoinositides in membrane traffic at the synapse. J Cell Sci 114:1041–1052
Cremona O, Di Paolo G, Wenk MR, Luthi A, Kim WT, Takei K, Daniell L, Nemoto Y, Shears SB, Flavell RA, McCormick DA, De Camilli P (1999) Essential role of phosphoinositide metabolism in synaptic vesicle recycling. Cell 99:179–188
Cuadra AE, Kuo SH, Kawasaki Y, Bredt DS, Chetkovich DM (2004) AMPA receptor synaptic targeting regulated by stargazin interactions with the Golgi-resident PDZ protein nPIST. J Neurosci 24:7491–7502
Dale LB, Bhattacharya M, Seachrist JL, Anborgh PH, Ferguson SS (2001) Agonist-stimulated and tonic internalization of metabotropic glutamate receptor 1a in human embryonic kidney 293 cells: agonist-stimulated endocytosis is beta-arrestin1 isoform-specific. Mol Pharmacol 60:1243–1253
Daw MI, Bortolotto ZA, Saulle E, Zaman S, Collingridge GL, Isaac JT (2002) Phosphatidylinositol 3 kinase regulates synapse specificity of hippocampal long-term depression. Nat Neurosci 5:835–836
De Heuvel E, Bell AW, Ramjaun AR, Wong K, Sossin WS, McPherson PS (1997) Identification of the major synaptojanin-binding proteins in brain. J Biol Chem 272:8710–8716
De Wit H, Walter AM, Milosevic I, Gulyas-Kovacs A, Riedel D, Sorensen JB, Verhage M (2009) Synaptotagmin-1 docks secretory vesicles to syntaxin-1/SNAP-25 acceptor complexes. Cell 138:935–946
Dell'Acqua ML, Faux MC, Thorburn J, Thorburn A, Scott JD (1998) Membrane-targeting sequences on AKAP79 bind phosphatidylinositol-4, 5-bisphosphate. EMBO J 17:2246–2260
DeWire SM, Ahn S, Lefkowitz RJ, Shenoy SK (2007) Beta-arrestins and cell signaling. Annu Rev Physiol 69:483–510
Di Paolo G, De Camilli P (2006) Phosphoinositides in cell regulation and membrane dynamics. Nature 443:651–657
Di Paolo G, Pellegrini L, Letinic K, Cestra G, Zoncu R, Voronov S, Chang S, Guo J, Wenk MR, De Camilli P (2002) Recruitment and regulation of phosphatidylinositol phosphate kinase type 1 gamma by the FERM domain of talin. Nature 420:85–89
Di Paolo G, Moskowitz HS, Gipson K, Wenk MR, Voronov S, Obayashi M, Flavell R, Fitzsimonds RM, Ryan TA, De Camilli P (2004) Impaired PtdIns(4,5)P_2 synthesis in nerve terminals produces defects in synaptic vesicle trafficking. Nature 431:415–422
Dickman DK, Horne JA, Meinertzhagen IA, Schwarz TL (2005) A slowed classical pathway rather than kiss-and-run mediates endocytosis at synapses lacking synaptojanin and endophilin. Cell 123:521–533
Dittman JS, Kaplan JM (2006) Factors regulating the abundance and localization of synaptobrevin in the plasma membrane. Proc Natl Acad Sci U S A 103:11399–11404
Dittman J, Ryan TA (2009) Molecular circuitry of endocytosis at nerve terminals. Annu Rev Cell Dev Biol 25:133–160
Divecha N, Roefs M, Halstead JR, D'Andrea S, Fernandez-Borga M, Oomen L, Saqib KM, Wakelam MJ, D'Santos C (2000) Interaction of the type Ialpha PIPkinase with phospholipase D: a role

for the local generation of phosphatidylinositol 4, 5-bisphosphate in the regulation of PLD2 activity. EMBO J 19:5440–5449

Dos Remedios CG, Chhabra D, Kekic M, Dedova IV, Tsubakihara M, Berry DA, Nosworthy NJ (2003) Actin binding proteins: regulation of cytoskeletal microfilaments. Physiol Rev 83:433–473

Doughman RL, Firestone AJ, Anderson RA (2003) Phosphatidylinositol phosphate kinases put PI4,5P(2) in its place. J Membr Biol 194:77–89

Eberhard DA, Cooper CL, Low MG, Holz RW (1990) Evidence that the inositol phospholipids are necessary for exocytosis. Loss of inositol phospholipids and inhibition of secretion in permeabilized cells caused by a bacterial phospholipase C and removal of ATP. Biochem J 268:15–25

Eisenberg E, Greene LE (2007) Multiple roles of auxilin and hsc70 in clathrin-mediated endocytosis. Traffic 8:640–646

Farsad K, Ringstad N, Takei K, Floyd SR, Rose K, De Camilli P (2001) Generation of high curvature membranes mediated by direct endophilin bilayer interactions. J Cell Biol 155:193–200

Ferguson SM, Brasnjo G, Hayashi M, Wolfel M, Collesi C, Giovedi S, Raimondi A, Gong LW, Ariel P, Paradise S, O'Toole E, Flavell R, Cremona O, Miesenbock G, Ryan TA, De Camilli P (2007) A selective activity-dependent requirement for dynamin 1 in synaptic vesicle endocytosis. Science 316:570–574

Ferraguti F, Crepaldi L, Nicoletti F (2008) Metabotropic glutamate 1 receptor: current concepts and perspectives. Pharmacol Rev 60:536–581

Fonseca R, Vabulas RM, Hartl FU, Bonhoeffer T, Nagerl UV (2006) A balance of protein synthesis and proteasome-dependent degradation determines the maintenance of LTP. Neuron 52:239–245

Ford MG, Mills IG, Peter BJ, Vallis Y, Praefcke GJ, Evans PR, McMahon HT (2002) Curvature of clathrin-coated pits driven by epsin. Nature 419:361–366

Frost A, Perera R, Roux A, Spasov K, Destaing O, Egelman EH, De Camilli P, Unger VM (2008) Structural basis of membrane invagination by F-BAR domains. Cell 132:807–817

Frost A, Unger VM, De Camilli P (2009) The BAR domain superfamily: membrane-molding macromolecules. Cell 137:191–196

Fukata Y, Tzingounis AV, Trinidad JC, Fukata M, Burlingame AL, Nicoll RA, Bredt DS (2005) Molecular constituents of neuronal AMPA receptors. J Cell Biol 169:399–404

Fukazawa Y, Saitoh Y, Ozawa F, Ohta Y, Mizuno K, Inokuchi K (2003) Hippocampal LTP is accompanied by enhanced F-actin content within the dendritic spine that is essential for late LTP maintenance in vivo. Neuron 38:447–460

Gad H, Ringstad N, Low P, Kjaerulff O, Gustafsson J, Wenk M, Di Paolo G, Nemoto Y, Crun J, Ellisman MH, De Camilli P, Shupliakov O, Brodin L (2000) Fission and uncoating of synaptic clathrin-coated vesicles are perturbed by disruption of interactions with the SH3 domain of endophilin. Neuron 27:301–312

Gaffield MA, Rizzoli SO, Betz WJ (2006) Mobility of synaptic vesicles in different pools in resting and stimulated frog motor nerve terminals. Neuron 51:317–325

Gaidarov I, Keen JH (1999) Phosphoinositide-AP-2 Interactions required for targeting to plasma membrane clathrin-coated pits. J Cell Biol 146:755–764

Gaidarov I, Krupnick JG, Falck JR, Benovic JL, Keen JH (1999) Arrestin function in G protein-coupled receptor endocytosis requires phosphoinositide binding. EMBO J 18:871–881

Gallop JL, Jao CC, Kent HM, Butler PJ, Evans PR, Langen R, McMahon HT (2006) Mechanism of endophilin N-BAR domain-mediated membrane curvature. EMBO J 25:2898–2910

Geering B, Cutillas PR, Nock G, Gharbi SI, Vanhaesebroeck B (2007) Class IA phosphoinositide 3-kinases are obligate p85-p110 heterodimers. Proc Natl Acad Sci U S A 104:7809–7814

Geppert M, Goda Y, Hammer RE, Li C, Rosahl TW, Stevens CF, Sudhof TC (1994) Synaptotagmin I: a major Ca^{2+} sensor for transmitter release at a central synapse. Cell 79:717–727

Giessel AJ, Sabatini BL (2010) M1 muscarinic receptors boost synaptic potentials and calcium influx in dendritic spines by inhibiting postsynaptic SK channels. Neuron 68:936–947

Giudici ML, Emson PC, Irvine RF (2004) A novel neuronal-specific splice variant of type I phosphatidylinositol 4-phosphate 5-kinase isoform gamma. Biochem J 379:489–496
Gomez LL, Alam S, Smith KE, Horne E, Dell'Acqua ML (2002) Regulation of A-kinase anchoring protein 79/150-cAMP-dependent protein kinase postsynaptic targeting by NMDA receptor activation of calcineurin and remodeling of dendritic actin. J Neurosci 22:7027–7044
Gong LW, De Camilli P (2008) Regulation of postsynaptic AMPA responses by synaptojanin 1. Proc Natl Acad Sci U S A 105:17561–17566
Gong LW, Di Paolo G, Diaz E, Cestra G, Diaz ME, Lindau M, De Camilli P, Toomre D (2005) Phosphatidylinositol phosphate kinase type I gamma regulates dynamics of large dense-core vesicle fusion. Proc Natl Acad Sci U S A 102:5204–5209
Gonzalez-Gaitan M, Jackle H (1997) Role of Drosophila alpha-adaptin in presynaptic vesicle recycling. Cell 88:767–776
Gracheva EO, Burdina AO, Touroutine D, Berthelot-Grosjean M, Parekh H, Richmond JE (2007) Tomosyn negatively regulates CAPS-dependent peptide release at Caenorhabditis elegans synapses. J Neurosci 27:10176–10184
Grigliatti TA, Hall L, Rosenbluth R, Suzuki DT (1973) Temperature-sensitive mutations in Drosophila melanogaster. XIV. A selection of immobile adults. Mol Gen Genet 120:107–114
Grishanin RN, Kowalchyk JA, Klenchin VA, Ann K, Earles CA, Chapman ER, Gerona RR, Martin TF (2004) CAPS acts at a prefusion step in dense-core vesicle exocytosis as a PIP2 binding protein. Neuron 43:551–562
Gu J, Lee CW, Fan Y, Komlos D, Tang X, Sun C, Yu K, Hartzell HC, Chen G, Bamburg JR, Zheng JQ (2010) ADF/cofilin-mediated actin dynamics regulate AMPA receptor trafficking during synaptic plasticity. Nat Neurosci 13:1208–1215
Gu M, Schuske K, Watanabe S, Liu Q, Baum P, Garriga G, Jorgensen EM (2008) Mu2 adaptin facilitates but is not essential for synaptic vesicle recycling in Caenorhabditis elegans. J Cell Biol 183:881–892
Guan R, Dai H, Harrison SC, Kirchhausen T (2010) Structure of the PTEN-like region of auxilin, a detector of clathrin-coated vesicle budding. Structure 18:1191–1198
Guo S, Stolz LE, Lemrow SM, York JD (1999) SAC1-like domains of yeast SAC1, INP52, and INP53 and of human synaptojanin encode polyphosphoinositide phosphatases. J Biol Chem 274:12990–12995
Guo J, Wenk MR, Pellegrini L, Onofri F, Benfenati F, De Camilli P (2003) Phosphatidylinositol 4-kinase type IIalpha is responsible for the phosphatidylinositol 4-kinase activity associated with synaptic vesicles. Proc Natl Acad Sci U S A 100:3995–4000
Haass C (2004) Take five–BACE and the gamma-secretase quartet conduct Alzheimer's amyloid beta-peptide generation. EMBO J 23:483–488
Haffner C, Takei K, Chen H, Ringstad N, Hudson A, Butler MH, Salcini AE, Di Fiore PP, De Camilli P (1997) Synaptojanin 1: localization on coated endocytic intermediates in nerve terminals and interaction of its 170 kDa isoform with Eps15. FEBS Lett 419:175–180
Haffner C, Di Paolo G, Rosenthal JA, De Camilli P (2000) Direct interaction of the 170kDa isoform of synaptojanin 1 with clathrin and with the clathrin adaptor AP-2. Curr Biol 10:471–474
Harel A, Wu F, Mattson MP, Morris CM, Yao PJ (2008) Evidence for CALM in directing VAMP2 trafficking. Traffic 9:417–429
Harris TW, Hartwieg E, Horvitz HR, Jorgensen EM (2000) Mutations in synaptojanin disrupt synaptic vesicle recycling. J Cell Biol 150:589–600
Hashimoto K, Fukaya M, Qiao X, Sakimura K, Watanabe M, Kano M (1999) Impairment of AMPA receptor function in cerebellar granule cells of ataxic mutant mouse stargazer. J Neurosci 19:6027–6036
Hawkins PT, Anderson KE, Davidson K, Stephens LR (2006) Signalling through Class I PI3Ks in mammalian cells. Biochem Soc Trans 34:647–662
Hay JC, Martin TF (1993) Phosphatidylinositol transfer protein required for ATP-dependent priming of Ca(2+)-activated secretion. Nature 366:572–575
Hay JC, Fisette PL, Jenkins GH, Fukami K, Takenawa T, Anderson RA, Martin TF (1995) ATP-dependent inositide phosphorylation required for Ca(2+)-activated secretion. Nature 374:173–177

Hayashi M, Raimondi A, O'Toole E, Paradise S, Collesi C, Cremona O, Ferguson SM, De Camilli P (2008) Cell- and stimulus-dependent heterogeneity of synaptic vesicle endocytic recycling mechanisms revealed by studies of dynamin 1-null neurons. Proc Natl Acad Sci U S A 105:2175–2180

Henne WM, Boucrot E, Meinecke M, Evergren E, Vallis Y, Mittal R, McMahon HT (2010) FCHo proteins are nucleators of clathrin-mediated endocytosis. Science 328:1281–1284

Hilgemann DW (2007) Local PIP(2) signals: when, where, and how? Pflugers Arch 455:55–67

Hoeffer CA, Tang W, Wong H, Santillan A, Patterson RJ, Martinez LA, Tejada-Simon MV, Paylor R, Hamilton SL, Klann E (2008) Removal of FKBP12 enhances mTOR-Raptor interactions, LTP, memory, and perseverative/repetitive behavior. Neuron 60:832–845

Horne EA, Dell'Acqua ML (2007) Phospholipase C is required for changes in postsynaptic structure and function associated with NMDA receptor-dependent long-term depression. J Neurosci 27:3523–3534

Hou L, Klann E (2004) Activation of the phosphoinositide 3-kinase-Akt-mammalian target of rapamycin signaling pathway is required for metabotropic glutamate receptor-dependent long-term depression. J Neurosci 24:6352–6361

Hsieh H, Boehm J, Sato C, Iwatsubo T, Tomita T, Sisodia S, Malinow R (2006) AMPAR removal underlies Abeta-induced synaptic depression and dendritic spine loss. Neuron 52:831–843

Hu H, Qin Y, Bochorishvili G, Zhu Y, Van Aelst L, Zhu JJ (2008) Ras signaling mechanisms underlying impaired GluR1-dependent plasticity associated with fragile X syndrome. J Neurosci 28:7847–7862

Huang YY, Nguyen PV, Abel T, Kandel ER (1996) Long-lasting forms of synaptic potentiation in the mammalian hippocampus. Learn Mem 3:74–85

Hughes WE, Cooke FT, Parker PJ (2000) Sac phosphatase domain proteins. Biochem J 350(Pt 2): 337–352

Hui E, Johnson CP, Yao J, Dunning FM, Chapman ER (2009) Synaptotagmin-mediated bending of the target membrane is a critical step in Ca(2+)-regulated fusion. Cell 138:709–721

Irie F, Okuno M, Pasquale EB, Yamaguchi Y (2005) EphrinB-EphB signalling regulates clathrin-mediated endocytosis through tyrosine phosphorylation of synaptojanin 1. Nat Cell Biol 7:501–509

Ishihara H, Shibasaki Y, Kizuki N, Katagiri H, Yazaki Y, Asano T, Oka Y (1996) Cloning of cDNAs encoding two isoforms of 68-kDa type I phosphatidylinositol-4-phosphate 5-kinase. J Biol Chem 271:23611–23614

Ishihara H, Shibasaki Y, Kizuki N, Wada T, Yazaki Y, Asano T, Oka Y (1998) Type I phosphatidylinositol-4-phosphate 5-kinases. Cloning of the third isoform and deletion/substitution analysis of members of this novel lipid kinase family. J Biol Chem 273:8741–8748

Itoh T, Koshiba S, Kigawa T, Kikuchi A, Yokoyama S, Takenawa T (2001) Role of the ENTH domain in phosphatidylinositol-4,5-bisphosphate binding and endocytosis. Science 291:1047–1051

Itoh T, Erdmann KS, Roux A, Habermann B, Werner H, De Camilli P (2005) Dynamin and the actin cytoskeleton cooperatively regulate plasma membrane invagination by BAR and F-BAR proteins. Dev Cell 9:791–804

Jackson C, Welch HC, Bellamy TC (2010a) Control of cerebellar long-term potentiation by P-Rex-family guanine-nucleotide exchange factors and phosphoinositide 3-kinase. PLoS One 5:e11962

Jackson LP, Kelly BT, McCoy AJ, Gaffry T, James LC, Collins BM, Honing S, Evans PR, Owen DJ (2010b) A large-scale conformational change couples membrane recruitment to cargo binding in the AP2 clathrin adaptor complex. Cell 141:1220–1229

Jakobsson J, Gad H, Andersson F, Low P, Shupliakov O, Brodin L (2008) Role of epsin 1 in synaptic vesicle endocytosis. Proc Natl Acad Sci U S A 105:6445–6450

James DJ, Khodthong C, Kowalchyk JA, Martin TF (2008) Phosphatidylinositol 4,5-bisphosphate regulates SNARE-dependent membrane fusion. J Cell Biol 182:355–366

Jarquin-Pardo M, Fitzpatrick A, Galiano FJ, First EA, Davis JN (2007) Phosphatidic acid regulates the affinity of the murine phosphatidylinositol 4-phosphate 5-kinase-Ibeta for phosphatidylinositol-4-phosphate. J Cell Biochem 100:112–128

Jaworski J, Spangler S, Seeburg DP, Hoogenraad CC, Sheng M (2005) Control of dendritic arborization by the phosphoinositide-3′-kinase-Akt-mammalian target of rapamycin pathway. J Neurosci 25:11300–11312

Jin W, Ge WP, Xu J, Cao M, Peng L, Yung W, Liao D, Duan S, Zhang M, Xia J (2006) Lipid binding regulates synaptic targeting of PICK1, AMPA receptor trafficking, and synaptic plasticity. J Neurosci 26:2380–2390

Jockusch WJ, Speidel D, Sigler A, Sorensen JB, Varoqueaux F, Rhee JS, Brose N (2007) CAPS-1 and CAPS-2 are essential synaptic vesicle priming proteins. Cell 131:796–808

Jorgensen EM, Hartwieg E, Schuske K, Nonet ML, Jin Y, Horvitz HR (1995) Defective recycling of synaptic vesicles in synaptotagmin mutants of Caenorhabditis elegans. Nature 378:196–199

Jost M, Simpson F, Kavran JM, Lemmon MA, Schmid SL (1998) Phosphatidylinositol-4,5-bisphosphate is required for endocytic coated vesicle formation. Curr Biol 8:1399–1402

Jurado S, Benoist M, Lario A, Knafo S, Petrok CN, Esteban JA (2010) PTEN is recruited to the postsynaptic terminal for NMDA receptor-dependent long-term depression. EMBO J 29:2827–2840

Kahlfeldt N, Vahedi-Faridi A, Koo SJ, Schafer JG, Krainer G, Keller S, Saenger W, Krauss M, Haucke V (2010) Molecular basis for association of PIPKI gamma-p90 with clathrin adaptor AP-2. J Biol Chem 285:2734–2749

Kastning K, Kukhtina V, Kittler JT, Chen G, Pechstein A, Enders S, Lee SH, Sheng M, Yan Z, Haucke V (2007) Molecular determinants for the interaction between AMPA receptors and the clathrin adaptor complex AP-2. Proc Natl Acad Sci U S A 104:2991–2996

Kelly BT, McCoy AJ, Spate K, Miller SE, Evans PR, Honing S, Owen DJ (2008) A structural explanation for the binding of endocytic dileucine motifs by the AP2 complex. Nature 456:976–979

Khuong TM, Habets RL, Slabbaert JR, Verstreken P (2010) WASP is activated by phosphatidylinositol-4,5-bisphosphate to restrict synapse growth in a pathway parallel to bone morphogenetic protein signaling. Proc Natl Acad Sci U S A 107:17379–17384

Kim SH, Ryan TA (2010) CDK5 serves as a major control point in neurotransmitter release. Neuron 67:797–809

Kim WT, Chang S, Daniell L, Cremona O, Di Paolo G, De Camilli P (2002) Delayed reentry of recycling vesicles into the fusion-competent synaptic vesicle pool in synaptojanin 1 knockout mice. Proc Natl Acad Sci U S A 99:17143–17148

Kirchhausen T (2000) Clathrin. Annu Rev Biochem 69:699–727

Koenig JH, Ikeda K (1999) Contribution of active zone subpopulation of vesicles to evoked and spontaneous release. J Neurophysiol 81:1495–1505

Krauss M, Haucke V (2005) Functional assay of effectors of ADP ribosylation factor 6 during clathrin/AP-2 coat recruitment to membranes. GTPases Regul Membr Dyn 404:388–398

Krauss M, Kinuta M, Wenk MR, De Camilli P, Takei K, Haucke V (2003) ARF6 stimulates clathrin/AP-2 recruitment to synaptic membranes by activating phosphatidylinositol phosphate kinase type Igamma. J Cell Biol 162:113–124

Krauss M, Kukhtina V, Pechstein A, Haucke V (2006) Stimulation of phosphatidylinositol kinase type I-mediated phosphatidylinositol (4,5)-bisphosphate synthesis by AP-2mu-cargo complexes. Proc Natl Acad Sci U S A 103:11934–11939

Kunz J, Wilson MP, Kisseleva M, Hurley JH, Majerus PW, Anderson RA (2000) The activation loop of phosphatidylinositol phosphate kinases determines signaling specificity. Mol Cell 5:1–11

Kusano K, Abe H, Obinata T (1999) Detection of a sequence involved in actin-binding and phosphoinositide-binding in the N-terminal side of cofilin. Mol Cell Biochem 190:133–141

Kwon CH, Luikart BW, Powell CM, Zhou J, Matheny SA, Zhang W, LI Y, Baker SJ, Parada LF (2006) Pten regulates neuronal arborization and social interaction in mice. Neuron 50:377–388

Lam AD, Tryoen-Toth P, Tsai B, Vitale N, Stuenkel EL (2008) SNARE-catalyzed fusion events are regulated by Syntaxin1A-lipid interactions. Mol Biol Cell 19:485–497

Landman N, Jeong SY, Shin SY, Voronov SV, Serban G, Kang MS, Park MK, Di Paolo G, Chung S, Kim TW (2006) Presenilin mutations linked to familial Alzheimer's disease cause an imbalance in phosphatidylinositol 4,5-bisphosphate metabolism. Proc Natl Acad Sci U S A 103:19524–19529

Lee A, Frank DW, Marks MS, Lemmon MA (1999) Dominant-negative inhibition of receptor-mediated endocytosis by a dynamin-1 mutant with a defective pleckstrin homology domain. Curr Biol 9:261–264

Lee SH, Simonetta A, Sheng M (2004a) Subunit rules governing the sorting of internalized AMPA receptors in hippocampal neurons. Neuron 43:221–236

Lee SY, Wenk MR, Kim Y, Nairn AC, De Camilli P (2004b) Regulation of synaptojanin 1 by cyclin-dependent kinase 5at synapses. Proc Natl Acad Sci U S A 101:546–551

Lee SY, Voronov S, Letinic K, Nairn AC, Di Paolo G, De Camilli P (2005) Regulation of the interaction between PIPKI gamma and talin by proline-directed protein kinases. J Cell Biol 168:789–799

Liao H, Ellena J, Liu L, Szabo G, Cafiso D, Castle D (2007) Secretory carrier membrane protein SCAMP2 and phosphatidylinositol 4,5-bisphosphate interactions in the regulation of dense core vesicle exocytosis. Biochemistry 46:10909–10920

Lin DT, Huganir RL (2007) PICK1 and phosphorylation of the glutamate receptor 2 (GluR2) AMPA receptor subunit regulates GluR2 recycling after NMDA receptor-induced internalization. J Neurosci 27:13903–13908

Lin CH, Yeh SH, Lu KT, Leu TH, Chang WC, Gean PW (2001) A role for the PI-3 kinase signaling pathway in fear conditioning and synaptic plasticity in the amygdala. Neuron 31:841–851

Ling K, Doughman RL, Firestone AJ, Bunce MW, Anderson RA (2002) Type I gamma phosphatidylinositol phosphate kinase targets and regulates focal adhesions. Nature 420:89–93

Liu J, Kaksonen M, Drubin DG, Oster G (2006) Endocytic vesicle scission by lipid phase boundary forces. Proc Natl Acad Sci U S A 103:10277–10282

Liu J, Sun Y, Oster GF, Drubin DG (2010) Mechanochemical crosstalk during endocytic vesicle formation. Curr Opin Cell Biol 22:36–43

Logothetis DE, Petrou VI, Adney SK, Mahajan R (2010) Channelopathies linked to plasma membrane phosphoinositides. Pflugers Arch 460:321–341

Lott IT, Head E (2001) Down syndrome and Alzheimer's disease: a link between development and aging. Ment Retard Dev Disabil Res Rev 7:172–178

Lou X, Korogod N, Brose N, Schneggenburger R (2008) Phorbol esters modulate spontaneous and Ca^{2+}-evoked transmitter release via acting on both Munc13 and protein kinase C. J Neurosci 28:8257–8267

Loyet KM, Kowalchyk JA, Chaudhary A, Chen J, Prestwich GD, Martin TF (1998) Specific binding of phosphatidylinositol 4,5-bisphosphate to calcium-dependent activator protein for secretion (CAPS), a potential phosphoinositide effector protein for regulated exocytosis. J Biol Chem 273:8337–8343

Lu J, Helton TD, Blanpied TA, Racz B, Newpher TM, Weinberg RJ, Ehlers MD (2007) Postsynaptic positioning of endocytic zones and AMPA receptor cycling by physical coupling of dynamin-3 to Homer. Neuron 55:874–889

Lu W, Shi Y, Jackson AC, Bjorgan K, During MJ, Sprengel R, Seeburg PH, Nicoll RA (2009) Subunit composition of synaptic AMPA receptors revealed by a single-cell genetic approach. Neuron 62:254–268

Lue LF, Kuo YM, Roher AE, Brachova L, Shen Y, Sue L, Beach T, Kurth JH, Rydel RE, Rogers J (1999) Soluble amyloid beta peptide concentration as a predictor of synaptic change in Alzheimer's disease. Am J Pathol 155:853–862

Luthi A, Di Paolo G, Cremona O, Daniell L, De Camilli P, McCormick DA (2001) Synaptojanin 1 contributes to maintaining the stability of GABAergic transmission in primary cultures of cortical neurons. J Neurosci 21:9101–9111

Maehama T, Dixon JE (1999) PTEN: a tumour suppressor that functions as a phospholipid phosphatase. Trends Cell Biol 9:125–128

Man HY, Wang Q, Lu WY, Ju W, Ahmadian G, Liu L, D'Souza S, Wong TP, Taghibiglou C, Lu J, Becker LE, Pei L, Liu F, Wymann MP, Macdonald JF, Wang YT (2003) Activation of PI3-kinase is required for AMPA receptor insertion during LTP of mEPSCs in cultured hippocampal neurons. Neuron 38:611–624

Mani M, Lee SY, Lucast L, Cremona O, Di Paolo G, De Camilli P, Ryan TA (2007) The dual phosphatase activity of synaptojanin1 is required for both efficient synaptic vesicle endocytosis and reavailability at nerve terminals. Neuron 56:1004–1018

Mao L, Takamiya K, Thomas G, Lin DT, Huganir RL (2010) GRIP1 and 2 regulate activity-dependent AMPA receptor recycling via exocyst complex interactions. Proc Natl Acad Sci U S A 107:19038–19043

Marks B, McMahon HT (1998) Calcium triggers calcineurin-dependent synaptic vesicle recycling in mammalian nerve terminals. Curr Biol 8:740–749

Marone R, Cmiljanovic V, Giese B, Wymann MP (2008) Targeting phosphoinositide 3-kinase: moving towards therapy. Biochim Biophys Acta 1784:159–185

Martens S, Kozlov MM, McMahon HT (2007) How synaptotagmin promotes membrane fusion. Science 316:1205–1208

Martin TF (1997) Phosphoinositides as spatial regulators of membrane traffic. Curr Opin Neurobiol 7:331–338

Martin TF (2003) Tuning exocytosis for speed: fast and slow modes. Biochim Biophys Acta 1641:157–165

Massol RH, Boll W, Griffin AM, Kirchhausen T (2006) A burst of auxilin recruitment determines the onset of clathrin-coated vesicle uncoating. Proc Natl Acad Sci U S A 103:10265–10270

Matthew WD, Tsavaler L, Reichardt LF (1981) Identification of a synaptic vesicle-specific membrane protein with a wide distribution in neuronal and neurosecretory tissue. J Cell Biol 91:257–269

Mattila PK, Pykalainen A, Saarikangas J, Paavilainen VO, Vihinen H, Jokitalo E, Lappalainen P (2007) Missing-in-metastasis and IRSp53 deform PI(4,5)P_2-rich membranes by an inverse BAR domain-like mechanism. J Cell Biol 176:953–964

McLaughlin S, Wang J, Gambhir A, Murray D (2002) PIP(2) and proteins: interactions, organization, and information flow. Annu Rev Biophys Biomol Struct 31:151–175

McPherson PS, Takei K, Schmid SL, De Camilli P (1994) p145, a major Grb2-binding protein in brain, is co-localized with dynamin in nerve terminals where it undergoes activity-dependent dephosphorylation. J Biol Chem 269:30132–30139

McPherson PS, Garcia EP, Slepnev VI, David C, Zhang X, Grabs D, Sossin WS, Bauerfeind R, Nemoto Y, De Camilli P (1996) A presynaptic inositol-5-phosphatase. Nature 379:353–357

Micheva KD, Kay BK, McPherson PS (1997) Synaptojanin forms two separate complexes in the nerve terminal. Interactions with endophilin and amphiphysin. J Biol Chem 272:27239–27245

Miesenbock G, De Angelis DA, Rothman JE (1998) Visualizing secretion and synaptic transmission with pH-sensitive green fluorescent proteins. Nature 394:192–195

Milosevic I, Sorensen JB, Lang T, Krauss M, Nagy G, Haucke V, Jahn R, Neher E (2005) Plasmalemmal phosphatidylinositol-4,5-bisphosphate level regulates the releasable vesicle pool size in chromaffin cells. J Neurosci 25:2557–2565

Milstein AD, Nicoll RA (2008) Regulation of AMPA receptor gating and pharmacology by TARP auxiliary subunits. Trends Pharmacol Sci 29:333–339

Morgan JR, Di Paolo G, Werner H, Shchedrina VA, Pypaert M, Pieribone VA, De Camilli P (2004) A role for talin in presynaptic function. J Cell Biol 167:43–50

Morishita W, Marie H, Malenka RC (2005) Distinct triggering and expression mechanisms underlie LTD of AMPA and NMDA synaptic responses. Nat Neurosci 8:1043–1050

Moritz A, De Graan PN, Gispen WH, Wirtz KW (1992) Phosphatidic acid is a specific activator of phosphatidylinositol-4-phosphate kinase. J Biol Chem 267:7207–7210

Mundell SJ, Matharu AL, Pula G, Roberts PJ, Kelly E (2001) Agonist-induced internalization of the metabotropic glutamate receptor 1a is arrestin- and dynamin-dependent. J Neurochem 78:546–551

Nakano-Kobayashi A, Yamazaki M, Unoki T, Hongu T, Murata C, Taguchi R, Katada T, Frohman MA, Yokozeki T, Kanaho Y (2007) Role of activation of PIP5Kgamma661 by AP-2 complex in synaptic vesicle endocytosis. EMBO J 26:1105–1116

Nelson CD, Perry SJ, Regier DS, Prescott SM, Topham MK, Lefkowitz RJ (2007) Targeting of diacylglycerol degradation to M1 muscarinic receptors by beta-arrestins. Science 315:663–666

Nelson CD, Kovacs JJ, Nobles KN, Whalen EJ, Lefkowitz RJ (2008) Beta-arrestin scaffolding of phosphatidylinositol 4-phosphate 5-kinase Ialpha promotes agonist-stimulated sequestration of the beta2-adrenergic receptor. J Biol Chem 283:21093–21101

Newpher TM, Ehlers MD (2008) Glutamate receptor dynamics in dendritic microdomains. Neuron 58:472–497

Newton PM, Messing RO (2006) Intracellular signaling pathways that regulate behavioral responses to ethanol. Pharmacol Ther 109:227–237

Ning K, Pei L, Liao M, Liu B, Zhang Y, Jiang W, Mielke JG, Li L, Chen Y, El-Hayek YH, Fehlings MG, Zhang X, Liu F, Eubanks J, Wan Q (2004) Dual neuroprotective signaling mediated by downregulating two distinct phosphatase activities of PTEN. J Neurosci 24:4052–4060

Niwa R, Nagata-Ohashi K, Takeichi M, Mizuno K, Uemura T (2002) Control of actin reorganization by Slingshot, a family of phosphatases that dephosphorylate ADF/cofilin. Cell 108:233–246

Nonet ML, Holgado AM, Brewer F, Serpe CJ, Norbeck BA, Holleran J, Wei L, Hartwieg E, Jorgensen EM, Alfonso A (1999) UNC-11, a caenorhabditis elegans AP180 homologue, regulates the size and protein composition of synaptic vesicles. Mol Biol Cell 10:2343–2360

Obar RA, Collins CA, Hammarback JA, Shpetner HS, Vallee RB (1990) Molecular cloning of the microtubule-associated mechanochemical enzyme dynamin reveals homology with a new family of GTP-binding proteins. Nature 347:256–261

Ono S (2007) Mechanism of depolymerization and severing of actin filaments and its significance in cytoskeletal dynamics. Int Rev Cytol 258:1–82

Opazo P, Choquet D (2010) A three-step model for the synaptic recruitment of AMPA receptors. Mol Cell Neurosci 46:1–8

Osborne SL, Wen PJ, Meunier FA (2006) Phosphoinositide regulation of neuroexocytosis: adding to the complexity. J Neurochem 98:336–342

Osten P, Khatri L, Perez JL, Kohr G, Giese G, Daly C, Schulz TW, Wensky A, Lee LM, Ziff EB (2000) Mutagenesis reveals a role for ABP/GRIP binding to GluR2 in synaptic surface accumulation of the AMPA receptor. Neuron 27:313–325

Page DT, Kuti OJ, Prestia C, Sur M (2009) Haploinsufficiency for Pten and Serotonin transporter cooperatively influences brain size and social behavior. Proc Natl Acad Sci U S A 106:1989–1994

Palop JJ, Mucke L (2010) Synaptic depression and aberrant excitatory network activity in Alzheimer's disease: two faces of the same coin? Neuromolecular Med 12:48–55

Palop JJ, Chin J, Roberson ED, Wang J, Thwin MT, Bien-Ly N, Yoo J, Ho KO, Yu GQ, Kreitzer A, Finkbeiner S, Noebels JL, Mucke L (2007) Aberrant excitatory neuronal activity and compensatory remodeling of inhibitory hippocampal circuits in mouse models of Alzheimer's disease. Neuron 55:697–711

Pan L, Wu H, Shen C, Shi Y, Jin W, Xia J, Zhang M (2007) Clustering and synaptic targeting of PICK1 requires direct interaction between the PDZ domain and lipid membranes. EMBO J 26:4576–4587

Pang ZP, Sudhof TC (2010) Cell biology of Ca^{2+} -triggered exocytosis. Curr Opin Cell Biol 22:496–505

Parrish WR, Stefan CJ, Emr SD (2004) Essential role for the myotubularin-related phosphatase Ymr1p and the synaptojanin-like phosphatases Sjl2p and Sjl3p in regulation of phosphatidylinositol 3-phosphate in yeast. Mol Biol Cell 15:3567–3579

Pearce LR, Komander D, Alessi DR (2010) The nuts and bolts of AGC protein kinases. Nat Rev Mol Cell Biol 11:9–22

Pechstein A, Shupliakov O, Haucke V (2010) Intersectin 1: a versatile actor in the synaptic vesicle cycle. Biochem Soc Trans 38:181–186

Peineau S, Taghibiglou C, Bradley C, Wong TP, Liu L, Lu J, Lo E, Wu D, Saule E, Bouschet T, Matthews P, Isaac JT, Bortolotto ZA, Wang YT, Collingridge GL (2007) LTP inhibits LTD in the hippocampus via regulation of GSK3beta. Neuron 53:703–717

Pendaries C, Tronchere H, Plantavid M, Payrastre B (2003) Phosphoinositide signaling disorders in human diseases. FEBS Lett 546:25–31

Perandones C, Costanzo RV, Kowaljow V, Pivetta OH, Carminatti H, Radrizzani M (2004) Correlation between synaptogenesis and the PTEN phosphatase expression in dendrites during postnatal brain development. Brain Res Mol Brain Res 128:8–19

Perera RM, Zoncu R, Lucast L, De Camilli P, Toomre D (2006) Two synaptojanin 1 isoforms are recruited to clathrin-coated pits at different stages. Proc Natl Acad Sci U S A 103:19332–19337

Perin MS, Brose N, Jahn R, Sudhof TC (1991) Domain structure of synaptotagmin (p65). J Biol Chem 266:623–629

Peter BJ, Kent HM, Mills IG, Vallis Y, Butler PJ, Evans PR, McMahon HT (2004) BAR domains as sensors of membrane curvature: the amphiphysin BAR structure. Science 303:495–499

Pucadyil TJ, Schmid SL (2008) Real-time visualization of dynamin-catalyzed membrane fission and vesicle release. Cell 135:1263–1275

Qin Y, Zhu Y, Baumgart JP, Stornetta RL, Seidenman K, Mack V, van Aelst L, Zhu JJ (2005) State-dependent Ras signaling and AMPA receptor trafficking. Genes Dev 19(17):2000–2015

Ramjaun AR, McPherson PS (1996) Tissue-specific alternative splicing generates two synaptojanin isoforms with differential membrane binding properties. J Biol Chem 271:24856–24861

Raymond CR, Redman SJ, Crouch MF (2002) The phosphoinositide 3-kinase and p70 S6 kinase regulate long-term potentiation in hippocampal neurons. Neuroscience 109:531–536

Rebecchi MJ, Pentyala SN (2000) Structure, function, and control of phosphoinositide-specific phospholipase C. Physiol Rev 80:1291–1335

Reeves RH, Irving NG, Moran TH, Wohn A, Kitt C, Sisodia SS, Schmidt C, Bronson RT, Davisson MT (1995) A mouse model for Down syndrome exhibits learning and behaviour deficits. Nat Genet 11:177–184

Renden R, Berwin B, Davis W, Ann K, Chin CT, Kreber R, Ganetzky B, Martin TF, Broadie K (2001) Drosophila CAPS is an essential gene that regulates dense-core vesicle release and synaptic vesicle fusion. Neuron 31:421–437

Reyes-Harde M, Stanton PK (1998) Postsynaptic phospholipase C activity is required for the induction of homosynaptic long-term depression in rat hippocampus. Neurosci Lett 252:155–158

Rhee JS, Betz A, Pyott S, Reim K, Varoqueaux F, Augustin I, Hesse D, Sudhof TC, Takahashi M, Rosenmund C, Brose N (2002) Beta phorbol ester- and diacylglycerol-induced augmentation of transmitter release is mediated by Munc13s and not by PKCs. Cell 108:121–133

Richards DA, Rizzoli SO, Betz WJ (2004) Effects of wortmannin and latrunculin A on slow endocytosis at the frog neuromuscular junction. J Physiol 557:77–91

Ringstad N, Nemoto Y, De Camilli P (1997) The SH3p4/Sh3p8/SH3p13 protein family: binding partners for synaptojanin and dynamin via a Grb2-like Src homology 3 domain. Proc Natl Acad Sci U S A 94:8569–8574

Robinson MS (2004) Adaptable adaptors for coated vesicles. Trends Cell Biol 14:167–174

Roizen NJ, Patterson D (2003) Down's syndrome. Lancet 361:1281–1289

Rong R, Ahn JY, Huang H, Nagata E, Kalman D, Kapp JA, Tu J, Worley PF, Snyder SH, Ye K (2003) PI3 kinase enhancer-Homer complex couples mGluRI to PI3 kinase, preventing neuronal apoptosis. Nat Neurosci 6:1153–1161

Rouach N, Byrd K, Petralia RS, Elias GM, Adesnik H, Tomita S, Karimzadegan S, Kealey C, Bredt DS, Nicoll RA (2005) TARP gamma-8 controls hippocampal AMPA receptor number, distribution and synaptic plasticity. Nat Neurosci 8:1525–1533

Roux A, Uyhazi K, Frost A, De Camilli P (2006) GTP-dependent twisting of dynamin implicates constriction and tension in membrane fission. Nature 441:528–531

Rust MB, Gurniak CB, Renner M, Vara H, Morando L, Gorlich A, Sassoe-Pognetto M, Banchaabouchi MA, Giustetto M, Triller A, Choquet D, Witke W (2010) Learning, AMPA receptor mobility and synaptic plasticity depend on n-cofilin-mediated actin dynamics. EMBO J 29:1889–1902

Saarikangas J, Zhao H, Lappalainen P (2008) Regulation of the actin cytoskeleton-plasma membrane interplay by phosphoinositides. Physiol Rev 90:259–289

Sadakata T, Kakegawa W, Mizoguchi A, Washida M, Katoh-Semba R, Shutoh F, Okamoto T, Nakashima H, Kimura K, Tanaka M, Sekine Y, Itohara S, Yuzaki M, Nagao S, Furuichi T (2007) Impaired cerebellar development and function in mice lacking CAPS2, a protein involved in neurotrophin release. J Neurosci 27:2472–2482

Sankaranarayanan S, Ryan TA (2000) Real-time measurements of vesicle-SNARE recycling in synapses of the central nervous system. Nat Cell Biol 2:197–204

Sankaranarayanan S, Atluri PP, Ryan TA (2003) Actin has a molecular scaffolding, not propulsive, role in presynaptic function. Nat Neurosci 6:127–135

Sanna PP, Cammalleri M, Berton F, Simpson C, Lutjens R, Bloom FE, Francesconi W (2002) Phosphatidylinositol 3-kinase is required for the expression but not for the induction or the maintenance of long-term potentiation in the hippocampal CA1 region. J Neurosci 22:3359–3365

Schafer DA (2002) Coupling actin dynamics and membrane dynamics during endocytosis. Curr Opin Cell Biol 14:76–81

Schiavo G, Gu QM, Prestwich GD, Sollner TH, Rothman JE (1996) Calcium-dependent switching of the specificity of phosphoinositide binding to synaptotagmin. Proc Natl Acad Sci U S A 93:13327–13332

Schneggenburger R, Neher E (2000) Intracellular calcium dependence of transmitter release rates at a fast central synapse. Nature 406:889–893

Schuske KR, Richmond JE, Matthies DS, Davis WS, Runz S, Rube DA, Van Der Bliek AM, Jorgensen EM (2003) Endophilin is required for synaptic vesicle endocytosis by localizing synaptojanin. Neuron 40:749–762

Seidenman KJ, Steinberg JP, Huganir R, Malinow R (2003) Glutamate receptor subunit 2 Serine 880 phosphorylation modulates synaptic transmission and mediates plasticity in CA1 pyramidal cells. J Neurosci 23:9220–9228

Seol GH, Ziburkus J, Huang S, Song L, Kim IT, Takamiya K, Huganir RL, Lee HK, Kirkwood A (2007) Neuromodulators control the polarity of spike-timing-dependent synaptic plasticity. Neuron 55:919–929

Seregaza Z, Roubertoux PL, Jamon M, Soumireu-Mourat B (2006) Mouse models of cognitive disorders in trisomy 21: a review. Behav Genet 36:387–404

Shih SC, Katzmann DJ, Schnell JD, Sutanto M, Emr SD, Hicke L (2002) Epsins and Vps27p/Hrs contain ubiquitin-binding domains that function in receptor endocytosis. Nat Cell Biol 4:389–393

Shimada A, Niwa H, Tsujita K, Suetsugu S, Nitta K, Hanawa-Suetsugu K, Akasaka R, Nishino Y, Toyama M, Chen L, Liu ZJ, Wang BC, Yamamoto M, Terada T, Miyazawa A, Tanaka A, Sugano S, Shirouzu M, Nagayama K, Takenawa T, Yokoyama S (2007) Curved EFC/F-BAR-domain dimers are joined end to end into a filament for membrane invagination in endocytosis. Cell 129:761–772

Shin OH, Lu J, Rhee JS, Tomchick DR, Pang ZP, Wojcik SM, Camacho-Perez M, Brose N, Machius M, Rizo J, Rosenmund C, Sudhof TC (2010) Munc13 C2B domain is an activity-dependent Ca^{2+} regulator of synaptic exocytosis. Nat Struct Mol Biol 17:280–288

Shpetner HS, Vallee RB (1989) Identification of dynamin, a novel mechanochemical enzyme that mediates interactions between microtubules. Cell 59:421–432

Shupliakov O, Bloom O, Gustafsson JS, Kjaerulff O, Low P, Tomilin N, Pieribone VA, Greengard P, Brodin L (2002) Impaired recycling of synaptic vesicles after acute perturbation of the presynaptic actin cytoskeleton. Proc Natl Acad Sci U S A 99:14476–14481

Siksou L, Triller A, Marty S (2009) An emerging view of presynaptic structure from electron microscopic studies. J Neurochem 108:1336–1342

Simpson F, Hussain NK, Qualmann B, Kelly RB, Kay BK, McPherson PS, Schmid SL (1999) SH3-domain-containing proteins function at distinct steps in clathrin-coated vesicle formation. Nat Cell Biol 1:119–124

Singer-Kruger B, Nemoto Y, Daniell L, Ferro-Novick S, De Camilli P (1998) Synaptojanin family members are implicated in endocytic membrane traffic in yeast. J Cell Sci 111(Pt 22):3347–3356
Slepnev VI, De Camilli P (2000) Accessory factors in clathrin-dependent synaptic vesicle endocytosis. Nat Rev Neurosci 1:161–172
Slepnev VI, Ochoa GC, Butler MH, Grabs D, De Camilli P (1998) Role of phosphorylation in regulation of the assembly of endocytic coat complexes. Science 281:821–824
Speese S, Petrie M, Schuske K, Ailion M, Ann K, Iwasaki K, Jorgensen EM, Martin TF (2007) UNC-31 (CAPS) is required for dense-core vesicle but not synaptic vesicle exocytosis in Caenorhabditis elegans. J Neurosci 27:6150–6162
Speidel D, Bruederle CE, Enk C, Voets T, Varoqueaux F, Reim K, Becherer U, Fornai F, Ruggieri S, Holighaus Y, Weihe E, Bruns D, Brose N, Rettig J (2005) CAPS1 regulates catecholamine loading of large dense-core vesicles. Neuron 46:75–88
Stefan CJ, Audhya A, Emr SD (2002) The yeast synaptojanin-like proteins control the cellular distribution of phosphatidylinositol (4,5)-bisphosphate. Mol Biol Cell 13:542–557
Stein A, Radhakrishnan A, Riedel D, Fasshauer D, Jahn R (2007) Synaptotagmin activates membrane fusion through a Ca^{2+}-dependent trans interaction with phospholipids. Nat Struct Mol Biol 14:904–911
Steinberg JP, Takamiya K, Shen Y, Xia J, Rubio ME, Yu S, Jin W, Thomas GM, Linden DJ, Huganir RL (2006) Targeted in vivo mutations of the AMPA receptor subunit GluR2 and its interacting protein PICK1 eliminate cerebellar long-term depression. Neuron 49:845–860
Sudhof TC, Malenka RC (2008) Understanding synapses: past, present, and future. Neuron 60:469–476
Suh BC, Hille B (2008) PIP2 is a necessary cofactor for ion channel function: how and why? Annu Rev Biophys 37:175–195
Sulis ML, Parsons R (2003) PTEN: from pathology to biology. Trends Cell Biol 13:478–483
Sumioka A, Yan D, Tomita S (2010) TARP phosphorylation regulates synaptic AMPA receptors through lipid bilayers. Neuron 66:755–767
Sundborger A, Soderblom C, Vorontsova O, Evergren E, Hinshaw JE, Shupliakov O (2010) An endophilin-dynamin complex promotes budding of clathrin-coated vesicles during synaptic vesicle recycling. J Cell Sci 124:133–143
Suzuki DT, Grigliatti T, Williamson R (1971) Temperature-sensitive mutations in Drosophila melanogaster. Proc Natl Acad Sci U S A 68:890–893
Takei K, McPherson PS, Schmid SL, De Camilli P (1995) Tubular membrane invaginations coated by dynamin rings are induced by GTP-gamma S in nerve terminals. Nature 374:186–190
Takei K, Slepnev VI, Haucke V, De Camilli P (1999) Functional partnership between amphiphysin and dynamin in clathrin-mediated endocytosis. Nat Cell Biol 1:33–39
Tandon A, Bannykh S, Kowalchyk JA, Banerjee A, Martin TF, Balch WE (1998) Differential regulation of exocytosis by calcium and CAPS in semi-intact synaptosomes. Neuron 21:147–154
Tang SJ, Reis G, Kang H, Gingras AC, Sonenberg N, Schuman EM (2002) A rapamycin-sensitive signaling pathway contributes to long-term synaptic plasticity in the hippocampus. Proc Natl Acad Sci U S A 99:467–472
Terashima A, Pelkey KA, Rah JC, Suh YH, Roche KW, Collingridge GL, McBain CJ, Isaac JT (2008) An essential role for PICK1 in NMDA receptor-dependent bidirectional synaptic plasticity. Neuron 57:872–882
Thieman JR, Mishra SK, Ling K, Doray B, Anderson RA, Traub LM (2009) Clathrin regulates the association of PIPKIgamma661 with the AP-2 adaptor beta2 appendage. J Biol Chem 284:13924–13939
Tsokas P, Ma T, Iyengar R, Landau EM, Blitzer RD (2007) Mitogen-activated protein kinase upregulates the dendritic translation machinery in long-term potentiation by controlling the mammalian target of rapamycin pathway. J Neurosci 27:5885–5894
Tsujita K, Suetsugu S, Sasaki N, Furutani M, Oikawa T, Takenawa T (2006) Coordination between the actin cytoskeleton and membrane deformation by a novel membrane tubulation domain of PCH proteins is involved in endocytosis. J Cell Biol 172:269–279

Tu JC, Xiao B, Naisbitt S, Yuan JP, Petralia RS, Brakeman P, Doan A, Aakalu VK, Lanahan AA, Sheng M, Worley PF (1999) Coupling of mGluR/Homer and PSD-95 complexes by the Shank family of postsynaptic density proteins. Neuron 23:583–592

Van Der Bliek AM, Meyerowitz EM (1991) Dynamin-like protein encoded by the Drosophila shibire gene associated with vesicular traffic. Nature 351:411–414

Van Epps HA, Hayashi M, Lucast L, Stearns GW, Hurley JB, De Camilli P, Brockerhoff SE (2004) The zebrafish nrc mutant reveals a role for the polyphosphoinositide phosphatase synaptojanin 1 in cone photoreceptor ribbon anchoring. J Neurosci 24:8641–8650

Varoqueaux F, Sigler A, Rhee JS, Brose N, Enk C, Reim K, Rosenmund C (2002) Total arrest of spontaneous and evoked synaptic transmission but normal synaptogenesis in the absence of Munc13-mediated vesicle priming. Proc Natl Acad Sci U S A 99:9037–9042

Venkitaramani DV, Chin J, Netzer WJ, Gouras GK, Lesne S, Malinow R, Lombroso PJ (2007) Beta-amyloid modulation of synaptic transmission and plasticity. J Neurosci 27:11832–11837

Verstreken P, Koh TW, Schulze KL, Zhai RG, Hiesinger PR, Zhou Y, Mehta SQ, Cao Y, Roos J, Bellen HJ (2003) Synaptojanin is recruited by endophilin to promote synaptic vesicle uncoating. Neuron 40:733–748

Verstreken P, Ohyama T, Haueter C, Habets RL, Lin YQ, Swan LE, Ly CV, Venken KJ, De Camilli P, Bellen HJ (2009) Tweek, an evolutionarily conserved protein, is required for synaptic vesicle recycling. Neuron 63:203–215

Volpicelli-Daley LA, Lucast L, Gong LW, Liu L, Sasaki J, Sasaki T, Abrams CS, Kanaho Y, De Camilli P (2010) Phosphatidylinositol-4-phosphate 5-kinases and phosphatidylinositol 4,5-bisphosphate synthesis in the brain. J Biol Chem 285:28708–28714

Voronov SV, Frere SG, Giovedi S, Pollina EA, Borel C, Zhang H, Schmidt C, Akeson EC, Wenk MR, Cimasoni L, Arancio O, Davisson MT, Antonarakis SE, Gardiner K, De Camilli P, Di Paolo G (2008) Synaptojanin 1-linked phosphoinositide dyshomeostasis and cognitive deficits in mouse models of down's syndrome. Proc Natl Acad Sci U S A 105:9415–9420

Walent JH, Porter BW, Martin TF (1992) A novel 145 kd brain cytosolic protein reconstitutes Ca(2+)-regulated secretion in permeable neuroendocrine cells. Cell 70:765–775

Walsh DM, Klyubin I, Fadeeva JV, Cullen WK, Anwyl R, Wolfe MS, Rowan MJ, Selkoe DJ (2002) Naturally secreted oligomers of amyloid beta protein potently inhibit hippocampal long-term potentiation in vivo. Nature 416:535–539

Weernink PA, Meletiadis K, Hommeltenberg S, Hinz M, Ishihara H, Schmidt M, Jakobs KH (2004) Activation of type I phosphatidylinositol 4-phosphate 5-kinase isoforms by the Rho GTPases, RhoA, Rac1, and Cdc42. J Biol Chem 279:7840–7849

Wegner AM, Nebhan CA, Hu L, Majumdar D, Meier KM, Weaver AM, Webb DJ (2008) N-wasp and the arp2/3 complex are critical regulators of actin in the development of dendritic spines and synapses. J Biol Chem 283:15912–15920

Weissenhorn W (2005) Crystal structure of the endophilin-A1 BAR domain. J Mol Biol 351:653–661

Wenk MR, Pellegrini L, Klenchin VA, Di Paolo G, Chang S, Daniell L, Arioka M, Martin TF, De Camilli P (2001) PIP kinase Igamma is the major PI(4,5)P(2) synthesizing enzyme at the synapse. Neuron 32:79–88

Wenthold RJ, Yokotani N, Doi K, Wada K (1992) Immunochemical characterization of the non-NMDA glutamate receptor using subunit-specific antibodies. Evidence for a hetero-oligomeric structure in rat brain. J Biol Chem 267:501–507

Wiedemann C, Schafer T, Burger MM (1996) Chromaffin granule-associated phosphatidylinositol 4-kinase activity is required for stimulated secretion. EMBO J 15:2094–2101

Wiedemann C, Schafer T, Burger MM, Sihra TS (1998) An essential role for a small synaptic vesicle-associated phosphatidylinositol 4-kinase in neurotransmitter release. J Neurosci 18:5594–5602

Yamabhai M, Hoffman NG, Hardison NL, McPherson PS, Castagnoli L, Cesareni G, Kay BK (1998) Intersectin, a novel adaptor protein with two Eps15 homology and five Src homology 3 domains. J Biol Chem 273:31401–31407

Yamada H, Padilla-Parra S, Park SJ, Itoh T, Chaineau M, Monaldi I, Cremona O, Benfenati F, De Camilli P, Coppey-Moisan M, Tramier M, Galli T, Takei K (2009) Dynamic interaction of amphiphysin with N-WASP regulates actin assembly. J Biol Chem 284:34244–34256
Yamashita T, Hige T, Takahashi T (2005) Vesicle endocytosis requires dynamin-dependent GTP hydrolysis at a fast CNS synapse. Science 307:124–127
Yang Y, Wang XB, Zhou Q (2010) Perisynaptic GluR2-lacking AMPA receptors control the reversibility of synaptic and spines modifications. Proc Natl Acad Sci U S A 107:11999–12004
Yarar D, Surka MC, Leonard MC, Schmid SL (2008) SNX9 activities are regulated by multiple phosphoinositides through both PX and BAR domains. Traffic 9:133–146
Yim YI, Sun T, Wu LG, Raimondi A, De Camilli P, Eisenberg E, Greene LE (2010) Endocytosis and clathrin-uncoating defects at synapses of auxilin knockout mice. Proc Natl Acad Sci U S A 107:4412–4417
Yin HL, Janmey PA (2003) Phosphoinositide regulation of the actin cytoskeleton. Annu Rev Physiol 65:761–789
Yonezawa N, Homma Y, Yahara I, Sakai H, Nishida E (1991) A short sequence responsible for both phosphoinositide binding and actin binding activities of cofilin. J Biol Chem 266:17218–17221
Yuen EY, Liu W, Kafri T, Van Praag H, Yan Z (2010) Regulation of AMPA receptor channels and synaptic plasticity by cofilin phosphatase Slingshot in cortical neurons. J Physiol 588:2361–2371
Zhang B, Koh YH, Beckstead RB, Budnik V, Ganetzky B, Bellen HJ (1998) Synaptic vesicle size and number are regulated by a clathrin adaptor protein required for endocytosis. Neuron 21:1465–1475
Zheng J, Cahill SM, Lemmon MA, Fushman D, Schlessinger J, Cowburn D (1996) Identification of the binding site for acidic phospholipids on the pH domain of dynamin: implications for stimulation of GTPase activity. J Mol Biol 255:14–21
Zhou Q, Homma KJ, Poo MM (2004) Shrinkage of dendritic spines associated with long-term depression of hippocampal synapses. Neuron 44:749–757

Chapter 6
Phosphatidylinositol 4, 5 Bisphosphate and the Actin Cytoskeleton

Li Zhang, Yuntao S. Mao, Paul A. Janmey and Helen L. Yin

Abstract Dynamic changes in PM PIP_2 have been implicated in the regulation of many processes that are dependent on actin polymerization and remodeling. PIP_2 is synthesized primarily by the type I phosphatidylinositol 4 phosphate 5 kinases (PIP5Ks), and there are three major isoforms, called α, β and γ. There is emerging evidence that these PIP5Ks have unique as well as overlapping functions. This review will focus on the isoform-specific roles of individual PIP5K as they relate to the regulation of the actin cytoskeleton. We will review recent advances that establish PIP_2 as a critical regulator of actin polymerization and cytoskeleton/membrane linkages, and show how binding of cytoskeletal proteins to membrane PIP_2 might alter lateral or transverse movement of lipids to affect raft formation or lipid asymmetry. The mechanisms for specifying localized increase in PIP_2 to regulate dynamic actin remodeling will also be discussed.

Keywords Phosphatidylinositol 4 phosphate 5 kinase (PIP5K) · PIP_2 · Actin cytoskeleton

6.1 Historical Perspective

The phosphoinositide (PPI) phosphatidylinositol 4, 5 bisphosphate [PtdIns(4,5)P_2; referred to as PIP_2 in this chapter] is particularly abundant at the plasma membrane (PM). Initially, the interest in PIP_2 was centered around its role as the immediate precursor to three pivotal second messengers, diacylglycerol (DAG), inositol (1,4,5)-trisphosphate ($InsP_3$), and phosphatidylinositol (3,4,5)-trisphosphate (PIP_3) (Divecha and Irvine 1995). Since then, PIP_2 has emerged as an important spatial and

H. L. Yin (✉) · L. Zhang
Department of Physiology, University of Texas Southwestern Medical Center,
5323 Harry Hines Blvd, Dallas TX 75390-9040, USA
e-mail: Helen.Yin@UTSouthwestern.edu

Y. S. Mao
Cold Spring Harbor Laboratory, 1 Bungtown Rd., Cold Spring Harbor NY 11727-2202, USA

P. A. Janmey
Institute for Medicine and Engineering, 1010 Vagelos Laboratories, University of Pennsylvania,
3340 Smith Walk, Philadelphia PA 19104, USA

T. Balla et al. (eds.), *Phosphoinositides II: The Diverse Biological Functions*,
Subcellular Biochemistry 59, DOI 10.1007/978-94-007-3015-1_6,

temporal regulator of cell signaling, especially in the regulation of cytoskeletal and membrane dynamics (Yin and Janmey 2003; Mao and Yin 2007; Janmey et al. 2009; van den Bout and Divecha 2009; Kwiatkowska 2010; Funakoshi et al. 2011). PIP_2 also regulates the activity of PM located transmembrane proteins, such as ion channels, transporters and signaling receptors (Michailidis et al. 2011; Xu et al. 2010; Klein et al. 2008; Pan et al. 2008). In addition, PIP_2 is an organelle marker that distinguishes the PM from internal membranes that have less abundant PIP_2. However, instead of being simply a passive platform or a source of second messengers, this phospholipid actively regulates PIP_2-binding proteins including actin binding proteins, clathrin adaptor proteins, some Rho family GTPases (Heo et al. 2006) and their effectors (Strochlic et al. 2010). The importance of PIP_2 in human biology is underscored by the expanding list of inherited human diseases that are linked to perturbations of the lipid kinases and phosphatases that regulate PIP_2 homeostasis (McCrea and De Camilli 2009; Chang et al. 2011).

6.1.1 PIP_2 Regulation of Actin Dependent Processes

Dynamic changes in PM PIP_2 have been implicated in the regulation of many processes that are dependent on dynamic actin polymerization and remodeling. These include the maintenance of cytoskeletal-PM linkage (Sheetz et al. 2006), substrate adhesion, mechanotransduction and integrin signaling (Parsons et al. 2010), cell migration during development, chemotaxis, morphogenesis and metastasis (Chao et al. 2010b; Kisseleva et al. 2005). PIP_2 is also a regulator of the endocytic and exocytic membrane trafficking machineries (Krauss and Haucke 2007; Martin 2001; Di Paolo and De Camilli 2006), phagocytosis (Botelho et al. 2000; Grinstein 2010; Mao et al. 2009b; Szymanska et al. 2008; Arora et al. 2005), epithelial cell morphogenesis (Ling et al. 2007; Xie et al. 2009; El Sayegh et al. 2007; Martin-Belmonte et al. 2007), cytokinesis (Logan and Mandato 2006), and responses to apoptotic (Mejillano et al. 2001; Halstead et al. 2006), oxidative (Halstead et al. 2006; Chen et al. 2009), inflammatory stresses (Kagan and Medzhitov 2006) and cytokinesis (Logan and Mandato 2006; Janetopoulos and Devreotes 2006).

PIP_2 regulates the organization and dynamics of the actin cytoskeleton at multiple levels (Fig. 6.1). First, it binds and regulates actin binding proteins that are critically involved in actin nucleation, filament end capping and severing, and reinforcement of the membrane cytoskeletal linkages (Yin and Janmey 2003; Mao and Yin 2007; Janmey et al. 2009; Saarikangas et al. 2010). Second, PIP_2 recruits scaffolding proteins such as cytoskeletal effector proteins, endocytic adaptors and their accessory proteins (Di Paolo and De Camilli 2006) to promote the interplay between the PM and the cytoskeleton. Third, PIP_2 regulates the activity of Rho family GTPases, predominantly through activation/inactivation of Rho GEFs and GAPs that bind PPIs (Audhya and Emr 2002, 2003; Russo et al. 2001). Recently, there are also emerging evidence that the PM recruitment of some Rho GTPases is mediated partly through binding to both PIP_2 and PI(3,4,5)P_3 (Heo et al. 2006). Furthermore, PIP5Ks bind Rac directly (Tolias et al. 1995; Halstead et al. 2010, Chao et al. 2010b), and

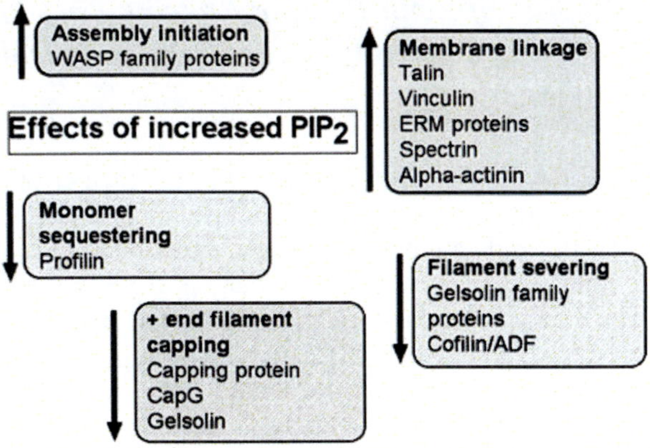

Fig. 6.1 Effects of increasing PIP_2 on the actin cytoskeleton. PIP_2 regulates the organization and dynamics of the actin cytoskeleton at multiple levels. It activates proteins involved in initiating actin filament assembly, reinforcing membrane cytoskeletal linkages, and inactivates monomer sequestering and filament severing proteins. It also dissociates filament + end capping proteins to promote nucleated actin assembly

are known in some cases to recruit/activate Rac1 and RhoA independently of PIP_2 synthesis (Chao et al. 2010b; Manes et al. 2010).

6.1.2 *PIP_2 Regulated Actin Binding Proteins*

Cytoskeletal proteins were among the first shown to be regulated by PIP_2 (Yin and Janmey 2003; Mao and Yin 2007; Janmey et al. 2009; van den Bout and Divecha 2009; Kwiatkowska 2010; Funakoshi et al. 2011). This began with reports that profilin (Malm et al. 1983), alpha-actinin (Burn et al. 1985), vinculin (Niggli et al. 1986; Ito et al. 1983) and components of the erythrocyte cytoskeleton (Anderson and Marchesi 1985) bound acidic phospholipids, and that PIP_2 dissociated complexes of profilin and actin (Lassing and Lindberg 1985). There is now strong biochemical evidence that many more cytoskeletal proteins bind PIP_2. Some are activated by PIP_2, while others are inhibited, at least in biochemical assays. These and subsequent findings suggested that increases in cellular PIP_2 would usually drive the polymerization of cytoskeletal actin and stabilize its interaction with the plasma membrane (Mao and Yin 2007; Janmey et al. 2009). Manipulations of PIP_2 homeostasis corroborate that at least a subset of these reactions occur physiologically within the cellular context, and reveal additional complexities such as the existence of distinct PIP_2 pools.

6.1.3 *The Majority of Actin Binding Proteins Bind PIP_2 Without Using Currently Recognized Canonical PPI Binding Motifs*

In the past decade, the number of PPI-binding proteins such as those involved in membrane trafficking, ion transport or spatial localization of signaling has increased enormously, and actin binding proteins are now a minority of the total ligands proposed for these lipids (Catimel et al. 2008). Many of the newly reported proteins were identified by their possession of well defined PPI-interacting modules, such as pleckstrin (PH), FYVE, PX, FERM or ENTH/ANTH domains and their lipid binding potential was confirmed thereafter *in vitro*. Among these well-defined PPI binding modules, only a few preferentially bind PIP_2 and they often bind exclusively to the lipid headgroup. These include the PLCδ1-PH domain (Balla 2009; Flannagan and Grinstein 2010), the Tubby domain (Szentpetery et al. 2009), and the FERM domain (band 4.1, ezrin, radixin, moesin) (Elliott et al. 2010) found in ezrin/radixin/moesin (ERM) (Pearson et al. 2000; Fehon et al. 2010) and talin (Elliott et al. 2010).

In contrast, the majority of actin regulatory proteins bind PIP_2 using less obviously structured motifs that contain clusters of basic/aromatic amino acids (Ciano-Oliveira et al. 2003; Janmey et al. 2009). Some examples are: the Wiskott Aldrich Syndrome protein (WASP) superfamily that promotes actin assembly by activating the nucleating Arp2/3 complex; capping protein/CapZ and gCap39 that uncap the filament (+) ends; cofilin, which severs actin filaments and accelerates actin treadmilling; gelsolin family proteins, which sever and cap actin filaments to promote dynamic actin reorganization; and vinculin, which regulates (FA) adhesion turnover (Fig. 6.1).

PIP_2's charged inositol headgroup and hydrophobic acyl chain are both required for gelsolin and profilin binding to PIP_2 (Janmey and Stossel 1987; Lassing and Lindberg 1985; Gorbatyuk et al. 2006). N-WASP simultaneously binds several PIP_2 via its polybasic domain and acts as a density sensor that responds to small changes in PIP_2 surface density (Papayannopoulos et al. 2005). The question of how cofilin binds PIP_2 is controversial. Initially, a NMR structure suggests that cofilin binds the acyl and head group of PIP_2 in a binding pocket (Gorbatyuk et al. 2006). However, another study reports that cofilin binds PIP_2 headgroups, but not acyl chains, in a multivalent, cooperative manner without hydrophobic involvement (Zhao et al. 2010). Thus, cofilin, like N-WASP, is a PIP_2 sensor that is responsive to small changes in PIP_2 density at the PM. In both cases, the PIP_2 signals are integrated with other inputs, such as Rho family GTPases and phosphorylation (Yin and Janmey 2003; Mao and Yin 2007; Janmey et al. 2009; van den Bout and Divecha 2009; Kwiatkowska 2010; Funakoshi et al. 2011).

6.1.4 *This Review's Goals*

This review will focus on recent advances that establish PIP_2 as a critical regulator of actin polymerization and cytoskeleton/membrane linkages, and show how binding of cytoskeletal proteins to membrane PIP_2 might alter lateral or transverse movement of lipids to affect raft formation or lipid asymmetry. It will review how PIP_2 regulates selected functions that involve the actin cytoskeleton. Since PIP_2 regulation of the actin cytoskeleton has been extensively reviewed recently (Mao and Yin 2007; Janmey et al. 2009; Saarikangas et al. 2010), we will focus instead on the isoform-specific roles of individual PIP5K that generate PIP_2. The mechanisms for specifying localized increase in PIP_2 to regulate dynamic actin remodeling will also be discussed.

6.2 PIP_2 Dynamics

There is now overwhelming evidence indicating that some pools of PIP_2 are generated in a spatially- and temporally-regulated manner (Balla and Varnai 2009; Grinstein 2010), and that downregulation of the PIP_2 signal is critically important for the cycling of almost all PIP_2-dependent processes (Scott et al. 2005; Brown et al. 2001; Cremona and De Camilli 2001). Localized dynamic PIP_2 cycles are required for actin regulation (Mao et al. 2009b; Coppolino et al. 2002) and endocytic/exocytic membrane trafficking (Mao et al. 2009a). Therefore, a precise understanding of how PIP_2 synthesis and metabolism/dissipation are regulated is of central importance to biology and medicine. We will therefore briefly review current knowledge about how PIP_2 homeostasis is maintained.

6.2.1 *PIP_2 Synthesis*

PIP_2 is generated primarily from phosphatidylinositol monophosphates (PIPs), through two distinct pathways: first, by the type I phosphatidylinositol phosphate kinases (PIPK) that phosphorylate PI(4)P on the D-5 position of the inositol ring (henceforth referred to as PIP5Ks), and second, by the type II PIP4Ks that phosphorylate PI(5)P at the D-4 position. These two types of PIPKs have distinct functions (Clarke et al. 2010). Since PI(4)P is much more abundant than PI(5)P (Toker and Cantley 1997), PIP5Ks are likely to be the major source of PIP_2. We will focus on PIP5Ks exclusively.

6.2.2 *PIP_2 Metabolism and Dissipation*

PIP_2 can be decreased in multiple ways. It is hydrolyzed by phosphatidylinositol specific phospholipase C (PLC) to generate $InsP_3$ and DAG and converted by the class I phosphoinositide 3 kinases (PI3Ks) to generate PIP_3. Additionally, PIP_2 synthesized locally will be dissipated by diffusion, unless there is continuous local PIP_2 generation to maintain a gradient and/or PIP_2 capture by scaffolding molecules at a site of synthesis (Corbett-Nelson et al. 2006). PIP_2 is also dephosphorylated on the

D4 and D5 phosphate groups in its inositol ring by PPI phosphatases (ptase) (Liu and Bankaitis 2010). Among these, synaptojanin 1 (SYNJ1), which has both 4- and 5-phosphatase activity, is particularly well characterized in terms of its effects on membrane trafficking.

Disruption of the *Synj1* gene in mice results in an accumulation of clathrin coated vesicles and polymerized actin at the endocytic zone of nerve terminals (Cremona et al. 1999). Furthermore, transgenic mice overexpressing SYNJ1 have deficits in brain functions and learning (Voronov et al. 2008). Since *SYNJ1* maps to human chromosome 21, and it is overexpressed in Down's syndrome, it has been postulated that PIP_2 dyshomeostasis may partially contribute to brain dysfunctions in Down's syndrome patients. Interesting, the Oculo-Cerebro-Renal syndrome of Lowe protein 1 (OCRL1), a 5-ptase that is linked to the developmental disease characterized by mental retardation, cataracts, and renal failure, has also been implicated in endosomal trafficking (Mao et al. 2009a) and actin regulation. Fibroblasts from Lowe syndrome patients have impaired migration, spreading, and fluid phase endocytosis (Suchy and Nussbaum 2002; Coon et al. 2009), suggesting that PIP_2 dyshomeostasis may contribute to their cytoskeletal defects.

6.2.3 *The Expanding PIP_2 Toolkit*

6.2.3.1 Tools to Monitor PIP_2 Dynamics

Static Snapshot

Anti-PIP_2 antibodies were the first reagents developed to visualize the intracellular localization of PIP_2 and several studies using anti-PIP_2 antibodies reveal localized concentrations of PIP_2 (Das et al. 1987; Gascard et al. 1991; Tran et al. 1993). Recently, an improved fixation and permeabilization protocol that optimizes the preservation of membrane lipids and labeling with anti-PIP_2 has been described (Hammond et al. 2009). In addition, a freeze fracture method that does not use chemical fixation has been developed (Fujita et al. 2009) to probe PIP_2 distribution at the nanoscale with recombinant pleckstrin homology (PH) of PLCδ, which binds PIP_2.

Spatial and Temporal Dynamics

PIP_2's dynamic behavior has been monitored by high resolution live cell imaging of (PH)-PLCδ-GFP (Balla and Varnai 2002; Balla 2009; Flannagan and Grinstein 2010). PH-PLCδ-GFP, when overexpressed at low level, accurately reports PIP_2 levels at the PM, as corroborated in fixed cells by using anti-PIP_2 antibody (Wang et al. 2003; Laux et al. 2000; Hammond et al. 2009). A caveat is that PH-PLCδ also binds $InsP_3$, a product of PIP_2 hydrolysis, with higher affinity than PIP_2 (Hirose et al. 1999). Therefore, in some cases, PH-PLCδ translocation from the PM can potentially reflect an increase in $InsP_3$, rather than a decrease in PM PIP_2 *per se*.

To circumvent this potential complication, a FRET biosensor that detects interaction of PH-PLCδ-YFP and -CFP has been developed (van Rheenen et al. 2005; Kalwa and Michel 2011). In addition, Tubby, a membrane associated transcription factor that binds PIP_2 through its carboxyl terminal "tubby domain", has recently gained popularity as an alternate PIP_2 sensor (Lee et al. 2010; Brown et al. 2007; Quinn et al. 2008; Nelson et al. 2008). Unlike PH-PLCδ, Tubby does not bind $InsP_3$ (Hughes et al. 2007). However, a side by side comparison shows that Tubby is less sensitive to small changes in PIP_2 transients than PH-PLCδ, presumably because it has a higher affinity for PIP_2 (Szentpetery et al. 2009).

Fluorescent derivatives of PIP_2 have also been developed to image trafficking of this lipid in cells. Fluorescent PIP_2 is delivered by patch pipette or microinjection of lipid micelles (Cho et al. 2005; Golebiewska et al. 2008).

At present there does not appear to be a universally accepted method by which to detect cellular PIP_2 without doubts about specificity or that the imaging technique induces or alters the distribution of cellular PIP_2. Anti-PIP_2 antibodies and PH domains often bind other possibly more abundant targets and might not bind to all forms of PIP_2. Labeled PIP_2 has the disadvantage that the label itself is nearly as large as the lipid and often electrostatically charged, and depending on where it is conjugated potentially alters the localization of the lipid in the cell.

6.2.3.2 Tools to Increase PIP_2

PIP5K Overexpression

Initially, transient PIP5K overexpression was used to explore the role of PIP_2 in cellular function. Overexpressed PIP5Ks induce a variety of abnormal actin structures such as actin-rich needles, N-WASP dependent comets (Rozelle et al. 2000), Rho- and Rho-kinase dependent stress fibers (Yamamoto et al. 2001), and the arrest of Arf 6 and actin dependent endosomes trafficking (Brown et al. 2001). The responses vary depending on the cell types used and most likely the extent of overexpression. Unfortunately, these overexpression studies were not able to distinguish between the isoform-specific roles of each PIP5K, probably because high level overexpression may have overwhelmed the normal mechanisms for specifying the PIP5K localization (van den Bout and Divecha 2009). This problem was further confounded by the fact that some of the putative kinase dead (KD) PIP5K mutants used in early studies are not actually completely KD and most are not dominant negative (Mao et al. 2009b; Mao and Yin 2007).

PIP_2 Shuttling

PIP_2 shuttling directly introduces PIP_2 into cells by using a histone or polyamine carrier to shield PIP_2's charged groups (Wang et al. 2003). PIP_2 is "shuttled" into the entire cell (Mao et al. 2009b), or selectively on the basal vs. apical side of polarized epithelial cells (Martin-Belmonte et al. 2007). This approach has been used to rescue PIP5K knockdown/knockout phenotype to establish its relation to loss of PIP_2 (Mao et al. 2009b).

6.2.3.3 Tools to Deplete PIP_2

PIP_2 Sequestration

PH-PLCδ has been used extensively to mask cellular PIP_2. It is introduced either by high level transient overexpression or transduction of the tat-recombinant fusion protein. The mutant PH-PLCδ that does not bind PIP_2 is used as a negative control. WT but not mutant PH-PLCδ has dramatic effects on cytoskeletal dependent processes, such as phagocytosis (Szymanska et al. 2008) and invadopodia formation (Yamaguchi et al. 2010). The cell permeant PIP_2 binding peptide (PBP10), which is derived from gelsolin's PIP_2 binding domain, has also been used to sequester PIP_2. It inhibits glucose transport in an actin-dependent manner (Funaki et al. 2006), induces rapid Ca^{2+}-dependent actin depolymerization (Finkelstein et al. 2010), and interferes with the actin- and gelsolin-dependent formation of N-cadherin junctions (El Sayegh et al. 2007).

Another approach is to microinject a PIP_2 specific monoclonal antibody to deplete PIP_2 in cells. Like PH-PLCδ overexpression, anti-PIP_2 inhibits invadopodia formation in cultured breast cancer cells (Yamaguchi et al. 2010).

5-ptase Overexpression

The soluble fragment of the yeast cytosolic 5-ptase (Inp54p) has been linked to a PM anchor to examine the direct effect of decreasing PM PIP_2.

Constitutive Targeting

Inp54p is targeted to the PM by attaching the lyn minimal PM targeting peptide. Transient overexpression of Insp54p linked to the lyn minimal target decreases membrane-cytoskeleton interactions, implicating PIP_2 in maintaining the cortical cytoskeleton (Raucher et al. 2000). Inp54p has also been targeted using the lyk and c-src minimal membrane anchors that reside predominantly in raft vs nonraft domains to explore the role of these PIP_2 pools in T cell functions (Johnson et al. 2008; Chichili et al. 2009). The important question of how PIP_2 is compartmentalized in raft vs nonraft microdomains and how it exerts different cytoskeletal effects in these domains will be discussed below.

Table 6.1 Mammalian PIP5K isoforms and major γ splice variants

Human isoforms and splice variants	Mouse equivalent	MW (kDa)	No. of residues in human (mouse)
hPIP5Kα (A)	mPIP5Kβ (A)	68	549 (546)
hPIP5Kβ (B)	mPIP5Kα (B)	68	540 (539)
hPIP5Kγ87/hPIP5Kγ640 (C)	mPIP5Kγ635 (γ_b;C)	87	640 (635)
hPIP5Kγ90/hPIP5Kγ668 (C)	mPIP5Kγ661(γ_a C)	90	668 (661)
*	mPIP5Kγ688 (γ_c C)	93	NA (688)
hPIP5Kγ700 (C)	*	100	700 (*)
hPIP4Kγ707 (C)	*	100	707 (*)

*Not found in human or mouse. The α, β, and γ isoforms are also referred to as A, B and C, respectively, and the γ splice variants as $\gamma_{a,b,c}$

Acute Targeting

Acute targeting is achieved by using the rapamycin-inducible FKBP and FRB dimerization system. The lyn minimal anchor (Suh et al. 2006; Hao et al. 2009) or the GAP43 palmitoylation motif (Varnai et al. 2006) are used as the PM membrane anchor. Since PIP_2 level is depleted acutely in the absence of the cascade of Ca^{2+}, DAG or $InsP_3$ signals that normally accompany PLC mediated PIP_2 signaling, PM PIP_2, and not the second messengers derived from PIP_2, is established as the direct regulator of transferrin and EGF receptor endocytosis (Varnai et al. 2006), clathrin and actin dynamics during endocytosis (Zoncu et al. 2007), ERM recruitment to the PM (Hao et al. 2009), the activation of the K^+ and Ca^{2+} channels (Suh et al. 2006, 2010), and receptor phosphorylation during Wnt signaling (Pan et al. 2008).

6.3 PIP5K Isoforms and Splice Variants

6.3.1 PIP5K Cloning and Nomenclature

In 1996, two different PIP5Ks were cloned simultaneously from human and mice (Ishihara et al. 1996; Loijens and Anderson 1996). These isoforms were independently named α and β, but unfortunately, the human and mouse isoform designations were reversed. That is, human PIP5Kα is equivalent to mouse β, and human PIP5Kβ is equivalent to mouse α. *In this review, we will use the human isoform designation*, in accordance with the recent NCBI guideline (Fig. 6.2 and Table 6.1). These isoforms are also referred to as A and B to avoid the α and β designations altogether (Schill and Anderson 2009; Funakoshi et al. 2011; Xu et al. 2010).

In 1998, a third isoform, named PIP5Kγ (also called C), was cloned (Ishihara et al. 1998). It was originally reported to have a 87 kDa short form and a longer 90 kDa carboxyl-terminal splice variant that has a short tail extension (Ishihara et al. 1998; Giudici et al. 2004). The "tail" has multiple binding partners, including talin, AP-2,

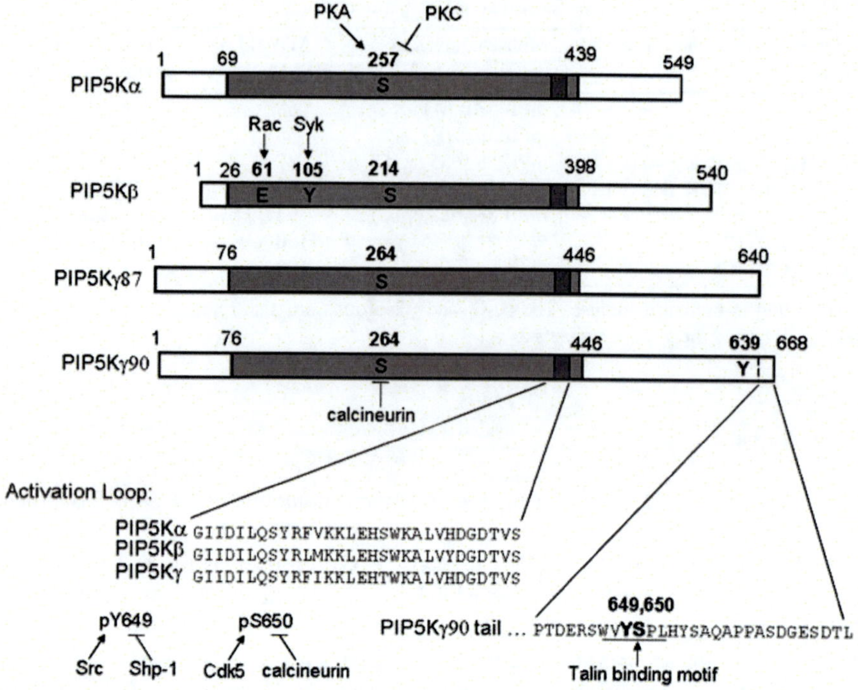

Fig. 6.2 The domain structure of human PIP5K isoforms and their phosphorylation sites. Lightly and darkly shaded regions indicate the conserved kinase homology domain and the activation loop. Two major PIP5Kγ splice variants, which differ by the absence or presence of a "tail" (28 or 26 residues in human and mouse, respectively), are shown. PIP5Kα, β and γ are inhibited by phosphorylation at a conserved ser residue under normal growth conditions. PIP5Kβ is phosphorylated at tyr105 resulting in inactivation and dissociation from the PM. The unique PIP5Kγ90 tail has two tandem phosphorylation sites (tyr649 and ser650/mouse 634,645) that activates and inhibits, respectively

and cadherins (Di Paolo et al. 2002; Ling et al. 2002, 2007; El Sayegh et al. 2007). These isoforms have been referred to by their molecular weights and the number of amino acids. The latter nomenclature is confusing, because the human and mouse homologs have different numbers of amino acids and multiple splice variants have been identified (Table 6.1). These splice variants are also designated as $\gamma_{a,b,c}$ (Powner et al. 2005), but there is a recent proposal to rename them according to the HUGO guideline for isoform designation (Schill and Anderson 2009). We will focus on the most well characterized splice variants, PIP5Kγ87 and 90 (or γ_b and $_a$) here.

6.3.2 PIP5K Domain Organization

The three PIP5K isoforms have a highly conserved central lipid kinase core and divergent N- and C-terminal extensions (Fig. 6.2). The kinase core contains an "activation loop" that resembles the activation loop found in PIP4K, for which a crystal structure has been solved (Kunz et al. 2000, 2002). Domain swapping and site-directed mutagenesis of the activation loops in these related enzymes establish that PIP5Ks have a conserved glu that recognizes PI4P, while PIP4Ks have a conserved ala that recognizes PI5P. The minimal targeting motif also includes two invariant lys residues used for electrostatic interaction, and two tandem basic residues at the C-terminus of the kinase domain (Arioka et al. 2004). These residues are conserved in all PIP5Ks throughout phylogeny, suggesting that they may specify PIP5Ks' universal PM targeting code.

Early *in vitro* studies have shown that, with few exceptions, all PIP5K isoforms appear at the biochemical level to be functionally similar. They have comparable enzyme kinetics and are all activated by phosphatidic acid (PA) (Stace et al. 2008; Ishihara et al. 1998), by ser/thr dephosphorylation (Yamamoto et al. 2006; Park et al. 2001), and by small GTPases such as RhoA (Chong et al. 1994), Rac1 (Weernink et al. 2004; Halstead et al. 2010), and Arf6 (Honda et al. 1999; Funakoshi et al. 2011). Nevertheless, there are now increasing evidence for isoform specific differences, in addition to overlapping functions (Mao and Yin 2007; van den Bout and Divecha 2009; Kwiatkowska 2010). These differences are likely to be specified by PIP5Ks' distinct N- and C-terminal extensions. Current evidence for isoform specific regulation of actin-dependent functions will be discussed in a subsequent section.

6.3.3 PIP5K Expression

PIP5Ks isoforms are ubiquitously expressed, although their pattern of expression varies between tissues and cell types. PIP5Kγs, particularly γ90, are particularly abundant in the adult brain (Wenk et al. 2001; Di Paolo et al. 2002; Volpicelli-Daley et al. 2010; Fairn et al. 2009). The prevailing hypothesis is that PIP5Kγ90 regulates synaptic vesicle trafficking (Di Paolo et al. 2004). Western blotting with an anti-PIP5Kγ antibody that recognizes all PIP5Kγ isoforms shows that PIP5Kγ level increases significantly during embryonic development and continues to increase postnatally (Volpicelli-Daley et al. 2010). In contrast, PIP5Kα, which is expressed at high level during mid-gestation, does not increase further after birth, while PIP5Kβ is not detectible until after birth. These results led to the hypothesis that PIP5Kγ and β may have specialized roles in the postnatal brain, while PIP5Kα may have a "housekeeping" function.

Compared with the adult brain, murine RAW 264.7 macrophage-like cells have much less PIP5Kγ protein, similar amounts of PIP5Kα and more PIP5Kβ (Fairn et al. 2009). Quantitative real time PCR shows that although nonneuronal cells have much less PIP5Kγ (Volpicelli-Daley et al. 2010) than the brain, PIP5Kγ mRNA is nevertheless more abundant than the other PIP5Ks (Mao et al. 2009b; Micucci et al. 2008; Wang et al. 2008a). Neutrophils have more PIP5Kγ90 than PIP5Kγ87, while bone marrow derived macrophages and HeLa cells have less (Mao et al. 2009b; Wang et al. 2004).

6.4 Lessons Learned from PIP5K Gene Silencing

The question of why mammals have so many PIP5Ks is now more amenable to analysis using genetic approaches. Initially, RNAi silencing was used (Padron et al. 2003; Heldwein et al. 2004; Wang et al. 2004). Recently, animals with disruption of individual or multiple PIP5K gene(s) have been generated. In addition, micro-RNAi based knockdown is beginning to be used (Lee et al. 2010). Here we will briefly review the effects of PIP5K KO and RNAi.

6.4.1 PIP5Kγ Knockouts (KO)

There are currently three PIP5Kγ disrupted mouse lines. Two have eliminated expression of all known PIP5Kγs (pan KO) (*PIP5Kγ-/-*) (Di Paolo et al. 2004; Wang et al. 2007), and the third line is *PIP5Kγ90-/-*, but *PIP5Kγ87+/+* (Wernimont et al. 2010).

6.4.1.1 A *PIP5Kγ-/-* Line (Pan KO Line 1) Generated by Homologous Recombination

These mice die within a day of birth (Di Paolo et al. 2004). Synaptosome prepared from their brains have 40% less PIP_2 than WT, and they fail to generate PIP_2 in response to K^+ depolarization. *In vitro* studies revealed that the PIP5Kγ-/- cells have multiple defects. Their cultured cortical neurons have severe defects in synaptic transmission, which correlate with abnormal exocytosis and defective clathrin-mediated endocytosis (Di Paolo et al. 2004). Their chromaffin cells have defective vesicle priming and fusion dynamics (Gong et al. 2005). Their mast cells have defective $InsP_3$ signaling (Vasudevan et al. 2009) (see below). Their macrophages have defective FcγR mediated phagocytosis because they have decreased binding to IgG opsonized particles (Mao et al. 2009b). Although the macrophages have normal basal PIP_2 level, they are unable to replenish PIP_2 depleted by FcγR ligation (Mao et al. 2009b). Their neutrophils chemotax normally *in vitro*, but have impaired polarization and decreased infiltration to inflammatory sites *in vivo* (Xu et al. 2010). Intravital microscopy shows that they have decreased adhesion to the endothelium, but no defect in rolling and rate of emigration *per se*. Rescue experiments show that PIP5Kγ90 corrected these defects, while PIP5Kγ87 restored some functions, but not polarization (Xu et al. 2010).

6.4.1.2 A *PIP5Kγ-/-* Line (Pan KO Line 2) Generated by Gene Trap

These KO animals die at embryonic day 11.5 and have neural tube closure and heart septation defects (Wang et al. 2007). Their megakaryocytes bleb abnormally

(Wang et al. 2008b). Taken together, these results suggest that there are underlying cytoskeletal defects that impact cell migration and cytoskeletal membrane linkages. The reason for the different phenotypes between lines 1 and 2 is not known, and may be due to differences in genetic background.

6.4.1.3 A *PIP5Kγ90-/-* Line Generated by Homologous Recombination

Since all known PIP5Kγs are knocked out in lines 1 and 2, it is not possible to distinguish between the relative contributions of individual PIP5Ks. PIP5Kγ90 binds talin, cadherin, AP-2 and AP-1 subunits (Di Paolo et al. 2002; Ling et al. 2002, 2007; El Sayegh et al. 2007), and therefore has been implicated in clathrin mediated endocytosis, cell migration, cell:matrix and cell:substrate attachments. It is assumed that a large subset of the pan KO phenotypes is due primarily to the loss of PIP5Kγ90. However, if this were the case, it is difficult to reconcile how the embryos in PIP5Kγ KO lines (particularly line 1) develop normally, since development is critically dependent on cell:matrix attachment, cell:cell interactions, and migration. This conundrum raises the possibility PIP5Kγ90 is not solely responsible for all the functions ascribed to them by *in vitro* studies, and/or there are compensation by other PIP5Ks after PIP5Kγ silencing. This is a possibility because *PIP5Kγ* KO brains expressed more PIP5Kα and β at the protein level (Volpicelli-Daley et al. 2010), and PIP5Kγ RNAi likewise increases their expression in nonneuronal cells (Padron et al. 2003).

To address the question of how PIP5Kγ90 contributes, the PIP5Kγ90 specific exon that specifies its unique C-terminal tail has been selectively targeted by RNAi (Wang et al. 2004) and by homologous recombination (Wernimont et al. 2010). Surprisingly, unlike pan *PIP5Kγ KO*, *PIP5Kγ90-/-* mice are viable, have no obvious defects, and survive to adulthood (Wernimont et al. 2010). Since *PIP5Kγ 90-/-* mice live while *PIP5Kγ-/-* mice die, the almost inevitable conclusion is that PIP5Kγ87 and 90 together are essential for life, while PIP5Kγ90 (in the presence of γ87) is not. PIP5Kγ87 has long been overshadowed by PIP5Kγ90, because less is known about the shorter isoform. Additional work will be required to establish how PIP5Kγ87 *per se* is necessary for late embryonic/postnatal survival, and to characterize the *PIP5Kγ 90-/-* mouse more completely. An important first question is whether PIP5Kγ90 KO is compensated for by increases in PIP5Kγ87 or the other PIP5Ks.

PIP5Kγ 90-/- T lymphocytes are hyper-responsive due to sustained β integrin:LFA-1 (lymphocyte function-associated antigen 1) coupling at the immunological synapse (Wernimont et al. 2010). Paradoxically, other studies using pan PIP5Kγ shRNAi report the opposite result in T cells (Bolomini-Vittori et al. 2009) and natural killer (NK) cells (Mace et al. 2010). These results raises the possibility that if PIP5Kγ90 were indeed a negative regulator of antigen induced T cell adhesion and activation (Wernimont et al. 2010), then the suppressed response in pan PIP5Kγ cells may be explained by assuming that PIP5Kγ87 is a positive regulator in this context.

6.4.2 PIP5Kα-/- Mice Generated by Gene Trap

*PIP5K*α-/- mice (mouse *PIP5K*β) bred less well than WT mice, but otherwise exhibit no apparent gross defects and survive to adulthood (Wang et al. 2008a). Their macrophages binds IgG opsonized particles normally (unlike PIP5Kγ KO) but are impaired in their ability to extend pseudopodia for particle ingestion (Mao et al. 2009b). Their platelets have a blunted response to thrombin stimulation, as manifested by decreased $InsP_3$ generation and aggregation *in vitro*. Defective thrombosis is corroborated *in vivo* in a carotid artery injury model (Wang et al. 2008a). Since platelet aggregation is dependent on actin remodeling, and both PLCβ and γ2 are activated (Elvers et al. 2010), these results strongly implicate PIP5Kα in actin regulation and PLC mediated signal transduction. Interestingly, although *PIP5K*β-/- platelets also have decreased $InsP_3$ generation, they do not have an aggregation defect (Wang et al. 2008a). *PIP5K*α-/- macrophages and platelets have a 20% and 40% decrease in total PIP_2, respectively.

6.4.3 PIP5Kβ-/- Mice (Mouse PIP5Kα-/-) Generated by Homologous Recombination

These mice are viable and develop normally (Sasaki et al. 2005). The only documented phenotype at the whole animal level is enhanced passive cutaneous and systematic anaphylaxis. Their mast cells have 35% less PIP_2 and less cortical actin (Sasaki et al. 2005). FcεRI stimulation induces more robust degranulation and Erk activation. These defects are corrected by jasplakinolide, which stabilizes the actin cytoskeleton. Thus, PIP5Kβ acts as a negative regulator that maintains the cortical actin barrier to dampen FcεR-mediated signaling and regulated exocytosis.

Notably, unlike in PIP5Kγ KO, PIP5Kα or β KO does not induce a compensatory increase in the expression of the nontargeted PIP5Ks in brains (Volpicelli-Daley et al. 2010) or platelets (Wang et al. 2008a). Taken together, these results suggest that PIP5Kα or β may be less critical than PIP5Kγ in the brain. Additional studies will be required to determine if there are compensatory changes in other tissues that rely more on PIP5Kα or β.

6.4.4 Double KO Lines

The relation between the PIP5Ks is examined by using double KO mice (Volpicelli-Daley et al. 2010). *PIP5K*α/β KO mice survive to adulthood, suggesting that PIP5Kγ alone is sufficient for sustaining normal development and life. *PIP5K*β/γ line 1 double KO develop normally *in utero*, but die within minutes, rather than hours after birth for PIP5Kγ KO only. Thus, PIP5Kβ and γ may have synergistic functions postnatally. No *PIP5K*α/γ double KO mouse was born, suggesting that in the absence of PIP5Kγ, the additional loss of α is more devastating than that of β. Since PIP5Kα is expressed early during development, it may have a housekeeping function that overlaps with γ during prenatal development.

In summary, these knockout (KO) mice establish that each PIP5K has unique as well as previously unanticipated overlapping/complementary functions at the organismal level, and that there are considerable tissue-specific differences. Evidence for PIP5K isoform specific regulation of several cytoskeletal related functions will be reviewed below.

6.4.5 Evidence for PIP5K Isoform Specific Ca^{2+} *Signaling*

PIP_2 is the obligatory substrate for PLC-mediated $InsP_3$ generation, and $InsP_3$ elicits Ca^{2+} release from the endoplasmic reticulum (ER) to activate multiple Ca^{2+} dependent responses, including actin remodeling. Ca^{2+} release from ER stores also activates store operated Ca^{2+} entry. PIP5K silencing studies show that the PIP5Ks are used differentially for Ca^{2+} signaling, and the dominant pools used are cell specific. PIP5Kγ KO and RNAi decrease $InsP_3$ release in mast cells (Vasudevan et al. 2009) and HeLa cells (Wang et al. 2004), suggesting that PIP5Kγ is primarily responsible for the PIP_2 pool used to generate $InsP_3$ in these cells. In contrast, PIP5Kβ silencing increases store-operated Ca^{2+} entry in mast cells (Vasudevan et al. 2009), suggesting that it is a negative regulator of Ca^{2+} in a step downstream of PIP5Kγ regulated Ca^{2+} release from ER stores. PIP5Kγ (and to a lesser extent PIP5Kα) is also implicated in Ca^{2+}-regulated exocytosis of lytic granules in NK cells (Micucci et al. 2008), although the effect of RNAi on $InsP_3$ *per se* was not examined (Micucci et al. 2008). On the other hand, PIP5Kγ KO (line 2) platelets have normal $InsP_3$ signaling, while PIP5Kα or β single KO has partially decreases $InsP_3$ signaling (Wang et al. 2008a) (see below). PIP5Kα and β supports two independent PIP_2 pools for $InsP_3$ generation, becausePIP5K α/β double KO platelets have almost no $InsP_3$ response (Volpicelli-Daley et al. 2010).

6.4.6 Evidence for PIP5K Isoform Specific Actin Regulation

Cell crawling involves the extension of the leading edge (also called lamellipodium, and is related to invadopodium or pseudopodium in other contexts), dynamic cycles of adhesion/detachment through FAs, and contraction at the rear of the cell. These motility components must be coordinated for efficient translocation, and PIP_2/PIP5Ks have been implicated in all aspects of cell crawling.

6.4.6.1 Cell Adhesion, Integrin Signaling and Focal Adhesion (FA) Dynamics

FAs are sites of actin filament attachment to the extracellular matrix through integrin receptors, and they are also mediators of bidirectional integrin signaling and mechanotransduction (Dubash et al. 2009). PIP_2 level increases transiently during cell attachment to the extracellular matrix, and PIP_2 activates several key FA components, including talin, vinculin, and α-actinin. Talin has a key role in activation of the integrin family of cell adhesion receptors and also directly links the cytoplasmic tail of β integrin to the actin cytoskeleton. It has a FERM domain that binds PIP_2 and is activated by PIP_2 to bind β integrin (Elliott et al. 2010; Goksoy et al. 2008).

Most studies suggest that PIP5Kγ90, which is concentrated in FA, is primarily responsible for the regulation of FA dynamics. PIP5Kγ90 binds talin through its COOH-terminal tail (Ling et al. 2002; Lee et al. 2005) (Fig. 6.2) to increase local PIP_2 synthesis and promote assembly of the nascent FA. After assembling the mature FA, PIP_2 production is dialed down because activated integrin displaces PIP5Kγ90 from talin (Elliott et al. 2010; Dubash et al. 2009). Self limiting of the PIP_2 signal promotes dynamic FA turnover.

Besides PIP5Kγ90, other PIP5Ks may regulate FA dynamics through other mechanisms. For example, PIP5Kγ87, which lacks PIP5Kγ90's COOH-terminal tail extension (Di Paolo et al. 2002; Ling et al. 2002), has also been implicated in integrin-mediated adhesion via a phospholipase D2-mediated mechanism (Powner et al. 2005). PIP5Kβ promotes β integrin endocytosis in a clathrin and dynamin dependent manner to facilitate FA disassembly (Chao et al. 2010a).

6.4.6.2 Membrane Ruffling and Cell Migration

Epidermal growth factor (EGF) and platelet derived growth factor (PDGF) induces membrane ruffling and cell migration (Sun et al. 2007; Chao et al. 2010b). Upon EGF stimulation of breast carcinoma cells, cofilin dissociates from the PM in response to PLCγ1 mediated PIP_2 hydrolysis (van Rheenen et al. 2007). A cofilin mutant that binds PIP_2 with above normal affinity is less readily released from the PM and impairs cell steering (Leyman et al. 2009). There is also evidence for crosstalk between cofilin phosphorylation and regulation by PIP2 (Meira et al. 2009).

EGF appears to regulate membrane ruffling and cell migration by activating different PIP5Ks. EGF promotes PIP5Kγ90 phosphorylation at tyr649, resulting in an increased interaction with talin as described above, to promote FA formation and turnover. PIP5Kγ's role is confirmed by the finding that PIP5Kγ RNAi in breast cancer cell lines (MDA-MB-231 and -435S) blocks cell migration, invasion, and proliferation (Sun et al. 2010). PIP5Kγ is however not the sole regulator in breast cancer cells, because PIP5Kα, which is most highly expressed in MDA-MB-231 cells, promotes invadopodia formation and gelatin degradation (Yamaguchi et al. 2010).

PIP5Kα's role in growth factor induced membrane ruffling has been studied extensively (Doughman et al. 2003). Overexpressed PIP5Kα promotes the formation of actin foci when Rac1 is inhibited, but stimulates ruffle formation when Rac1 is activated. These results suggest that PIP5Kα promotes actin assembly, and that additional inputs from Rac1 are required to generate active ruffles. The LIM protein ajuba, which is a component of the integrin-mediated adhesive complex and a Rac activator, is a potential effector (Kisseleva et al. 2005). It promotes PIP5Kα localization to membrane ruffles and leading lamellipodia. The relation between PIP5Kα and Rac is complex, because recent evidence suggests that during PDGF stimulation of membrane ruffling, PIP5Kα has a scaffolding function upstream of Rac (Chao et al. 2010b). This relation will be discussed further below.

6.4.6.3 Leukocyte Polarity and Phagocytosis

Polarization

Leukocytes are polarized during chemotaxis and formation of the immunological synapse, and polarization depends on the coordinated regulation of the frontness and backness signaling cascades (Wang 2009). We have already reviewed briefly how PIP5Ks and PIP_2 are involved in the leading edge (frontness). The backness signal is orchestrated in the uropod, a PM protrusion found at the rear of motile leukocytes. It is generated by segregation of proteins and lipids, and local recruitment and activation of components of the actin cytoskeleton. The uropod has been implicated in intracellular signaling, detachment from the substrate and extravasation (Sanchez-Madrid and Serrador 2009; Ludwig et al. 2010). It is enriched with ERM and some PIP5K isoforms.

ERMs

ERMs provide structural reinforcements for the cell cortex and are particularly concentrated in uropods. They maintain lymphocyte cortical rigidity (Hao et al. 2009) and are activated by phosphorylation and by PIP_2. Ezrin binds PIP_2 through its FERM domain to relieve the autoinhibited state (Pearson et al. 2000; Fehon et al. 2010). The importance of the PIP_2 binding for membrane localization and regulation of ERM proteins has become increasingly clear in recent studies. Mutation of four basic residues in ezrin's PIP_2 binding domain prevents ezrin localization to actin-rich membrane structures (Barret et al. 2000). Chemokine activation of B lymphocytes inactivates ERM by releasing them from the PM and decreasing phosphorylation (Hao et al. 2009). Release is initiated by PLC mediated PIP_2 hydrolysis. ERM has also been implicated as a Rho A downstream effector. However, this relation is complicated by the recent findings that ERM may also act upstream of RhoA both positively and negatively, by interaction with Rho GAPS, GEFs and GDIs (Fehon et al. 2010).

PIP5Kβ and PIP5Kγ90

PIP5Kβ and PIP5Kγ90 are both enriched in uropods. PIP5Kβ's uropod localization in the neutrophil-like HL-60 cells is independent of its lipid kinase activity and is mediated through binding of the last four residues in its unique 83 amino acid C-terminal extension (Lacalle et al. 2007) (Fig. 6.2) to the PDZ domain of ERM-binding phosphoprotein 50. The complex in turn binds ERM and activates RhoA (Manes et al. 2010). These interactions provide a venue for PIP5Kβ to function as a scaffold to coordinate rear signaling during leukocyte migration. PIP5Kβ silencing impairs RhoA activation during neutrophil polarization, suggesting that it acts upstream of RhoA (see further description below).

PIP5Kγ90 is also enriched in uropods of lymphocytes and neutrophils, while PIP5Kγ87 is uniformly distributed. Neutrophils crawling on fibrinogen coated substrate in the absence of chemotactic peptides already have more PIP5Kγ90 at the rear than the front, and rear recruitment is dependent on engagement of β2 integrins (Xu et al. 2010). Chemoattractants induce further recruitment of other components to the rear of the cell to form the uropod (Xu et al. 2010). The roles of PIP5Kγ90 and 87 isoforms has been elucidated using PIP5Kγ pan KO (line 1) and rescue. These studies show that PIP5Kγ90 is required for PIP_2 generation at uropods and the establishment of "backness" signaling during chemotaxis (Lokuta et al. 2007; Xu et al. 2010).

Phagocytosis

Fcγ receptor (FcγR)-mediated phagocytosis is orchestrated by highly coordinated PIP_2, PIP_3 and actin transients (Botelho et al. 2000; Grinstein 2010) that regulate the particle attachment, engulfment, and phagosome maturation steps (Groves et al. 2008). PIP_2 promotes actin assembly by recruiting WASP family proteins to induce Arp2/3 dependent actin nucleation, and inhibits gelsolin and cofilin to prevent actin severing during the ingestion phase (Arora et al. 2005; Adachi et al. 2002). PIP_2 also activates ERM to promote phagosome maturation (Erwig et al. 2006).

PIP5Kγ and α are both recruited to the nascent phagocytic cup but they regulate different steps in phagocytosis (Botelho et al. 2000; Coppolino et al. 2002; Mao et al. 2009b). PIP5Kγ KO macrophages have a particle attachment defect and a RhoA/Rac 1 imbalance that generates an abnormally stable actin cytoskeleton (Mao et al. 2009b). This impedes FcγR microclustering that is promotes high avidity ligand binding to secure stable particle attachment. PIP5Kα macrophages bind opsonized particles normally, but are not able to extend pseudopodia to ingest the attached particles. The pseudopodia extension defect is due to defective activation of WASP to induce Arp2/3-dependent actin polymerization at the nascent phagocytic cup. In addition, PIP5Kγ is rapidly and transiently activated by the spleen tyrosine kinase (syk), while PIP5Kα is not. These findings establish that PIP5Kγ and α orchestrate different types of actin remodeling at sequential stages of phagocytosis, and that they are differentially regulated (Mao et al. 2009b).

The contribution of PIP5Kβ to FcγR mediated phagocytosis has not been examined, but it has been implicated in integrin-mediated *Yersinia* phagocytosis (Wong and Isberg 2003).

6.4.6.4 Context-dependent Interaction of PIP_2 with Cytoskeletal Proteins

Overview

Considering the number of proteins affected by PIP_2 and the different ways in which protein function can be affected, there is a striking pattern to the functions that are activated or inhibited by these lipids (Fig. 6.1). All PIP_2-sensitive actin monomer binding proteins (e.g. profilin), and proteins that sever actin filaments (e.g. gelsolin and cofilin) are inactivated by PIP_2. In contrast, proteins that promote actin assembly (e.g. WASP family proteins) or that link F-actin to the membrane (e.g. ERM, talin and vinculin) are activated by PIP_2. In addition some motor proteins, such as myosin I (Komaba and Coluccio 2010), are targeted to PIP_2-enriched membranes. Filament bundling proteins (e.g. alpha-actinin) are generally inactivated by PIP_2. The overall effect in the cell, assuming that all activities are equally impacted is increased actin polymerization and linkage to the membrane, with dissociation of actin bundles to promote more open network structures as actin assembles at the membrane interface. When PIP_2 levels decreases, actin polymerizing is inhibited and polymerized actin is actively severed/depolymerized while factors that stabilize bundling become active. This constellation of *in vitro* activities is consistent with the effects of increasing or decreasing PIP_2 levels in the cell either by various manipulations described above or by gene KO and RNAi. However, there are few if any studies to determine which of the many possible PIP_2-mediated reactions is preferentially stimulated under different cellular conditions.

The Puzzle

Perhaps the main challenge in understanding how PPIs regulate cytoskeletal or other proteins is the sheer number of PPI (usually PIP_2) binding proteins that have been reported and generally well characterized biochemically as specific and high affinity ligands for these lipids. When only a handful of proteins, mostly actin binding proteins, were shown to be regulated by PIP_2, it was plausible to propose that such proteins with μM concentrations in the cell could be all be regulated by PIP_2, present at 10's to 100's of μM concentrations, when specific signals were initiated. Currently however, over a hundred proteins are reported to bind PIP_2 with similar affinity and there are very few studies of how different PIP_2-binding proteins might compete with each other. There is also good evidence that proteins such as annexin 2, which bind a single PIP_2 molecule, can prevent the binding of multiple PIP_2s in a lipid bilayer with their protein ligands, either by occluding their access to soluble proteins, or by changing the structure of the membrane. The emerging evidence that different PIP5Ks, which all produce the same lipid have very different cellular effects, also suggests that access of PIP_2 to its multiple potential targets depends strongly on the location and environment in which the PIP_2 is placed.

As a result, a major unresolved question is how PIP_2 distributes laterally within the PM and whether all PIP_2 molecules within a membrane are equally effective at binding their targets. Two critical issues, for which there are conflicting reports and no consensus, are the relation of PIP_2 to formation of cholesterol-dependent lipids rafts, and whether PIP_2 can self-associate to form clusters independent of or at least not requiring cholesterol. A challenge for future studies is to delineate more clearly the localization and diffusivity of PIP_2 in the different membranes within the cell, and to evaluate how the large number of PIP_2 ligands are selectively regulated when changes in PIP_2 levels or localization are generated *in vivo*.

6.5 Regulation of PIP5K Localization

PIP5Ks are predominantly cytosolic proteins that are recruited to the PM in response to extracellular signals (Wang et al. 2005). Some PIP5Ks are recruited to the PM in the same part of the cell, to different cell regions, and in some cases, to raft vs nonraft domains. It is hypothesized that PIP5Ks are targeted to the PM using common targeting cues and to specific regions by isoform specific cues. The use of multiple combinatorial targeting determinants ensures specificity and amplification of spatially and temporally defined PIP_2 signaling.

6.5.1 Electrostatic Interactions with PI4P and PIP_2

6.5.1.1 PIP5K Interaction with PI4P

The crystal structure of PIP4K show that PI4Ks dimerize to form a flattened surface that docks on the lipid bilayer, and the activation domain specifies PIP4K's preference for PI(5)P (Rao et al. 1998). PIP5Ks are likely to have a similar overall organization (Fairn et al. 2009) and bind PI4P in a similar manner. Recently, it has also become clear that substrate recognition is necessary but not sufficient for specifying PIP5K localization at the PM. For example, although the trans *Golgi* network has much more PI(4)P than the PM (Wang et al. 2003), it has less PIP5K/PIP_2. In contrast, PIP_2 is particularly enriched at the PM, making it more attractive as a PIP5K PM recruiter.

6.5.1.2 PIP5K Interaction with PIP_2 in a Bistable Electrostatic Switch Model (Fairn et al. (2009))

This model is based on several recent discoveries. A PIP5Kα fragment (human aa374-440) within the activation loop (Fig. 6.2) binds PIP_2 preferentially compared with PI4P, even though the full length kinase slightly prefers PI4P (Szymanska et al. 2008). Overexpression of this fragment decreases PIP5Kα association with the PM, decreases PIP_2 level, disrupts the actin cytoskeleton, and inhibits phagocytosis. Modeling of the PIP5K structure shows that, like the related PIP4K, all PIP5K isoforms have a flat and highly positively charged surface that can dock electrostatically onto negatively charged groups on the inner leaflet of the PM (Fairn et al. 2009). The importance of electrostatic targeting is again confirmed using the phagocytosis model (Fairn et al. 2009).

The bistable electrostatic switch model is attractive for several reasons. First, it provides a concerted detection process to ensure specificity of PIP5K targeting to the PM and local amplification of PIP_2 production (Fairn et al. 2009). Since PIP_2 contributes to PM's anionic charge and regulates PIP5K targeting, local PIP_2 increase would recruit more PIP5K to the PM to amplify PIP_2 generation. Conversely, a decrease in PIP_2 (either from 5-ptase or conversion to other metabolites) would displace PIP5K from the PM, to rapidly switch off further PIP_2 production. Second, it also explains why some type of PIP5K phosphorylation decreases PIP5K association with the PM (Yamamoto et al. 2006; Halstead et al. 2006; Chen et al. 2009), and suggests how the anionic phosphatidic acid activates PIP5Ks (Jarquin-Pardo et al. 2007).

6.5.2 Regulation by Phosphorylation

PIP5K phosphorylation regulates PM association and lipid kinase activity. We will first review evidence for all PIP5Ks, and discuss regulation of PIP5Kγ90's COOH-terminal tails separately.

6.5.2.1 Inactivating ser/thr Phosphorylation

An unknown fraction of each PIP5K is partially phosphorylated on ser/thr residues under normal tissue culture conditions (Itoh et al. 2000). Phosphorylation decreases PIP5Ks' lipid kinase activity. The cAMP-dependent protein kinase A phosphorylates PIP5Kβ *in vitro* and ser214 is identified as a phosphorylation site (Park et al. 2001) (Fig. 6.2). Stimuli that increase PIP_2 and actin polymerization promote PIP5Kβ dephosphorylation. These include lysophosphatidic acid (Park et al. 2001) and hypertonic stress (Yamamoto et al. 2006). PIP5Kα and γ are not dephosphorylated by hypertonicity, establishing that there are isoform specific responses.

6.5.2.2 Activating and Inactivating Tyr Phosphorylation

It is known for some time that pervanadate, a potent tyr phosphatase inhibitor, increases PIP_2 in HEK293 and REF52 cells and increases actin comet formation (Rumenapp et al. 1998; Rozelle et al. 2000). However, paradoxically, oxidative stress, which activates multiple tyr kinases, decreases PIP_2 and disrupts actin filaments in HeLa cells (Halstead et al. 2006; Chen et al. 2009). This dichotomy can be explained by the finding that tyrosine phosphorylation inhibits PIP5Kβ but activates PIP5Kγ. Thus, the overall impact on PIP_2 homeostasis depends on the PIP5K expression profile in a particular cell type.

Oxidative Stress

Oxidative stress decreases PIP_2 and induces actin depolymerization in HeLa cells. Oxidants promotes PIP5Kβ tyrosine phosphorylation and dissociation from the PM in a syk and src dependent manner (Halstead et al. 2006; Chen et al. 2009). PIP5Kβ coimmunoprecipitates with syk and the syk phosphorylation site has been mapped to human tyr105 (Fig. 6.1). The phosphomimetic PIP5KβY105E phosphomimetic is catalytically inactive and cytosolic, while the nonphosphorylatable PIP5KβY105F has elevated kinase activity and is constitutively PM association (Chen et al. 2009). The large decrease in PM PIP_2 in oxidant-stressed cells may be an early trigger for apoptosis, because PIP5Kβ overexpression, which prevents oxidant-induced PIP_2 decrease, protects cells from apoptosis (Halstead et al. 2006).

Other PIP5Ks have also been implicated in protection against apoptosis. These include PIP5Kα, which is cleaved by caspase 3 during apoptosis (Mejillano et al. 2001). Protection is dependent on PIP_2 generation, because KD PIP5Kα shows no protection. Recently, PIP5Kα is reported to promote cell survival in the MIN6B1 beta cell line (Tomas et al. 2010).

Phagocytosis

PIP5Kβ is also phosphorylated by syk during FcγR mediated phagocytosis (Mao et al. 2009b). In addition, PIP5Kγ87 and 90 are transiently tyrosine phosphorylated as well, but at an earlier time point compared with PIP5Kβ (Mao et al. 2009b). PIP5Kγ is activated by syk dependent tyrosine phosphorylation, while PIP5Kβ is inhibited. Additional studies will be required to identify the syk phosphorylation site PIP5Kγ and how it inhibits activity. Paradoxically, PIP5Kα, which is not tyrosine phosphorylated during phagocytosis, can nonetheless be phosphorylated by syk *in vitro* (Mao et al. 2009b). Since PIP5Kα is recruited to the phagosome with a similar time frame as PIP5Kγ, this result suggests that PIP5Kα and PIP5Kγ may differentially accessible to syk even though both are located in the narrow confines of the phagosome.

PIP5Kγ90 Specific tyr and ser Phosphorylations

The unique PIP5Kγ90 tail has two tandem phosphorylation sites (tyr649 and ser650 in WVYSPL), and phosphorylation of these residues have opposite effects and are mutually exclusive. Tyr phosphorylation activates the tail to bind talin and an AP-2 subunit, while ser phosphorylation decreases binding. Tyr649 is phosphorylated by src (Ling et al. 2003) or EGFR (Sun et al. 2007) and dephosphorylated by Shp-1 (Bairstow et al. 2005). In neurons, ser650 is phosphorylated by Cdk5, and dephosphorylated by calcineurin in response to K^+-induced depolarization (Nakano-Kobayashi et al. 2007; Lee et al. 2005).

6.5.3 Recruitment and Regulation by Small GTPases

The bistable electronic switch model cannot completely explain how PIP5K isoforms are differentially targeted to different parts of the cell, such as the leading edge vs uropod (Manes et al. 2010; Xu et al. 2010; Sun et al. 2011) and to raft vs nonraft membrane microdomains (Johnson et al. 2008; Szymanska et al. 2009; Mao et al. 2009b). Targeting at these levels is likely to rely on additional cues, such as association with binding partners. These include Rac1 and RhoA, which are important regulators of the actin cytoskeleton and operate in an antagonistic manner to dictate the ultimate cytoskeletal outcome. We will focus on their role to highlight new information about their relation to PIP5Ks.

6.5.3.1 PIP5Ks Downstream of Rac/Rho

It has been shown previously that RhoA and Rac1 act upstream of PIP5Ks (Ren et al. 1996; Chong et al. 1994; Hartwig et al. 1995; Tolias and Carpenter 2000; Weernink et al. 2004). The path from Rac1 to PIP5K is recently confirmed in a study on lysophosphatidic acid induced neurite remodeling (Halstead et al. 2010), which shows that PIP5Kβ is recruited to the PM by Rac1. PIP5Kβ binds Rac1 but not RhoA, and binding is mediated through an N-terminal sequence. Mutation of glu61 to leu in this region (Fig. 6.2) decreases PM recruitment and blunts neurite extension, in spite of normal lipid kinase activity. PIP5Kα and γ also interact with Rac using residues equivalent to PIP5Kβ. Vinculin, which binds PIP_2, is implicated downstream of PIP5Kβ because it dissociates from neurite attachment sites during neurite retraction, and vinculin mutants that are defective in PIP_2 binding attenuate neurite retraction (Halstead et al. 2010). Thus, this study clearly establishes that all PIP5Ks act downstream of Rac to induce neurite retraction.

This study also contributes to the evolving understanding of how PIP_2 regulates vinculin at FA. Originally it was proposed that vinculin is recruited to FA by binding PIP_2 (Gilmore and Burridge 1996; Weekes et al. 1996). However, unactivated vinculin binds PIP_2 weakly (Johnson et al. 1998; Bakolitsa et al. 2004) and vinculin PIP_2 mutants used are still recruited to FA but they impede FA turnover (Chandrasekar et al. 2005; Saunders et al. 2006). Thus, these studies support the hypothesis that transient increases in PIP_2 displace vinculin from actin filaments to facilitate the FA turnover.

6.5.3.2 PIP5Ks Upstream of Rac/Rho

There is now also emerging evidence that places each of the PIP5Ks upstream of Rho family GTPases. PIP5KγKO macrophages, which have a particle attachment defect during FcγR mediated phagocytosis and an abnormally stable actin cytoskeleton, have abnormally high RhoA activation and decreased Rac1 activation (Mao et al. 2009b). Inhibition of RhoA or activation of Rac1 corrects both defects, suggesting that PIP5Kγ acts upstream of Rac and Rho. Another study shows that PIP5Kα promotes Rac1 translocation/activation at the PM during integrin induced migration (Chao et al. 2010b). This PIP5Kα specific function requires physical interaction of PIP5Kα with the Rac1 polybasic domain and is independent of PIP5Kα's catalytic activity. Thus, PIP5Kα acts upstream of Rac1 as a scaffolding protein in this scenario. These interactions provide a venue for PIP5Kβ to function as a scaffold to coordinate rear signaling during leukocyte migration. PIP5Kβ silencing impairs RhoA activation during neutrophil polarization, suggesting that it acts upstream of RhoA (Manes et al. 2010).

6.6 PIP_2 and PIP5K Partitioning in Raft vs Nonraft Microdomains

The increasing evidence for PIP5K isoform specific differences strongly suggests that they generate distinct PIP_2 pools to regulate different functions. The questions of whether PIP_2 exists in heterogeneous microdomains, and whether these domains are formed by PIP_2 in cholesterol rich rafts by interaction with proteins, or by a combination of both, need to be answered. They hold the key to understanding how PIP_2 is regulated spatially and temporally, and how the PIP_2 pools generated by the PIP5Ks are functionally and possibly physically segregated. Some aspects of how different PIP_2 pools are created and maintained in the PM have been reviewed recently (Kwiatkowska 2010). We will focus on evidence for the existence of PIP_2 microdomains, speculate on how these domains are maintained in spite of diffusivity of lipids, and discuss emerging evidence that PIP5Ks may be differentially localized in these domains as well.

6.6.1 Lateral Distribution and Diffusivity of PPIs in Cell and Model Membranes

An important feature of PIP_2 signaling is potential partitioning of PIP_2 within regions of the membrane with different composition and dynamics and the possibility that PIP_2 is targeted to different proteins by this membrane localization. Evidence for different pools of PIP_2 in the cell originates from the discovery of fractions of PIP_2 in the PM that remained inaccessible to PLC hydrolysis following activation of that enzyme (Berridge 1983; Agranoff et al. 1983; Haeffner 1993). Consistent with the existence of different pools of PIP_2, immunofluorescence microscopy studies of various PIP_2-binding domains (Laux et al. 2000; Pike and Miller 1998), labeled

PIP_2 (Cho et al. 2005) and antibodies (Varnai et al. 2002; Huang et al. 2004) have reported a non-homogeneous distribution of PIP_2 in the PM.

Multiple mechanisms have been proposed for segregation of PIP_2 in different membrane microenvironments. These include corralling by the cortical actin fence (Johnson et al. 2008; Chichili et al. 2009; Morone et al. 2006; Cho et al. 2005) or microtubules (Golub and Caroni 2005), selective aggregation by peripheral membrane binding proteins, sequestration into liquid disordered membrane domains, and direct attractive interactions among PIP_2 headgroups or with other lipids. In addition, lipid diffusion may also be restricted by extreme membrane curvature as found in the phagocytic cup. During phagocytosis, PIP_2 persists in the phagocytic cup for minutes without diffusing away and the persistence of PIP_2 in the cup is not dependent on rafts or the actin cytoskeleton (Corbett-Nelson et al. 2006).

6.6.1.1 PIP_2 Sequestration by Proteins

A protein-based mechanism for PIP_2 oligomerization has been proposed, as the result of electrostatic interactions between several neighboring lipids and an unstructured polybasic protein domain such as that in the MARCKS protein (Laux et al. 2000; McLaughlin et al. 2002) and GAP-43, which sequesters PIP_2 primarily in rafts (Tong et al. 2008). In this model, the protein is required for PIP_2 clustering and serves to prevent interaction of PIP_2 with other potential protein targets until MARCKS or analogous proteins dissociate from PIP_2 as the result of their phosphorylation by enzymes such as protein kinase C. MARCKS sequesters PIP_2 under basal conditions, and is induced by agonist signaling to release PIP_2 for interaction with other PIP_2 targets (McLaughlin and Murray 2005). The physiological relevance of this model is confirmed by examining the effects of MARCKS knockdown on the response of endothelial cells to insulin (Kalwa and Michel 2011). MARCKS silencing attenuates endothelial wound healing, induces derangement of the actin cytoskeleton, and reduces the amount of PIP_2 in the light membrane fractions (caveolae/rafts) where MARCKS is also located. It interferes with insulin induced PIP_2 increase in lipid rafts and N-WASP/Arp2/3 activation. These results suggest that the PIP_2 required for N-WASP activation is likely to be exclusively supplied by release from MARCKS.

6.6.1.2 PIP_2 Sequestration by Counter Ions and Hydrogen Bonds

An alternative or complementary mechanism for formation of lateral PIP_2 aggregates in lamellar membranes, without a need for protein binding, is suggested by the following evidence: PIP_2 headgroups, unlike those of nearly all other phospholipids, can form extensive hydrogen-bonded networks (Redfern and Gericke 2004, 2005; Levental et al. 2008b; Liepina et al. 2003); divalent cations reduce the electrostatic repulsion between the anionic PIP_2 headgroups and also act as bridges between two adjacent lipids (Levental et al. 2009a; Christian et al. 2009; Lorenz et al. 2008).

In nearly all studies that consider the distribution of PIP_2 in the lipid bilayer, the self-interaction between PIP_2 molecules is assumed to be entirely repulsive due to their large net negative charge, approximately -3.5 at physiological ionic strength, and to a lesser extent because of the large steric size of the headgroup. Both experimental and modeling studies strongly suggest that hydrogen bonding between the multiple hydroxyl and phosphoester moieties on the inositol ring overcome to a large extent the electrostatic repulsions (Levental et al. 2008b; Redfern and Gericke 2005; Lorenz et al. 2008).

The experimentally measured surface pressure of pure PIP_2 monolayers is significantly less than the theoretically predicted values using a conservative model for purely electrostatic repulsion, and it is strongly increased by soluble agents that disrupt water structure, suggesting further bridging between PIP_2 mediated by interactions with solvent (Levental et al. 2008a, 2008b). Physiologic concentrations of divalent cations, especially Ca^{2+} during its influx from transmembrane channels, also potentiates formation of nano-scale clusters of PIP_2 (Levental et al. 2009a).

6.6.2 *The Controversy and Potential Resolution*

In recent years, the existence of cholesterol rich rafts *per se* has been intensely debated (Shaw 2006) and the relationship between PIP_2 domains and cholesterol-enriched rafts has been even more difficult to reconcile. Initial studies showed that PIP_2 cosedimented with cholesterol in low temperature detergent-insoluble fractions of cell membranes, suggesting that it concentrated in cholesterol-enriched rafts. This finding has been challenged by suggestions that detergent extraction *per se* induces artifactual clustering (van Rheenen et al. 2005), and optical measurements give mixed results, with only some data supporting the existence of a less mobile lipid population (Cho et al. 2005).

Colocalization of PIP_2 with cholesterol in solid-like or liquid ordered domains is also difficult to reconcile with the presence of the highly disordered arachidonate chain at the S3 position of natural PIP_2, and with studies o PIP_2 distribution in cholesterol-containing model membranes. In lipid monolayers compressed to the density of a bilayer, cholesterol-depended formation of liquid ordered (Lo) and liquid disordered (Ld) phases strongly partitions PIP_2 into the Ld phase, whereas cholesterol is presumed to be enriched in the Lo phase (Levental et al. 2009a, 2009b) Thus, even if PIP_2 and cholesterol do not co-segregate, the formation of Lo/Ld phase separation would nevertheless strongly alter the spatial distribution of PIP_2 in the cell membrane and contribute to its global and local partitioning in different pools (Levental et al. 2009a). Therefore, in spite of the above concerns, there is a compelling case for cholesterol's contribution to PM structure and dynamics that can alter the manner in which PIP_2 interacts with its protein targets.

6.6.3 Evidence for Raft and Nonraft Associated PIP_2 Pools

It has been reported based on biochemical evidence that approximately half of the cell's PIP_2 is synthesized preferentially in cholesterol/sphingolipid enriched caveolae light membrane fractions ("rafts") (Golub and Caroni 2005; Pike and Miller 1998; Morris et al. 2006), and that these PIP_2-enriched microdomains exhibit locally regulated PIP_2 turnover and restricted diffusion-mediated exchanges with their environment (Golub and Caroni 2005). PIP_2 in caveolar rafts is required for the budding of respiratory syncytial virus (Yeo et al. 2009). There are also reports that PIP_2 is enriched in noncaveolar raft microdomains that are the staging platforms for choreographing signaling and cytoskeletal dynamics.

6.6.3.1 PIP_2 Microdomains

The existing data suggest that there are multiple functionally independent pools of PIP_2, and there have been many attempts to show that these pools exist physically in the PM.

Immunofluorescence Microscopy

These studies show that PM PIP_2 microdomains are heterogeneous; some contain conventional raft markers, while others are enriched for syntaxin, which is involved in Ca^{2+}-mediated exocytosis and mostly excluded from the low density raft fraction (Aoyagi et al. 2005; Milosevic et al. 2005). PIP_2 clustering proteins, such as MARCKS (McLaughlin and Murray 2005; Kalwa and Michel 2010) and GAP-43 (Tong et al. 2008) have also been identified.

A study using fluorescent lipids delivered locally by a patch pipette shows that whereas labeled PI and PI(4)P diffuse freely throughout the PM from the site of delivery, the mobility of labeled PIP_2 is much more restricted, suggestive of its recruitment into membrane domains (Cho et al. 2005). When the actin network is destabilized using cytochalasin D, labeled PIP_2 diffuses freely, suggesting that part of the PIP_2-sequestration mechanism requires an intact cytoskeleton. Delivery of labeled (Bodipy TMR-) PIP_2 by microinjection of lipid micelles followed by fluorescence correlation spectroscopy also reveals a low diffusion rate compared to that of the labeled PIP_2 in simple bilayer vesicles, again suggesting that much of the PIP_2 in the cell is sequestered by proteins or other cellular structures (Golebiewska et al. 2008).

Electron Microscopy

There have been many attempts to identify PIP_2 pools by electron microscopy, but most of the techniques previously used lack sufficient resolution to distinguish between raft *vs* nonraft PIP_2 unequivocally. Recently, a freeze fracture method that does not use chemical fixation has been developed to probe PIP_2 distribution with GST-PH-PLCδ PIP_2 (Fujita et al. 2009). It shows that PIP_2 is highly enriched at the rim of caveolae and in coated pits, and is less clustered in other parts of the PM. When human dermal fibroblasts were stimulated with angiotensin II, the caveolar PIP_2 decreases by 80% in 40 s, while the nonraft pool decreases by 60% within 10 s. but recovers to normal level thereafter. The coated pit associated pool decreases only by 30% within this time frame.

Differential Targeting of 5-ptase

A recent report uses raft-targeted (L_{10}-) and non-raft-targeted (S_{15}-) Inp54p, respectively to probe their effects on T cell functions (Johnson et al. 2008). L_{10}-Inp54p targeting in T cells generates in a smooth PM phenotype that is devoid of membrane ruffles and filopodia. PIP_2 depletion is confirmed by dot-blot of light membrane fractions with an anti-PIP_2 antibody. Unexpectedly, S_{15}-Inp54 targeting increases raft PIP_2 and promotes filopodia formation and cell spreading. Although the mechanism for this increase is not understood, these results raise the possibility that the raft and nonraft PIP_2 pools regulate different cytoskeletal functions.

6.6.3.2 Differential PIP5K Localization

There had been many early attempts using floatation on density gradients to determine if PIP5Ks are raft or nonraft associated. Many of these early studies use *in vitro* lipid kinase activity or Western blotting with antibodies that cannot distinguish between the three isoforms. These studies failed to establish that PIP5K protein or activity is enriched in the raft fractions in PC12 cells and platelets (Yang et al. 2004; Aoyagi et al. 2005). In addition, the single yeast PIP5K ortholog, Mss4p, is not present in rafts, although its association with membrane is sphingolipid dependent (Kobayashi et al. 2005). Notwithstanding these early results, there is now new evidence to show that PIP5Kα, which is partially raft associated in the resting cell, is further recruited to rafts following B cell receptor (Saito et al. 2003) and FcγRIIA activation (Szymanska et al. 2009). These results raise the possibility that PIP_2 are generated *in situ* in rafts. It will be important to determine if the other PIP5Ks are raft associated and their response to receptor ligation.

References

Adachi R, Takeuchi K, Suzuki K (2002) Antisense oligonucleotide to cofilin enhances respiratory burst and phagocytosis in opsonized zymosan-stimulated mouse macrophage j774.1 cells. J Biol Chem 277:45566–45571

Agranoff BW, Murthy P, Seguin EB (1983) Thrombin-induced phosphodiesteratic cleavage of phosphatidylinositol bisphosphate in human platelets. J Biol Chem 258:2076–2078

Anderson RA, Marchesi VT (1985) Regulation of the association of membrane skeletal protein 4.1 with glycophorin by a polyphosphoinositide. Nature 318:295–298

Aoyagi K, Sugaya T, Umeda M, Yamamoto S, Terakawa S, Takahashi M (2005) The activation of exocytotic sites by the formation of phosphatidylinositol 4,5-bisphosphate microdomains at syntaxin clusters. J Biol Chem 280:17346–17352

Arioka M, Nakashima S, Shibasaki Y, Kitamoto K (2004) Dibasic amino acid residues at the carboxy-terminal end of kinase homology domain participate in the plasma membrane localization and function of phosphatidylinositol 5-kinase [gamma]. Biochem Biophys Res Commun 319:456–463

Arora PD, Chan MWC, Anderson RA, Janmey PA, McCulloch CA (2005) Separate functions of gelsolin mediate sequential steps of collagen phagocytosis. Mol Biol Cell 16:5175–5190

Audhya A, Emr SD (2002) Stt4 pi 4-kinase localizes to the plasma membrane and functions in the pkc1-mediated map kinase cascade. Dev Cell 2:593–605

Audhya A, Emr SD (2003) Regulation of pi4,5p2 synthesis by nuclear-cytoplasmic shuttling of the mss4 lipid kinase. EMBO J 22:4223–4236

Bairstow SF, Ling K, Anderson RA (2005) Phosphatidylinositol phosphate kinase type i{gamma} directly associates with and regulates shp-1 tyrosine phosphatase. J Biol Chem 280:23884–23891

Bakolitsa C, Cohen DM, Bankston LA, Bobkov AA, Cadwell GW, Jennings L, Critchley DR, Craig SW, Liddington RC (2004) Structural basis for vinculin activation at sites of cell adhesion. Nature 430:583–586

Balla T (2009) Green light to illuminate signal transduction events. Trends Cell Biol 19:575–586

Balla T, Varnai P (2002) Visualizing cellular phosphoinositide pools with gfp-fused protein-modules. Sci STKE 2002:PL3

Balla T, Varnai P (2009) Visualization of cellular phosphoinositide pools with gfp-fused protein-domains. Curr Protoc Cell Biol Chapter 24:Unit 24 4

Barret C, Roy C, Montcourrier P, Mangeat P, Niggli V (2000) Mutagenesis of the phosphatidylinositol 4,5-bisphosphate (pip(2)) binding site in the nh(2)-terminal domain of ezrin correlates with its altered cellular distribution. J Cell Biol 151:1067–1080

Berridge MJ (1983) Rapid accumulation of inositol trisphosphate reveals that agonists hydrolyse polyphosphoinositides instead of phosphatidylinositol. Biochem J 212:849–858

Bolomini-Vittori M, Montresor A, Giagulli C, Staunton D, Rossi B, Martinello M, Constantin G, Laudanna C (2009) Regulation of conformer-specific activation of the integrin lfa-1 by a chemokine-triggered rho signaling module. Nat Immunol 10:185–194

Botelho RJ, Teruel M, Dierckman R, Anderson R, Wells A, York JD, Meyer T, Grinstein S (2000) Localized biphasic changes in phosphatidylinositol-4,5-bisphosphate at sites of phagocytosis. J Cell Biol 151:1353–1368

Brown FD, Rozelle AL, Yin HL, Balla T, Donaldson JG (2001) Phosphatidylinositol 4,5-bisphosphate and arf6-regulated membrane traffic. J Cell Biol 154:1007–1017

Brown DA, Hughes SA, Marsh SJ, Tinker A (2007) Regulation of m(kv7.2/7.3) channels in neurons by pip(2) and products of pip(2) hydrolysis: significance for receptor-mediated inhibition. J Physiol 582:917–925

Burn P, Rotman A, Meyer RK, Burger MM (1985) Diacylglycerol in large alpha-actinin/actin complexes and in the cytoskeleton of activated platelets. Nature 314:469–472

Catimel B, Schieber C, Condron M, Patsiouras H, Connolly L, Catimel J, Nice EC, Burgess AW, Holmes AB (2008) The pi(3,5)p2 and pi(4,5)p2 interactomes. J Proteome Res 7:5295–5313

Chandrasekar I, Stradal TEB, Holt MR, Entschladen F, Jockusch BM, Ziegler WH (2005) Vinculin acts as a sensor in lipid regulation of adhesion-site turnover. J Cell Sci 118:1461–1472

Chang LC, Huang TH, Chang CS, Tsai YR, Lin RH, Lee PW, Hsu MF, Huang LJ, Wang JP (2011) Signaling mechanisms of inhibition of phospholipase d activation by chs-111 in formyl peptide-stimulated neutrophils. Biochem Pharmacol 81:269–278

Chao WT, Ashcroft F, Daquinag AC, Vadakkan T, Wei Z, Zhang P, Dickinson ME, Kunz J (2010a) Type i phosphatidylinositol phosphate kinase beta regulates focal adhesion disassembly by promoting beta1 integrin endocytosis. Mol Cell Biol 30:4463–4479

Chao WT, Daquinag AC, Ashcroft F, Kunz J (2010b) Type i pipk-alpha regulates directed cell migration by modulating rac1 plasma membrane targeting and activation. J Cell Biol 190:247–262

Chen MZ, Zhu X, Sun HQ, Mao YS, Wei Y, Yamamoto M, Yin HL (2009) Oxidative stress decreases phosphatidylinositol 4,5-bisphosphate levels by deactivating phosphatidylinositol- 4-phosphate 5-kinase beta in a syk-dependent manner. J Biol Chem 284:23743–23753

Chichili GR, Westmuckett AD, Rodgers W (2009) T cell signal regulation by the actin cytoskeleton. J Biol Chem 285:14737–14746

Cho H, Kim YA, Yoon J-Y, Lee D, Kim JH, Lee SH, Ho W-K (2005) Low mobility of phosphatidylinositol 4,5-bisphosphate underlies receptor specificity of gq-mediated ion channel regulation in atrial myocytes. PNAS 102:15241–15246

Chong LD, Traynor-Kaplan A, Bokoch GM, Schwartz MA (1994) The small gtp-binding protein rho regulates a phosphatidylinositol 4-phosphate 5-kinase in mammalian cells. Cell 79:507–513

Christian DA, Tian A, Ellenbroek WG, Levental I, Rajagopal K, Janmey PA, Liu AJ, Baumgart T, Discher DE (2009) Spotted vesicles, striped micelles and janus assemblies induced by ligand binding. Nat Mater 8:843–849

Ciano-Oliveira CD, Sirokmany G, Szaszi K, Arthur WT, Masszi A, Peterson M, Rotstein OD, Kapus A (2003) Hyperosmotic stress activates rho: differential involvement in rho kinase-dependent mlc phosphorylation and nkcc activation. Am J Physiol Cell Physiol 285:C555–C566

Clarke JH, Wang M, Irvine RF (2010) Localization, regulation and function of type ii phosphatidylinositol 5-phosphate 4-kinases. Adv Enzyme Regul 50:12–18

Coon BG, Mukherjee D, Hanna CB, Riese DJ 2nd, Lowe M, Aguilar RC (2009) Lowe syndrome patient fibroblasts display ocrl1-specific cell migration defects that cannot be rescued by the homologous inpp5b phosphatase. Hum Mol Genet 18:4478–4491

Coppolino MG, Dierckman R, Loijens J, Collins RF, Pouladi M, Jongstra-Bilen J, Schreiber AD, Trimble WS, Anderson R, Grinstein S (2002) Inhibition of phosphatidylinositol-4-phosphate 5-kinase ialpha impairs localized actin remodeling and suppresses phagocytosis. J Biol Chem 277:43849–43857

Corbett-Nelson EF, Mason D, Marshall JG, Collette Y, Grinstein S (2006) Signaling-dependent immobilization of acylated proteins in the inner monolayer of the plasma membrane. J Cell Biol 174:255–265

Cremona O, De Camilli P (2001) Phosphoinositides in membrane traffic at the synapse. J Cell Sci 114:1041–1052

Cremona O, Di Paolo G, Wenk MR, Luthi A, Kim WT, Takei K, Daniell L, Nemoto Y, Shears SB, Flavell RA, McCormick DA, De Camilli P (1999) Essential role of phosphoinositide metabolism in synaptic vesicle recycling. Cell 99:179–188

Das ND, Yoshioka T, Samuelson D, Shichi H (1987) Immunohistochemical localization of phosphatidylinositol-4,5-bisphosphate in the rat lens. Ophthalmic Res 19:57–60

Di Paolo G, De Camilli P (2006) Phosphoinositides in cell regulation and membrane dynamics. Nature 443:651–657

Di Paolo G, Pellegrini L, Letinic K, Cestra G, Zoncu R, Voronov S, Chang S, Guo J, Wenk MR, De Camilli P (2002) Recruitment and regulation of phosphatidylinositol phosphate kinase type 1 gamma by the ferm domain of talin. Nature 420:85–89

Di Paolo G, Moskowitz HS, Gipson K, Wenk MR, Voronov S, Obayashi M, Flavell R, Fitzsimonds RM, Ryan TA, De Camilli P (2004) Impaired ptdins(4,5)p2 synthesis in nerve terminals produces defects in synaptic vesicle trafficking. Nature 431:415–422

Divecha N, Irvine RF (1995) Phospholipid signaling. Cell 80:269–278

Doughman RL, Firestone AJ, Wojtasiak ML, Bunce MW, Anderson RA (2003) Membrane ruffling requires coordination between type ialpha phosphatidylinositol phosphate kinase and rac signaling. J Biol Chem 278:23036–23045

Dubash AD, Menold MM, Samson T, Boulter E, Garcia-Mata R, Doughman R, Burridge K (2009) Focal adhesions: new angles on an old structure. Int Rev Cell Mol Biol 277:1–65
El Sayegh TY, Arora PD, Ling K, Laschinger C, Janmey PA, Anderson RA, McCulloch CA (2007) Phosphatidylinositol-4,5 bisphosphate produced by pip5kigamma regulates gelsolin, actin assembly, and adhesion strength of n-cadherin junctions. Mol Biol Cell 18:3026–3038
Elliott PR, Goult BT, Kopp PM, Bate N, Grossmann JG, Roberts GC, Critchley DR, Barsukov IL (2010) The structure of the talin head reveals a novel extended conformation of the ferm domain. Structure 18:1289–1299
Elvers M, Pozgaj R, Pleines I, May F, Kuijpers MJ, Heemskerk JM, Yu P, Nieswandt B (2010) Platelet hyperreactivity and a prothrombotic phenotype in mice with a gain-of-function mutation in phospholipase cgamma2. J Thromb Haemost 8:1353–1363
Erwig L-P, McPhilips KA, Wynes MW, Ivetic A, Ridley AJ, Henson PM (2006) Differential regulation of phagosome maturation in macrophages and dendritic cells mediated by rho gtpases and ezrin-radixin-moesin (erm) proteins. Proc Natl Acad Sci 103:12825–12830
Fairn GD, Ogata K, Botelho RJ, Stahl PD, Anderson RA, De Camilli P, Meyer T, Wodak S, Grinstein S (2009) An electrostatic switch displaces phosphatidylinositol phosphate kinases from the membrane during phagocytosis. J Cell Biol 187:701–714
Fehon RG, McClatchey AI, Bretscher A (2010) Organizing the cell cortex: the role of erm proteins. Nat Rev Mol Cell Biol 11:276–287
Finkelstein M, Etkovitz N, Breitbart H (2010) Role and regulation of sperm gelsolin prior to fertilization. J Biol Chem 285:39702–39709
Flannagan RS, Grinstein S (2010) The application of fluorescent probes for the analysis of lipid dynamics during phagocytosis. Methods Mol Biol 591:121–134
Fujita A, Cheng J, Tauchi-Sato K, Takenawa T, Fujimoto T (2009) A distinct pool of phosphatidylinositol 4,5-bisphosphate in caveolae revealed by a nanoscale labeling technique. Proc Natl Acad Sci U S A 106:9256–9261
Funaki M, DiFransico L, Janmey PA (2006) Pi 4,5-p2 stimulates glucose transport activity of glut4 in the plasma membrane of 3t3-l1 adipocytes. Biochim Biophys Acta (BBA) Mol Cell Res 1763:889–899
Funakoshi Y, Hasegawa H, Kanaho Y (2011) Regulation of pip5k activity by arf6 and its physiological significance. J Cell Physiol 226:888–895
Gascard P, Tran D, Sauvage M, Sulpice JC, Fukami K, Takenawa T, Claret M, Giraud F (1991) Asymmetric distribution of phosphoinositides and phosphatidic acid in the human erythrocyte membrane. Biochim Biophys Acta 1069:27–36
Gilmore AP, Burridge K (1996) Regulation of vinculin binding to talin and actin by phosphatidylinositol-4-5-bisphosphate. Nature 381:531–535
Giudici M-L, Emson PC, Irvine RF (2004) A novel neuronal-specific splice variant of type i phosphatidylinositol 4-phosphate 5-kinase isoform gamma. Biochem J 379:489–496
Goksoy E, Ma Y-Q, Wang X, Kong X, Perera D, Plow EF, Qin J (2008) Structural basis for the autoinhibition of talin in regulating integrin activation. Mol Cell 31:124–133
Golebiewska U, Nyako M, Woturski W, Zaitseva I, McLaughlin S (2008) Diffusion coefficient of fluorescent phosphatidylinositol 4,5-bisphosphate in the plasma membrane of cells. Mol Biol Cell 19:1663–1669
Golub T, Caroni P (2005) Pi(4,5)p2-dependent microdomain assemblies capture microtubules to promote and control leading edge motility. J Cell Biol 169:151–165
Gong L-W, Di Paolo G, Diaz E, Cestra G, Diaz M-E, Lindau M, De Camilli P, Toomre D (2005) Phosphatidylinositol phosphate kinase type i{gamma} regulates dynamics of large dense-core vesicle fusion. PNAS 102:5204–5209
Gorbatyuk VY, Nosworthy NJ, Robson SA, Bains NPS, Maciejewski MW, dos Remedios CG, King GF (2006) Mapping the phosphoinositide-binding site on chick cofilin explains how pip2 regulates the cofilin-actin interaction. Mol Cell 24:511–522
Grinstein S (2010) Imaging signal transduction during phagocytosis: phospholipids, surface charge, and electrostatic interactions. Am J Physiol Cell Physiol 299:C876–C881
Groves E, Dart AE, Covarelli V, Caron E (2008) Molecular mechanisms of phagocytic uptake in mammalian cells. Cell Mol Life Sci 65:1957–1976

Haeffner EW (1993) Transient temporal relationship between 1-oleoyl-2-acetyl-sn-glycerol (oag)-activated synthesis and hydrolysis of polyphosphoinositides: desensitization of phospholipase c and the inositol lipid kinases upon long-term treatment of ascites cells by exogenous oag. J Lipid Mediat 7:239–252

Halstead JR, Rheenen J van, Snel Mireille HJ, Meeuws S, Mohammed S, D'Santos CS, Heck Albert J, Jalink K, Divecha N (2006) A role for ptdins(4,5)p2 and pip5k[alpha] in regulating stress-induced apoptosis. Curr Biol 16:1850–1856

Halstead JR, Savaskan NE, Bout I van den, Van Horck F, Hajdo-Milasinovic A, Snell M, Keune WJ, Ten Klooster JP, Hordijk PL, Divecha N (2010) Rac controls pip5k localisation and ptdins(4,5)p synthesis, which modulates vinculin localisation and neurite dynamics. J Cell Sci 123:3535–3546

Hammond GR, Schiavo G, Irvine RF (2009) Immunocytochemical techniques reveal multiple, distinct cellular pools of ptdins4p and ptdins(4,5)p(2). Biochem J 422:23–35

Hao JJ, Liu Y, Kruhlak M, Debell KE, Rellahan BL, Shaw S (2009) Phospholipase c-mediated hydrolysis of pip2 releases erm proteins from lymphocyte membrane. J Cell Biol 184:451–462

Hartwig JH, Bokoch GM, Carpenter CL, Janmey PA, Taylor LA, Toker A, Stossel TP (1995) Thrombin receptor ligation and activated rac uncap actin filament barbed ends through phosphoinositide synthesis in permeabilized human platelets. Cell 82:643–653

Heldwein EE, Macia E, Wang J, Yin HL, Kirchhausen T, Harrison SC (2004) Crystal structure of the clathrin adaptor protein 1 core. PNAS 101:14108–14113

Heo WD, Inoue T, Park WS, Kim ML, Park BO, Wandless TJ, Meyer T (2006) Pi(3,4,5)p3 and pi(4,5)p2 lipids target proteins with polybasic clusters to the plasma membrane. Science 314:1458–1461

Hirose K, Kadowaki S, Tanabe M, Takeshima H, Iino M (1999) Spatiotemporal dynamics of inositol 1,4,5-trisphosphate that underlies complex ca2 + mobilization patterns. Science 284:1527–1530

Honda A, Nogami M, Yokozeki T, Yamazaki M, Nakamura H, Watanabe H, Kawamoto K, Nakayama K, Morris AJ, Frohman MA, Kanaho Y (1999) Phosphatidylinositol 4-phosphate 5-kinase [alpha] is a downstream effector of the small g protein arf6 in membrane ruffle formation. Cell 99:521–532

Huang S, Lifshitz L, Patki-Kamath V, Tuft R, Fogarty K, Czech MP (2004) Phosphatidylinositol-4,5-bisphosphate-rich plasma membrane patches organize active zones of endocytosis and ruffling in cultured adipocytes. Mol Cell Biol 24:9102–9123

Hughes S, Marsh SJ, Tinker A, Brown DA (2007) Pip(2)-dependent inhibition of m-type (kv7.2/7.3) potassium channels: direct on-line assessment of pip(2) depletion by gq-coupled receptors in single living neurons. Pflugers Arch 455:115–124

Ishihara H, Shibasaki Y, Kizuki N, Katagiri H, Yazaki Y, Asano T, Oka Y (1996) Cloning of cdnas encoding two isoforms of 68-kda type i phosphatidylinositol4-phosphate 5-kinase. J Biol Chem 271:23611–23614

Ishihara H, Shibasaki Y, Kizuki N, Wada T, Yazaki Y, Asano T, Oka Y (1998) Type i phosphatidylinositol-4-phosphate 5-kinases. Cloning of the third isoform and deletion/substitution analysis of members of this novel lipid kinase family. J Biol Chem 273:8741–8748

Ito S, Werth DK, Richert ND, Pastan I (1983) Vinculin phosphorylation by the src kinase. Interaction of vinculin with phospholipid vesicles. J Biol Chem 258:14626–14631

Itoh T, Ishihara H, Shibasaki Y, Oka Y, Takenawa T (2000) Autophosphorylation of type i phosphatidylinositol phosphate kinase regulates its lipid kinase activity. J Biol Chem 275:19389–19394

Janetopoulos C, Devreotes P (2006) Phosphoinositide signaling plays a key role in cytokinesis. J Cell Biol 174:485–490

Janmey PA, Stossel TP (1987) Modulation of gelsolin function by phosphatidylinositol 4,5-bisphosphate. Nature 325:362–364

Janmey PA, Bucki R, Yin HL (2009) Phosphoinositides and actin cytoskeletal rearrangement. In: Bradshaw T, Dennis E (eds) The handbook of cell signaling, 2nd edn. Elsevier, New York

Jarquin-Pardo M, Fitzpatrick A, Galiano FJ, First EA, Davis JN (2007) Phosphatidic acid regulates the affinity of the murine phosphatidylinositol 4-phosphate 5-kinase-ibeta for phosphatidylinositol-4-phosphate. J Cell Biochem 100:112–128

Johnson RP, Niggli V, Durrer P, Craig SW (1998) A conserved motif in the tail domain of vinculin mediates association with and insertion into acidic phospholipid bilayers. Biochemistry 37:10211–10222

Johnson CM, Chichili GR, Rodgers W (2008) Compartmentalization of phosphatidylinositol 4,5-bisphosphate signaling evidenced using targeted phosphatases. J Biol Chem 283:29920–29928

Kagan JC, Medzhitov R (2006) Phosphoinositide-mediated adaptor recruitment controls toll-like receptor signaling. Cell 125:943–955

Kalwa H, Michel T (2010) The marcks protein plays a critical role in phosphatidylinositol 4,5-bisphosphate metabolism and directed cell movement in vascular endothelial cells. J Biol Chem 286:2320–2330

Kalwa H, Michel T (2011) The marcks protein plays a critical role in phosphatidylinositol 4,5-bisphosphate metabolism and directed cell movement in vascular endothelial cells. J Biol Chem 286:2320–2330

Kisseleva M, Feng Y, Ward M, Song C, Anderson RA, Longmore GD (2005) The lim protein ajuba regulates phosphatidylinositol 4,5-bisphosphate levels in migrating cells through an interaction with and activation of pipki{alpha}. Mol Cell Biol 25:3956–3966

Klein RM, Spofford LS, Abel EV, Ortiz A, Aplin AE (2008) B-raf regulation of rnd3 participates in actin cytoskeletal and focal adhesion organization. Mol Biol Cell 19:498–508

Kobayashi T, Takematsu H, Yamaji T, Hiramoto S, Kozutsumi Y (2005) Disturbance of sphingolipid biosynthesis abrogates the signaling of mss4, phosphatidylinositol-4-phosphate 5-kinase, in yeast. J Biol Chem 280:18087–18094

Komaba S, Coluccio LM (2010) Localization of myosin 1b to actin protrusions requires phosphoinositide binding. J Biol Chem 285:27686–27693

Krauss M, Haucke V (2007) Phosphoinositide-metabolizing enzymes at the interface between membrane traffic and cell signalling. EMBO Rep 8:241–246

Kunz J, Wilson MP, Kisseleva M, Hurley JH, Majerus PW, Anderson RA (2000) The activation loop of phosphatidylinositol phosphate kinases determines signaling specificity. Mol Cell 5:1–11

Kunz J, Fuelling A, Kolbe L, Anderson RA (2002) Stereo-specific substrate recognition by phosphatidylinositol phosphate kinases is swapped by changing a single amino acid residue. J Biol Chem 277:5611–5619

Kwiatkowska K (2010) One lipid, multiple functions: how various pools of pi(4,5)p(2) are created in the plasma membrane. Cell Mol Life Sci 67:3927–3946

Lacalle RA, Peregil RM, Albar JP, Merino E, Martinez AC, Merida I, Manes S (2007) Type i phosphatidylinositol 4-phosphate 5-kinase controls neutrophil polarity and directional movement. J Cell Biol 179:1539–1553

Lassing I, Lindberg U (1985) Specific interaction between phosphatidylinositol 4,5-bisphosphate and profilactin. Nature 314:472–474

Laux T, Fukami K, Thelen M, Golub T, Frey D, Caroni P (2000) Gap43, marcks, and cap23 modulate pi(4,5)p2 at plasmalemmal rafts, and regulate cell cortex actin dynamics through a common mechanism. J Cell Biol 149:1455–1472

Lee SY, Voronov S, Letinic K, Nairn AC, Di Paolo G, De Camilli P (2005) Regulation of the interaction between pipki{gamma} and talin by proline-directed protein kinases. J Cell Biol 168:789–799

Lee SY, Kim B, Yoon S, Kim YJ, Liu T, Woo JH, Chwae YJ, Joe EH, Jou I (2010) Phosphatidylinositol 4-phosphate 5-kinase alpha is induced in ganglioside-stimulated brain astrocytes and contributes to inflammatory responses. Exp Mol Med 42:662–673

Levental I, Cebers A, Janmey PA (2008a) Combined electrostatics and hydrogen bonding determine intermolecular interactions between polyphosphoinositides. J Am Chem Soc 130:9025–9030

Levental I, Janmey PA, Cebers A (2008b) Electrostatic contribution to the surface pressure of charged monolayers containing polyphosphoinositides. Biophys J 95:1199–1205

Levental I, Byfield FJ, Chowdhury P, Gai F, Baumgart T, Janmey PA (2009a) Cholesterol-dependent phase separation in cell-derived giant plasma-membrane vesicles. Biochem J 424:163–167

Levental I, Christian DA, Wang YH, Madara JJ, Discher DE, Janmey PA (2009b) Calcium-dependent lateral organization in phosphatidylinositol 4,5-bisphosphate (pip2)- and cholesterol-containing monolayers. Biochemistry 48:8241–8248

Leyman S, Sidani M, Ritsma L, Waterschoot D, Eddy R, Dewitte D, Debeir O, Decaestecker C, Vandekerckhove J, Rheenen J van, Ampe C, Condeelis J, Van Troys M (2009) Unbalancing the phosphatidylinositol-4,5-bisphosphate-cofilin interaction impairs cell steering. Mol Biol Cell 20:4509–4523

Liepina I, Czaplewski C, Janmey P, Liwo A (2003) Molecular dynamics study of a gelsolin-derived peptide binding to a lipid bilayer containing phosphatidylinositol 4,5-bisphosphate. Biopolymers 71:49–70

Ling K, Doughman RL, Firestone AJ, Bunce MW, Anderson RA (2002) Type i[gamma] phosphatidylinositol phosphate kinase targets and regulates focal adhesions. Nature 420:89–93

Ling K, Doughman RL, Iyer VV, Firestone AJ, Bairstow SF, Mosher DF, Schaller MD, Anderson RA (2003) Tyrosine phosphorylation of type i{gamma} phosphatidylinositol phosphate kinase by src regulates an integrin-talin switch. J Cell Biol 163:1339–1349

Ling K, Bairstow SF, Carbonara C, Turbin DA, Huntsman DG, Anderson RA (2007) Type igamma phosphatidylinositol phosphate kinase modulates adherens junction and e-cadherin trafficking via a direct interaction with mu 1b adaptin. J Cell Biol 176:343–353

Liu Y, Bankaitis VA (2010) Phosphoinositide phosphatases in cell biology and disease. Prog Lipid Res 49:201–217

Logan MR, Mandato CA (2006) Regulation of the actin cytoskeleton by pip2 in cytokinesis. Biol Cell 98:377–388

Loijens JC, Anderson RA (1996) Type i phosphatidylinositol-4-phosphate 5-kinases are distinct members of this novel lipid kinase family. J Biol Chem 271:32937–32943

Lokuta MA, Senetar MA, Bennin DA, Nuzzi PA, Chan KT, Ott VL, Huttenlocher A (2007) Type igamma pip kinase is a novel uropod component that regulates rear retraction during neutrophil chemotaxis. Mol Biol Cell 18:5069–5080

Lorenz CD, Faraudo J, Travesset A (2008) Hydrogen bonding and binding of polybasic residues with negatively charged mixed lipid monolayers. Langmuir 24:1654–1658

Ludwig A, Otto GP, Riento K, Hams E, Fallon PG, Nichols BJ (2010) Flotillin microdomains interact with the cortical cytoskeleton to control uropod formation and neutrophil recruitment. J Cell Biol 191:771–781

Mace EM, Zhang J, Siminovitch KA, Takei F (2010) Elucidation of the integrin lfa-1-mediated signaling pathway of actin polarization in natural killer cells. Blood 116:1272–1279

Malm B, Larsson H, Lindberg U (1983) The profilin–actin complex: further characterization of profilin and studies on the stability of the complex. J Muscle Res Cell Motil 4:569–588

Manes S, Fuentes G, Peregil RM, Rojas AM, Lacalle RA (2010) An isoform-specific pdz-binding motif targets type i pip5 kinase beta to the uropod and controls polarization of neutrophil-like hl60 cells. FASEB J 24:3381–3392

Mao YS, Yin HL (2007) Regulation of the actin cytoskeleton by phosphatidylinositol 4-phosphate 5 kinases. Pflugers Arch 455:5–18

Mao Y, Balkin DM, Zoncu R, Erdmann KS, Tomasini L, Hu F, Jin MM, Hodsdon ME, De Camilli P (2009a) A ph domain within ocrl bridges clathrin-mediated membrane trafficking to phosphoinositide metabolism. EMBO J 28:1831–1842

Mao YS, Yamaga M, Zhu X, Wei Y, Sun HQ, Wang J, Yun M, Wang Y, Di Paolo G, Bennett M, Mellman I, Abrams CS, De Camilli P, Lu CY, Yin HL (2009b) Essential and unique roles of pip5k-gamma and -alpha in fcgamma receptor-mediated phagocytosis. J Cell Biol 184:281–296

Martin TFJ (2001) Pi(4,5)p2 regulation of surface membrane traffic. Curr Opin Cell Biol 13:493–499

Martin-Belmonte F, Gassama A, Datta A, Yu W, Rescher U, Gerke V, Mostov K (2007) Pten-mediated apical segregation of phosphoinositides controls epithelial morphogenesis through cdc42. Cell 128:383–397

McCrea HJ, De Camilli P (2009) Mutations in phosphoinositide metabolizing enzymes and human disease. Physiology (Bethesda) 24:8–16

McLaughlin S, Murray D (2005) Plasma membrane phosphoinositide organization by protein electrostatics. Nature 438:605–611

McLaughlin S, Wang J, Gambhir A, Murray D (2002) Pip2 and proteins: interactions, organization, and information flow. Annu Rev Biophys Biomol Struct 31:151–175

Meira M, Masson R, Stagljar I, Lienhard S, Maurer F, Boulay A, Hynes NE (2009) Memo is a cofilin-interacting protein that influences plc{gamma}1 and cofilin activities, and is essential for maintaining directionality during erbb2-induced tumor-cell migration. J Cell Sci 122:787–797

Mejillano M, Yamamoto M, Rozelle AL, Sun H-Q, Wang X, Yin HL (2001) Regulation of apoptosis by phosphatidylinositol 4,5-bisphosphate inhibition of caspases, and caspase inactivation of phosphatidylinositol phosphate 5-kinases. J Biol Chem 276:1865–1872

Michailidis IE, Rusinova R, Georgakopoulos A, Chen Y, Iyengar R, Robakis NK, Logothetis DE, Baki L (2011) Phosphatidylinositol-4,5-bisphosphate regulates epidermal growth factor receptor activation. Pflugers Arch 461:387–397

Micucci F, Capuano C, Marchetti E, Piccoli M, Frati L, Santoni A, Galandrini R (2008) Pi5ki-dependent signals are critical regulators of the cytolytic secretory pathway. Blood 111:4165–4172

Milosevic I, Sorensen JB, Lang T, Krauss M, Nagy G, Haucke V, Jahn R, Neher E (2005) Plasmalemmal phosphatidylinositol-4,5-bisphosphate level regulates the releasable vesicle pool size in chromaffin cells. J Neurosci 25:2557–2565

Morone N, Fujiwara T, Murase K, Kasai RS, Ike H, Yuasa S, Usukura J, Kusumi A (2006) Three-dimensional reconstruction of the membrane skeleton at the plasma membrane interface by electron tomography. J Cell Biol 174:851–862

Morris JB, Huynh H, Vasilevski O, Woodcock EA (2006) Alpha1-adrenergic receptor signaling is localized to caveolae in neonatal rat cardiomyocytes. J Mol Cell Cardiol 41:17–25

Nakano-Kobayashi A, Yamazaki M, Unoki T, Hongu T, Murata C, Taguchi R, Katada T, Frohman MA, Yokozeki T, Kanaho Y (2007) Role of activation of pip5kgamma661 by ap-2 complex in synaptic vesicle endocytosis. EMBO J 26:1105–1116

Nelson CD, Kovacs JJ, Nobles KN, Whalen EJ, Lefkowitz RJ (2008) Beta-arrestin scaffolding of phosphatidylinositol 4-phosphate 5-kinase ialpha promotes agonist-stimulated sequestration of the beta2-adrenergic receptor. J Biol Chem 283:21093–21101

Niggli V, Dimitrov DP, Brunner J, Burger MM (1986) Interaction of the cytoskeletal component vinculin with bilayer structures analyzed with a photoactivatable phospholipid. J Biol Chem 261:6912–6918

Padron D, Wang YJ, Yamamoto M, Yin H, Roth MG (2003) Phosphatidylinositol phosphate 5-kinase ibeta recruits ap-2 to the plasma membrane and regulates rates of constitutive endocytosis. J Cell Biol 162:693–701

Pan W, Choi SC, Wang H, Qin Y, Volpicelli-Daley L, Swan L, Lucast L, Khoo C, Zhang X, Li L, Abrams CS, Sokol SY, Wu D (2008) Wnt3a-mediated formation of phosphatidylinositol 4,5-bisphosphate regulates lrp6 phosphorylation. Science 321:1350–1353

Papayannopoulos V, Co C, Prehoda KE, Snapper S, Taunton J, Lim WA (2005) A polybasic motif allows n-wasp to act as a sensor of pip(2) density. Mol Cell 17:181–191

Park SJ, Itoh T, Takenawa T (2001) Phosphatidylinositol 4-phosphate 5-kinase type i is regulated through phosphorylation response by extracellular stimuli. J Biol Chem 276:4781–4787

Parsons JT, Horwitz AR, Schwartz MA (2010) Cell adhesion: integrating cytoskeletal dynamics and cellular tension. Nat Rev Mol Cell Biol 11:633–643

Pearson MA, Reczek D, Bretscher A, Karplus PA (2000) Structure of the erm protein moesin reveals the ferm domain fold masked by an extended actin binding tail domain. Cell 101:259–270

Pike LJ, Miller JM (1998) Cholesterol depletion delocalizes phosphatidylinositol bisphosphate and inhibits hormone-stimulated phosphatidylinositol turnover. J Biol Chem 273:22298–22304

Powner DJ, Payne RM, Pettitt TR, Giudici ML, Irvine RF, Wakelam MJO (2005) Phospholipase d2 stimulates integrin-mediated adhesion via phosphatidylinositol 4-phosphate 5-kinase i{gamma}b. J Cell Sci 118:2975–2986

Quinn K, Behe P, Tinker A (2008) Monitoring changes in membrane phosphatidylinositol 4,5-bisphosphate in living cells using a domain from the transcription factor tubby. J Physiol 586:2855–2871

Rao VD, Misra S, Boronenkov IV, Anderson RA, Hurley JH (1998) Structure of type ii[beta] phosphatidylinositol phosphate kinase: a protein kinase fold flattened for interfacial phosphorylation. Cell 94:829–839

Raucher D, Stauffer T, Chen W, Shen K, Guo S, York JD, Sheetz MP, Meyer T (2000) Phosphatidylinositol 4,5-bisphosphate functions as a second messenger that regulates cytoskeleton-plasma membrane adhesion. Cell 100:221–228

Redfern DA, Gericke A (2004) Domain formation in phosphatidylinositol monophosphate/phosphatidylcholine mixed vesicles. Biophys J 86:2980–2992

Redfern DA, Gericke A (2005) Ph-dependent domain formation in phosphatidylinositol polyphosphate/phosphatidylcholine mixed vesicles. J Lipid Res 46:504–515

Ren XD, Bokoch GM, Traynor-Kaplan A, Jenkins GH, Anderson RA, Schwartz MA (1996) Physical association of the small gtpase rho with a 68-kda phosphatidylinositol 4-phosphate 5-kinase in swiss 3t3 cells. Mol Biol Cell 7:435–442

Rozelle AL, Machesky LM, Yamamoto M, Driessens MHE, Insall RH, Roth MG, Luby-Phelps K, Marriott G, Hall A, Yin HL (2000) Phosphatidylinositol 4,5-bisphosphate induces actin-based movement of raft-enriched vesicles through wasp-arp2/3. Curr Biol 10:311–320

Rumenapp U, Schmidt M, Olesch S, Ott S, Eichel-Streiber CV, Jakobs KH (1998) Tyrosine-phosphorylation-dependent and rho-protein-mediated control of cellular phosphatidylinositol 4,5-bisphosphate levels. Biochem J 334(Pt 3):625–631

Russo C, Gao Y, Mancini P, Vanni C, Porotto M, Falasca M, Torrisi MR, Zheng Y, Eva A (2001) Modulation of oncogenic dbl activity by phosphoinositol phosphate binding to pleckstrin homology domain. J Biol Chem 276:19524–19531

Saarikangas J, Zhao H, Lappalainen P (2010) Regulation of the actin cytoskeleton-plasma membrane interplay by phosphoinositides. Physiol Rev 90:259–289

Saito K, Tolias KF, Saci A, Koon HB, Humphries LA, Scharenberg A, Rawlings DJ, Kinet J-P, Carpenter CL (2003) Btk regulates ptdins-4,5-p2 synthesis: importance for calcium signaling and pi3k activity. Immunity 19:669–677

Sanchez-Madrid F, Serrador JM (2009) Bringing up the rear: defining the roles of the uropod. Nat Rev Mol Cell Biol 10:353–359

Sasaki J, Sasaki T, Yamazaki M, Matsuoka K, Taya C, Shitara H, Takasuga S, Nishio M, Mizuno K, Wada T, Miyazaki H, Watanabe H, Iizuka R, Kubo S, Murata S, Chiba T, Maehama T, Hamada K, Kishimoto H, Frohman MA, Tanaka K, Penninger JM, Yonekawa H, Suzuki A, Kanaho Y (2005) Regulation of anaphylactic responses by phosphatidylinositol phosphate kinase type i {alpha}. J Exp Med 201:859–870

Saunders RM, Holt MR, Jennings L, Sutton DH, Barsukov IL, Bobkov A, Liddington RC, Adamson EA, Dunn GA, Critchley DR (2006) Role of vinculin in regulating focal adhesion turnover. Eur J Cell Biol 85:487–500

Schill NJ, Anderson RA (2009) Two novel phosphatidylinositol-4-phosphate 5-kinase type igamma splice variants expressed in human cells display distinctive cellular targeting. Biochem J 422:473–482

Scott CC, Dobson W, Botelho RJ, Coady-Osberg N, Chavrier P, Knecht DA, Heath C, Stahl P, Grinstein S (2005) Phosphatidylinositol-4,5-bisphosphate hydrolysis directs actin remodeling during phagocytosis. J Cell Biol 169:139–149

Shaw AS (2006) Lipid rafts: now you see them, now you don't. Nat Immunol 7:1139–1142

Sheetz MP, Sable JE, Dobereiner HG (2006) Continuous membrane-cytoskeleton adhesion requires continuous accommodation to lipid and cytoskeleton dynamics. Annu Rev Biophys Biomol Struct 35:417–434

Stace C, Manifava M, Delon C, Coadwell J, Cockcroft S, Ktistakis NT (2008) Pa binding of phosphatidylinositol 4-phosphate 5-kinase. Adv Enzyme Regul 48:55–72
Strochlic TI, Viaud J, Rennefahrt UEE, Anastassiadis T, Peterson JR (2010) Phosphoinositides are essential coactivators for p21-activated kinase 1. Mol Cell 40:493–500
Suchy SF, Nussbaum RL (2002) The deficiency of pip2 5-phosphatase in lowe syndrome affects actin polymerization. Am J Hum Genet 71:1420–1427
Suh B-C, Inoue T, Meyer T, Hille B (2006) Rapid chemically induced changes of ptdins(4,5)p2 gate kcnq ion channels. Science 314:1454–1457
Suh BC, Leal K, Hille B (2010) Modulation of high-voltage activated ca(2 +) channels by membrane phosphatidylinositol 4,5-bisphosphate. Neuron 67:224–238
Sun Y, Dandekar RD, Mao YS, Yin HL, Wulfing C (2011) Phosphatidylinositol (4,5) bisphosphate controls T cell activation by regulating T cell rigidity and organization. PLoS One 6:e27227
Sun Y, Ling K, Wagoner MP, Anderson RA (2007) Type i gamma phosphatidylinositol phosphate kinase is required for egf-stimulated directional cell migration. J Cell Biol 178:297–308
Sun Y, Turbin D, Ling K, Thapa N, Leung S, Huntsman D, Anderson R (2010) Type i gamma phosphatidylinositol phosphate kinase modulates invasion and proliferation and its expression correlates with poor prognosis in breast cancer. Breast Cancer Res 12:R6
Szentpetery Z, Balla A, Kim Y, Lemmon M, Balla T (2009) Live cell imaging with protein domains capable of recognizing phosphatidylinositol 4,5-bisphosphate; a comparative study. BMC Cell Biol 10:67
Szymanska E, Sobota A, Czurylo E, Kwiatkowska K (2008) Expression of pi(4,5)p2-binding proteins lowers the pi(4,5)p2level and inhibits fcgammariia-mediated cell spreading and phagocytosis. Eur J Immunol 38:260–272
Szymanska E, Korzeniowski M, Raynal P, Sobota A, Kwiatkowska K (2009) Contribution of pip-5 kinase i[alpha] to raft-based fc[gamma]riia signaling. Exp Cell Res 315:981–995
Toker A, Cantley LC (1997) Signalling through the lipid products of phosphoinositide-3-oh kinase. Nature 387:673–676
Tolias K, Carpenter CL (2000) In vitro interaction of phosphoinositide-4-phosphate 5-kinases with rac. Methods Enzymol 325:190–200
Tolias KF, Cantley LC, Carpenter CL (1995) Rho family gtpases bind to phosphoinositide kinases. J Biol Chem 270:17656–17659
Tomas A, Yermen B, Regazzi R, Pessin JE, Halban PA (2010) Regulation of insulin secretion by phosphatidylinositol-4,5-bisphosphate. Traffic 11:123–137
Tong J, Nguyen L, Vidal A, Simon SA, Skene JH, McIntosh TJ (2008) Role of gap-43 in sequestering phosphatidylinositol 4,5-bisphosphate to raft bilayers. Biophys J 94:125–133
Tran D, Gascard P, Berthon B, Fukami K, Takenawa T, Giraud F, Claret M (1993) Cellular distribution of polyphosphoinositides in rat hepatocytes. Cell Signal 5:565–581
van den Bout I, Divecha N (2009) Pip5k-driven ptdins(4,5)p2 synthesis: regulation and cellular functions. J Cell Sci 122:3837–3850
Rheenen J van, Achame EM, Janssen H, Calafat J, Jalink K (2005) Pip2 signaling in lipid domains: a critical re-evaluation. EMBO J 24:1664–1673
Rheenen J van, Song X, Roosmalen W van, Cammer M, Chen X, Desmarais V, Yip SC, Backer JM, Eddy RJ, Condeelis JS (2007) Egf-induced pip2 hydrolysis releases and activates cofilin locally in carcinoma cells. J Cell Biol 179:1247–1259
Varnai P, Lin X, Lee S, Tuymetova G, Bondeva T, Spat A, Rhee S, Hajnoczky G, Balla T (2002) Inositol lipid binding and membrane localization of isolated pleckstrin homology (ph) domains. Studies on the ph domains of phospholipase c delta 1 and p130. J Biol Chem 277:27412–27422
Varnai P, Thyagarajan B, Rohacs T, Balla T (2006) Rapidly inducible changes in phosphatidylinositol 4,5-bisphosphate levels influence multiple regulatory functions of the lipid in intact living cells. J Cell Biol 175:377–382
Vasudevan L, Jeromin A, Volpicelli-Daley L, De Camilli P, Holowka D, Baird B (2009) The beta- and gamma-isoforms of type i pip5k regulate distinct stages of ca2 + signaling in mast cells. J Cell Sci 122:2567–2574

Volpicelli-Daley LA, Lucast L, Gong LW, Liu L, Sasaki J, Sasaki T, Abrams CS, Kanaho Y, De Camilli P (2010) Phosphatidylinositol-4-phosphate 5-kinases and phosphatidylinositol 4,5-bisphosphate synthesis in the brain. J Biol Chem 285:28708–28714

Voronov SV, Frere SG, Giovedi S, Pollina EA, Borel C, Zhang H, Schmidt C, Akeson EC, Wenk MR, Cimasoni L, Arancio O, Davisson MT, Antonarakis SE, Gardiner K, De Camilli P, Di Paolo G (2008) Synaptojanin 1-linked phosphoinositide dyshomeostasis and cognitive deficits in mouse models of down's syndrome. Proc Natl Acad Sci U S A 105:9415–9420

Wang F (2009) The signaling mechanisms underlying cell polarity and chemotaxis. Cold Spring Harb Perspect Biol 1:a002980

Wang YJ, Wang J, Sun HQ, Martinez M, Sun YX, Macia E, Kirchhausen T, Albanesi JP, Roth MG, Yin HL (2003) Phosphatidylinositol 4 phosphate regulates targeting of clathrin adaptor ap-1 complexes to the golgi. Cell 114:299–310

Wang Y, Li W, Wang J, Xu K, Dong P, Luo X, Yin H (2004) Critical role of pip5ki{gamma}87 in insp3-mediated ca(2+) signaling. J Cell Biol 167:1005–1010

Wang L, Li G, Sugita S (2005) A central kinase domain of type i phosphatidylinositol phosphate kinases is sufficient to prime exocytosis: isoform specificity and its underlying mechanism. J Biol Chem 280:16522–16527

Wang Y, Lian L, Golden JA, Morrisey EE, Abrams CS (2007) Pip5ki gamma is required for cardiovascular and neuronal development. Proc Natl Acad Sci U S A 104:11748–11753

Wang Y, Chen X, Lian L, Tang T, Stalker TJ, Sasaki T, Brass LF, Choi JK, Hartwig JH, Abrams CS (2008a) Loss of pip5kibeta demonstrates that pip5ki isoform-specific pip2 synthesis is required for ip3 formation. Proc Natl Acad Sci U S A 105:14064–14069

Wang Y, Litvinov RI, Chen X, Bach TL, Lian L, Petrich BG, Monkley SJ, Critchley DR, Sasaki T, Birnbaum MJ, Weisel JW, Hartwig J, Abrams CS (2008b) Loss of pip5kigamma, unlike other pip5ki isoforms, impairs the integrity of the membrane cytoskeleton in murine megakaryocytes. J Clin Invest 118:812–819

Weekes J, Barry ST, Critchley DR (1996) Acidic phospholipids inhibit the intramolecular association between the N- and C-terminal regions of vinculin, exposing actin-binding and protein kinase C phosphorylation sites. Biochem J 314(Pt 3):827–832

Weernink PAO, Meletiadis K, Hommeltenberg S, Hinz M, Ishihara H, Schmidt M, Jakobs KH (2004) Activation of type i phosphatidylinositol 4-phosphate 5-kinase isoforms by the rho gtpases, rhoa, rac1, and cdc42. J Biol Chem 279:7840–7849

Wenk MR, Pellegrini L, Klenchin VA, Di Paolo G, Chang S, Daniell L, Arioka M, Martin TF, De Camilli P (2001) Pip kinase i[gamma] is the major pi(4,5)p2 synthesizing enzyme at the synapse. Neuron 32:79–88

Wernimont SA, Legate KR, Simonson WT, Fassler R, Huttenlocher A (2010) Pipki gamma 90 negatively regulates lfa-1-mediated adhesion and activation in antigen-induced cd4+t cells. J Immunol 185:4714–4723

Wong K-W, Isberg RR (2003) Arf6 and phosphoinositol-4-phosphate-5-kinase activities permit bypass of the rac1 requirement for {beta}1 integrin-mediated bacterial uptake. J Exp Med 198:603–614

Xie Z, Chang SM, Pennypacker SD, Liao EY, Bikle DD (2009) Phosphatidylinositol 4-phosphate 5-kinase 1 alpha mediates extracellular calcium-induced keratinocyte differentiation. Mol Biol Cell 20:1695–1704

Xu W, Wang P, Petri B, Zhang Y, Tang W, Sun L, Kress H, Mann T, Shi Y, Kubes P, Wu D (2010) Integrin-induced pip5k1c kinase polarization regulates neutrophil polarization, directionality, and in vivo infiltration. Immunity 33:340–350

Yamaguchi H, Yoshida S, Muroi E, Kawamura M, Kouchi Z, Nakamura Y, Sakai R, Fukami K (2010) Phosphatidylinositol 4,5-bisphosphate and pip5-kinase ialpha are required for invadopodia formation in human breast cancer cells. Cancer Sci 101:1632–1638

Yamamoto M, Hilgemann DH, Feng S, Bito H, Ishihara H, Shibasaki Y, Yin HL (2001) Phosphatidylinositol 4,5-bisphosphate induces actin stress-fiber formation and inhibits membrane ruffling in cv1 cells. J Cell Biol 152:867–876

Yamamoto M, Chen MZ, Wang YJ, Sun HQ, Wei Y, Martinez M, Yin HL (2006) Hypertonic stress increases phosphatidylinositol 4,5-bisphosphate levels by activating pip5kibeta. J Biol Chem 281:32630–32638

Yang SA, Carpenter CL, Abrams CS (2004) Rho and rho-kinase mediate thrombin-induced phosphatidylinositol 4-phosphate 5-kinase trafficking in platelets. J Biol Chem 279:42331–42336

Yeo DS, Chan R, Brown G, Ying L, Sutejo R, Aitken J, Tan BH, Wenk MR, Sugrue RJ (2009) Evidence that selective changes in the lipid composition of raft-membranes occur during respiratory syncytial virus infection. Virology 86:168–182

Yin HL, Janmey PA (2003) Phosphoinositide regulation of the actin cytoskeleton. Annu Rev Physiol 65:761–789

Zhao H, Hakala M, Lappalainen P (2010) Adf/cofilin binds phosphoinositides in a multivalent manner to act as a pip(2)-density sensor. Biophys J 98:2327–2336

Zoncu R, Perera RM, Sebastian R, Nakatsu F, Chen H, Balla T, Ayala G, Toomre D, De Camilli PV (2007) Loss of endocytic clathrin-coated pits upon acute depletion of phosphatidylinositol 4,5-bisphosphate. Proc Natl Acad Sci U S A 104:3793–3798

Chapter 7
Phosphoinositides in Chemotaxis

Michael C. Weiger and Carole A. Parent

Abstract Phosphatidylinositol lipids generated through the action of phosphinositide 3-kinase (PI3K) are key mediators of a wide array of biological responses. In particular, their role in the regulation of cell migration has been extensively studied and extends to amoeboid as well as mesenchymal migration. Through the emergence of fluorescent probes that target PI3K products as well as the use of specific inhibitors and knockout technologies, the spatio-temporal distribution of PI3K products in chemotaxing cells has been shown to represent a key anterior polarity signal that targets downstream effectors to actin polymerization. In addition, through intricate cross-talk networks PI3K products have been shown to regulate signals that control posterior effectors. Yet, in more complex environments or in conditions where chemoattractant gradients are steep, a variety of cell types can still chemotax in the absence of PI3K signals. Indeed, parallel signal transduction pathways have been shown to coordinately regulate cell polarity and directed movement. In this chapter, we will review the current role PI3K products play in the regulation of directed cell migration in various cell types, highlight the importance of mathematical modeling in the study of chemotaxis, and end with a brief overview of other signaling cascades known to also regulate chemotaxis.

Keywords Cell migration · Chemotaxis · *Dictyostelium discoideum* · Fibroblasts · Neutrophils · Phosphoinositide 3-kinase

7.1 Introduction

Scientists have had a long-standing fascination with the intricate migration patterns exhibited by a wide array of cells. Depending on the cell type and environmental context, cells can also migrate alone as individuals or collectively in a group of cells. Single-cell migration has been classified into two types: amoeboid and mesenchymal migration (see Fig. 7.1). The morphological and motility characteristics of amoeboid

C. A. Parent (✉) · M. C. Weiger
Laboratory of Cellular and Molecular Biology,
National Cancer Institute, National Institutes of Health, 37 Convent Drive,
Bldg.37/Rm2066, Bethesda MD 20892-4256, USA
e-mail: parentc@mail.nih.gov

T. Balla et al. (eds.), *Phosphoinositides II: The Diverse Biological Functions*,
Subcellular Biochemistry 59, DOI 10.1007/978-94-007-3015-1_7,

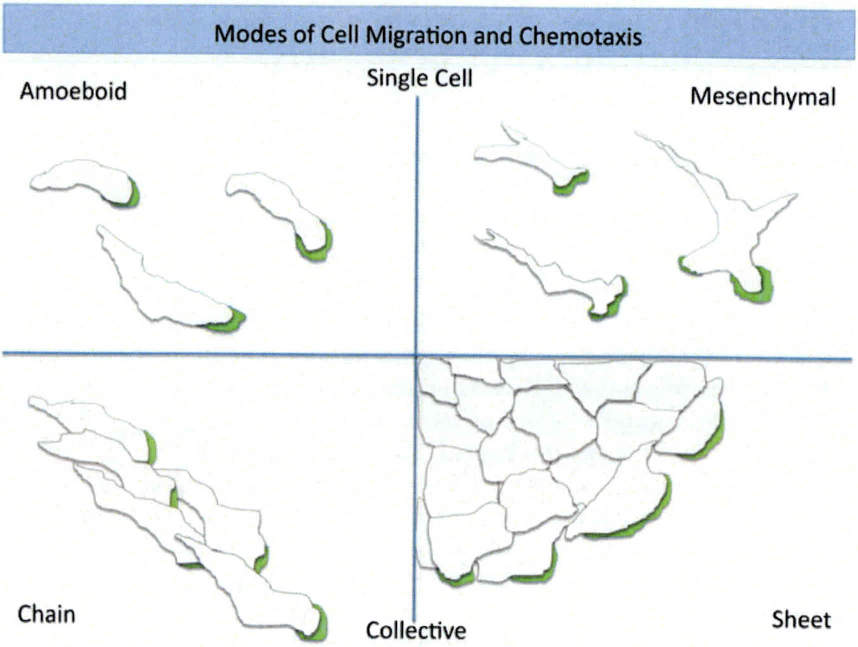

Fig. 7.1 Modes of cell migration and chemotaxis. During both random and directed migration (chemotaxis), cells can migrate as single cells or collectively. Single cell migration can be divided into two primary modes: amoeboid and mesenchymal. Cells with amoeboid characteristics appear elliptical, migrate at rapid rates (10 μm/min) and exhibit low adherence to their substrate. Cells exhibiting mesenchymal migration display more spindle and fibroblast-like morphologies, form strong adhesions with their substratum and migrate at much slower rates (0.1–1 μm/min). Collective migration is composed of chain and sheet migration. These forms of migration involve cell-cell adhesions and cell-cell signaling through various signaling molecules. During chemotaxis, cells show polarized morphologies as well as asymmetric activation of signaling pathways such as PI3K and 3' PI accumulation (*green*)

migration have been established through studies of the soil amoeba, *Dictyostelium discoideum* and mammalian leukocytes (Haston et al. 1982; Devreotes and Zigmond 1988). Cells exhibiting amoeboid migration appear ellipsoid in shape and translocate via cycles of protrusive membrane expansion (pseudopodia) and contraction of the cell rear, giving the impression that the cells are 'gliding' over the substratum. Cells acquire this 'gliding' behavior by exerting low adherence to their substrate, being either completely independent or nearly independent of integrin mediate adhesions (Yumura et al. 1984; Devreotes and Zigmond 1988; Fey et al. 2002). Due to exceptional deformability, amoeboid migration allows cells to rapidly (10 μm/min) maneuver through barriers posed by networks of extracellular matrix (EMC) and connective tissues as elegantly demonstrated by activated neutrophils. Several carcinoma cells display this type of motility and it is thought to mediate cancer metastasis from the primary tumor (Wood 1958; Verschueren et al. 1991; Farina et al. 1998; Rintoul and Sethi 2001).

Single-cell, mesenchymal migration is a much slower process (0.1–2 μm/min) (Abercrombie and Heaysman 1953; Abercrombie 1978; Friedl et al. 1998) that is dependent on integrins forming strong adhesions (nascent and focal) with the underlying substratum. Cells that undergo mesenchymal migration have fibroblast-like, spindle-shaped morphologies and cycle through discrete phases to generate forward motion. First, a cell sends out protrusions in the form of filopodia (thin, spike-like structures) and/or lamellipodia (broad, sheet-like structures). The advancing membranes eventually establish the leading edge of the cell and form new adhesions with the surface. Next, motor proteins mediate contraction through a network of actin filaments coupled to integrin adhesion sites, or focal adhesion. These cell adhesions mature in the leading edge and dissemble in the trailing edge causing the cell body to move forward. Once complete, this cycle is repeated giving rise to cell translocation (Lauffenburger and Horwitz 1996) Mesenchymal migration is predominantly found in cultured fibroblasts (Abercrombie et al. 1970; Gail and Boone 1970; Albrecht-Buehler 1979; Chen 1979; Dunn 1980; Sheetz et al. 1998; Munevar et al. 2001) and in cells from connective-tissue tumors such as fibrosarcomas, gliomas and dedifferentiated epithelial cancers (Paulus et al. 1996; Polette et al. 1998; Tester et al. 2000; Wolf et al. 2003).

In both physiological and cell culture conditions, cells have also been shown to migrate collectively either as a sheet or in chains. The ability of cells to migrate collectively as a sheet is a common phenomenon that occurs during embryological development (Davidson and Keller 1999; Klinowska et al. 1999; Simian et al. 2001), the formation of new blood vessels, ducts, and glands, and also during cancer invasion (Alexander et al. 2008; Friedl and Gilmour 2009). Strong cell-cell adhesions contribute to the ability of a group of cells to move as a functional unit (Hegerfeldt et al. 2002). Cells at the front of the collective body often exhibit a high degree of mobility and appear to drag the rest of the cell sheet (Vaughan and Trinkaus 1966). However, recent data has shown that cells at the front as well as those well within the sheet generate traction forces necessary to drive sheet migration, raising the possibility that cells in sheets can communicate with each other over long distances and coordinate their movements (Trepat et al. 2009).

Cells migrating in chains can communicate via both physical linkages (cell-cell and cell-ECM) and/or secreted signals. For instance, upon starvation and subsequent aggregation, *Dictyostelium* amoebae display an elegant form of chain migration, called streaming. Randomly dispersed *Dictyostelium* cells communicate with one another by locally producing, secreting, and responding to cAMP. As a result, these cells organize in head-to-tail chains and chemotax toward aggregation centers. Subsequent studies of this streaming process introduced the concept of signal relay, a chemotactic-based communication system that can regulate long-range coordinated movements of groups of cells (Dormann et al. 2002; Kriebel et al. 2003). In mammalian cells, chain migration occurs in neural precursor cells (Jacques et al. 1998), myoblasts (El Fahime et al. 2000) and melanomas (Friedl et al. 1997). Chain migration is also observed in breast carcinoma (Page and Anderson 1987; Pitts et al.

1991), ovarian cancer and melanoma of a vascular-type pattern where single tumor cells align between stromal fibers (Friedl and Wolf 2003). It has also been observed that macrophages can stimulate mammary tumor cells to follow them, via paracrine signaling, toward a blood vessel and at perivascular regions promote tumor intravasation (Goswami et al. 2005; Wyckoff et al. 2007).

At the core of all of these migration modes is a requirement for individual cells to acquire morphological polarity, where actin-based forces are primarily found at the leading edge and myosin II-based contractions are concentrated at the back. The acquisition of such polarized responses is regulated by complex signal transduction cascades that ultimately control where and when cytoskeletal networks are activated and assembled. For example, in the context of chemotaxis, where cells move directionally in response to external chemical gradients, polarized responses are generated through the amplification of external signals. In this chapter, we will review how specific phosphoinositides are involved in cell migration directed by such extracellular cues.

7.2 PI3K Isoforms

In recent years, numerous researchers have sought to define the role phosphoinositide 3-kinases (PI3K) and their lipid products in regulating chemotaxis under various conditions. The PI3Ks are a family of lipid kinases that regulate several cellular processes including differentiation, proliferation, cell survival, as well as cell growth and migration (Leevers et al. 1999). Activated PI3Ks can phosphorylate a wide array of lipid substrates, including phosphatidylinositol (PI), phosphatidylinositol 4-phosphate (PI(4)P) and phosphatidylinositol 4,5-bisphosphate (PI(4,5)P_2), at the 3' position of their inositol ring to generate specific phosphorylated PIs (see Fig. 7.2). In particular, phosphatidylinositol 3,4,5-trisphosphate (PI(3,4,5)P_3) and phosphatidylinositol 3,4-bisphosphate (PI(3,4)P_2) have been implicated in regulating downstream signaling events that control cell migration (Vanhaesebroeck et al. 2001). For the remainder of the chapter, we will refer to PI(3,4,5)P_3 and PI(3,4)P_2 as 3' PIs.

PI3Ks are classified into three groups according to their substrate specificity and sequence homology: class I, class II, and class III (see Fig. 7.3). The class I PI3Ks are further divided into sub-groups A and B. Structurally, sub-groups A and B are closely related and both exist as stable heterodimers comprising a highly homologous 110 kDa catalytic subunit and a smaller tightly associated regulatory subunit. Subgroup A is composed of three catalytic isoforms, α and β, which have broad tissue distribution, and δ, whose expression pattern is more restricted. These catalytic subunits can bind to one of five regulatory subunits (p85α, p85β, p55α, p55γ, and p50α). The class IA PI3Ks are activated by receptor tyrosine kinases (RTKs). Activation results in the recruitment of PI3Ks (class IA) from the cytosol to the plasma membrane where they associate with phosphorylated tyrosine residues on the cytoplasmic tail(s) of RTKs via their Src-homology 2 (SH2) domains. This localization brings the enzymes within close proximity to their lipid substrates. PI(4,5)P_2 is the preferred

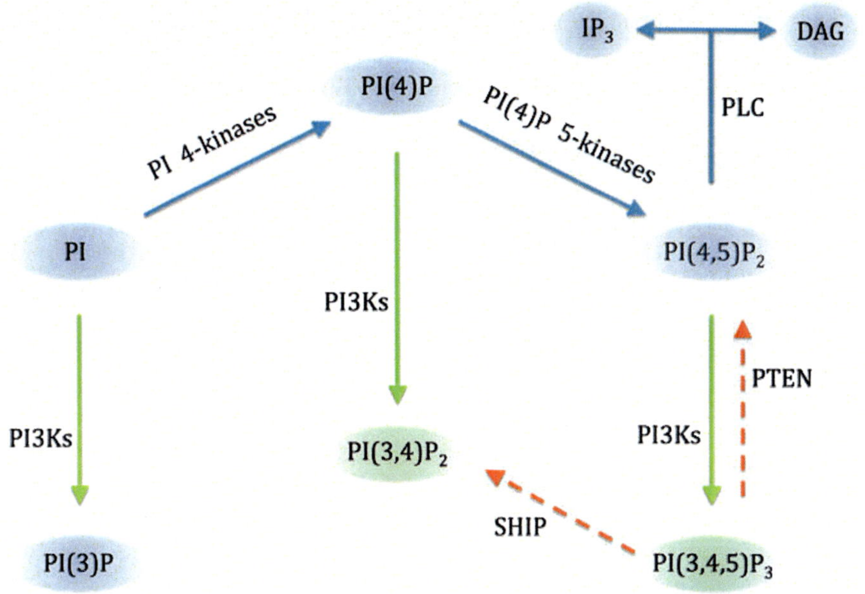

Fig. 7.2 The phosphoinositide cycle during chemotaxis. After chemoattractant stimulation, active PI3Ks primarily convert $PI(4,5)P_2$ into $PI(3,4,5)P_3$ by phosphorylating the 3' position of the inositol ring. The level and spatial distribution of $PI(3,4,5)P_3$ is regulated by the actions of two phosphatases, PTEN and SHIP, which dephosphorylate $PI(3,4,5)P_3$ yielding $PI(4,5)P_2$ and $PI(3,4,)P_2$, respectively. Chemoattractant stimulation also leads to activation of PLC which subsequently cleaves the PI3K substrate $PI(4,5)P_2$ into IP_3 and DAG

substrate for class IA and class IB PI3Ks *in vivo* (Cain and Ridley 2009). Class IB PI3Ks are composed of a p110γ catalytic isoform bound to either a p101 or p84 regulatory subunit (Suire et al. 2005) and, unlike class IA, they are primarily activated by G-protein-coupled receptors (GPCRs) and Ras—the catalytic subunits of class I PI3Ks contain a Ras-binding domain (RBD) that can regulate their activity.

There are three isoforms of class II PI3Ks in mammals (PI3K-C2α, PI3K-C2β, PI3K-C2γ). Currently, this class of PI3K is thought to not require a regulatory subunit and it has been reported that these enzymes function in the context of multi-protein complexes. There is some evidence that class II PI3Ks regulate cell adhesion and actin reorganization during cell migration and wound healing in non-immune cell systems (Domin et al. 2005; Maffucci et al. 2005). However, further studies have proven difficult due to the limited availability pharmacological inhibitors to these isoforms (Virbasius et al. 1996; Domin et al. 1997). Class III PI3Ks contain only one mammalian member, Vps34 (vacuolar protein sorting 34), which requires a regulatory subunit, p150, for its activity. It is involved in multiple steps of membrane trafficking, but currently no role in migration has been reported (Backer 2008).

While our understanding of the physiological role of class II and III PI3Ks continues to mature, class I PI3Ks have clearly been implicated in the control of directional

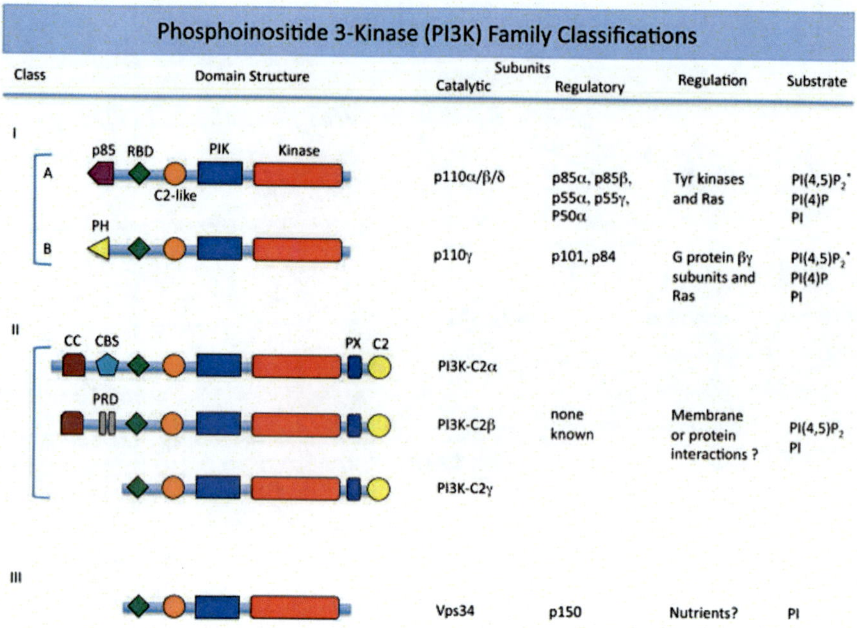

Fig. 7.3 The PI3K family of enzymes. PI3Ks are categorized into three classes. The domain structures of the catalytic subunit, isoforms of the catalytic and regulator subunits, primary method of regulation, and substrates (*preferred *in vivo*) are given, if known. Abbreviations: p85: p85-binding domain; RBD: Ras-binding domain, C2-like: PKC homology 2-like domain; CC: coiled-coil domain; CBS: clatherin-binding site; PX: phox domain; PRD: proline-rich domain

migration and cell polarity for well over a decade. An established signaling scheme linking PI3K to chemotaxis has indeed been extensively studied in numerous physiological and pathological contexts (Schneider and Haugh 2006; Stephens et al. 2008).

7.3 Probing 3' PIs and PI3K Function and Activity

Our understanding of the role of PI3K and 3' PIs during chemotaxis is in large part due to the availability of various tools to measure and inhibit PI3K activity as well as to visualize the cellular distribution of PI3K products. It is fitting to begin the discussion with the pharmacological inhibitors used to study the biological processes that require PI3K signaling, especially since these agents have proven invaluable in the context of chemotaxis. The fungal metabolite wortmannin was among the first widely available agents found to inhibit PI3K (Arcaro and Wymann 1993). This cell permeable and water soluble, small molecule was found to covalently modify a lysine in the p110 catalytic domain that is responsible for binding ATP, thereby effectively abolishing class I, II and III PI3K activity, with the exception of PI3K-C2α which is less sensitive (Virbasius et al. 1996; Wymann et al. 1996; Domin et al. 1997; Vanhaesebroeck et al. 2001). Another small molecule inhibitor of PI3Ks is the quercetin derivative LY294002, which competitively targets the ATP-binding site of the kinase. Unlike wortmannin, LY294002 is a reversible inhibitor, providing experimental flexibility compared to Wortmannin (Vlahos et al. 1994). More recently, isoform specific inhibitors have been developed for class I PI3Ks, which has tremendously facilitated the investigation of the specific contributions of Class I PI3Kα, β, δ, and γ isoforms to cellular processes (Bilancio et al. 2006; Knight et al. 2006). One drawback to small molecule inhibitors is cross-reactivity with other proteins, which typically depends on the inhibitor concentration used (Fedorov et al. 2007). For example, the myosin light chain kinase, DNA-dependent protein kinase, PI-4 kinase, mTOR, and NADPH oxidase in neutrophils are among several molecules also affected by wortmannin treatment (Nakanishi et al. 1992; Vlahos et al. 1994; Brunn et al. 1996; Banin et al. 1998; Izzard et al. 1999). LY294002 also has specificity issues. It has been show to inhibit DNA-dependent protein kinase and even mTOR at concentrations necessary to effectively inhibit PI3K (Brunn et al. 1996; Izzard et al. 1999).

Another technique often utilized to inhibit PI3K signaling is the introduction of a dominant negative form of the protein. Typically, a mutated form of the regulatory subunit (Δp85) that cannot interact with its receptor (like the platelet-derived growth factor (PDGF) receptor) is over-expressed in cells via transient transfection or viral infection. This mutant effectively sequesters the majority of the catalytic domains and prevents their translocation to and activation by the receptor (Wennstrom et al. 1994a; Dhand et al. 1994). However, this technique is limited to growth factor receptor signaling (RTKs). PI3K inhibition can also be accomplished by generating cells that do not express PI3K proteins. Using various genetic approaches, this is typically done through deletion (knockout) of the gene that encodes a PI3K isoform(s) or by introducing (knock-in) a gene that encodes for a kinase-dead mutant of the protein. Gene-targeted mice have been generated to all four class I PI3K catalytic isoforms (Bi et al. 1999, 2002; Sasaki et al. 2000; Hirsch et al. 2000; Li et al. 2000; Graupera et al. 2008) and both knockout and kinase-dead knock-in strategies have been used to assess the role of regulatory and catalytic class I PI3K subunits (Okkenhaug et al. 2002; Patrucco et al. 2004; Ali et al. 2004). It is important to mention that results obtained from these types of studies are sometimes difficult to interpret as altered expression of one subunit can affect the expression of the other subunit (Vanhaesebroeck et al. 2001; Okkenhaug and Vanhaesebroeck 2001). Moreover, these genetic manipulations can yield non-viable species. For example, p110α and p110β knockout mice are embryonic lethal (Bi et al. 1999, 2002). Cre-Lox recombination is often employed to circumvent this drawback (Jia et al. 2008). In addition, RNA interference (RNAi) to reduce the levels of specific proteins has also become prevalent in chemotaxis research. It has been used as a means of inhibiting PI3K activity by targeting either the regulatory or catalytic subunits of class I and class II PI3Ks (Lee et al. 2004; Maffucci et al. 2005; Nakhaei-Nejad et al. 2007). Although powerful, this technique also has its limitation in that great care must be taken to insure specific RNA targeting. Finally, instead of reducing 3' PI lipid production, a constitutively active form of the protein that is insensitive to intracellular regulatory mechanism can be used (Shioi et al. 2000; Costa et al. 2007; Attoub et al. 2008; Gan et al. 2011). Often pathological conditions result from the over or unregulated activation of PI3K and expression of a constitutively active PI3K can provide further insights into downstream signaling mechanisms. Alternatively, the prospect of using engineered molecules (chimeras) and light to control the local activity of signaling proteins has emerged as a powerful tool to explore how spatial segregation of signaling pathways regulate cell orientation and migration (Inoue and Meyer 2008; Wu et al. 2009).

While the level of 3' PIs can be readily measured biochemically using thin layer chromatography and even more accurately by high-pressure liquid chromatography (Traynor-Kaplan et al. 1989; Carpenter and Cantley 1990), the use of high-resolution imaging of fluorescently tagged molecules that specifically target PI3K products has provided a unique technical advance to the PI3K field. Indeed, the generation of intracellular probes consisting of green fluorescent protein (GFP) fused to the pleckstrin homology (PH) domain of signaling proteins known to interact with specific PIs has provided an unprecedented view of PI3K activity in live, migrating cells. Meyer and colleagues first demonstrated the utility of one such GFP probe when they showed the dynamic dissociation of the PH domain of PLCδ (which binds PI(4,5)P_2) from the plasma membrane to the cytoplasm of rat basophilic leukemia cells following stimulation with platelet activating factor (PAF) (Stauffer et al. 1998). This technology has since then been applied to PH domains that specifically bind 3' PIs as well as to specific proteins including PI3K and 3' phosphatases (Parent et al. 1998; Meili et al. 1999; Rickert et al. 2000; Servant et al. 2000; Haugh et al. 2000; Funamoto et al. 2001; Merlot and Firtel 2003; Janetopoulos et al. 2004; Sasaki et al. 2004; Ma et al. 2004). Importantly, PH domains can be somewhat promiscuous and bind with different affinity to more than one PI. The PH domain of Akt, for example, binds with highest affinity to both PI(3,4,5)P_3 and PI(3,4)P_2 and therefore represents a readout of the localization and levels of both lipids (Kavran et al. 1998; Lemmon 2007). Furthermore, with these types of probes, it is important that an appropriate fluorescent membrane marker be used to ensure that any observed local increase in the probe is not due to changes in membrane density, which occurs during membrane ruffling and folding. The extent of probe translocation can also be quantified by computing the difference between the intensities at the membrane and cytosol and then the ratio of this quantity and the cytosolic intensity as a final measurement (Teruel and Meyer 2000; Meyer and Oancea 2000).

7.4 The Chemotaxis Paradigm: PI3K and 3' PIs at the Leading Edge

How do PI3Ks and their 3' PI lipid products regulate chemotaxis? It all has to do with location, location, location. PI3K is recruited to the plasma membrane by activated chemoattractant receptors, either RTKs or GPCRs. Once brought to the membrane and activated, PI3K locally produces PI(3,4,5)P_3 from PI(4,5)P_2 on the inner leaflet of the plasma membrane (Rameh and Cantley 1999). Furthermore, the small GTPase Ras has been shown to interact with and activate PI3K providing another link between RTK and GPCR to PI3K signaling during chemotaxis (Rodriguez-Viciana et al. 1994;

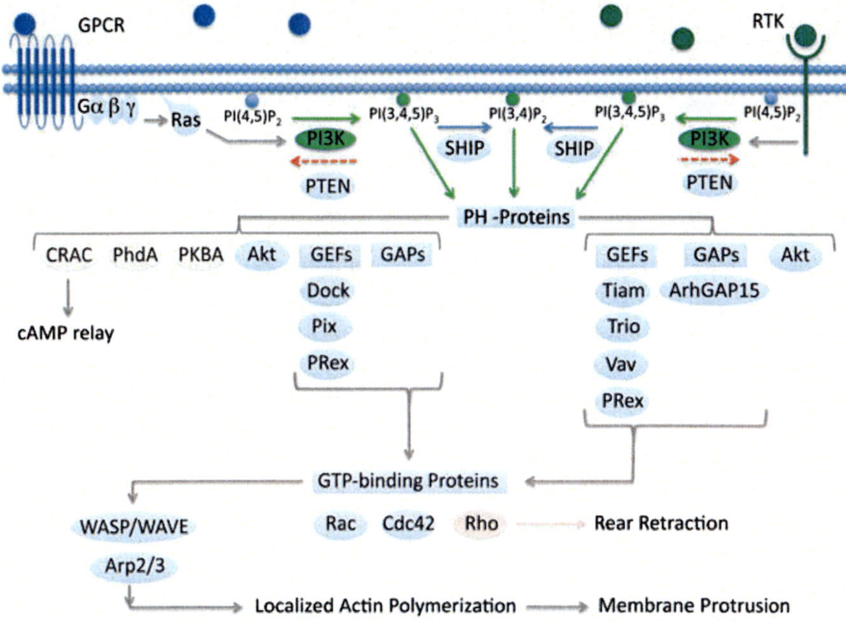

Fig. 7.4 PI3K-dependent signaling during chemotaxis. Depending on the chemoattractant stimulation, PI3K is recruited to and activated by ligand-bound G-protein coupled receptors (GPCRs) and/or receptor tyrosine kinases (RTKs). Once at the plasma membrane, PI3K (*green*) has access to PI(4,5)P_2, which it phosphorylates producing PI(3,4,5)P_3 (*green*). The lipid phosphatases PTEN and SHIP either convert PI(3,4,5)P_3 back to PI(4,5)$_2$ or produce PI(3,4)P_2 (*green*), respectively. In *Dicytostelium* and mammalian cells, a spatially confined pool of 3' PIs (PI(3,4,5)P_3 and PI(3,4)P_2) recruits pleckstrin homology (PH) domain containing proteins (PH-proteins) to the membrane, often facilitating their activation. A subset of mammalian PH-proteins, including GEFs and GAPs for the GTP-binding proteins (*blue ellipses*) is locally activated by 3' PIs, thus promoting local activation of Rac and Cdc42 and spatially confined actin polymerization and protrusion via WASP/WAVE and Arp/3 dependent mechanisms. The GTP-binding protein Rho (*pink ellipse*) regulates cell-rear retraction and contraction. PH-proteins in *Dictyostelium* (tan ellipse) regulate signal relay (CRAC) and cell polarity (PKBA). Mammalian Akt (*blue ellipse*) is also implicated in regulating actin polymerization. (Horizontal arrows: *red dashed* and *blue* denote dephosphorylation reactions; *green* denotes phosphorylation reactions; *grey* denotes molecular interactions)

Suire et al. 2006; Charest and Firtel 2006). Remarkably, when exposed to a chemoattractant gradient, PI3K and its lipid product PI(3,4,5)P_3 are exclusively distributed at the leading edge of cells. The asymmetric distribution of PI(3,4,5)P_3 is confined spatially by the actions of lipid phosphatases: namely the 3' phosphatase, phosphatase and tensin homolog on chromosome ten (PTEN), which converts PI(3,4,5)P_3 back to PI(4,5)P_2, and is enriched at the sides and back of chemotaxing cells. The Src homology phosphatases (SHIP1 and SHIP2) are also active within cells and can locally produce PI(3,4)P_2 by dephosphorylating PI(3,4,5)P_3 at the 5' position, thereby adjusting the local 3' PI composition (see Figs. 7.2 and 7.4). Together, the mutually exclusive cellular distribution of PI3K and PTEN leads to the strong enrichment of 3' PIs at the leading edge of cells and a strong amplification of the external chemical gradient. Due to the low abundance of 3' PIs under basal conditions ($\sim 0.25\%$ of the total membrane lipid content) and their relatively slow diffusion (~ 0.1–$1\ \mu m^2/s$), 3' PIs afford an ideal regulatory switch for gradient sensing (Rameh and Cantley 1999; Vanhaesebroeck et al. 2001; Schneider and Haugh 2006).

Once the 3' PI gradient is established, it provides a zip code that specifically and spatially targets effectors to the front of migrating cells (see Fig. 7.4). The 3' PI lipid products of PI3K, PI(3,4,5)P_3 and its dephosphorylated relative PI(3,4)P_2, mediate the recruitment of cytosolic signaling molecules containing PH domains. These domains represent the 11th most common in the human genome and are best known for their ability to target cellular membranes by binding specifically to phosphoinositides (Lemmon 2007). One of the most well known PH domain containing molecules recruited to the membrane by 3' PIs is the serine/threonine kinase Akt (also called protein kinase B or PKB). Akt is an important mediator of cell survival and growth (Klippel et al. 1997) while reports also link it to cell migration. It has also been shown to regulate actin dynamics via its direct binding to and phosphorylation of actin (Cenni et al. 2003; Vandermoere et al. 2007) or via phosphorylation of the actin-binding protein Girdin/APE (Akt phosphorylation enhancer) (Enomoto et al. 2005). The connection between Akt and cell motility is further supported with evidence from *Dictyostelium* in which cells lacking PKB and the PKB related protein 1 (PKBR1) exhibit motility defects (Meili et al. 2000; Funamoto et al. 2001; Kamimura et al. 2008).

Evidence also suggests that polarized 3' PIs mediate the local recruitment and activation of specific guanine-nucleotide-exchange factors (GEFs) and GTPase activating proteins (GAPs) via their PH domains (see Fig. 7.4). Locally activated GEFs and GAPs can then activate or inhibit members of the Rho family of small GTP binding proteins (Cdc42, Rac, and Rho) by promoting either the GTP (active) or GDP (inactive) bound forms. In their active GTP-bound state, Rho GTP binding proteins are responsible for regulating leading edge membrane protrusion and ruffling as well as tail retraction during chemotaxis. One such GEF whose activity is dependent on PI3K during chemotaxis is the Rac GEF, Vav. This GEF was shown to be important during macrophage-colony stimulating factor (M-CSF)-stimulated chemotaxis of mouse macrophages and that its activity is PI3K dependent fashion (Han et al. 1998; Vedham et al. 2005). Similarly, PC12 cells stimulated with nerve growth factor exhibit protrusions with local enrichment of PI(3,4,5)P_3, which mediates recruitment of Vav2/3. Moreover, loss of Vav2/3 suppresses PI(3,4,5)P_3 production, suggesting that Rac and Rac GEFs provide a positive feedback loop to PI3K (Aoki et al. 2005)—a process that has been shown during both neutrophil and *Dictyostelium* chemotaxis (Weiner et al. 2002; Park et al. 2004). New evidence suggests that a related positive-feedback loop between Rac and PI3K also exists in integrin-mediated migration. In this case, integrin engagement and clustering is sufficient to recruit both PI3K and Rac1 to focal adhesions. Newly synthesized PI(3,4,5)P_3 and integrins then activate Rac1 GEFs leading to Rac1 activation and further accumulation of PI3K (Smerling et al. 2007). The Dock180 and related proteins (Dock2, Dock7, and MBC) are other Rac GEFs that have been extensively studied in the context of cell migration and chemotaxis. These proteins have been shown to bind PI(3,4,5)P_3, which is important for both their localization and activity. The adaptor protein Elmo forms a complex with Dock180 and is either required for or enhances Rac activation at the plasma membrane (Cote and Vuori 2007). Neutrophils lacking the Dock 180-related protein Dock2 accumulate less PI(3,4,5)P_3 in response to the chemoattractant N-formylmethionyl-leucyl-phenylalanine (fMLP), suggesting a role for Dock2 in the stabilization of PI(3,4,5)P_3 or in the activation of PI3K (Kunisaki et al. 2006). Other Rac GEFs and GAPs have been shown to interact with 3' PIs, but their role in chemotaxis is complex and still not completely understood. For example, the activity of the Rac GEF, P-Rex1 is regulated by PI(3,4,5)P_3, although neutrophils lacking P-Rex1 show only slight reductions in chemotaxis and cell speed (Welch et al. 2002; Welch et al. 2005; Dong et al. 2005). Interestingly, the Rac GAP ArhGAP15 also appears to be influenced by PI(3,4,5)P_3 (Welch et al. 2002; Costa et al. 2007), possibly serving to limit the extent of Rac activity at the leading edge. On the other hand, the p21-activated kinase (PAK)-interacting exchange proteins (α and β PIX) serve as GEFs for both Rac and Cdc42. The regulation of Rac and Cdc42 by α and β PIX is complex and appears to be depend on protein-protein interactions with molecules such as PAK and GPCR-kinase-interacting protein (GIT), despite the fact that they contain PH domains (Frank and Hansen 2008). Moreover, αPIX has been shown to associate with the p85 subunit of PI3K and PI3K enhances αPIX GEF activity, which introduces the potential for more cross-talk between signaling pathways regulating directed migration (Yoshii et al. 1999). Taken together, these data implicate that 3' PIs and PI3K coordinate the spatial localization and activation of GEF and GAPs, which are posed to confine the activity of Rho GTP binding proteins.

An important function of Rho GTP binding proteins is to regulate the Arp2/3 complex, an actin nucleating protein, that functions to promote new sites of actin polymerization on existing actin filaments leading to highly branched cytoskeletal networks (Pollard and Borisy 2003). The Wiskott-Aldrich Syndrome Protein family (WASP, N-WASP, and Wave/Scar) coordinate actin reorganization by coupling Rho GTP binding protein signaling to the mobilization of the Arp2/3 complex thereby promoting actin polymerization and membrane protrusion (Takenawa and Suetsugu 2007; Pollitt and Insall 2009). Cdc42, which contributes to efficient chemotaxis, has been shown to bind WASP and N-WASP and promote actin filament formation (Symons et al. 1996; Miki et al. 1998). Similarly, Wave proteins act downstream of Rac, PI(3,4,5)P_3 and other PIs to mediate the formation of leading edge lamellipodia and dorsal ruffles (Suetsugu et al. 2003; Yamazaki et al. 2003). Activated Rho on the other hand signals to Rho-associated kinase (ROCK), which promotes both stress fiber formation by increasing acto-myosin contraction and focal adhesion assembly as well as myosin II filament assembly (Matsui et al. 1996; Narumiya et al. 2009). Interestingly, it now appears that the maintenance of asymmetric signaling during chemotaxis requires extensive front to back regulation. In mouse neutrophils, RhoA along with ROCK and Cdc42 can regulate the spatial localization and activity of PTEN in response to chemoattractant stimulation (Li et al. 2005). In macrophages under CSF-1 stimulation, PI3K (p110δ) activity inhibits RhoA activation, which limits ROCK's ability to activate PTEN (Papakonstanti et al. 2007). Furthermore, PI(3,4,5)P_3 induced activation of Cdc42 at the front of cells is thought to be connected to RhoA activation at the back (Xu et al. 2003). These findings highlight the importance of PI3K signaling in the proper spatial localization of many signaling components.

7.5 Cell Context Specificity of PI3Ks

The PI3K family actively contributes to the regulation of cell migration and chemotaxis in many different cell types. In eukaryotic cells, PI3K signaling in the context of chemotaxis has been extensively characterized in the soil amoeba *Dictyostelium discoideum* and neutrophils. Their similarities and unique differences help frame the complexities of PI3K-and PI-dependent signaling during chemotaxis. In addition, studies in mesenchymal cells such as fibroblasts have also shown that both PI3K and 3' PIs are important for efficient chemotaxis.

7.5.1 Dictyostelium Discoideum

Dictyostelium was discovered over 75 years ago and has had a major impact on the field of cytokinesis, cell motility, phagocytosis, signal transduction, and cell differentiation (Raper 1935; Maeda and Firtel 1997; Kay and Williams 1999; Daunderer et al. 1999; King and Insall 2009). It has proven to be an ideal model system for studying cell migration and chemotaxis because of its short life cycle and accessible genetics (Egelhoff et al. 1991; Kay and Williams 1999; Eichinger et al. 1999, 2005; Eichinger and Noegel 2003). Chemotaxis is an intricate part of the life cycle of *Dictyostelium*. In the presence of nutrients, the cells live independently of each other. However, under starvation conditions, the cells enter a developmental program where they interact with each other to form an aggregate that eventually differentiates into a multicellular structure composed of a spore head atop a stalk of vacuolated cells. In this process, up to 100,000 single cells signal to each other by secreting cyclic adenosine monophosphate (cAMP) and responding chemotactically to it. The chemotactic response begins at the receptor level, when cAMP binds to the specific GPCR cAMP receptor 1 (cAR1) (Klein et al. 1988; Kumagai et al. 1989;, Sun and Devreotes 1991; Johnson et al. 1992; Alberts 2002), which leads to the activation of a wide array of effectors that ultimately give rise to changes in gene expression, the regulation of chemotaxis, and the synthesis and secretion of cAMP or signal relay (Parent and Devreotes 1996; Aubry and Firtel 1999; Kimmel and Parent 2003; Firtel et al. 1989; Kim et al. 1998). The pathway that leads to PI3K activation requires inputs from Gβγ and Ras (Insall et al. 1996; Lim et al. 2001; Huang et al. 2003). There are five class I-like PI3Ks in *Dicytostelium*; all contain Ras binding domains (RBD) (Zhou et al. 1995; Meili et al. 1999, 2000; Funamoto et al. 2001, 2002). Unlike mammalian PI3Ks, which are associated with a regulatory subunit, PI3Ks in *Dictyostelium* have not been shown to bind to a regulatory subunit. Of the five PI3Ks, PI3K1, PI3K2, and PI3K3 are thought to be the major PI3Ks implicated in the production of $PI(3,4,5)P_3$ and the regulation of efficient chemotaxis (Funamoto et al. 2001; Takeda et al. 2007; Huang et al. 2003). Indeed, cells lacking all three PI3K isoforms move more slowly, exhibit a less polarized morphology and show reduced direct movement during chemotaxis compared to wild-type cells (Takeda et al. 2007).

Stimulating *Dictyostelium* cells with a uniform dose of cAMP, elicits a rapid (< 5–10 s) increase in 3' PI levels at the cell periphery. Further, when cAMP is presented as a gradient, PI3K and 3' PIs are spatially confined and elevated in regions of the cell membrane closest to the cAMP source (Parent et al. 1998). This leads to an intracellular gradient of 3' PIs that is amplified to approximately three to seven times that of an external cAMP gradient, and it was proposed that this polarized signal promotes local pseudopod formation during chemotaxis (Parent and Devreotes 1999; Funamoto et al. 2002; Janetopoulos et al. 2004). This increase in 3' PI levels is either transient (following a uniform stimulation) or sustained (when cells are placed in a gradient) due to the action of a PTEN-like 3-phosphatase (Iijima and Devreotes 2002; Funamoto et al. 2002; Huang et al. 2003). In resting cells, PTEN is uniformly associated with the plasma membrane. Upon a uniform stimulus, PTEN transiently redistributes to the cytoplasm. In contrast, in cells exposed to a gradient of cAMP, PTEN is exclusively associated with the back and sides of chemotaxing cells. The membrane localization of PTEN is mediated in part by an interaction with $PI(4,5)P_2$ (Rahdar et al. 2009). Indeed, data suggest that PLC can regulate the spatial distribution of PTEN by preferentially degrading $PI(4,5)P_2$ at the cell front toward the gradient source (van Haastert et al. 2007; Kortholt et al. 2007). PTEN was confirmed to be a critical regulator of cell polarization and migration as the loss of the *Dictyostelium* homologue to the mammalian 5-phosphatase SHIP, Dd5P2, only has minor effects on cell migration (Loovers et al. 2003) and, most importantly, cells lacking PTEN show a broad front with multiple pseudopods and significant chemotaxis defects (Iijima and Devreotes 2002; Wessels et al. 2007). Interestingly, it has been reported that the creation of a local F-actin network enhances the accumulation of 3' PIs through the ability of class I PI3Ks to bind to F-actin and thus promote a positive feedback loop with PI3K and locally enhancing Ras activity (Sasaki et al. 2004, 2007; Park et al. 2004). Yet, 3' PI polarization is clearly observed in cells treated with actin polymerization inhibitors, showing that gradient sensing occurs independently of cytoskeletal rearrangement and morphological polarization (Parent et al. 1998; Janetopoulos et al. 2004; Xu et al. 2005). This finding also established that eukaryotic cells are able to spatially sense gradients.

In *Dictyostelium*, three 3' PI effectors have been extensively studied: cytosolic regulator of adenylate cyclase (CRAC), PH domain-containing protein A (PhdA), and a PKB-related kinase (PKBA) (Parent et al. 1998; Meili et al. 1999; Funamoto et al. 2001; Huang et al. 2003). CRAC is required for the cAMP-mediated activation of the adenylyl cyclase ACA, which synthesizes cAMP during signal relay (Insall et al. 1994; Kriebel et al. 2003; Comer et al. 2005; Comer and Parent 2006), and for chemotaxis. Cells lacking CRAC do not produce cAMP in response to cAMP addition and do not spontaneously aggregate when starved. Loss of PhdA or PKBA results in reductions in chemoattractant-mediated actin polymerization and defects in cell polarization and chemotaxis (Meili et al. 1999, 2000; Funamoto et al. 2001, 2002). Yet, these cells can still chemotax. While numerous $PI(3,4,5)P_3$ effectors exists (Park et al. 2008; Swaney et al. 2010; Zhang et al. 2010), the fact that cells lacking two (PI3K1–2), three (PI3K1–3), or multiple (PI3K1–5) PI3K isoforms can still chemotax, albeit at reduced speed (Loovers et al. 2006; Takeda et al. 2007; Hoeller and Kay 2007; Kolsch et al. 2008), suggest that other pathways are involved. It has been suggested that PI3K activity controls the rate at which pseudopodia are randomly generated, but not the direction of the pseudopod formation, and PTEN suppresses lateral pseudopod formation thereby allowing cells to maintain their course of migration (Wessels et al. 2007). In addition, it is now well accepted that PI3K is important in regulating cell speed and locomotion in shallow, linear gradients (Gruver et al. 2008; Bosgraaf et al. 2008). In contrast, in steep exponential gradients (generated by a micropipette) PI3K appears to be dispensable as other pathways are sufficient to promote efficient chemotactic movement (Kolsch et al. 2008). Thus, in *Dictyostelium* it is now appreciated that 3' PIs are not essential for chemotaxis, but do play an integral role.

7.5.2 Leukocytes

Sites of infection or injury elicit strong chemotactic responses from the innate immune system. Both host cells and foreign entities such as bacteria, release chemoattractants and chemokines that stimulate phagocytic leukocytes like neutrophils and macrophages to migrate towards the source of infection or injury. The process of immune surveillance is the most prominent type of single cell migration *in vivo* and 3' PIs are critical in regulating leukocyte motility and chemotaxis.

Like *Dictyostelium*, neutrophils and macrophages migrate very rapidly (10–20 μm/min) using GPCR signaling cascades to transduce chemoattractant signals. However, in contrast to *Dictyostelium*, which primarily respond to cAMP, these cells can recognize a host of chemoattractants such as complement component 5a (C5a), fMLP, and chemokines, although all of them couple to pertussis toxin sensitive G protein, Gi (Weingarten and Bokoch 1990). Upon receptor activation, the class IB PI3Kγ becomes activated in a Gβγ and Ras dependent fashion giving rise to the bulk of the $PI(3,4,5)P_3$ generated (Vanhaesebroeck et al. 1997; Stephens et al. 1997; Dekker and Segal 2000; Hirsch et al. 2000; Fruman and Cantley 2002; Brock et al. 2003; Suire et al. 2006). It has been known that 3' PIs influence leukocyte movement as neutrophils treated with synthetic $PI(3,4,5)P_3$ rapidly polarize and accumulate F-actin and PH_{Akt}-GFP at their leading edge (Niggli 2000; Weiner et al. 2002)—a process that is thought to be further increased via a positive feedback loop that is dependent on Rac and F-actin networks (much like in *Dictyostelium*) (Xu et al. 2003; Wang et al. 2002; Weiner et al. 2002). Macrophages and neutrophils isolated from PI3Kγ-deficient mice show reduced chemotaxis toward various stimuli *in vitro*; and *in vivo* PI3Kγ-deficient mice show less accumulation of macrophages and neutrophils at inflammatory sites (Hirsch et al. 2000; Li et al. 2000; Sasaki et al. 2000;

Patrucco et al. 2004). Curiously, the class IA PI3Kδ has also been implicated as a regulator of neutrophil motility and chemotaxis, potentially via the Gαi-dependent activation of Src and Hck kinases (Stephens et al. 1993; Ptasznik et al. 1995; Sasaki et al. 2000; Li et al. 2000, Patrucco et al. 2004; Puri et al. 2004; Jou et al. 2002; Clayton et al. 2002; Sadhu et al. 2003; Hannigan et al. 2002). For instance, neutrophils treated with a specific PI3Kδ inhibitor and macrophages isolated from p110δ kinase dead knock-in mice do not show enriched PI(3,4,5)P_3 at their leading edges and exhibit polarity defects (Sadhu et al. 2003; Papakonstanti et al. 2007). In addition, in differentiated PLB-985 cells, a pluripotent hematopoietic cell line, the production of PI(3,4,5)P_3 in response to fMLP is biphasic, with the initial 30 s response dependent on PI3Kγ and a later 5-min phase solely dependent on PI3Kδ (Boulven et al. 2006). Similar, temporally staggered signaling dependences have been observed *in vivo* during CXC chemokine-induced leukocyte recruitment, where PI3Kγ was reported to be important in early (1.5 h) CXC chemokine emigration responses while later (4–5 h) emigration responses were mediated by PI3Kδ (Liu et al. 2007).

In mouse neutrophils it is well established that PI3Kγ localizes to the plasma membrane upon stimulation with fMLP, leading to an increase in 3' PI formation, which can be abolished by wortmannin and LY294002 treatment (Weiner et al. 2002; Wang et al. 2002; Condliffe et al. 2005; Ferguson et al. 2007). However, unlike what is observed in *Dictyostelium*, the role of PTEN in the localized accumulation of PI(3,4,5)P_3 is not clear. While it has been reported that neutrophils isolated from $PTEN^{-/-}$ mice exhibit elevated levels of PI(3,4,5)P_3, P-Akt, and actin polymerization (Subramanian et al. 2007), others have reported that neutrophils isolated from $PTEN^{-/-}$ mice do not exhibit elevated PI(3,4,5)P_3 levels and that SHIP is required for the polarization of PI(3,4,5)P_3 (Nishio et al. 2007). SHIP-1-deficient neutrophils exhibit elevated and diffuse PI(3,4,5)P_3 distribution and a weakly polarized morphology. Cells deficient in PTEN however, show normal polarization of PH_{Akt}-GFP (Nishio et al. 2007). The complete mechanistic details have not yet been fully resolved, but a prevailing view is that SHIP-1 is the primary phosphatase regulating PI(3,4,5)P_3 levels in neutrophils. Nevertheless, PTEN has been shown to co-localize with active RhoA at the posterior of chemotaxing neutrophils isolated from the bone marrow of mice (Li et al. 2005). This distribution is disrupted by inhibition of the RhoA effector ROCK or in neutrophils lacking αPIX (Li et al. 2005). These observations suggest that PI3K is involved in the asymmetric polarization of αPIX-Cdc42 and RhoA, which is coupled to the subsequent posterior localization and activation of PTEN, helping maintain molecular polarity. Indeed, differentiated HL-60 neutrophil-like cells with suppressed PTEN expression show reduced chemotaxis (Li et al. 2005).

Increasing evidence suggest that the role of PI3Ks in leukocyte directional sensing depends on environmental conditions and cell type (Nishio et al. 2007). Ferguson and colleagues reported that although accumulation of PI(3,4,5)P_3 and F-actin at the leading edge of stimulated neutrophil isolated from $p110\gamma^{-/-}$ mice is abolished, cells are still able to carry out chemotaxis normally on certain surfaces. Further, priming neutrophils with lipopolysaccharide (LPS) reduces their requirement for p110γ (Ferguson et al. 2007). It was therefore suggested that PI3Kγ plays a role in the amplification and stabilization of the polarization response, rather than its initiation (Ferguson et al. 2007). Others, investigating leukocyte (T and B cell, and natural killer cell) chemotaxis and homing, have reported that while both PI3Kγ and δ are important, typically one isoform predominantly contributes over the other in one cell type versus the other (Nombela-Arrieta et al. 2004; Reif et al. 2004; Saudemont et al. 2009). Even more intriguing is a report that in natural killer (NK) cells, GPCR may function to activate PI3Kδ in addition to PI3Kγ and that both isoforms are necessary for NK cell chemotaxis toward CXCL12 and CCL3 as well as to an inflamed peritoneum (Saudemont et al. 2009). More recently, elegant *in vivo* studies in wounded zebrafish embryos demonstrated for the first time that PI3Kγ activity is indeed required for neutrophil chemotaxis, where 3' PIs concentrate on dominant pseudopods (Yoo et al. 2010). From these studies, it appears that the role of PI3Ks and 3' PIs in leukocyte chemotaxis is far more complex than the simple 'molecular compass' hypothesis initially proposed (Rickert et al. 2000) and prompts further studies that directly address the role of environmental and cellular conditions are needed to determine the precise role of PI3K in chemotaxis.

7.5.3 Fibroblasts

In a response that lasts several days, dermal fibroblasts surrounding a wound site are stimulated to proliferate and directionally migrate into the provisional matrix of the fibrin clot. These cells engage in producing, remodeling, and later contracting new ECM to reform the lost and damaged tissue (Martin 1997; Singer and Clark 1999). Within the provisional matrix, blood platelets release PDGF, which acts as a potent chemoattractant for fibroblasts (Seppa et al. 1982; Deuel et al. 1991; Heldin and Westermark 1999). Other stimuli such as transforming growth factor β (TGF-β), insulin-like growth factor-1, and epidermal growth factor (EGF) also contribute to the proliferation and migration of fibroblasts. Activated macrophages within the wound further add to the pool of PDGF and TGF-β through secretion. While many factors regulate fibroblasts, PDGF is distinct in its capacity to accelerate fibroblast infiltration (Pierce et al. 1989, 1991; Deuel et al. 1991).

In the context of wound healing, fibroblasts transduce PDGF signals through PDGF receptors, a well-characterized class of RTKs. Ligand-induced receptor dimerization leads to the recruitment and activation of class IA PI3K and to the subsequent generation of 3' PIs at the plasma membrane. Elevated levels of 3' PI coincide with membrane ruffling and are important for chemotaxis (Hawkins et al. 1992; Jackson et al. 1992; Wennstrom et al. 1994a, b). It was also shown that pharmacological inhibition of PI3K blocks membrane protrusions and chemotaxis (Wymann and Arcaro 1994; Hawkins et al. 1995; Hooshmand-Rad et al. 1997). Furthermore, much like in neutrophils, the treatment of fibroblasts with exogenous forms of PI(3,4,5)P_3 elicits membrane ruffling and chemotaxis (Derman et al. 1997). With advances in fluorescent fusion biosensors, it was possible to observe local 3' PI accumulation in fibroblasts during PDGF stimulation (Haugh et al. 2000), providing further support that this pathway is important during the chemotactic response of highly adherent, slower mesenchymal cells (Worth and Parsons 2008).

Cell adhesion via integrin engagement also contributes to activation of PI3K and may mediate the chemotactic response of fibroblasts and other mesenchymal cells (Ginsberg et al. 2005). Remarkably, recent data in fibroblasts suggest that PI3K can be activated during cell spreading on adhesive surfaces seemingly uncoupled from integrin-mediated pathways (Weiger et al. 2009). Further, fibroblasts undergoing spreading and random migration exhibit dynamic 3' PI localization to regions of active membrane protrusions and these regions correlate with both cell direction and speed. Indeed, during random fibroblast migration, the characteristic lifetime of 3' PI localization corresponds to cell migration persistence and their stochastic dynamics appear coupled, suggesting that depletion of a limiting cytosol signaling component (PI3K) maybe responsible for these observed dynamics. When this fragile equilibrium is tipped, 3' PI dynamics spatially reorient favoring local protrusion and cell turning in a different region of the cell (Weiger et al. 2009, 2010).

The precise role of PI3K during mesenchymal chemotaxis cells is still murky with reports that inhibition of PI3K in smooth muscle cells and Swiss 3T3 cells does not significantly impact their ability to chemotax (Higaki et al. 1996). Despite these findings though, others have demonstrated that PI3K signaling is indispensable for PDGF stimulated chemotaxis in both aortic endothelial cells and NIH 3T3 cells (Kundra et al. 1994; Wennstrom et al. 1994b). In addition to PI3K, several other signaling molecules such as PLC and Ras are reported to be active downstream of RTKs and contribute to the mesenchymal chemotactic response adding further complexity in assessing the precise role of PI3K and 3' PIs (Anand-Apte and Zetter 1997; Condeelis et al. 2005).

7.6 PI3Ks Signaling Through the Eyes of Mathematic Modeling

Even before the biochemical and mechanistic details of the PI3K pathway were being uncovered, physicists and mathematicians had begun to describe the observation of pattern formation in various biological processes. Gierer and Meinhart's introduced a mathematical guideline for the construction of molecular models that account for the non-linear interactions in biological processes. They reasoned that pattern formation would require an autocatalytic, self-enhancing activation combined with inhibitory or depletion effects of wider range (Gierer and Meinhardt 1972). Their concepts would inspire the formalization of many models to explain the polarization behavior of PI3K signaling during chemoattractant stimulation.

Early models described leukocyte motility and chemotaxis in terms of quantitative metrics such as persistence time and directionality and also incorporated biochemical dynamics of chemoattractant/receptor binding (Tranquillo et al. 1988). As more and more biochemical details of the chemotactic process and other intracellular processes were revealed, mathematical descriptions were assembled and applied to numerous systems involving receptor binding, trafficking and downstream signaling (Lauffenburger 1993). However, the discovery that PH domain-containing proteins specifically translocate to the leading edge of chemotaxing *Dictyostelium* cells (Parent et al. 1998) sparked the generation of a wave of new models attempting to explain this unique behavior. A phenomenological model proposed by Hans Meinhardt became a foundation upon which many subsequent models were constructed. According to his model, polarization was achieved by a saturating, self-enhancing reaction (a species that signals for extension of filopods and lamellipods) and is coupled with two antagonistic reactions (inhibitors), one that acts locally and the other over the whole cell (Meinhardt 1999). Another more mechanistic model based on the PI cycle, which transfers PIs between the membrane and endoplasmic reticulum, was later proposed. By accounting for receptor desensitization and the reaction-diffusion processes of the PI cycle, this model was able to emulate many of the experimentally observed dynamics (Narang et al. 2001).

Many more mathematical models have since been introduced to provide further explanation of the experimental observations made in *Dictyostelium*, neutrophils and fibroblasts with the basic goal of identifying the molecular mechanisms that adequately explain the complex responses observed during chemoattractant stimulation (Meinhardt 1999; Postma and Van Haastert 2001; Levchenko and Iglesias 2002; Rappel et al. 2002; Ma et al. 2004; Subramanian and Narang 2004; Schneider and Haugh 2005; Skupsky et al. 2005; Gamba et al. 2005; Arrieumerlou and Meyer 2005). Probably one of the most well known models is the *l*ocal *e*xcitation *g*lobal *i*nhibition (LEGI) model and variations thereof (Ma et al. 2004; Levine et al. 2006). In this model, gradient sensing is achieved through receptor occupancy and subsequent generation of localized regions of an active species that is at the plasma membrane, and by a rapidly diffusing species (a global inhibitor) in the cytosol that negatively regulates the active species (Levchenko and Iglesias 2002; Kutscher et al. 2004). Under gradient stimulation, the active species (PI3K) and inhibitory species (PTEN or other lipid phosphatases) produce a polarized response (3' PI), thereby defining the front and rear of the cell in terms of spatial sensing.

A mathematical model is not solely intended to be the end all explanation, it is a valuable tool to examine, perturb, and identify important elements of a signaling system and guide new hypotheses and experiments (Schneider and Haugh 2006; Mogilner et al. 2006). Indeed, refined models that better recapitulate experimental observations continue to be introduced. For example, several of the LEGI-based models predicted a low response level at the rear of the cell (measured from the localization of fluorescently label PH domain containing probes), a response not observed experimentally in *Dictyostelium* (Janetopoulos et al. 2004). To address this issue, a model based on the LEGI mechanism introduced cytosolic and membrane bound inhibitors that are mutually antagonistic to the response and are produced at the same rate as the active species. Under this framework, a distinct, spatially segregated response (i.e. PI(3,4,5)P_3, PI3K or Ras) captured gradient sensing behavior similar to that seen experimentally (Levine et al. 2006). Other key features of this model include its ability to reverse internal asymmetry of the response and establish internal direction over a large range of gradient strengths and background concentration levels.

The observation that cells can chemotax in extremely shallow gradients has prompted the introduction of models employing positive feedback loops to recapitulate the large amplification of a second messenger (3' PI/PH domain binding sites) compared to the level of the extracellular chemoattractant, either gradient or uniform (Iglesias and Devreotes 2008). Similar to the LEGI mechanism, many of these models incorporate a second messenger response that is locally regulated by receptor occupancy and include a diffusing inhibitor, which is also locally activated/produced. The concept of local production of an inhibitory process was later supported experimentally (Xu et al. 2007). These models are unique in how they elicit a positive feedback. The elevated second messenger response (3' PI) augments its own production through mechanisms such as autocatalytic effects (Meier-Schellersheim et al. 2006), substrate delivery (Skupsky et al. 2005), or slowing its rate of degradation (Gamba et al. 2005). While these models generate responses with enhanced amplification compared to that obtained with the basic LEGI mechanism, they exhibit response shapes that are almost insensitive to extracellular stimulus dose (Iglesias and Devreotes 2008). In addition, other models do not evoke a global inhibitor but achieve similar gradient sensing and polarization behavior through substrate depletion descriptors (Postma and Van Haastert 2001; Onsum and Rao 2007).

Studies that couple both experiments and mathematical modeling have also been conducted in fibroblast during PDGF simulation. An extensive and elegant analysis that characterizes the kinetics, dose responsiveness and spatial regulation of PDGF gradient sensing in fibroblasts suggest that the mechanism of gradient sensing is fundamentally distinct from neutrophils and *Dictyostelium*, exhibiting neither significant amplification nor adaptation. It was also demonstrated that fibroblasts have lower sensitivity, and are more dependent on the midpoint PDGF concentration in a gradient compared to fast moving amoeboid cells (Park et al. 2003; Haugh and Schneider 2004; Schneider and Haugh 2004, 2005). Further, optimal gradient sensing is observed in a relatively narrow range of PDGF concentrations. Although it is reasonable to think that a feedback mechanism might exist in fibroblasts similar to those proposed in neutrophils and *Dictyostelium*, this doesn't appear to be necessary in fibroblasts during PDGF-mediated gradient sensing. In fact, the morphological polarity of fibroblasts may cause enhanced PI3K signaling at the leading edge possibly through local (not global) regulation of PI3-phosphatases or other 3' PI degradation pathways (Schneider et al. 2005). Remarkably, the gradient sensing mechanism derived from single cell observations has been cast into the larger physiological context of the wound healing process. Using a mathematical model that incorporates course-grained receptor dynamics and intracellular processes, fibroblasts are predicted to shape a gradient of PDGF through receptor endocytosis and degradation, allowing the optimal mid point concentration for migration to be maintained as the cells invade a wound (Haugh 2006). This model was later expanded upon and treated fibroblasts as individual entities having heterogenous PI3K signaling properties (protein expression levels) to mimic natural cell-to-cell variability (Monine and Haugh 2008), further demonstrating the utility of mathematical models.

Models continue to gain wide spread traction as a vehicle for testing mechanistic hypotheses and guiding experimental design. A freely accessible modeling analysis platform, known as Virtual Cell (University of Connecticut Heath Center, NRCAM and NCRR), has been developed and is freely available to the scientific community (Loew and Schaff 2001). This software package allows users to construct a molecular pathway that is subsequently converted by the software into a corresponding mathematical system of differential equations. Virtual cell allows users to define compartmental topology and geometry, molecular characteristics, and relevant interaction parameters. Numerous models have already been developed, implemented and tested, including the LEGI model for PI3K-dependent chemotactic response (Ma et al. 2004; Yang 2009).

As the number of new molecular players implicated in regulating chemotaxis increases, more scientists are turning to modeling to assist them in interpreting the complex and interconnected signaling networks controlling directed migration. For example, recent models have investigated the interplay between PI3K signaling and other downstream molecules in the regulation of directed migration in amoeboid cells (Dawes and Edelstein-Keshet 2007; Onsum and Rao 2007). Models can also make interesting and sometimes provocative predictions that stimulate debate. For example, Stephanou and colleagues developed a model that predicts that the persistence of mesenchymal cell migration may be affected by the relative life times of different adhesion complexes (Stephanou et al. 2008). As live-cell imaging methods and experimental approaches yield more molecular and quantitative details, analytical methods and mathematical models will undoubtedly play an increasingly important role in providing mechanistic interpretations of data (Welf and Haugh 2011).

7.7 Other Signaling Pathways Regulating Chemotaxis

An emerging consensus is beginning to cast 3' PIs, not as the sole regulators, but as an integral signaling component that acts with multiple other pathways to coordinate cell polarity and directed movement; serving to optimize polarization and subsequent migration under various conditions (see Fig. 7.5) (Loovers et al. 2006; Hoeller and Kay 2007; Kolsch et al. 2008). Many other pathways involving enzymes that cleave PIs have been implicated as chemotactic regulators. One such pathway involves phospholipase Cγ (PLCγ) and cofilin. Cofilin is an actin severing protein that is essential for the localized formation of actin barbed-ends, which are sites for new actin polymerization (Condeelis 2001; DesMarais et al. 2005). Cofilin is thought to be locally released and activated through the hydrolysis of $PI(4,5)P_2$ by PLCγ and, simultaneously, inactivated globally via phosphorylation by LIM kinase (LIMK) (Hitchcock-Degregori 2006; Mouneimne et al. 2006). In adenocarcinoma cells, EGF stimulation induces two peaks of actin polymerization; the first peak is dependent on PLCγ and cofilin (Chan et al. 1998, 2000; Mouneimne et al. 2004), while the second relies on PI3K activity (Hill et al. 2000; Chen et al. 2003). When PLCγ is inhibited in carcinoma cells, defects in gradient sensing are observed. On the other hand, inhibition of PI3K does not affect gradient sensing (Mouneimne et al. 2006). However, full lamellipod extension is dependent on PI3K activity since inhibition of PI3K abrogates the second peak of actin polymerization during EFG

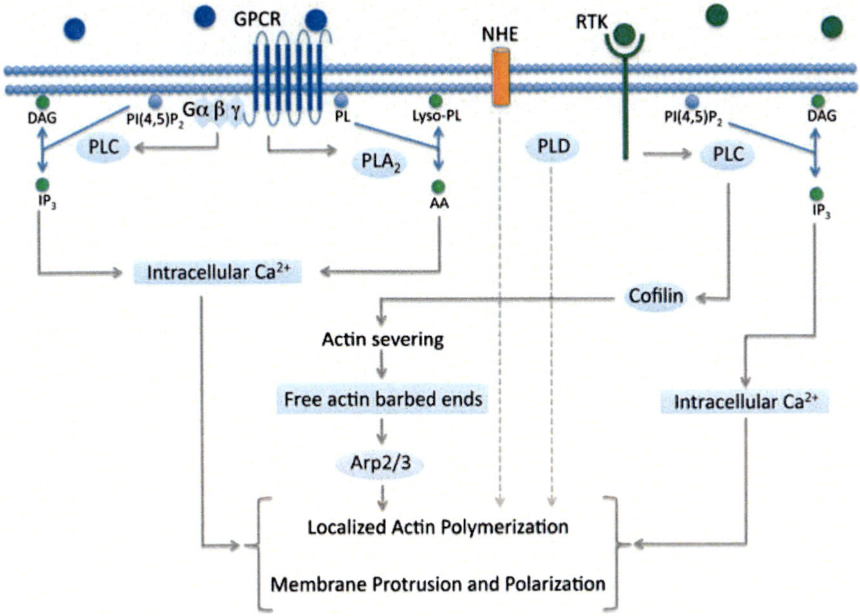

Fig. 7.5 Emerging pathways that regulate chemotaxis apart from 3' PIs. Several alternate pathways have been identified as potential regulators of local actin polymerization and membrane protrusion. Pathways involving PLC, PLA_2, and their second messengers as well as PLD have been shown to be important regulators of chemotaxis. A pathway involving PLC and cofilin has also been shown to mediate chemotactic responses under growth factor stimulation. Even small molecules like Ca^{2+} and the Na^+-H^+ exchanger NHE, which regulates intracellular pH (H^+ levels), regulate certain facets of chemotaxis

stimulation (Hill et al. 2000; Mouneimne et al. 2004). Thus, the PLCγ/cofilin and PI3K/Arp2/3 signaling pathways seem to cooperate in chemotactic gradient sensing and efficient lamellipod generation in response to EGF stimulation (Ghosh et al. 2004; Mouneimne et al. 2006). To underscore the importance of cell context, studies in *Dictyostelium* have shown that although cofilin localizes to the leading edge during chemotaxis there is no evidence that it regulates gradient sensing (Aizawa et al. 1995, 1997). However, treatment of human neutrophils with PLC inhibitors hinders chemotactic responses to fMLP, IL8, and leukotriene B_4 (LTB_4) (Hou et al. 2004; Liu et al. 2010).

Other phospholipase signaling pathways provide important regulatory inputs during chemotaxis. Two independent studies in *Dictyostelium* established that phospholipase A_2 (PLA_2), which cleaves the second acyl chain of phospholipids to predominantly produce the second messengers arachidonic acid and lyso-phospholipid, regulates chemotaxis in parallel with PI3K (Chen et al. 2007; van Haastert et al. 2007). It was determined that in shallow chemoattractant gradients, inhibition of either PI3K or PLA_2 inhibits chemotaxis, yet inhibition of both pathways is required to disrupt chemotaxis in a steep gradient. In monocytes, the Ca^{2+}-insensitive PLA_2

($iPLA_2\beta$) and the cytosolic PLA_2 ($cPLA_2\alpha$) localize to distinct cellular compartments and selectively affect both cell directionality and speed, respectively (Mishra et al. 2008). Interestingly, the inhibition of both PLC and PLA_2 in *Dictyostelium* completely inhibits the cAMP mediated PI(3,4,5)P_3 response and causes drastic defects in chemotaxis (van Haastert et al. 2007).

In an effort to identify other pathways that regulate chemotaxis, scientist screened a series of *Dictyostelium* mutants and identified the soluble guanylyl cyclase (sGC), which produces guanosine 3', 5'-monophosphate (cGMP), as being a key regulator (Roelofs et al. 2001). Interestingly, in *Dictyostelium* sGC co-localizes with newly polymerized actin at the leading edge (Veltman et al. 2005). Cells expressing sGC with no catalytic activity exhibit slower migration and elevated number of lateral pseudopods, but normal gradient sensing compared to wild-type cell. On the other hand, if the ability of sGC to localize to F-actin is abolished, cells show impaired directionality (Veltman and van Haastert 2008). It remains unclear if cGMP is important for chemotaxis in other cell types although some studies have linked nitric oxide production to sGC activity (Kaplan et al. 1989; Shibata et al. 2001) and shown that disruption of nitric oxide and cGMP pathways impair neutrophil chemotaxis (Wanikiat et al. 1997).

While phospholipase D (PLD) produces the second messenger phosphatidic acid (PA) and choline by hydrolyzing the phosphodiester bond of phospatidylcholine (Oude Weernink et al. 2007), it does not appear to be a direct regulator of chemotaxis; however, an increasing number of studies indicate that it can impact chemotaxis. In *Dictyostelium*, inhibition of PLD causes a dramatic decrease in PI(4,5)P_2 synthesis, resulting in severe defects in actin-based motility (Zouwail et al. 2005). Similarly, PLD binds to actin in human mast cells, which appears to be important for the regulation of PLD1b activity (Farquhar et al. 2007). During chemotaxis, PI(4,5)P_2 serves as a major substrate for PLC and simultaneously influences the subcellular localization and activity of PLD. More recently, reports in differentiated HL-60 cells, show that active PLD1 mediates chemotactic responses toward IL-8 and fMLP and that silencing both PLD1 and PLD2 leads to migration arrest (Lehman et al. 2006). Furthermore, in both macrophages and fibroblasts, over-expressing and silencing PLD2 results in enhanced and decreased chemotaxis, respectively (Knapek et al. 2010). Further studies will perhaps reveal even more evidence linking PLD to chemotaxis.

It is well established that calcium has a multifunctional role in directed migration, influencing directional sensing, cytoskeletal reorganization, traction force generation, and relocation of focal adhesions. It has been reported that intracellular calcium displays a rear-to-front gradient, with the lowest concentration being at the front of migrating cells (Brundage et al. 1991). Transient increases in calcium levels have been observed, albeit infrequently, in the cell rear and could mediate intermittent tail retraction (Lee et al. 1999). At the leading edge of macrophages, biochemical analyses have shown that calcium is essential for maintaining ruffling structures, actin polymerization and PI(3,4,5)P_3 signaling (Evans and Falke 2007). Calcium levels are also elevated after PDGF stimulation in fibroblasts, leading to activation of Ca^{2+}/calmodulin kinase II (CaMKII), an activator of the Rac1-specific GEF Tiam1 that causes Rac1-dependent membrane ruffling (Buchanan et al. 2000). In

Dictyostelium and neutrophils, Ca^{2+} signaling mainly regulates myosin-dependent contraction at the cell rear (Yumura et al. 1996; Nebl and Fisher 1997). Another study found *Dictyostelium* lacking the IP_3 receptor-like gene (ipIA), exhibit no changes in intracellular Ca^{2+} in response to cAMP and also showed no significant chemotaxis defects (Traynor et al. 2000). Interestingly, upon stimulation with chemotactic factors, neutrophils exhibit an increase in intracellular Ca^{2+}, which appears to modulate integrin cell adhesion molecules and regulate neutrophils migration on adhesive substrates (Niggli 2003). More recently, short-lived high-calcium microdomains ('calcium flickers') were observed to concentrate predominantly at the leading edge of migrating fibroblasts and were not present in stationary cells. When presented with a gradient acting perpendicular to cell movement, these 'calcium flickers' developed in greater proportion in the cell region facing a PDGF source and correlated with cell turning behavior (Wei et al. 2009). This flickering behavior appears to be coupled to membrane tension (stretch activated Ca^{2+} channel) and chemoattractant signal transduction (type 2 IP_3 receptors).

The notion that discrete, spatiotemporal organization of an ion can impact chemotaxis prompted further investigation into other ion-channels that might serve as candidates for chemotactic regulators. The Na^+-H^+ exchanger (NHE) has become a front-runner. In *Dictyostelium*, NHE1 localizes to the leading edge of polarized cells and is necessary for pH homeostasis. Cell lacking NHE1 can still chemotax but fail to attain a polarized morphology while still showing localized $PI(3,4,5)P_3$, suggesting that it is important for efficient chemotaxis (Patel and Barber 2005). Other studies have suggested that Rac1 and RhoA1 regulation of NHE1 activity induces morphological and cytoskeletal changes in metastatic cells (Paradiso et al. 2004). In fibroblasts, data suggest that NHE1 participates in a positive feedback loop with Cdc42 (Frantz et al. 2007). Also in fibroblasts, expression of a NHE1 mutant that prevents H^+ efflux abolished PDGF-induced biphasic accumulation of actin free barbed ends, where the first phase is cofilin dependent and the later is PI3K dependent. Interestingly, the cofilin dependent phase was recovered by mutating a pH sensitive region in cofilin (Frantz et al. 2008). Attention is also being given to NHE regulatory factors that appear able to regulate the duration of PI3K signaling under PDGF stimulation (Demoulin et al. 2003; Takahashi et al. 2006). All these additional signaling pathways along with PI3K and 3' PI signaling contribute to the chemotactic response of numerous cell types during multiple biological functions. The specific contexts, time scales, and extent in which each pathway contributes to chemotaxis remain largely unanswered and open to further exploration.

7.8 Conclusions

While significantly more work needs to be done, the advancements in our current understanding of the role of PI3K-mediated signaling pathways in the control of chemotaxis have been substantial. With the implementation of new technologies, mathematical models and well-conceived experiments, we will continue to peel away the complexities governing chemotaxis and finally establish the many contributions of 3' PI signaling in this fascinating process.

Acknowledgements We thank Drs. Philippe Afonso, Satarupa Das, and Vidya Vedham for their critical input. This work was supported by the Intramural Research Program of the Center for Cancer Research, NCI, National Institutes of Health.

References

Abercrombie M (1978) Fibroblasts. J Clin Pathol Suppl (R Coll Pathol) 12:1–6

Abercrombie M, Heaysman JE (1953) Observations on the social behaviour of cells in tissue culture. I. Speed of movement of chick heart fibroblasts in relation to their mutual contacts. Exp Cell Res 5:111–131

Abercrombie M, Heaysman JE, Pegrum SM (1970) The locomotion of fibroblasts in culture. I. Movements of the leading edge. Exp Cell Res 59:393–398

Aizawa H, Sutoh K, Tsubuki S, Kawashima S, Ishii A, Yahara I (1995) Identification, characterization, and intracellular distribution of cofilin in Dictyostelium discoideum. J Biol Chem 270:10923–10932

Aizawa H, Fukui Y, Yahara I (1997) Live dynamics of Dictyostelium cofilin suggests a role in remodeling actin latticework into bundles. J Cell Sci 110(Pt 19):2333–2344

Alberts B, Johnson A, Lewis J, Raff M, Roberts K, Walter P (2002) Molecular biology of the cell. Garland Science, New York

Albrecht-Buehler G (1979) The angular distribution of directional changes of guided 3T3 cells. J Cell Biol 80:53–60

Alexander S, Koehl GE, Hirschberg M, Geissler EK, Friedl P (2008) Dynamic imaging of cancer growth and invasion: a modified skin-fold chamber model. Histochem Cell Biol 130:1147–1154

Ali K, Bilancio A, Thomas M, Pearce W, Gilfillan AM, Tkaczyk C, Kuehn N, Gray A, Giddings J, Peskett E, Fox R, Bruce I, Walker C, Sawyer C, Okkenhaug K, Finan P, Vanhaesebroeck B (2004) Essential role for the p110delta phosphoinositide 3-kinase in the allergic response. Nature 431:1007–1011

Anand-Apte B, Zetter B (1997) Signaling mechanisms in growth factor-stimulated cell motility. Stem Cells 15:259–267

Aoki K, Nakamura T, Fujikawa K, Matsuda M (2005) Local phosphatidylinositol 3,4,5-trisphosphate accumulation recruits Vav2 and Vav3 to activate Rac1/Cdc42 and initiate neurite outgrowth in nerve growth factor-stimulated PC12 cells. Mol Biol Cell 16:2207–2217

Arcaro A, Wymann MP (1993) Wortmannin is a potent phosphatidylinositol 3-kinase inhibitor: the role of phosphatidylinositol 3,4,5-trisphosphate in neutrophil responses. Biochem J 296(Pt 2):297–301

Arrieumerlou C, Meyer T (2005) A local coupling model and compass parameter for eukaryotic chemotaxis. Dev Cell 8:215–227

Attoub S, De Wever O, Bruyneel E, Mareel M, Gespach C (2008) The transforming functions of PI3-kinase-gamma are linked to disruption of intercellular adhesion and promotion of cancer cell invasion. Ann N Y Acad Sci 1138:204–213

Aubry L, Firtel R (1999) Integration of signaling networks that regulate Dictyostelium differentiation. Annu Rev Cell Dev Biol 15:469–517

Backer JM (2008) The regulation and function of Class III PI3Ks: novel roles for Vps34. Biochem J 410:1–17

Banin S, Moyal L, Shieh S, Taya Y, Anderson CW, Chessa L, Smorodinsky NI, Prives C, Reiss Y, Shiloh Y, Ziv Y (1998) Enhanced phosphorylation of p53 by ATM in response to DNA damage. Science 281:1674–1677

Bi L, Okabe I, Bernard DJ, Wynshaw-Boris A, Nussbaum RL (1999) Proliferative defect and embryonic lethality in mice homozygous for a deletion in the p110alpha subunit of phosphoinositide 3-kinase. J Biol Chem 274:10963–10968

Bi L, Okabe I, Bernard DJ, Nussbaum RL (2002) Early embryonic lethality in mice deficient in the p110beta catalytic subunit of PI 3-kinase. Mamm Genome 13:169–172

Bilancio A, Okkenhaug K, Camps M, Emery JL, Ruckle T, Rommel C, Vanhaesebroeck B (2006) Key role of the p110delta isoform of PI3K in B-cell antigen and IL-4 receptor signaling: comparative analysis of genetic and pharmacologic interference with p110delta function in B cells. Blood 107:642–650

Bosgraaf L, Keizer-Gunnink I, Van Haastert PJ (2008) PI3-kinase signaling contributes to orientation in shallow gradients and enhances speed in steep chemoattractant gradients. J Cell Sci 121:3589–3597

Boulven I, Levasseur S, Marois S, Pare G, Rollet-Labelle E, Naccache PH (2006) Class IA phosphatidylinositide 3-kinases, rather than p110 gamma, regulate formyl-methionyl-leucyl-phenylalanine-stimulated chemotaxis and superoxide production in differentiated neutrophil-like PLB-985 cells. J Immunol 176:7621–7627

Brock C, Schaefer M, Reusch HP, Czupalla C, Michalke M, Spicher K, Schultz G, Nurnberg B (2003) Roles of G beta gamma in membrane recruitment and activation of p110 gamma/p101 phosphoinositide 3-kinase gamma. J Cell Biol 160:89–99

Brundage RA, Fogarty KE, Tuft RA, Fay FS (1991) Calcium gradients underlying polarization and chemotaxis of eosinophils. Science 254:703–706

Brunn GJ, Williams J, Sabers C, Wiederrecht G, Lawrence JC Jr, Abraham RT (1996) Direct inhibition of the signaling functions of the mammalian target of rapamycin by the phosphoinositide 3-kinase inhibitors, wortmannin and LY294002. EMBO J 15:5256–5267

Buchanan FG, Elliot CM, Gibbs M, Exton JH (2000) Translocation of the Rac1 guanine nucleotide exchange factor Tiam1 induced by platelet-derived growth factor and lysophosphatidic acid. J Biol Chem 275:9742–9748

Cain RJ, Ridley AJ (2009) Phosphoinositide 3-kinases in cell migration. Biol Cell 101:13–29

Carpenter CL, Cantley LC (1990) Phosphoinositide kinases. Biochemistry 29:11147–11156

Cenni V, Sirri A, Riccio M, Lattanzi G, Santi S, De Pol A, Maraldi NM, Marmiroli S (2003) Targeting of the Akt/PKB kinase to the actin skeleton. Cell Mol Life Sci 60:2710–2720

Chan AY, Raft S, Bailly M, Wyckoff JB, Segall JE, Condeelis JS (1998) EGF stimulates an increase in actin nucleation and filament number at the leading edge of the lamellipod in mammary adenocarcinoma cells. J Cell Sci 111(Pt 2):199–211

Chan AY, Bailly M, Zebda N, Segall JE, Condeelis JS (2000) Role of cofilin in epidermal growth factor-stimulated actin polymerization and lamellipod protrusion. J Cell Biol 148:531–542

Charest PG, Firtel RA (2006) Feedback signaling controls leading-edge formation during chemotaxis. Curr Opin Genet Dev 16:339–347

Chen WT (1979) Induction of spreading during fibroblast movement. J Cell Biol 81:684–691

Chen L, Janetopoulos C, Huang YE, Iijima M, Borleis J, Devreotes PN (2003) Two phases of actin polymerization display different dependencies on PI(3,4,5)P3 accumulation and have unique roles during chemotaxis. Mol Biol Cell 14:5028–5037

Chen L, Iijima M, Tang M, Landree MA, Huang YE, Xiong Y, Iglesias PA, Devreotes PN (2007) PLA2 and PI3K/PTEN pathways act in parallel to mediate chemotaxis. Dev Cell 12:603–614

Clayton E, Bardi G, Bell SE, Chantry D, Downes CP, Gray A, Humphries LA, Rawlings D, Reynolds H, Vigorito E, Turner M (2002) A crucial role for the p110delta subunit of phosphatidylinositol 3-kinase in B cell development and activation. J Exp Med 196:753–763

Comer FI, Parent CA (2006) Phosphoinositide 3-kinase activity controls the chemoattractant-mediated activation and adaptation of adenylyl cyclase. Mol Biol Cell 17:357–366

Comer FI, Lippincott CK, Masbad JJ, Parent CA (2005) The PI3K-mediated activation of CRAC independently regulates adenylyl cyclase activation and chemotaxis. Curr Biol 15:134–139

Condeelis J (2001) How is actin polymerization nucleated in vivo? Trends Cell Biol 11:288–293

Condeelis J, Singer RH, Segall JE (2005) The great escape: when cancer cells hijack the genes for chemotaxis and motility. Annu Rev Cell Dev Biol 21:695–718

Condliffe AM, Davidson K, Anderson KE, Ellson CD, Crabbe T, Okkenhaug K, Vanhaesebroeck B, Turner M, Webb L, Wymann MP, Hirsch E, Ruckle T, Camps M, Rommel C, Jackson SP, Chilvers ER, Stephens LR, Hawkins PT (2005) Sequential activation of class IB and class IA PI3K is important for the primed respiratory burst of human but not murine neutrophils. Blood 106:1432–1440

Costa C, Barberis L, Ambrogio C, Manazza AD, Patrucco E, Azzolino O, Neilsen PO, Ciraolo E, Altruda F, Prestwich GD, Chiarle R, Wymann M, Ridley A, Hirsch E (2007) Negative feedback regulation of Rac in leukocytes from mice expressing a constitutively active phosphatidylinositol 3-kinase gamma. Proc Natl Acad Sci U S A 104:14354–14359

Cote JF, Vuori K (2007) GEF what? Dock180 and related proteins help Rac to polarize cells in new ways. Trends Cell Biol 17:383–393

Daunderer C, Schliwa M, Graf R (1999) Dictyostelium discoideum: a promising centrosome model system. Biol Cell 91:313–320

Davidson LA, Keller RE (1999) Neural tube closure in Xenopus laevis involves medial migration, directed protrusive activity, cell intercalation and convergent extension. Development 126:4547–4556

Dawes AT, Edelstein-Keshet L (2007) Phosphoinositides and Rho proteins spatially regulate actin polymerization to initiate and maintain directed movement in a one-dimensional model of a motile cell. Biophys J 92:744–768

Dekker LV, Segal AW (2000) Perspectives: signal transduction. Signals to move cells. Science 287:982–983, 985

Demoulin JB, Seo JK, Ekman S, Grapengiesser E, Hellman U, Ronnstrand L, Heldin CH (2003) Ligand-induced recruitment of Na^+/H^+-exchanger regulatory factor to the PDGF (platelet-derived growth factor) receptor regulates actin cytoskeleton reorganization by PDGF. Biochem J 376:505–510

Derman MP, Toker A, Hartwig JH, Spokes K, Falck JR, Chen CS, Cantley LC, Cantley LG (1997) The lipid products of phosphoinositide 3-kinase increase cell motility through protein kinase C. J Biol Chem 272:6465–6470

Desmarais V, Ghosh M, Eddy R, Condeelis J (2005) Cofilin takes the lead. J Cell Sci 118:19–26

Deuel TF, Kawahara RS, Mustoe TA, Pierce AF (1991) Growth factors and wound healing: platelet-derived growth factor as a model cytokine. Annu Rev Med 42:567–584

Devreotes PN, Zigmond SH (1988) Chemotaxis in eukaryotic cells: a focus on leukocytes and Dictyostelium. Annu Rev Cell Biol 4:649–686

Dhand R, Hara K, Hiles I, Bax B, Gout I, Panayotou G, Fry MJ, Yonezawa K, Kasuga M, Waterfield MD (1994) PI 3-kinase: structural and functional analysis of intersubunit interactions. EMBO J 13:511–521

Domin J, Pages F, Volinia S, Rittenhouse SE, Zvelebil MJ, Stein RC, Waterfield MD (1997) Cloning of a human phosphoinositide 3-kinase with a C2 domain that displays reduced sensitivity to the inhibitor wortmannin. Biochem J 326(Pt 1):139–147

Domin J, Harper L, Aubyn D, Wheeler M, Florey O, Haskard D, Yuan M, Zicha D (2005) The class II phosphoinositide 3-kinase PI3K-C2beta regulates cell migration by a PtdIns3P dependent mechanism. J Cell Physiol 205:452–462

Dong X, Mo Z, Bokoch G, Guo C, Li Z, Wu D (2005) P-Rex1 is a primary Rac2 guanine nucleotide exchange factor in mouse neutrophils. Curr Biol 15:1874–1879

Dormann D, Weijer G, Parent CA, Devreotes PN, Weijer CJ (2002) Visualizing PI3 kinase-mediated cell-cell signaling during Dictyostelium development. Curr Biol 12:1178–1188

Dunn GA (1980) The locomotory machinery of fibroblasts. Eur J Cancer 16:6–8

Egelhoff TT, Titus MA, Manstein DJ, Ruppel KM, Spudich JA (1991) Molecular genetic tools for study of the cytoskeleton in Dictyostelium. Methods Enzymol 196:319–334

Eichinger L, Noegel AA (2003) Crawling into a new era-the Dictyostelium genome project. EMBO J 22:1941–1946

Eichinger L, Lee SS, Schleicher M (1999) Dictyostelium as model system for studies of the actin cytoskeleton by molecular genetics. Microsc Res Tech 47:124–134

Eichinger L, Pachebat JA, Glockner G, Rajandream MA, Sucgang R, Berriman M, Song J, Olsen R, Szafranski K, Xu Q, Tunggal B, Kummerfeld S, Madera M, Konfortov BA, Rivero F, Bankier AT, Lehmann R, Hamlin N, Davies R, Gaudet P, Fey P, Pilcher K, Chen G, Saunders D, Sodergren E, Davis P, Kerhornou A, Nie X, Hall N, Anjard C, Hemphill L, Bason N, Farbrother P, Desany B, Just E, Morio T, Rost R, Churcher C, Cooper J, Haydock S, Van Driessche N, Cronin A, Goodhead I, Muzny D, Mourier T, Pain A, Lu M, Harper D, Lindsay R, Hauser H, James K, Quiles M, Madan Babu M, Saito T, Buchrieser C, Wardroper A, Felder M, Thangavelu M, Johnson D, Knights A, Loulseged H, Mungall K, Oliver K, Price C, Quail MA, Urushihara H, Hernandez J, Rabbinowitsch E, Steffen D, Sanders M, Ma J, Kohara Y, Sharp S, Simmonds M, Spiegler S, Tivey A, Sugano S, White B, Walker D, Woodward J, Winckler T, Tanaka Y, Shaulsky G, Schleicher M, Weinstock G, Rosenthal A, Cox EC, Chisholm RL, Gibbs R, Loomis WF, Platzer M, Kay RR, Williams J, Dear PH, Noegel AA, Barrell B, Kuspa A (2005) The genome of the social amoeba Dictyostelium discoideum. Nature 435:43–57

El Fahime E, Torrente Y, Caron NJ, Bresolin MD, Tremblay JP (2000) In vivo migration of transplanted myoblasts requires matrix metalloproteinase activity. Exp Cell Res 258:279–287

Enomoto A, Murakami H, Asai N, Morone N, Watanabe T, Kawai K, Murakumo Y, Usukura J, Kaibuchi K, Takahashi M (2005) Akt/PKB regulates actin organization and cell motility via Girdin/APE. Dev Cell 9:389–402

Evans JH, Falke JJ (2007) Ca^{2+} influx is an essential component of the positive-feedback loop that maintains leading-edge structure and activity in macrophages. Proc Natl Acad Sci U S A 104:16176–16181

Farina KL, Wyckoff JB, Rivera J, Lee H, Segall JE, Condeelis JS, Jones JG (1998) Cell motility of tumor cells visualized in living intact primary tumors using green fluorescent protein. Cancer Res 58:2528–2532

Farquhar MJ, Powner DJ, Levine BA, Wright MH, Ladds G, Hodgkin MN (2007) Interaction of PLD1b with actin in antigen-stimulated mast cells. Cell Signal 19:349–358

Fedorov O, Marsden B, Pogacic V, Rellos P, Muller S, Bullock AN, Schwaller J, Sundstrom M, Knapp S (2007) A systematic interaction map of validated kinase inhibitors with Ser/Thr kinases. Proc Natl Acad Sci U S A 104:20523–20528

Ferguson GJ, Milne L, Kulkarni S, Sasaki T, Walker S, Andrews S, Crabbe T, Finan P, Jones G, Jackson S, Camps M, Rommel C, Wymann M, Hirsch E, Hawkins P, Stephens L (2007) PI(3)Kgamma has an important context-dependent role in neutrophil chemokinesis. Nat Cell Biol 9:86–91

Fey P, Stephens S, Titus MA, Chisholm RL (2002) SadA, a novel adhesion receptor in Dictyostelium. J Cell Biol 159:1109–1119

Firtel RA, Van Haastert PJ, Kimmel AR, Devreotes PN (1989) G protein linked signal transduction pathways in development: Dictyostelium as an experimental system. Cell 58:235–239

Frank SR, Hansen SH (2008) The PIX-GIT complex: a G protein signaling cassette in control of cell shape. Semin Cell Dev Biol 19:234–244

Frantz C, Karydis A, Nalbant P, Hahn KM, Barber DL (2007) Positive feedback between Cdc42 activity and H^+ efflux by the Na-H exchanger NHE1 for polarity of migrating cells. J Cell Biol 179:403–410

Frantz C, Barreiro G, Dominguez L, Chen X, Eddy R, Condeelis J, Kelly MJ, Jacobson MP, Barber DL (2008) Cofilin is a pH sensor for actin free barbed end formation: role of phosphoinositide binding. J Cell Biol 183:865–879

Friedl P, Gilmour D (2009) Collective cell migration in morphogenesis, regeneration and cancer. Nat Rev Mol Cell Biol 10:445–457

Friedl P, Wolf K (2003) Tumour-cell invasion and migration: diversity and escape mechanisms. Nat Rev Cancer 3:362–374

Friedl P, Maaser K, Klein CE, Niggemann B, Krohne G, Zanker KS (1997) Migration of highly aggressive MV3 melanoma cells in 3-dimensional collagen lattices results in local matrix reorganization and shedding of alpha2 and beta1 integrins and CD44. Cancer Res 57:2061–2070

Friedl P, Zanker KS, Brocker EB (1998) Cell migration strategies in 3-D extracellular matrix: differences in morphology, cell matrix interactions, and integrin function. Microsc Res Tech 43:369–378

Fruman DA, Cantley LC (2002) Phosphoinositide 3-kinase in immunological systems. Semin Immunol 14:7–18

Funamoto S, Milan K, Meili R, Firtel RA (2001) Role of phosphatidylinositol 3' kinase and a downstream pleckstrin homology domain-containing protein in controlling chemotaxis in Dictyostelium. J Cell Biol 153:795–810

Funamoto S, Meili R, Lee S, Parry L, Firtel RA (2002) Spatial and temporal regulation of 3-phosphoinositides by PI 3-kinase and PTEN mediates chemotaxis. Cell 109:611–623

Gail MH, Boone CW (1970) The locomotion of mouse fibroblasts in tissue culture. Biophys J 10:980–993

Gamba A, De Candia A, Di Talia S, Coniglio A, Bussolino F, Serini G (2005) Diffusion-limited phase separation in eukaryotic chemotaxis. Proc Natl Acad Sci U S A 102:16927–16932

Gan X, Wang J, Su B, Wu D (2011) Evidence for direct activation of mTORC2 kinase activity by phosphatidylinositol 3,4,5-trisphosphate. J Biol Chem 286:10998–11002

Ghosh M, Song X, Mouneimne G, Sidani M, Lawrence DS, Condeelis JS (2004) Cofilin promotes actin polymerization and defines the direction of cell motility. Science 304:743–746

Gierer A, Meinhardt H (1972) A theory of biological pattern formation. Kybernetik 12:30–39

Ginsberg MH, Partridge A, Shattil SJ (2005) Integrin regulation. Curr Opin Cell Biol 17:509–516

Goswami S, Sahai E, Wyckoff JB, Cammer M, Cox D, Pixley FJ, Stanley ER, Segall JE, Condeelis JS (2005) Macrophages promote the invasion of breast carcinoma cells via a colony-stimulating factor-1/epidermal growth factor paracrine loop. Cancer Res 65:5278–5283

Graupera M, Guillermet-Guibert J, Foukas LC, Phng LK, Cain RJ, Salpekar A, Pearce W, Meek S, Millan J, Cutillas PR, Smith AJ, Ridley AJ, Ruhrberg C, Gerhardt H, Vanhaesebroeck B (2008) Angiogenesis selectively requires the p110alpha isoform of PI3K to control endothelial cell migration. Nature 453:662–666

Gruver JS, Wikswo JP, Chung CY (2008) 3'-phosphoinositides regulate the coordination of speed and accuracy during chemotaxis. Biophys J 95:4057–4067

Han J, Luby-Phelps K, Das B, Shu X, Xia Y, Mosteller RD, Krishna UM, Falck JR, White MA, Broek D (1998) Role of substrates and products of PI 3-kinase in regulating activation of Rac-related guanosine triphosphatases by Vav. Science 279:558–560

Hannigan M, Zhan L, Li Z, Ai Y, Wu D, Huang CK (2002) Neutrophils lacking phosphoinositide 3-kinase gamma show loss of directionality during N-formyl-Met-Leu-Phe-induced chemotaxis. Proc Natl Acad Sci U S A 99:3603–3608

Haston WS, Shields JM, Wilkinson PC (1982) Lymphocyte locomotion and attachment on two-dimensional surfaces and in three-dimensional matrices. J Cell Biol 92:747–752

Haugh JM (2006) Deterministic model of dermal wound invasion incorporating receptor-mediated signal transduction and spatial gradient sensing. Biophys J 90:2297–2308

Haugh JM, Schneider IC (2004) Spatial analysis of 3' phosphoinositide signaling in living fibroblasts: I. Uniform stimulation model and bounds on dimensionless groups. Biophys J 86:589–598

Haugh JM, Codazzi F, Teruel M, Meyer T (2000) Spatial sensing in fibroblasts mediated by 3' phosphoinositides. J Cell Biol 151:1269–1280

Hawkins PT, Jackson TR, Stephens LR (1992) Platelet-derived growth factor stimulates synthesis of PtdIns(3,4,5)P3 by activating a PtdIns(4,5)P2 3-OH kinase. Nature 358:157–159

Hawkins PT, Eguinoa A, Qiu RG, Stokoe D, Cooke FT, Walters R, Wennstrom S, Claesson-Welsh L, Evans T, Symons M et al (1995) PDGF stimulates an increase in GTP-Rac via activation of phosphoinositide 3-kinase. Curr Biol 5:393–403

Hegerfeldt Y, Tusch M, Brocker EB, Friedl P (2002) Collective cell movement in primary melanoma explants: plasticity of cell-cell interaction, beta1-integrin function, and migration strategies. Cancer Res 62:2125–2130

Heldin CH, Westermark B (1999) Mechanism of action and in vivo role of platelet-derived growth factor. Physiol Rev 79:1283–1316

Higaki M, Sakaue H, Ogawa W, Kasuga M, Shimokado K (1996) Phosphatidylinositol 3-kinase-independent signal transduction pathway for platelet-derived growth factor-induced chemotaxis. J Biol Chem 271:29342–29346

Hill K, Welti S, Yu J, Murray JT, Yip SC, Condeelis JS, Segall JE, Backer JM (2000) Specific requirement for the p85-p110alpha phosphatidylinositol 3-kinase during epidermal growth factor-stimulated actin nucleation in breast cancer cells. J Biol Chem 275:3741–3744

Hirsch E, Katanaev VL, Garlanda C, Azzolino O, Pirola L, Silengo L, Sozzani S, Mantovani A, Altruda F, Wymann MP (2000) Central role for G protein-coupled phosphoinositide 3-kinase gamma in inflammation. Science 287:1049–1053

Hitchcock-Degregori SE (2006) Chemotaxis: cofilin in the driver's seat. Curr Biol 16:R1030–R1032

Hoeller O, Kay RR (2007) Chemotaxis in the absence of PIP3 gradients. Curr Biol 17:813–817

Hooshmand-Rad R, Claesson-Welsh L, Wennstrom S, Yokote K, Siegbahn A, Heldin CH (1997) Involvement of phosphatidylinositide 3'-kinase and Rac in platelet-derived growth factor-induced actin reorganization and chemotaxis. Exp Cell Res 234:434–441

Hou C, Kirchner T, Singer M, Matheis M, Argentieri D, Cavender D (2004) In vivo activity of a phospholipase C inhibitor 1-(6-((17beta-3-methoxyestra-1,3,5(10)-trien-17-yl)amino)hexyl)-1H-pyrrole -2,5-dione (U73122), in acute and chronic inflammatory reactions. J Pharmacol Exp Ther 309:697–704

Huang YE, Iijima M, Parent CA, Funamoto S, Firtel RA, Devreotes P (2003) Receptor-mediated regulation of PI3Ks confines PI(3,4,5)P3 to the leading edge of chemotaxing cells. Mol Biol Cell 14:1913–1922

Iglesias PA, Devreotes PN (2008) Navigating through models of chemotaxis. Curr Opin Cell Biol 20:35–40

Iijima M, Devreotes P (2002) Tumor suppressor PTEN mediates sensing of chemoattractant gradients. Cell 109:599–610

Inoue T, Meyer T (2008) Synthetic activation of endogenous PI3K and Rac identifies an AND-gate switch for cell polarization and migration. PLoS One 3:e3068

Insall R, Kuspa A, Lilly PJ, Shaulsky G, Levin LR, Loomis WF, Devreotes P (1994) CRAC, a cytosolic protein containing a pleckstrin homology domain, is required for receptor and G protein-mediated activation of adenylyl cyclase in Dictyostelium. J Cell Biol 126:1537–1545

Insall RH, Borleis J, Devreotes PN (1996) The aimless RasGEF is required for processing of chemotactic signals through G-protein-coupled receptors in Dictyostelium. Curr Biol 6:719–729

Izzard RA, Jackson SP, Smith GC (1999) Competitive and noncompetitive inhibition of the DNA-dependent protein kinase. Cancer Res 59:2581–2586

Jackson TR, Stephens LR, Hawkins PT (1992) Receptor specificity of growth factor-stimulated synthesis of 3-phosphorylated inositol lipids in Swiss 3T3 cells. J Biol Chem 267:16627–16636

Jacques TS, Relvas JB, Nishimura S, Pytela R, Edwards GM, Streuli CH, Ffrench-Constant C (1998) Neural precursor cell chain migration and division are regulated through different beta1 integrins. Development 125:3167–3177

Janetopoulos C, Ma L, Devreotes PN, Iglesias PA (2004) Chemoattractant-induced phosphatidylinositol 3,4,5-trisphosphate accumulation is spatially amplified and adapts, independent of the actin cytoskeleton. Proc Natl Acad Sci U S A 101:8951–8956

Jia S, Liu Z, Zhang S, Liu P, Zhang L, Lee SH, Zhang J, Signoretti S, Loda M, Roberts TM, Zhao JJ (2008) Essential roles of PI(3)K-p110beta in cell growth, metabolism and tumorigenesis. Nature 454:776–779

Johnson RL, Van Haastert PJ, Kimmel AR, Saxe CL 3rd, Jastorff B, Devreotes PN (1992) The cyclic nucleotide specificity of three cAMP receptors in Dictyostelium. J Biol Chem 267:4600–4607

Jou ST, Carpino N, Takahashi Y, Piekorz R, Chao JR, Wang D, Ihle JN (2002) Essential, nonredundant role for the phosphoinositide 3-kinase p110delta in signaling by the B-cell receptor complex. Mol Cell Biol 22:8580–8591

Kamimura Y, Xiong Y, Iglesias PA, Hoeller O, Bolourani P, Devreotes PN (2008) PIP3-independent activation of TorC2 and PKB at the cell's leading edge mediates chemotaxis. Curr Biol 18:1034–1043

Kaplan SS, Billiar T, Curran RD, Zdziarski UE, Simmons RL, Basford RE (1989) Inhibition of chemotaxis Ng-monomethyl-L-arginine: a role for cyclic GMP. Blood 74:1885–1887

Kavran JM, Klein DE, Lee A, Falasca M, Isakoff SJ, Skolnik EY, Lemmon MA (1998) Specificity and promiscuity in phosphoinositide binding by pleckstrin homology domains. J Biol Chem 273:30497–30508

Kay RR, Williams JG (1999) The Dictyostelium genome project: an invitation to species hopping. Trends Genet 15:294–297

Kim JY, Borleis JA, Devreotes PN (1998) Switching of chemoattractant receptors programs development and morphogenesis in Dictyostelium: receptor subtypes activate common responses at different agonist concentrations. Dev Biol 197:117–128

Kimmel AR, Parent CA (2003) The signal to move: D. discoideum go orienteering. Science 300:1525–1527

King JS, Insall RH (2009) Chemotaxis: finding the way forward with Dictyostelium. Trends Cell Biol 19:523–530

Klein PS, Sun TJ, Saxe CL 3rd, Kimmel AR, Johnson RL, Devreotes PN (1988) A chemoattractant receptor controls development in Dictyostelium discoideum. Science 241:1467–1472

Klinowska TC, Soriano JV, Edwards GM, Oliver JM, Valentijn AJ, Montesano R, Streuli CH (1999) Laminin and beta1 integrins are crucial for normal mammary gland development in the mouse. Dev Biol 215:13–32

Klippel A, Kavanaugh WM, Pot D, Williams LT (1997) A specific product of phosphatidylinositol 3-kinase directly activates the protein kinase Akt through its pleckstrin homology domain. Mol Cell Biol 17:338–344

Knapek K, Frondorf K, Post J, Short S, Cox D, Gomez-Cambronero J (2010) The molecular basis of phospholipase D2-induced chemotaxis: elucidation of differential pathways in macrophages and fibroblasts. Mol Cell Biol 30:4492–4506

Knight ZA, Gonzalez B, Feldman ME, Zunder ER, Goldenberg DD, Williams O, Loewith R, Stokoe D, Balla A, Toth B, Balla T, Weiss WA, Williams RL, Shokat KM (2006) A pharmacological map of the PI3-K family defines a role for p110alpha in insulin signaling. Cell 125:733–747

Kolsch V, Charest PG, Firtel RA (2008) The regulation of cell motility and chemotaxis by phospholipid signaling. J Cell Sci 121:551–559

Kortholt A, King JS, Keizer-Gunnink I, Harwood AJ, Van Haastert PJ (2007) Phospholipase C regulation of phosphatidylinositol 3,4,5-trisphosphate-mediated chemotaxis. Mol Biol Cell 18:4772–4779

Kriebel PW, Barr VA, Parent CA (2003) Adenylyl cyclase localization regulates streaming during chemotaxis. Cell 112:549–560

Kumagai A, Pupillo M, Gundersen R, Miake-Lye R, Devreotes PN, Firtel RA (1989) Regulation and function of G alpha protein subunits in Dictyostelium. Cell 57:265–275

Kundra V, Escobedo JA, Kazlauskas A, Kim HK, Rhee SG, Williams LT, Zetter BR (1994) Regulation of chemotaxis by the platelet-derived growth factor receptor-beta. Nature 367:474–476

Kunisaki Y, Nishikimi A, Tanaka Y, Takii R, Noda M, Inayoshi A, Watanabe K, Sanematsu F, Sasazuki T, Sasaki T, Fukui Y (2006) DOCK2 is a Rac activator that regulates motility and polarity during neutrophil chemotaxis. J Cell Biol 174:647–652

Kutscher B, Devreotes P, Iglesias PA (2004) Local excitation, global inhibition mechanism for gradient sensing: an interactive applet. Sci STKE 2004:pl3

Lauffenburger DA, Horwitz AF (1996) Cell migration: a physically integrated molecular process. Cell 84:359–369

Lauffenburger DL, Linderman JJ (1993) Receptors: models for binding, trafficking, and signaling. Oxford University Press, New York

Lee J, Ishihara A, Oxford G, Johnson B, Jacobson K (1999) Regulation of cell movement is mediated by stretch-activated calcium channels. Nature 400:382–386

Lee JS, Hmama Z, Mui A, Reiner NE (2004) Stable gene silencing in human monocytic cell lines using lentiviral-delivered small interference RNA. Silencing of the p110alpha isoform of phosphoinositide 3-kinase reveals differential regulation of adherence induced by 1alpha,25-dihydroxycholecalciferol and bacterial lipopolysaccharide. J Biol Chem 279:9379–9388

Leevers SJ, Vanhaesebroeck B, Waterfield MD (1999) Signalling through phosphoinositide 3-kinases: the lipids take centre stage. Curr Opin Cell Biol 11:219–225

Lehman N, Di Fulvio M, McCray N, Campos I, Tabatabaian F, Gomez-Cambronero J (2006) Phagocyte cell migration is mediated by phospholipases PLD1 and PLD2. Blood 108:3564–3572

Lemmon MA (2007) Pleckstrin homology (PH) domains and phosphoinositides. Biochem Soc Symp 81–93

Levchenko A, Iglesias PA (2002) Models of eukaryotic gradient sensing: application to chemotaxis of amoebae and neutrophils. Biophys J 82:50–63

Levine H, Kessler DA, Rappel WJ (2006) Directional sensing in eukaryotic chemotaxis: a balanced inactivation model. Proc Natl Acad Sci U S A 103:9761–9766

Li Z, Jiang H, Xie W, Zhang Z, Smrcka AV, Wu D (2000) Roles of PLC-beta2 and -beta3 and PI3Kgamma in chemoattractant-mediated signal transduction. Science 287:1046–1049

Li Z, Dong X, Wang Z, Liu W, Deng N, Ding Y, Tang L, Hla T, Zeng R, Li L, Wu D (2005) Regulation of PTEN by Rho small GTPases. Nat Cell Biol 7:399–404

Lim CJ, Spiegelman GB, Weeks G (2001) RasC is required for optimal activation of adenylyl cyclase and Akt/PKB during aggregation. EMBO J 20:4490–4499

Liu L, Puri KD, Penninger JM, Kubes P (2007) Leukocyte PI3Kgamma and PI3Kdelta have temporally distinct roles for leukocyte recruitment in vivo. Blood 110:1191–1198

Liu L, Das S, Losert W, Parent CA (2010) mTORC2 regulates neutrophil chemotaxis in a cAMP- and RhoA-dependent fashion. Dev Cell 19:845–857

Loew LM, Schaff JC (2001) The virtual cell: a software environment for computational cell biology. Trends Biotechnol 19:401–406

Loovers HM, Veenstra K, Snippe H, Pesesse X, Erneux C, Van Haastert PJ (2003) A diverse family of inositol 5-phosphatases playing a role in growth and development in Dictyostelium discoideum. J Biol Chem 278:5652–5658

Loovers HM, Postma M, Keizer-Gunnink I, Huang YE, Devreotes PN, Van Haastert PJ (2006) Distinct roles of PI(3,4,5)P3 during chemoattractant signaling in Dictyostelium: a quantitative in vivo analysis by inhibition of PI3-kinase. Mol Biol Cell 17:1503–1513

Ma L, Janetopoulos C, Yang L, Devreotes PN, Iglesias PA (2004) Two complementary, local excitation, global inhibition mechanisms acting in parallel can explain the chemoattractant-induced regulation of PI(3,4,5)P3 response in Dictyostelium cells. Biophys J 87:3764–3774

Maeda M, Firtel RA (1997) Activation of the mitogen-activated protein kinase ERK2 by the chemoattractant folic acid in Dictyostelium. J Biol Chem 272:23690–23695

Maffucci T, Cooke FT, Foster FM, Traer CJ, Fry MJ, Falasca M (2005) Class II phosphoinositide 3-kinase defines a novel signaling pathway in cell migration. J Cell Biol 169:789–799

Martin P (1997) Wound healing–aiming for perfect skin regeneration. Science 276:75–81

Matsui T, Amano M, Yamamoto T, Chihara K, Nakafuku M, Ito M, Nakano T, Okawa K, Iwamatsu A, Kaibuchi K (1996) Rho-associated kinase, a novel serine/threonine kinase, as a putative target for small GTP binding protein Rho. EMBO J 15:2208–2216

Meier-Schellersheim M, Xu X, Angermann B, Kunkel EJ, Jin T, Germain RN (2006) Key role of local regulation in chemosensing revealed by a new molecular interaction-based modeling method. PLoS Comput Biol 2:e82

Meili R, Ellsworth C, Lee S, Reddy TB, Ma H, Firtel RA (1999) Chemoattractant-mediated transient activation and membrane localization of Akt/PKB is required for efficient chemotaxis to cAMP in Dictyostelium. EMBO J 18:2092–2105

Meili R, Ellsworth C, Firtel RA (2000) A novel Akt/PKB-related kinase is essential for morphogenesis in Dictyostelium. Curr Biol 10:708–717

Meinhardt H (1999) Orientation of chemotactic cells and growth cones: models and mechanisms. J Cell Sci 112(Pt 17):2867–2874

Merlot S, Firtel RA (2003) Leading the way: directional sensing through phosphatidylinositol 3-kinase and other signaling pathways. J Cell Sci 116:3471–3478

Meyer T, Oancea E (2000) Studies of signal transduction events using chimeras to green fluorescent protein. Methods Enzymol 327:500–513

Miki H, Sasaki T, Takai Y, Takenawa T (1998) Induction of filopodium formation by a WASP-related actin-depolymerizing protein N-WASP. Nature 391:93–96

Mishra RS, Carnevale KA, Cathcart MK (2008) iPLA2beta: front and center in human monocyte chemotaxis to MCP-1. J Exp Med 205:347–359

Mogilner A, Wollman R, Marshall WF (2006) Quantitative modeling in cell biology: what is it good for? Dev Cell 11:279–287
Monine MI, Haugh JM (2008) Cell population-based model of dermal wound invasion with heterogeneous intracellular signaling properties. Cell Adh Migr 2:137–146
Mouneimne G, Soon L, Desmarais V, Sidani M, Song X, Yip SC, Ghosh M, Eddy R, Backer JM, Condeelis J (2004) Phospholipase C and cofilin are required for carcinoma cell directionality in response to EGF stimulation. J Cell Biol 166:697–708
Mouneimne G, Desmarais V, Sidani M, Scemes E, Wang W, Song X, Eddy R, Condeelis J (2006) Spatial and temporal control of cofilin activity is required for directional sensing during chemotaxis. Curr Biol 16:2193–2205
Munevar S, Wang YL, Dembo M (2001) Distinct roles of frontal and rear cell-substrate adhesions in fibroblast migration. Mol Biol Cell 12:3947–3954
Nakanishi S, Kakita S, Takahashi I, Kawahara K, Tsukuda E, Sano T, Yamada K, Yoshida M, Kase H, Matsuda Y et al (1992) Wortmannin, a microbial product inhibitor of myosin light chain kinase. J Biol Chem 267:2157–2163
Nakhaei-Nejad M, Hussain AM, Zhang QX, Murray AG (2007) Endothelial PI 3-kinase activity regulates lymphocyte diapedesis. Am J Physiol Heart Circ Physiol 293:H3608–H3616
Narang A, Subramanian KK, Lauffenburger DA (2001) A mathematical model for chemoattractant gradient sensing based on receptor-regulated membrane phospholipid signaling dynamics. Ann Biomed Eng 29:677–691
Narumiya S, Tanji M, Ishizaki T (2009) Rho signaling, ROCK and mDia1, in transformation, metastasis and invasion. Cancer Metastasis Rev 28:65–76
Nebl T, Fisher PR (1997) Intracellular Ca^{2+} signals in Dictyostelium chemotaxis are mediated exclusively by Ca^{2+} influx. J Cell Sci 110(Pt 22):2845–2853
Niggli V (2000) A membrane-permeant ester of phosphatidylinositol 3,4:5-trisphosphate (PIP(3)) is an activator of human neutrophil migration. FEBS Lett 473:217–221
Niggli V (2003) Signaling to migration in neutrophils: importance of localized pathways. Int J Biochem Cell Biol 35:1619–1638
Nishio M, Watanabe K, Sasaki J, Taya C, Takasuga S, Iizuka R, Balla T, Yamazaki M, Watanabe H, Itoh R, Kuroda S, Horie Y, Forster I, Mak TW, Yonekawa H, Penninger JM, Kanaho Y, Suzuki A, Sasaki T (2007) Control of cell polarity and motility by the PtdIns(3,4,5)P3 phosphatase SHIP1. Nat Cell Biol 9:36–44
Nombela-Arrieta C, Lacalle RA, Montoya MC, Kunisaki Y, Megias D, Marques M, Carrera AC, Manes S, Fukui Y, Martinez AC, Stein JV (2004) Differential requirements for DOCK2 and phosphoinositide-3-kinase gamma during T and B lymphocyte homing. Immunity 21:429–441
Okkenhaug K, Vanhaesebroeck B (2001) New responsibilities for the PI3K regulatory subunit p85 alpha. Sci STKE 2001:pe1
Okkenhaug K, Bilancio A, Farjot G, Priddle H, Sancho S, Peskett E, Pearce W, Meek SE, Salpekar A, Waterfield MD, Smith AJ, Vanhaesebroeck B (2002) Impaired B and T cell antigen receptor signaling in p110delta PI 3-kinase mutant mice. Science 297:1031–1034
Onsum M, Rao CV (2007) A mathematical model for neutrophil gradient sensing and polarization. PLoS Comput Biol 3:e36
Oude Weernink PA, Lopez De Jesus M, Schmidt M (2007) Phospholipase D signaling: orchestration by PIP2 and small GTPases. Naunyn Schmiedebergs Arch Pharmacol 374:399–411
Page DL, Anderson TJ (1987) Diagnostic histopathology of the breast. Chruchill Livingstone, New York
Papakonstanti EA, Ridley AJ, Vanhaesebroeck B (2007) The p110delta isoform of PI 3-kinase negatively controls RhoA and PTEN. EMBO J 26:3050–3061
Paradiso A, Cardone RA, Bellizzi A, Bagorda A, Guerra L, Tommasino M, Casavola V, Reshkin SJ (2004) The Na^+-H^+exchanger-1 induces cytoskeletal changes involving reciprocal RhoA and Rac1 signaling, resulting in motility and invasion in MDA-MB-435 cells. Breast Cancer Res 6:R616–R628
Parent CA, Devreotes PN (1996) Molecular genetics of signal transduction in Dictyostelium. Annu Rev Biochem 65:411–440

Parent CA, Devreotes PN (1999) A cell's sense of direction. Science 284:765–770
Parent CA, Blacklock BJ, Froehlich WM, Murphy DB, Devreotes PN (1998) G protein signaling events are activated at the leading edge of chemotactic cells. Cell 95:81–91
Park CS, Schneider IC, Haugh JM (2003) Kinetic analysis of platelet-derived growth factor receptor/phosphoinositide 3-kinase/Akt signaling in fibroblasts. J Biol Chem 278:37064–37072
Park KC, Rivero F, Meili R, Lee S, Apone F, Firtel RA (2004) Rac regulation of chemotaxis and morphogenesis in Dictyostelium. EMBO J 23:4177–4189
Park WS, Heo WD, Whalen JH, O'Rourke NA, Bryan HM, Meyer T, Teruel MN (2008) Comprehensive identification of PIP3-regulated PH domains from C. elegans to H. sapiens by model prediction and live imaging. Mol Cell 30:381–392
Patel H, Barber DL (2005) A developmentally regulated Na-H exchanger in Dictyostelium discoideum is necessary for cell polarity during chemotaxis. J Cell Biol 169:321–329
Patrucco E, Notte A, Barberis L, Selvetella G, Maffei A, Brancaccio M, Marengo S, Russo G, Azzolino O, Rybalkin SD, Silengo L, Altruda F, Wetzker R, Wymann MP, Lembo G, Hirsch E (2004) PI3Kgamma modulates the cardiac response to chronic pressure overload by distinct kinase-dependent and -independent effects. Cell 118:375–387
Paulus W, Baur I, Beutler AS, Reeves SA (1996) Diffuse brain invasion of glioma cells requires beta 1 integrins. Lab Invest 75:819–826
Pierce GF, Mustoe TA, Lingelbach J, Masakowski VR, Griffin GL, Senior RM, Deuel TF (1989) Platelet-derived growth factor and transforming growth factor-beta enhance tissue repair activities by unique mechanisms. J Cell Biol 109:429–440
Pierce GF, Mustoe TA, Altrock BW, Deuel TF, Thomason A (1991) Role of platelet-derived growth factor in wound healing. J Cell Biochem 45:319–326
Pitts WC, Rojas VA, Gaffey MJ, Rouse RV, Esteban J, Frierson HF, Kempson RL, Weiss LM (1991) Carcinomas with metaplasia and sarcomas of the breast. Am J Clin Pathol 95:623–632
Polette M, Gilles C, De Bentzmann S, Gruenert D, Tournier JM, Birembaut P (1998) Association of fibroblastoid features with the invasive phenotype in human bronchial cancer cell lines. Clin Exp Metastasis 16:105–112
Pollard TD, Borisy GG (2003) Cellular motility driven by assembly and disassembly of actin filaments. Cell 112:453–465
Pollitt AY, Insall RH (2009) WASP and SCAR/WAVE proteins: the drivers of actin assembly. J Cell Sci 122:2575–2578
Postma M, Van Haastert PJ (2001) A diffusion-translocation model for gradient sensing by chemotactic cells. Biophys J 81:1314–1323
Ptasznik A, Traynor-Kaplan A, Bokoch GM (1995) G protein-coupled chemoattractant receptors regulate Lyn tyrosine kinase. Shc adapter protein signaling complexes. J Biol Chem 270:19969–19973
Puri KD, Doggett TA, Douangpanya J, Hou Y, Tino WT, Wilson T, Graf T, Clayton E, Turner M, Hayflick JS, Diacovo TG (2004) Mechanisms and implications of phosphoinositide 3-kinase delta in promoting neutrophil trafficking into inflamed tissue. Blood 103:3448–3456
Rahdar M, Inoue T, Meyer T, Zhang J, Vazquez F, Devreotes PN (2009) A phosphorylation-dependent intramolecular interaction regulates the membrane association and activity of the tumor suppressor PTEN. Proc Natl Acad Sci U S A 106:480–485
Rameh LE, Cantley LC (1999) The role of phosphoinositide 3-kinase lipid products in cell function. J Biol Chem 274:8347–8350
Raper KB (1935) Dictyostelium discoideum, a new species of slime mold from decaying forest leaves. J Agric Res 50:135–147
Rappel WJ, Thomas PJ, Levine H, Loomis WF (2002) Establishing direction during chemotaxis in eukaryotic cells. Biophys J 83:1361–1367
Reif K, Okkenhaug K, Sasaki T, Penninger JM, Vanhaesebroeck B, Cyster JG (2004) Cutting edge: differential roles for phosphoinositide 3-kinases, p110gamma and p110delta, in lymphocyte chemotaxis and homing. J Immunol 173:2236–2240

Rickert P, Weiner OD, Wang F, Bourne HR, Servant G (2000) Leukocytes navigate by compass: roles of PI3Kgamma and its lipid products. Trends Cell Biol 10:466–473

Rintoul RC, Sethi T (2001) The role of extracellular matrix in small-cell lung cancer. Lancet Oncol 2:437–442

Rodriguez-Viciana P, Warne PH, Dhand R, Vanhaesebroeck B, Gout I, Fry MJ, Waterfield MD, Downward J (1994) Phosphatidylinositol-3-OH kinase as a direct target of Ras. Nature 370:527–532

Roelofs J, Meima M, Schaap P, Van Haastert PJ (2001) The Dictyostelium homologue of mammalian soluble adenylyl cyclase encodes a guanylyl cyclase. EMBO J 20:4341–4348

Sadhu C, Masinovsky B, Dick K, Sowell CG, Staunton DE (2003) Essential role of phosphoinositide 3-kinase delta in neutrophil directional movement. J Immunol 170:2647–2654

Sasaki T, Irie-Sasaki J, Jones RG, Oliveira-Dos-Santos AJ, Stanford WL, Bolon B, Wakeham A, Itie A, Bouchard D, Kozieradzki I, Joza N, Mak TW, Ohashi PS, Suzuki A, Penninger JM (2000) Function of PI3Kgamma in thymocyte development, T cell activation, and neutrophil migration. Science 287:1040–1046

Sasaki AT, Chun C, Takeda K, Firtel RA (2004) Localized Ras signaling at the leading edge regulates PI3K, cell polarity, and directional cell movement. J Cell Biol 167:505–518

Sasaki AT, Janetopoulos C, Lee S, Charest PG, Takeda K, Sundheimer LW, Meili R, Devreotes PN, Firtel RA (2007) G protein-independent Ras/PI3K/F-actin circuit regulates basic cell motility. J Cell Biol 178:185–191

Saudemont A, Garcon F, Yadi H, Roche-Molina M, Kim N, Segonds-Pichon A, Martin-Fontecha A, Okkenhaug K, Colucci F (2009) p110gamma and p110delta isoforms of phosphoinositide 3-kinase differentially regulate natural killer cell migration in health and disease. Proc Natl Acad Sci U S A 106:5795–5800

Schneider IC, Haugh JM (2004) Spatial analysis of 3' phosphoinositide signaling in living fibroblasts: II. Parameter estimates for individual cells from experiments. Biophys J 86:599–608

Schneider IC, Haugh JM (2005) Quantitative elucidation of a distinct spatial gradient-sensing mechanism in fibroblasts. J Cell Biol 171:883–892

Schneider IC, Haugh JM (2006) Mechanisms of gradient sensing and chemotaxis: conserved pathways, diverse regulation. Cell Cycle 5:1130–1134

Schneider IC, Parrish EM, Haugh JM (2005) Spatial analysis of 3' phosphoinositide signaling in living fibroblasts, III: influence of cell morphology and morphological Polarity. Biophys J 89:1420–1430

Seppa H, Grotendorst G, Seppa S, Schiffmann E, Martin GR (1982) Platelet-derived growth factor in chemotactic for fibroblasts. J Cell Biol 92:584–588

Servant G, Weiner OD, Herzmark P, Balla T, Sedat JW, Bourne HR (2000) Polarization of chemoattractant receptor signaling during neutrophil chemotaxis. Science 287:1037–1040

Sheetz MP, Felsenfeld DP, Galbraith CG (1998) Cell migration: regulation of force on extracellular-matrix-integrin complexes. Trends Cell Biol 8:51–54

Shibata K, Warbington ML, Gordon BJ, Kurihara H, Van Dyke TE (2001) Nitric oxide synthase activity in neutrophils from patients with localized aggressive periodontitis. J Periodontol 72:1052–1058

Shioi T, Kang PM, Douglas PS, Hampe J, Yballe CM, Lawitts J, Cantley LC, Izumo S (2000) The conserved phosphoinositide 3-kinase pathway determines heart size in mice. EMBO J 19:2537–2548

Simian M, Hirai Y, Navre M, Werb Z, Lochter A, Bissell MJ (2001) The interplay of matrix metalloproteinases, morphogens and growth factors is necessary for branching of mammary epithelial cells. Development 128:3117–3131

Singer AJ, Clark RA (1999) Cutaneous wound healing. N Engl J Med 341:738–746

Skupsky R, Losert W, Nossal RJ (2005) Distinguishing modes of eukaryotic gradient sensing. Biophys J 89:2806–2823

Smerling C, Tang K, Hofmann W, Danker K (2007) Role of the alpha(1) integrin cytoplasmic tail in the formation of focal complexes, actin organization, and in the control of cell migration. Exp Cell Res 313:3153–3165

Stauffer TP, Ahn S, Meyer T (1998) Receptor-induced transient reduction in plasma membrane PtdIns(4,5)P2 concentration monitored in living cells. Curr Biol 8:343–346

Stephanou A, Mylona E, Chaplain M, Tracqui P (2008) A computational model of cell migration coupling the growth of focal adhesions with oscillatory cell protrusions. J Theor Biol 253:701–716

Stephens L, Eguinoa A, Corey S, Jackson T, Hawkins PT (1993) Receptor stimulated accumulation of phosphatidylinositol (3,4,5)-trisphosphate by G-protein mediated pathways in human myeloid derived cells. EMBO J 12:2265–2273

Stephens LR, Eguinoa A, Erdjument-Bromage H, Lui M, Cooke F, Coadwell J, Smrcka AS, Thelen M, Cadwallader K, Tempst P, Hawkins PT (1997) The G beta gamma sensitivity of a PI3K is dependent upon a tightly associated adaptor, p101. Cell 89:105–114

Stephens L, Milne L, Hawkins P (2008) Moving towards a better understanding of chemotaxis. Curr Biol 18:R485–R494

Subramanian KK, Narang A (2004) A mechanistic model for eukaryotic gradient sensing: spontaneous and induced phosphoinositide polarization. J Theor Biol 231:49–67

Subramanian KK, Jia Y, Zhu D, Simms BT, Jo H, Hattori H, You J, Mizgerd JP, Luo HR (2007) Tumor suppressor PTEN is a physiologic suppressor of chemoattractant-mediated neutrophil functions. Blood 109:4028–4037

Suetsugu S, Yamazaki D, Kurisu S, Takenawa T (2003) Differential roles of WAVE1 and WAVE2 in dorsal and peripheral ruffle formation for fibroblast cell migration. Dev Cell 5:595–609

Suire S, Coadwell J, Ferguson GJ, Davidson K, Hawkins P, Stephens L (2005) p84, a new Gbetagamma-activated regulatory subunit of the type IB phosphoinositide 3-kinase p110gamma. Curr Biol 15:566–570

Suire S, Condliffe AM, Ferguson GJ, Ellson CD, Guillou H, Davidson K, Welch H, Coadwell J, Turner M, Chilvers ER, Hawkins PT, Stephens L (2006) Gbetagammas and the Ras binding domain of p110gamma are both important regulators of PI(3)Kgamma signalling in neutrophils. Nat Cell Biol 8:1303–1309

Sun TJ, Devreotes PN (1991) Gene targeting of the aggregation stage cAMP receptor cAR1 in Dictyostelium. Genes Dev 5:572–582

Swaney KF, Huang CH, Devreotes PN (2010) Eukaryotic chemotaxis: a network of signaling pathways controls motility, directional sensing, and polarity. Annu Rev Biophys 39:265–289

Symons M, Derry JM, Karlak B, Jiang S, Lemahieu V, McCormick F, Francke U, Abo A (1996) Wiskott-Aldrich syndrome protein, a novel effector for the GTPase CDC42Hs, is implicated in actin polymerization. Cell 84:723–734

Takahashi Y, Morales FC, Kreimann EL, Georgescu MM (2006) PTEN tumor suppressor associates with NHERF proteins to attenuate PDGF receptor signaling. EMBO J 25:910–920

Takeda K, Sasaki AT, Ha H, Seung HA, Firtel RA (2007) Role of phosphatidylinositol 3-kinases in chemotaxis in Dictyostelium. J Biol Chem 282:11874–11884

Takenawa T, Suetsugu S (2007) The WASP-WAVE protein network: connecting the membrane to the cytoskeleton. Nat Rev Mol Cell Biol 8:37–48

Teruel MN, Meyer T (2000) Translocation and reversible localization of signaling proteins: a dynamic future for signal transduction. Cell 103:181–184

Tester AM, Ruangpanit N, Anderson RL, Thompson EW (2000) MMP-9 secretion and MMP-2 activation distinguish invasive and metastatic sublines of a mouse mammary carcinoma system showing epithelial-mesenchymal transition traits. Clin Exp Metastasis 18:553–560

Tranquillo RT, Lauffenburger DA, Zigmond SH (1988) A stochastic model for leukocyte random motility and chemotaxis based on receptor binding fluctuations. J Cell Biol 106:303–309

Traynor D, Milne JL, Insall RH, Kay RR (2000) Ca(2+) signalling is not required for chemotaxis in Dictyostelium. EMBO J 19:4846–4854

Traynor-Kaplan AE, Thompson BL, Harris AL, Taylor P, Omann GM, Sklar LA (1989) Transient increase in phosphatidylinositol 3,4-bisphosphate and phosphatidylinositol trisphosphate during activation of human neutrophils. J Biol Chem 264:15668–15673

Trepat X, Wasserman MR, Angelini TE, Millet E, Weitz DA, Butler JP, Fredberg JJ (2009) Physical forces during collective cell migration. Nat Phys 5:426–430

Van Haastert PJ, Keizer-Gunnink I, Kortholt A (2007) Essential role of PI3-kinase and phospholipase A2 in Dictyostelium discoideum chemotaxis. J Cell Biol 177:809–816

Vandermoere F, El Yazidi-Belkoura I, Demont Y, Slomianny C, Antol J, Lemoine J, Hondermarck H (2007) Proteomics exploration reveals that actin is a signaling target of the kinase Akt. Mol Cell Proteomics 6:114–124

Vanhaesebroeck B, Welham MJ, Kotani K, Stein R, Warne PH, Zvelebil MJ, Higashi K, Volinia S, Downward J, Waterfield MD (1997) P110delta, a novel phosphoinositide 3-kinase in leukocytes. Proc Natl Acad Sci U S A 94:4330–4335

Vanhaesebroeck B, Leevers SJ, Ahmadi K, Timms J, Katso R, Driscoll PC, Woscholski R, Parker PJ, Waterfield MD (2001) Synthesis and function of 3-phosphorylated inositol lipids. Annu Rev Biochem 70:535–602

Vaughan RB, Trinkaus JP (1966) Movements of epithelial cell sheets in vitro. J Cell Sci 1:407–413

Vedham V, Phee H, Coggeshall KM (2005) Vav activation and function as a rac guanine nucleotide exchange factor in macrophage colony-stimulating factor-induced macrophage chemotaxis. Mol Cell Biol 25:4211–4220

Veltman DM, Van Haastert PJ (2008) The role of cGMP and the rear of the cell in Dictyostelium chemotaxis and cell streaming. J Cell Sci 121:120–127

Veltman DM, Roelofs J, Engel R, Visser AJ, Van Haastert PJ (2005) Activation of soluble guanylyl cyclase at the leading edge during Dictyostelium chemotaxis. Mol Biol Cell 16:976–983

Verschueren H, De Baetselier P, Bereiter-Hahn J (1991) Dynamic morphology of metastatic mouse T-lymphoma cells invading through monolayers of 10T1/2 cells. Cell Motil Cytoskeleton 20:203–214

Virbasius JV, Guilherme A, Czech MP (1996) Mouse p170 is a novel phosphatidylinositol 3-kinase containing a C2 domain. J Biol Chem 271:13304–13307

Vlahos CJ, Matter WF, Hui KY, Brown RF (1994) A specific inhibitor of phosphatidylinositol 3-kinase 2-(4-morpholinyl)-8-phenyl-4H-1-benzopyran-4-one (LY294002). J Biol Chem 269:5241–5248

Wang F, Herzmark P, Weiner OD, Srinivasan S, Servant G, Bourne HR (2002) Lipid products of PI(3)Ks maintain persistent cell polarity and directed motility in neutrophils. Nat Cell Biol 4:513–518

Wanikiat P, Woodward DF, Armstrong RA (1997) Investigation of the role of nitric oxide and cyclic GMP in both the activation and inhibition of human neutrophils. Br J Pharmacol 122:1135–1145

Wei C, Wang X, Chen M, Ouyang K, Song LS, Cheng H (2009) Calcium flickers steer cell migration. Nature 457:901–905

Weiger MC, Wang CC, Krajcovic M, Melvin AT, Rhoden JJ, Haugh JM (2009) Spontaneous phosphoinositide 3-kinase signaling dynamics drive spreading and random migration of fibroblasts. J Cell Sci 122:313–323

Weiger MC, Ahmed S, Welf ES, Haugh JM (2010) Directional persistence of cell migration coincides with stability of asymmetric intracellular signaling. Biophys J 98:67–75

Weiner OD, Neilsen PO, Prestwich GD, Kirschner MW, Cantley LC, Bourne HR (2002) A PtdInsP(3)- and Rho GTPase-mediated positive feedback loop regulates neutrophil polarity. Nat Cell Biol 4:509–513

Weingarten R, Bokoch GM (1990) GTP binding proteins and signal transduction in the human neutrophil. Immunol Lett 26:1–6

Welch HC, Coadwell WJ, Ellson CD, Ferguson GJ, Andrews SR, Erdjument-Bromage H, Tempst P, Hawkins PT, Stephens LR (2002) P-Rex1, a PtdIns(3,4,5)P3- and Gbetagamma-regulated guanine-nucleotide exchange factor for Rac. Cell 108:809–821

Welch HC, Condliffe AM, Milne LJ, Ferguson GJ, Hill K, Webb LM, Okkenhaug K, Coadwell WJ, Andrews SR, Thelen M, Jones GE, Hawkins PT, Stephens LR (2005) P-Rex1 regulates neutrophil function. Curr Biol 15:1867–1873

Welf ES, Haugh JM (2011) Signaling pathways that control cell migration: models and analysis. Wiley Interdiscip Rev Syst Biol Med 3:231–240

Wennstrom S, Hawkins P, Cooke F, Hara K, Yonezawa K, Kasuga M, Jackson T, Claesson-Welsh L, Stephens L (1994a) Activation of phosphoinositide 3-kinase is required for PDGF-stimulated membrane ruffling. Curr Biol 4:385–393

Wennstrom S, Siegbahn A, Yokote K, Arvidsson AK, Heldin CH, Mori S, Claesson-Welsh L (1994b) Membrane ruffling and chemotaxis transduced by the PDGF beta-receptor require the binding site for phosphatidylinositol 3' kinase. Oncogene 9:651–660

Wessels D, Lusche DF, Kuhl S, Heid P, Soll DR (2007) PTEN plays a role in the suppression of lateral pseudopod formation during Dictyostelium motility and chemotaxis. J Cell Sci 120:2517–2531

Wolf K, Mazo I, Leung H, Engelke K, Von Andrian UH, Deryugina EI, Strongin AY, Brocker EB, Friedl P (2003) Compensation mechanism in tumor cell migration: mesenchymal-amoeboid transition after blocking of pericellular proteolysis. J Cell Biol 160:267–277

Wood S Jr (1958) Pathogenesis of metastasis formation observed in vivo in the rabbit ear chamber. AMA Arch Pathol 66:550–568

Worth DC, Parsons M (2008) Adhesion dynamics: mechanisms and measurements. Int J Biochem Cell Biol 40:2397–2409

Wu YI, Frey D, Lungu OI, Jaehrig A, Schlichting I, Kuhlman B, Hahn KM (2009) A genetically encoded photoactivatable Rac controls the motility of living cells. Nature 461:104–108

Wyckoff JB, Wang Y, Lin EY, Li JF, Goswami S, Stanley ER, Segall JE, Pollard JW, Condeelis J (2007) Direct visualization of macrophage-assisted tumor cell intravasation in mammary tumors. Cancer Res 67:2649–2656

Wymann M, Arcaro A (1994) Platelet-derived growth factor-induced phosphatidylinositol 3-kinase activation mediates actin rearrangements in fibroblasts. Biochem J 298(Pt 3):517–520

Wymann MP, Bulgarelli-Leva G, Zvelebil MJ, Pirola L, Vanhaesebroeck B, Waterfield MD, Panayotou G (1996) Wortmannin inactivates phosphoinositide 3-kinase by covalent modification of Lys-802, a residue involved in the phosphate transfer reaction. Mol Cell Biol 16:1722–1733

Xu J, Wang F, Van Keymeulen A, Herzmark P, Straight A, Kelly K, Takuwa Y, Sugimoto N, Mitchison T, Bourne HR (2003) Divergent signals and cytoskeletal assemblies regulate self-organizing polarity in neutrophils. Cell 114:201–214

Xu X, Meier-Schellersheim M, Jiao X, Nelson LE, Jin T (2005) Quantitative imaging of single live cells reveals spatiotemporal dynamics of multistep signaling events of chemoattractant gradient sensing in Dictyostelium. Mol Biol Cell 16:676–688

Xu X, Meier-Schellersheim M, Yan J, Jin T (2007) Locally controlled inhibitory mechanisms are involved in eukaryotic GPCR-mediated chemosensing. J Cell Biol 178:141–153

Yamazaki D, Suetsugu S, Miki H, Kataoka Y, Nishikawa S, Fujiwara T, Yoshida N, Takenawa T (2003) WAVE2 is required for directed cell migration and cardiovascular development. Nature 424:452–456

Yang L, Iglesias PA (2009) Modeling spatial and temporal dynamics of chemotactic networks. In: Hereld D and Tian J (eds) Chemotaxis: methods and protocols. Springer Protocols

Yoo SK, Deng Q, Cavnar PJ, Wu YI, Hahn KM, Huttenlocher A (2010) Differential regulation of protrusion and polarity by PI3K during neutrophil motility in live zebrafish. Dev Cell 18:226–236

Yoshii S, Tanaka M, Otsuki Y, Wang DY, Guo RJ, Zhu Y, Takeda R, Hanai H, Kaneko E, Sugimura H (1999) alphaPIX nucleotide exchange factor is activated by interaction with phosphatidylinositol 3-kinase. Oncogene 18:5680–5690

Yumura S, Mori H, Fukui Y (1984) Localization of actin and myosin for the study of ameboid movement in Dictyostelium using improved immunofluorescence. J Cell Biol 99:894–899

Yumura S, Furuya K, Takeuchi I (1996) Intracellular free calcium responses during chemotaxis of Dictyostelium cells. J Cell Sci 109(Pt 11):2673–2678

Zhang P, Wang Y, Sesaki H, Iijima M (2010) Proteomic identification of phosphatidylinositol (3,4,5) triphosphate-binding proteins in Dictyostelium discoideum. Proc Natl Acad Sci U S A 107:11829–11834

Zhou K, Takegawa K, Emr SD, Firtel RA (1995) A phosphatidylinositol (PI) kinase gene family in Dictyostelium discoideum: biological roles of putative mammalian p110 and yeast Vps34p PI 3-kinase homologs during growth and development. Mol Cell Biol 15:5645–5656

Zouwail S, Pettitt TR, Dove SK, Chibalina MV, Powner DJ, Haynes L, Wakelam MJ, Insall RH (2005) Phospholipase D activity is essential for actin localization and actin-based motility in Dictyostelium. Biochem J 389:207–214

Chapter 8
Phosphoinositides in Golgi Complex Function

Giovanni D'Angelo, Mariella Vicinanza, Cathal Wilson and Maria Antonietta De Matteis

Abstract The Golgi complex is a ribbon-like organelle composed of stacks of flat cisternae interconnected by tubular junctions. It occupies a central position in the endomembrane system as proteins and lipids that are synthesized in the endoplasmic reticulum (ER) pass through the Golgi complex to undergo biosynthetic modification (mainly glycosylation) and to be sorted to their final destinations. In addition the Golgi complex possesses a number of activities, apparently not directly connected with its main role in trafficking and sorting, which have been recently reviewed in Wilson et al. 2011. In spite of the constant massive flux of material the Golgi complex maintains its identity and phosphoinositides (PIs), among other factors, play a central role in this process. The active metabolism of PIs at the Golgi is necessary for the proper functioning of the organelle both in terms of membrane trafficking/sorting and its manifold metabolic and signalling activities. Phosphatidylinositol 4-phosphate (PtdIns4*P*), in particular, is responsible for the recruitment of numerous cytosolic proteins that recognise and bind PtdIns4*P* via specific lipid-binding domains. In this chapter we will summarize the findings that have contributed to our current understanding of the role of PIs in the biology of the Golgi complex in terms of the regulation of PI metabolism and the functional roles and regulation of PtdIns4*P* effectors.

Keywords Golgi · Phosphatidylinositol 4-kinase · Phosphoinositide phosphatase · Arf1 · Trans Golgi network

M. A. De Matteis (✉) · G. D'Angelo · M. Vicinanza · C. Wilson
Telethon Institute of Genetics and Medicine, Via Pietro Castellino 111, 80131 Napoli, Italy
e-mail: dematteis@tigem.it

G. D'Angelo
e-mail: gdangelo@tigem.it

M. Vicinanza
e-mail: vicinanza@tigem.it

C. Wilson
e-mail: wilson@tigem.it

T. Balla et al. (eds.), *Phosphoinositides II: The Diverse Biological Functions*,
Subcellular Biochemistry 59, DOI 10.1007/978-94-007-3015-1_8,

8.1 Phosphoinositide Metabolism at the Golgi Complex

The Golgi complex hosts a number of enzymatic activities of the PI metabolic cycle (see Fig. 8.1). The net effect of these activities is that the predominant PI species on the Golgi membranes is PtdIns4*P*, with minor pools of PtdIns4,5P_2 and PtdIns3,4,5P_3 being also detectable (D'Angelo et al. 2008a; Watt et al. 2002; Low et al. 2010).

8.1.1 *PI Kinases*

PtdIns-4-kinases play crucial roles in Golgi function as their inhibition leads to pleiotropic defects in membrane trafficking and Golgi-localized sugar and lipid metabolism (D'Angelo et al. 2008a). In mammalian cells, all four PtdIns-4-kinases (PI4KIIα, PI4KIIβ, PI4KIIIα, and PI4KIIIβ) have been shown to localize and function, at least partially, at the Golgi complex (D'Angelo et al. 2008a), with PI4KIIα and PI4KIIIβ accounting for the bulk of Golgi PtdIns4*P* production (D'Angelo et al. 2008a).

PI4KIIIβ is a cytosolic protein with conserved multidomain organization between the yeast (Pik1p) and the mammalian (PI4KIIIβ) homologues. PI4KIIIβ consists of a N-terminal lipid kinase unique (LKU) domain (predicted to be a helical domain), a central Hom2 domain (identified as a Rab-binding site) and a C-terminal catalytic domain (de Graaf et al. 2004). The catalytic activity of PI4KIIIβ is sensitive to wortmannin (a well-established PtdIns 3-kinase inhibitor) due to a high degree of homology of the PI4KIIIβ catalytic domain with that of PtdIns 3-kinases. Adjacent to these domains, additional binding and phosphorylation sites were mapped in both Pik1p and PI4KIIIβ. These include the Frequenin-binding site at the C-terminus of the LKU domain, a Ser-rich domain containing phosphorylation sites for PKD at the N-terminus of the Hom2 domain and a proline-rich domain at the N-terminus of the protein (Hausser et al. 2005). Finally basic and Leu-rich stretches are proposed to contribute to the nucleo-cytoplasmic shuttling of PI4KIIIβ (Heilmeyer et al. 2003).

Recruitment of PI4KIIIβ to the Golgi is controlled by a direct interaction with the GTP-bound form of the small GTPase Arf1 (Godi et al. 1999). Efficient membrane association of PI4KIIIβ requires, in addition, the formation of a complex between PI4KIIIβ and Arf1 with NCS-1 (neuronal calcium sensor-1, an N-myristoylated Ca2+-binding protein (Haynes et al. 2005). The direct interaction between NCS-1 and Arf1 negatively controls PI4KIIIβ activation, thus providing a feedback control loop for the PI4KIIIβ enzymatic machinery (Haynes et al. 2005). A similar mechanism is also found in yeast where the Golgi recruitment of the yeast ortholog of PI4KIIIβ, Pik1p, requires the N-myristoylated cofactor Frq1p, which is highly homologous to NCS-1 (Hendricks et al. 1999).

In addition, protein kinase D (both PKD1 and PKD2) regulates the activity of PI4KIIIβ on the Golgi (Hausser et al. 2005). These kinases phosphorylate PI4KIIIβ on a specific motif that is then recognized by 14-3-3 proteins leading to a stabilization of the activated form of PI4KIIIβ (Hausser et al. 2006). Similarly, phosphorylation

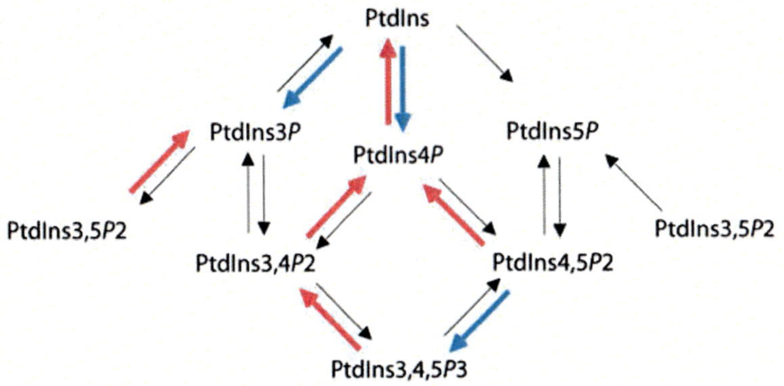

	enzyme	activity	substrate	yeast homologues	localization	regulators	gene location	disease
PI-kinases	PI4KIIα	4-kinase	PtdIns	Lsb6p	TGN/endosome	cholesterol	10q24.2	
	PI4KIIβ	4-kinase	PtdIns	Lsb6p	TGN/endosome		4p15.2	
	PI4KIIIα	4-kinase	PtdIns	Stt4p	ER/Golgi	Rab1b	22q11.21	
	PI4KIIIβ	4-kinase	PtdIns	Pik1p	Golgi/nucleus	ARF1, PKD, NCS-1, 14-3-3	1q21.1-q21.3	
	PI3Kδ	3-kinase	PtdIns4,5P2		Golgi	LPS	1p36.22	
	VPS34	3-kinase	PtdIns	VPS34p	Golgi/endosome		18q12.3	
	PI3K-C2α	3-kinase	PtdIns4,5P2		Golgi/endosome		11p15.1	
	PI3K-C2γ	3-kinase	PtdIns4,5P2		Golgi		12p12.3	
PI-phosphatases	OCRL1	5-ptase	PtdIns4,5P2		TGN/endosome	Rabs	Xq26.1	Lowe syndrome Dent2 disease
	INPP5B	5-ptase	PtdIns4,5P2		ERGIC	Rabs	1p34	
	INPP5E	5-ptase	PtdIns3,4,5P3 PtdIns3,5P2		Golgi/cilia		9q34.3	Joubert syndrome 1
	Sac1	4-ptase	PtdIns4P	Sac1p	ER/Golgi	growth factors, MAPKs	3p21.3	
	INPP4B	4-ptase	PtdIns3,4,5P3		Golgi/cytosol		4q31.21	
	TPIPα,β,γ	3-ptase	PtdIns3,4,5P3		ER/Golgi		13q12.11	
	PTEN1,2	3-ptase	PtdIns3,4,5P3 PtdIns3,4,P2	Tep1p	plasma membrane Golgi, nucleus		10q23.31	genetic cancer

Fig. 8.1 Schematic representation of the PI metabolic cycle (*upper panel*). *Blue* and *pink arrows* represent phosphorylation and dephosphorylation reactions, respectively, reported to happen on Golgi membranes. List of the PI metabolizing enzymes (*lower panel*) reported to localize (at least partially) to the Golgi complex

of the yeast Pik1p regulates its binding to 14-3-3 proteins. However, in this case 14-3-3 binding removes Pik1p from Golgi membranes and regulates its distribution between the nucleus and the Golgi in response to cell growth conditions (see below Demmel et al. 2008).

PI4KIIα is a tightly membrane-associated 4-kinase differing considerably from PI4KIIIβ, being smaller in size and simpler in structural organization. PI4KIIα contains an N-terminal Pro-rich domain and a kinase domain split into two stretches, with the central catalytic domain containing a conserved Cys-rich domain. This Cys-rich domain is palmitoylated and has been shown to be required for critical properties of the kinase, including catalytic activity, intracellular localization (mainly targeting to the trans Golgi network (TGN)) and membrane association (association with lipid rafts) (Barylko et al. 2009).

PI4KIIα activity may respond to cholesterol levels, function in sterol-dependent modulation of non-vesicular ceramide transport (Minogue et al. 2006; Banerji et al. 2010) and be required for cargo sorting via the AP-1/clathrin-dependent pathway (Wang et al. 2003). However, how PI4KIIα is recruited to specific Golgi regions has not been defined and it also associates with several other organelles, such as the plasma membrane, endosomes and the ER.

The other two mammalian PtdIns-4-kinases, PI4KIIIα and PI4KIIβ, are not primarily localized on Golgi membranes. PI4KIIβ, indeed, localizes to endosomal and perinuclear membranes under resting conditions (D'Angelo et al. 2008a). Recently a pool of PtdIns4*P* produced by PI4KIIIα has been reported to recruit the Arf1 guanine nucleotide exchange factor GBF1 to the Golgi complex (Dumaresq-Doiron et al. 2010). Indeed, Dumaresq-Doiron et al. proposed that Rab1b temporally and spatially regulates the GBF1 interaction with Golgi membranes activating the production of PtdIns4*P* by PI4KIIIα (Dumaresq-Doiron et al. 2010).

As mentioned above, there is evidence that a minor pool of PtdIns4,5P_2, which is mainly present at the plasma membrane, localizes to Golgi membranes (Watt et al. 2002). Indeed, Arf1-dependent recruitment of PtdIns-4-phosphate-5-kinase (PIP5K) activity to Golgi membranes in vitro was reported (Godi et al. 1999), yet the molecular identity of the PIP5K isoform that is responsible for generating Golgi PtdIns4,5P_2 has not been resolved.

In a recent report the type I PtdIns-3-kinase ∂ isoform has been shown to localize to Golgi membranes in LPS-activated macrophages where it is required for the selective trafficking of cytokines by participating in the fissioning of TGN-generated membrane carriers (Low et al. 2010). This suggests a role for a 3-phosphorylated PI (most probably PtdIns3,4,5P_3) in this process and, indeed, a PtdIns3,4,5P_3 pool has been visualized at the Golgi complex in macrophages following LPS stimulation (Low et al. 2010). However, the specific PtdIns3,4,5P_3 effectors at the Golgi and the role this species might play in non-specialized cells remain to be investigated.

8.1.2 PI Phosphatases

A number of PI phosphatases have been found to localize to the Golgi complex. These are OCRL1 and INPP5B, which are polyphosphoinositide-5-phosphatases with preference for PtdIns4,5P_2, and the 4-phosphatase Sac1.

The PtdIns4,5P_2-specific 5-phosphatase OCRL1 is a multidomain protein with a split N-terminal pleckstrin-homology (PH) domain, a central 5-phosphatase catalytic domain, an ASPM-SPD-2-Hydin (ASH) domain, a C-terminal inactive Rho GTPase activating protein (GAP) domain, and multiple clathrin-binding motifs and Rab-binding regions. OCRL1 has been found to associate with the Golgi, and in particular with the TGN, in different tissue culture cells (Vicinanza et al. 2008), but additional pools of OCRL1 have also been observed at endosomal membranes (Vicinanza et al. 2008). Recruitment of OCRL1 to Golgi and endosomal membranes appears to be mediated by small GTPases of the Rab family, with Rab1 and Rab6 controlling its association with the Golgi and Rab5 with the endosomes (Vicinanza et al. 2008). Mutations in OCRL1 cause Lowe syndrome, a rare genetic condition that leads to congenital cataracts, mental retardation and Fanconi syndrome of the proximal renal tubules, (thus also known as oculo-cerebro-renal (OCRL) syndrome). Determining the molecular mechanisms that underlie the physiological role of OCRL1 will be an important step for understanding the pathophysiology of Lowe Syndrome.

OCRL1 has a similar domain structure and shows 45% amino acid sequence identity to INPP5B, which also localizes to the Golgi (Williams et al. 2007). Similarly to OCRL1, INPP5B interacts with Rab5 and is targeted to early endosomes. Additionally, INPP5B has been found to interact with Rab proteins such as Rab1, Rab2, Rab6 and Rab9. While OCRL1 is mainly associated with the TGN, INPP5B also associates with earlier Golgi compartments, including the ERGIC (ER-Golgi Intermediate Compartment).

Another lipid-5-phosphatase (INPP5E) that prefers PtdIns3,4,5P_3 and PtdIns3,5P_2 as substrates co-localizes on Golgi membranes with the small GTPase Rab20 (Fukuda et al. 2008). Mutations of the INPP5E gene cause Joubert syndrome and destabilize pre-formed primary cilia in response to stimulation (Bielas et al. 2009). INPP5E has also been linked to the regulation of macrophage phagocytosis and insulin signaling (Ooms et al. 2009). Understanding the precise role of INPP5E on Golgi membranes remains to be addressed.

The Golgi complex also hosts phosphoinositide 4-phosphatase activities of which the best characterized is that dependent on the phosphatase Sac1. Sac1 is an integral transmembrane protein residing in the ER and the Golgi complex both in yeast and mammals depending on growth conditions (Blagoveshchenskaya and Mayinger 2009 and see below).

The phosphatase activity associated with the Sac phosphatase domain is mainly directed towards PtdIns4P, PtdIns3P and PtdIns3,5P_2 *in vitro*. However, yeast *Sac1* mutants have very high levels of PtdIns4P, thus indicating that *in vivo* Sac1p has a preference for PtdIns4P over the other two substrates. Yeast strains with mutations in the *Sac1* gene show an array of phenotypes, including inositol auxotrophy (Whitters et al. 1993), secretory defects in chitin deposition, disorganisation of the actin cytoskeleton, and impairment of ATP uptake and protein translocation to the ER (Mayinger et al. 1995). Moreover, mutations in *Sac1* can bypass the essential requirement for *Sec14* (responsible for the major yeast phosphatidylinositol/phosphatidylcholine transfer protein) in protein transport from the Golgi complex to the PM. The bypass of the sec14-1ts defect by sac1-22 has been shown to involve an eightfold increase in PtdIns4P rather than sphingolipid biosynthesis and increase in the diacylglycerol levels at the Golgi complex (Mayinger et al. 1995).

The human homologue of *Sac1* has been cloned and it behaves as the yeast isoform in terms of its substrate specificity and its localisation to the ER and Golgi complex (Rohde et al. 2003). Sac1-depletion in mammalian cells causes manifest disorganization of *cis*-, medial and *trans* Golgi compartments that however do not correlate with overt membrane trafficking defects (Liu et al. 2008).

8.2 PtdIns4*P* Effectors at the Golgi Complex

PtdIns4*P* has been shown to be essential for proper Golgi functioning and a growing number of PtdIns4*P* effectors are being found to associate to the Golgi complex (see Fig. 8.2). Three functionally distinct classes of Golgi PtdIns4*P* effectors can be distinguished: coat adaptors, lipid transfer proteins, and G-protein regulators.

The association of these effectors is mediated by protein modules which bind PtdIns4*P* usually with low affinity. These modules may be either PH-domains (such as in FAPPs, CERT, OSBP), ENTH/ANTH domains (such as in epsinR) or "simply" exposed patches or pockets containing basic residues that can interact with the phosphates of PtdIns4*P* and aromatic residues (usually a tryptophan) that can pack against the inositol ring (as in Vps74/Golph3, GGAs, Drs2p and others).

8.2.1 Coat Adaptors

The coat adaptors that have been shown to be under PtdIns4*P* control are the clathrin adaptor protein complex 1 (AP-1) (Wang et al. 2003), the Golgi-localised, gamma-ear containing, ADP ribosylation factor-binding proteins (GGAs) (Wang et al. 2007), EpsinR (Mills et al. 2003), and the retromer adaptor Vps74/GOLPH3 (Dippold et al. 2009; Wood et al. 2009).

The AP-1 complex and GGAs are clathrin-associated adaptor proteins that function in the trafficking of cargo from the TGN to endosomes and lysosomes. In both cases, recruitment to the Golgi requires detection of activated Arf1 and PtdIns4*P* (Wang et al. 2003, 2007). Purified AP-1 binds PtdIns4*P* and mutation of the gamma chain of AP-1 impairs PtdIns4*P* binding and recruitment to the Golgi (Heldwein et al. 2004). Decreasing Golgi PtdIns4*P* levels by RNAi interference of PI4KIIα blocks the recruitment of AP-1 to the Golgi and inhibits AP-1-associated functions (Wang et al. 2003).

Similarly, reducing PtdIns4*P* levels also affects the recruitment of GGAs to the TGN (Wang et al. 2007). Mutation of residues within the GAT domain of GGAs that are required for binding PtdIns4*P* reduce GGA recruitment to the Golgi. PtdIns4*P* also binds to a motif in the N-terminal domain of the yeast GGA2 protein (Demmel et al. 2008). In addition to participating in the membrane recruitment of GGAs, PtdIns4*P* promotes the recognition of ubiquitin-sorting signals by mammalian and yeast GGAs (Wang et al. 2007) and thus is involved in ubiquitin-dependent sorting at the TGN.

EpsinR is yet another adaptor protein for clathrin-coated vesicles at the TGN-endosome boundary. It binds to PtdIns4*P*, AP-1 adaptors and to clathrin (Mills et al. 2003; Hirst et al. 2003).

GOLPH3/Vps74 interacts with components of the retromer complex (e.g. VPS35) (Scott et al. 2009), a coat involved in retrograde endosome-to-Golgi transport, and with the cytosolic tails of Golgi glycosyltransferases. Of note, the GOLPH3/Vps74 gene is located in a chromosome region that is frequently amplified

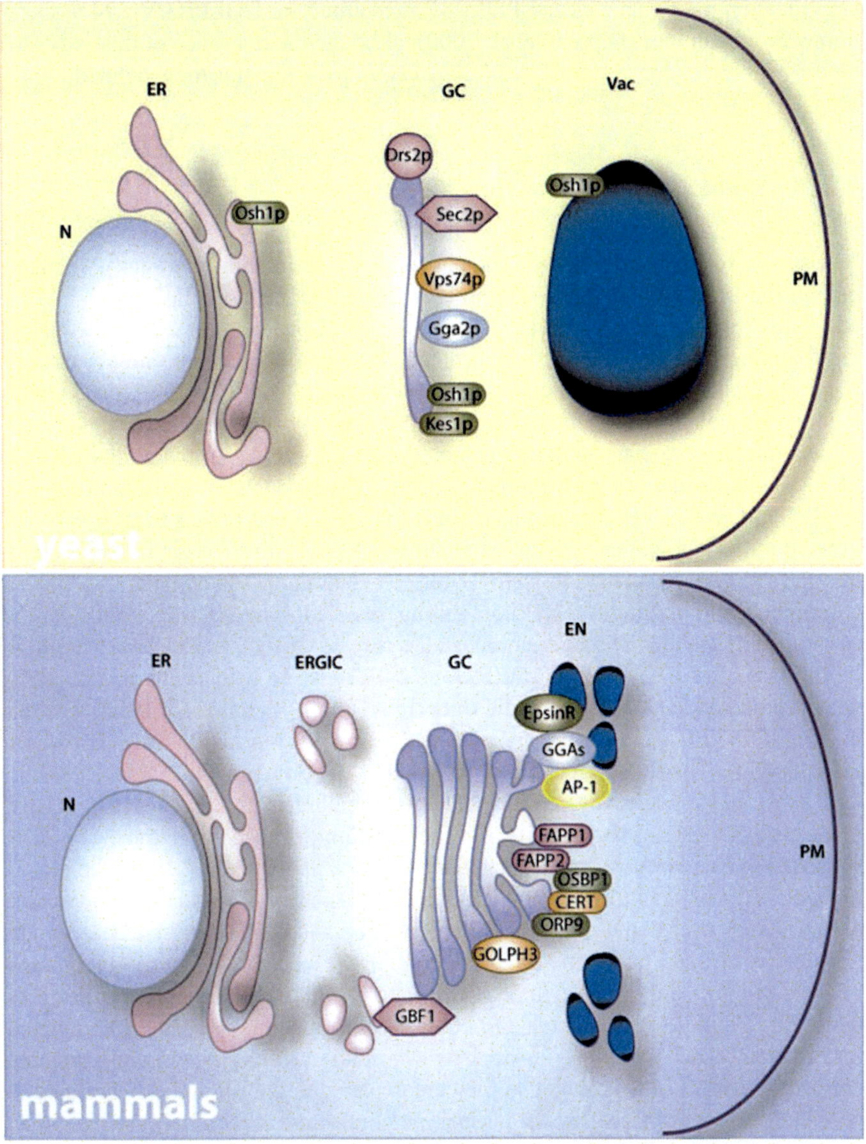

Fig. 8.2 Schematic representation of PtdIns4*P* effectors localised to Golgi complex in yeast (*upper panel*) and mammalian (*lower panel*) cells. *N* nucleus; *ER* endoplasmic reticulum; *ERGIC* ER-Golgi intermediate compartment; *GC* Golgi complex; *EN* endosomes; *VAC* vacuole; *PM* plasma membrane

in multiple solid tumours (Scott et al. 2009). Functional analysis of the genes belonging to this region has uncovered a direct role for GOLPH3/Vps74 in the positive control of cell proliferation via the Akt/mTOR pathway (Scott et al. 2009). This has thus defined GOLPH3/Vps74 as a new oncogene that is commonly targeted for amplification in human cancers. PtdIns4*P* is required for GOLPH3/Vps74 association with membranes (Dippold et al. 2009), although the relevance of this PtdIns4*P* binding for the oncogenic activity of GOLPH3/Vps74 remains to be defined.

8.2.2 Lipid Transfer Proteins-flippases

In a screen for the evaluation of the phosphinositide binding specificity of mammalian Plekstrin Homology (PH) domains, the PH domain of the protein FAPP1 was found to specifically interact with PtdIns4*P* (Dowler et al. 2000). FAPP1 is the founding member of a class of proteins sharing a similar PH domain in their N-terminal portion that also includes the proteins CERT, FAPP2, OSBP1, and ORP9 (D'Angelo et al. 2008b). All the members of this class of proteins have been shown to dynamically associate with Golgi membranes depending on the ability of their PH domains to interact with PtdIns4*P* (D'Angelo et al. 2008b). In addition, the OSBP1 and FAPP1 PH domains also bind Arf1, which contributes to their association with the Golgi complex (Godi et al. 2004). A distinctive feature of members of this class of proteins is that, in addition to their N-terminal PH domain, they have a C-terminal lipid binding/transfer domain that is able to interact with sterols (in the case of OSBP1 and ORP9), ceramide (in the case of CERT), and glucosylceramide (in the case of FAPP2). OSBP1, ORP9, and CERT also contain an amino acid motif with two phenylalanines in an acidic tract (FFAT motif) that is responsible for their binding to the ER resident transmembrane proteins VAP-A and VAP-B (D'Angelo et al. 2008b). Thanks to their FFAT motifs and PH domains OSBP1, ORP9, and CERT are able to cycle between ER and Golgi membranes in a process that, at least in the case of CERT, is regulated by phosphorylation/dephosphorylation cycles. The supposed molecular function associated with the members of this class of proteins is to regulate the intracellular distribution of the lipid species to which they bind thanks to their C-terminal domain. OSBP1/ORP9, CERT, and FAPP2, are, indeed, able to transfer sterols, ceramide, and glucosylceramide, respectively, between membrane bilayers *in vitro*, and thus they can be considered as intracellular lipid transfer proteins. The synthesis of ceramide and cholesterol in mammalian cells is completed on ER membranes while the steady state distribution of cholesterol and ceramide derivatives within cells is such that they are depleted in the ER and concentrated in post-Golgi compartments (i.e. the plasma membrane and endosomal system). Thus, a possible role for CERT and OSBP1/ORP9 is to foster the supply of these lipids to the Golgi membranes from where they can be delivered to their final subcellular destinations via conventional vesicular trafficking (Hanada et al. 2003; Ngo and Ridgway 2009). In the case of CERT, such a model is corroborated by the evidence that ER-produced ceramide is converted into sphingomyelin at the Golgi complex in a process that requires CERT as a cofactor for full efficiency (Hanada et al. 2003). A similar paradigm can be applied to FAPP2 and glucosylceramide as glucosylceramide is synthesized on the cytosolic leaflet of the cis-Golgi membranes and metabolized to complex glycosphingolipids in the late Golgi in a process that requires FAPP2 as a cofactor (D'Angelo et al. 2007). Nevertheless, in the case of FAPP2, the directionality of glucosylceramide transfer and the route followed by this lipid to its site of conversion remain to be fully understood (D'Angelo et al. 2007; Halter et al. 2007).

In addition to their lipid transfer activity, CERT, FAPP2, and OSBP1/ORP9 have been shown to be involved in membrane trafficking events at ER-Golgi and Golgi-plasma membrane interfaces (see below), thus suggesting a scenario where non-vesicular lipid transfer and membrane trafficking intersect to determine the intracellular membrane composition in a PtdIns4*P*-regulated manner.

A further PtdIns4*P* effector at the Golgi complex that transfers lipids is the yeast phospholipid translocase (flippase) Drs2p. Drs2p is required for efficient vesicle-mediated protein transport from the TGN to endosomes, possibly by producing membrane curvature at sites of concentrated Arf1-GTP, AP-1 and clathrin localization. A region of basic amino acids within Drs2p is required for direct binding to PtdIns4*P* and this binding is required for Drs2p activity. Interestingly, via the same domain, Drs2p interacts with the Arf-GEF Gea2p. Interestingly the Arf-GEF Gea2p and PtdIns4*P* synergistically stimulate the flippase activity of Drs2p (Natarajan et al. 2009).

8.2.3 G-protein Regulators

The interactions between PtdIns4*P* and small GTPases at the Golgi complex are manifold in kind and function. On the one hand, the ability of some protein domains to bind both PtdIns4*P* and small GTPases (see above the PH domain of FAPP1 and OSBP and see below the example of Sec2p) mediates the coincident detection of these two key components of Golgi membranes and represents a versatile targeting signal to sub-domains of this organelle. On the other hand, not only is the production of PtdIns4*P* at the Golgi complex under control of the small GTPase Arf1, (see above), but PtdIns4*P* can in turn control the activity of this and other small GTPases at the Golgi, either directly by binding G-protein regulatory proteins or indirectly through PtdIns4*P* effectors that, in turn, act on small GTPases.

At the TGN, PtdIns4*P* (in cooperation with the Rab protein, Ypt32p) can bind directly to the yeast guanine exchange factor (GEF) Sec2p (Mizuno-Yamasaki et al. 2010), which activates the Rab/Ypt protein Sec4p. PtdIns4*P*, Sec2p and Ypt32p can form a ternary complex where PtdIns4*P* inhibits binding of Sec2p to Sec15p, a member of the Exocyst complex and a Sec4p effector. A drop in the levels of PtdIns4*P* in secretory vesicles budded from the TGN would relieve this inhibition, allowing the interaction of Sec15p with Sec2p, and leading to a displacement of Ypt32p by Sec15p. At this point Sec2p can efficiently activate Sec4p that acts on its effector Sec15p. Therefore, the levels of PtdIns4*P* may regulate the relative affinity of Sec2p for different protein partners, thus providing a temporal-spatial control for the activation of Sec4p and ensuring the efficiency of Golgi-to-PM transport.

With respect to the indirect control of small GTPases through PtdIns4*P* effectors, two interesting examples are those of the above-mentioned Arf exchange factor

Gea2p that interacts with the Drs2p flippase that binds PtdIns4*P* and the property of the PH domains of FAPP1 and OSBP to interact with Arf and to compete with the activity of the Arf-GAP1 (Godi et al. 2004).

8.3 Regulation of the Activities of the Golgi Complex by PtdIns4*P*

Thanks to its ability to recruit/activate the above-described effectors, PtdIns4*P* regulates key steps in membrane trafficking, protein glycosylation and sphingolipid metabolism at the Golgi complex.

8.3.1 PtdIns4P and Membrane Trafficking

PtdIns4*P* has a well-documented role in the regulation of several membrane trafficking events at the Golgi complex. An acute reduction of PtdIns4*P* levels at the Golgi inhibited transport to the lysosome and trafficking of VSV-G from the Golgi to the PM (Szentpetery et al. 2010). PtdIns4*P*, indeed, regulates the trafficking between Golgi-endosomal/lysosomal compartments via the recruitment of the clathrin adaptors AP-1, GGAs, and EpsinR (see above). The exit from the Golgi of cargo proteins destined to different plasma membrane domains depends on the ability of PtdIns4*P* to recruit a number of effectors to the Golgi that are involved in different steps of post-Golgi carrier formation, which are membrane bending/tubulation, membrane fission and translocation of carriers via motors acting on cytoskeletal tracks (De Matteis and Luini 2008). As mentioned above, FAPP1 and FAPP2 localize to the TGN and interact with PtdIns4*P* and Arf1 (Godi et al. 2004). FAPP2 knockdown has been shown to inhibit protein transport to the plasma membrane (Godi et al. 2004) both in polarized and non-polarized cells (Vieira et al. 2005). The FAPP PH domain has been demonstrated to induce membrane tubulation both *in vitro* and in culture cells (Godi et al. 2004; Lenoir et al. 2010) suggesting a role for these proteins in carrier budding from the TGN (Godi et al. 2004). A role in the subsequent carrier formation step (i.e. linking tubular carriers to actomyosin) has been invoked for the PtdIns4*P* effector GOLPH3/Vps74, which is able to link Golgi membranes to the unconventional myosin MYO18A and thus to F-actin (Dippold et al. 2009). Moreover, both GOLPH3 and MYO18A are required for carrier formation at the Golgi complex (Dippold et al. 2009). On the other hand, the yeast homologue of GOLPH3, Vps74p, is required for the retention of glycosylating enzymes in the Golgi and thus for efficient protein glycosylation (see below, Wood et al. 2009). The final step in post-Golgi carrier formation is represented by the fission of the budded membranes from the donor compartment (De Matteis and Luini 2008). This process is dependent on the activity of protein kinases of the PKD family (Liljedahl et al. 2001). PKDs phosphorylate and activate PI4KIIIβ leading to increased PtdIns4*P* production (Hausser et al. 2005) while they also phosphorylate CERT, diminishing its ceramide transfer activity. CERT-mediated sphingomyelin synthesis is coupled to the production of diacylglycerol, which recruits and activates PKDs on Golgi membranes (Fugmann et al. 2007). Thus, the fission of post-Golgi plasma membrane-directed carriers relies on a tightly controlled machinery that contains PKDs, PtdIns4*P*-producing enzymes and PtdIns4*P* effectors such as CERT. CERT silencing, indeed, results in reduced protein transport to the plasma membrane (Fugmann et al. 2007).

8.3.2 PtdIns4P and Protein Glycosylation

The enzymes responsible for protein glycosylation at the Golgi complex are a group of type II integral membrane proteins with a C-terminal, lumenally oriented, enzymatic domain, a stem region, and a transmembrane domain followed by an N-terminal cytosolic domain (Wilson et al. 2011). The order of enzymatic reactions operated by Golgi glycosyltransferases is mirrored by the sub-Golgi distribution of the processing enzymes (Wilson et al. 2011) with early reactions being catalyzed by enzymes located in the proximal and distal aspects of the Golgi complex, respectively. The molecular determinants regulating the peculiar sub-Golgi distribution of the different enzymes are multiple. Nevertheless, a central role of PtdIns4*P* in regulating the efficiency and fidelity of protein glycosylation has been demonstrated both in yeast and in mammalian cells, mainly due to the control of the distribution/trafficking/retention of glycosyltransferases.

In a first report in yeast, Faulhammer et al. [2005] showed that the PtdIns4-phosphatase Sac1p localises at the ER in exponentially growing S. cerevisiae cells thanks to its interaction with the dolichol phosphate mannose (Dol-P-Man) synthase Dpm1p, which is a key enzyme for protein glycosylation reactions in the ER lumen. The deletion of the *Sac1* gene or its substitution with an enzymatically-dead mutant results in defects in glycosylation in yeast (Faulhammer et al. 2005). In addition, Sac1 downregulation in mammalian cells results in specific protein glycosylation defects possibly linked to the mislocalization of some Golgi glycosyltransferases following the altered balance of PtdIns4*P* levels in different Golgi sub-compartments (Cheong et al. 2010).

Interestingly, the yeast protein Vps74p (the yeast homolog of GOLPH3) has been demonstrated to recognize a pentameric cytoplasmic amino acid motif in the majority of yeast Golgi glycosyltranferases that maintains their steady-state localization in the Golgi (Schmitz et al. 2008; Tu et al. 2008). As seen above, Vps74p localizes to Golgi membranes thanks to its binding to PtdIns4*P* and this Vps74p-PtdIns4*P* interaction is needed to maintain glycosyltransferase localization to the Golgi complex (Tu et al. 2008).

8.3.3 PtdIns4P and Sphingolipid Metabolism

Sphingolipid synthesis starts with the production of ceramide on ER membranes. When transported to the Golgi complex, ceramide is the common precursor for sphingomyelin synthesis in the late Golgi or for glucosylceramide synthesis on the cytosolic leaflet of the *cis*-Golgi. Glucosylceramide is then a substrate for further glycosylation leading to the synthesis of different classes of complex glycosphingolipids after being relocated in the luminal side of late Golgi compartments (D'Angelo et al. 2008b). In this context, CERT has been shown to mediate ceramide non-vesicular transport from its site of production (the ER) to the late Golgi where it is consumed to specifically produce sphingomyelin (Hanada et al. 2003). As a consequence, sphingomyelin synthesis depends on the ability of CERT to cycle between the ER and Golgi membranes and thus on the interactions of CERT with Golgi-located PtdIns4*P* and the ER proteins VAP-A and B (see above). The pool of ceramide escaping CERT-mediated transport would, in turn, be inserted into membrane-bound carriers and transported to the *cis*-Golgi where it serves as a substrate for glucosylceramide synthesis. In a manner similar to ceramide, glucosylceramide is subjected to non-vesicular transport mediated by FAPP2 (see above), which fosters complex glycosphingolipid production (D'Angelo et al. 2007; Halter et al. 2007). A similar mechanism can also be envisaged for the two cholesterol transfer proteins OSBP1 and ORP9, which are thought to mediate cholesterol transport from the ER directly to the *trans*-Golgi complex, and, by doing so, to contribute to the enrichment of cholesterol in post-Golgi complex compartments. Interestingly, the fact that sphingomyelin and glycosphingolipid synthesis are, together with cholesterol distribution, under the common control of PtdIns4*P* suggests a certain degree of cross-talk among these different metabolic pathways (D'Angelo et al. 2008b). It has been demonstrated that sphingomyelin synthesis is stimulated by sterols in a process that depends on CERT and is regulated by OSBP1 (Perry and Ridgway 2006).

8.4 Integration of the Activities of the Golgi Complex with Cell Growth and Metabolism by PtdIns4*P*

While PIs in the Golgi regulate biosynthetic activity, the levels of the PIs are in turn regulated by the metabolic and growth conditions of the cell. As mentioned above, the yeast phosphoinositide 4-phosphatase Sac1p associates with the Golgi complex depending on the growth conditions: it translocates from the ER to the Golgi complex under cell-starvation conditions (Faulhammer et al. 2007). After translocation, Sac1p eliminates the Pik1p-generated pool of PtdIns4*P*. In addition, the Pik1p/Frq1p complex is released from the Golgi complex upon glucose depletion, suggesting the existence of a synergistic loop between the PtdIns-4-kinases and PtdIns-4-phosphatases that are responsible for the growth-dependent control of the Golgi PIs (Faulhammer et al. 2007). Re-addition of glucose to starved cells leads to the relocation of Sac1p to the ER and Pik1p to the Golgi, resulting in an increase in PtdIns4*P* levels at the Golgi complex and a resumption of transport. Along similar lines, it has been shown recently that mammalian Sac1 oligomerizes and translocates from the ER to the Golgi complex in quiescent cells, where it consumes the Golgi-complex-produced PtdIns4*P*, and by doing so, down-regulates anterograde trafficking. After stimulation by mitogens, Sac1 is relocated to the ER resulting in an increase in PtdIns4*P* levels at the Golgi complex and consequently in the promotion of constitutive secretion (Blagoveshchenskaya et al. 2008). The molecular mechanisms underlying the ER-Golgi shift of Sac1 have been clarified and involve the transition from an oligomeric to a monomeric state of Sac1 induced by mitogens. As a monomer Sac1 is retrieved to the ER via the coatomer protein-I (COPI)-mediated pathway. Indeed, hSac1 interacts with the COPI complex and mutation of a putative COPI-binding motif (KXKXX) abolishes this interaction and results in the accumulation of hSac1 in the Golgi complex (Rohde et al. 2003).

In conclusion, the above reports indicate that the cell can modulate the PI composition of the Golgi complex in response to growth conditions and this in turn can affect the biosynthetic activity of the cell.

References

Banerji S, Ngo M, Lane CF, Robinson CA, Minogue S, Ridgway ND (2010) Oxysterol binding protein-dependent activation of sphingomyelin synthesis in the golgi apparatus requires phosphatidylinositol 4-kinase IIalpha. Mol Biol Cell 21:4141–4150

Barylko B, Mao YS, Wlodarski P, Jung G, Binns DD, Sun HQ, Yin HL, Albanesi JP (2009) Palmitoylation controls the catalytic activity and subcellular distribution of phosphatidylinositol 4-kinase II{alpha}. J Biol Chem 284:9994–10003

Bielas SL, Silhavy JL, Brancati F, Kisseleva MV, Al-Gazali L, Sztriha L, Bayoumi RA, Zaki MS, Abdel-Aleem A, Rosti RO, Kayserili H, Swistun D, Scott LC, Bertini E, Boltshauser E, Fazzi E, Travaglini L, Field SJ, Gayral S, Jacoby M, Schurmans S, Dallapiccola B, Majerus PW, Valente EM, Gleeson JG (2009) Mutations in INPP5E, encoding inositol polyphosphate-5-phosphatase E, link phosphatidyl inositol signaling to the ciliopathies. Nat Genet 41:1032–1036

Blagoveshchenskaya A, Mayinger P (2009) SAC1 lipid phosphatase and growth control of the secretory pathway. Mol Biosyst 5:36–42

Blagoveshchenskaya A, Cheong FY, Rohde HM, Glover G, Knodler A, Nicolson T, Boehmelt G, Mayinger P (2008) Integration of Golgi trafficking and growth factor signaling by the lipid phosphatase SAC1. J Cell Biol 180:803–812

Cheong FY, Sharma V, Blagoveshchenskaya A, Oorschot VM, Brankatschk B, Klumperman J, Freeze HH, Mayinger P (2010) Spatial regulation of Golgi phosphatidylinositol-4-phosphate is required for enzyme localization and glycosylation fidelity. Traffic 11:1180–1190

D'Angelo G, Polishchuk E, Di Tullio G, Santoro M, Di Campli A, Godi A, West G, Bielawski J, Chuang CC, Spoel AC van der, Platt FM, Hannun YA, Polishchuk R, Mattjus P, De Matteis MA (2007) Glycosphingolipid synthesis requires FAPP2 transfer of glucosylceramide. Nature 449:62–67

D'Angelo G, Vicinanza M, Di Campli A, De Matteis MA (2008a) The multiple roles of PtdIns(4)P—not just the precursor of PtdIns(4,5)P2. J Cell Sci 121:1955–1963

D'Angelo G, Vicinanza M, De Matteis MA (2008b) Lipid-transfer proteins in biosynthetic pathways. Curr Opin Cell Biol 20:360–370

Demmel L, Gravert M, Ercan E, Habermann B, Muller-Reichert T, Kukhtina V, Haucke V, Baust T, Sohrmann M, Kalaidzidis Y, Klose C, Beck M, Peter M, Walch-Solimena C (2008) The clathrin adaptor Gga2p is a phosphatidylinositol 4-phosphate effector at the Golgi exit. Mol Biol Cell 19:1991–2002

Graaf P de, Zwart WT, Dijken RA van, Deneka M, Schulz TK, Geijsen N, Coffer PJ, Gadella BM, Verkleij AJ, Sluijs P van der, Bergen en Henegouwen PM van (2004) Phosphatidylinositol

4-kinasebeta is critical for functional association of rab11 with the Golgi complex. Mol Biol Cell 15:2038–2047

De Matteis MA, Luini A (2008) Exiting the Golgi complex. Nat Rev Mol Cell Biol 9:273–284

Dippold HC, Ng MM, Farber-Katz SE, Lee SK, Kerr ML, Peterman MC, Sim R, Wiharto PA, Galbraith KA, Madhavarapu S, Fuchs GJ, Meerloo T, Farquhar MG, Zhou H, Field SJ (2009) GOLPH3 bridges phosphatidylinositol-4-phosphate and actomyosin to stretch and shape the Golgi to promote budding. Cell 139:337–351

Dowler S, Currie RA, Campbell DG, Deak M, Kular G, Downes CP, Alessi DR (2000) Identification of pleckstrin-homology-domain-containing proteins with novel phosphoinositide-binding specificities. Biochem J 351:19–31

Dumaresq-Doiron K, Savard MF, Akam S, Costantino S, Lefrancois S (2010) The phosphatidylinositol 4-kinase PI4KIIIalpha is required for the recruitment of GBF1 to Golgi membranes. J Cell Sci 123:2273–2280

Faulhammer F, Konrad G, Brankatschk B, Tahirovic S, Knodler A, Mayinger P (2005) Cell growth-dependent coordination of lipid signaling and glycosylation is mediated by interactions between Sac1p and Dpm1p. J Cell Biol 168:185–191

Faulhammer F, Kanjilal-Kolar S, Knodler A, Lo J, Lee Y, Konrad G, Mayinger P (2007) Growth control of Golgi phosphoinositides by reciprocal localization of Sac1 lipid phosphatase and Pik1 4-Kinase. Traffic 8(11):1554–1567

Fugmann T, Hausser A, Schoffler P, Schmid S, Pfizenmaier K, Olayioye MA (2007) Regulation of secretory transport by protein kinase D-mediated phosphorylation of the ceramide transfer protein. J Cell Biol 178:15–22

Fukuda M, Kanno E, Ishibashi K, Itoh T (2008) Large scale screening for novel rab effectors reveals unexpected broad Rab binding specificity. Mol Cell Proteomics 7:1031–1042

Godi A, Pertile P, Meyers R, Marra P, Di Tullio G, Iurisci C, Luini A, Corda D, De Matteis MA (1999) ARF mediates recruitment of PtdIns-4-OH kinase-beta and stimulates synthesis of PtdIns(4,5)P2 on the Golgi complex. Nat Cell Biol 1:280–287

Godi A, Di Campli A, Konstantakopoulos A, Di Tullio G, Alessi DR, Kular GS, Daniele T, Marra P, Lucocq JM, De Matteis MA (2004) FAPPs control Golgi-to-cell-surface membrane traffic by binding to ARF and PtdIns(4)P. Nat Cell Biol 6:393–404

Halter D, Neumann S, Dijk SM van, Wolthoorn J, Maziere AM de, Vieira OV, Mattjus P, Klumperman J, Meer G van, Sprong H (2007) Pre- and post-Golgi translocation of glucosylceramide in glycosphingolipid synthesis. J Cell Biol 179:101–115

Hanada K, Kumagai K, Yasuda S, Miura Y, Kawano M, Fukasawa M, Nishijima M (2003) Molecular machinery for non-vesicular trafficking of ceramide. Nature 426:803–809

Hausser A, Storz P, Martens S, Link G, Toker A, Pfizenmaier K (2005) Protein kinase D regulates vesicular transport by phosphorylating and activating phosphatidylinositol-4 kinase IIIbeta at the Golgi complex. Nat Cell Biol 7:880–886

Hausser A, Link G, Hoene M, Russo C, Selchow O, Pfizenmaier K (2006) Phospho-specific binding of 14-3-3 proteins to phosphatidylinositol 4-kinase III beta protects from dephosphorylation and stabilizes lipid kinase activity. J Cell Sci 119:3613–3621

Haynes LP, Thomas GM, Burgoyne RD (2005) Interaction of neuronal calcium sensor-1 and ADP-ribosylation factor 1 allows bidirectional control of phosphatidylinositol 4-kinase beta and trans-Golgi network-plasma membrane traffic. J Biol Chem 280:6047–6054

Heilmeyer LM Jr, Vereb G Jr, Vereb G, Kakuk A, Szivak I (2003) Mammalian phosphatidylinositol 4-kinases. IUBMB Life 55:59–65

Heldwein EE, Macia E, Wang J, Yin HL, Kirchhausen T, Harrison SC (2004) Crystal structure of the clathrin adaptor protein 1 core. Proc Natl Acad Sci U S A 101:14108–14113

Hendricks KB, Wang BQ, Schnieders EA, Thorner J (1999) Yeast homologue of neuronal frequenin is a regulator of phosphatidylinositol-4-OH kinase. Nat Cell Biol 1:234–241

Hirst J, Motley A, Harasaki K, Peak Chew SY, Robinson MS (2003) EpsinR: an ENTH domain-containing protein that interacts with AP-1. Mol Biol Cell 14:625–641

Lenoir M, Coskun U, Grzybek M, Cao X, Buschhorn SB, James J, Simons K, Overduin M (2010) Structural basis of wedging the Golgi membrane by FAPP pleckstrin homology domains. EMBO Rep 11:279–284

Liljedahl M, Maeda Y, Colanzi A, Ayala I, Van Lint J, Malhotra V (2001) Protein kinase D regulates the fission of cell surface destined transport carriers from the trans-Golgi network. Cell 104:409–420

Liu Y, Boukhelifa M, Tribble E, Morin-Kensicki E, Uetrecht A, Bear JE, Bankaitis VA (2008) The Sac1 phosphoinositide phosphatase regulates Golgi membrane morphology and mitotic spindle organization in mammals. Mol Biol Cell 19:3080–3096

Low PC, Misaki R, Schroder K, Stanley AC, Sweet MJ, Teasdale RD, Vanhaesebroeck B, Meunier FA, Taguchi T, Stow JL (2010) Phosphoinositide 3-kinase delta regulates membrane fission of Golgi carriers for selective cytokine secretion. J Cell Biol 190:1053–1065

Mayinger P, Bankaitis VA, Meyer DI (1995) Sac1p mediates the adenosine triphosphate transport into yeast endoplasmic reticulum that is required for protein translocation. J Cell Biol 131:1377–1386

Mills IG, Praefcke GJ, Vallis Y, Peter BJ, Olesen LE, Gallop JL, Butler PJ, Evans PR, McMahon HT (2003) EpsinR: an AP1/clathrin interacting protein involved in vesicle trafficking. J Cell Biol 160:213–222

Minogue S, Chu KM, Westover EJ, Covey DF, Hsuan JJ, Waugh MG (2006) Relationship between phosphatidylinositol 4-phosphate synthesis, membrane organization, and lateral diffusion of PI4KIIalpha at the trans-Golgi network. J Lipid Res 51:2314–2324

Mizuno-Yamasaki E, Medkova M, Coleman J, Novick P (2010) Phosphatidylinositol 4-phosphate controls both membrane recruitment and a regulatory switch of the Rab GEF Sec2p. Dev Cell 18:828–840

Natarajan P, Liu K, Patil DV, Sciorra VA, Jackson CL, Graham TR (2009) Regulation of a Golgi flippase by phosphoinositides and an ArfGEF. Nat Cell Biol 11:1421–1426

Ngo M, Ridgway ND (2009) Oxysterol binding protein-related Protein 9 (ORP9) is a cholesterol transfer protein that regulates Golgi structure and function. Mol Biol Cell 20:1388–1399

Ooms LM, Horan KA, Rahman P, Seaton G, Gurung R, Kethesparan DS, Mitchell CA (2009) The role of the inositol polyphosphate 5-phosphatases in cellular function and human disease. Biochem J 419:29–49

Perry RJ, Ridgway ND (2006) Oxysterol-binding protein and vesicle-associated membrane protein-associated protein are required for sterol-dependent activation of the ceramide transport protein. Mol Biol Cell 17:2604–2616

Rohde HM, Cheong FY, Konrad G, Paiha K, Mayinger P, Boehmelt G (2003) The human phosphatidylinositol phosphatase SAC1 interacts with the coatomer I complex. J Biol Chem 278:52689–52699

Schmitz KR, Liu J, Li S, Setty TG, Wood CS, Burd CG, Ferguson KM (2008) Golgi localization of glycosyltransferases requires a Vps74p oligomer. Dev Cell 14:523–534

Scott KL, Kabbarah O, Liang MC, Ivanova E, Anagnostou V, Wu J, Dhakal S, Wu M, Chen S, Feinberg T, Huang J, Saci A, Widlund HR, Fisher DE, Xiao Y, Rimm DL, Protopopov A, Wong KK, Chin L (2009) GOLPH3 modulates mTOR signalling and rapamycin sensitivity in cancer. Nature 459:1085–1090

Szentpetery Z, Varnai P, Balla T (2010) Acute manipulation of Golgi phosphoinositides to assess their importance in cellular trafficking and signaling. Proc Natl Acad Sci U S A 107:8225–8230

Tu L, Tai WC, Chen L, Banfield DK (2008) Signal-mediated dynamic retention of glycosyltransferases in the Golgi. Science 321:404–407

Vicinanza M, D'Angelo G, Di Campli A, De Matteis MA (2008) Function and dysfunction of the PI system in membrane trafficking. EMBO J 27:2457–2470

Vieira OV, Verkade P, Manninen A, Simons K (2005) FAPP2 is involved in the transport of apical cargo in polarized MDCK cells. J Cell Biol 170:521–526

Wang YJ, Wang J, Sun HQ, Martinez M, Sun YX, Macia E, Kirchhausen T, Albanesi JP, Roth MG, Yin HL (2003) Phosphatidylinositol 4 phosphate regulates targeting of clathrin adaptor AP-1 complexes to the Golgi. Cell 114:299–310

Wang J, Sun HQ, Macia E, Kirchhausen T, Watson H, Bonifacino JS, Yin HL (2007) PI4P promotes the recruitment of the GGA adaptor proteins to the trans-Golgi network and regulates their recognition of the ubiquitin sorting signal. Mol Biol Cell 18:2646–2655

Watt SA, Kular G, Fleming IN, Downes CP, Lucocq JM (2002) Subcellular localization of phosphatidylinositol 4,5-bisphosphate using the pleckstrin homology domain of phospholipase C delta1. Biochem J 363:657–666

Whitters EA, Cleves AE, McGee TP, Skinner HB, Bankaitis VA (1993) SAC1p is an integral membrane protein that influences the cellular requirement for phospholipid transfer protein function and inositol in yeast. J Cell Biol 122:79–94

Wilson C, Venditti R, Rega LR, Colanzi A, D'Angelo G, De Matteis MA (2011) The Golgi apparatus: an organelle with multiple complex functions. Biochem J 433:1–9

Williams C, Choudhury R, McKenzie E, Lowe M (2007) Targeting of the type II inositol polyphosphate 5-phosphatase INPP5B to the early secretory pathway. J Cell Sci 120:3941–3951

Wood CS, Schmitz KR, Bessman NJ, Setty TG, Ferguson KM, Burd CG (2009) PtdIns4P recognition by Vps74/GOLPH3 links PtdIns 4-kinase signaling to retrograde Golgi trafficking. J Cell Biol 187:967–975

Chapter 9
Sec14 Like PITPs Couple Lipid Metabolism with Phosphoinositide Synthesis to Regulate Golgi Functionality

Carl J. Mousley, James M. Davison and Vytas A. Bankaitis

Abstract An interface coordinating lipid metabolism with proteins that regulate membrane trafficking is necessary to regulate Golgi morphology and dynamics. Such an interface facilitates the membrane deformations required for vesicularization, forms platforms for protein recruitment and assembly on appropriate sites on a membrane surface and provides lipid co-factors for optimal protein activity in the proper spatio-temporally regulated manner. Importantly, Sec14 and Sec14-like proteins are a unique superfamily of proteins that sense specific aspects of lipid metabolism, employing this information to potentiate phosphoinositide production. Therefore, Sec14 and Sec14 like proteins form central conduits to integrate multiple aspects of lipid metabolism with productive phosphoinositide signaling.

Keywords: Phosphatidylinositol · PI-transfer protein · Sec14 domain · Golgi · Phosphatidylinositol 4-kinase

9.1 Lipid Metabolism and the Golgi System

The functional integrity of the Golgi network requires cooperation between protein function and lipid metabolism. Lipid metabolism interfaces with the proteinaceous membrane trafficking machinery in three primary ways. First, it facilitates the membrane deformations that accompany vesicle budding, fusion and tubulation. Second, lipid metabolism creates platforms for protein recruitment to appropriate sites on a membrane surface. Third, it produces lipids which serve as co-factors for optimal protein activity in a spatio-temporally regulated manner. Thus, the interface of lipid

C. J. Mousley (✉) · J. M. Davison · V. A. Bankaitis
Department of Cell & Developmental Biology, Lineberger Comprehensive Cancer Center, University of North Carolina School of Medicine, Chapel Hill, NC 27599-7090, USA
e-mail: mousley@email.unc.edu

V. A. Bankaitis
e-mail: vytas@med.unc.edu

T. Balla et al. (eds.), *Phosphoinositides II: The Diverse Biological Functions*,
Subcellular Biochemistry 59, DOI 10.1007/978-94-007-3015-1_9,

metabolism with mechanisms of membrane trafficking is complex, and this interface is a major factor in controlling Golgi morphology and dynamics.

9.2 PtdIns-4-Phosphate and Golgi Function

Phosphoinositides (PIPs) are phosphorylated in all combinations on either the D-3, D-4 or D-5 hydroxyl moieties of the inositol headgroup of phosphatidylinositol (PtdIns). The observation that PtdIns and phosphoinositides regulate membrane trafficking describes the first established case for lipids playing an active role in regulating membrane trafficking reactions. Herein, we focus on the functions of PtdIns-4-phosphate within the Golgi system.

Two distinct classes of PtdIns 4-OH kinases, Pik1 and Lsb6 in yeast or PI4KIIIβ and PI4KIIα in mammals respectively, associate with the Golgi network (Balla and Balla 2006; Strahl and Thorner 2007). The PI4KIIIβ enzymes have been more extensively studied and provide the focus herein. PI4KIIIβ activity is stimulated by heterodimerization with NCS1 (frequenin in yeast), a small myristoylated Ca^{2+}-binding protein (Hendricks et al. 1999; Zhao et al. 2001). Upon association with the Golgi membrane PI4KIIIβ enzymes have been found to directly interact with components of the vesicular trafficking machinery, PI4KIIIβ interacts with the GTP-bound form of Arf1 on the Golgi in mammalian cells (Godi et al. 1999; Haynes et al. 2005). In contrast to mammalian PI4KIIIβ, yeast Pik1 does not directly interact with Arf1, instead it interacts with the ARF1 specific GTP exchange factor, Sec7 (Gloor et al. 2010).

Membrane associated PtdIns 4-OH kinases employs dual activities to regulate Golgi secretory function. Firstly, PI4KIIIβ directly binds to and regulates the activity of the small Rab GTPase, Rab11 (the mammalian orthologue of yeast Ypt31). Therefore, one critical activity of PI4KIIIβ kinases is to function as a scaffold for the recruitment of key components of the membrane trafficking machinery (Polevoy et al. 2009). Secondly, the catalytic activity of PtdIns 4-OH kinase is necessary for Golgi function. Acute inactivation of yeast Pik1 kinase activity (Hama et al. 1999; Walch-Solimena and Novick 1999), or evoked recruitment of the Sac1 phosphoinositide phosphatase catalytic domain which degrades PtdIns-4-phosphate to PtdIns (Szentpetery et al. 2010) to the Golgi induces membrane trafficking defects through this organelle. In addition, inactivation of PtdIns-binding proteins (e.g. Sec14; see below), which potentiate the PtdIns 4-OH kinase of these enzymes, also compromises protein trafficking through the Golgi complex (Bankaitis et al. 1990; Schaaf et al. 2008).

PtdIns-4-phosphate regulates Golgi secretory functions by several mechanisms. PtdIns-4-phosphate serves as an adaptor required to recruit peripheral membrane proteins required for vesicle biogenesis. These include adaptor proteins for clathrin binding (i.e. AP-1; Carlton and Cullen 2005; Wang et al. 2003), GGA proteins which potentiate Arf1-GTP activity (Wang et al. 2007; Demmel et al. 2008), Rab GTPases and their concomitant guanidine nucleotide exchange factors (de Graaf et al. 2004; Sciorra et al. 2005; Mizuno-Yamasaki et al. 2010a), oxysterol binding proteins which interface with phosphoinositide-4-phosphate signaling (Li et al. 2002a; Litvak et al.

2005; Stefan et al. 2011), and other lipid binding/transfer proteins which further remodel Golgi membrane lipid composition (see below). Second, PtdIns-4-phosphate directly regulates the activities of other Golgi resident proteins. One example is the amino-phospholipid flippase Drs2, a Type-IV integral membrane ATPase (Natarajan et al. 2004; Muthusamy et al. 2009), which transfers phosphatidylserine (PtdSer) and phosphatidylethanolamine (PtdEtn) from the luminal leaflet to the cytosolic face of Golgi/endosomal membranes. Drs2 executes significant pro-secretory functions in these compartments, and its flippase activity is stimulated by binding to both PtdIns-4-phosphate and to a GTPase exchange factor for Arf1 (Chantalat et al. 2004). How Drs2 dependent aminophospholipid flippase activity regulates membrane trafficking pathways remains unknown.

Another requirement for PtdIns-4-phosphate is to maintain the integrity of the Golgi apparatus. Sec2, the major Sec4 Rab GTPase exchange factor, binds to PtdIns-4-phosphate, which inhibits Sec2 binding to the Sec15 component of the exocyst complex. Therefore, PtdIns-4-phosphate inhibits the inappropriate assembly of the exocyst complex on Golgi membranes (Mizuno-Yamasaki et al. 2010b). The exocyst marks secretory vesicles for fusion to the plasma membrane, therefore PtdIns-4-phosphate inhibits the premature recruitment of the exocyst to Golgi membranes (i.e. membranes from which secretory vesicles form)—an event which could potentially confuse the distinction between secretory vesicles and the Golgi system.

In addition to PtdIns-4-phosphate promoting anterograde membrane trafficking pathways from the Golgi, recent studies show that PtdIns-4-phosphate also regulates retrograde membrane trafficking from endosomes back to the Golgi (Mousley et al. 2008; Wood et al. 2009). Specifically, the yeast Vps74 protein, which facilitates the sorting of escaped Golgi resident glycosyltransferases to retrograde carriers for retrieval back to the Golgi system, is a PtdIns-4-phosphate binding protein. Vps74 binds directly to the cytosolic tails of the cargo glycosyltransferases as well as to the COP1 subunit of the coatomer complex whose assembly is regulated by Arf1-GTP (Wood et al. 2009). Vps74 is the yeast functional ortholog of the mammalian GOLPH3 protein that has been implicated in MYO18A-dependent control of Golgi morphology (Wood et al. 2009). GOLPH3 is reported to interact with at least one subunit of the retromer complex which functions in retrograde trafficking in the endosomal pathway (Vergés et al. 2006) therefore GOLPH3 may execute Vps74-like functions in cargo sorting and retrieval in mammals.

9.3 Diacylglycerol and Golgi Function

Diacylglycerol (DAG) is a neutral lipid which regulates vesicle budding at multiple steps in the exocytic pathway including transport from the TGN (Litvak et al. 2005; Antonny et al. 1997; Kearns et al. 1997; Yanagisawa et al. 2002; Baron and Malhotra 2002) and the formation of COP1-vesicles for retrograde trafficking from early Golgi cisternae back to the ER. It regulates vesicular budding in two general ways. First, DAG is known to directly regulate protein components of the membrane trafficking machinery. It stimulates the activities of Arf-GTPase activating proteins in both yeast and mammalian systems (Antonny et al. 1997; Yanagisawa et al. 2002). Also, in mammalian cells, DAG is necessary and sufficient to recruit protein kinase D (PKD) isoforms to Golgi membranes. Upon activation, PKD coordinates DAG metabolism and signaling to downstream lipid metabolic events that optimize membrane trafficking from the TGN (Liljedahl et al. 2001; Bard and Malhotra 2006; Bossard et al. 2007). A prominent feature of this circuit is that PI4KIIIβ is recruited to the Golgi, the consequence of which is described above. DAG recruits a number of other signaling proteins to Golgi membranes as well including various protein kinases C isoforms and Ras guanosine nucleotide release proteins (Lehel et al. 1995; Maissel et al. 2006; Wang et al. 1999; Caloca et al. 2003). Included in this cohort is PKC$\eta\prime$, which phosphorylates and activates PKD (Díaz Ael and Malhotra 2005).

The second regulatory mechanism utilizes the unique topological properties of DAG to orchestrate the nucleation and/or propagation of the membrane deformations necessary for vesicle budding and scission (Chernomordik et al. 1995; Burger 2000). The extreme inverted-cone shapes assumed by DAG species (due to their small headgroup to acyl chain axial area ratios) are compatible with non-bilayer lipid arrangements. The biophysical properties and signaling capacity of DAG is likely coupled. DAG regulates at least two steps in COP1-dependent vesicle biogenesis. At an early point in vesicle formation, it enables membrane deformation required to generate buds/tubules. Later, DAG regulates the scission event required to release the newly formed vesicle from its donor membrane (Asp et al. 2009). Budding/tubulation does not require the involvement of the DAG-activated ArfGAP1 and is posited to act primarily as a topological regulator of membrane curvature. The DAG involvement at the scission step does require ArfGAP1 activity—suggesting that DAG potentiates scission both by activating ArfGAP1 and by promoting formation of the non-bilayer membrane structures which characterize terminal fission intermediates (Asp et al. 2009). The DAG-activated PKD involvements in membrane trafficking from the TGN also display scission-step execution points (Liljedahl et al. 2001; Bard and Malhotra 2006).

9.4 Lipid Transfer Proteins

The extensive involvement of lipids in regulating the functional integrity of the Golgi complex requires coordination between both lipid metabolism and lipid signaling. Lipid transfer proteins have been found to perform such a task and PtdIns/PtdCho transfer proteins (PITPs) represents a suitable example of this coordination. Sec14, the major yeast PITP, regulates an essential interface between lipid metabolism and membrane trafficking from the Golgi network. Historically, it has been proposed that Sec14 utilizes its intrinsic ability to transfer PtdIns or PtdCho monomers in-between bilayers to deliver lipids from the endoplasmic reticulum to the Golgi. Genetic studies in yeast have proven invaluable in determining what biological activities are regulated by Sec14p and other Sec14-like PITPs; the findings of which are not consistent with classical transfer mechanisms for Sec14p function. Intracellular Sec14p levels are approximately 100-fold in excess of the levels needed for cell viability (Salama et al. 1990; Cleves et al. 1991), thus, for a transfer model to fit biologically, cells must demonstrate an extremely low threshold for PtdIns transfer. Second, disruption of the CDP-choline pathway for PtdCho biosynthesis or Sac1 dependent PtdIns-4-phosphate catabolism bypasses the essential Sec14p requirement for yeast viability (Rivas et al. 1999; Xie et al. 1998; Phillips et al. 1999). Thus, Sec14p regulates lipid metabolism rather than PtdIns supply/transport. Third, vectoral phospholipid transfer models predict that increased affinity of a PITP for PtdIns versus PtdCho is an important functional property; yet, Sec14p activity is surprisingly insensitive to specific reductions in PtdIns binding affinity (Salama et al. 1990; Sha et al. 1998). Finally, transfer models predict that alternative supply of PtdIns to membranes bypass essential Sec14p requirement. Contrary to this observation, manipulating the yeast lipidome by increasing PtdIns to 40 mol% of total glycerophospholipid mass (which should solve all PtdIns supply demands) fails to relieve cells of the essential Sec14p requirement (Rivas et al. 1999).

A description for Sec14 activity is better served where it functions primarily as a lipid sensor that instructs specific enzymes when and where to execute biochemical reactions. Such an activity provides a mechanism of coincidence detection that integrates multiple aspects of lipid metabolism with PIP signaling. Ultimately, this provides additional layers by which the membrane trafficking machinery is regulated. These alternative possibilities suggest concepts that might translate into functional mechanisms for other LTPs. Utilization of this class of proteins for linking specific channels of metabolic information with the action of interfacial lipid-modifying enzymes represents a novel theme in cell signaling.

9.5 The Sec14 Superfamily of PITPs

The Sec14 domain (SMART entry: smart00516), for which the yeast Sec14p is the prototype (Sha et al. 1998), represents an ancient and versatile structural unit restricted to eukaryotes. To date 1551 Sec14 domains, representing 1550 proteins, are annotated in the NCBI database (http://www.ncbi.nlm.nih.gov). Even simple eukaryotes express multiple Sec14 family members. Saccharomyces cerevisiae expresses five Sec14-like proteins in addition to Sec14p (Li et al. 2002b), and Homo sapiens, Mus musculus, Drosophila melanogaster, Caenorhabditis elegans and Aradidopsis thaliana each possess >20 individual genes that encode distinct Sec14 superfamily members.

As expected from the diversity of the superfamily, proteins containing a Sec14 domain interface with a multitude of cellular activities. Studies in yeast and plants demonstrate multiple roles for Sec14-like PITPs in regulating: housekeeping membrane-trafficking pathways (Phillips et al. 1999; Wu et al. 2000; Carmen-Lopez et al. 1994), developmental membrane-trafficking circuits for dimorphic growth and sporulation in yeast (Carmen-Lopez et al. 1994; Nakase et al. 2001; Rudge et al.

2004), and root hair biogenesis in plants (Vincent et al. 2005). In addition, Sec14 domains are frequently observed in proteins that regulate activities of small GTPases of the Ras, Rho and Rac families. Examples include the Ras-GAP neurofibromins NF1 and NF2 (Aravind et al. 1999), Rho-GAPs and Cdc42-GAPs of the BCH and BNIP families (Shang et al. 2003; Tcherkezian and Lamarche-Vane 2007; Sirokmany et al. 2005), and the Rho-GEFs Trio, Dbl and Duo (Ueda et al. 2004; Debant et al. 1996). The PTP-MEG2 protein tyrosine phosphatase also harbors a Sec14 domain (Gu et al. 1992).

The Sec14 domain is also associated with uncharacterized modules that include nodulin domains (in higher plants) (Vincent et al. 2005; Kapranov et al. 2001), Golgi dynamics (GOLD) domains in metazoans and higher plants (Anantharaman and Aravind 2002), metazoan-specific PRELI domains (thought to function as mitochondrial targeting motifs) (Dee and Moffat 2005), and a GTPase motif of uncertain function (Habermehl et al. 2004). Stand-alone Sec14 domain proteins are more prevalent in simpler eukaryotes. Higher eukaryotes often couple Sec14 domains with more complex arrangements in addition to expressing stand-alone Sec14 domains. Examples of mammalian stand-alone Sec14 domain proteins include cellular retinaldehyde binding protein (CRALBP) (Liu et al. 2005), caytaxin (Bomar et al. 2003), and α-tocopherol and retinaldehyde transfer proteins (Gotoda et al. 1995; Hentati et al. 1996; Golovleva et al. 2003). The biological importance of Sec14 superfamily members in mammals is demonstrated by the linkage of human diseases to dysfunction of Sec14-like proteins and of Sec14 domains in multidomain proteins. Such diseases include autosomal-dominant cancers attributed to neurofibromin insufficiencies (Cichowski and Jacks 2001), ataxia with vitamin E deficiency arising from diminished α-TTP function (Ouachi et al. 1995), ataxia caused by loss of caytaxin function (Hentati et al. 1996), and retinal degeneration syndromes elicited by CRALBP dysfunction (Maw et al. 1997; Fishman et al. 2004).

How is the diversity of the Sec14 superfamily employed in the context of the lipid signaling circuitry of eukaryotic cells? Together it is proposed that the Sec14-protein superfamily has been engineered for sensing specific aspects of lipid metabolism and for transduction of sensing information to an activity that employs PIP signaling for action. Below, we describe a physical picture of how the Sec14 superfamily generates a productive 'crosstalk' between the larger lipidome, PIP signaling, and membrane trafficking control.

9.6 Ligand Binding by Sec14-like Proteins

Available crystal structures of Sec14-like proteins include detergent-bound forms of Sec14p, several phospholipid-bound forms of yeast Sfh1p, ligand-bound and unbound versions of α-tocopherol transfer protein (α-TTP), the mammalian Sec14-GOLD protein Sec14L2, and detergent-bound and phospholipid-bound forms of the neurofibromin Sec14-like domain (Schaaf et al. 2008; Sha et al. 1998; Min et al. 2003; Meier et al. 2003; Stocker and Baumann 2003; D'Angelo et al. 2006; Welti et al. 2007). These structures show that the Sec14 fold is structurally conserved

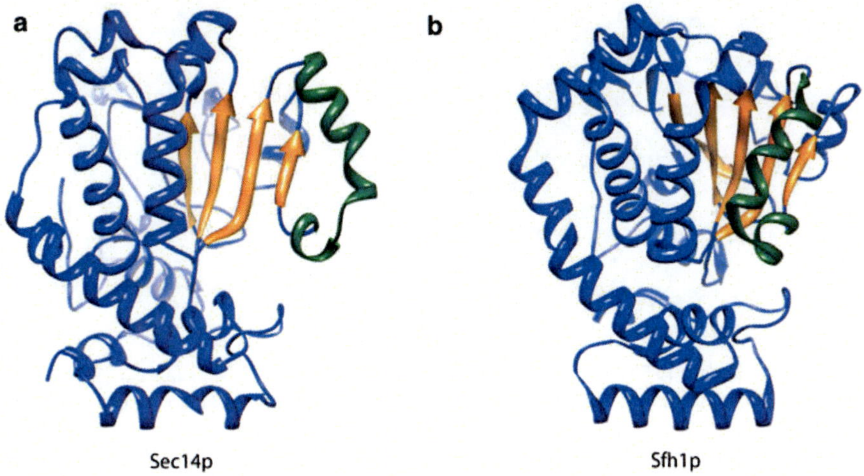

Fig. 9.1 A helical substructure of the Sec14-like proteins regulates entrance of the ligand into the hydrophobic binding pocket. Ribbon diagrams of (**a**) open Sec14p (1AUA, open conformer bound to the detergent β-octylglucoside) and (**b**) closed Sfh1p (3B7N, closed conformer bound to PtdIns) are shown. Comparisons of open versus closed conformers provide insight into how the helical substructure (*green*) regulates entry into the binding pocket. α-helices and turns are depicted in *blue* and the binding cavity floor β-sheets are depicted in *yellow*

comprising approximately 280-residue two-lobed globular structure. In the case of apo-Sec14p, the amino-terminal region consists of four anti-parallel α-helices whereas the carboxy-terminal lobe (also referred to as the CRAL_TRIO domain) forms the phospholipid binding pocket (Fig. 9.1). The Sec14p lipid binding cavity consists of five parallel β-strands comprising a β-sheet that is sandwiched by two long α-helices on one side and two short α-helices plus one 310-helix on the other. Molecular dynamics simulations (Ryan et al. 2007) and comparisons of apo-Sec14p and holo-Sfh1p structures (Schaaf et al. 2008; Sha et al. 1998) demonstrate that access to the hydrophobic lipid-binding cavity is regulated by a helical gate (Fig. 9.1). This gate is closed when Sec14 is bound to ligand. Transitions from open to closed conformations bring about rigid body motions which displaces the helical gate. These conformational dynamics are coupled to interfacial phospholipid exchange reactions on the surface of biological membranes. Other members of the Sec14 superfamily probably undergo similar conformational dynamics during the course of ligand loading/unloading reactions. A regulatory substructure, the approximately 20-residue gating module, regulates an H-bond network that transduces conformational information to the helical gate (Ryan et al. 2007), probably in response to conformational changes initiated by membrane docking. Examples of human disease missense mutations that map to the gating modules of Sec14 superfamily proteins confirm the functional importance of this substructure (Ryan et al. 2007).

A notable property of Sec14 is that the hydrophobic gradient within the lipid binding domain closely matches that found in the cytosolic leaflet of a membrane.

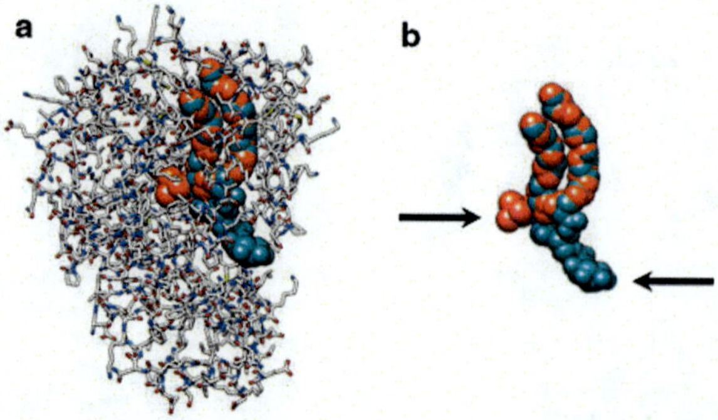

Fig. 9.2 Sec 14-like PITPs bind phospholipid head groups at two different sites. (**a**) 3B7Z Bound PtdIns (*Aqua spheres*) and PtdCho (*Orange spheres*) is shown in a complex with the Sec14-like PITP, Sfh1p (*Grey*, *red* and *blue wire*). (**b**) The Sfh1 bound phospholipid configurations are depicted without the PITP. The *arrows* indicate the lipid headgroups. Extraordinarily, the acyl chains of these two different lipid species occupy the same hydrophobic cavity of the transfer protein, but their headgroups bind at distinct sites with a clear physical barrier between binding regions. We propose that this physical barrier between the headgroup binding sites in Sec14p molecule forms the basis for how heterotypic exchange reactions present a PtdIns headgroup to the lipid kinase

This similarity implies that lipid binding/release by Sec14p, and presumably other Sec14-like proteins, is driven by simple substrate partitioning from one aprotic environment to another (Smirnova et al. 2007). How Sec14 selects lipid ligands on the membrane surface, and how these are subsequently configured to enable abstraction from a membrane bilayer evoke questions for future analysis. The local membrane deformations that must proceed to enable these selection/configuration processes also remain unknown.

The most surprising feature of both Sfh1p and Sec14p is that the headgroups of PtdCho and PtdIns are bound at distinct sites within the lipid binding pocket (Schaaf et al. 2008, Fig. 9.2). The inositol binding site is situated near the protein surface (Schaaf et al. 2008, Fig. 9.2). By contrast, the PtdCho headgroup is buried within the interior of the hydrophobic cleft, an interaction stabilized by tyrosine-mediated cation–π interactions (Schaaf et al. 2008). Another unusual feature of Sfh1p and Sec14p is their ability to accommodate phospholipid molecules with different volumes (cavity volumes for Sfh1p–PtdIns and Sfh1p–PtdCho are 4050.6 Å3 and 3068.7 Å3, respectively) without significantly effecting the shape of the protein surface. This is accomplished in part by loading the unoccupied PtdCho headgroup binding site with ordered water molecules in the Sec14–PtdIns complex and, reciprocally, loading the unoccupied PtdIns headgroup binding site with ordered water in the Sec14–PtdCho complex (Schaaf et al. 2008). The flux of water into and from the hydrophobic pocket during heterotypic exchange reactions is a major factor in overcoming the differences in Sec14p relative binding affinities for PtdIns and PtdCho so that heterotypic exchange reactions (PtdIns for PtdCho or PtdCho for PtdIns) can actually take place. That Sfh1, a protein that shares 64% sequence identity with

Sec14, is functionally distinct to Sec14 is proposed to be attributed to specific reconfigurations in atomic interactions between amino acid side chains and ordered water molecules within the lipid binding cavity (Schaaf et al. 2011). Such altered dynamics reconstitutes a functional gating module that communicates conformational energy from within the hydrophobic pocket to the helical structure that gates access to the pocket (Schaaf et al. 2011). It is predicted that a consequence of this is that the rates of phospholipid cycling into and out of the Sfh1 and Sec14 hydrophobic pocket differs such that Sfh1 cannot substitute for Sec14 activity. This remains open for future investigation.

9.7 Coincidence Sensors that Couple Lipid Metabolic Inputs to PIP Synthesis

Both PtdIns- and PtdCho-binding activities must reside on the same Sec14p molecule to generate a biologically functional protein able to stimulate PtdIns 4-OH kinase activity (Schaaf et al. 2008). Thus, heterotypic exchange reactions are required for Sec14p-mediated stimulation of PtdIns kinases (and PIP synthesis) in vivo. This indicates that Sec14p cannot stimulate PtdIns 4-OH kinases in cells unless sufficient amounts of PtdCho are present to facilitate heterotypic exchange reactions necessary to activate PtdIns kinases (Fig. 9.3). Together this connects Sec14p as a PtdCho sensor which transmits PtdCho metabolic information to PIP synthesis. This activity is consistent with the apparent coupling between the cytidine diphosphate (CDP)–choline pathway for PtdCho biosynthesis and membrane trafficking control (Cleves et al. 1991; Skinner et al. 1993).

The Sec14p requirement for coordinating the PtdCho biosynthesis/membrane trafficking interface ensures that the DAG pools necessary for TGN/endosomal trafficking are not exhausted by the CDP–choline pathway for PtdCho biosynthesis. Thus Sec14p senses PtdCho as a readout for DAG consumption (Skinner et al. 1995). As PtdCho levels increase from synthesis through the CDP–choline pathway, Sec14p is activated for heterotypic PtdIns/PtdCho exchange, stimulating PIP production by PtdIns 4-OH kinases (Fig. 9.3). As a consequence, PtdIns-4-phosphate synthesis would serve to activate downstream effectors that promote vesicle budding from TGN/endosomes. In addition, PtdIns-4-phosphate might also inhibit the cholinephosphate cytidylyltransferase, the rate-determining enzyme of the CDP–choline pathway (Fig. 9.3).

Do these concepts hold true for other members of the Sec14 superfamily? Bioinformatic analyses have identified primary sequence 'bar codes' for PtdIns and PtdCho binding (Schaaf et al. 2008). It is apparent that PtdCho binding is not a conserved feature of Sec14-like proteins. However, the holo-Sec14L2 and α-TTP structures, and the biochemical properties of CRALBP and related proteins, show that members of the Sec14 superfamily lacking key PtdCho-binding residues are able to bind to alternative hydrophobic ligands (Schaaf et al. 2008; Min et al. 2003; Meier et al. 2003; Stocker and Baumann 2003; D'Angelo et al. 2006; Welti et al. 2007). In contrast to the PtdCho binding motif, bioinformatics identifies the PtdIns-binding 'bar

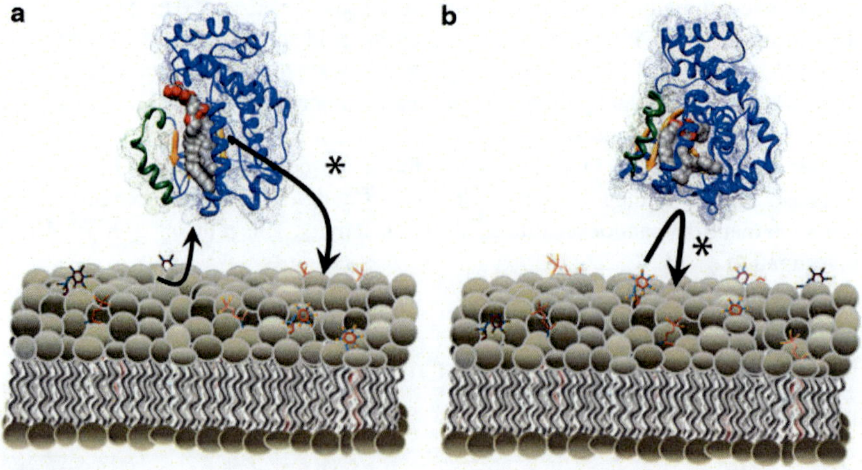

Fig. 9.3 Heterotypic exchange promotes PtdIns presentation. Heterotypic exchange reactions can support PtdIns presentation by two different models. **a** PtdCho vectorial displaces a Sec14p bound PtdIns in a head-first manner. The displaced PtdIns exits the binding pocket through a portal distinct from the portal through which PtdCho invades. PtdIns4-OH Kinase (not shown) modifies the exiting PtdIns during this exchange. **b** A second mode by which heterotypic exchange promotes PtdIns presentation is by frustration of an invading PtdIns. In this mode, PtdIns attempts to invade into the hydrophobic pocket of a PtdCho bound Sec14p. The bound PtdCho frustrates the PtdIns, allowing PtdIns 4-OH kinase to modify its substrate. Both models satisfy the requirement that Ptd-4-phosphate cannot collapse back into the hydrophobic pocket, as this complex results in locked Sec14p-PIP species that cannot be reversed by phospholipid exchange

code' to be ubiquitous to the superfamily. It is attractive to propose that a two-ligand PITP-mediated mechanism for PtdIns kinase activation (analogous to that described for Sec14p and Sfh1p) might be broadly utilized by the Sec14 superfamily members. Together, the Sec14 superfamily of proteins link diverse aspects of the lipid metabolome with PIP signaling.

9.8 Instructive Mechanisms for Regulating PtdIns-kinase Activity

That Sec14p and Sec14-like PITPs are required physiologically to potentiate yeast PtdIns 4-OH kinase activity suggests that these enzymes, alone, inadequately engage membrane-incorporated PtdIns. Infact, to circumvent the poor interfacial utilization of PtdIns as a substrate by PtdIns 4-OH kinase in an in vitro PtdIns kinase assay PtdIns is presented to the kinase in the context of mixed phospholipid-detergent micelles. Thus, it is attractive to speculate that PtdIns 4-OH kinases have evolved to be inefficient interfacial enzymes to enable opportunities for regulating lipid signaling. That is, Sec14-like PITPs solve the inadequacies of PtdIns 4-OH kinases in engaging microsomal PtdIns, in turn providing significant mechanisms to coordinate lipid metabolism with a plethora of intracellular signaling pathways.

Insights into Sec14 functionality predict that the priming of Sec14-like PITPs through association with their cognate sensory ligand (PtdCho in the case of Sec14) triggers, through heterotypic exchange, the ability to present PtdIns headgroups to PtdIns 4-OH kinases in a conformation readily susceptible to phosphorylation. In this model, the efficiency with which PtdIns 4-OH kinase modifies PtdIns is low, a consequence attributed to the short window of opportunity that a PtdIns kinase has to phosphorylate its substrate. However, these exchange reactions occur rapidly and independent of ATP consumption, therefore, the high frequency with which a PITP presents PtdIns to a kinase compensates for this inefficient system.

The term 'heterotypic exchange' implies complete transition, in one cycle, of a holo-Sec14p–PtdCho complex to a holo-Sec14p–PtdIns complex or vice versa. This definition might be too strict. The effect of heterotypic exchange reactions on interfacial presentation of PtdIns to PtdIns 4-OH kinases could reflect consequences of abortive heterotypic exchange reactions (Schaaf et al. 2008). Assuming that the Sec14p–PtdCho state represents the 'primed' Sec14p, sequestration of PtdCho deep within the hydrophobic pocket might result in a 'slow' exit of this phospholipid from the pocket. Thus, a deep ligand obstructs the pocket from an invading PtdIns molecule. During abortive heterotypic exchange, where PtdCho binding is a prerequisite to PtdIns presentation, the invading lipid is neither embedded in the membrane nor sequestered by the engaged Sec14p enabling the lipid headgroup to become highly susceptible to by PtdIns 4-OH kinase mediated phosphorylation. Multiple abortive PtdIns exchange events could theoretically accompany a complete heterotypic exchange event, particularly if the rate of PtdIns invasion exceeds the rate of PtdCho egress. In support of this idea the rates of PtdIns transfer are approximately 20-fold faster than that of PtdCho. In an alternative scenario, however, an invading PtdCho molecule would drive the ejection of a pre-bound PtdIns ligand from the PITP lipid binding pocket. Thus, during its egress, the PtdIns headgroup would become susceptible to PtdIns 4-OH dependent modification. These alternative mechanisms might not be mutually exclusive and make distinguishing experimental predictions of lipid trajectories during heterotypic exchange. Molecular dynamic approaches will be required to investigate these mechanisms further.

9.9 Definition of Sensing Territories

Spatial and temporal restriction of PITP activity can be achieved by multiple paths. Protein–protein or protein–lipid interactions that involve the Sec14-like PITP surface (as opposed to the hydrophobic binding pocket) can restrict localization to membrane subdomains. Indeed, several yeast Sec14-like PITPs stably target specific intracellular locations. The identities of proteins (or lipids) that impart specific localization to PITPs have yet to be described. PITP receptors are of interest, as such proteins (or lipid platforms) define 'sensing' territories. In that regard, the PtdIns kinases themselves represent obvious candidates for PITP receptors. Is there a requirement for dedicated physical interactions between PITPs and the kinases? This requirement is unlikely to be the case as Sec14p defects are rescued by the expression in yeast of mammalian class 1 PITPs (i.e. proteins with no primary sequence homology or structural similarity to Sec14p).

Sec14 membrane association is transient yet it is still sufficient to productively activate PtdIns kinases. Presentation mechanisms remain plausible given that the rate with which Sec14 dissociates from membranes (or any given PITP) is much slower than the rate of heterotypic exchange. The residence time of a single Sec14 molecule is unknown. However, reasonable estimates suggest that a Sec14 molecule could execute up to 10 exchange cycles on a membrane surface in a 2 s residency. Direct measurements of the time scales for lipid exchange by individual PITP molecules, especially when considered relative to PITP membrane dwell times, define important future directions essential for assessing nanoreactor versus lipid transfer models.

In this regard, Sec14 domains in modular proteins (e.g. Ras/Rho GAPs and GEFs and the MEG2 PTP) (Aravind et al. 1999; Shang et al. 2003; Tcherkezian and Lamarche-Vane 2007; Sirokmany et al. 2005; Ueda et al. 2004; Debant et al. 2004; Gu et al. 1992) are well-engineered to act as intrinsic lipid sensing units for instructing local changes in PIP environment in response to metabolic cues. Both GEF and GAP activities are responsive to PIPs, therefore it is attractive to postulate that the Sec14 domain 'senses' the lipid environment, then directs PtdIns kinases for 'on-demand' PIP synthesis, and finally, recruits GEF/GAP/PTP catalytic domains to the newly synthesized pool of PIPs (e.g. as in Dbl and Dbs). Some authors suggesting that lipid binding by the Sec14 domain recruits Ras/Rho-GEFs to membrane surfaces (Sirokmany et al. 2005; Debant et al. 1996; Kostenko et al. 2004). However, it is more likely that such membrane-targeting activities probably involve protein-lipid interactions on the Sec14 domain surface rather than the hydrophobic pocket.

Similar designs might also apply to Sec14-nodulin proteins of higher plants (Aravind et al. 1999; Anantharaman and Aravind 2002). The AtSfh1 Sec14-nodulin protein expressed in Arabidopsis root hair cells both stimulates PtdIns-4,5-P2 synthesis and distribution in growing root hairs. In this manner, an AtSfh1-PtdIns-kinase-PIP-kinase axis enables polarized programs for membrane trafficking, actin organization and calcium signaling in growing root hairs (Vincent et al. 2005). AtSfh1 might help develop the PIP landscape via a sensing role for the Sec14 domain and consequently signals and potentiates PIP synthesis by neighboring PtdIns 4-OH and PtdIns-4-phosphate 5-OH kinases (Preuss et al. 2006; Stenzel et al. 2008).

Nodulin domains represent membrane association elements, and the chemical properties of the AtSfh1 nodulin domain and of other nodulin domains suggest additional layers of regulation. These units present basic carboxy-terminal tails (Vincent et al. 2005; Kapranov et al. 2001), thereby resembling known PIP-binding motifs that operate via membrane surface electrostatics (e.g. the MARCKS peptide (McLaughlin and Murray 2005). It is a testable proposition that AtSfh1 functionally specifies the production and organization of dedicated PIP pools. Regulated electrostatic interactions suggest mechanisms for imprinting 'caged' PIP patterns for subsequent and regulated PIP release in a spatially organized program of downstream signaling

(McLaughlin and Murray 2005). The multiplicity of plant Sec14-nodulin proteins forecasts a large diversity in such a strategy for coupling distinct Sec14-like modules of multidomain proteins to developmental pathways for membrane morphogenesis (Vincent et al. 2005). Sec14-GOLD domain proteins that bind PIPs (e.g. PATELLIN 1) might also operate similarly (Peterman et al. 2004).

9.10 Concluding Remarks

New lines of research are providing novel understandings into the mechanisms of how lipid metabolism plays a central role in regulating membrane trafficking and signaling. Recognition of Sec14-like PITPs' integral function in lipid metabolism and PIP signaling, and the recent advances that demonstrate a physical appreciation of how such integration could work, present the trafficking and signaling fields with new questions to address. Although we propose that PITPs primarily function as nanoreactores for regulating lipid metabolism, as opposed to the classic "lipid carrier" model, the mechanism will vary from case to case. For instance, the metazoan-specific PITPs (e.g. Drosophila RdgBα and mammalian NIR2) are structurally unrelated to the Sec14-like PITPs, yet bind PtdIns and PtdCho in the same binding siter. Do these PITPs function as signaling nanoreactors by potentiating PIP synthesis or do they simply carry lipids from one membrane to another? Furthermore, the nanoreactor/lipid-carrier dichotomy may be applied to other putative transfer proteins that operate in cells (e.g. oxysterol binding proteins and ceramide transfer protein. These competing conceptual frameworks can now be experimentally examined, thereby providing an escape from the circular arguments that have plagued the lipid transfer protein field since its inception.

A unique way of regulating signal transduction pathways maybe the use of non-enzymatic protein biosensors, such as Sec14-like PITPs, to couple metabolic cues with the action of interfacial lipid modifying enzymes. If Sec14p can bind other inositol-phospholipids in addition to PtdIns, then the versatility of Sec14p as a biosensor might be more than previously anticipated. There is no structural rationale to exclude the possibility that Sec14p may also bind inositol-phosphoceramides in vivo, a property that could additionally allow Sec14p the ability to transmit PtdCho metabolic information to inositol sphingolipid metabolism.

Currently, our understanding of spatial and temporal regulation of lipid metabolism is inadequate for a complete biological evaluation. Technical complications will be a major challenge in the future and impede scientific growth and information. Understanding when and where Sec14-like PITPs are active is a central component to understanding lipid metabolism. Thus, engineering reliable conformational biosensors will be required to elucidate the roles of Sec14-like PITPs at the interface between membrane trafficking and signaling in living cells.

References

Anantharaman V, Aravind L (2002) The GOLD domain, a novel protein module involved in Golgi function and secretion. Genome Biol 3:0023

Antonny B et al (1997) Activation of ADP-ribosylation factor 1 GTPase-activating protein by phosphatidylcholine-derived diacylglycerols. J Biol Chem 272:30848–30851

Aravind L et al (1999) Sec14p-like domains in NF1 and Dbl-like proteins indicate lipid regulation of Ras and Rho signaling. Curr Biol 9:R195–R197

Asp L et al (2009) Early stages of Golgi vesicle and tubule formation require diacylglycerol. Mol Biol Cell 20:780–790

Balla A, Balla T (2006) Phosphatidylinositol 4-kinases: old enzymes with emerging functions. Trends Cell Biol 16:351–361

Bankaitis VA et al (1990) An essential role for a phospholipid transfer protein in yeast Golgi function. Nature 347:561–562

Bard F, Malhotra V (2006) The formation of TGN-to-plasma-membrane transport carriers. Annu Rev Cell Dev Biol 22:439–455

Baron CL, Malhotra V (2002) Role of diacylglycerol in PKD recruitment to the TGN and protein transport to the plasma membrane. Science 295:325–328

Bomar JM et al (2003) Mutations in a novel gene encoding a CRAL-TRIO domain cause human Cayman ataxia and ataxia/dystonia in the jittery mouse. Nat Genet 35:264–269

Bossard C et al (2007) Dimeric PKD regulates membrane fission to form transport carriers at the TGN. J Cell Biol 179:1123–1131

Burger KNJ (2000) Greasing membrane fusion and fission machineries. Traffic 1:605–613

Caloca MJ et al (2003) Exchange factors of the RasGRP family mediate Ras activation in the Golgi. J Biol Chem 278:33465–33473

Carlton JG, Cullen PJ (2005) Coincidence detection in phosphoinositide signaling. Trends Cell Biol 15:540–547

Carmen-Lopez M et al (1994) A phosphatidylinositol/phosphatidylcholine transfer protein is required for differentiation of the dimorphic yeast Yarrowia lipolytica from the yeast to the mycelial form. J Cell Biol 124:113–127

Chantalat S et al (2004) The Arf activator Gea2p and the P-type ATPase Drs2p interact at the Golgi in Saccharomyces cerevisiae. J Cell Sci 117:711–722

Chernomordik L et al (1995) Lipids in biological membrane fusion. J Membr Biol 146:1–14

Cichowski K, Jacks T (2001) NF1 tumor suppressor gene function: narrowing the GAP. Cell 104:593–604

Cleves AE et al (1991) Mutations in the CDP-choline pathway for phospholipid biosynthesis bypass the requirement for an essential phospholipid transfer protein. Cell 64:789–800

D'Angelo I et al (2006) A novel bipartite phospholipid-binding module in the neurofibromatosis type 1 protein. EMBO Rep 7:174–179

Debant A et al (1996) The multidomain protein Trio binds the LAR transmembrane tyrosine phosphatase, contains a protein kinase domain, and has separate rac-specific and rho-specific guanine nucleotide exchange factor domains. Proc Natl Acad Sci U S A 93:5466–5471

Dee CT, Moffat KG (2005) A novel family of mitochondrial proteins is represented by the Drosophila genes slmo, preli-like and real-time. Dev Genes Evol 215:248–254

Demmel L et al (2008) The clathrin adaptor Gga2p is a phosphatidylinositol-4-phosphate effector at the Golgi exit. Mol Biol Cell 19:1991–2002

Díaz Ael AM, Malhotra V (2005) PKCeta is required for $\beta 1\gamma 2/\beta 3\gamma 2$ and PKD mediated transport to the cell surface and the organization of the Golgi apparatus. J Cell Biol 169:83–91

Fishman GA et al (2004) Vovel mutations in the cellular retinaldehyde-binding protein gene (RLBP1) associated with retinitis punctata albescens: evidence of interfamilial genetic heterogeneity and fundus changes in heterozygotes. Arch Opthamol 122:70–75

Gloor Y et al (2010) Interaction between Sec7p and Pik1p: the first clue for the regulation of a coincidence detection signal. Eur J Cell Biol 89:575–583

Godi A et al (1999) ARF mediates recruitment of PtdIns 4-OH kinase beta and stimulates synthesis of PtdIns(4,5)P2 on the Golgi complex. Nat Cell Biol 1:280–287

Golovleva I et al (2003) Disease causing mutations in the cellular retinaldehyde binding protein tighten and abolish ligand interactions. J Biol Chem 278:12397–12402

Gotoda T et al (1995) Adult-onset spinocerebellar dysfunction caused by a mutation in the gene for alpha-tocopherol transfer protein. N Engl J Med 333:1313–1318

de Graaf P et al (2004) Phosphatidylinositol 4-kinase beta is critical for functional association of rab11 with the Golgi complex. Mol Biol Cell 15:2038–2047

Gu M et al (1992) Cloning and expression of a cytosolic megakaryocyte protein-tyrosine-phosphatase with sequence homology to retinaldehyde-binding protein and yeast SEC14p. Proc Natl Acad Sci U S A 89:2980–2984

Habermehl D et al (2004) Recombinant SEC14-like proteins (TAP) possess GTPase activity. Biochem Biophys Res Commun 326:254–259

Hama H et al (1999) Direct involvement of phosphatidylinositol-4-phosphate in secretion in the yeast Saccharomyces cerevisiae. J Biol Chem 274:34294–34301

Haynes LP et al (2005) Interaction of neuronal calcium sensor-1 (NCS-1) and ADP-ribosylation factor 1 allows bidirectional control of phosphatidylinositol-4-kinase beta and trans-Golgi network-plasma membrane traffic. J Biol Chem 280:6047–6054

Hendricks KB et al (1999) Yeast homologue of neuronal frequenin is a regulator of phosphatidylinositol-4-OH kinase. Nat Cell Biol 1:234–241

Hentati A et al (1996) Human alpha-tocopherol transfer protein: gene structure and mutations in familial vitamin E deficiency. Ann Neurol 39:295–300

Kapranov P et al (2001) Nodule-specific regulation of phosphatidylinositol transfer protein expression in Lotus japonicus. Plant Cell 13:1369–1382

Kearns BG et al (1997) Essential role for diacylglycerol in protein transport from the yeast Golgi complex. Nature 387:101–105

Kostenko EV et al (2004) The Sec14 homology domain regulates the cellular distribution and transforming activity of the Rho-specific guanine nucleotide exchange factor Dbs. J Biol Chem 280:2807–2817

Lehel C et al (1995) Protein kinase C epsilon subcellular localization domains and proteolytic degradation sites. A model for protein kinase C conformational changes. J Biol Chem 270:19651–19658

Li X et al (2002a) Analysis of oxysterol binding protein homologue Kes1p function in regulation of Sec14p-dependent protein transport from the yeast Golgi complex. J Cell Biol 157:63–77

Li X et al (2002b) Identification of a novel family of nonclassic yeast phosphatidylinositol transfer proteins whose function modulates phospholipase D activity and Sec14p-independent cell growth. Mol Biol Cell 11:1989–2005

Liljedahl M et al (2001) Protein kinase D regulates the fission of cell surface destined transport carriers from the trans-Golgi network. Cell 104:409–420

Litvak V et al (2005) Maintenance of the diacylglycerol level in the Golgi apparatus by the Nir2 protein is critical for Golgi secretory function. Nat Cell Biol 7:225–234

Liu T et al (2005) Structural insights into the cellular retinaldehyde-binding protein (CRALBP). Proteins Struct Funct Bioinform 61:412–422

Maissel A et al (2006) PKCeta is localized in the Golgi, ER and nuclear envelope and translocates to the nuclear envelope upon PMA activation and serum-starvation: C1b domain and the pseudo-substrate containing fragmenttarget PKCeta to the Golgi and the nuclear envelope. Cell Signal 18:1127–1139

Maw MA et al (1997) Mutation of the gene encoding cellular retinaldehyde-binding protein in autosomal recessive retinitis pigmentosa. Nat Genet 17:198–200

McLaughlin S, Murray D (2005) Plasma membrane phosphoinositide organization by protein electrostatics. Nature 438:605–611

Meier R et al (2003) The molecular basis of vitamin E retention: structure of human alpha-tocopherol transfer protein. J Mol Biol 331:725–734

Min KC et al (2003) Crystal structure of α-tocopherol transfer protein bound to its ligand: Implications for ataxia with vitamin E deficiency. Proc Natl Acad Sci U S A 100:14713–14718
Mizuno-Yamasaki E et al (2010a) Phosphatidylinositol-4-phosphate controls both membrane recruitment and a regulatory switch of the Rab GEF Sec2. Dev Cell 18:828–840
Mizuno-Yamasaki E et al (2010b) Phosphatidylinositol 4-phosphate controls both membrane recruitment and a regulatory switch of the Rab GEF Sec2p. Dev Cell 18(5):828–840
Mousley CJ et al (2008) Trans-Golgi network and endosome dynamics connect ceramide homeostasis with regulation of the unfolded protein response and TOR signaling in yeast. Mol Biol Cell 19:4785–4803
Muthusamy B-P et al (2009) Linking phospholipid flippases to vesicle-mediated transport. Biochim Biophys Acta 179:612–619
Nakase Y et al (2001) The Schizosaccharomyces pombe spo20(+) gene encoding a homologue of Saccharomyces cerevisiae Sec14 plays an important role in forespore membrane formation. Mol Biol Cell 4:901–917
Natarajan P et al (2004) Regulation of a golgi flippase by phosphoinositides and an Arf-GEF. Proc Natl Acad Sci U S A 101:10614–10619
Ouachi K et al (1995) Ataxia with vitamin E deficiency is caused by mutations in the α-tocopherol transfer protein. Nat Genet 9:141–145
Peterman TK et al (2004) Patellin1, a novel Sec14-like protein, localizes to the cell plate and binds phosphoinositides. Plant Physiol 136:3080–3094
Phillips SE et al (1999) Yeast Sec14p deficient in phosphatidylinositol transfer activity is functional in vivo. Mol Cell 4:187–197
Polevoy G et al (2009) Dual roles for the Drosophila PI 4-kinase four wheel drive in localizing Rab11 during cytokinesis. J Cell Biol 187:847–858
Preuss ML et al (2006) A role for the RabA4b effector protein PI-4Kbeta1 in polarized expansion of root hairs in Arabidopsis thaliana. J Cell Biol 172:261–268
Rivas MP et al (1999) Relationship between altered phospholipid metabolism, DAG, 'bypass Sec14p', and the inositol auxotrophy of yeast sac1 mutants. Mol Biol Cell 10:2235–2250
Routt SM et al (2005) Nonclassical PITPs activate phospholipase D via an Stt4p-dependent pathway and modulate function of late stages of the secretory pathway in vegetative yeast cells. Traffic 6:1157–1172
Rudge SA et al (2004) Roles of phosphoinositides and of Spo14p (phospholipase D)-generated phosphatidic acid during yeast sporulation. Mol Biol Cell 15:207–218
Ryan MM et al (2007) Conformational dynamics of the major yeast phosphatidylinositol transfer protein Sec14: Insights into the mechanisms of PL exchange and diseases of Sec14-like protein deficiencies. Mol Biol Cell 18:1928–1942
Salama SR et al (1990) Cloning and characterization of the Kluyveromyces lactis SEC14: A gene whose product stimulates Golgi secretory function in S. cerevisiae. J Bacteriol 172:4510–4521
Schaaf G et al (2008) The functional anatomy of phospholipid binding and regulation of phosphoinositide homeostasis by proteins of the Sec14-superfamily. Mol Cell 29:191–206
Schaaf G et al (2011) Resurrection of a functional phosphatidylinositol transfer protein from a pseudo-Sec14 scaffold by directed evolution. Mol Biol Cell 22(6):892–905
Sciorra V et al (2005) Synthetic gene array analysis of the PtdIns 4-kinase Pik1p identifies components in a Golgi specific Ypt31/rab-GTPase signaling pathway. Mol Biol Cell 15:2038–2047
Sha B et al (1998) Crystal structure of the Saccharomyces cerevisiae phosphatidylinositol transfer protein Sec14. Nature 391:506–510
Shang X et al (2003) Concerted regulation of cell dynamics by BNIP-2 and Cdc42GAP homology/Sec14p-like, proline-rich, and GTPase-activating protein domains of a novel rhoGTPase-activating protein, BPGAP1. J Biol Chem 278:45903–45914
Sirokmany G et al (2005) Sec14 homology domain targets p50RhoGAP to endosomes and provides a link between Rab- and Rho GTPases. J Biol Chem 281:6096–6105

Skinner HB et al (1993) Phospholipid transfer activity is relevant to but not sufficient for the essential function of the yeast SEC14 gene product. EMBO J 12:4775–4784

Skinner HB et al (1995) Phosphatidylinositol transfer protein stimulates yeast Golgi secretory function by inhibiting choline-phosphate cytidylyltransferase activity. Proc Natl Acad Sci U S A 92:112–116

Smirnova T et al (2007) Local polarity and hydrogen bonding inside the Sec14 PL-binding cavity: high-field multifrequency studies. Biophys J 92:3686–3695

Stefan CJ, Manford AG, Baird D, Yamada-Hanff J, Mao Y, Emr SD (2011) Osh proteins regulate phosphoinositide metabolism at ER-plasma membrane contact sites. Cell 144:389–401

Stenzel I et al (2008) The type B phosphatidylinositol-4-phosphate 5-kinase 3 is essential for root hair formation in Arabidopsis thaliana. Plant Cell 20:124–141

Stocker A, Baumann U (2003) Supernatant protein factor in complex with RRR-alpha-tocopherylquinone: a link between oxidized vitamin E and cholesterol biosynthesis. J Mol Biol 332:759–765

Strahl T, Thorner J (2007) Synthesis and function of membrane phosphoinositides in budding yeast, Saccharomyces cerevisiae. Biochim Biophys Acta 1771:353–404

Szentpetery Z et al (2010) Acute manipulation of Golgi phosphoinositides to assess their importance in membrane trafficking and signaling. Proc Natl Acad Sci U S A 107:8225–8230

Tcherkezian J, Lamarche-Vane N (2007) Current knowledge of the large RhoGAP family of proteins. Biol Cell 26:67–86

Ueda S et al (2004) Role of the Sec14-like domain of Dbl family exchange factors in the regulation of Rho family GTPases in different subcellular sites. Cell Signal 16:826–906

Vergés M et al (2006) The mammalian retromer regulates transcytosis of the polymeric immunoglobulin receptor. Nat Cell Biol 6:763–769

Vincent P et al (2005) A Sec14p-nodulin domain phosphatidylinositol transfer protein polarizes membrane growth of Arabidopsis thaliana root hairs. J Cell Biol 168:801–812

Walch-Solimena C, Novick P (1999) The yeast phosphatidylinositol-4-OH kinase Pik1 regulates secretion at the Golgi. Nat Cell Biol 1:523–555

Wang QJ et al (1999) Differential localization of protein kinase C delta by phorbol esters and related compounds using a fusion protein with green fluorescent protein. J Biol Chem 274:37233–37239

Wang J et al (2003) Phosphatidylinositol-4-phosphate regulates targeting of clathrin adaptor AP-1 complexes to the Golgi. Cell 114:299–310

Wang J et al (2007) PI4P promotes the recruitment of the GGA adaptor proteins to the trans-Golgi network and regulates their recognition of the ubiquitin sorting signal. Mol Biol Cell 18:2646–2655

Welti S et al (2007) The sec14 homology module of neurofibromin binds cellular glycerophospholipids: mass spectrometry and structure of a lipid complex. J Mol Biol 366:551–562

Wood CS et al (2009) PtdIns4P recognition by Vps74/GOLPH3 links PtdIns 4-kinase signaling to retrograde Golgi trafficking. J Cell Biol 187:967–975

Wu WI et al (2000) A new gene involved in the transport-dependent metabolism of phosphatidylserine, PSTB2/PDR17, shares sequence similarity with the gene encoding the phosphatidylinositol/phosphatidylcholine transfer protein, SEC14. J Biol Chem 275:14446–14456

Xie Z et al (1998) Phospholipase D activity is required for suppression of yeast phosphatidylinositol transfer protein defects. Proc Natl Acad Sci U S A 95:12346–12351

Yanagisawa L et al (2002) Activity of specific lipid-regulated ARFGAPs is required for Sec14p-dependent Golgi secretory function in yeast. Mol Biol Cell 13:2193–2206

Zhao X et al (2001) Interaction of neuronal calcium sensor-1 (NCS-1) with phosphatidylinositol-4-kinase beta stimulates lipid kinase activity and affects membrane trafficking in COS-7 cells. J Biol Chem 276:40183–40189

Chapter 10
Phosphoinositide Sensitivity of Ion Channels, a Functional Perspective

Nikita Gamper and Tibor Rohacs

Abstract Phosphoinositides, especially phosphatidylinositol 4,5-bisphosphate [PtdIns(4,5)P_2] are required for the activity of many different ion channels. This chapter will highlight various aspects of this paradigm, by discussing current knowledge on four different ion channel families: inwardly rectifying K^+ (Kir) channels, KCNQ voltage gated K^+ channels, voltage gated Ca^{2+} (VGCC) channels and Transient Receptor Potential (TRP) channels. Our main focus is to discuss functional aspects of this regulation, i.e. how changes in the concentration of PtdIns(4,5)P_2 in the plasma membrane upon phospholipase C activation may modulate the activity of ion channels, and what are the major determinants of this regulation. We also discuss how channels act as coincidence detectors sensing phosphoinositide levels and other signalling molecules. We also briefly discuss the available methods to study phosphoinositide regulation of ion channels, and structural aspects of interaction of ion channel proteins with these phospholipids. Finally, in several cases the effect of PtdIns(4,5)P_2 is more complex than a simple dependence of ion channel activity on the lipid, and we will discuss some these complexities.

Keywords PIP_2 · PtdIns(4,5)P_2 · Ion channel · Phospholipase C · G protein coupled receptor

10.1 Introduction

Membrane phosphoinositides play a multitude of roles in a variety of biological processes. Phosphatidylinositol 4,5-bisphosphate [PtdIns(4,5)P_2], generally referred to

---

N. Gamper (✉)
Institute of Membrane and Systems Biology, Faculty of Biological Sciences, University of Leeds, LS2 9JT, Leeds, UK
e-mail: n.gamper@leeds.ac.uk

T. Rohacs (✉)
Department of Pharmacology and Physiology, University of Medicine and Dentistry of New Jersey—New Jersey Medical School, Newark, 07103 NJ, USA
e-mail: tibor.rohacs@umdnj.edu

T. Balla et al. (eds.), *Phosphoinositides II: The Diverse Biological Functions*, Subcellular Biochemistry 59, DOI 10.1007/978-94-007-3015-1_10,

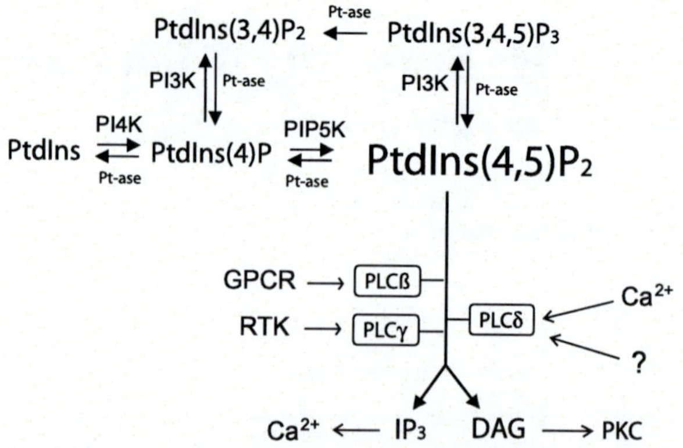

Fig. 10.1 Phosphoinositide metabolism. PtdIns(4,5)P_2 is generated from its precursor PtdIns by two consecutive phosphorylation steps by phosphatidylinositol 4-kinases (PI4K) and phosphatidylinositol 4-phosphate 5 kinases (PIP5K). The reversibility of these processes is ensured by various phosphatases (Ptase). Phosphoinositide 3 kinases (PI3K) generate PtdIns(3,4)P_2 and PtdInse(3,4,5)P_3. Various PLC isoforms, activated by G Protein Coupled Receptors (GPCR), receptor tyrosine kinases (RTK) and other factors, hydrolyse PtdIns(4,5)P_2, and generate inositol 1,4,5 trisphopshate (IP_3) and diacylglycerol (DAG). IP_3 liberates Ca^{2+} from intracellular stores, and DAG activates protein kinase C (PKC). (From Rohacs (2007) with permission)

as PIP_2, is the substrate for phospholipase C (PLC) (Fig. 10.1) and constitutes up to 1% of the phospholipids in the plasma membrane, where it is localized in the inner, cytoplasmic leaflet. Its immediate precursor PtdIns(4)P is found at comparable quantities, whereas phosphatidylinositol is more abundant, but it is usually not efficient in regulating ion channels. Other phosphoinositides, such as PtdIns(3,4)P_2 and PtdIns(3,4,5)P_3, the products of PI3 Kinases are found in much smaller quantities even in stimulated cells than PtdIns(4,5)P_2. PtdIns(4,5)P_2 is the physiologically most important regulator of most studied ion channels, because it is more abundant, and/or more active than other phosphoinositides.

Tremendous progress has been made in the last one and a half decades in understanding the regulation of transporters and ion channels by these lipids. Several sporadic early studies reported the modulation of various membrane transporters by phosphoinositides, such as the plasma membrane Ca^{2+} ATP-ase; these studies have been thoroughly reviewed recently (Huang 2007). It was, however, Hilgeman's seminal paper in 1996 (Hilgemann and Ball 1996) reporting PIP_2 dependence of two cardiac ion transport proteins in excised patches, that sparked a new era of research on this field. Shortly after this discovery, a surprising number and variety of ion channels have been reported to be regulated by PIP_2. It currently seems that dependence of activity on PIP_2 is a property of a large number, if not the majority, of mammalian plasma membrane ion channel proteins (Gamper and Shapiro 2007a; Suh and Hille 2008). This chapter will discuss this topic using the example of four different groups of ion channels, all of which share the property of being regulated by membrane phosphoinositides (chiefly PIP_2) and illustrate various features of this paradigm.

Our first two examples are the inwardly-rectifying K^+ (Kir) channels and the M-type K^+ channels (Kv7 or KCNQ) for which the regulation by phosphoinositides is widely studied and relatively well understood. It is clear that all members of both families require the presence of PIP_2 for activity and depletion of the lipid in the plasma membrane inhibits them; there is also a consensus for the physiological role of the apparent affinity for PIP_2 in the regulation of these channels. The next example is the family of voltage-gated Ca^{2+} channels (VGCC) which also require PIP_2 for their activity although whether PIP_2 is a physiological regulator of VGCC activity in vivo remains under discussion. Finally our last example is the Transient Receptor Potential (TRP) channel family; these channels are in the focus of intense research, however, their phosphoinositide regulation is complex in many cases, not very well understood, and even controversial.

10.2 Tools to Study Phosphoinositide Regulation of Ion Channels

A powerful toolkit for studying regulation of ion channels by phosphoinositides has been developed over the recent years. Since a similar combination of methods and approaches has been applied to investigation of regulation by phosphoinositides of all channel families discussed here, we will briefly summarize them in this section.

(i) *Excised patch recordings.* In this experimental paradigm an inside-out excised patch configuration of the patch clamp technique is used. After the excision of the membrane patch, the activity of PIP_2 dependent ion channels decreases, a phenomenon referred to as run-down. For many PIP_2-sensitive channels the mechanism underlying run-down is the decrease in PIP_2 concentration in the patch due to the action of membrane-bound lipid phosphatases present in the patch membrane. Mg^{2+} applied to the intracellular surface of the patch accelerates channel run-down by providing a cofactor for lipid phosphatases (Huang et al. 1998) and also by the direct screening of the negative charges of the phosphoinositides (Suh and Hille 2007; Piron et al. 2010). MgATP can prevent current run-down, or re-activate the channels after run-down (Hilgemann 1997; Sui et al. 1998) by providing substrate for lipid kinases that regenerate PtdIns(4,5)P_2 from its precursors. Run-down can be slowed down with an inhibitory cocktail against lipid phosphatases (Hilgemann and Ball 1996; Zhang et al. 1999). This implies that most major kinases and phosphatases remain associated with the patch membrane even after excision. The velocity of the run-down of the activity of a given channel generally correlates with its apparent affinity for PIP_2; channels with higher PIP_2 affinity run down slower than channels with lower affinity. One can also apply PIP_2 chelating agents, such as PIP_2 antibody (Huang et al. 1998) or poly-Lysine (Lopes et al. 2002) to excised patches to accelerate run-down. Poly-Lys is less selective than the antibody but it works more reliably.

Perhaps the most direct way to study the effects of PIP_2 on ion channels is to apply phsophoinositides directly to the cytoplasmic surface of excised inside-out patches after current rundown. Phosphoinositides with various lipid side chains are available for these experiments; PIP_2 from natural sources has mainly arachidonyl-stearyl side chains, while synthetic PIP_2 usually contains two palmitoyl side chains. These long acyl chain lipids accumulate in the patch membrane, thus it is difficult to control their effective concentration. After activation with these PIP_2 analogues most ion channels run down quite slowly upon cessation of the application of the lipid, making repeated application of the these analogues impractical (Rohacs et al. 2002). Short acyl chain, water soluble (e.g. DiC_8) analogues of PIP_2 on the other hand activate most ion channels quickly and reversibly, presumably because they diffuse out the membrane easily upon washout (Rohacs et al. 1999). DiC_8 phosphoinositides are water soluble, whereas long acyl chain phosphoinositides are found in micelles in aqueous solutions; these solutions often need to be sonicated in order to prevent aggregation of the lipids. It is important to keep in mind that the patch membrane contains significant amount of PIP_2 in the cell-attached configuration.

(ii) *Biochemical approaches to study phosphoinositide binding to ion channels.* Several techniques have been used to measure direct biochemical binding of phosphoinositides to ion channels (Huang et al. 1998; Soom et al. 2001; MacGregor et al. 2002). Most of these studies were performed with truncated cytoplasmic segments of ion channels. The advantage of this approach is that it measures direct association of phosphoinositides with ion channels. On the other hand, it is possible that the binding to these isolated channel fragments does not correspond to the biologically important interactions. In several cases mutations that affected PIP_2 binding were reintroduced into the full length channel and functional effects were shown on phosphoinositide sensitivity (Huang et al. 1998). This is a strong argument for direct activation of a channel by PIP_2. A perhaps even stronger evidence for direct activation is the demonstration of the effect of the lipid on a purified channel reconstituted in the artificial membranes of known composition. Recently activation by PIP_2 of purified TRPM8 channels in lipid bilayers (Zakharian et al. 2009, 2010) and Kir channels in liposomes (D'Avanzo et al. 2010) have been demonstrated.

(iii) *Manipulations of phosphoinositide levels in living cells.* Only limited pharmacological tools are available to inhibit various enzymes involved in PIP_2 metabolism, and they are not very specific. At relatively low concentrations wortmannin (10–100 nM) and LY294002 (10 μM) are widely used as PI3K inhibitors. At higher concentrations (> 5 μM for wortmannin and > 100 μM for LY294002) they also inhibit PI4K isoforms (Balla 2001), thus slowly depleting PIP_2 and also preventing the recovery of PIP_2 levels after the PLC-mediated hydrolysis. PLC can be inhibited by U73122 and edelfosine, but these drugs have a number of side effects (Horowitz et al. 2005). In intact cells PIP_2 levels can be modified using a variety of tools. PIP_2 levels can be decreased by activating PLC via G-protein coupled receptors (PLCβ), receptor tyrosine kinases (PLCγ) or Ca^{2+} influx (probably PLCδ). However, PLCs not only hydrolyze PIP_2 but concurrently release IP_3 and DAG which, in turn, trigger their own signalling cascades (e.g. release of Ca^{2+} from the IP_3-sensitive stores, activation of PKC, arachidonic acid release etc.; Fig. 10.1) which often complicates interpretation of results. An alternative approach to modify PIP_2 levels in living cells is by over-expression of various lipid kinases and phosphatases. The first generation of such tools contained constitutively active enzymes; for instance a widely used approach to tonically deplete PIP_2 in cells is to overexpress construct contained PIP_2-specific 5' phosphotases of INP family from yeast (Stolz et al. 1998) tagged with GFP and a membrane localization sequence from tyrosine kinase Lyn. When overexpressed in cells such constructs localize to the plasma membrane and tonically deplete PIP_2 by converting it into PIP. Similarly, overexpression of PI4- and PIP5-kinases is used to tonically increase membrane PIP_2 levels. In a new generation of such probes, a phenomenon of chemically-induced dimeryzation (CID) has been utilized to make the lipid 5' phosphatase or kinase activity to become acutely inducible in living cell. These constructs were independently developed in the labs of Tobias Meyer (Suh et al. 2006) and Tamas Balla (Varnai et al. 2006). The PIP_2 depleting CID system uses two different proteins with high affinity to immunosuppressant rapamycin: the FRB domain of the mammalian target of rapamycin (mTOR) and the FK506 binding protein FKBP12. In a study by Suh and colleagues the rapamycin-binding domain of FRB was fused to the membrane-localisation tag of Lyn kinase while the rapamycin-binding domain of FKBP was attached to the 5' phosphatase Inp54p and CFP. When these constructs were co-transfected together with Kv7.2/Kv7.3 channels into ts-A cells, acute addition of the rapamycin analogue induced rapid recruitment of Inp54p to plasma membrane, dephosphorylation of PIP_2 and virtually complete inhibition of M channel activity (Suh et al. 2006). Similarly, TRPM8 channel activity was almost completely inhibited by rapamycin-induced PIP_2 depletion (Varnai et al. 2006). Another type of inducible lipid phosphatase which is used to study PIP_2-sensitivity of ion channels is voltage-sensitive phosphatase VSP which contains a voltage sensor domain homologous to those of voltage-gated ion channels and a 5'-phosphatase domain homologous to PTEN (Iwasaki et al. 2008). The phosphatase domain of VSP is inactive at potentials below 0 mV but can be activated by strong depolarization pulses (e.g. to above +40 mV) (Iwasaki et al. 2008), thus strong depolarizing pulses to above +40 mV induce rapid, reversible depletion of PIP_2 in cells overexpressing VSP; this, in turn, was shown to inhibit Kir and Kv7 channels (Murata and Okamura 2007; Falkenburger et al. 2010c).

(iv) *Optical probes for phosphoinositide metabolism.* Attachment of fluorescent proteins to different phosphoinositide-binding domains has been used for live monitoring of phosphoinositide levels in living cells. Thus, the pleckstrin homology (PH) domain of PLCδ1 fused with GFP (PLC-PH-GFP) has been widely used to monitor PIP_2 hydrolysis by PLC or dephosphorylation by Inp phosphatases. At basal conditions this probe localizes to the inner leaflet of the plasma membrane where it binds to PIP_2; when PIP_2 concentration in the membrane decreases (e.g. due to its hydrolysis to IP_3 and DAG by PLC), the probe translocates to the cytoplasm, which can be easily monitored using confocal microscope. Similar fluorescence resonance energy transfer

(FRET)-based variants of this probe have also been developed (van der Wal et al. 2001). These probes have been widely used for correlating kinetics of PIP_2 hydrolysis with the activity of ion channel of interest (Mitchell et al. 1996; Hsuan et al. 1998). The disadvantage of the PLC-PH-GFP probe is that it has higher affinity to IP_3 in vitro as compared to PIP_2 (Hirose et al. 1999; Varnai and Balla 1998) thus, the interpretation of its translocation in terms of PIP_2 levels is not straightforward (Varnai and Balla 2006; Liu et al. 2010). Recently a new PIP_2 probe that does not bind IP_3 has been developed (Quinn et al. 2008), it is based on the PIP_2 affinity of the transcription factor tubby. The probe is probably a better reporter of the membrane PIP_2 levels, although it is less sensitive an indicator of PLC activity compared to PLC-PH-GFP (Szentpetery et al. 2009; Liu et al. 2010). In addition to PIP_2, similar optical probes have been developed for other lipids, such as PtdIns(3,4)P_2, PtdIns(3,4,5)P_3, PtdIns(4)P, DAG and others (Balla and Varnai 2009; Balla 2009).

10.3 Inwardly Rectifying K^+ (Kir) Channels

Kir channels are K^+ selective ion channels that conduct more current in the inward than in the outward direction if measured through a range of voltages in patch clamp experiments. They have two transmembrane domains per subunit, four of which form the functional homo or heterotetrameric channels. Most of them are open near the resting membrane potential and they conduct outward currents in most cases, but they close down upon major depolarizations, thus they allow the development of the action potentials. The mammalian family has 15 members, divided into various subfamilies, numbered 1–7 (Hibino et al. 2010). Most Kir channels are constitutively active, with two exceptions. G-protein activated inwardly rectifying K^+ (GIRK) channels are members of the Kir3.x subfamily, while ATP inhibited K^+ (K_{ATP}) channels are members of the Kir6.x family. GIRK channels are activated by the $\beta\gamma$ subunits of heterotrimeric G-proteins and play roles in processes such as the regulation of heart rate by muscarinic stimulation and in the analgesic effects of opioids. K_{ATP} channels are inhibited by cytoplasmic ATP, and they are best known for their role in glucose-induced stimulation of insulin secretion. It has been shown that all members of the mammalian Kir channel family require PIP_2 for activity (Rohacs et al. 2003; Du et al. 2004). Mutations in Kir channel genes may cause a variety of disease, such as diabetes, hyperinsulinemia, Andersen's syndrome, Bartter's syndrome, and vitreoretinal degeneration (Hibino et al. 2010).

10.3.1 How Does PIP_2 Interact with Ion Channels?

This question is most thoroughly studied in Kir channels, and our understanding of how PIP_2 interacts with channels is probably the most comprehensive here. The generally accepted view is that the negatively charged head group of PIP_2 interacts with positively charged amino acid residues in the cytoplasmic domains of ion

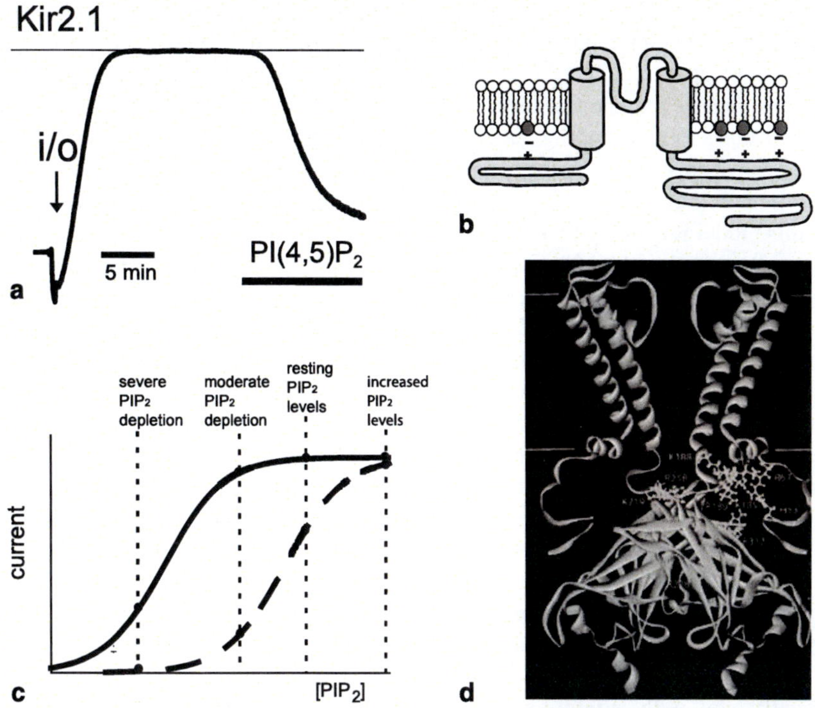

Fig. 10.2 Phosphoinositide regulation of Kir channels. **a** Kir2.1 activity in an excised macropatch from a Xenopus oocyte at −80 mV. Current activity runs down after establishment of the inside out configuration (i/o). The channels are re-activated by the application of PIP_2. (Modified from Rohacs et al. (1999) with permission). **b** Simplified cartoon showing that PIP_2 interacts with a Kir channel through binding of the negatively charged head-group of the lipid to positively charged residues in the C- and N-termini of the channel protein. **c** The effect of the apparent affinity of the channel for PIP_2. (From Rohacs (2007) with permission). Hypothetical concentration dependence of the effect of PIP_2 on a channel with high (*solid line*), and low (*dashed line*) apparent affinity for PIP_2. **d** A 3D model of Kir2.1 channels with putative PIP_2 interacting residues, based on partial crystal structures of various Kir channels. (From Logothetis et al. (2007a) with permission)

channels (Fig. 10.2b, d). Early work identified residues in the proximal C-terminus, close to the pore- forming second transmembrane domain that are involved in PIP_2 interactions (Fan and Makielski 1997; Hilgemann and Ball 1996). Later several studies systematically mutated conserved positively charged residues in Kir channels to identify additional PIP_2 interacting residues. As a general rule neutralizing a PIP_2 interacting residue decreases Kir channel apparent PIP_2 affinity (Fig. 10.2c), which is manifested in the decrease of channel open probability, decreased macroscopic current amplitude, increased speed of run down in excised patches, and increased inhibition by depletion of PIP_2 (see below). The putative PIP_2 interacting residues identified this way were located in various places in the linear sequence, including more distal regions in the C-terminus, and residues in the N-terminus (Lopes et al. 2002).

When the partial crystal structures of various Kir channels were published, most of the putative PIP_2 interacting residues identified earlier by mutagenesis (Lopes et al. 2002) lined up on the interface of the channel with the membrane, compatible with the idea that they are part of the PIP_2 binding site. This shows that the mutagenesis approach is useful in finding putative PIP_2 interacting residues. A homology model based on partial crystal structures of various Kir channels has been proposed to depict PIP_2 interacting residues in Kir channels (Logothetis et al. 2007b) (Fig. 10.2d). Based on this model it is likely that positively charged residues from different parts of the channel come close together in 3 dimensions to form a PIP_2 binding pocket.

Even though this relatively simple model is quite widely accepted, the real picture may be somewhat more complex, thus, several points of caution need to be made. First, the phosphoinositide specificity profile of Kir channels is variable, some channels, such as K_{ATP} are activated equally well by PtdIns(4,5)P_2, PtdIns(3,4)P_2 and PtdIns(3,4,5,)P_3, whereas others such as Kir2.1, are activated specifically by PtdIns (4,5)P_2, but not by the other two lipids (Rohacs et al. 2003). This stereospecific activation is difficult to explain with a purely charge mediated binding. Second, crystal structures of known phosphoinositide interacting soluble proteins show that in addition to positively charged amino acids, non-charged residues also invariably contribute to phosphoinositide binding (Rosenhouse-Dantsker and Logothetis 2007). Virtually no effort has been made so far to identify non-charged residues that interact with phosphoinositides in ion channels. Third, even when a positively charged residue is identified, mutation of which alters channel activation by PIP_2, it is very difficult to tell with certainty, based on mutagenesis data whether this residue interacts directly with PIP_2, or its mutation alters PIP_2 interactions indirectly (Colquhoun 1998). Even when the crystal structure of the channel is solved without the interacting lipid, it is not trivial to dock PIP_2 and tell which residue it is in contact with.

10.3.2 *The Role of Apparent Affinity for PIP_2 and Relationship to Other Regulators*

Under resting conditions the inner leaflet of the plasma membrane contains significant amounts of PIP_2. Whether this is enough to keep a particular PIP_2 sensitive channel maximally open, depends on the apparent affinity of the channel for the lipid (Fig. 10.2c). Channels with high affinity for PIP_2 cannot be further activated by excess PIP_2 because resting PIP_2 levels are saturating them. On the other hand, channels with lower PIP_2 affinity can be theoretically further activated by increased PIP_2 levels. Conversely, and more importantly, channels with lower PIP_2 affinity can be easily inhibited by depletion of PIP_2, whereas channels with high PIP_2 affinity are not inhibited significantly by moderate (physiological) PIP_2 depletions (Fig. 10.2c). However, even high affinity channels can be inhibited by complete depletion of PIP_2, such as by applying a PIP_2 chelator in excised patches.

Mutation of PIP_2 interacting residues may convert a high affinity channel to a low affinity one, shifting its PIP_2 dose-response to the right and rendering it more sensitive to inhibition by PIP_2 depletion. This phenomenon is utilized in mutation studies aiming at locating putative PIP_2 interacting residues (Lopes et al. 2002). In several Kir channels mutations are also found that strengthen PIP_2 interactions, leading to decreased sensitivity for PIP_2 depletion (Zhang et al. 1999; Du et al. 2004).

Even in the absence of mutations, the apparent affinity for PIP_2 is not necessarily static. A classic example of PIP_2 affinity modulated by a channel ligand is GIRK channels, where it was proposed that both $G_{\beta\gamma}$ and Na^+ open the channels by stabilizing its interaction with PIP_2 (Huang et al. 1998; Zhang et al. 1999). This would manifest as a left shift in the PIP_2 dose-response, and increased channel activity at a constant PIP_2 concentration. Consistently with this, GIRK channel currents are inhibited less by PLC induced PIP_2 depletion in the presence of excess $G_{\beta\gamma}$ (Keselman et al. 2007). Intracellular Na^+, another activator of several GIRK channels has a similar effect (Zhang et al. 1999). For Na^+, a compelling mechanistic model was proposed recently to explain how it increases PIP_2 sensitivity of Kir3.4 channels: Na^+ binds to an Asp and His that triggers a structural switch that frees a crucial Arg enabling it to interact with PIP_2 (Rosenhouse-Dantsker et al. 2008). A similar model was proposed later for Kir3.2, based on crystallographic studies (Inanobe et al. 2010). Many other modulators of Kir channels, such as protein kinase C, intracellular Mg^{2+} and pH have also been reported to affect channel PIP_2 interactions (Du et al. 2004; Keselman et al. 2007).

10.3.3 Metabolic Regulation and Phosphoinositides—K_{ATP} Channels

Kir6.x channels are the pore-forming subunits of K_{ATP} channels. They are considered to be metabolic sensors directly inhibited by cytoplasmic ATP and they open in conditions when cytoplasmic ATP concentrations decrease. Functional K_{ATP} channels have an auxiliary subunit, the sulfonylurea receptor (SUR). ATP inhibits the channel through direct binding to the pore-forming Kir6.2 subunit and the SUR subunit modifies this effect. ADP on the other hand activates the channels through the SUR subunit, and the K_{ATP} channels are considered to be sensors of cellular ATP/ADP ratio. The two best characterized combinations are Kir6.2–SUR1, the K_{ATP} channel in insulin secreting pancreatic beta cells and the cardiac Kir6.2–SUR2A, found in ventricular cardiomyocytes. The physiological function of the pancreatic K_{ATP} channel is very well established, these cells respond to physiological changes in extracellular glucose concentrations by changes in intracellular ATP levels. An increase in extracellular glucose increases the ATP/ADP ratio inside these cells, leading to the closing of K_{ATP} channels, depolarizing the membrane potential and the consequential opening of voltage gated Ca^{2+} channels which, in turn, stimulates insulin secretion. In other cell types such as cardiomyocytes, physiological changes in cellular metabolism are not expected to change cellular ATP levels. K_{ATP} channels there are likely to act as brakes on cellular metabolism under severe metabolic conditions, such as ischemia, when they open, hyperpolarize the cell, and thus limit further activity.

K_{ATP} channels require PIP_2 for activity, and their phosphoinositide regulation is intimately related to their metabolic regulation. In excised patches ATP sensitivity of these channels show a marked reduction after application of phosphoinositides (Shyng and Nichols 1998; Baukrowitz et al. 1998) and it was proposed that different phosphoinositide levels among different cells may underlie the well known variability of ATP sensitivity in excised patches. Both ATP and the head-group of PIP_2 are highly negatively charged, and binding of both molecules to K_{ATP} channels is thought to involve positively charged residues. Mechanistically, it is possible that PIP_2 and ATP bind to overlapping binding sites, and binding of ATP displaces the activating lipids head-group. Another important activator of K_{ATP} channels is long acyl chain coenzyme-A (LC-CoA) (Tucker and Baukrowitz 2008; Shumilina et al. 2006). Even though originally it was proposed that PIP_2 and LC-CoA activates K_{ATP} channels via different mechanisms (Gribble et al. 1998), there has been a growing consensus that the negatively charged LC-CoA acts through the phosphoinositide binding site of Kir6.2 (Tucker and Baukrowitz 2008), based mainly on the following data. Most Kir channels show some level of isomer specificity among various phosphoinositides, with PtdIns(4,5)P_2 being the most active, and are inhibited by LC-CoA (Rohacs et al. 2003; Rapedius et al. 2005). Kir6.2 channels on the other hand show no isomer selectivity among phosphoinositides, and are activated by LC-CoA (Rohacs et al. 2003). When Kir2.1 and Kir7.1 channels were engineered to be less selective among various isomers of PIP_2, they were activated by LC-CoA (Rohacs et al. 2003). Furthermore, it was demonstrated that PIP_2 binding to the C-terminus of Kir6.2 and Kir2.1 is antagonized by LC-CoA (Rapedius et al. 2005).

10.4 M-type (Kv7, KCNQ) Channels

Following Kir channels, the Kv7 K^+ channel family gives another example of 'classical' PIP_2-sensitive channels for which phosphoinositide binding site has been suggested and physiological role of the channel-PIP_2 interaction has been confirmed. The direct PIP_2-dependency of Kv7 open probability even allowed some researchers to use Kv7 channels as biosensors of plasma membrane PIP_2 levels (much like for the case of Kir channels; see e.g. (Suh et al. 2006; Murata and Okamura 2007)).

In mammals there are five KCNQ genes (KCNQ1-5) coding for five Kv7 α-subunits (Kv7.1–Kv7.5) which give rise to several physiologically important potassium currents. In the mammalian central and peripheral nervous systems Kv7.2, Kv7.3 and Kv7.5 form the so-called 'M-type channels' underlying neuronal M current, an important cellular instrument for stabilizing neuronal resting membrane potential, setting the threshold for action potential firing and controlling firing frequency (Wang et al. 1998; Shapiro et al. 2000; Selyanko et al. 2002); reviewed in (Delmas and Brown 2005; Brown and Passmore 2009). The M current was discovered some 30 years ago by David Brown and colleagues (Brown and Adams 1980) as a specific K^+ current fraction in sympathetic neurons which is characterized by slow kinetics of activation and inactivation, very negative (negative to −60 mV) threshold for activation and no inactivation under physiological conditions (Fig. 10.3a). In the original study (Brown and Adams 1980) this current fraction was eliminated by stimulation of muscarinic acetylcholine receptors (mAchR), hence, it received the name 'M current'. Distinctive biophysical properties of neuronal M channels bestow them a strong control over neuronal excitability. Thus, the negative threshold for activation and no inactivation allows a fraction of M channels to be open near the resting membrane potential of a neuron whilst slow kinetics of activation confers a role in the accommodation (wearing-off) within the bursts of action potentials (Delmas and Brown 2005; Brown and Passmore 2009). The importance of M currents in mammalian CNS is exemplified by the fact that loss-of-function mutations within principal M channel genes KCNQ2 and KCNQ3 often result in a form of epilepsy, benign familial neonatal convulsions (BFNC) and even mutations causing as little as 25% of the M current reduction are sufficient to cause a disease (Maljevic et al. 2010); genetic deletion of KCNQ2 in mice is lethal (Watanabe et al. 2000). The general 'rule of thumb' is that neurons expressing high levels of M-current-forming Kv7 channels are 'phasic' neurons with high threshold for action potential firing (Jia et al. 2008); acute pharmacological or receptor-induced inhibition of M current (Jia et al. 2008; Liu et al. 2010) or genetic downregulation of KCNQ expression (Mucha et al. 2010) can switch these neurons into highly excitable, constantly firing ('tonic') phenotype which explains KCNQ-associated seizures as well as recently reported role of M channels in pain (Linley et al. 2008; Liu et al. 2010; Mucha et al. 2010).

Another important K^+ current conducted by a member of Kv7 family, Kv7.1, can be found in the heart. In cardiomyocytes Kv7.1 multimerizes with its auxiliary subunit, KCNE1 to produce the slow component of the cardiac delayed rectifier current (I_K), I_{Ks} (Barhanin et al. 1996; Sanguinetti et al. 1996; Wang et al. 1996). The I_{Ks} is responsible for the repolarization of the cardiac action potential and for the control of action potential duration (reviewed in Charpentier et al. (2010)). The loss-of-function mutations within KCNQ1 gene often result in the group of cardiac arrhythmias called inherited long QT syndrome form 1 (e.g. the autosomal dominant Romano-Ward syndrome and the autosomal recessive Jervel and Lange-Nielsen syndrome); the gain-of-function KCNQ1 mutations have also been reported and these result in familial atrial fibrillation and another form of arrhythmia—short QT syndrome (reviewed in Charpentier et al. (2010)).

Kv7.4 is a Kv7 subunit which is predominantly expressed in the auditory pathways and loss-of-function mutations within the KCNQ4 result in DFNA2 nonsyndromic hearing loss (Kubisch et al. 1999). KCNQ4 is abundantly expressed in the inner ear, particularly in the outer hair cells (OHCs) of the organ of Corti (Kubisch et al. 1999) as well as in several nuclei and tracts of the auditory pathways in the brainstem (Kharkovets et al. 2000). In OHCs Kv7.4 localizes to the basal membrane and might provide a pathway for the extrusion of potassium entering OHCs through the mechanosensitive channels at the apical membrane (Kharkovets et al. 2000, 2006).

In addition to the neuronal and cardiac roles, several Kv7 subunits are expressed in smooth (Greenwood and Ohya 2009) and skeletal (Iannotti et al. 2010) muscles

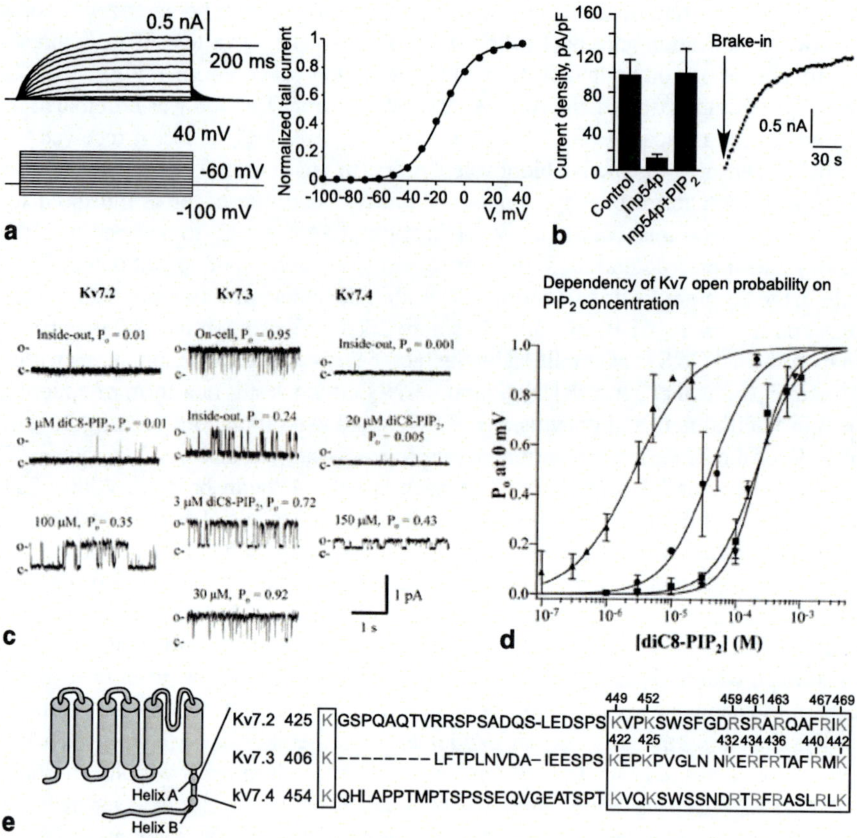

Fig. 10.3 PIP_2 sensitivity of the Kv7 channels. **a** Whole-cell current traces elicited in CHO cells overexpressing Kv7.5 by the train of voltage pulses depicted underneath. Activation curve (normalized tail current amplitudes plotted against voltage) is given on the right. **b** Inhibition of Kv7.2/Kv7.3 current in CHO cells by the PIP_2 depletion with Inp54p phosphatase. Cells were transfected with plasmids coding for KCNQ2, KCNQ3 with or without membrane-targeted Inp54p (*Inp54p*). Bars represent Kv7.2/7.3 current densities recorded in the whole-cell patch clamp mode from the cells transfected with KCNQ2, KCNQ3 only (Control) or with KCNQ2, KCNQ3 and Inp54p (Inp54p) or from KCNQ2, KCNQ3, Inp54p transfected cells with 100 μM DiC_8-PIP_2 added to the pipette solution. Trace on the left represent the time course of Kv7.2/7.3 current recovery in Inp54p-overexpressing cell upon breaking into whole cell with a pipette solution containing 100 μM DiC_8-PIP_2. (Modified from Linley et al. (2008) with permission). **c** PIP_2 sensitivity of the Kv7 channels studied using singe-channel recordings. Individual traces recorded in the cell-attached ('on-cell') or inside-out configurations (as indicated) from patches of CHO cell membrane containing single Kv7.2 (*left*), Kv7.3 (*middle*) or Kv7.4 (*right*) channels in the presence of the indicated concentrations of DiC_8-PIP_2. (From Li et al. (2005) with permission). **d** A dependency of the P_o of Kv7.2 (*squares*), Kv7.3 (*triangles*), Kv7.4 (*inverted triangles*) and Kv7.2/Kv7.3 (*circles*) from the DiC_8-PIP_2 concentration in inside-out patches. (From Li et al. (2005) with permission). **e** Alignment of the putative PIP_2-interacive domains of Kv7.2, Kv7.3 and Kv7.4 channels. The *large* and *small boxes* enclose a cluster of positively charged residues and a critical conserved basic residue, respectively. (Based on the data from Hernandez et al. (2008a))

and in epithelia (Vallon et al. 2005) where they may contribute to the control of contractility (Greenwood and Ohya 2009), skeletal muscle proliferation (Iannotti et al. 2010), transepithelial transport (Vallon et al. 2005) and cell volume regulation (Piron et al. 2010).

As discussed above, a number of important physiological roles for Kv7 channels have been identified, accordingly, mutations or other impairments of Kv7 channels often result in severe disorders (e.g. seizures, pain, arrhythmias and deafness), therefore Kv7 channel regulation (to which PIP_2 plays one of the key roles) has attracted some high-profile research.

10.4.1 M Channel Modulation: Focus on PIP_2

The M current has been discovered as a neuronal K^+ current fraction inhibited by mAchRs. Later it became apparent that not only mAchRs (more precisely, M_1, M_3 and M_5 mAchR isoforms), but also receptors for bradykinin, angiotensin II, histamine, protease activated receptor-2 (PAR-2), P2Y receptors and potentially any other GPCR that is coupled to the $G_{q/11}$ subtype of G proteins can inhibit M channels (reviewed in (Delmas and Brown 2005; Gamper and Shapiro 2007a; Linley et al. 2010)). Deciphering the signalling pathways linking GPCR and M channels took some time though.

Before the KCNQ genes have been cloned, most of the M current research has been performed on sympathetic neurons which express robust M currents. It was soon discovered that in these neurons the muscarinic inhibition of M current is mediated by M_1 mAchR (Marrion et al. 1989) and require G_q or G_{11} type of G_α subunits (Haley et al. 1998, 2000) and their usual downstream effector, PLCβ. An important experiment by A. Selyanko in David Brown's group demonstrated that external application of muscarinic agonist inhibited M channels isolated in cell-attached patches from the superior cervical ganglion (SCG) sympathetic neurons. Based on the distinction first made by Soejima and Noma (1984) between "membrane-delimited" and "diffusible messenger" signalling (Soejima and Noma 1984), it was concluded that muscarinic M current inhibition must be mediated by a diffusible intracellular second messenger (Selyanko et al. 1992). It required an additional 10 years of concentrated effort before this elusive "mystery" messenger has been identified and, as it often happens, it turned out to be not quite what everyone has been looking for.

After all 'usual suspects' of the PLC signalling cascade (Fig. 10.1) have been exhaustively probed and failed to satisfy the experimental data (see e.g. (Robbins et al. 1993; Hille 1994; Marrion 1997)), complementary studies by the Hille and Logothetis groups came to a suggestion that the actual mediator of PLC-induced M current inhibition may not be a downstream product of PIP_2 hydrolysis but the PIP_2 hydrolysis itself. Indeed, the recovery of M current amplitude from the mAchR-mediated inhibition was shown to be prevented by blocking PIP_2 resynthesis with a PI4-kinase inhibitor (Suh and Hille 2002; Zhang et al. 2003). Furthermore, the application of standard phosphoinositide research toolkit soon revealed that Kv7/M channels (both in the expression systems and in SCG neurons) are indeed highly PIP_2 sensitive. Thus, currents from cloned Kv7.2/Kv7.3 channels expressed in *Xenopus* oocytes ran-down upon patch excision and this run-down was successfully prevented or attenuated by addition of PIP_2 (or an analog) to the inner leaflet of the plasma membrane; Kv7 current in excised patches was promptly inhibited by PIP_2 scavengers such as anti-PIP_2 antibody and polylysine (Zhang et al. 2003). Overexpression of Inp54p phosphotase tonically inhibited Kv7.2/7.3 currents in CHO cells while perfusion of DiC_8-PIP_2 through the patch pipette rapidly recovered current amplitude in the Inp54p-overexpressing cells (Fig. 10.3b; (Li et al. 2005; Linley et al. 2008)). Similar experiments were later repeated for Kv7.1 and IKs channels (Loussouarn et al. 2003; Piron et al. 2010) and a similar PIP_2 requirement for channel activity has been seen (with the exception of voltage dependency of the PIP_2 effect seen with Kv7.1 channels, see below).

Development of new optical and biochemical tools for monitoring and manipulating PIP_2 levels in living cells allowed researchers to further probe the relationships between receptor-mediated PIP_2 hydrolysis and Kv7 channel activity. Thus the PLCδ–PH probes have been extensively used to correlate the PIP_2 hydrolysis by PLC with the Kv7/M current inhibition in the expression systems and neurons (e.g. Horowitz et al. 2005; Winks et al. 2005; Falkenburger et al. 2010c); indeed, these experiments found a good correlation between the kinetics of both processes (especially in the expression systems) and allowed to compose a detailed kinetic models taking into account affinities, abundances and kinetics of interaction for different key players of $G_{q/11}$ signalling (such as receptors, G_{α}, $G_{\beta\gamma}$, PLC, etc.), PIP_2 and Kv7 subunits (Hernandez et al. 2009; Falkenburger et al. 2010b).

A significant complication of PLC signalling, which for some time casted a cloud of doubt over the 'PIP_2 hypothesis', is the fact that the PIP_2 hydrolysis is always accompanied by the concurrent release of several second messengers (Fig. 10.1) and some of them were also shown to cause M current inhibition (e.g. Ca^{2+} (Selyanko and Brown 1996; Cruzblanca et al. 1998; Gamper and Shapiro 2003; Gamper et al. 2005) and DAG/PKC (Hoshi et al. 2003)). Thus, while PIP_2 sensitivity of Kv7 channels *per se* was convincingly demonstrated, it was not clear whether receptor-mediated Kv7/M current inhibition in native cells can be solely mediated by the PIP_2 depletion. To a certain degree this question has been clarified with the use of inducible 5'-phosphatases (VSPs and rapamycin-inducible CID system). Unlike PLC, these phosphatases convert PtdIns(4,5)P_2 to PtdIns(4)P without the release of any relevant second messengers; nevertheless, both type of inducible phosphatases were shown to be able to inhibit Kv7/M channel activity almost completely (Suh et al. 2006; Murata and Okamura 2007; Falkenburger et al. 2010c). There were also other types of experiments which solidified the 'PIP_2 hypothesis'. Thus, application of highly-basic palmitoilated PIP_2-binding peptides reduced M current in SCG neurons and sensitized the current to depression by muscarinic stimulation (Robbins et al. 2006). In contrast, when tonic membrane PIP_2 levels were elevated by over-expression of PIP5-kinase, the tonic amplitude of overexpressed Kv7.2/7.3 channels dramatically increased (Li et al. 2005). A similar maneuver reduced the extent of muscarinic suppression of M current in sympathetic neurons (Winks et al. 2005).

10.4.2 Structure-functional Aspects of Kv7 Channel Sensitivity to PIP_2

Single channel recordings from the cells expressing cloned Kv7 channels revealed that homomeric channels assembled from the individual Kv7 subunits have distinct and highly-variable maximal open probability (P_o) in cell attached patches (Selyanko et al. 2001; Li et al. 2004). Thus, the tonic P_o (at saturating voltages) of Kv7.2 and Kv7.4 is very low (~0.1), the P_o of Kv7.3 is near unity and the P_o of Kv7.2/7.3 heteromultimers is in the range of 0.3 (Selyanko et al. 2001; Li et al. 2004, 2005). In a series of inside-out single channel experiments Li et al. (2005) found out that the P_o of Kv7 channels tested (Kv7.2, Kv7.3, heteromeric Kv7.2/Kv7.3 and Kv7.4) can be interpreted as a Hill function of the DiC_8-PIP_2 concentration (with Hill coefficients between 1 and 1.7; Fig. 10.3c, d). Interestingly, these experiments also revealed that Kv7 channels with low tonic $P_o$max (Kv7.2 and Kv7.4) had approximately 100 times lower apparent PIP_2 affinity as compared to Kv7.3 which tonic P_o max ≈ 1 (DiC_8-PIP_2 $EC_{50} \sim 200\,\mu M$ *vs.* $\sim 2\,\mu M$); Kv7.2/7.3 heteromultimers had an intermediate values for both the PIP_2 affinity ($EC_{50} \sim 40\,\mu M$) and $P_o$max (~0.3), in accord with them being heteromeric channels containing subunits with both high and low PIP_2 affinity. Thus, it has been suggested that the tonic activity of Kv7 subunits depends directly on their apparent PIP_2 affinity and on the tonic concentration of PIP_2 in the plasma membrane. Furthermore, it was hypothesized that the P_o of Kv7 channels is directly governed by membrane PIP_2 abundance.

The different intrinsic affinity for PIP_2 of different Kv7 channels implies that M channels assembled from different Kv7 subunits should respond to muscarinic stimulation with different sensitivities, and indeed, this is what has been observed: the concentration-dependency of the inhibition of homomeric Kv7.3 and Kv7.4 channels by M_1 mAChR agonist Oxotremorin-M had IC_{50} of 1 μM and 66 nM respectively (Hernandez et al. 2009).

The single channel analysis of Kv7 channel PIP_2 dependence has been extended in the further work by the Shapiro group which used chimeras between high- and low-PIP_2-affinity Kv7 isoforms (Kv7.3 and Kv7.4) to pin-point a site of PIP_2 binding within the Kv7 channels. This chimeric approach in combination with point-mutations, homology modeling and energy minimization analysis revealed a cluster of positively-charged amino acids within the linker between the first two (out of four) helical domains of Kv7 carboxy-termini (helixes A and B) as such PIP_2 binding site (Fig. 10.3e (Hernandez et al. 2008a)). The motif identified in Kv7 channels contained conserved K/R residues at the positions (in Kv7.2) 425, 452, 459, 461, 463 and 467 (Fig. 10.3e) which were suggested to play a key role in the channel interaction with PIP_2. Homology modelling based on the solved structure of the PIP_2 binding sites of Kir2.1 (Pegan et al. 2005) implied that Kv7 channels may have PIP_2-binding modules which structurally are similar to Kir2.1. In another study (Zhang et al. 2003) a more proximal positively charged residue of the C-terminus (H328 of Kv7.2) has been suggested to participate in the channel interaction with PIP_2. It is thus conceivable that while chimeric approach did identify some core PIP_2 binding residues within Kv7 channel, there may be some other regions within Kv7 channel proteins that participate in the interactions with PIP_2; in addition a caution needs to be taken regarding the modelling of Kv7 channels on the basis Kir channel structure as the homology between these two channel families is not close.

Interestingly, while swapping the A-B linker between Kv7.4 and Kv7.3 does invert PIP_2 sensitivity of the channels, the suggested PIP_2-binding K/R residues are conserved among the Kv7.2-Kv7.5 channels (but not in Kv7.1, see below). Thus, it is still unclear if the strikingly different apparent PIP_2 affinities of individual Kv7 subunits (e.g. ~100-fold difference in DiC_8-PIP_2 EC_{50} between Kv7.4 and Kv7.3) arise from the different biochemical *binding affinities* of individual PIP_2 binding sites or from the divergent *coupling efficiencies* between the PIP_2-binding domains and the gating machinery of the channel (Hernandez et al. 2008a).

Kv7.1 is a member of Kv7 family which in many structural and functional aspects stands apart from the rest of the family (e.g. it is the only Kv7 channel that inactivates, it is not inhibited by Ca^{2+} etc.); the part of the Kv7.1 C-terminus which is homologous to the putative PIP_2 binding motif of Kv7.2–Kv7.5 carries much less similarity with the rest of the family. Accordingly, the putative PIP_2-interacting residues that were identified within the Kv7.1 are distributed more diffusely. Among three putative PIP_2-interacting residues identified within the Kv7.1 two belong to the helix B (R539 and R555) of the C-terminus, another putative site was identified as an arginine within the S4-S5 linker (R243) (Park et al. 2005).

Interestingly, PIP_2 dependency of the Kv7.1 channel gating shows a noticeable dissimilarity from that of other Kv7s: the action of PIP_2 on Kv7.2–Kv7.4 comprises of the voltage-independent increase in channel P_o while voltage-dependence and kinetics of channel activation and deactivation is not affected (Li et al. 2005; Delmas et al. 2005); in contrast, binding of PIP_2 to Kv7.1 induces negative shift in voltage-dependence and slows deactivation (Loussouarn et al. 2003; Piron et al. 2010). While it is accepted that for all Kv7s PIP_2 acts to stabilise the open state of the channel, the difference in the effect of PIP_2 on channel gating further highlights likely structural dissimilarity in PIP_2 action on KV7.1 and the rest of Kv7 family.

10.4.3 PIP_2 Depletion by GPCR Activation in Neurons—Is It Really Happening?

While experiments with inducible phosphatases did unambiguously prove that Kv7 channels can be acutely inhibited by PIP_2 depletion in living cells, what these experiments did not prove is whether physiological M channel inhibition by the PLC-coupled GPCR in vivo is indeed mediated by PIP_2 depletion or, to put it differently, whether GPCR activation in neurons can produce enough PIP_2 depletion to inhibit native M current without the need for other second messengers. For the muscarinic suppression of M current in sympathetic SCG neurons the answer is most likely 'yes' as the other second messengers produced by the PLC hydrolysis were ruled out by exhaustive experimentation (reviewed in Delmas et al. (2005);

Gamper and Shapiro (2007b)), moreover, M_1 receptors in SCG do not couple to IP_3-sensitive Ca^{2+} stores and do not release Ca^{2+}, which is another potent inhibitor of M channels (Selyanko and Brown 1996; Cruzblanca et al. 1998; Gamper and Shapiro 2003; Zaika et al. 2007). However, in more general terms the answer is probably "no" or "not quite" as even the pioneers of 'PIP_2 hypothesis', Hillgeman (Hilgemann et al. 2001) and Hille (Falkenburger et al. 2010a) acknowledge that it is very difficult to envision a PLC-mediated PIP_2 depletion as a specific signalling mechanism on its own as strong PIP_2 depletion would simultaneously 'shut down' too many membrane proteins. Accordingly, a closer look at the modulation of native M currents by other $G_{q/11}$-PLC-coupled receptors revealed more complex nature of this signalling cascade. Thus, endogenous B_2 (bradykinin) and P2Y (purinergic) receptors in SCG neurons do induce IP_3-mediated rises in cytosolic Ca^{2+} and only weakly suppress M current if intracellular Ca^{2+} is held constant, IP_3 receptors are blocked, Ca^{2+} stores are depleted or when an IP_3 phosphatase or an IP_3 sponge is over-expressed (Shapiro et al. 1994; Cruzblanca et al. 1998; Delmas et al. 2002; Gamper and Shapiro 2003; Zaika et al. 2006, 2007). Similarly, in sensory neurons from dorsal root ganglia (DRG), bradykinin B_2 (Liu et al. 2010) and PAR-2 receptors (Linley et al. 2008) robustly inhibit native M current but mostly via Ca^{2+}-mediated mechanism while saturation of the plasma membrane with the excess of DiC_8-PIP_2 by the intracellular dialysis only marginally reduces such inhibition (Linley et al. 2008). Moreover, study by Liu and colleagues suggested that the degree of membrane PIP_2 depletion estimated with the optical probes based on the PH domain of PLCδ (hitherto a major PIP_2 probe used by many labs) is likely to be overestimated as this probe has higher affinity for IP_3 than for PIP_2 (Hirose et al. 1999). Indeed, translocation of the probe from membrane to the cytosol may not necessarily indicate a significant drop in membrane PIP_2 levels, as IP_3, the hydrolysis product of PIP_2, may also cause the probe to translocate (Gamper et al. 2004; Liu et al. 2010). Accordingly, in DRG neurons PLCδ-PH probe robustly translocated to cytosol in response to bradykinin stimulation but another PIP_2 probe, YFP-tubby, which does not bind IP_3 (Quinn et al. 2008), did not translocate unless exogenous B_2 receptors are overexpressed (Liu et al. 2010). These observations suggest that in DRG neurons bradykinin induces enough PIP_2 hydrolysis to produce IP_3 necessary for Ca^{2+} release from the stores and to cause PLCδ-PH probe to translocate but the overall drop in the membrane PIP_2 level is not sufficient to cause YFP-tubby probe translocation or to significantly inhibit M current. Thus it is likely that for the many PLC-mediated signalling pathways PIP_2 depletion is a contributing factor but not a sole mediator of M current inhibition. In a most likely scenario, activation of PLC by a GPCR concomitantly triggers three different signals that modulate M channel activity in a cumulative way: (i) some drop in membrane PIP_2 (probably localized, although see (Gamper and Shapiro 2007b) for discussion of problems with local PIP_2 depletion); (ii) release of Ca^{2+} from intracellular stores, Ca^{2+}-bound calmodulin then inhibits M channels (Gamper and Shapiro 2003; Gamper et al. 2005); (iii) activation of PKC and AKAP-dependent phosphorylation of M channel protein (Hoshi et al. 2003; Bal et al. 2010). These concurrent pathways are ultimately interrelated as phosphorylation of Kv7.2 by PKC increases the sensitivity of this M channel to muscarinic inhibition (presumably by decreasing channel affinity to PIP_2) (Hoshi et al. 2003; Bal et al. 2010), likewise, since the suggested calmodulin- and PIP_2 binding sites are in close proximity or overlap (Yus-Najera et al. 2002; Gamper and Shapiro 2003; Hernandez et al. 2008a), calmodulin binding to M channel could compete PIP_2 off the channel and thus reduce channel PIP_2 affinity (the opposite should also hold true: PIP_2 depletion should increase the affinity for calmodulin binding) (Gamper and Shapiro 2007a). This putative 'coincidence detection' mechanism may insure the fidelity and specificity of PLC-mediated regulation of M channels (see more on this issue in our recent reviews (Gamper and Shapiro 2007a, b)).

10.4.4 Physiological Significance of PLC-mediated M Channel Inhibition

This topic has been discussed at length in many recent reviews (e.g. Delmas and Brown 2005; Gamper and Shapiro 2007b; Hernandez et al. 2008b; Brown and Passmore 2009) therefore here we will just briefly outline the major concepts: (i) PLC-mediated M-type channel inhibition underlies the excitatory action of neurotransmitters (acetylcholine) and neuropeptides (e.g. bradykinin and angiotensin II; reviewed in Delmas and Brown (2005)); (ii) muscarinic inhibition of presinaptic M currents has been suggested to facilitate neurotransmitter release (Hernandez et al. 2008b; Kubista et al. 2009); (iii) in the PNS inhibition of M channels in nociceptive sensory fibers by the inflammatory mediators bradykinin and proteases mediates acute inflammatory pain (Linley et al. 2008, 2010; Liu et al. 2010).

10.5 Voltage-gated Ca^{2+} Channels

Although PIP_2 sensitivity of voltage-gated Ca^{2+} channels is less understood than that of Kv7s, the research in both fields historically paralleled each other in many ways as both VGCC and M channels are inhibited by M_1 AchR stimulation in SCG neurons and a common second ('mystery') messenger has been suggested (Bernheim et al. 1991; Mathie et al. 1992; Hille 1994).

VGCC form a large family of voltage gated ion channels which are selectively permeable to Ca^{2+}. VGCC are expressed in all types of excitable cells where they mediate release of neurotransmitters from synaptic terminals, secretion of neuromediators and hormones by neurons and neuroendocrine cells, excitation-contraction coupling and gene expression (see (Catterall 2000) for review). The family contains 'high-voltage-activated' channels (L-, N-, P/Q- and R-type) which are activated by strong depolarizations (above $\sim$$-30$ mV), and 'low-voltage-activated' T-type channels which are activated at more negative voltages (threshold voltage $\sim$$-50$ mV). The assembly of a VGCC is quite complex, it contains a pore-forming subunit $\alpha 1$, which has 24 transmembrane domains (TMD) organized in four 6-TMD repeats (with each 6-TMD repeat being analogous to a single α subunit of a voltage-gated K^+ channel). One $\alpha 1$ subunit is sufficient to provide a pore-forming channel core, however, functional VGCC are usually assembled with auxiliary subunits: β, α2δ and, in some cases, γ (with the exception of the T-type channels α1 subunit, which is sufficient to form functional channel). There are three groups of α1 subunits: Cav1.1–Cav1.4 are pore-forming subunits of L-type channels; Cav2.1 form P/Q channels, Cav2.2 form N-type channels, Cav2.3 form R-type channels and Cav3.1–Cav3.3 form T-type channels.

VGCC are very important mediators of Ca^{2+} influx and, thus, these channels are targeted by multiple and complex regulatory and modulatory signalling cascades. Particularly well researched is the regulation of VGCC by GPCR. There are two major pathways of such regulation. (i) The 'fast' pathway is voltage-dependent, membrane delimited and is mediated by the $G_{\beta\gamma}$ subunits; this pathway is mediated by the Pertussis Toxin-sensitive, $G_{o/i}$-coupled GPCR and is understood as direct voltage-dependent interaction of the channel with $G_{\beta\gamma}$ subunits (Bean 1989; Lipscombe et al. 1989; Herlitze et al. 1996; Zamponi and Snutch 1998). (ii) The 'slow' pathway encompass a group of mechanisms which share some common features such as voltage independence, often lack of sensitivity to PTX, and much slower (10 s seconds) kinetics as compared to direct $G_{\beta\gamma}$ inhibition (100 s ms). One of such slow pathways is initiated by the G_q-coupled receptors (Bernheim et al. 1991; Delmas and Brown 2005; Michailidis et al. 2007; Roberts-Crowley et al. 2009) and it has been suggested that this G_q-mediated slow pathway may require the same second messenger as the M channel inhibition (Bernheim et al. 1991; Mathie et al. 1992; Hille 1994). A growing body of evidence suggest that indeed, as in the case of Kv7/M channels, many types of VGCC are PIP_2 sensitive and receptor-mediated PIP_2 depletion is, again, a plausible candidate for a mediator of VGCC inhibition by some $G_{q/11}$ receptor agonists.

10.5.1 Experimental Evidence for PIP_2 Sensitivity of N-, P/Q- and L-type Channels

The first indications that some VGCC may require PIP_2 for their activity were published around the same time as that for the M channels: Wu et al discovered that the run-down of cloned P/Q-type Ca^{2+} channels in inside-out macropatches can be reversed by application of PIP_2 to the inner leaflet of the plasma membrane (Wu et al. 2002). Later, these findings were extended to N- (Gamper et al. 2004) and L-type (Michailidis et al. 2007) channels (Fig. 10.4a). The same set of techniques used to study PIP_2 sensitivity of Kir and Kv7 channels has since been applied to VGCC. Thus, anti-PIP_2 antibodies accelerated the run-down of P/Q-type channels, an effect reversed by direct application of PIP_2 (Wu et al. 2002). Chelation of membrane PIP_2 by the overexpression of PLCδ-PH construct (Gamper et al. 2004; Suh et al. 2010) or tonic depletion of PIP_2 by overexpression of Inp54p 5' phosphatase reduced current density of native VGCC (mostly N-type) in SCG neurons and reduced the amount of VGCC inhibition by M_1 AchR triggering (Gamper et al. 2004). Furthermore, it has been found that recovery of N-type current from the muscarinic inhibition was

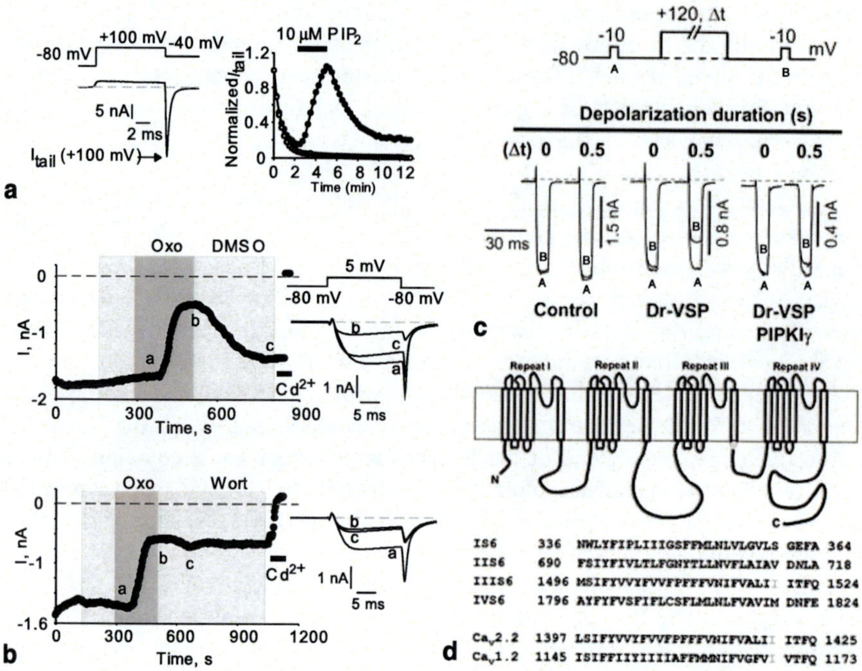

Fig. 10.4 PIP$_2$ sensitivity of VGCC. **a** Reactivation of N-type Ca^{2+} channels by PIP$_2$ in inside-out macropatches. Shown on the right is a current trace evoked by the voltage step from −80 to +100 mV (voltage protocol is given above the trace) in the macropatch from *Xenopus laevis* oocyte overexpressing N-type Ca^{2+} channels. Shown on the right is a time course of rundown and reactivation by PIP$_2$ application of the tail current induced by stepping from +100 to −40 mV. (From Gamper et al. (2004) with permission). **b** Inhibition of PIP$_2$ resynthesis with wortmannin in cultured sympathetic neurons prevents recovery of VGCC from muscarinic modulation. Plotted are the amplitudes of inward Ca^{2+} currents evoked by 15 ms depolarizing voltage pulses given every 3 s from a holding potential of −80 to +5 mV recorded in the perforated patch configuration of the patch-clamp technique. Top panel shows a control experiment in which the neuron was treated with vehicle (0.1% DMSO). DMSO, Oxotremorine (Oxo, 10 μM), and $CdCl_2$ (Cd^{2+}; 100 μM) were applied during the periods indicated by the shaded areas. *Insets* on the right depict current traces recorded at times indicated. Lower panel depicts similar experiment but wortmannin (Wort; 50 μM) was applied instead of DMSO. (From Gamper et al. (2004) with permission). **c** Inhibition of CaV1.3 L-type Ca^{2+} channels in tsA cells by voltage-sensitive phosphatase. Typical current traces before and after activation of Dr-VSP by depolarizations to +120 mV. Cells without Dr-VSP (Control), cells transfected with Dr-VSP, or cells transfected with Dr-VSP and PI5-K received a 10 ms test pulse to −10 mV and then were depolarized to +120 mV for zero or 0.5 s followed by a second test pulse (voltage protocol is depicted above). The currents before (A) and after (B) the +120 mV-depolarizing pulse are superimposed. (From Suh et al. (2010) with permission). **d** Putative PIP$_2$-interactive residues within VGCC. Shown are amino acid sequences of the S6 TMDs in the four repeats of Cav2.1 α subunit and the S6 segment in the third repeat of Cav2.2 and Cav1.2 α subunits. I1520 in Cav2.1 and homologous residues in Cav2.2 and Cav1.2 are shown in *grey*. Cartoon depicting transmembrane topology of the α1 subunit of voltage-gated Ca^{2+} channels is shown above; *grey dot* at the intracellular end of S6 within the third repeat indicates the location of I1520. (From Zhen et al. (2006) with permission)

abolished by PI4 kinase blockade with wortmannin (Fig. 10.4b) whereas the inhibition itself was attenuated by dialysis of DiC$_8$-PIP$_2$ via the patch pipette (Gamper et al. 2004). Hille's group used the inducible phosphatase approach to probe if PLC-independent PIP$_2$ depletion can inhibit heterologousely expressed VGCC and also to screen for PIP$_2$-sensitive VGCC isoforms (Suh et al. 2010). This study has confirmed major original findings of Jiang's and Shapiro's groups (Wu et al. 2002; Gamper et al. 2004) and brought several important conclusions in support of the 'PIP$_2$ hypothesis' for VGCC: (i) inducible enzymatic depletion of membrane PIP$_2$ without any GPCR or PLC activation and without co-release of any relevant signalling molecules can inhibit N-, P/Q- and L-type (Cav1.2 and Cav1.3) channels (Fig. 10.4c); (ii) inducible enzymatic PIP$_2$ depletion prevented (Cav1.3) or dramatically reduced (Cav2.2) subsequent muscarinic inhibition of Ca^{2+} currents; (iii) kinetics of PIP$_2$-sensitive VGCC inhibition and recovery follows the kinetics of enzymatic PIP$_2$ depletion and recovery closely (especially true for Cav1.3 channels although not so true for Cav2.2; see below).

Not all experiments on PIP$_2$ dependency of VGCC are coherent, thus, cloned P/Q- (Wu et al. 2002) and N-type (Gamper et al. 2004) VGCC expressed in oocytes appear to display a bi-modal sensitivity to PIP$_2$: low concentrations of PIP$_2$ produced a voltage-independent stabilizing effect, whereas higher concentrations induced a positive shift of channel voltage-dependence reminiscent of the transition from 'willing' to 'reluctant' (terms used to denote either free or $G_{\beta\gamma}$-bound channels in a $G_{o/i}$-coupled receptor modulation paradigm (Bean 1989; Ikeda 1996; Herlitze et al. 1997)) states of VGCC. Accordingly, a model has been proposed in which P/Q- and N-type channels have two distinct PIP$_2$ binding sites: a higher-affinity site that binds PIP$_2$ to maintain channel activity and a lower affinity site which, when PIP$_2$ is bound, shifts the channel into the 'reluctant' mode (Wu et al. 2002; Michailidis et al. 2007). However, the voltage-dependent action of PIP$_2$ on these channels was not observed in the whole cell experiments in SCG neurons (Gamper et al. 2004), likewise, little evidence for voltage dependence of PIP$_2$ effect was found in the whole cell experiments on the L- and N-type channels overexpressed in ts-A cells (Suh et al. 2010). It has been hypothesized that channel phosphorylation or possibly some cytosolic factor that modifies VGCC sensitivity to PIP$_2$ can be lost in excised-patch experiments (Suh et al. 2010). Despite of this slight discrepancy, the excised-patch and the whole-cell experiments do suggest that the 'slow' pathway of $G_{q/11}$-coupled receptor-induced VGCC inhibition in neurons can be mediated (at least in part) by a high-affinity, voltage independent action of PIP$_2$.

PIP$_2$-binding site(s) within VGCC remain elusive. Low-specificity interactions between the Cav2.1 (P/Q-type) subunit C-terminus and several phosphoinositide species have been reported (Rousset et al. 2004). In addition, a substitution of single isoleucin (I1520) by histidine or aspartate in the cytosolic loop after S6 in the third 6-TMD repeat significantly attenuated the run-down of recombinant P/Q channels in inside-out patches and prevented channel inhibition by PIP$_2$-scavenging MARCKS peptide (Fig. 10.4d); similar effects were seen after substitution of homologous residues in N- and L-type channels (Zhen et al. 2006). These effects were attributed to the changes in channel-PIP$_2$ interaction. Mutagenesis experiments described above are suggestive but further work is needed to characterize PIP_2 binding sites within VGCC.

10.5.2 PIP_2 vs. Arachidonic Acid

There is a competing hypothesis for the 'slow' $G_{q/11}$-mediated inhibition of L-, N- and P/Q channels according to which the main second messenger is the arachidonic acid (AA). AA is a frequent constituent of phospholipids, including PIP_2 as it is covalently attached to the C2 (*sn*-2) carbon atom of the glycerol backbone of phospholipids; estimated 80% of PIP_2 has AA in the *sn*-2 position (Wenk et al. 2003; Roberts-Crowley et al. 2009). Phospholipase A2 group IVa ($cPLA_2$) selectivity cleaves AA at the *sn*-2 position of phospholipids (Leslie 2004). $cPLA_2$ can bind to PIP_2 via its C2 domain and $G_{q/11}$ receptor stimulation can acutely activate $cPLA_2$ via the ERK1/2-dependent phosphorylation (Roberts-Crowley et al. 2009). Thus, the same receptors that trigger PIP_2 hydrolysis can cause concurrent release of the AA which, according to the 'AA hypothesis', is the main signal mediating VGCC inhibition. In support of this hypothesis, exogenously applied AA inhibits currents of native and recombinant VGCC of major subtypes with IC_{50} in the range of 1–10 μM, which is considered as a physiologically relevant range (Xiao et al. 1997; Vellani et al. 2000; Zhang et al. 2000; Liu et al. 2001; Talavera et al. 2004; Liu 2007). In contrast to PIP_2 which stabilise the open state of the channels, AA was suggested to stabilize the closed state (Roberts-Crowley et al. 2009). Several other experiments, mostly by the Rittenhouse group, suggested involvement of the $cPLA_2$ in the G_q-mediated inhibition of N- and L-type VGCC in SCG neurons. Thus, muscarinic stimulation of SCG neurons was shown to induce phosphorylation of the $cPLA_2$ protein (Liu et al. 2006), moreover, pharmacological inhibition of PLA reduced N-type Ca^{2+} current inhibition by Oxo-M (Liu and Rittenhouse 2003). Likewise, L-type channel inhibition by Oxo-M was lost in neurons from $cPLA_2^{-/-}$ mice (no change in M current inhibition by Oxo-M in such neurons was noticed) (Liu et al. 2006) (for further discussion of regulation of VGCC by AA see the excellent recent review (Roberts-Crowley et al. 2009)).

As for the case of PIP_2, the evidence for the sensitivity of VGCC to the exogenously applied AA is sound but whether AA is a second messenger of the receptor-mediated physiological signals regulating the VGCC in native neurons is much harder to prove due to the plethora of second messengers released by GPCR. In addition, some labs were unable to find evidence in support for the requirement of PLA for the $G_{q/11}$-mdiated inhibition of N- and L-type channels (Bannister et al. 2002; Gamper et al. 2004; Lechner et al. 2005). An experimental design which would allow enzymatic release of AA without concurrent production of other second messengers (similar to inducible phosphotases developed to probe the PIP_2 sensitivity of channels) would help to further support the 'AA hypothesis'.

As in the case for M channels, attempts to unify the 'PIP_2' and the 'AA' hypotheses into a 'coincidence detection' mechanism has been made (e.g. Gamper and Shapiro 2007a; Roberts-Crowley et al. 2009) with the most comprehensive model proposed by the Rittenhouse group. In this hypothesis it is suggested that PIP_2 is docked within a VGCC channel complex in such a way that its inositol head group binds to one site within the channel while its AA tail interacts with another binding site. Stimulation of a $G_{q/11}$-coupled receptor simultaneously (or in a rapid succession) activates PLC, PLA_2 and DAG lipase which, in turn, comprehensively degrade PIP_2 molecule into IP_3, glycerol and free fatty acids. This full degradation is needed for maximal destabilization of the open state of the channel (Roberts-Crowley et al. 2009). The hypothesis is very attractive as it accounts for many conflicting evidence from both 'PIP_2' and 'AA' hypotheses and also provides some way of specificity for the $G_{q/11}$ signalling towards the VGCC channels as it ensures that simple PIP_2 hydrolysis is not enough to produce maximal inhibition of VGCC. However, new data from the inducible phosphatase experiments (Suh et al. 2010), which suggest that L-, N- and P/Q-type VGCC can be significantly inhibited by the conversion of PtdIns(4,5)P_2 into PtdIns(4)P (without AA release) pose some difficulty here. Nevertheless, at least for the N-type channels, there is an additional small component of inhibition induced by Oxo-M which is not prevented by PIP_2 dephosphorylation (Suh et al. 2010). This may indicate a need for a cofactor such as AA. Clearly further research is needed to develop inclusive model for the slow pathway of VGCC inhibition.

10.5.3 Possible Physiological Implications

The physiological significance of VGCC regulation is hard to overestimate since the activity of these channels control synaptic transmission, muscle contraction and gene expression. Accordingly, dysfunctions of VGCC cause severe human conditions ranging from movement disorders, arrhythmias and hypertension to neurological disorders, epilepsy and migraine (Gribkoff 2006). Emerging evidence suggest that PIP_2 stabilizes activity of P/Q-, N- and L-type VGCC and receptor-mediated PIP_2 depletion underlies (or at least contributes to) inhibition of these channels. Therefore, it is straightforward to suggest that PIP_2 sensitivity of VGCC may provide one of the core mechanisms for control over the physiological processes which are regulated through VGCC.

10.6 TRP Channels

Transient Receptor Potential (TRP) channels are distant relatives of the voltage gated ion channel superfamily (Yu and Catterall 2004). They have six transmembrane domains per subunit and four subunits form the functional channel (Ramsey et al. 2006). Most TRP channels are non-selective, Ca^{2+} permeable cation channels, and display outward rectification. Based on sequence homology, mammalian TRP channels are subdivided into six groups: TRPC (Classical or Canonical), TRPV (Vanilloid), TRPM (Melastatin), TRPP (Polycystin), TRPML (Mucolipin) and TRPA

(Ankyrin). They play essential roles in a wide variety of physiological processes, such as thermosensation, mechanosensation, nociception, taste, vision, fertilization, intra- and extracellular Ca^{2+} and Mg^{2+} homeostasis (Clapham et al. 2001; Montell et al. 2002; Minke and Cook 2002). Mutations in TRP channels and TRP related proteins cause various diseases such as hypomagnesemia (Walder et al. 2002), polycystic kidney disease (Wilson 2004), familial focal segmental glomerulosclerosis (Winn et al. 2005) and mucolipidosis (Raychowdhury et al. 2004), reviewed in Nilius et al. (2005); Nilius and Owsianik (2010). In congruence with the variety of functions they are involved in, their activation mechanisms are also quite diverse; these include temperature, mechanical stimuli, pH, and various signalling pathways and chemical ligands.

Despite the high diversity of activation mechanisms and physiological functions, most, if not all TRP channels are regulated by phosphoinositides (Rohacs 2007, 2009; Nilius et al. 2008). However, their regulation by phosphoinositides is quite complex. All ion channel families discussed so far were activated by PIP_2, in other words their activity depended on the presence of the lipid. Dependence of activity on phosphoinositides have been described for many TRP family members as well; it is possible that this is a conserved feature of this ion channel family, but inhibition by phosphoinositides have also been described for many of them. Table 10.1 summarizes current knowledge based on the primary literature on phosphoinositide effects on TRP channels. Here we discuss the literature on a selected few channels. Two of our examples (TRPM8 and TRPV5/6) are similar to the channels discussed so far, their activity depends on PIP_2, whereas our other two examples, TRPCs and TRPV1 are channels where the regulation by PIP_2 is quite complex and controversial, and our understanding is limited.

10.6.1 TRPM Channels

TRPMs are the functionally most diverse group in the TRP channel superfamily with eight mammalian members. Most TRPMs are non-selective Ca^{2+} permeable cation channels, similar to other TRP-s; exceptions are TRPM4 and TRPM5, which conduct monovalent cations, but not Ca^{2+}. PIP_2 regulation has been reported for 4 members of this group, in all cases PIP_2 activated the respective channel (Table 10.1) and thus PIP_2 dependence is probably a common feature of TRPM channels. Here we discuss the literature on TRPM8, which, with respect to phosphoinositide regulation, is the most thoroughly studied member of this family.

TRPM8 is an ion channel activated by cold temperatures and cooling agents such as menthol or icilin in sensory neurons (McKemy et al. 2002; Peier et al. 2002). Genetic deletion of TRPM8 in mice convincingly demonstrated the involvement of this channel in sensing cold temperatures (Dhaka et al. 2007; Colburn et al. 2007; Bautista et al. 2007). TRPM8 has also been proposed to be involved in mediating the analgesic effects of moderate cold and menthol (Proudfoot et al. 2006).

TRPM8 clearly requires PIP_2 for activity. Its activity runs down in excised patches, and application of PIP_2 reactivates the channel (Liu and Qin 2005; Rohacs et al.

Table 10.1 Regulation of TRP channels by PIP_2

Name	Regulation/function	Regulation by PIP_2, reference
Classical TRPs		
dTRPL	Activated downstream of PLC, drosophila vision	PIP_2 inhibits in excised patches (Estacion et al. 2001) PIP2 activates, but PIP and PI inhibits in excised patches (Huang et al. 2010)
TRPC1	Activated downstream of PLC	PIP_2 activates in excised patches, native cells (Saleh et al. 2008, 2009)
TRPC3	Activated downstream of PLC, DAG activates	PIP_2 activates in excised patches, expression system (Lemonnier et al. 2007)
TRPC4α	Activated downstream of PLC	TRPC4α but not TRPC4β is inhibited by PIP_2, whole cell patch clamp (Otsuguro et al. 2008)
TRPC5	Activated downstream of PLC,	PIP_2 activates in excised patches, but inhibits in whole cell, PIP_2 depletion may inhibit or activate it (Trebak et al. 2008) PIP_2 inhibits desensitization in whole-cell patch clamp (Kim et al. 2008b)
TRPC6	Activated downstream of PLC, DAG activates	PIP_2 activates in excised patches (expression system) (Lemonnier et al. 2007) PIP_2 inhibits in excised patches (native smooth muscle cells) (Albert et al. 2008; Ju et al. 2010) Extracellularly applied PIP_2 enhances it in platelets (Jardin et al. 2008) Calmodulin inhibits by displacing PIP_3 (Kwon et al. 2007)
TRPC7	Activated downstream of PLC, DAG activates	PIP_2 activates in excised patches, expression system (Lemonnier et al. 2007) PIP_2 inhibits in excised patches, native channels (Ju et al. 2010)
Vanilloid TRPs		
TRPV1	Heat, capsaicin, low pH, involved in nociception	PIP_2 may partially inhibit in intact cells (Chuang et al. 2001; Prescott and Julius 2003; Lukacs et al. 2007) PIP_2 activates in excised patches (Stein et al. 2006; Lukacs et al. 2007; Klein et al. 2008) PIP_2 inhibits desensitization in intact cells (Liu et al. 2005; Lukacs et al. 2007; Lishko et al. 2007)
TRPV5	Constitutively active epithelial Ca^{2+} channel	PIP_2 activates in excised patches (Rohacs et al. 2005; Lee et al. 2005)
TRPV6	Constitutively active epithelial Ca^{2+} channel	PIP_2 activates in excised patches, PIP2 depletion inhibits (Thyagarajan et al. 2008)
Melastatin TRPs		
TRPM4	Intracellular Ca^{2+} activates	PIP_2 activates in excised patches (Zhang et al. 2005)
TRPM5	Intracellular Ca^{2+} activates	PIP_2 activates in excised patches (Liu and Liman 2003; Nilius et al. 2006)

Table 10.1 (Continued)

Name	Regulation/function	Regulation by PIP_2, reference
TRPM7	cAMP, shear stress, Mg^{2+} transport, Mg^{2+} inhibits	PIP_2 activates in excised patches (Runnels et al. 2002) Role of PIP2 depletion has been challenged (Takezawa et al. 2004; Langeslag et al. 2007)
TRPM8	Cold, menthol	PIP_2 activates in excised patches, and lipid bilayers, PIP_2 depletion inhibits (Liu and Qin 2005; Rohacs et al. 2005; Varnai et al. 2006; Daniels et al. 2008; Zakharian et al. 2009, 2010)
Other TRPs		
TRPA1	Mustard oil and other noxious chemicals Noxious cold	PIP_2 inhibits heterologous desensitization by capsaicin (Akopian et al. 2007) PIP_2 activates in excised patches, inhibits desensitization in whole cell (Karashima et al. 2008) PIP_2 inhibits sensitization by PAR2 in whole-cell (Dai et al. 2007) PIP_2 inhibits in excised patches in the presence of PPPi, no effect w/o PPPi (Kim and Cavanaugh 2007; Kim et al. 2008c) Depletion of PIP_2 with rapamycin-inducible phosphatase have no effect (Wang et al. 2008)
TRPP2	Mutated in polycystic kidney disease, Mechanosensor?	PIP_2 inhibits, depletion of PIP_2 by EGF activates (Ma et al. 2005)
TRPML1	Intracellular channels Mutation causes mucolipidosis	Specifically activated by PtdIns(3,5)P_2 (Dong et al. 2010)

2005). The activating effect is isomer specific; PtdIns(4,5)P_2, is more effective than PtdIns(3,4)P_2, PtdIns(3,4,5)P_3 or PtdIns(4)P (Rohacs et al. 2005). PIP_2 chelating agents, such as PIP_2 antibody, or poly-Lysine also inhibit TRPM8 in excised patches (Liu and Qin 2005; Rohacs et al. 2005). The activity of the purified TRPM8 reconstituted into lipid bilayers depends on the presence of PIP_2 with a similar phosphoinositide specificity profile as in excised patches, providing a strong evidence for direct activation of the channel by PIP_2 (Zakharian et al. 2009, 2010), see also Fig. 10.5a, 10.5b. Activation of PLC via cell surface receptors (Liu and Qin 2005; Rohacs et al. 2005), by Ca^{2+} influx through TRPM8 (Rohacs et al. 2005; Daniels et al. 2008) or pharmacologically with m-3M3FBS (Daniels et al. 2008) inhibits TRPM8. PLC independent depletion of PIP_2 using a rapamycin-inducible phosphatase (Varnai et al. 2006; Wang et al. 2008; Daniels et al. 2008) or high concentrations of wortmannin (Liu and Qin 2005; Rohacs et al. 2005) also inhibits TRPM8 further supporting its dependence on PIP_2.

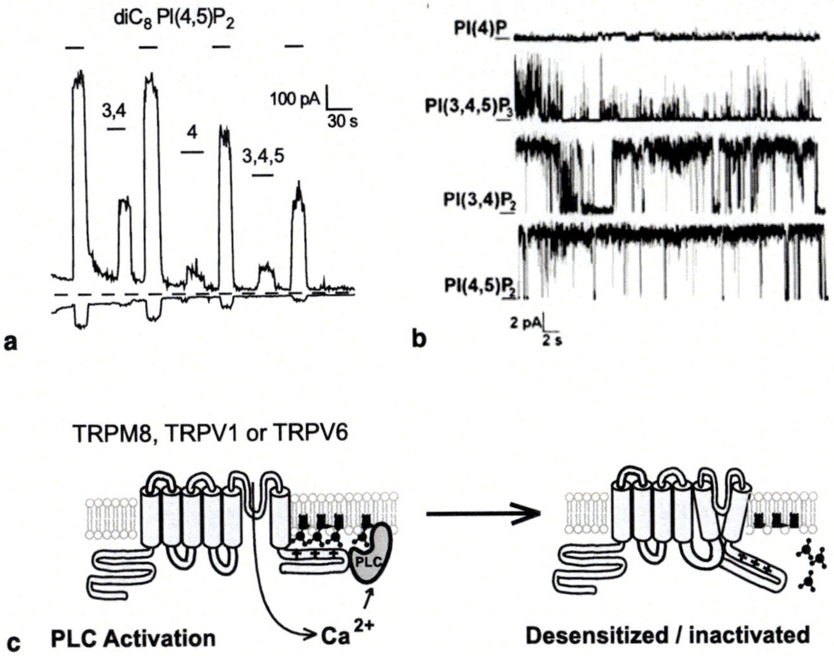

Fig. 10.5 **a** PIP_2 regulation of TRP channels. Activation of TRPM8 currents in an excised inside out macropatch in Xenopus oocytes. Currents are show after run-down, and the effects of various diC_8 phosphoinositides are shown. (From Rohacs et al. (2005) with permission). **b** Activation of the purified TRPM8 by various diC_8 phosphoinositides in planar lipid bilayers. (From Zakharian et al. (2010) with permission). **c** A model for Ca^{2+} induced inactivation/desensitization for various TRP channels. (Modified from Rohacs (2009) with permission)

In addition to being important for channel activity, PIP_2 is also likely to be involved in desensitization of TRPM8. TRPM8 currents activated by menthol (McKemy et al. 2002; Rohacs et al. 2005; Daniels et al. 2008) or cold (Reid et al. 2002; Daniels et al. 2008) gradually diminish in the presence of extracellular Ca^{2+}, a process called desensitization or adaptation. This effect has been reported both in expression systems (McKemy et al. 2002; Rohacs et al. 2005) and in native sensory neurons (Reid et al. 2002). Similarly, physiological responses to cold (Darian-Smith et al. 1973) and menthol (Eccles 1994) have been shown to desensitize. It was proposed that the mechanism of desensitization is the Ca^{2+}-induced activation of PLC and the ensuing depletion of PIP_2 (Rohacs et al. 2005) (Fig. 10.5c). This idea is based on the following findings. As mentioned earlier, PIP_2 activates TRPM8 in excised patches and depletion of the lipid inhibits the channel in intact cells. Ca^{2+} influx through TRPM8 leads to activation of PLC and the depletion of PIP_2, (Rohacs et al. 2005; Daniels et al. 2008). TRPM8 desensitization is slowed down by co-expressing PIP5K that synthesizes PIP_2, and accelerated by co-expressing the highly Ca^{2+} sensitive PLC isoform PLCδ1 (Rohacs et al. 2005).

How does PIP_2 activate TRPM8? Two questions will be discussed here briefly: what is the relationship of PIP_2 to other regulators of TRPM8, and where are the

PIP_2 interacting residues? TRPM8 is activated by cold and cooling agents, such as menthol. Cooling agents shift the activation threshold of the channel to warmer temperatures (McKemy et al. 2002). It was shown that both cold and menthol increase sensitivity of TRPM8 to PIP_2, i.e. shift PIP_2 dose-response curves to the left. Concurrently, the channel becomes less sensitive to PIP_2 depletion in the presence of menthol (Rohacs et al. 2005). This is similar to the effect of $G_{\beta\gamma}$ on Kir channels, as discussed earlier.

TRP channels are also thought to be directly activated by PIP_2 through binding to positively charged residues, but only limited efforts have been made to identify those residues in TRP channels so far. Mutation of positively charged residues in the highly conserved TRP domain of TRPM8 substantially decreased the apparent affinity of the channel for PIP_2 (Rohacs et al. 2005). The same mutations rendered the channel more sensitive to inhibition by depletion of PIP_2. This is compatible with the idea that these residues are part of a PIP_2 binding site. However, the R1008Q mutation that had the most dramatic effect on PIP_2 sensitivity also affected menthol and cold sensitivity. PIP_2 sensitivity of this mutant was however still much less than that of the wild-type channel when examined at lower temperatures and higher menthol concentrations arguing for a primary effect on PIP_2 sensitivity. Nevertheless, as discussed earlier, it cannot be excluded that these mutations affect PIP_2 sensitivity indirectly. Two of the three TRP domain mutants only moderately affected PIP_2 sensitivity, thus it is unlikely that this domain is solely responsible for PIP_2 sensitivity of TRPM8. It is likely that other parts of the channel also contribute to PIP_2 binding, a notion further supported by the fact that mutation of equivalent TRP domain residues did not affect PIP_2 sensitivity of TRPM4 (Nilius et al. 2006).

10.6.2 TRPV Channels

The TRPV family has 6 mammalian members. They can be separated into 2 groups. TRPV1-4 are sensory channels, all are activated by heat with various thresholds. Most of these channels are expressed in sensory neurons, or keratinocytes in the skin. TRPV4, in addition to being activated by heat, is also a mechanosensitive channel. TRPV5 and 6 on the other hand are epithelial Ca^{2+} channels involved in organism level Ca^{2+} homeostasis. PIP_2 regulation was reported for three members of this family: TRPV1, TRPV5 and TRPV6 (Table 10.1). All three of these channels are activated by PIP_2 in excised patches but for TRPV1 an additional indirect inhibitory effect of the lipid in intact cells may complicate the picture.

10.6.2.1 TRPV5 and TRPV6

TRPV5 and TRPV6 are Ca^{2+} selective channels, located on the apical membrane of epithelial cells that are responsible for active transcellular Ca^{2+} transport (Hoenderop et al. 2005). They share high homology to each other, but much less to the other members of the TRPV family. Unlike all other TRP channels, TRPV5 and 6 show inward rectification and are selective for calcium and other divalent cations (Hoenderop et al. 2005). TRPV5 is expressed in the kidney, in the late distal convoluted and the connecting tubules, whereas TRPV6 is mainly expressed in the duodenum. TRPV6 is regulated at the transcriptional level by active vitamin D3 (calcitriol). Genetic deletion of either of these channels results in disturbances in calcium homeostasis in mice (Bianco et al. 2006; Hoenderop et al. 2003). The rate of TRPV6 protein evolution was shown to be accelerated in the human lineage (Akey et al. 2006) and its ancestral overactive variant was shown to be associated with increased prevalence of kidney stones in humans, presumably by increased intestinal Ca^{2+} absorption and compensatory hypercalciuria (Suzuki et al. 2008). Both TRPV5 and TRPV6 undergo Ca^{2+}-induced inactivation, which presumably protects the cells form toxic Ca^{2+} levels and limits epithelial Ca^{2+} transport.

Both TRPV5 and TRPV6 require PIP_2 for activity; their activity runs down in excised patches, which is accelerated by poly-Lysine (Rohacs et al. 2005) and they are reactivated by application of PIP_2 (Lee et al. 2005; Thyagarajan et al. 2008). We have proposed that Ca^{2+}-induced inactivation of TRPV6 proceeds through PLC activation and the resulting depletion of PIP_2 (Thyagarajan et al. 2008, 2009), similarly to TRPM8 (Fig. 10.5b). This model is based on the following findings. TRPV6 is activated in excised patches by PIP_2 but not PIP. Ca^{2+}-induced inactivation is inhibited by dialyzing PtdIns(4,5)P_2, but not PtdIns(4)P through the patch pipette in whole-cell patch clamp experiments. Ca^{2+} influx through TRPV6 leads to depletion of PIP_2 and formation of IP_3, indicating activation of PLC. PLC independent depletion of PIP_2 with the rapamycin-inducible PIP_2 phosphatase, or high concentrations of wortmannin inhibited TRPV6 (Thyagarajan et al. 2008). Both PIP_2 depletion and Ca^{2+}-induced inactivation of TRPV6 were inhibited by PLC inhibitors (Thyagarajan et al. 2009).

The calcium sensor calmodulin has also been proposed to play a role in Ca^{2+}-induced inactivation of TRPV6 (Derler et al. 2006; Niemeyer et al. 2001). Again, just like in other cases, it is possible that both mechanisms contribute to Ca^{2+}-induced inactivation. Competition of CaM with PIP_2, as proposed for other TRP channels (Kwon et al. 2007) and Kv7 channels (see above) is a feasible mechanism that would integrate CaM and PIP_2 regulation, but it has not been experimentally tested on TRPV6.

10.6.2.2 TRPV1

TRPV1 was the first non-canonical mammalian TRP channel to be cloned (Caterina et al. 1997). Its major activators are heat, capsaicin (the pungent compound in hot peppers), and tissue acidosis. This channel is involved in nociception and there are many other factors that activate or regulate it (Pingle et al. 2007). TRPV1 was also the first mammalian TRP channel that was reported to be regulated by PIP_2. It was proposed that PIP_2 tonically inhibits TRPV1, and depletion of this lipid by pro-inflammatory agents, such as bradykinin, relieves this inhibition, and potentiates

TRPV1 activity at low stimulation levels (Chuang et al. 2001). This potentiation is thought to underlie thermal hyperalgesia, the increased sensitivity of inflamed areas to heat. Later however, several laboratories reported that in contradiction to this model, PIP_2 and other phosphoinositides activate the channel in excised patches (Stein et al. 2006; Lukacs et al. 2007; Klein et al. 2008; Kim et al. 2008c). Agents that chelate PIP_2 (such as poly-Lysine) inhibit TRPV1 in excised patches, thus supporting the activating effect of the lipid (Stein et al. 2006; Lukacs et al. 2007). This apparent controversy is similar to that seen with TRPC5, see later.

What is the functional role of the activating effect of PIP_2? It is likely that depletion of the lipid plays a role in the Ca^{2+}-dependent desensitization of TRPV1, similarly to several other TRP channels, such as TRPM8 (Rohacs et al. 2005), TRPM4 (Nilius et al. 2006) and TRPV6 (Thyagarajan et al. 2008). The model is simple: when Ca^{2+} enters a cell through TRPV1, it activates a Ca^{2+} sensitive PLC, and the resulting PIP_2 depletion leads/contributes to decreased channel activity (Fig. 10.5c). This model is based on the following data. (i) As already mentioned, TRPV1 requires PIP_2 for activity in excised patches. (ii) Application of capsaicin in the presence of extracellular Ca^{2+} leads to hydrolysis of PIP_2 (Liu et al. 2005; Lukacs et al. 2007; Akopian et al. 2007; Yao and Qin 2009). (iii) Recovery from desensitization depends on the ability of the cell to resynthesize PIP_2 (Liu et al. 2005). (iv) PLC inhibitors reduce desensitization (Lukacs et al. 2007; Lishko et al. 2007). (v) Supplying excess PtdIns(4,5)P_2 or PtdIns(4)P through the patch pipette in whole-cell patch clamp experiments reduces desensitization (Lukacs et al. 2007; Lishko et al. 2007). PtdIns(4)P also activates TRPV1 in excised patches, and it is also depleted upon PLC activation (Lukacs et al. 2007). As the concentration of PtdIns(4)P is thought to be comparable to that of PIP_2, it may also play a role, together with PIP_2, in keeping TRPV1 open.

PIP_2 depletion is unlikely to be the mechanism solely responsible for desensitization of TRPV1, as both PLC inhibition and supplying excess PIP_2 only partially inhibited desensitization. Also in one study supplying PIP_2 through the patch pipette in whole-cell experiments only moderately reduced capsaicin-induced desensitization (Akopian et al. 2007). The ubiquitous Ca^{2+} sensor calmodulin has also been proposed to play a role in desensitization, both acting on the channel directly (Numazaki et al. 2003; Rosenbaum et al. 2004; Lishko et al. 2007), and by activating calcineurin (Docherty et al. 1996; Mohapatra and Nau 2005), and thus inducing dephosphorylation of the channel.

There seems to be a general agreement on the role of PIP_2 in activating TRPV1, and the originally proposed tonic inhibitory effect of PIP_2 is somewhat debated. Is it possible that PIP_2 has both inhibitory effects and is required for channel activity, similar to what was proposed for VGCC (Wu et al. 2002). It was found that depletion of the lipid with the rapamycin-inducible PIP_2 phosphatase system (Varnai et al. 2006) leads to further activation when the channel is only moderately stimulated by capsaicin or heat (Lukacs et al. 2007). This finding suggests a partial inhibition by PIP_2 in intact cells, in addition to its activating effect. Importantly, potentiation by PIP_2 depletion was only seen when the channel was stimulated by low concentration of capsaicin, or moderate heating, conditions where PLC mediated sensitization also occurs. When the channel was maximally stimulated by high capsaicin concentrations, neither activation, nor inhibition by the inducible phosphatase was observed (Lukacs et al. 2007). The lack of inhibition at high capsaicin concentrations was explained with PtdIns(4)P keeping the channel open under such conditions. PtdIns(4)P is not depleted by the phosphatase, indeed it is expected that its level increases when PIP_2 is converted to PtdIns(4)P. Conversely, when we over-expressed the PIP5K enzyme, generating excess of PIP_2, TRPV1 activity was inhibited at low, but not at high capsaicin concentrations (Lukacs et al. 2007). This finding is also compatible with a partial inhibitory effect of PIP_2 at moderate stimulation levels. This inhibitory effect, however, is likely to be indirect, because it is not detectable in excised patches.

Another article, on the other hand found that the rapamycin-inducible PIP_2 phosphatase inhibited TRPV1 both high and low concentrations of capsaicin (Klein et al. 2008). This is compatible with the activating effect of PIP_2 in excised patches, and argues against an inhibitory effect of the lipid. It is hard to tell what causes the discrepancies between the two studies (Lukacs et al. 2007; Klein et al. 2008). There are a number of differences in the experimental conditions including, the cell-type, the rapamycin analogue, the concentrations of capsaicin used, and the origin of the rapamycin-phosphatase system (Suh et al. (2006) vs. Varnai et al. (2006)). Some of these differences may explain the opposing findings of the two studies. It is worth noting however, that the same two articles reached very similar conclusions on the effects of the phosphoinositides PtdIns(4,5)P_2, PtdIns(4)P and PtdIns(3,4,5)P_3 in excised patches, despite several differences in experimental conditions (Lukacs et al. 2007; Klein et al. 2008).

A recent addition to the complexity of phosphoinositide regulation of TRPV1 is the discovery of Pirt (Kim et al. 2008a). Pirt is a two transmembrane domain protein, specifically expressed in DRG neurons and it interacts both with TRPV1 and phosphoinositides. It was proposed that phosphoinositides activate TRPV1 through binding to Pirt. PIP_2 however activates TRPV1 in excised patches in expression systems (Lukacs et al. 2007), where Pirt is unlikely to be present. It is unlikely that Pirt is an obligatory subunit for TRPV1 modulation by phosphoinositides, but it is present in the native environment of TRPV1; it interacts with the channel and modulates its function. Thus it is probably an important modulator of native TRPV1 channels, but clarifying its exact role in phosphoinositide regulation of these channels will require further experimental work.

In conclusion, TRPV1 clearly requires phosphoinositides for activity; PIP_2 reproducibly activates the channel in excised patches. There also seems to be an agreement that depletion of the lipid contributes to Ca^{2+}-induced desensitization. If there is a partial inhibition by PIP_2 in intact cells, it is likely to depend on a factor lost upon patch excision (indirect effect) because several laboratories found no evidence of it in excised patches using a variety of tools (Lukacs et al. 2007; Klein et al. 2008). PIP_2 regulation of TRPV1 has recently been reviewed with a discussion of ideas to integrate the activating and the possible inhibitory effects of PIP_2 in the PLC mediated regulation of TRPV1 (Rohacs et al. 2008).

10.6.3 TRPC Channels

TRPC channels are activated downstream of PLC, and mediate Ca^{2+} influx and presumably depolarization. The exact mechanism by which they are activated by PLC is not clear in most cases (Trebak et al. 2007). TRPC3, 6 and 7 has been shown to be activated by DAG, the downstream product of PLC activation (Hofmann et al. 1999), but the other TRPC isoforms are generally thought to be insensitive to DAG. Many TRPC isoforms have been shown to be inhibited by PIP_2 (table), and relief from tonic inhibition by PIP_2 upon PLC activation has been proposed as a mechanism for TRPC channel activation. As we will see, this mechanism may play a role in certain cases, but it is unlikely to be a general paradigm among TRPCs.

TRPCs are the closest mammalian homologues of the drosophila TRP and TRPL channels. In the drosophila eye the TRP/TRPL complex is activated by light in a PLC dependent manner, thus generating the receptor potential (Hardie and Raghu 2001). The TRPL channel was shown to be inhibited by PIP_2 in excised patches in an expression system (Estacion et al. 2001). The same channel was also activated by DAG analogues. Later studies showed that activation of the TRP/TRPL complex by PIP_2 depletion is unlikely to be the major mechanism to generate the receptor potential in the drosophila eye, even though it may play some auxiliary role (Hardie 2007). Confounding these observations, a recent report, found activation of heterologously expressed TRPL by PIP_2 in excised patches. The two precursors, PtdIns and PtdIns(4)P, on the other hand inhibited TRPL in excised patches (Huang et al. 2010).

The mammalian TRPC4 splice variant TRPC4α, but not TRPC4β is inhibited when PIP_2 is dialyzed through the patch pipette in whole-cell patch clamp experiments (Otsuguro et al. 2008). PIP_2 was shown to bind to the C-terminus of TRPC4α, but not TRPC4β. The inhibition by PIP_2 could be disrupted with the cytoskeletal inhibitor cytochalasin D or by deleting the C-terminal PDZ binding motif from TRPC4α. PIP_2 depletion, however, was not sufficient in itself to open the channels. The effects of PIP_2 in excised patches were not examined in this study (Otsuguro et al. 2008).

Another article showed that TRPC5 can be moderately activated by depleting PIP_2 using two inhibitors of PI4K, wortmannin and LY294002 (Trebak et al. 2008). Activation by wortmannin was inhibited by dialyzing PIP_2 through the patch pipette. Interestingly, depletion of PIP_2 using a rapamycin-inducible PIP_2 phosphatase inhibited TRPC5 when the channel was activated by a low concentration of carbachol (Trebak et al. 2008). When PIP_2 was tested in excised patches, however, it activated TRPC5 (Trebak et al. 2008), similarly to TRPC3, 6, and 7 (Lemonnier et al. 2007). These data suggest that PIP_2 has both activating and inhibitory effects on TRPC5. The inhibitory effect of PIP_2 is likely to be an indirect effect, because it is not detected in excised patches.

Presently there is a controversy whether TRPC6 and TRPC7 channels are activated or inhibited by PIP_2. One study showed that in an expression system, PIP_2 activates TRPC6 and TRPC7 in inside-out patches (Lemonnier et al. 2007). Another study found that in vascular smooth muscle cells PIP_2 inhibits native TRPC6 channels in excised patches. (Albert et al. 2008; Ju et al. 2010). The same study also showed that dialyzing PIP_2 through the whole-cell patch pipette inhibited activation of TRPC6 by angiotensin II and DAG. Collectively, the regulation of TRPC channels by PIP_2 is likely to be quite complex and not yet fully understood (Table 10.1).

In conclusion, the activity of many TRP channels depends on the presence of PIP_2 in the plasma membrane; in this respect, these channels are similar to Kir and KCNQ channels. Some TRP channels, such as TRPM8 and TRPV5 and 6, behave very similar to classical PIP_2 sensitive channels. Some other TRP channels however, are also reported to be both activated and inhibited by PIP_2. The difference between whether PIP_2 activates or inhibits was either the experimental setting, i.e. intact cells versus excised patches or endogenous versus heterologously expressed channels. Given the sheer prevalence of this "dual regulation", it is hard to dismiss it as an artifact, or unreliable data. Differences between native vs. expressed channels can be explained by different cellular components expressed in these cell in addition to the channel, whereas difference between excised patch and whole-cell measurements can be explained with lost cellular components in the latter. Altogether, regulation of many TRP channels by PIP_2 is quite complex and its understanding requires further investigation.

10.7 Conclusions

A large number and variety of ion channels are modulated by plasma membrane phosphoinositides. In most cases, the activity of the channels depend on the presence of PtdIns(4,5)P_2, and the depletion of the lipid inhibits them. In the last 10–15 years we have seen an explosion in the number of PIP_2-sensitive ion channels and transporters; in addition to the ones discussed here the list of PIP_2-sensitive ion channels now includes K2P, HERG, CNG, ENaC, CFTR, P2X—to name a few, but also many others. Importantly, we have also seen a tremendous progress in the development of tools and approaches to study this phenomenon; this progress gives hope that in the near future we will see further insights into the mechanisms and significance of ion channel interaction with phosphoinositides. Indeed, there are many intriguing yet unanswered questions ahead. One of such questions is why so many ion channel proteins display sensitivity to phosphoinositides? One hypothesis suggests that for many plasma membrane ion channels requirement for PIP_2 provides a mechanism for silencing these channels until they reach plasma membrane (Hilgemann et al. 2001). Indeed, during their life cycle, plasma membrane ion channels travel trough the various membranous organelles (ER, Golgi, endosomes etc.) but in most cases it is only plasma membrane where their activity is needed. Accordingly, in contrast to the plasma membrane, intracellular membranes usually contain very little PIP_2 and for the majority of PIP_2-sensitive ion channels the requirement for PIP_2 is permissive. Thus, at least for the channels with high PIP_2 affinity, the PIP_2-dependence may serve to ensure that their activity is 'turned off' until they reach their designated cellular localization. Ion channels with moderate and low PIP_2 affinity are however likely to be modulated by physiological fluctuations in plasma membrane PIP_2 abundance. The next 'hot' question therefore is how the specificity of PIP_2 signalling is achieved? One possible mechanism for such specificity is a local PIP_2 depletion which would affect only those PIP_2-sensitive membrane proteins that are in close spatial juxtaposition to a PIP_2-depleting activity (e.g. GPCR coupled to PLC). The idea of local PIP_2 depletion is attractive but is not easily reconcilable with the suggested fast lateral diffusion of PIP_2 in the biological membranes (Yaradanakul and Hilgemann 2007) nor with the experimental data in neurons demonstrating that extracellular application of GPCR agonists can inhibit PIP_2-sensitive ion channels (e.g. Kv7) within the isolated membrane patch during the cell-attached patch clamp recording (Selyanko et al. 1992). Nevertheless local PIP_2 depletion hypothesis may work for some type of cells (e.g. in cardiomyocytes; (Cho et al. 2005)) or in neurons with long processes. Another mechanism for specificity for the PIP_2-mediated signalling may arise from the coincidence detection (as discussed above). Indeed, if PIP_2 depletion requires a set of cofactors in order to mediate modulation of a given ion channel, and different ion channels require different sets of cofactors, than the functional outcome of the receptor-mediated PIP_2 depletion will be defined by the availability and timing of the cofactor release (or withdrawal). All these interesting questions and theories require further research, which is well warranted given the fundamental nature of the phosphoinositide sensitivity of ion channels. A further focus on the interactions of the ion channels and phosphoinositides is brought about by the increasing evidence that mutations within the ion channel genes that disrupt channel interaction with phosphoinositides may underlie severe disorders in humans. Thus, three arrhytmogenic mutations within the Kv7.1 channel (Park et al. 2005) were suggested to impair cardiac I_{Ks} current by reducing apparent PIP_2 affinity of Kv7.1. Likewise, mutations affecting channel-phosphoinositide interactions within several Kir channel genes were linked to diseases such as Andersen–Tawil syndrome (ATS), hyperprostaglandin E syndrome (HPS) and congenital hyperinsulinism (CHI), reviewed in Logothetis et al. (2010). Therefore comprehensive future studies of ion channel sensitivity to and regulation by phosphoinositides are necessary for elucidation of basic principles of membrane-associated cellular signalling in health and disease.

References

Akey JM, Swanson WJ, Madeoy J, Eberle M, Shriver MD (2006) TRPV6 exhibits unusual patterns of polymorphism and divergence in worldwide populations. Hum Mol Genet 15:2106–2113

Akopian AN, Ruparel NB, Jeske NA, Hargreaves KM (2007) Transient receptor potential TRPA1 channel desensitization in sensory neurons is agonist dependent and regulated by TRPV1-directed internalization. J Physiol 583:175–193

Albert AP, Saleh SN, Large WA (2008) Inhibition of native TRPC6 channel activity by phosphatidylinositol 4,5-bisphosphate in mesenteric artery myocytes. J Physiol 586:3087–3095

Bal M, Zhang J, Hernandez CC, Zaika O, Shapiro MS (2010) Ca^{2+}/calmodulin disrupts AKAP79/150 interactions with KCNQ (M-Type) K^+ channels. J Neurosci 30:2311–2323

Balla T (2001) Pharmacology of phosphoinositides, regulators of multiple cellular functions. Curr Pharm Des 7:475–507

Balla T (2009) Green light to illuminate signal transduction events. Trends Cell Biol 19:575–586

Balla T, Varnai P (2009) Visualization of cellular phosphoinositide pools with GFP-fused protein-domains. Curr Protoc Cell Biol Chapter 24:Unit 24.4

Bannister RA, Melliti K, Adams BA (2002) Reconstituted slow muscarinic inhibition of neuronal (Ca(v)1.2c) L-type Ca^{2+} channels. Biophys J 83:3256–3267

Barhanin J, Lesage F, Guillemare E, Fink M, Lazdunski M, Romey G (1996) K(V)LQT1 and lsK (minK) proteins associate to form the I(Ks) cardiac potassium current. Nature 384:78–80

Baukrowitz T, Schulte U, Oliver D, Herlitze S, Krauter T, Tucker SJ, Ruppersberg JP, Fakler B (1998) PIP_2 and PIP as determinants for ATP inhibition of KATP channels. Science 1141–1144

Bautista DM, Siemens J, Glazer JM, Tsuruda PR, Basbaum AI, Stucky CL, Jordt SE, Julius D (2007) The menthol receptor TRPM8 is the principal detector of environmental cold. Nature 448:204–208

Bean BP (1989) Neurotransmitter inhibition of neuronal calcium currents by changes in channel voltage dependence. Nature 340:153–156

Bernheim L, Beech DJ, Hille B (1991) A diffusible second messenger mediates one of the pathways coupling receptors to calcium channels in rat sympathetic neurons. Neuron 6:859–867

Bianco SD, Peng JB, Takanaga H, Suzuki Y, Crescenzi A, Kos CH, Zhuang L, Freeman MR, Gouveia CH, Wu J, Luo H, Mauro T, Brown EM, Hediger MA (2006) Marked disturbance of calcium homeostasis in mice with targeted disruption of the Trpv6 calcium channel gene. J Bone Miner Res 22:274–285

Brown DA, Adams PR (1980) Muscarinic suppression of a novel voltage-sensitive K^+ current in a vertebrate neurone. Nature 283:673–676

Brown DA, Passmore GM (2009) Neural KCNQ (Kv7) channels. Br J Pharmacol 156:1185–1195

Caterina MJ, Schumacher MA, Tominaga M, Rosen TA, Levine JD, Julius D (1997) The capsaicin receptor: a heat-activated ion channel in the pain pathway. Nature 389:816–824

Catterall WA (2000) Structure and regulation of voltage-gated Ca^{2+} channels. Annu Rev Cell Dev Biol 16:521–555

Charpentier F, Merot J, Loussouarn G, Baro I (2010) Delayed rectifier K^+ currents and cardiac repolarization. J Mol Cell Cardiol 48:37–44

Cho H, Kim YA, Yoon JY, Lee D, Kim JH, Lee SH, Ho WK (2005) Low mobility of phosphatidylinositol 4,5-bisphosphate underlies receptor specificity of Gq-mediated ion channel regulation in atrial myocytes. Proc Natl Acad Sci U S A 102:15241–15246

Chuang HH, Prescott ED, Kong H, Shields S, Jordt SE, Basbaum AI, Chao MV, Julius D (2001) Bradykinin and nerve growth factor release the capsaicin receptor from PtdIns(4,5)P_2 mediated inhibition. Nature 411:957–962

Clapham DE, Runnels LW, Strubing C (2001) The TRP ion channel family. Nat Rev Neurosci 2:387–396

Colburn RW, Lubin ML, Stone DJ Jr, Wang Y, Lawrence D, D'Andrea MR, Brandt MR, Liu Y, Flores CM, Qin N (2007) Attenuated cold sensitivity in TRPM8 null mice. Neuron 54:379–386

Colquhoun D (1998) Binding, gating, affinity and efficacy: the interpretation of structure-activity relationships for agonists and of the effects of mutating receptors. Br J Pharmacol 125:924–947

Cruzblanca H, Koh DS, Hille B (1998) Bradykinin inhibits M current via phospholipase C and Ca^{2+} release from PI_3-sensitive Ca^{2+} stores in rat sympathetic neurons. Proc Natl Acad Sci U S A 95:7151–7156

D'Avanzo N, Cheng WW, Doyle DA, Nichols CG (2010) Direct and specific activation of human inward rectifier K^+ channels by membrane phosphatidylinositol 4,5-bisphosphate. J Biol Chem

Dai Y, Wang S, Tominaga M, Yamamoto S, Fukuoka T, Higashi T, Kobayashi K, Obata K, Yamanaka H, Noguchi K (2007) Sensitization of TRPA1 by PAR2 contributes to the sensation of inflammatory pain. J Clin Invest 117:1979–1987

Daniels RL, Takashima Y, McKemy DD (2008) The activity of the neuronal cold sensor TRPM8 is regulated by phospholipase C via the phospholipid phosphoinositol-4,5-bisphosphate. J Biol Chem 284:1570–1582

Darian-Smith I, Johnson KO, Dykes R (1973) "Cold" fiber population innervating palmar and digital skin of the monkey: responses to cooling pulses. J Neurophysiol 36:325–346

Delmas P, Brown DA (2005) Pathways modulating neural KCNQ/M (Kv7) potassium channels. Nat Rev Neurosci 6:850–862

Delmas P, Wanaverbecq N, Abogadie FC, Mistry M, Brown DA (2002) Signaling microdomains define the specificity of receptor-mediated $InsP_3$ pathways in neurons. Neuron 34:209–220

Delmas P, Coste B, Gamper N, Shapiro MS (2005) Phosphoinositide lipid second messengers: new paradigms for calcium channel modulation. Neuron 47:179–182

Derler I, Hofbauer M, Kahr H, Fritsch R, Muik M, Kepplinger K, Hack ME, Moritz S, Schindl R, Groschner K, Romanin C (2006) Dynamic but not constitutive association of calmodulin with rat TRPV6 channels enables fine tuning of Ca^{2+}-dependent inactivation. J Physiol 577:31–44

Dhaka A, Murray AN, Mathur J, Earley TJ, Petrus MJ, Patapoutian A (2007) TRPM8 is required for cold sensation in mice. Neuron 54:371–378

Docherty RJ, Yeats JC, Bevan S, Boddeke HW (1996) Inhibition of calcineurin inhibits the desensitization of capsaicin-evoked currents in cultured dorsal root ganglion neurones from adult rats. Pflugers Arch 431:828–837

Dong XP, Shen D, Wang X, Dawson T, Li X, Zhang Q, Cheng X, Zhang Y, Weisman LS, Delling M, Xu H (2010) $PI(3,5)P_2$ controls membrane traffic by direct activation of mucolipin Ca release channels in the endolysosome. Nat Commun 1

Du X, Zhang H, Lopes CM, Mirshahi T, Rohacs T, Logothetis DE (2004) Characteristic interactions with phosphatidylinositol 4,5-bisphosphate determine regulation of Kir channels by diverse modulators. J Biol Chem 279:37271–37281

Eccles R (1994) Menthol and related cooling compounds. J Pharm Pharmacol 46:618–630

Estacion M, Sinkins WG, Schilling WP (2001) Regulation of Drosophila transient receptor potential-like (TrpL) channels by phospholipase C-dependent mechanisms. J Physiol 530:1–19

Falkenburger BH, Jensen JB, Dickson EJ, Suh BC, Hille B (2010a) Phosphoinositides: lipid regulators of membrane proteins. J Physiol 588:3179–3185

Falkenburger BH, Jensen JB, Hille B (2010b) Kinetics of M1 muscarinic receptor and G protein signaling to phospholipase C in living cells. J Gen Physiol 135:81–97

Falkenburger BH, Jensen JB, Hille B (2010c) Kinetics of PIP_2 metabolism and KCNQ2/3 channel regulation studied with a voltage-sensitive phosphatase in living cells. J Gen Physiol 135:99–114

Fan Z, Makielski JC (1997) Anionic phospholipids activate ATP-sensitive potassium channels. J Biol Chem 272:5388–5395

Gamper N, Shapiro MS (2003) Calmodulin mediates Ca^{2+}-dependent modulation of M-type K^+ channels. J Gen Physiol 122:17–31

Gamper N, Shapiro MS (2007a) Regulation of ion transport proteins by membrane phosphoinositides. Nat Rev Neurosci 8:921–934

Gamper N, Shapiro MS (2007b) Target-specific PIP_2 signalling: how might it work? J Physiol 582:967–975

Gamper N, Reznikov V, Yamada Y, Yang J, Shapiro MS (2004) Phosphotidylinositol 4,5-bisphosphate signals underlie receptor-specific $G_{q/11}$-mediated modulation of N-type Ca^{2+} channels. J Neurosci 24:10980–10992

Gamper N, Li Y, Shapiro MS (2005) Structural requirements for differential sensitivity of KCNQ K^+ channels to modulation by Ca^{2+}/calmodulin. Mol Biol Cell 16:3538–3551

Greenwood IA, Ohya S (2009) New tricks for old dogs: KCNQ expression and role in smooth muscle. Br J Pharmacol 156:1196–1203

Gribble FM, Proks P, Corkey BE, Ashcroft FM (1998) Mechanism of cloned ATP-sensitive potassium channel activation by oleoyl-CoA. J Biol Chem 273:26383–26387

Gribkoff VK (2006) The role of voltage-gated calcium channels in pain and nociception. Semin Cell Dev Biol 17:555–564

Haley JE, Abogadie FC, Delmas P, Dayrell M, Vallis Y, Milligan G, Caulfield MP, Brown DA, Buckley NJ (1998) The alpha subunit of Gq contributes to muscarinic inhibition of the M-type potassium current in sympathetic neurons. J Neurosci 18:4521–4531

Haley JE, Delmas P, Offermanns S, Abogadie FC, Simon MI, Buckley NJ, Brown DA (2000) Muscarinic inhibition of calcium current and M current in Galpha q-deficient mice. J Neurosci 20:3973–3979

Hardie RC (2007) TRP channels and lipids: from Drosophila to mammalian physiology. J Physiol 578:9–24

Hardie RC, Raghu P (2001) Visual transduction in Drosophila. Nature 413:186–193

Herlitze S, Garcia DE, Mackie K, Hille B, Scheuer T, Catterall WA (1996) Modulation of Ca^{2+} channels by G-protein beta gamma subunits. Nature 380:258–262

Herlitze S, Hockerman GH, Scheuer T, Catterall WA (1997) Molecular determinants of inactivation and G protein modulation in the intracellular loop connecting domains I and II of the calcium channel alpha1A subunit. Proc Natl Acad Sci U S A 94:1512–1516

Hernandez CC, Zaika O, Shapiro MS (2008a) A carboxy-terminal inter-helix linker as the site of phosphatidylinositol 4,5-bisphosphate action on Kv7 (M-type) K^+ channels. J Gen Physiol 132:361–381

Hernandez CC, Zaika O, Tolstykh GP, Shapiro MS (2008b) Regulation of neural KCNQ channels: signalling pathways, structural motifs and functional implications. J Physiol 586:1811–1821

Hernandez CC, Falkenburger B, Shapiro MS (2009) Affinity for phosphatidylinositol 4,5-bisphosphate determines muscarinic agonist sensitivity of $Kv7K^+$ channels. J Gen Physiol 134:437–448

Hibino H, Inanobe A, Furutani K, Murakami S, Findlay I, Kurachi Y (2010) Inwardly rectifying potassium channels: their structure, function, and physiological roles. Physiol Rev 90:291–366

Hilgemann DW (1997) Cytoplasmic ATP-dependent regulation of ion transporters and channels: mechanisms and messengers. Annu Rev Physiol 59:193–220

Hilgemann DW, Ball R (1996) Regulation of cardiac Na^+/Ca^{2+} exchange and K_{ATP} potassium channels by PIP_2. Science 273:956–959

Hilgemann DW, Feng S, Nasuhoglu C (2001) The complex and intriguing lives of PIP_2 with ion channels and transporters. Sci STKE 2001:RE19

Hille B (1994) Modulation of ion-channel function by G-protein-coupled receptors. Trends Neurosci 17:531–536

Hirose K, Kadowaki S, Tanabe M, Takeshima H, Iino M (1999) Spatiotemporal dynamics of inositol 1,4,5-trisphosphate that underlies complex Ca^{2+} mobilization patterns. Science 284:1527–1530

Hoenderop JG, Leeuwen JP van, Eerden BC van der, Kersten FF, Kemp AW van der, Merillat AM, Waarsing JH, Rossier BC, Vallon V, Hummler E, Bindels RJ (2003) Renal Ca^{2+} wasting, hyperabsorption, and reduced bone thickness in mice lacking TRPV5. J Clin Invest 112:1906–1914

Hoenderop JG, Nilius B, Bindels RJ (2005) Calcium absorption across epithelia. Physiol Rev 85:373–422

Hofmann T, Obukhov AG, Schaefer M, Harteneck C, Gudermann T, Schultz G (1999) Direct activation of human TRPC6 and TRPC3 channels by diacylglycerol. Nature 397:259–263

Horowitz LF, Hirdes W, Suh BC, Hilgemann DW, Mackie K, Hille B (2005) Phospholipase C in living cells: activation, inhibition, Ca^{2+} requirement, and regulation of M current. J Gen Physiol 126:243–262

Hoshi N, Zhang JS, Omaki M, Takeuchi T, Yokoyama S, Wanaverbecq N, Langeberg LK, Yoneda Y, Scott JD, Brown DA, Higashida H (2003) AKAP150 signaling complex promotes suppression of the M-current by muscarinic agonists. Nat Neurosci 6:564–571

Hsuan JJ, Minogue S, dos Santos M (1998) Phosphoinositide 4- and 5-kinases and the cellular roles of phosphatidylinositol 4,5-bisphosphate. Adv Cancer Res 74:167–216

Huang CL (2007) Complex roles of PIP_2 in the regulation of ion channels and transporters. Am J Physiol Renal Physiol 293:F1761–F1765

Huang CL, Feng S, Hilgemann DW (1998) Direct activation of inward rectifier potassium channels by PIP_2 and its stabilization by $G_{\beta\gamma}$. Nature 391:803–806

Huang J, Liu CH, Hughes SA, Postma M, Schwiening CJ, Hardie RC (2010) Activation of TRP channels by protons and phosphoinositide depletion in Drosophila photoreceptors. Curr Biol 20:189–197

Iannotti FA, Panza E, Barrese V, Viggiano D, Soldovieri MV, Taglialatela M (2010) Expression, localization, and pharmacological role of Kv7 potassium channels in skeletal muscle proliferation, differentiation, and survival after myotoxic insults. J Pharmacol Exp Ther 332:811–820

Ikeda SR (1996) Voltage-dependent modulation of N-type calcium channels by G-protein beta gamma subunits. Nature 380:255–258

Inanobe A, Nakagawa A, Matsuura T, Kurachi Y (2010) A structural determinant for the control of PIP_2-sensitivity in G protein-gated inward rectifier K^+ channels. J Biol Chem

Iwasaki H, Murata Y, Kim Y, Hossain MI, Worby CA, Dixon JE, McCormack T, Sasaki T, Okamura Y (2008) A voltage-sensing phosphatase, Ci-VSP, which shares sequence identity with PTEN, dephosphorylates phosphatidylinositol 4,5-bisphosphate. Proc Natl Acad Sci U S A 105:7970–7975

Jardin I, Redondo PC, Salido GM, Rosado JA (2008) Phosphatidylinositol 4,5-bisphosphate enhances store-operated calcium entry through hTRPC6 channel in human platelets. Biochim Biophys Acta 1783:84–97

Jia Z, Bei J, Rodat-Despoix L, Liu B, Jia Q, Delmas P, Zhang H (2008) NGF inhibits M/KCNQ currents and selectively alters neuronal excitability in subsets of sympathetic neurons depending on their M/KCNQ current background. J Gen Physiol 131:575–587

Ju M, Shi J, Saleh SN, Albert AP, Large WA (2010) Ins(1,4,5)P_3 interacts with PIP_2 to regulate activation of TRPC6/C7 channels by diacylglycerol in native vascular myocytes. J Physiol 588:1419–1433

Karashima Y, Prenen J, Meseguer V, Owsianik G, Voets T, Nilius B (2008) Modulation of the transient receptor potential channel TRPA1 by phosphatidylinositol 4,5-bisphosphate manipulators. Pflugers Arch 457:77–89

Keselman I, Fribourg M, Felsenfeld DP, Logothetis DE (2007) Mechanism of PLC-mediated Kir3 current inhibition. Channels (Austin) 1:113–123

Kharkovets T, Hardelin JP, Safieddine S, Schweizer M, El Amraoui A, Petit C, Jentsch TJ (2000) KCNQ4, a K^+ channel mutated in a form of dominant deafness, is expressed in the inner ear and the central auditory pathway. Proc Natl Acad Sci U S A 97:4333–4338

Kharkovets T, Dedek K, Maier H, Schweizer M, Khimich D, Nouvian R, Vardanyan V, Leuwer R, Moser T, Jentsch TJ (2006) Mice with altered KCNQ4 K^+ channels implicate sensory outer hair cells in human progressive deafness. EMBO J 25:642–652

Kim D, Cavanaugh EJ (2007) Requirement of a soluble intracellular factor for activation of transient receptor potential A1 by pungent chemicals: role of inorganic polyphosphates. J Neurosci 27:6500–6509

Kim AY, Tang Z, Liu Q, Patel KN, Maag D, Geng Y, Dong X (2008a) Pirt, a phosphoinositide-binding protein, functions as a regulatory subunit of TRPV1. Cell 133:475–485

Kim BJ, Kim MT, Jeon JH, Kim SJ, So I (2008b) Involvement of phosphatidylinositol 4,5-bisphosphate in the desensitization of canonical transient receptor potential 5. Biol Pharm Bull 31:1733–1738

Kim D, Cavanaugh EJ, Simkin D (2008c) Inhibition of transient receptor potential A1 channel by phosphatidylinositol-4,5-bisphosphate. Am J Physiol Cell Physiol 295:C92–C99

Klein RM, Ufret-Vincenty CA, Hua L, Gordon SE (2008) Determinants of molecular specificity in phosphoinositide regulation: PI(4,5)P_2 is the endogenous lipid regulating TRPV1. J Biol Chem 283:26208–26216

Kubisch C, Schroeder BC, Friedrich T, Lutjohann B, El Amraoui A, Marlin S, Petit C, Jentsch TJ (1999) KCNQ4, a novel potassium channel expressed in sensory outer hair cells, is mutated in dominant deafness. Cell 96:437–446

Kubista H, Kosenburger K, Mahlknecht P, Drobny H, Boehm S (2009) Inhibition of transmitter release from rat sympathetic neurons via presynaptic M(1) muscarinic acetylcholine receptors. Br J Pharmacol 156:1342–1352

Kwon Y, Hofmann T, Montell C (2007) Integration of phosphoinositide- and calmodulin-mediated regulation of TRPC6. Mol Cell 25:491–503

Langeslag M, Clark K, Moolenaar WH, Leeuwen FN van, Jalink K (2007) Activation of TRPM7 channels by PLC-coupled receptor agonists. J Biol Chem 282:232–239

Lechner SG, Hussl S, Schicker KW, Drobny H, Boehm S (2005) Presynaptic inhibition via a phospholipase C- and phosphatidylinositol bisphosphate-dependent regulation of neuronal Ca^{2+} channels. Mol Pharmacol 68:1387–1396

Lee J, Cha SK, Sun TJ, Huang C-L (2005) PIP_2 activates TRPV5 and releases its inhibition by intracellular Mg^{2+}. J Gen Physiol 126:439–451

Lemonnier L, Trebak M, Putney JW Jr (2007) Complex regulation of the TRPC3, 6 and 7 channel subfamily by diacylglycerol and phosphatidylinositol-4,5-bisphosphate. Cell Calcium 43: 506–514

Leslie CC (2004) Regulation of the specific release of arachidonic acid by cytosolic phospholipase A2. Prostaglandins Leukot Essent Fatty Acids 70:373–376

Li Y, Gamper N, Shapiro MS (2004) Single-channel analysis of KCNQ K^+ channels reveals the mechanism of augmentation by a cysteine-modifying reagent. J Neurosci 24:5079–5090

Li Y, Gamper N, Hilgemann DW, Shapiro MS (2005) Regulation of Kv7 (KCNQ) K^+ channel open probability by phosphatidylinositol 4,5-bisphosphate. J Neurosci 25:9825–9835

Linley JE, Rose K, Patil M, Robertson B, Akopian AN, Gamper N (2008) Inhibition of M current in sensory neurons by exogenous proteases: a signaling pathway mediating inflammatory nociception. J Neurosci 28:11240–11249

Linley JE, Rose K, Ooi L, Gamper N (2010) Understanding inflammatory pain: ion channels contributing to acute and chronic nociception. Pflugers Arch 459:657–669

Lipscombe D, Kongsamut S, Tsien RW (1989) Alpha-adrenergic inhibition of sympathetic neurotransmitter release mediated by modulation of N-type calcium-channel gating. Nature 340:639–642

Lishko PV, Procko E, Jin X, Phelps CB, Gaudet R (2007) The ankyrin repeats of TRPV1 bind multiple ligands and modulate channel sensitivity. Neuron 54:905–918

Liu SJ (2007) Inhibition of L-type Ca^{2+} channel current and negative inotropy induced by arachidonic acid in adult rat ventricular myocytes. Am J Physiol Cell Physiol 293:C1594–C1604

Liu D, Liman ER (2003) Intracellular Ca^{2+} and the phospholipid PIP_2 regulate the taste transduction ion channel TRPM5. Proc Natl Acad Sci U S A 100:15160–15165

Liu L, Rittenhouse AR (2003) Arachidonic acid mediates muscarinic inhibition and enhancement of N-type Ca^{2+} current in sympathetic neurons. Proc Natl Acad Sci U S A 100:295–300

Liu B, Qin F (2005) Functional control of cold- and menthol-sensitive TRPM8 ion channels by phosphatidylinositol 4,5-bisphosphate. J Neurosci 25:1674–1681

Liu L, Barrett CF, Rittenhouse AR (2001) Arachidonic acid both inhibits and enhances whole cell calcium currents in rat sympathetic neurons. Am J Physiol Cell Physiol 280:C1293–C1305

Liu B, Zhang C, Qin F (2005) Functional recovery from desensitization of vanilloid receptor TRPV1 requires resynthesis of phosphatidylinositol 4,5-bisphosphate. J Neurosci 25:4835–4843

Liu L, Zhao R, Bai Y, Stanish LF, Evans JE, Sanderson MJ, Bonventre JV, Rittenhouse AR (2006) M1 muscarinic receptors inhibit L-type Ca^{2+} current and M-current by divergent signal transduction cascades. J Neurosci 26:11588–11598

Liu B, Linley JE, Du X, Zhang X, Ooi L, Zhang H, Gamper N (2010) The acute nociceptive signals induced by bradykinin in rat sensory neurons are mediated by inhibition of M-type K^+ channels and activation of Ca^{2+}-activated Cl^- channels. J Clin Invest 120:1240–1252

Logothetis DE, Jin T, Lupyan D, Rosenhouse-Dantsker A (2007a) Phosphoinositide-mediated gating of inwardly rectifying K^+ channels. Pflugers Arch 455:83–95

Logothetis DE, Lupyan D, Rosenhouse-Dantsker A (2007b) Diverse Kir modulators act in close proximity to residues implicated in phosphoinositide binding. J Physiol 582:953–965

Logothetis DE, Petrou VI, Adney SK, Mahajan R (2010) Channelopathies linked to plasma membrane phosphoinositides. Pflugers Arch 460:321–341

Lopes CMB, Zhang H, Rohacs T, Jin T, Logothetis DE (2002) Alterations in Conserved Kir Channel-PIP2 Interactions Underlie Channelopathies. Neuron 34:933–944

Loussouarn G, Park KH, Bellocq C, Baro I, Charpentier F, Escande D (2003) Phosphatidylinositol-4,5-bisphosphate, PIP_2, controls KCNQ1/KCNE1 voltage-gated potassium channels: a functional homology between voltage-gated and inward rectifier K^+ channels. EMBO J 22:5412–5421

Lukacs V, Thyagarajan B, Balla A, Varnai P, Balla T, Rohacs T (2007) Dual regulation of TRPV1 by phosphoinositides. J Neurosci 27:7070–7080

Ma R, Li WP, Rundle D, Kong J, Akbarali HI, Tsiokas L (2005) PKD2 functions as an epidermal growth factor-activated plasma membrane channel. Mol Cell Biol 25:8285–8298

MacGregor GG, Dong K, Vanoye CG, Tang L, Giebisch G, Hebert SC (2002) Nucleotides and phospholipids compete for binding to the C terminus of KATP channels. Proc Natl Acad Sci U S A 99:2726–2731

Maljevic S, Wuttke TV, Seebohm G, Lerche H (2010) KV7 channelopathies. Pflugers Arch 460:277–288

Marrion NV (1997) Control of M-current. Annu Rev Physiol 59:483–504

Marrion NV, Smart TG, Marsh SJ, Brown DA (1989) Muscarinic suppression of the M-current in the rat sympathetic ganglion is mediated by receptors of the M1-subtype. Br J Pharmacol 98:557–573

Mathie A, Bernheim L, Hille B (1992) Inhibition of N- and L-type calcium channels by muscarinic receptor activation in rat sympathetic neurons. Neuron 8:907–914

McKemy DD, Neuhausser WM, Julius D (2002) Identification of a cold receptor reveals a general role for TRP channels in thermosensation. Nature 416:52–58

Michailidis IE, Zhang Y, Yang J (2007) The lipid connection-regulation of voltage-gated Ca^{2+} channels by phosphoinositides. Pflugers Arch 455:147–155

Minke B, Cook B (2002) TRP channel proteins and signal transduction. Physiol Rev 82:429–472

Mitchell CA, Brown S, Campbell JK, Munday AD, Speed CJ (1996) Regulation of second messengers by the inositol polyphosphate 5- phosphatases. Biochem Soc Trans 24:994–1000

Mohapatra DP, Nau C (2005) Regulation of Ca^{2+}-dependent desensitization in the vanilloid receptor TRPV1 by calcineurin and cAMP-dependent protein kinase. J Biol Chem 280:13424–13432

Montell C, Birnbaumer L, Flockerzi V (2002) The TRP channels, a remarkably functional family. Cell 108:595–598

Mucha M, Ooi L, Linley JE, Mordaka P, Dalle C, Robertson B, Gamper N, Wood IC (2010) Transcriptional control of KCNQ channel genes and the regulation of neuronal excitability. J Neurosci 30:13235–13245

Murata Y, Okamura Y (2007) Depolarization activates the phosphoinositide phosphatase Ci-VSP, as detected in Xenopus oocytes coexpressing sensors of PIP_2. J Physiol 583:875–889

Niemeyer BA, Bergs C, Wissenbach U, Flockerzi V, Trost C (2001) Competitive regulation of CaT-like-mediated Ca^{2+} entry by protein kinase C and calmodulin. Proc Natl Acad Sci U S A 98:3600–3605

Nilius B, Owsianik G (2010) Channelopathies converge on TRPV4. Nat Genet 42:98–100

Nilius B, Voets T, Peters J (2005) TRP channels in disease. Sci STKE 2005:re8

Nilius B, Mahieu F, Prenen J, Janssens A, Owsianik G, Vennekens R, Voets T (2006) The Ca^{2+}-activated cation channel TRPM4 is regulated by phosphatidylinositol 4,5-biphosphate. EMBO J 25:467–478

Nilius B, Owsianik G, Voets T (2008) Transient receptor potential channels meet phosphoinositides. EMBO J 27:2809–2816

Numazaki M, Tominaga T, Takeuchi K, Murayama N, Toyooka H, Tominaga M (2003) Structural determinant of TRPV1 desensitization interacts with calmodulin. Proc Natl Acad Sci U S A 100:8002–8006

Otsuguro KI, Tang J, Tang Y, Xiao R, Freichel M, Tsvilovskyy V, Ito S, Flockerzi V, Zhu MX, Zholos AV (2008) Isoform-specific inhibition of TRPC4 channel by phosphatidylinositol 4,5-bisphosphate. J Biol Chem 283:10026–10036

Park KH, Piron J, Dahimene S, Merot J, Baro I, Escande D, Loussouarn G (2005) Impaired KCNQ1-KCNE1 and phosphatidylinositol-4,5-bisphosphate interaction underlies the long QT syndrome. Circ Res 96:730–739

Pegan S, Arrabit C, Zhou W, Kwiatkowski W, Collins A, Slesinger PA, Choe S (2005) Cytoplasmic domain structures of Kir2.1 and Kir3.1 show sites for modulating gating and rectification. Nat Neurosci 8:279–287

Peier AM, Moqrich A, Hergarden AC, Reeve AJ, Andersson DA, Story GM, Earley TJ, Dragoni I, McIntyre P, Bevan S, Patapoutian A (2002) A TRP channel that senses cold stimuli and menthol. Cell 108:705–715

Pingle SC, Matta JA, Ahern GP (2007) Capsaicin receptor: TRPV1 a promiscuous TRP channel. Handb Exp Pharmacol 155–171

Piron J, Choveau FS, Amarouch MY, Rodriguez N, Charpentier F, Merot J, Baro I, Loussouarn G (2010) KCNE1-KCNQ1 osmoregulation by interaction of phosphatidylinositol-4,5-bisphosphate with Mg^{2+} and polyamines. J Physiol 588:3471–3483

Prescott ED, Julius D (2003) A modular PIP_2 binding site as a determinant of capsaicin receptor sensitivity. Science 300:1284–1288

Proudfoot CJ, Garry EM, Cottrell DF, Rosie R, Anderson H, Robertson DC, Fleetwood-Walker SM, Mitchell R (2006) Analgesia mediated by the TRPM8 cold receptor in chronic neuropathic pain. Curr Biol 16:1591–1605

Quinn KV, Behe P, Tinker A (2008) Monitoring changes in membrane phosphatidylinositol 4,5-bisphosphate in living cells using a domain from the transcription factor tubby. J Physiol 586:2855–2871

Ramsey IS, Delling M, Clapham DE (2006) An introduction to TRP channels. Annu Rev Physiol 68:619–647

Rapedius M, Soom M, Shumilina E, Schulze D, Schonherr R, Kirsch C, Lang F, Tucker SJ, Baukrowitz T (2005) Long chain CoA esters as competitive antagonists of phosphatidylinositol 4,5-bisphosphate activation in Kir channels. J Biol Chem 280:30760–30767

Raychowdhury MK, Gonzalez-Perrett S, Montalbetti N, Timpanaro GA, Chasan B, Goldmann WH, Stahl S, Cooney A, Goldin E, Cantiello HF (2004) Molecular pathophysiology of mucolipidosis type IV: pH dysregulation of the mucolipin-1 cation channel. Hum Mol Genet 13:617–627

Reid G, Babes A, Pluteanu F (2002) A cold- and menthol-activated current in rat dorsal root ganglion neurones: properties and role in cold transduction. J Physiol 545:595–614

Robbins J, Marsh SJ, Brown DA (1993) On the mechanism of M-current inhibition by muscarinic m1 receptors in DNA-transfected rodent neuroblastoma x glioma cells. J Physiol 469:153–178

Robbins J, Marsh SJ, Brown DA (2006) Probing the regulation of M (Kv7) potassium channels in intact neurons with membrane-targeted peptides. J Neurosci 26:7950–7961

Roberts-Crowley ML, Mitra-Ganguli T, Liu L, Rittenhouse AR (2009) Regulation of voltage-gated Ca^{2+} channels by lipids. Cell Calcium 45:589–601

Rohacs T (2007) Regulation of TRP channels by PIP_2. Pflugers Arch 453:753–762

Rohacs T (2009) Phosphoinositide regulation of non-canonical transient receptor potential channels. Cell Calcium 45:554–565

Rohacs T, Chen J, Prestwich GD, Logothetis DE (1999) Distinct specificities of inwardly rectifying K^+ channels for phosphoinositides. J Biol Chem 274:36065–36072

Rohacs T, Lopes C, Mirshahi T, Jin T, Zhang H, Logothetis DE (2002) Assaying phosphatidylinositol bisphosphate regulation of potassium channels. Methods Enzymol 345:71–92

Rohacs T, Lopes CM, Jin T, Ramdya PP, Molnar Z, Logothetis DE (2003) Specificity of activation by phosphoinositides determines lipid regulation of Kir channels. Proc Natl Acad Sci U S A 100:745–750

Rohacs T, Lopes CMB, Michailidis I, Logothetis DE (2005) $PI(4,5)P_2$ regulates the activation and desensitization of TRPM8 channels through the TRP domain. Nat Neurosci 8:626–634

Rohacs T, Thyagarajan B, Lukacs V (2008) Phospholipase C mediated modulation of TRPV1 channels. Mol Neurobiol 37:153–163

Rosenbaum T, Gordon-Shaag A, Munari M, Gordon SE (2004) Ca^{2+}/calmodulin modulates TRPV1 activation by capsaicin. J Gen Physiol 123:53–62

Rosenhouse-Dantsker A, Logothetis DE (2007) Molecular characteristics of phosphoinositide binding. Pflugers Arch 455:45–53

Rosenhouse-Dantsker A, Sui JL, Zhao Q, Rusinova R, Rodriguez-Menchaca AA, Zhang Z, Logothetis DE (2008) A sodium-mediated structural switch that controls the sensitivity of Kir channels to PtdIns(4,5)P(2). Nat Chem Biol 4:624–631

Rousset M, Cens T, Gouin-Charnet A, Scamps F, Charnet P (2004) Ca^{2+} and phosphatidylinositol 4,5-bisphosphate stabilize a $G_{\beta\gamma}$-sensitive state of Cav2 Ca $^{2+}$ channels. J Biol Chem 279: 14619–14630

Runnels LW, Yue L, Clapham DE (2002) The TRPM7 channel is inactivated by PIP2 hydrolysis. Nat Cell Biol 4:329–336

Saleh SN, Albert AP, Large WA (2008) Obligatory role for phosphatidylinositol-4, 5-bisphosphate in activation of native TRPC1 store-operated channels in vascular myocytes. J Physiol 587:531–540

Saleh SN, Albert AP, Large WA (2009) Activation of native TRPC1/C5/C6 channels by endothelin-1 is mediated by both PIP3 and PIP2 in rabbit coronary artery myocytes. J Physiol 587:5361–5375

Sanguinetti MC, Curran ME, Zou A, Shen J, Spector PS, Atkinson DL, Keating MT (1996) Coassembly of K_VLQT1 and minK (IsK) proteins to form cardiac I_{Ks} potassium channel. Nature 384:80–83

Selyanko AA, Brown DA (1996) Intracellular calcium directly inhibits potassium M channels in excised membrane patches from rat sympathetic neurons. Neuron 16:151–162

Selyanko AA, Stansfeld CE, Brown DA (1992) Closure of potassium M-channels by muscarinic acetylcholine-receptor stimulants requires a diffusible messenger. Proc Biol Sci 250:119–125

Selyanko AA, Hadley JK, Brown DA (2001) Properties of single M-type KCNQ2/KCNQ3 potassium channels expressed in mammalian cells. J Physiol 534:15–24

Selyanko AA, Delmas P, Hadley JK, Tatulian L, Wood IC, Mistry M, London B, Brown DA (2002) Dominant-negative subunits reveal potassium channel families that contribute to M-like potassium currents. J Neurosci 22:RC212

Shapiro MS, Wollmuth LP, Hille B (1994) Angiotensin II inhibits calcium and M current channels in rat sympathetic neurons via G proteins. Neuron 12:1319–1329

Shapiro MS, Roche JP, Kaftan EJ, Cruzblanca H, Mackie K, Hille B (2000) Reconstitution of muscarinic modulation of the KCNQ2/KCNQ3 K^+ channels that underlie the neuronal M current. J Neurosci 20:1710–1721

Shumilina E, Klocker N, Korniychuk G, Rapedius M, Lang F, Baukrowitz T (2006) Cytoplasmic accumulation of long-chain coenzyme A esters activates K_{ATP} and inhibits Kir2.1 channels. J Physiol 575:433–442

Shyng SL, Nichols CG (1998) Membrane phospholipid control of nucleotide sensitivity of K_{ATP} channels. Science 1138–1141

Soejima M, Noma A (1984) Mode of regulation of the ACh-sensitive K-channel by the muscarinic receptor in rabbit atrial cells. Pflugers Arch 400:424–431

Soom M, Schonherr R, Kubo Y, Kirsch C, Klinger R, Heinemann SH (2001) Multiple PIP_2 binding sites in Kir2.1 inwardly rectifying potassium channels. FEBS Lett 490:49–53

Stein AT, Ufret-Vincenty CA, Hua L, Santana LF, Gordon SE (2006) Phosphoinositide 3-Kinase Binds to TRPV1 and Mediates NGF-stimulated TRPV1 Trafficking to the Plasma Membrane. J Gen Physiol 128:509–522

Stolz LE, Huynh CV, Thorner J, York JD (1998) Identification and characterization of an essential family of inositol polyphosphate 5-phosphatases (INP51, INP52 and INP53 gene products) in the yeast Saccharomyces cerevisiae. Genetics 148:1715–1729

Suh BC, Hille B (2002) Recovery from muscarinic modulation of M current channels requires phosphatidylinositol 4,5-bisphosphate synthesis. Neuron 35:507–520

Suh BC, Hille B (2007) Electrostatic interaction of internal Mg^{2+} with membrane PIP_2 Seen with KCNQ K^+ channels. J Gen Physiol 130:241–256
Suh BC, Hille B (2008) PIP_2 is a necessary cofactor for ion channel function: how and why? Annu Rev Biophys 37:175–195
Suh BC, Inoue T, Meyer T, Hille B (2006) Rapid chemically induced changes of PtdIns(4,5)P_2 gate KCNQ ion channels. Science 314:1454–1457
Suh BC, Leal K, Hille B (2010) Modulation of high-voltage activated Ca^{2+} channels by membrane phosphatidylinositol 4,5-bisphosphate. Neuron 67:224–238
Sui JL, Petit Jacques J, Logothetis DE (1998) Activation of the atrial K_{ACh} channel by the betagamma subunits of G proteins or intracellular Na^+ ions depends on the presence of phosphatidylinositol phosphates. Proc Natl Acad Sci U S A 95:1307–1312
Suzuki Y, Pasch A, Bonny O, Mohaupt MG, Hediger MA, Frey FJ (2008) Gain-of-function haplotype in the epithelial calcium channel TRPV6 is a risk factor for renal calcium stone formation. Hum Mol Genet 17:1613–1618
Szentpetery Z, Balla A, Kim YJ, Lemmon MA, Balla T (2009) Live cell imaging with protein domains capable of recognizing phosphatidylinositol 4,5-bisphosphate; a comparative study. BMC Cell Biol 10:67
Takezawa R, Schmitz C, Demeuse P, Scharenberg AM, Penner R, Fleig A (2004) Receptor-mediated regulation of the TRPM7 channel through its endogenous protein kinase domain. Proc Natl Acad Sci U S A 101:6009–6014
Talavera K, Staes M, Janssens A, Droogmans G, Nilius B (2004) Mechanism of arachidonic acid modulation of the T-type Ca^{2+} channel alpha1G. J Gen Physiol 124:225–238
Thyagarajan B, Lukacs V, Rohacs T (2008) Hydrolysis of phosphatidylinositol 4,5-bisphosphate mediates calcium induced inactivation of TRPV6 channels. J Biol Chem 283:14980–14987
Thyagarajan B, Benn BS, Christakos S, Rohacs T (2009) Phospholipase C mediated regulation of TRPV6 channels: implications in active intestinal Ca^{2+} transport. Mol Pharmacol 75:608–616
Trebak M, Lemonnier L, Smyth JT, Vazquez G, Putney JW Jr (2007) Phospholipase C-coupled receptors and activation of TRPC channels. Handb Exp Pharmacol 593–614
Trebak M, Lemonnier L, Dehaven WI, Wedel BJ, Bird GS, Putney JW Jr (2008) Complex functions of phosphatidylinositol 4,5-bisphosphate in regulation of TRPC5 cation channels. Pflugers Arch
Tucker SJ, Baukrowitz T (2008) How highly charged anionic lipids bind and regulate ion channels. J Gen Physiol 131:431–438
Vallon V, Grahammer F, Volkl H, Sandu CD, Richter K, Rexhepaj R, Gerlach U, Rong Q, Pfeifer K, Lang F (2005) KCNQ1-dependent transport in renal and gastrointestinal epithelia. Proc Natl Acad Sci U S A 102:17864–17869
Wal J van der, Habets R, Varnai P, Balla T, Jalink K (2001) Monitoring agonist-induced phospholipase C activation in live cells by fluorescence resonance energy transfer. J Biol Chem 276:15337–15344
Varnai P, Balla T (1998) Visualization of phosphoinositides that bind pleckstrin homology domains: calcium- and agonist-induced dynamic changes and relationship to myo-[3H]inositol-labeled phosphoinositide pools. J Cell Biol 143:501–510
Varnai P, Balla T (2006) Live cell imaging of phosphoinositide dynamics with fluorescent protein domains. Biochim Biophys Acta 1761:957–967
Varnai P, Thyagarajan B, Rohacs T, Balla T (2006) Rapidly inducible changes in phosphatidylinositol 4,5-bisphosphate levels influence multiple regulatory functions of the lipid in intact cells. J Cell Biol 175:377–382
Vellani V, Reynolds AM, McNaughton PA (2000) Modulation of the synaptic Ca^{2+} current in salamander photoreceptors by polyunsaturated fatty acids and retinoids. J Physiol 529:333–344
Walder RY, Landau D, Meyer P, Shalev H, Tsolia M, Borochowitz Z, Boettger MB, Beck GE, Englehardt RK, Carmi R, Sheffield VC (2002) Mutation of TRPM6 causes familial hypomagnesemia with secondary hypocalcemia. Nat Genet 31:171–174

Wang Q, Curran ME, Splawski I, Burn TC, Millholland JM, VanRaay TJ, Shen J, Timothy KW, Vincent GM, Jager T de, Schwartz PJ, Toubin JA, Moss AJ, Atkinson DL, Landes GM, Connors TD, Keating MT (1996) Positional cloning of a novel potassium channel gene: K_VLQT1 mutations cause cardiac arrhythmias. Nat Genet 12:17–23
Wang HS, Pan Z, Shi W, Brown BS, Wymore RS, Cohen IS, Dixon JE, McKinnon D (1998) KCNQ2 and KCNQ3 potassium channel subunits: molecular correlates of the M-channel. Science 282:1890–1893
Wang YY, Chang RB, Waters HN, McKemy DD, Liman ER (2008) The nociceptor ion channel TRPA1 is potentiated and inactivated by permeating calcium ions. J Biol Chem 283:32691–32703
Watanabe H, Nagata E, Kosakai A, Nakamura M, Yokoyama M, Tanaka K, Sasai H (2000) Disruption of the epilepsy KCNQ2 gene results in neural hyperexcitability. J Neurochem 75:28–33
Wenk MR, Lucast L, Di Paolo G, Romanelli AJ, Suchy SF, Nussbaum RL, Cline GW, Shulman GI, McMurray W, De Camilli P (2003) Phosphoinositide profiling in complex lipid mixtures using electrospray ionization mass spectrometry. Nat Biotechnol 21:813–817
Wilson PD (2004) Polycystic kidney disease. N Engl J Med 350:151–164
Winks JS, Hughes S, Filippov AK, Tatulian L, Abogadie FC, Brown DA, Marsh SJ (2005) Relationship between membrane phosphatidylinositol-4,5-bisphosphate and receptor-mediated inhibition of native neuronal M channels. J Neurosci 25:3400–3413
Winn MP, Conlon PJ, Lynn KL, Farrington MK, Creazzo T, Hawkins AF, Daskalakis N, Kwan SY, Ebersviller S, Burchette JL, Pericak-Vance MA, Howell DN, Vance JM, Rosenberg PB (2005) A mutation in the TRPC6 cation channel causes familial focal segmental glomerulosclerosis. Science
Wu L, Bauer CS, Zhen XG, Xie C, Yang J (2002) Dual regulation of voltage-gated calcium channels by PtdIns(4,5)P_2. Nature 419:947–952
Xiao YF, Gomez AM, Morgan JP, Lederer WJ, Leaf A (1997) Suppression of voltage-gated L-type Ca^{2+} currents by polyunsaturated fatty acids in adult and neonatal rat ventricular myocytes. Proc Natl Acad Sci U S A 94:4182–4187
Yao J, Qin F (2009) Interaction with phosphoinositides confers adaptation onto the TRPV1 pain receptor. PLoS Biol 7:e46
Yaradanakul A, Hilgemann DW (2007) Unrestricted diffusion of exogenous and endogenous PIP_2 in baby hamster kidney and Chinese hamster ovary cell plasmalemma. J Membr Biol 220:53–67
Yu FH, Catterall WA (2004) The VGL-chanome: a protein superfamily specialized for electrical signaling and ionic homeostasis. Sci STKE 2004:re15
Yus-Najera E, Santana-Castro I, Villarroel A (2002) The identification and characterization of a noncontinuous calmodulin-binding site in noninactivating voltage-dependent KCNQ potassium channels. J Biol Chem 277:28545–28553
Zaika O, Lara LS, Gamper N, Hilgemann DW, Jaffe DB, Shapiro MS (2006) Angiotensin II regulates neuronal excitability via phosphatidylinositol 4,5-bisphosphate-dependent modulation of Kv7 (M-type) K^+ channels. J Physiol 575:49–67
Zaika O, Tolstykh GP, Jaffe DB, Shapiro MS (2007) Inositol triphosphate-mediated Ca^{2+} signals direct purinergic P2Y receptor regulation of neuronal ion channels. J Neurosci 27:8914–8926
Zakharian E, Thyagarajan B, French RJ, Pavlov E, Rohacs T (2009) Inorganic polyphosphate modulates TRPM8 channels. PLoS One 4:e5404
Zakharian E, Cao C, Rohacs T (2010) Gating of transient receptor potential melastatin 8 (TRPM8) channels activated by cold and chemical agonists in planar lipid bilayers. J Neurosci 30:12526–12534
Zamponi GW, Snutch TP (1998) Decay of prepulse facilitation of N type calcium channels during G protein inhibition is consistent with binding of a single Gbeta subunit. Proc Natl Acad Sci U S A 95:4035–4039
Zhang H, He C, Yan X, Mirshahi T, Logothetis DE (1999) Activation of inwardly rectifying K^+ channels by distinct PtdIns(4,5)P_2 interactions. Nat Cell Biol 1:183–188

Zhang Y, Cribbs LL, Satin J (2000) Arachidonic acid modulation of alpha1H, a cloned human T-type calcium channel. Am J Physiol Heart Circ Physiol 278:H184–H193

Zhang H, Craciun LC, Mirshahi T, Rohacs T, Lopes CMB, Jin T, Logothetis DE (2003) PIP_2 activates KCNQ channels and its hydrolysis underlies receptor-mediated inhibition of M currents. Neuron 37:963–975

Zhang Z, Okawa H, Wang Y, Liman ER (2005) Phosphatidylinositol 4,5-bisphosphate rescues TRPM4 channels from desensitization. J Biol Chem 280:39185–39192

Zhen XG, Xie C, Yamada Y, Zhang Y, Doyle C, Yang J (2006) A single amino acid mutation attenuates rundown of voltage-gated calcium channels. FEBS Lett 580:5733–5738

Chapter 11
Nuclear Phosphoinositides: Location, Regulation and Function

Roberta Fiume, Willem Jan Keune, Irene Faenza, Yvette Bultsma, Giulia Ramazzotti, David R. Jones, Alberto M. Martelli, Lilly Somner, Matilde Y. Follo, Nullin Divecha and Lucio Cocco

Abstract Lipid signalling in human disease is an important field of investigation and stems from the fact that phosphoinositide signalling has been implicated in the control of nearly all the important cellular pathways including metabolism, cell cycle control, membrane trafficking, apoptosis and neuronal conduction. A distinct nuclear inositide signalling metabolism has been identified, thus defining a new role for inositides in the nucleus, which are now considered essential co-factors for several nuclear processes, including DNA repair, transcription regulation, and RNA dynamics. Deregulation of phoshoinositide metabolism within the nuclear compartment may contribute to disease progression in several disorders, such as chronic inflammation, cancer, metabolic, and degenerative syndromes. In order to utilize these very druggable pathways for human benefit there is a need to identify how nuclear inositides are regulated specifically within this compartment and what downstream nuclear effectors process and integrate inositide signalling cascades in order to specifically control nuclear function. Here we describe some of the facets of nuclear inositide metabolism with a focus on their relationship to cell cycle control and differentiation.

Keywords Nucleus · Phosphoinositides · Phospholipase C · PtdIns5*P* · Myelodysplastic Syndrome

R. Fiume (✉)
Cellular Signalling Laboratory, Department of Human Anatomical Sciences,
University of Bologna, Bologna, Italy
e-mail: roberta.fiume@unibo.it

N. Divecha
CRUK Inositide Laboratory, The Paterson Institute for Cancer Research, M204BX, United Kingdom
e-mail: ndivecha@picr.man.ac.uk

L. Cocco
Cellular Signalling Laboratory, Department of Human Anatomical Sciences,
University of Bologna, Bologna, Italy
e-mail: lucio.cocco@unibo.it

T. Balla et al. (eds.), *Phosphoinositides II: The Diverse Biological Functions*,
Subcellular Biochemistry 59, DOI 10.1007/978-94-007-3015-1_11,

11.1 Nuclear Phosphoinositide Signalling

Phosphorylation at the 3,4, or 5, position of the inositol head group of phosphatidylinositol generates seven different phosphoinositides that form the basis of a ubiquitous membrane signalling system. An array of tightly regulated phosphoinositide kinases and phosphatases, ultimately control the subcellular profile of phosphoinositides (Irvine 2005), which can regulate protein localisation, ion channel function and protein enzymatic activity and impact on cellular processes including vesicle transport, cytoskeletal dynamics, cell proliferation and survival, gene transcription, cell polarity and migration (McCrea and De Camilli 2009). Phosphoinositides are tethered tightly into the membrane and can recruit and localise proteins to specific subcellular membrane domains through specific phosphoinositide interacting domains (PID) (Lemmon 2003). Because the membrane can be considered more akin to a two dimensional system, membrane interaction is analogous to inducing protein/protein interactions and acts to concentrate upstream regulators and downstream targets together leading to enhanced downstream signalling and specificity. Phosphoinositide signalling occurs on many different intracellular membranes including the inner surface of the plasma-membrane, the Golgi, the endoplasmic reticulum and on membrane vesicles that move between these compartments and their deregulation has been implicated in an array of human diseases (McCrea and De Camilli 2009). Phosphoinositide metabolism also occurs within the nucleus. When isolated nuclei are incubated with radiolabeled ^{32}P-ATP, radioactivity is incorporated into Phosphatidylinositol phosphate (PtdIns*P*), Phosphatidylinositol bisphosphate (PtdInsP_2) and Phosphatidic acid (PtdOH) (Smith and Wells 1983a, b, 1984a, b). As the nuclei were intact and not disrupted, phosphoinositides must be present in nuclei and the kinases that can make them are also present and located at the same sites.

11.1.1 The Location of Phosphoinositide Signalling in the Nucleus

Phosphoinositides are normally presented within the context of a membrane. Phosphoinositides contain two long hydrophobic fatty acyl tails linked to a glycerol group, which is coupled via a phosphodiester linkage to the phosphorylated inositol head group. This chemical structure is ideally suited to form the interface between the hydrophobic membrane, through insertion of the fatty acyl tails, and the cytosol. The nucleus is an organelle that is bounded by a double bilayer membrane, the outer part being contiguous with the endoplasmic reticulum. While one would imagine that inositide signalling in the nucleus might occur on the inner surface of this double bilayer, the first clues that this may not be the case came from studies in Murine erthythroleukemia (MEL) cells (Cocco et al. 1987). In this case nuclei were isolated from control MEL cells or from cells that had been differentiated down the erythroid pathway. The isolated nuclei were then incubated with radiolabelled

ATP, which became incorporated into phosphoinositides. What was fascinating and previously undocumented, was that upon differentiation there were changes in the amount of radiolabelled phosphoinositide present in the nuclei. This suggested that phosphoinositides are dynamically regulated in response to extracellular signals and that phosphoinositides in the nucleus may constitute a signalling pathway that could specifically control nuclear functions. The physiochemical nature of phoshoinositides within the nucleus is still not clear. Experiments utilising detergents to remove the nuclear membrane prevented neither the radiolabelling of nuclear phosphoinositides in the control conditions nor the changes in phosphoinositide labelling observed upon differentiation (Cocco et al. 1987). In a more detailed analysis, we prepared nuclei from rat liver, which have a beautiful intact nuclear envelope after isolation, and used increasing concentrations of detergent to remove the envelope, which was analysed by electron microscopy. Radiolabelling of nuclear phosphoinositides and the mass of various phosphoinositides and phospholipids were also measured (Vann et al. 1997). The data clearly demonstrated that removal of the nuclear envelope correlated with loss of phospholipids such as phosphatidylcholine, but did not correlate with either the removal of phosphoinositides or phosphoinositide kinases. These data suggest that the nuclear phosphoinositide pools that are involved in regulating nuclear processes, are present within the nucleus rather than in the nuclear membrane envelope (Divecha et al. 1991; Banfic et al. 1993; Cocco et al. 1987, 1988; Payrastre et al. 1992). Using a specific PID or antibodies that interact specifically with the phosphoinositide phosphatidylinositol(4,5)bisphosphate (PtdIns(4,5)P_2), it appears that PtdIns(4,5)P_2, and by inference, other phosphoinositides, are clustered in nuclear structures called interchromatin granules (Watt et al. 2002; Boronenkov et al. 1998; Mellman et al. 2008). These structures are also nuclear regions that are highly enriched in factors used for splicing mRNA. Although this would suggest a role for PtdIns(4,5)P_2 in splicing, it is not clear whether these regions are where splicing occurs or where splicing components are stored. However immunodepletion of PtdIns(4,5)P_2 from nuclear extracts attenuates in vitro splicing (Osborne et al. 2001). The exact chemical nature of how phosphoinositides are presented in these structures is far from clear but likely, phosphoinositides are sequestered by proteins that interact with and hide their hydrophobic tails but are able to present the inositol head group for further phosphorylation or phospholipase C mediated cleavage.

11.1.2 PtdIns(4,5)P_2 Synthesis and Signalling in the Nucleus

Within the nucleus Phosphatidylinositol(4,5)bisphosphate (PtdIns(4,5)P_2) is central to phosphoinositide signalling being a second messenger itself (van den Bout and Divecha 2009) and a substrate for both phosphatidylinositol-3-kinase (PtdIns-3-kinase) and phospholipase C (PLC).

For PtdIns(4,5)P_2 to function as a specific second messenger within the nucleus we expect that PtdIns(4,5)P_2 levels would be controlled by nuclear specific factors and in turn they would regulate nuclear specific downstream targets.

PtdIns(4,5)P_2 can be synthesised by two different enzyme families that are highly related. Phosphatidylinositol-4-phosphate (PtdIns4*P*)-5-kinases (PIP5Ks) phosphorylate PtdIns4*P* (Loijens et al. 1996) on the 5-position while phosphatidylinositol-5-phosphate (PtdIns5*P*)-4-kinases (PIP4Ks) phosphorylate PtdIns5*P* on the 4-position (van den Bout and Divecha 2009). So which family is responsible for the synthesis of PtdIns(4,5)P_2 in the nucleus? Isoforms of both families are present in the nucleus (Ciruela et al. 2000; Boronenkov et al. 1998; Mellman et al. 2008), however, the mass level of PtdIns4*P* is at least 20-fold higher than the level of PtdIns5*P*. This suggests that PtdIns(4,5)P_2 is synthesised primarily through the PIP5K pathway. In order to further analyse this, we incubated isolated nuclei with radiolabelled ATP for short time periods and then isolated the PtdIns(4,5)P_2 and determined on which position the label was incorporated. We found that the relative labelling ratio of the 5 to the 4 position was approximately 1.8 (Vann et al. 1997). There are two possible interpretations to these experimental data. The first is that approximately two times more PtdIns(4,5)P_2 is synthesised through the PIP5K than the PIP4K pathway (we cannot determine if the radiolabel was on the same molecule or on different molecules of PtdIns(4,5)P_2). The second is that the labelling of the 4-position occurs because of new synthesis of PtdIns4*P* from PtdIns phosphorylation, which is passed on to PIP5K for synthesis of PtdIns(4,5)P_2. In order to differentiate between these possibilities, we undertook a similar nuclear labelling experiment in the presence of inhibitors of PI4K that synthesise PtdIns4*P*. High concentrations of wortmannin inhibit the PI4KIII family of enzymes while adenosine is a specific inhibitor of the PI4KII family (Balla and Balla 2006). Interestingly, both inhibitors blocked PtdIns4*P* synthesis to about 50% each in isolated intact nuclei and to approximately 90% when incubated in combination. When we determined the ratio of radiolabelling of the 5 to the 4 position of PtdIns(4,5)P_2 labelled in the presence of the inhibitors we found that treatment with wortmannin increased the ratio to 10:1 while adenosine had no effect. Neither wortmannin nor adenosine had any effect on the in vitro activity of the PIP4K enzymes. These simple and elegant in vitro studies suggest that in nuclei there are two families of enzymes that synthesise PtdIns4*P*, but that only the wortmannin sensitive enzymes provide PtdIns4*P* that is used by the PIP5K to generate PtdIns(4,5)P_2. The adenosine sensitive pool of PtdIns4*P* may be involved in direct signalling. Furthermore it would appear that at least 90% of nuclear PtdIns(4,5)P_2 is derived from the PIP5K pathway with the PIP4K pathway possibly providing a small minority of the nuclear PtdIns(4,5)P_2. The data also suggest that the role of the PIP4K may not be related to their ability to generate PtdIns(4,5)P_2 but that they may have a more specialised function in the nucleus (see later).

11.1.3 Nuclear Specific Regulators of PtdIns(4,5)P_2 Synthesis

It is still unclear which isoforms of PIP5K are present in the nucleus. This is in part a consequence of the lack of suitable antibodies and because when overexpressed in cells, PIP5K generally localise to the plasma-membrane. However, PIP5Kα (Mellman et al. 2008; Boronenkov et al. 1998) and two splice variants of PIP5Kγ have been shown to be present in the nucleus (Schill and Anderson 2009). How their localisation is regulated is not clear. However, two nuclear specific regulators of PIP5K have been defined. We initially demonstrated that the Retinoblastoma protein (pRB) interacts with all isoforms of PIP5K (Divecha et al. 2002). pRB is a nuclear localised master regulator of differentiation, cell survival and progression through the cell cycle. Moreover the pRB pathway is deregulated in nearly all human tumours. pRB interacts with and stimulates the activity of PIP5K and pRB activity can control the synthesis of nuclear PtdIns(4,5)P_2. In fact we have shown that pRB acts as a scaffold protein for a number of different enzymes involved in phosphoinositide regulation including PIP4K and a Diacylglycerol kinase (DGK) (Los et al. 2006). Interestingly, DGKζ, but not DGKα or DGKθ, interacts with pRB in vitro and in vivo, and acts in vivo as a downstream effector of pRB to regulate nuclear levels of diacylglycerol and phosphatidic acid, and cell cycle progression in response to DNA damage induced by γ-irradiation (Los et al. 2006). DGKζ also localized mainly to the nucleus in C2C12 cells and its overexpression (but not of a kinase dead mutant or of a mutant that did not enter the nucleus), blocked C2C12 cells in the G_1 phase of the cell cycle (Evangelisti et al. 2007). In contrast, the down-regulation of endogenous DGKζ by siRNA increased the number of cells both in S and G_2/M phases of the cell cycle. The cell cycle arrest of cells overexpressing wild-type DGKζ was accompanied by decreased levels of pRB phosphorylated on Ser-807/811. pRB also interacts with the p55 regulatory subunit of PI-3-kinase (Xia et al. 2003) furthering the idea that pRB may act to scaffold nuclear inositide metabolising enzymes.

Another well characterised regulator of PIP5K has emerged from the Anderson laboratory. Using yeast two hybrid analysis, Star-PAP was identified as an interactor with PIP5Kα (Mellman et al. 2008). Interaction regulates the localisation of PIP5Kα to nuclear speckles, where PtdIns(4,5)P_2, presumably synthesised by PIP5Kα, regulates the activity of Star-PAP. Star-PAP is a poly(A) polymerase that regulates the length of the poly-A tail of a select set of mRNAs, some of which are involved in regulating responses to oxidative stress. RNAi mediated suppression of PIP5Kα leads to a decrease in a similar set of mRNAs that are also regulated by Star-PAP. In vitro, Star-PAP activity is dramatically stimulated by PtdIns(4,5)P_2, suggesting that it is also a downstream target for nuclear PtdIns(4,5)P_2 signalling. The data suggest that the Star-PAP complex acts as a hub for nuclear PtdIns(4,5)P_2 signalling to control the response to oxidative stress. pRB is also critical for responses to oxidative damage and thus may impinge on the Star-PAP pathway through regulation of PtdIns(4,5)P_2 synthesis. Whether Star-PAP is directly regulated by PtdIns(4,5)P_2 is not clear as star-PAP is also regulated by phosphorylation by Caesin kinase 1, an enzyme that is also regulated by PtdIns(4,5)P_2 (Gonzales et al. 2008).

Besides Star-PAP, few nuclear specific PtdIns(4,5)P_2 interactors have been identified. The BAF complex is a chromating regulating complex which is able to interact with PtdIns(4,5)P_2 (Zhao et al. 1998; Rando et al. 2002) although it is not clear which component of the complex interacts with this phosphoinostide nor what this interaction can do to the function of the BAF complex. PtdIns(4,5)P_2 also interacts with histone H1 and disrupts its ability to suppress basal transcription by RNA polymerase in vitro. PtdIns(4,5)P_2 interaction is abolished upon Protein Kinase C (PKC) mediated phosphorylation of the H1 (Yu et al. 1998). Furthermore the drosophila PIP5K homologue, skittles, interacts with ASH2, a component of a chromating remodelling

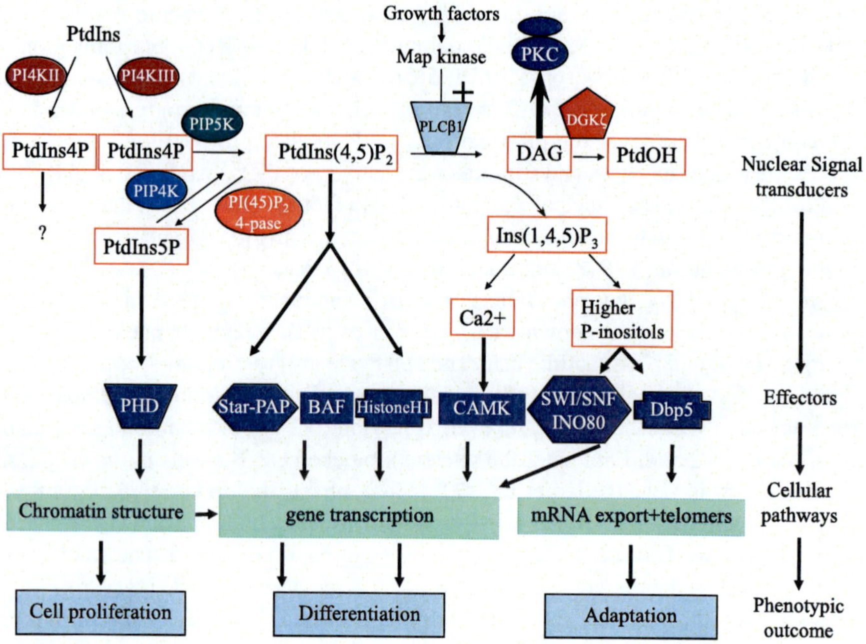

Fig. 11.1 Nuclear phosphoinositides are metabolized by phosphatases, kinases and phospholipases (*colored shapes* next to *arrows*) to generate a profile of inositides and their derivatives within the nucleus. These nuclear signal transducers interact with and regulate the function of a set of nuclear effectors (*dark blue shapes*) in order to modulate nuclear specific pathways (*green boxes*). The modulation of these nuclear pathways eventually leads to the modulation of cell or organismal behavior (*light blue boxes*)

complex and its functional complex modulates histone H1 hyperphosphorylation in vivo (Cheng and Shearn 2004). These data suggest that PtdIns(4,5)P_2 may play direct role in modulating chromatin assembly and regulating transcription. Using a number of different strategies to enrich and purify PtdIns(4,5)P_2 interacting proteins from the nuclei of MEL cells, we have identified an enrichment of proteins that are part of the pre-mRNA and mRNA splicing complexes and proteins involved in regulating DNA damage responses. While the vast majority of these contained lysine/arginine-rich patches with the following motif, K/R-(Xn = 3-7)-K-X-K/R-K/R, we also identified a smaller subset of known phosphoinositide-binding proteins containing pleckstrin homology (PH) or plant homeodomain (PHD) modules (Lewis et al. 2011). Proteins with no prior history of phosphoinositide interaction were also identified, some of which have functional roles in chromatin assembly. DNA topology was exemplar amongst these with the identification of topoisomerase IIα (TopoIIα). Biochemical assays validated our proteomic data supporting a direct interaction between PtdIns(4,5)P_2 and DNA TopoIIα. Furthermore, we also showed that in vitro, phosphoinositides could modulate TopoIIα decatenation activity (Lewis et al. 2011). Clearly further definition of PIP5K and PtdIns(4,5)P_2 interactors will be critical for understanding the complexity of PtdIns(4,5)P_2 synthesis in the nucleus. A snapshot of phosphoinositide signaling is shown in Fig. 11.1.

11.2 Nuclear PLC and Cell Cycle Regulation

The study of nuclear inositide signalling has been fraught with difficulties, mainly due to the problems of obtaining intact nuclei, deprived of the outer membrane (that could carry endoplasmic reticulum (ER) contamination) combined with the complexity of isolating cells in a precise phase of the cell cycle.

Among the nuclear PI-metabolising enzymes, the inositide specific PLC has been one of the most extensively studied. The activation of nuclear PLC was first demonstrated in two "founder" reports, showing that insulin-like growth factor (IGF)-1 stimulation of Swiss 3T3 mouse fibroblasts produced a decrease in PtdIns4*P* and PtdIns(4,5)P_2 and a concomitant increase in diacylglycerol (DAG) levels in membrane-stripped nuclei (Divecha et al. 1991; Cocco et al. 1989). On the contrary, no changes in PtdIns4*P*, PtdIns(4,5)P_2, and DAG amount were detected in whole cell homogenates or in nuclei in which the envelope was maintained. Moreover, bombesin, which is another strong mitogen, stimulated inositide metabolism at the plasma membrane, but not in the nucleus, suggesting the existence of a nuclear polyphosphoinositide signalling system entirely distinct from the one at the plasma membrane (Divecha et al. 1991). It was also shown that PKC translocates from the cytoplasm to the nucleus in response to increased nuclear DAG levels (Divecha et al. 1991). PLCβ1 was also shown to be present in nuclei of Swiss 3T3, and PLCβ1 activity was up-regulated in response to IGF-1 stimulation (Martelli et al. 1992). The presence and activity of PLCβ1 and PtdIns4*P*-5-Kinase was subsequently confirmed in rat liver nuclei (Divecha et al. 1993).

The regulation of nuclear PLCβ1 has been investigated extensively. At the plasma membrane, PLCβ1 is activated by both Gαq/α_{11} and Gβγ subunits of heterotrimeric G-proteins. Activation of nuclear PLCβ1 appears to involve an entirely different mechanism, involving mitogen-activated protein kinases (MAPK) phosphorylation (Fig. 11.2). Following IGF-1 stimulation of quiescent Swiss 3T3 mouse fibroblasts, activated p42/44 MAPK translocates to the nucleus where it phosphorylates Ser 982 in the C-terminal tail of PLCβ1 (Xu et al. 2001). This phosphorylation is inhibited by the MAPK inhibitor PD098059. Phosphorylation of PLCβ1 by MAPK also occurred in vitro using recombinant PLCβ1 and MAPK proteins. However, phosphorylation of Ser 982 *per se* does not increase PLCβ1 activity, as seen previously after IGF-1 stimulation (Martelli et al. 1992). Thus phosphorylation of Ser 982 in vivo might cause the recruitment of other components which stimulate PLCβ1 activity. Nonetheless, Swiss 3T3 mouse fibroblasts, stably transfected with PLCβ1 harboring a Ser 982 Gly mutation, showed a significant loss in mitogenesis in response to IGF-1 (Xu et al. 2001) similar to the one obtained through the down regulation of PLCβ1 by anti-sense RNA (Manzoli et al. 1997). Recent data also suggest the involvement of the subunits Gq/11 in regulating PLCβ1 in nuclei of striatal neurons (Kumar et al. 2008) and activation of nuclear PtdIns(4,5)P_2 hydrolysis in rat hepatocytes in response to insulin requires translocation of the insulin receptor to the nucleus (Rodrigues et al. 2008).

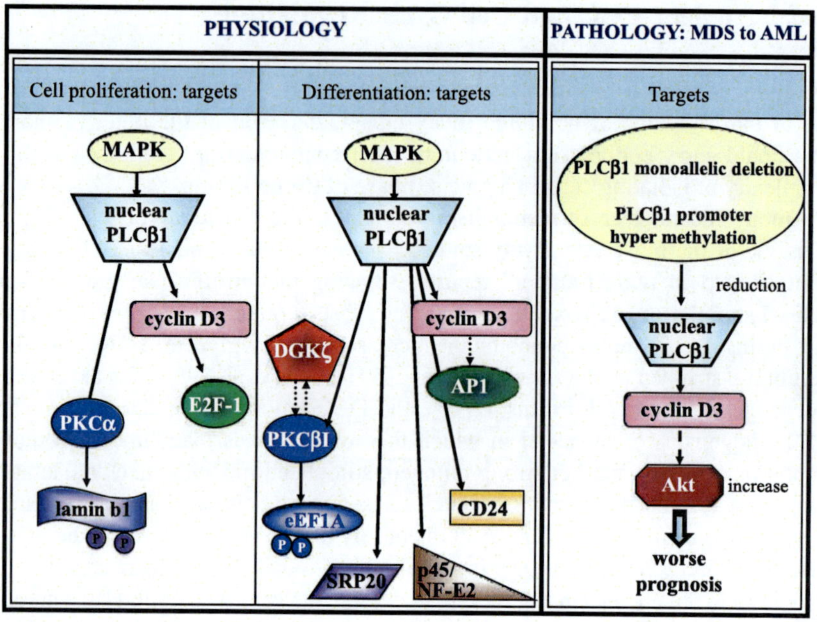

Fig. 11.2 Targets of PLCβ1 in physiological and pathological conditions. Under physiological conditions PLCβ1 activates a variety of downstream targets which ultimately regulate cell proliferation or differentiation. During MDS progression to AML, reduction in PLCβ1 expression correlates with an increase in Akt which results in a worse prognosis. Cyclin D3 is a critical downstream transducer of PLCβ1 common to both the physiological and pathological state

In synchronized HL-60 cells two peaks of PLCβ1 nuclear activity were observed 1 and 8,5 h after the release from a nocodazole block, that correlated with G_2/M and late G_1 phases of the cell cycle (Lukinovic-Skudar et al. 2005). The mechanism of PLCβ1 activation involved an extracellular signal-regulated kinases (MEK) inhibitor sensitive to phosphorylation on a serine, similar to the one occurring in Swiss 3T3 cells (Lukinovic-Skudar et al. 2005). These studies were confirmed and extended in serum starved HL-60 cells progressing through the G_1 phase by re-addition of serum (Lukinovic-Skudar et al. 2007). Also in this case two temporally distinct waves of nuclear PLCβ1 activity occurred in cells mitogenically stimulated: one important for the G_1/S and the other for the G_2/M transition (Lukinovic-Skudar et al. 2007) and reviewed in (Visnjic and Banfic 2007). However, the mechanisms of interaction between PLCβ1 products, DAG and inositol 1,4,5-triphosphate (Ins(1,4,5)P_3) with the cell cycle machinery are still unclear. Pioneering work by Fields and co-workers demonstrated that DAG levels rise to a peak in nuclei coincident with the G_2/M transition, and this increase was sufficient to selectively stimulate PKCβII translocation to the nucleus, where it directly phosphorylates lamin B in the nuclear envelope, leading to nuclear lamina disassembly and mitosis progression (Hocevar and Fields 1991; Hocevar et al. 1993; Goss et al. 1994; Walker et al. 1995; Sun et al. 1997; Gokmen-Polar and Fields 1998). These authors also demonstrated that the generation of nuclear DAG at the G_2/M transition was dependent on PLC activation. In MEL cells, PLCβ1 is required for the activation of PKC and the phosphorylation of lamin B in G_2/M (Fiume et al. 2009). MAPKs, in particular Jun N-terminal Kinase (JNK), can be activated by serum stimulation, and then it can translocate to the nucleus, where it mediates PLCβ1 activation. These events can regulate PKCα-dependent phosphorylation of lamin B, nuclear envelope disassembly and thus cell cycle progression (Fiume et al. 2009). Nuclear PLCβ1 also appears to be important in the resumption of meiosis in mouse oocyte (Avazeri et al. 2000). PLCβ1 translocates to the nucleus, apparently to perichromatin and interchromatin granules, and this is followed by a later shift to the nucleoplasm, as demonstrated by immuno-electron microscopy analysis. Importantly, microinjection into the nucleus of an antibody to PLCβ1 blocked germinal vesicle breakdown (Avazeri et al. 2000).

Nuclear phospholipid metabolism is also particularly important during G_1/S transition and S phase. Nuclear specific phosphatidylinositol (PtdIns) lipid breakdown occurs during S phase, releasing nuclear inositol phosphates, including inositol(1,4)-bisphosphate (Ins(1,4)P_2), which may function to stimulate DNA polymerase activity (York and Majerus 1994; Sylvia et al. 1988, 1989). In the G_1/S transition PLCδ1 accumulates in the nucleus by binding to PtdIns(4,5)P_2 (Stallings et al. 2005), and the suppression of PLCδ1 alters S phase progression and inhibits cell proliferation, possibly through a block in S-phase exit. PLCδ1 suppression also increased cyclin E level (Stallings et al. 2008) which has also been shown to attenuate S-phase exit. PLCβ1 overexpression in MEL cell nuclei (Faenza et al. 2000), leads to an upregulation of cyclin D3, along with its kinase (cdk4), even in cells that are serum-starved. As a consequence of increased cyclin D3 levels, retinoblastoma protein (pRB) is phosphorylated and this leads to the activation of the E2F-1 transcription factor (Faenza et al. 2000) (Fig. 11.2).

The contribution of PtdIns(4,5)P_2 in S phase entry has been linked also to cyclin A2. In Swiss 3T3 cells nuclear PtdIns(4,5)P_2 down-regulation may cause a delay in phorbol ester-induced S phase entry and this was at least in part channeled through cyclin A2 at the transcriptional level, thus identifying cyclin A2 as a downstream effector of the nuclear PtdIns(4,5)P_2 signalling network (Nelson et al. 2008).

All in all, the above mentioned studies clearly demonstrate that nuclear inositide signalling is able to integrate cellular signals to control important master regulators of the cell cycle in order to impinge on progression through the cell cycle.

11.3 NuclearPLCβ1 During Cell Differentiation

PLCβ1 has been implicated in the control of differentiation. For instance, nuclear PI metabolism changes during dimethyl sulfoxide (DMSO)-induced erythroid differentiation of MEL cells (Cocco et al. 1987). We have also demonstrated that the DMSO-induced differentiation of these cells is accompanied by an accumulation of nuclear PtdIns(4,5)P_2 (Manzoli et al. 1989), concomitant with a decrease of PLCβ1 in the nucleus (Martelli et al. 1994) and a decrease in nuclear DAG level (Divecha et al. 1995). The nuclear localization of PLCβ1 was shown to be crucial for the differentiation of MEL cells and targets a reduction in the expression of the transcription factor p45/NF-E2 which is required for the expression of the β-globin gene (Matteucci et al. 1998) (Fig. 11.2). p45/NF-E2 is a highly specific target for PLCβ1 signalling as other transcription factors involved in erythroid differentiation of MEL cells, such as members of the GATA family, are not regulated (Faenza et al. 2002). A proteomic approach also identified SRp20, a member of the highly conserved serine/arginine-rich splicing factor (SR) family of splicing regulators as a target of nuclear PLCβ1 in MEL cells. In addition, by immunoprecipitation and subcellular fractioning, it has been shown that endogenous PLCβ1 and SRp20 physically interact in the nucleus (Bavelloni et al. 2006). Nuclear PLCβ1 also up-regulates the expression of CD24 in MEL cells. CD24 is an antigen involved in differentiation and haematopoiesis, is overexpressed in a number of leukemias and is considered as a critical molecule in the metastasizing ability of solid tumors. When PLCβ1 expression is reduced by RNAi, CD24 expression is also down-regulated. The regulation of PLCβ1 on CD24 is mediated at the transcriptional level at least in part, since PLCβ1 affects the promoter activity of CD24 (Fiume et al. 2005).

In general nuclear PLCβ1 correlates with increased replicative capacity of the cell. However in some models of differentiation nuclear PLCβ1 activity appear to be required for differentiation. Nuclear PLCβ1 is increased upon insulin induced differentiation of C2C12 mouse myoblasts (Faenza et al. 2003). Skeletal muscle differentiation is characterized by terminal withdrawal from the cell cycle, the activation of muscle-specific gene expression, and morphological changes including myoblast alignment, elongation, and fusion of mononucleated myotubes. These events are coordinated by a family of four muscle-specific basic helix-loop-helix transcription factors: MyoD1, Myf5, myogenin, and Mrf4, termed the muscle regulatory factors (MRFs) (Lassar et al. 1994). An imbalance of nuclear and cytoplasmic PLCβ1 suppresses myogenesis as the overexpression of a cytoplasmic PLCβ1 mutant that lacks a nuclear localization sequence suppresses the differentiation of C2C12 myoblasts (Faenza et al. 2003), while the expression of the wild type PLC1β1 or PLCγ1 induces C2C12 differentiation. Upon differentiation PLCβ1 becomes highly concentrated in the nuclei, while PLCγ1 increases in the cytosol, suggesting that they may target a common pathway but in an independent manner (Faenza et al. 2004). In fact both PLCγ1 and β1 activate transcription of cyclin D3, however they appear to do this in different ways. PLCβ1 targets the activation of the AP1/Jun pathway while the regulation of cylin D3 transcription by PLCγ1 is not clear (Ramazzotti et al. 2008). Increased cyclin D3 levels are known to play an important role in regulating myocyte differentiation (Kiess et al. 1995; Cenciarelli et al. 1999; Chu and Lim 2000).

A common theme of the role of PLCβ1 in differentiation of MEL cells and C2C12 cells is the activation of cyclin D3. Nuclear PLCβ1 activates cyclin D3 in both systems. Cyclin D3, however, has opposite effect in the two cell types, stimulating the progression through G1 phase of the cell cycle in the case of MEL cells (Faenza et al. 2005; Cocco et al. 2009) and promoting the differentiation of myoblasts to myotubes in the case of C2C12 cells. Also during 3T3-L1 adipocyte differentiation nuclear PLCβ1 regulates the expression of cyclin D3. During 3T3-L1 adipocyte differentiation there are two phases of PLCβ1 activity; the first occurs within 5 min of treatment with differentiation medium, does not require translocation of PLCβ1 to the nucleus but is regulated by ERK and PKCα. The second phase occurs from day 2 of differentiation, requires translocation of PLCβ1 to the nucleus and is independent of regulation by ERK and PKCα. Over-expression of PLC mutants, which either lack the ERK phosphorylation site or the nuclear localization sequence, revealed that both phases of PLCβ1 activity are required for terminal differentiation to occur. Inhibition of PLCβ1 activity prevents the upregulation of cyclin D3 and cdk4 protein, suggesting that PLCβ1 plays a role in the control of the cell cycle during differentiation (O'Carroll et al. 2009). How PLCβ1 controls differentiation is not clear, however, using a combination of proteomics, immunocytochemistry and molecular biology, we identified a functional signaling cascade elicited by PLCβ1 in the nucleus during C2C12 myogenic differentiation. DAG generation from PLC-mediated PtdIns(4,5)P_2 hydrolysis results in the activation of nuclear PKCbI and the subsequent phosphorylation of the eukaryotic elongation factor 1A (eEF1A) on Ser 53 (Piazzi et al. 2010). PLCβ1 also co-localizes and interacts with DGKζ, in nuclear speckles in C2C12 cells. Like PLCβ1, nuclear DGKζ also increases during myoblast differentiation, and impairment of DGKζ upregulation markedly inhibits differentiation (Evangelisti et al. 2006). Furthermore over expression of DGKζ facilitates differentiation, although this appears to be through the inhibition of cyclin D1 transcription. These data would suggest that there may also be a role for nuclear phosphatidic acid in the regulation of myogenesis.

Nuclear PLC signalling therefore targets the activity of numerous proteins involved in cell cycle machinery in order to regulate cell specific fate such as proliferation or differentiation.

11.4 PLC and the Regulation of Nuclear Inositol Phosphates

Phospholipase C mediated cleavage of PtdIns(4,5)P_2 also generates DAG and inositol(1,4,5)trisphosphate (Ins(1,4,5)P_3). DAG is a potent activator of PKC (Nishizuka 1984), which translocates to the nucleus in response to IGF-1 stimulation (Divecha et al. 1991; Banfic et al. 1993; Martelli et al. 1991). There are many nuclear substrates of PKC, however, if any of them regulate proliferation in response to IGF-1 stimulation is not clear. The other second messenger, Ins(1,4,5)P_3 can regulate a number of pathways. Ins(1,4,5)P_3 receptors that regulate calcium flux have been found on the inner nuclear envelope (Malviya et al. 1990; Humbert et al. 1996) and recent studies have suggested that nuclear Ins(1,4,5)P_3 may specifically mediate increases in nuclear calcium (Rodrigues et al. 2007, 2008, 2009; Gomes et al. 2008). Increased nuclear calcium could potentially regulate an array of transcriptional regulators to modulate cell behaviour (Bading et al. 1997; Hardingham et al. 1997). Ins(1,4,5)P_3 can also be further phosphorylated in the nucleus to generate a number of highly phosphorylated inositols. Indeed the inositol polyphosphate multikinase (IPMK), which phosphorylates Ins(1,4,5)P_3 to generate higher phosphorylated inositols, localizes in the nucleus and regulates transcription (Resnick et al. 2005; Resnick and Saiardi 2008). The recent characterization of IPMK knockout mice demonstrates critical roles for IPMK in embryogenesis and central nervous system development (Frederick et al. 2005). Highly phosphorylated inositols are water soluble second messengers, which have been implicated in the control of chromatin remodelling, mRNA export and telomere function (Tsui and York 2010). In a similar manner to phosphoinositides, inositol phosphates also regulate protein function by specifically interacting with protein domains. In this case they are unlikely to regulate localisation, but are more likely to modulate protein conformation which in turn regulates their activity and function.

11.4.1 Class I PI-3-Kinase

PtdIns(4,5)P_2 can also be converted to PtdIns(3,4,5)P_3 by a nuclear PI-3-kinase. While PI-3-kinase has been shown to be present in the nucleus (Bacqueville et al. 2001; Deleris et al. 2006; Metjian et al. 1999), is not clear whether PtdIns(3,4,5)P_3 is actually synthesized in the nucleus. The most convincing data showing the presence of PtdIns(3,4,5)P_3 in nuclei has come from Lindsay et al. (2006). In this study the authors used the PH domain from GRP1, which shows high specificity towards PtdIns(3,4,5)P_3. This phosphoinositide probe was used as an affinity probe on electron microscopy sections of control cells and of those stimulated with PDGF. PDGF stimulated approximately a twofold increase in the PtdIns(3,4,5)P_3 signal seen in nuclei. Interestingly the increase in PtdIns(3,4,5)P_3 signal was sensitive to pretreatment with the PI-3-kinase inhibitor wortmannin, however, it was not sensitive to expression of a nuclear targeted PTEN (a PtdIns(3,4,5)P_3 phosphatase), although the increase in PtdIns(3,4,5)P_3 seen in the plasma membrane was sensitive to both. This may suggest that the microenvironment of PtdIns(3,4,5)P_3 within the nucleus is not conducive to PTEN mediated dephosphorylation. Interestingly a FRET probe developed to visualise PtdIns(3,4,5)P_3 in vivo was also unable to detect a significant pool of nuclear PIP_3 although the probe could detect hormone activated PtdIns(3,4,5)P_3 signalling at the plasmamembrane (Ananthanarayanan et al. 2005). There are two class I PI-3kinases, that convert PtdIns(4,5)P_2 into PtdIns(3,4,5)P_3 and both have been found in the nucleus. Class IA PI-3-kinase can be regulated by PIKE (PI-3-kinase enhancer), which localizes in the nucleus and can bind GTP. PLCγ1 interacts directly with PIKE and stimulates binding of GTP and the ability of PIKE to interact with and activate PI-3-kinase (Ahn et al. 2004; Ahn and Ye 2005; Ye 2005).

There are few targets that have been well characterized as acting downstream of nuclear PI-3-kinase signalling. Nucleophosmin interacts with PtdIns(3,4,5)P_3 and with CAD (caspase activated DNase) and the trimeric complex appears to be important in inhibiting DNA fragmentation (Ahn and Ye 2005). Akt/PKB is a well characterized PtdIns(3,4,5)P_3-regulated target at the plasma membrane, which also plays a role in the nucleus. However, whether PKB is activated by nuclear PIP_3 or is activated at the plasmamembrane and then translocated into the nucleus is not clear. There are many potential targets for nuclear Akt activity however few of them have been well characterized. Using a proteomic screening procedure for novel substrates of Akt in C2C12 myoblasts, lamin A/C was found to be a bone fide nuclear substrate of Akt (Cenni et al. 2008). Endogenous lamin A/C and Akt proteins interact, and lamin A/C is phosphorylated by Akt not only in vitro but also in vivo in response to insulin stimulation. By mass spectrometry and mutagenesis, Akt was shown to phosphorylate lamin A at Ser404, in the evolutionary conserved RSRGRASSH Akt motif. Since arginine at -3 is a prerequisite for Akt phosphorylation, these data suggest why lamin A/C mutated at Arg401, which is found in primary EDMD-2 (Emery-Dreifuss muscular dystrophy-2) cells, is not phosphorylated in vitro by recombinant Akt. Moreover, in primary myoblasts transfected with lamin S404A, the presence of misshapen nuclei and nuclear abnormalities, such as nuclear envelope breaches, blebs, and honeycomb lamina structures together with concentrated foci of lamin A in the nucleoplasm was observed, which are hallmarks of the EDMD-2 phenotype (Cenni et al. 2008; Marmiroli et al. 2009). Also Akt2 localizes in the nucleus of the differentiated myoblasts and plays a specific role in the commitment of myoblasts to differentiation (Vandromme et al. 2001). HL-60 cells differentiate into monocyte-like cells following exposure to interferon-Q (IFN-Q) and vitamin D3. All-trans-retinoic acid (ATRA) and DMSO induce maturation along neutrophilic pathways and phorbol 12-myristate 13-acetate (PMA) causes the cells to differentiate into a macrophage-like phenotype (Collins 1987). PI3K activity also progressively increases in the nuclei of ATRA-treated HL-60 cells, and wortmannin, a PI3K inhibitor, prevented ATRA mediated antiproliferative and differentiative effects (Bertagnolo et al. 2004). Moreover, the level of active nuclear Akt increases in both HL-60 and NB4 (Matkovic et al. 2006) cells after 96 h of ATRA-treatment.

11.4.2 Class II PI-3-Kinase

Several studies demonstrated the presence of Class II PI-3-kinase in the nucleus. Class II enzymes predominantly convert PtdIns to PtdIns3*P* and PtdIns3*P* has been shown to be present in the nuclei. Phosphatidylinositol 3-kinase C2alpha (PI3K-C2α) was found to be present in nuclei of HeLa cells, where it localised to nuclear speckles. Inhibition of RNA polymerase II activity led to phosphorylation of PI3K-C2α suggesting that PI3K-C2α may play a role in transcription or splicing (Didichenko and Thelen 2001). Human promyelocytic leukemia HL-60 cells have been extensively studied as an experimental model for leukemic and myelocytic differentiation. The activity of nuclear phosphoinositide 3-kinase C2β (PI3K-C2β) was investigated in HL-60 cells induced to differentiate along granulocytic or monocytic lineages. Visnjic et al. demonstrated a significant increase in the activity of PI3K-C2β immunoprecipitated from both the nuclei and the nuclear envelopes of ATRA differentiated HL-60 cells. They also showed an increased level of tyrosine phosphorylation of the enzyme and a parallel increase in the level of nuclear PtdIns3*P*, suggesting that the enzyme may be activated by tyrosine phosphorylation (Visnjic et al. 2002; Visnjic and Banfic 2007). An increase in nuclear PI3K-C2β was also demonstrated during cell cycle progression in HL-60 cells (Visnjic et al. 2003) and in vivo during compensatory hepatic growth after partial hepatectomy (Sindic et al. 2006). How increased nuclear PI3K-C2 activity regulates nuclear functions and which cellular pathways are regulated are not clear.

11.4.3 PtdIns5P Signalling in the Nucleus

Using specific assays we showed that PtdIns5*P* is present in the nucleus and its levels are regulated in response to cellular stressors such as oxidative imbalance or UV irradiation through the activation of the p38 stress activated kinase (Jones et al. 2006). So how are PtdIns5*P* levels regulated in response to stress activation? We demonstrated that in *C.Elegans* and in *Drosophila*, knockout of the single PIP4K enzyme leads to increased levels of PtdIns5*P*, without significant changes in the levels of PtdIns(4,5)P_2. These data suggest that in vivo the role of PIP4K is to regulate PtdIns5*P* levels. There are three isoforms of PIP4K, α, β and γ and we showed that the α isoform was predominantly cytosolic, while the β isoform was cytosolic and nuclear (Ciruela et al. 2000) and the γ isoform was found on intracellular membranes. We showed that in response to UV-irradiation PIP4Kβ was directly phosphorylated by p38 at serine 326 and that this phosphorylation led to a decrease in PIP4K activity associated with PIP4Kβ. To demonstrate that PIP4Kβ controls nuclear PtdIns5*P* levels, we showed that overexpression of PIP4Kβ decreased, while RNAi mediated suppression of PIP4Kβ increased nuclear PtdIns5*P* (Jones et al. 2006). Interestingly, detailed analysis of the difference in the activities of the three isoforms of PIP4K showed that PIP4Kβ has 2000 times less PIP4K activity compared to PIP4Kα (Bultsma et al. 2010; Wang et al. 2010). Therefore, how can PIP4Kβ, which has very little PIP4K activity, regulate nuclear PtdIns5*P*? To begin to understand this we immunoprecipitated PIP4Kβ from cells and identified associated proteins by using mass spectrometry. Interestingly, PIP4Kβ associates with PIP4Kα. We then carried out a series of experiments to demonstrate that in vivo the majority of PIP4K activity in a PIP4Kβ immunoprecipitate was actually derived from its association with PIP4Kα (Bultsma et al. 2010). Our previous data demonstrated that PIP4Kα was actually a cytosolic enzyme while PIP4Kβ was a nuclear enzyme. So how does PIP4Kβ regulate nuclear PtdIns5*P* levels? We found that when co-overexpressed PIP4Kβ was able to target the activity of PIP4Kα to the nucleus (Bultsma et al. 2010), where it can presumably regulate the levels of nuclear PtdIns5*P*. How phosphorylation by the p38 pathway regulates PIP4K activity associated with PIP4Kβ is not clear, but it may regulate the association between PIP4Kα and PIP4Kβ.

While PIP4Ks are able to regulate PtdIns5*P* levels by phosphorylating it to PtdIns(4,5)P_2 what is really unclear is how PtdIns5*P* is synthesised. There are other enzymatic activities, present in the nucleus, which could synthesise PtdIns5*P*. The PIP5K family can synthesise PtdIns5*P* from PtdIns, albeit very inefficiently.

PtdIns5*P* can also be generated by dephosphorylation of PtdIns(4,5)P_2 and a PtdIns(4,5)P_2-4-phosphatase has been characterised in mammalian cells that translocates to the nucleus upon stress induction (Ungewickell et al. 2005; Zou et al. 2007). Finally myotubularins can dephosphorylate PtdIns(3,5)P_2 to generate PtdIns5*P* (Coronas et al. 2008; Walker et al. 2001) although PtdIns(3,5)P_2 has not been demonstrated in the nucleus. Alternatively, and perhaps more interesting, there may be a novel enzymatic activity that synthesises nuclear PtdIns5*P*.

11.4.4 Targets for Nuclear PtdIns5P Signalling

The level of nuclear PtdIns5*P* is increased in response to oxidative stress and UV treatment and this occurs downstream of the activation of the stress activated p38 pathway (Jones et al. 2006). So what are the consequences of increased PtdIns5*P* in the nucleus? A seminal paper from Gozani et al (Gozani et al. 2003) demonstrated that the PHD finger of the growth inhibitory protein 2 (ING2) was able to interact with phosphoinositides and in particular PtdIns5*P*. Of interest, PHD fingers are generally only found in nuclear proteins many of which are involved in regulating gene transcription through the modulation of chromatin structure. ING2 also regulates the level of acetylation of the tumour suppressor p53 and increases in PtdIns5*P* induced acetylation and activation of p53 in a stress dependent manner. P53 is a master regulator of cell proliferation and is highly mutated and inactivated in human tumours. These data therefore link stress-activated modulation of nuclear PtdIns5*P* to the function of an important human tumour suppressor gene. To determine how common PHD interaction with phosphoinositides is, we have cloned over 30 of them and have assessed their interaction with phosphoinositides. We have found that among these, 10 PHD fingers interact strongly with phospholipids. Some PHD fingers also interact with trimethylated lysine 4 of histone H3 (H3K4me3) suggesting that they can also translate the histone code (Shi et al. 2006; Wysocka et al. 2006). For example TAF3 is a component of the basal transcription complex and interaction between the PHD finger of TAF3 and H3K4me3 stimulates transcription (Vermeulen et al. 2007). However, of the 30 PHD fingers that we have cloned only four showed interaction with peptides containing modified or unmodified sequences of the histone H3 tail. A subset of PHD fingers interact both with phosphoinositides and with modified histones, however it is not clear if phosphoinositide interaction can modulate or compete with histone interaction.

11.4.5 PtdIns5P Levels Can Regulate Histone Modification

In vivo, DNA is wrapped around histone octamers to form nucleosomes, which can be further packaged to form dense arrays of nucleosomes. Regulating the packing of these arrays and the actual positioning of the histone octamers are important in the regulation of transcription and gene expression. In all species, histones are highly conserved and residues within their N-terminal tails are targets for a number of different post-translational modifications, such as methylation, acetylation, ubiquitination and phosphorylation. Post translational modifications of histone proteins control gene expression by recruiting protein complexes that are able to modulate nucleosomal packing and positioning and modulate transcription directly (Ruthenburg et al. 2007). Interestingly, many of the proteins that contain PHD fingers are also present in these complexes (Musselman and Kutateladze 2009). The PHD finger of ATX1, a plant homologue of the mammalian trithorax proteins, does not interact with H3K4me3but shows exquisite preference for interaction with PtdIns5*P* (Alvarez-Venegas et al. 2006). ATX1 is a master controller of plant development and flowering (Alvarez-Venegas et al. 2003; Pien et al. 2008) and also plays a role in the regulation of gene transcription in response to environmental stress. ATX1 contains a SET domain that can trimethylate lysine 4 of histone H3, a histone tail modification that has been shown to be present at the promoter of genes that are being actively transcribed. Using expression arrays the WRKY70 gene was identified as a gene that was regulated both by ATX1 and by increased levels of drought stress. The expression of WRKY70 and the level of H3K4me3 on nucleosomes around its promoter was used to study how changing PtdIns5*P* modulated ATX1 activity in vivo. We showed that drought stress induced an increase in total levels of cellular PtdIns5*P*, that was mediated by the plant homologue of myotubularin, a PtdIns(3,5)P_2 3-phosphatase. The increase in PtdIns5*P* led to a decrease in the presence of H3K4me3 at the promoter of WRKY70. Using CHIP analysis we found that increased PtdIns5*P* also led to a decrease in the levels of ATX1 associated with promoters. In fact using immunofluorescence microscopy we showed that increased cellular PtdIns5*P* led to a change in the localisation of ATX1 from the nucleus to the cytoplasm which was dependent on the integrity of the PHD finger. Thus cellular PtdIns5*P* levels directly impinge on the activity of an important SET-domain containing protein to regulate the levels of H3K4me3 at specific promoters (Ndamukong et al. 2010) (Fig. 11.3). Other PHD fingers also interact with a different subset of histone marks including acetylated histones (Zeng et al. 2010; Matsuyama et al. 2010) and again according to our own data some of these are also able to interact with phosphoinositides. The data suggest that nuclear PtdIns5*P* levels may have an important role in modulating where and to what extend PHD finger containing proteins are activated and how they then impinge on chromatin structure and gene expression.

11.4.6 Nuclear Phosphoinositides and Human Disease

As nuclear PLCβ1 and nuclear inositides are involved in key steps of cell growth and differentiation, it is likely that they also play a role in disease development. Myelodysplastic syndrome (MDS) are a heterogeneous group of bone marrow disorders characterized by an impaired stem cell differentiation leading to progressive cytopenia and an increased, although variable, risk of evolution to acute myeloid leukemia (AML) transformation (Scott and Deeg 2010). The MDS diagnosis is

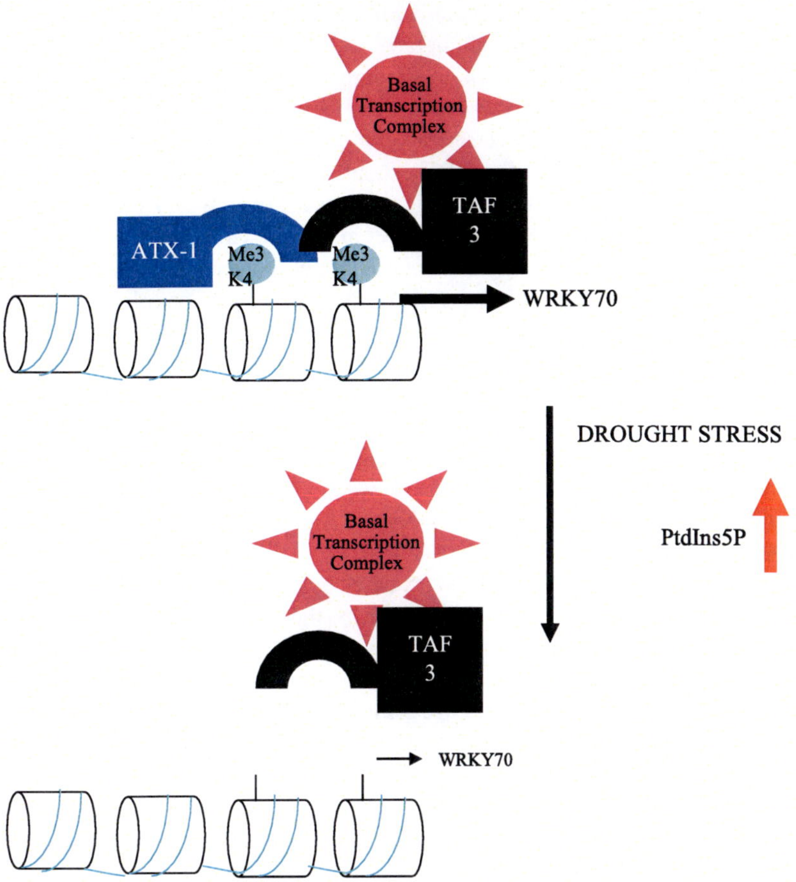

Fig. 11.3 ATX-1 is a plant trithorax homologue that tri-methylates lysine 4 of histone H3, which is often present at actively transcribed promoters. H3K4me3 can recruit the TAF complex protein TAF3 to increase transcription of genes such as WRKY70. Upon drought stress, PtdIns5*P* levels increase as a consequence of the activation of myotubularin. Increased PtdIns5*P* leads to decrease in the localisation of ATX-1 and H3K4me3 at promoters. This may lead to a decrease in TAF3 at the promoter and hence a decrease in WRKY70 transcription

currently based on morphological evaluations, according to either the French-American-British (FAB) (Bennett et al. 1982) or World Health Organization (WHO) (Vardiman et al. 2009) classification, as well as following two more complex systems, based on the percentage of marrow blasts, number of cytopenias, and bone marrow cytogenetic findings, useful for the assessment of the risk of evolution into AML, i.e. the International Prognostic Scoring System (IPSS) (Greenberg et al. 1997) and/or the WHO classification-based Scoring System (WPSS) (Malcovati et al. 2007). The identification of factors that define the risk of progression of MDS to AML or those that identify population that are beneficially effected by certain therapeutic regimes is essential in dictating the best overall therapeutic approach for individual patients.

Aberrations in DNA methylation, often affecting critical cancer-related signalling pathways, can lead to gene inactivation and contribute to tumorigenesis. However, DNA hypermethylation can be reversed by demethylating treatments, and indeed the current therapeutic approach for MDS patients is based on azacitidine, a DNA methyltransferase (DNMT) inhibitor which increases the overall survival and delays the MDS progression towards AML (Fenaux et al. 2009). However, the molecular mechanisms underlying this therapy are not completely understood, although they do lead to increased expression of genes involved in the control of cell cycle, such as p15/INK4B, p21WAF/Cip1 and p73 (Daskalakis et al. 2002; Raj et al. 2007). Furthermore, it is not clear why some patients benefit from azacitidine treatment, while others do not.

Among the enzymes of the nuclear PI cycle, PLCβ1 appears to play a fundamental role in MDS as it is deregulated at both a genetic and an epigenetic level. Fluorescence in situ hybridization (FISH) analyses showed that the PLCβ1 gene mapped to chromosome 20p12 (Peruzzi et al. 2000). Recent reports demonstrated that MDS patients can show a specific, cryptic and interstitial mono-allelic gene deletion of PLCβ1. In fact, in a study involving 80 MDS cases belonging to all of the IPSS risk groups, about 30% of all of the MDS patients showed the PLCβ1 gene deletion (Follo et al. 2009a). Interestingly, MDS patients bearing this mono-allelic deletion rapidly evolved to AML, suggesting that the alteration of the PLCβ1 signalling can be linked to a higher risk of MDS progression. Furthermore, PLCβ1 is aberrantly expressed in high-risk MDS, as compared to healthy donors (Follo et al. 2006). Interestingly, cells from MDS patients always expressed higher levels of the PLCβ1b splicing variant, which is localized predominantly in the nucleus compared to PLCβ1a, localised both in the nucleus and in the cytoplasm, hinting that an imbalance in nuclear versus cytoplasmatic PLCβ1 signalling could be important in the MDS phenotype. PLCβ1 promoter methylation and gene expression were quantified in high-risk MDS patients during azacitidine administration and compared to the expression of patients treated with only best supportive care as well as healthy subjects (Follo et al. 2009b). Strikingly, PLCβ1 promoter methylation was decreased upon azacitidine treatment and this correlated with an increase in PLCβ1 gene expression. Importantly, changes in PLCβ1 expression induced by azacitidine treatment correlated with and anticipated the clinical outcome. The variation in PLCβ1 expression, increase or decrease, in response to azacitidine were detectable much earlier compared to the clinical improvement or worsening, respectively, suggesting that monitoring PLCβ1 levels could lead therapeutic decisions (Follo et al. 2009b). Moreover, the combination of azacitidine (AZA) and valproic acid (VPA) treatment, in high-risk myelodysplastic patients, leads to a synergistic increase in PLCβ1 and cyclin D3 expression suggesting a potential for the combined activity of AZA and VPA in inducing PLCβ1 signalling and positively affecting clinical outcome (Follo et al. 2011). Our recent demonstration that there is an inverse correlation between PLCβ1 expression and Akt activation (Follo et al. 2008, 2009b) and that MDS patients showing a positive response had reduced levels of activated Akt (Follo et al. 2008), may underlie the importance of PLCβ1 expression levels in determining clinical outcome (Fig. 11.2). All in all, these data hint at a possible role for nuclear PLCβ1 in the pathogenesis of myeloid malignancies, and offer new tools for both diagnosis and prognosis of human MDS.

Many PHD finger containing proteins have also been implicated in the development of human immunodeficiency disease, neuronal dysfunction and cancer; moreover disease associated mutations often target the integrity of the PHD finger. In some cases, the disease associated mutations are linked to the ability of the PHD finger to associate with histone tails (both modified and unmodified), however, a number of mutations also lie outside of the PHD finger and in many cases the function of the PHD finger is unknown. For example somatic mutations in inhibitor of growth 1 (ING1) are linked to the development of breast cancer, melanoma, squamous cell carcinoma, head and neck cancer, and many of the mutations target the interaction of the PHD finger with H3K4me3. However there is a disease associated mutation in ING1 (260stop), which inserts a stop codon leaving an intact PHD finger. The stop codon however, truncates the final polybasic region and we presume from studies on ING2 (a close isoform of ING1), that this will attenuate its interaction with phosphoinositides. In the case of Borjeson-Forssman-Lehmann syndrome, disease associated mutations occur within the PHD fingers of PHF6, however no function has been associated with the PHD finger. Interestingly, we have found that one of the PHD fingers from PHF6 interacts with phosphoinositides. Ours and others studies (Shi et al. 2006) have shown that it is possible to generate mutants that dissociate the interaction of PHD fingers with histones from their ability to interact with phosphoinositides. These mutants will enable us to discern the role of the interaction between PHD fingers and various ligands in the development of human diseases.

11.5 Summary

The existence and function of inositide signalling in the nucleus is well documented and now has been linked to the regulation of nuclear specific functions to control proliferation, differentiation and cellular responses to environmental stressors. Although we are now beginning to understand some of the mechanisms that control nuclear phosphoinositide metabolism, more regulators and downstream targets need to be identified in order to exploit this independent signalling pathway for therapeutic purposes. We and others have focused on PLCβ1 which is the most extensively investigated PLC isoform in the nuclear compartment. PLCβ1 is a key player in the regulation of nuclear inositol lipid signalling, however it is neither clear how nuclear PLCβ1 activity is transduced into changes in nuclear function nor whether other isoforms of PLC are also important in regulating nuclear inositide metabolism. It is possible and highly likely that phoshorylation of Ins(1,4,5)P_3 to generate higher phosphorylated derivatives such as IP_6 and IP_7 will be important in regulating nuclear function. In yeast, IP_6 and IP_7 have been implicated in the regulation of mRNA export (York et al. 1999) and chromatin regulation (Odom et al. 2000;

Lee et al. 2007) and it is clear from knockout studies that they have important function in mammalian physiology (Tsui and York 2010). Phosphoinositides themselves through their interaction with specific chromatin remodelling complexes can also regulate chromatin conformation and gene transcription. The evidence, in a number of patients with myelodysplastic syndromes, that the mono-allelic deletion of PLCβ1 is associated with an increased risk of developing acute myeloid leukemia paves the way for an entirely new field of investigation. Indeed the genetic defect evidenced, in addition to being a useful prognostic tool, also suggests that altered expression of this enzyme could have a role in the pathogenesis of this disease by causing an imbalance between proliferation and apoptosis. Targeting the enzymes that control nuclear phosphoinositides will likely yield novel therapeutic targets.

Acknowledgements This work was supported by funding from the Italian MIUR-FIRB Human Proteome Net, Italian CARISBO Foundation, and Celgene Corp. Work in the laboratory of N. Divecha was funded by the Dutch Cancer Society and by CRUK.

References

Ahn JY, Ye K (2005) PIKE GTPase signaling and function. Int J Biol Sci 1:44–50

Ahn JY Rong R, Liu X, Ye K (2004) PIKE/nuclear PI 3-kinase signaling mediates the antiapoptotic actions of NGF in the nucleus. EMBO J 23:3995–4006

Alvarez-Venegas R, Pien S, Sadder M, Witmer X, Grossniklaus U, Avramova Z (2003) ATX-1, an Arabidopsis homolog of trithorax, activates flower homeotic genes. Curr Biol 13:627–637

Alvarez-Venegas R, Sadder M, Hlavacka A, Baluska F, Xia Y, Lu G, Firsov A, Sarath G, Moriyama H, Dubrovsky JG, Avramova Z (2006) The Arabidopsis homolog of trithorax, ATX1, binds phosphatidylinositol 5-phosphate, and the two regulate a common set of target genes. Proc Natl Acad Sci U S A 103:6049–6054

Ananthanarayanan B, Ni Q, Zhang J (2005) Signal propagation from membrane messengers to nuclear effectors revealed by reporters of phosphoinositide dynamics and Akt activity. Proc Natl Acad Sci U S A 102:15081–15086

Avazeri N, Courtot AM, Pesty A, Duquenne C, Lefevre B (2000) Cytoplasmic and nuclear phospholipase C-beta 1 relocation: role in resumption of meiosis in the mouse oocyte. Mol Biol Cell 11:4369–4380

Bacqueville D, Deleris P, Mendre C, Pieraggi MT, Chap H, Guillon G, Perret B, Breton-Douillon M (2001) Characterization of a G protein-activated phosphoinositide 3-kinase in vascular smooth muscle cell nuclei. J Biol Chem 276:22170–22176

Bading H, Hardingham GE, Johnson CM, Chawla S (1997) Gene regulation by nuclear and cytoplasmic calcium signals. Biochem Biophys Res Commun 236:541–543

Balla A, Balla T (2006) Phosphatidylinositol 4-kinases: old enzymes with emerging functions. Trends Cell Biol 16:351–361

Banfic H, Zizak M, Divecha N, Irvine RF (1993) Nuclear diacylglycerol is increased during cell proliferation in vivo. Biochem J 290(Pt 3):633–636

Bavelloni A, Faenza I, Cioffi G, Piazzi M, Parisi D, Matic I, Maraldi NM, Cocco L (2006) Proteomic-based analysis of nuclear signaling: PLCbeta1 affects the expression of the splicing factor SRp20 in Friend erythroleukemia cells. Proteomics 6:5725–5734

Bennett JM, Catovsky D, Daniel MT, Flandrin G, Galton DA, Gralnick HR, Sultan C (1982) Proposals for the classification of the myelodysplastic syndromes. Br J Haematol 51:189–199

Bertagnolo V, Brugnoli F, Marchisio M, Capitani S (2004) Inositide-modifying enzymes: a cooperative role in regulating nuclear morphology during differentiation of myeloid cells. J Biol Regul Homeost Agents 18:381–386

Boronenkov IV, Loijens JC, Umeda M, Anderson RA (1998) Phosphoinositide signaling pathways in nuclei are associated with nuclear speckles containing pre-mRNA processing factors. Mol Biol Cell 9:3547–3560

Bultsma Y, Keune WJ, Divecha N (2010) PIP4Kbeta interacts with and modulates nuclear localization of the high-activity PtdIns5P-4-kinase isoform PIP4Kalpha. Biochem J 430:223–235

Cenciarelli C, De Santa F, Puri PL, Mattei E, Ricci L, Bucci F, Felsani A, Caruso M (1999) Critical role played by cyclin D3 in the MyoD-mediated arrest of cell cycle during myoblast differentiation. Mol Cell Biol 19:5203–5217

Cenni V, Bertacchini J, Beretti F, Lattanzi G, Bavelloni A, Riccio M, Ruzzene M, Marin O, Arrigoni G, Parnaik V, Wehnert M, Maraldi NM, De Pol A, Cocco L, Marmiroli S (2008) Lamin A Ser404 is a nuclear target of Akt phosphorylation in C2C12 cells. J Proteome Res 7:4727–4735

Cheng MK, Shearn A (2004) The direct interaction between ASH2, a Drosophila trithorax group protein, and SKTL, a nuclear phosphatidylinositol 4-phosphate 5-kinase, implies a role for phosphatidylinositol 4,5-bisphosphate in maintaining transcriptionally active chromatin. Genetics 167:1213–1223

Chu CY, Lim RW (2000) Involvement of p27(kip1) and cyclin D3 in the regulation of cdk2 activity during skeletal muscle differentiation. Biochim Biophys Acta 1497:175–185

Ciruela A, Hinchliffe KA, Divecha N, Irvine RF (2000) Nuclear targeting of the beta isoform of type II phosphatidylinositol phosphate kinase (phosphatidylinositol 5-phosphate 4-kinase) by its alpha-helix 7. Biochem J 346(Pt 3):587–591

Cocco L, Gilmour RS, Ognibene A, Letcher AJ, Manzoli FA, Irvine RF (1987) Synthesis of polyphosphoinositides in nuclei of Friend cells. Evidence for polyphosphoinositide metabolism inside the nucleus which changes with cell differentiation. Biochem J 248:765–770

Cocco L, Martelli AM, Gilmour RS, Ognibene A, Manzoli FA, Irvine RF (1988) Rapid changes in phospholipid metabolism in the nuclei of Swiss 3T3 cells induced by treatment of the cells with insulin-like growth factor I. Biochem Biophys Res Commun 154:1266–1272

Cocco L, Martelli AM, Gilmour RS, Ognibene A, Manzoli FA, Irvine RF (1989) Changes in nuclear inositol phospholipids induced in intact cells by insulin-like growth factor I. Biochem Biophys Res Commun 159:720–725

Cocco L, Faenza I, Follo MY, Billi AM, Ramazzotti G, Papa V, Martelli AM, Manzoli L (2009) Nuclear inositides: PI-PLC signaling in cell growth, differentiation and pathology. Adv Enzyme Regul 49:2–10

Collins SJ (1987) The HL-60 promyelocytic leukemia cell line: proliferation, differentiation, and cellular oncogene expression. Blood 70:1233–1244

Coronas S, Lagarrigue F, Ramel D, Chicanne G, Delsol G, Payrastre B, Tronchere H (2008) Elevated levels of PtdIns5P in NPM-ALK transformed cells: implication of PIKfyve. Biochem Biophys Res Commun 372:351–355

Daskalakis M, Nguyen TT, Nguyen C, Guldberg P, Kohler G, Wijermans P, Jones PA, Lubbert M (2002) Demethylation of a hypermethylated P15/INK4B gene in patients with myelodysplastic syndrome by 5-Aza-2′-deoxycytidine (decitabine) treatment. Blood 100:2957–2964

Deleris P, Gayral S, Breton-Douillon M (2006) Nuclear Ptdlns(3,4,5)P_3 signaling: an ongoing story. J Cell Biochem 98:469–485

Didichenko SA, Thelen M (2001) Phosphatidylinositol 3-kinase c2alpha contains a nuclear localization sequence and associates with nuclear speckles. J Biol Chem 276:48135–48142

Divecha N, Banfic H, Irvine RF (1991) The polyphosphoinositide cycle exists in the nuclei of Swiss 3T3 cells under the control of a receptor (for IGF-I) in the plasma membrane, and stimulation of the cycle increases nuclear diacylglycerol and apparently induces translocation of protein kinase C to the nucleus. EMBO J 10:3207–3214

Divecha N, Rhee SG, Letcher AJ, Irvine RF (1993) Phosphoinositide signalling enzymes in rat liver nuclei: phosphoinositidase C isoform beta 1 is specifically, but not predominantly, located in the nucleus. Biochem J 289(Pt 3):617–620

Divecha N, Letcher AJ, Banfic HH, Rhee SG, Irvine RF (1995) Changes in the components of a nuclear inositide cycle during differentiation in murine erythroleukaemia cells. Biochem J 312(Pt 1):63–67

Divecha N, Roefs M, Los A, Halstead J, Bannister A, D'Santos C (2002) Type I PIPkinases interact with and are regulated by the retinoblastoma susceptibility gene product-pRB. Curr Biol 12:582–587

Evangelisti C, Riccio M, Faenza I, Zini N, Hozumi Y, Goto K, Cocco L, Martelli AM (2006) Subnuclear localization and differentiation-dependent increased expression of DGK-zeta in C2C12 mouse myoblasts. J Cell Physiol 209:370–378

Evangelisti C, Tazzari PL, Riccio M, Fiume R, Hozumi Y, Fala F, Goto K, Manzoli L, Cocco L, Martelli AM (2007) Nuclear diacylglycerol kinase-zeta is a negative regulator of cell cycle progression in C2C12 mouse myoblasts. FASEB J 21:3297–3307

Faenza I, Matteucci A, Manzoli L, Billi AM, Aluigi M, Peruzzi D, Vitale M, Castorina S, Suh PG, Cocco L (2000) A role for nuclear phospholipase Cbeta 1 in cell cycle control. J Biol Chem 275:30520–30524

Faenza I, Matteucci A, Bavelloni A, Marmiroli S, Martelli AM, Gilmour RS, Suh PG, Manzoli L, Cocco L (2002) Nuclear PLCbeta(1) acts as a negative regulator of p45/NF-E2 expression levels in friend erythroleukemia cells. Biochim Biophys Acta 1589:305–310

Faenza I, Bavelloni A, Fiume R, Lattanzi G, Maraldi NM, Gilmour RS, Martelli AM, Suh PG, Billi AM, Cocco L (2003) Up-regulation of nuclear PLCbeta1 in myogenic differentiation. J Cell Physiol 195:446–452

Faenza I, Bavelloni A, Fiume R, Santi P, Martelli AM, Maria Billi A, Lo Vasco VR, Manzoli L, Cocco L (2004) Expression of phospholipase C beta family isoenzymes in C2C12 myoblasts during terminal differentiation. J Cell Physiol 200:291–296

Faenza I, Billi AM, Follo MY, Fiume R, Martelli AM, Cocco L, Manzoli L (2005) Nuclear phospholipase C signaling through type 1 IGF receptor and its involvement in cell growth and differentiation. Anticancer Res 25:2039–2041

Fenaux P, Mufti GJ, Hellstrom-Lindberg E, Santini V, Finelli C, Giagounidis A, Schoch R, Gattermann N, Sanz G, List A, Gore SD, Seymour JF, Bennett JM, Byrd J, Backstrom J, Zimmerman L, McKenzie D, Beach C, Silverman LR (2009) Efficacy of azacitidine compared with that of conventional care regimens in the treatment of higher-risk myelodysplastic syndromes: a randomised, open-label, phase III study. Lancet Oncol 10:223–232

Fiume R, Faenza I, Matteucci A, Astolfi A, Vitale M, Martelli AM, Cocco L (2005) Nuclear phospholipase C beta1 (PLCbeta1) affects CD24 expression in murine erythroleukemia cells. J Biol Chem 280:24221–24226

Fiume R, Ramazzotti G, Teti G, Chiarini F, Faenza I, Mazzotti G, Billi AM, Cocco L (2009) Involvement of nuclear PLC{beta}1 in lamin B1 phosphorylation and G2/M cell cycle progression. FASEB J 23:957–966

Follo MY, Bosi C, Finelli C, Fiume R, Faenza I, Ramazzotti G, Gaboardi GC, Manzoli L, Cocco L (2006) Real-time PCR as a tool for quantitative analysis of PI-PLCbeta1 gene expression in myelodysplastic syndrome. Int J Mol Med 18:267–271

Follo MY, Finelli C, Bosi C, Martinelli G, Mongiorgi S, Baccarani M, Manzoli L, Blalock WL, Martelli AM, Cocco L (2008) PI-PLCbeta-1 and activated akt levels are linked to azacitidine responsiveness in high-risk myelodysplastic syndromes. Leukemia 22:198–200

Follo MY, Finelli C, Clissa C, Mongiorgi S, Bosi C, Martinelli G, Baccarani M, Manzoli L, Martelli AM, Cocco L (2009a) Phosphoinositide-phospholipase C beta1 mono-allelic deletion is associated with myelodysplastic syndromes evolution into acute myeloid leukemia. J Clin Oncol 27:782–790

Follo MY, Finelli C, Mongiorgi S, Clissa C, Bosi C, Testoni N, Chiarini F, Ramazzotti G, Baccarani M, Martelli AM, Manzoli L, Martinelli G, Cocco L (2009b) Reduction of phosphoinositide-phospholipase C beta1 methylation predicts the responsiveness to azacitidine in high-risk MDS. Proc Natl Acad Sci U S A 106:16811–16816

Follo MY, Finelli C, Mongiorgi S, Clissa C, Chiarini F, Ramazzotti G, Paolini S, Martinelli G, Martelli AM, Cocco L (2011) Synergistic induction of PI-PLCbeta1 signaling by azacitidine and valproic acid in high-risk myelodysplastic syndromes. Leukemia 25(2):271–280

Frederick JP, Mattiske D, Wofford JA, Megosh LC, Drake LY, Chiou ST, Hogan BL, York JD (2005) An essential role for an inositol polyphosphate multikinase, Ipk2, in mouse embryogenesis and second messenger production. Proc Natl Acad Sci U S A 102:8454–8459

Gokmen-Polar Y, Fields AP (1998) Mapping of a molecular determinant for protein kinase C betaII isozyme function. J Biol Chem 273:20261–20266

Gomes DA, Rodrigues MA, Leite MF, Gomez MV, Varnai P, Balla T, Bennett AM, Nathanson MH (2008) c-Met must translocate to the nucleus to initiate calcium signals. J Biol Chem 283:4344–4351

Gonzales ML, Mellman DL, Anderson RA (2008) CKIalpha is associated with and phosphorylates star-PAP and is also required for expression of select star-PAP target messenger RNAs. J Biol Chem 283:12665–12673

Goss VL, Hocevar BA, Thompson LJ, Stratton CA, Burns DJ, Fields AP (1994) Identification of nuclear beta II protein kinase C as a mitotic lamin kinase. J Biol Chem 269:19074–19080

Gozani O, Karuman P, Jones DR, Ivanov D, Cha J, Lugovskoy AA, Baird CL, Zhu H, Field SJ, Lessnick SL, Villasenor J, Mehrotra B, Chen J, Rao VR, Brugge JS, Ferguson CG, Payrastre B, Myszka DG, Cantley LC, Wagner G, Divecha N, Prestwich GD, Yuan J (2003) The PHD finger of the chromatin-associated protein ING2 functions as a nuclear phosphoinositide receptor. Cell 114:99–111

Greenberg P, Cox C, Lebeau MM, Fenaux P, Morel P, Sanz G, Sanz M, Vallespi T, Hamblin T, Oscier D, Ohyashiki K, Toyama K, Aul C, Mufti G, Bennett J (1997) International scoring system for evaluating prognosis in myelodysplastic syndromes. Blood 89:2079–2088

Hardingham GE, Chawla S, Johnson CM, Bading H (1997) Distinct functions of nuclear and cytoplasmic calcium in the control of gene expression. Nature 385:260–265

Hocevar BA, Fields AP (1991) Selective translocation of beta II-protein kinase C to the nucleus of human promyelocytic (HL60) leukemia cells. J Biol Chem 266:28–33

Hocevar BA, Burns DJ, Fields AP (1993) Identification of protein kinase C (PKC) phosphorylation sites on human lamin B. Potential role of PKC in nuclear lamina structural dynamics. J Biol Chem 268:7545–7552

Humbert JP, Matter N, Artault JC, Koppler P, Malviya AN (1996) Inositol 1,4,5-trisphosphate receptor is located to the inner nuclear membrane vindicating regulation of nuclear calcium signaling by inositol 1,4,5-trisphosphate. Discrete distribution of inositol phosphate receptors to inner and outer nuclear membranes. J Biol Chem 271:478–485

Irvine RF (2005) Inositide evolution—towards turtle domination? J Physiol 566:295–300

Jones DR, Bultsma Y, Keune WJ, Halstead JR, Elouarrat D, Mohammed S, Heck AJ, D'Santos CS, Divecha N (2006) Nuclear PtdIns5P as a transducer of stress signaling: an in vivo role for PIP4Kbeta. Mol Cell 23:685–695

Kiess M, Gill RM, Hamel PA (1995) Expression of the positive regulator of cell cycle progression, cyclin D3, is induced during differentiation of myoblasts into quiescent myotubes. Oncogene 10:159–166

Kumar V, Jong YJ, O'Malley KL (2008) Activated nuclear metabotropic glutamate receptor mGlu5 couples to nuclear Gq/11 proteins to generate inositol 1,4,5-trisphosphate-mediated nuclear Ca^{2+} release. J Biol Chem 283:14072–14083

Lassar AB, Skapek SX, Novitch B (1994) Regulatory mechanisms that coordinate skeletal muscle differentiation and cell cycle withdrawal. Curr Opin Cell Biol 6:788–794

Lee YS, Mulugu S, York JD, O'Shea EK (2007) Regulation of a cyclin-CDK-CDK inhibitor complex by inositol pyrophosphates. Science 316:109–112

Lemmon MA (2003) Phosphoinositide recognition domains. Traffic 4:201–213

Lewis AE, Sommer L, Arntzen MO, Strahm Y, Morrice NA, Divecha N, D'Santos CS (2011) Identification of nuclear phosphatidylinositol 4,5-bisphosphate-interacting proteins by neomycin extraction. Mol Cell Proteomics 10(2):M110.003376

Lindsay Y, McCoull D, Davidson L, Leslie NR, Fairservice A, Gray A, Lucocq J, Downes CP (2006) Localization of agonist-sensitive PtdIns(3,4,5)P_3 reveals a nuclear pool that is insensitive to PTEN expression. J Cell Sci 119:5160–5168

Loijens JC, Boronenkov IV, Parker GJ, Anderson RA (1996) The phosphatidylinositol 4-phosphate 5-kinase family. Adv Enzyme Regul 36:115–140

Los AP, Vinke FP, De Widt J, Topham MK, Van Blitterswijk WJ, Divecha N (2006) The retinoblastoma family proteins bind to and activate diacylglycerol kinase zeta. J Biol Chem 281:858–866

Lukinovic-Skudar V, Donlagic L, Banfic H, Visnjic D (2005) Nuclear phospholipase C-beta1b activation during G2/M and late G1 phase in nocodazole-synchronized HL-60 cells. Biochim Biophys Acta 1733:148–156

Lukinovic-Skudar V, Matkovic K, Banfic H, Visnjic D (2007) Two waves of the nuclear phospholipase C activity in serum-stimulated HL-60 cells during G(1) phase of the cell cycle. Biochim Biophys Acta 1771:514–521

Malcovati L, Germing U, Kuendgen A, Della Porta MG, Pascutto C, Invernizzi R, Giagounidis A, Hildebrandt B, Bernasconi P, Knipp S, Strupp C, Lazzarino M, Aul C, Cazzola M (2007) Time-dependent prognostic scoring system for predicting survival and leukemic evolution in myelodysplastic syndromes. J Clin Oncol 25:3503–3510

Malviya AN, Rogue P, Vincendon G (1990) Stereospecific inositol 1,4,5-[32P]trisphosphate binding to isolated rat liver nuclei: evidence for inositol trisphosphate receptor-mediated calcium release from the nucleus. Proc Natl Acad Sci U S A 87:9270–9274

Manzoli FA, Martelli AM, Capitani S, Maraldi NM, Rizzoli R, Barnabei O, Cocco L (1989) Nuclear polyphosphoinositides during cell growth and differentiation. Adv Enzyme Regul 28:25–34

Manzoli L, Billi AM, Rubbini S, Bavelloni A, Faenza I, Gilmour RS, Rhee SG, Cocco L (1997) Essential role for nuclear phospholipase C beta1 in insulin-like growth factor I-induced mitogenesis. Cancer Res 57:2137–2139

Marmiroli S, Bertacchini J, Beretti F, Cenni V, Guida M, De Pol A, Maraldi NM, Lattanzi G (2009) A-type lamins and signaling: the PI 3-kinase/Akt pathway moves forward. J Cell Physiol 220:553–561

Martelli AM, Gilmour RS, Neri LM, Manzoli L, Corps AN, Cocco L (1991) Mitogen-stimulated events in nuclei of Swiss 3T3 cells. Evidence for a direct link between changes of inositol lipids, protein kinase C requirement and the onset of DNA synthesis. FEBS Lett 283:243–246

Martelli AM, Gilmour RS, Bertagnolo V, Neri LM, Manzoli L, Cocco L (1992) Nuclear localization and signalling activity of phosphoinositidase C beta in Swiss 3T3 cells. Nature 358:242–245

Martelli AM, Billi AM, Gilmour RS, Neri LM, Manzoli L, Ognibene A, Cocco L (1994) Phosphoinositide signaling in nuclei of Friend cells: phospholipase C beta down-regulation is related to cell differentiation. Cancer Res 54:2536–2540

Matkovic K, Brugnoli F, Bertagnolo V, Banfic H, Visnjic D (2006) The role of the nuclear Akt activation and Akt inhibitors in all-trans-retinoic acid-differentiated HL-60 cells. Leukemia 20:941–951

Matsuyama R, Takada I, Yokoyama A, Fujiyma-Nakamura S, Tsuji N, Kitagawa H, Fujiki R, Kim M, Kouzu-Fujita M, Yano T, Kato S (2010) Double PHD fingers protein DPF2 recognizes acetylated histones and suppresses the function of estrogen-related receptor alpha through histone deacetylase 1. J Biol Chem 285:18166–18176

Matteucci A, Faenza I, Gilmour RS, Manzoli L, Billi AM, Peruzzi D, Bavelloni A, Rhee SG, Cocco L (1998) Nuclear but not cytoplasmic phospholipase C beta 1 inhibits differentiation of erythroleukemia cells. Cancer Res 58:5057–5060

McCrea HJ, De Camilli P (2009) Mutations in phosphoinositide metabolizing enzymes and human disease. Physiology (Bethesda) 24:8–16

Mellman DL, Gonzales ML, Song C, Barlow CA, Wang P, Kendziorski C, Anderson RA (2008) A PtdIns4,5P2-regulated nuclear poly(A) polymerase controls expression of select mRNAs. Nature 451:1013–1017

Metjian A, Roll RL, Ma AD, Abrams CS (1999) Agonists cause nuclear translocation of phosphatidylinositol 3-kinase gamma. A Gbetagamma-dependent pathway that requires the p110gamma amino terminus. J Biol Chem 274:27943–27947

Musselman CA, Kutateladze TG (2009) PHD fingers: epigenetic effectors and potential drug targets. Mol Interv 9:314–323

Ndamukong I, Jones DR, Lapko H, Divecha N, Avramova Z (2010) Phosphatidylinositol 5-phosphate links dehydration stress to the activity of ARABIDOPSIS TRITHORAX-LIKE factor ATX1. PLoS One 5:e13396

Nelson TJ, Sun MK, Hongpaisan J, Alkon DL (2008) Insulin, PKC signaling pathways and synaptic remodeling during memory storage and neuronal repair. Eur J Pharmacol 585:76–87

Nishizuka Y (1984) The role of protein kinase C in cell surface signal transduction and tumour promotion. Nature 308:693–698

O'Carroll SJ, Mitchell MD, Faenza I, Cocco L, Gilmour RS (2009) Nuclear PLCbeta1 is required for 3T3-L1 adipocyte differentiation and regulates expression of the cyclin D3-cdk4 complex. Cell Signal 21:926–935

Odom AR, Stahlberg A, Wente SR, York JD (2000) A role for nuclear inositol 1,4,5-trisphosphate kinase in transcriptional control. Science 287:2026–2029

Osborne SL, Thomas CL, Gschmeissner S, Schiavo G (2001) Nuclear PtdIns(4,5)P2 assembles in a mitotically regulated particle involved in pre-mRNA splicing. J Cell Sci 114:2501–2511

Payrastre B, Nievers M, Boonstra J, Breton M, Verkleij AJ, Van Bergen En Henegouwen PM (1992) A differential location of phosphoinositide kinases, diacylglycerol kinase, and phospholipase C in the nuclear matrix. J Biol Chem 267:5078–5084

Peruzzi D, Calabrese G, Faenza I, Manzoli L, Matteucci A, Gianfrancesco F, Billi AM, Stuppia L, Palka G, Cocco L (2000) Identification and chromosomal localisation by fluorescence in situ hybridisation of human gene of phosphoinositide-specific phospholipase C beta(1). Biochim Biophys Acta 1484:175–182

Piazzi M, Bavelloni A, Faenza I, Blalock W, Urbani A, D'Aguanno S, Fiume R, Ramazzotti G, Maraldi NM, Cocco L (2010) eEF1A phosphorylation in the nucleus of insulin-stimulated C2C12 myoblasts: Ser(3) is a novel substrate for protein kinase C betaI. Mol Cell Proteomics 9:2719–2728

Pien S, Fleury D, Mylne JS, Crevillen P, Inze D, Avramova Z, Dean C, Grossniklaus U (2008) ARABIDOPSIS TRITHORAX1 dynamically regulates FLOWERING LOCUS C activation via histone 3 lysine 4 trimethylation. Plant Cell 20:580–588

Raj K, John A, Ho A, Chronis C, Khan S, Samuel J, Pomplun S, Thomas NS, Mufti GJ (2007) CDKN2B methylation status and isolated chromosome 7 abnormalities predict responses to treatment with 5-azacytidine. Leukemia 21:1937–1944

Ramazzotti G, Faenza I, Gaboardi GC, Piazzi M, Bavelloni A, Fiume R, Manzoli L, Martelli AM, Cocco L (2008) Catalytic activity of nuclear PLC-beta(1) is required for its signalling function during C2C12 differentiation. Cell Signal 20:2013–221

Rando OJ, Zhao K, Janmey P, Crabtree GR (2002) Phosphatidylinositol-dependent actin filament binding by the SWI/SNF-like BAF chromatin remodeling complex. Proc Natl Acad Sci U S A 99:2824–2829

Resnick AC, Saiardi A (2008) Inositol polyphosphate multikinase: metabolic architect of nuclear inositides. Front Biosci 13:856–866

Resnick AC, Snowman AM, Kang BN, Hurt KJ, Snyder SH, Saiardi A (2005) Inositol polyphosphate multikinase is a nuclear PI3-kinase with transcriptional regulatory activity. Proc Natl Acad Sci U S A 102:12783–12788

Rodrigues MA, Gomes DA, Leite MF, Grant W, Zhang L, Lam W, Cheng YC, Bennett AM, Nathanson MH (2007) Nucleoplasmic calcium is required for cell proliferation. J Biol Chem 282:17061–17068

Rodrigues MA, Gomes DA, Andrade VA, Leite MF, Nathanson MH (2008) Insulin induces calcium signals in the nucleus of rat hepatocytes. Hepatology 48:1621–1631

Rodrigues MA, Gomes DA, Nathanson MH, Leite MF (2009) Nuclear calcium signaling: a cell within a cell. Braz J Med Biol Res 42:17–20

Ruthenburg AJ, Li H, Patel DJ, Allis CD (2007) Multivalent engagement of chromatin modifications by linked binding modules. Nat Rev Mol Cell Biol 8:983–994

Schill NJ, Anderson RA (2009) Two novel phosphatidylinositol-4-phosphate 5-kinase type Igamma splice variants expressed in human cells display distinctive cellular targeting. Biochem J 422:473–482

Scott BL, Deeg HJ (2010) Myelodysplastic syndromes. Annu Rev Med 61:345–358

Shi X, Hong T, Walter KL, Ewalt M, Michishita E, Hung T, Carney D, Pena P, Lan F, Kaadige MR, Lacoste N, Cayrou C, Davrazou F, Saha A, Cairns BR, Ayer DE, Kutateladze TG, Shi Y, Cote J, Chua KF, Gozani O (2006) ING2 PHD domain links histone H3 lysine 4 methylation to active gene repression. Nature 442:96–99

Sindic A, Crljen V, Matkovic K, Lukinovic-SKUDAR V, Visnjic D, Banfic H (2006) Activation of phosphoinositide 3-kinase C2 beta in the nuclear matrix during compensatory liver growth. Adv Enzyme Regul 46:280–287

Smith CD, Wells WW (1983a) Phosphorylation of rat liver nuclear envelopes. I. Characterization of in vitro protein phosphorylation. J Biol Chem 258:9360–9367

Smith CD, Wells WW (1983b) Phosphorylation of rat liver nuclear envelopes. II. Characterization of in vitro lipid phosphorylation. J Biol Chem 258:9368–9373

Smith CD, Wells WW (1984a) Characterization of a phosphatidylinositol 4-phosphate-specific phosphomonoesterase in rat liver nuclear envelopes. Arch Biochem Biophys 235:529–537

Smith CD, Wells WW (1984b) Solubilization and reconstitution of a nuclear envelope-associated ATPase. Synergistic activation by RNA and polyphosphoinositides. J Biol Chem 259:11890–11894

Stallings JD, Tall EG, Pentyala S, Rebecchi MJ (2005) Nuclear translocation of phospholipase C-delta1 is linked to the cell cycle and nuclear phosphatidylinositol 4,5-bisphosphate. J Biol Chem 280:22060–22069

Stallings JD, Zeng YX, Narvaez F, Rebecchi MJ (2008) Phospholipase C-delta1 expression is linked to proliferation, DNA synthesis, and cyclin E levels. J Biol Chem 283:13992–14001

Sun B, Murray NR, Fields AP (1997) A role for nuclear phosphatidylinositol-specific phospholipase C in the G2/M phase transition. J Biol Chem 272:26313–26317

Sylvia V, Curtin G, Norman J, Stec J, Busbee D (1988) Activation of a low specific activity form of DNA polymerase alpha by inositol-1,4-bisphosphate. Cell 54:651–658

Sylvia VL, Joe CO, Norman JO, Curtin GM, Tilley RD, Busbee DL (1989) Interaction of phosphatidylinositol-4-monophosphate with a low activity form of DNA polymerase alpha: a potential mechanism for enzyme activation. Int J Biochem 21:347–353

Tsui MM, York JD (2010) Roles of inositol phosphates and inositol pyrophosphates in development, cell signaling and nuclear processes. Adv Enzyme Regul 50:324–337

Ungewickell A, Hugge C, Kisseleva M, Chang SC, Zou J, Feng Y, Galyov EE, Wilson M, Majerus PW (2005) The identification and characterization of two phosphatidylinositol-4,5-bisphosphate 4-phosphatases. Proc Natl Acad Sci U S A 102:18854–18859

van den Bout I, Divecha N (2009) PIP5K-driven PtdIns(4,5)P2 synthesis: regulation and cellular functions. J Cell Sci 122:3837–3850

Vandromme M, Rochat A, Meier R, Carnac G, Besser D, Hemmings BA, Fernandez A, Lamb NJ (2001) Protein kinase B beta/Akt2 plays a specific role in muscle differentiation. J Biol Chem 276:8173–8179

Vann LR, Wooding FB, Irvine RF, Divecha N (1997) Metabolism and possible compartmentalization of inositol lipids in isolated rat-liver nuclei. Biochem J 327(Pt 2):569–576

Vardiman JW, Thiele J, Arber DA, Brunning RD, Borowitz MJ, Porwit A, Harris NL, Le BEAU MM, Hellstrom-Lindberg E, Tefferi A, Bloomfield CD (2009) The 2008 revision of the World Health Organization (WHO) classification of myeloid neoplasms and acute leukemia: rationale and important changes. Blood 114:937–951

Vermeulen M, Mulder KW, Denissov S, Pijnappel WW, Van Schaik FM, Varier RA, Baltissen MP, Stunnenberg HG, Mann M, Timmers HT (2007) Selective anchoring of TFIID to nucleosomes by trimethylation of histone H3 lysine 4. Cell 131:58–69

Visnjic D, Banfic H (2007) Nuclear phospholipid signaling: phosphatidylinositol-specific phospholipase C and phosphoinositide 3-kinase. Pflugers Arch 455:19–30

Visnjic D, Crljen V, Curic J, Batinic D, Volinia S, Banfic H (2002) The activation of nuclear phosphoinositide 3-kinase C2beta in all-trans-retinoic acid-differentiated HL-60 cells. FEBS Lett 529:268–274

Visnjic D, Curic J, Crljen V, Batinic D, Volinia S, Banfic H (2003) Nuclear phosphoinositide 3-kinase C2beta activation during G2/M phase of the cell cycle in HL-60 cells. Biochim Biophys Acta 1631:61–71

Walker SD, Murray NR, Burns DJ, Fields AP (1995) Protein kinase C chimeras: catalytic domains of alpha and beta II protein kinase C contain determinants for isotype-specific function. Proc Natl Acad Sci U S A 92:9156–9160

Walker DM, Urbe S, Dove SK, Tenza D, Raposo G, Clague MJ (2001) Characterization of MTMR3. An inositol lipid 3-phosphatase with novel substrate specificity. Curr Biol 11:1600–1605

Wang M, Bond NJ, Letcher AJ, Richardson JP, Lilley KS, Irvine RF, Clarke JH (2010) Genomic tagging reveals a random association of endogenous PtdIns5P 4-kinases IIalpha and IIbeta and a partial nuclear localization of the IIalpha isoform. Biochem J 430:215–221

Watt SA, Kular G, Fleming IN, Downes CP, Lucocq JM (2002) Subcellular localization of phosphatidylinositol 4,5-bisphosphate using the pleckstrin homology domain of phospholipase C delta1. Biochem J 363:657–666

Wysocka J, Swigut T, Xiao H, Milne TA, Kwon SY, Landry J, Kauer M, Tackett AJ, Chait BT, Badenhorst P, Wu C, Allis CD (2006) A PHD finger of NURF couples histone H3 lysine 4 trimethylation with chromatin remodelling. Nature 442:86–90

Xia X, Cheng A, Akinmade D, Hamburger AW (2003) The N-terminal 24 amino acids of the p55 gamma regulatory subunit of phosphoinositide 3-kinase binds Rb and induces cell cycle arrest. Mol Cell Biol 23:1717–1725

Xu A, Suh PG, Marmy-Conus N, Pearson RB, Seok OY, Cocco L, Gilmour RS (2001) Phosphorylation of nuclear phospholipase C beta1 by extracellular signal-regulated kinase mediates the mitogenic action of insulin-like growth factor I. Mol Cell Biol 21:2981–2990

Ye K (2005) PIKE/nuclear PI 3-kinase signaling in preventing programmed cell death. J Cell Biochem 96:463–472

York JD, Majerus PW (1994) Nuclear phosphatidylinositols decrease during S-phase of the cell cycle in HeLa cells. J Biol Chem 269:7847–7850

York JD, Odom AR, Murphy R, Ives EB, Wente SR (1999) A phospholipase C-dependent inositol polyphosphate kinase pathway required for efficient messenger RNA export. Science 285:96–100

Yu H, Fukami K, Watanabe Y, Ozaki C, Takenawa T (1998) Phosphatidylinositol 4,5-bisphosphate reverses the inhibition of RNA transcription caused by histone H1. Eur J Biochem 251:281–287

Zeng L, Zhang Q, Li S, Plotnikov AN, Walsh MJ, Zhou MM (2010) Mechanism and regulation of acetylated histone binding by the tandem PHD finger of DPF3b. Nature 466:258–262

Zhao K, Wang W, Rando OJ, Xue Y, Swiderek K, Kuo A, Crabtree GR (1998) Rapid and phosphoinositol-dependent binding of the SWI/SNF-like BAF complex to chromatin after T lymphocyte receptor signaling. Cell 95:625–636

Zou J, Marjanovic J, Kisseleva MV, Wilson M, Majerus PW (2007) Type I phosphatidylinositol-4,5-bisphosphate 4-phosphatase regulates stress-induced apoptosis. Proc Natl Acad Sci U S A 104:16834–16839

Chapter 12
Phosphoinositides and Cellular Pathogens

Bernard Payrastre, Frédérique Gaits-Iacovoni, Philippe Sansonetti, and Hélène Tronchère

Abstract Phosphoinositides are considered as highly dynamic players in the spatiotemporal organization of key signaling pathways, actin cytoskeleton rearrangements, establishment of cell polarity and intracellular vesicle trafficking. Their metabolism is accurately controlled and mutations in several phosphoinositide metabolizing enzymes take part in the development of human pathologies. Interestingly, evidence is accumulating that modulation of the phosphoinositide metabolism is critical for pathogenicity and virulence of many human pathogens. Given the importance of phosphoinositides, which link membrane and cytoskeleton dynamics to cell responses, it is not surprising that many invasive pathogens hijack their metabolism as part of their strategies to establish infection. In fact, according to their lifestyle, cellular pathogens use the phosphoinositide metabolism in order to trigger their uptake in nonphagocytic cells and/or modulate the maturation of the pathogen-containing vacuole to establish their replicative niche or escape in the cytosol and promote host cell survival. The last two decades have been marked by the discovery of different tactics used by cellular pathogens to modulate the phosphoinositide metabolism as part of their strategies to survive, proliferate and disseminate in a hostile environment.

Keywords Phosphoinositides · Kinases and phosphatases · Cellular pathogens · Infection · Trafficking

12.1 Introduction

Phosphatidylinositol (PtdIns), the most abundant phosphoinositide, can be reversibly phosphorylated at the 3, 4 and 5 positions of the inositol ring by specific kinases to generate a family of seven polyphosphoinositides, which are quantitatively minor

B. Payrastre (✉) · F. Gaits-Iacovoni · H. Tronchère
Inserm, U1048, Université Toulouse 3, I2MC, Hôpital de Toulouse,
BP 84225 Avenue Jean Poulhès, 31432 Toulouse Cedex 04, France
e-mail: bernard.payrastre@inserm.fr

P. Sansonetti
Inserm, U786, Unité de pathogénie microbienne moléculaire,
Institut Pasteur, Paris, France

T. Balla et al. (eds.), *Phosphoinositides II: The Diverse Biological Functions*,
Subcellular Biochemistry 59, DOI 10.1007/978-94-007-3015-1_12,

membrane lipids playing important roles in the control of a variety of intracellular mechanisms in eukaryotic cells (Sasaki et al. 2009; Di Paolo and De Camilli 2006; Payrastre et al. 2001). The classical pathway involving 4 and 5-kinases and -phosphatases produces phosphatidylinositol 4, 5 bisphosphate (PtdIns(4,5)P_2), which can be hydrolyzed by phospholipases C (PLC) to generate the well known second messengers diacylglycerol (DAG), an activator of protein kinase C (PKC), and inositol 1,4,5 *tris*phosphate ($InsP_3$), initiating calcium release from the endoplasmic reticulum. PtdIns(4,5)P_2 is also the substrate of class I phosphoinositide 3-kinases (PI 3-kinases) which generate the second messenger PtdIns(3,4,5)P_3. Furthermore, PtdIns(4,5)P_2 4-phosphatases can generate PtdIns5P, a lipid emerging as an important player in cell regulation which can also be produced through the action of 3-phosphatases of the myotubularin family targeting the PtdIns3P/PtdIns(3,5)P_2 pathway known to control vesicular trafficking. Finally, PtdIns(4,5)P_2 can act as a signaling molecule on its own by directly and specifically interacting with proteins involved in cytoskeleton reorganization and signal transduction mechanisms (Van den Bout and Divecha 2009). In fact, recent discoveries have suggested that probably all polyphosphoinositides have distinct biological roles. They can be rapidly synthesized and degraded in discrete membrane domains through different metabolic pathways involving specific 3, 4 or 5-kinases and -phosphatases (Sasaki et al. 2009). All polyphosphoinositides can interact with protein domains including PH, FYVE, PX, PHD or ENTH (Lemmon 2003; Carlton and Cullen 2005) allowing specific relocalization, activation or changes in protein conformation. Thus, phosphoinositides are considered as highly dynamic actors in the spatiotemporal organization of key signaling pathways, actin cytoskeleton rearrangements, establishment of cell polarity and intracellular vesicle trafficking (Sasaki et al. 2009). Their metabolism is accurately controlled and mutations in several phosphoinositide metabolizing enzymes take part in the development of human pathologies including cancer and genetic diseases (Pendaries et al. 2003; McCrea and De Camilli 2009). Accordingly, gene knock-out/knock-in studies in mice have shown disease-related phenotypes in many cases.

Interestingly, evidence is also accumulating that modulation of the phosphoinositide metabolism is important for pathogenicity and virulence of many human pathogens (DeVinney et al. 2000; Pizarro-Cerdà and Cossart 2004; Hilbi 2006). Given the importance of phosphoinositides in membrane and cytoskeleton dynamics, it is not surprising that many invasive pathogens divert the phosphoinositide metabolism in order to trigger their uptake in nonphagocytic cells and to modulate the maturation of the pathogen-containing vacuole to establish their replicative niche and to avoid the bactericidal lysosomal compartment (Brumel and Grinstein 2003; Weber et al. 2009b). In fact, according to their lifestyle, cellular pathogens have evolved many different strategies to subvert the phosphoinositide metabolism. This can be achieved at different locations along the infection process to ensure critical steps such as entry into eukaryotic cells, initiation of a replicative vacuole, escape to the cytosol and promotion of host cell survival. It is becoming clear that using these strategies is essential for many pathogens to survive, proliferate and disseminate in a hostile environment. The last two decades have been marked by the discovery of different tactics used by cellular pathogens to modulate the phosphoinositide metabolism. They include at least four general mechanisms: (i) activating host cell membrane receptors that mediate their action *via* the phosphoinositide metabolism, (ii) translocation of effectors with phosphoinositide-metabolizing functions, (iii) translocation of effectors that recruit and modulate host cell phosphoinositide metabolizing enzymes or (iv) translocation or secretion of effectors with phosphoinositide binding capacity.

This chapter covers several aspects of our current knowledge about common as well as unique mechanisms evolved by cellular pathogens to interfere with the phosphoinositide metabolism as part of their strategies to establish infection.

12.2 Subvertion of Host Cell Receptor-mediated Phosphoinositide Signaling

Pathogens have evolved a wide array of mechanisms to hijack the host cell membrane components of non-phagocytic cells to achieve infection. They target not only membrane proteins, but also phospholipids and sugars that are present on the external surface of mammalian cells. In many cases these interactions will induce bacterial intake, but they can also elicit cellular innate immune responses to fight the infection. Bacterial-induced phagocytosis involves complex interplay between the invading pathogen and the host cell (Cossart and Sansonetti 2004; Pizarro-Cerdà and Cossart 2004; Hilbi 2006). The following paragraph describes representative examples of subversion of plasma membrane receptors by pathogens to colonize target cells. One of the first reports showing that a bacterial pathogen can trigger a specific receptor that induces changes in phosphoinositide metabolism to initiate phagocytosis is *Listeria monocytogenes* (Ireton et al. 1996).

Listeria monocytogenes is a food-borne Gram-negative enterobacterium that invades host cells by receptor-mediated phagocytosis. It is responsible for the human disease, Listeriosis. The intestinal phase of the disease usually remains subclinical. However, in vulnerable individuals (i.e.: immunocompromised patients, pregnant women and newborns), bacteria that have crossed the intestinal epithelial barrier can disseminate systematically and cross other barriers like the blood brain barrier to cause meningitis, or the placenta barrier to cause severe granulomatous infections in newborns. *L. monocytogenes* expresses two main invasion proteins that activate uptake by the host cell, internalin A and B (InlA and InlB) (Hamon et al. 2006). After endocytosis *via* a so-called zipper mechanism, bacteria reside transiently in a phagosome, which maturation is delayed by maintaining an early endocytic phenotype (Rab7, EEA1 and PtdIns3P-positive vacuoles), and then lyse the vacuole and escape into the host cell cytoplasm where they multiply.

Several reports have demonstrated that InlB targeted the Met tyrosine kinase receptor for the hepatocyte growth factor (HGF) (Shen et al. 2000). The N-terminal part of InlB has a leucine-rich repeat (LRR) concave domain that forms homodimers

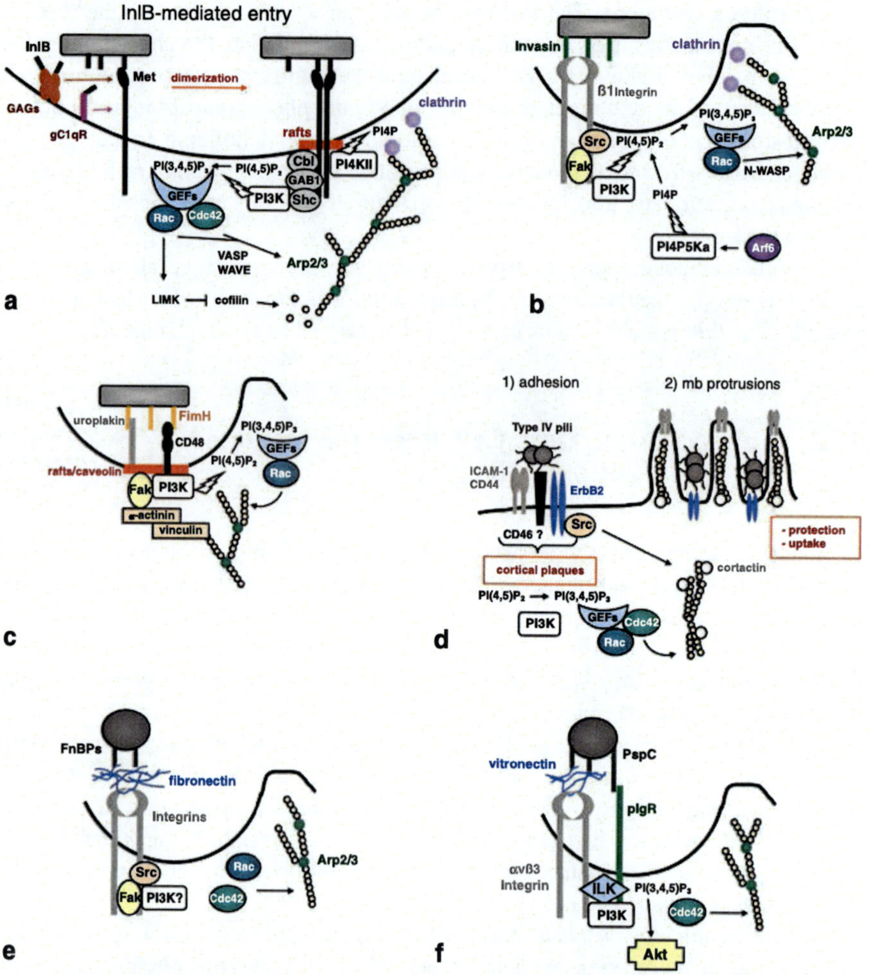

Fig. 12.1 Receptor-mediated pathways utilized by bacterial pathogens to modulate the phosphoinositide metabolism. **a** Invasion by *L. monocytogenes via* the Met receptor. Free InlB binds to membrane GAGs and gC1qR that cooperate with binding of bacterial InlB to the Met receptor, causing its dimerization and activation. Lipid rafts also play a role in this process. Met then recruits Cbl, Gab1, Shc and PI 3-kinase (PI3K) gets activated and produces PtdIn(3,4,5)P_3 (PI(3,4,5)P_3), leading to activation of Rac1 and Cdc42. The VASP/VAWE GTPases effectors stimulate the Arp2/3 complex and actin polymerization. PtdIns 4-kinase II is also recruited and generates PtdIns4P (PI4P) that binds to a clathrin binding protein and helps in the organization of actin coated pits. Ultimately, F-actin gets dissociated through the action of Rac on the LIMK/cofilin module. **b** Invasins of the bacterial surface of *Yersinia* bind β1 integrins, inducing relocation of Src, FAK (Fak), PI 3-kinase and PtdIns4P 5-kinase α (PI4P5Kα) at the entry site. PtdIns4P 5-kinase α is activated by Arf6 and produces PtdIns(4,5)P_2 (PI(4,5)P_2) that will participate in increasing the local levels of PtdIns(3,4,5)P_3. Then, GEFs and Rac get recruited and activate actin polymerization. **c** UPEC entry into bladder epithelial cells. FimH binds to uroplakin Ia and the CD48 receptor resulting in a rafts/caveolin-dependent recruitment of FAK and PI 3-kinase. Actin polymerization and binding to the entry site is linked to the subsequent Rac pathway activation and to α-actinin/vinculin association. **d** Entry of *N. meningitis* into endothelial cells. (1) Type IV pili allow interaction of

capable of activating Met, probably by forcing dimerization of the receptor, which they bind on a different site than the natural ligand (Marino et al. 1999; Niemann et al. 2007). Binding of HGF to Met is known to activate survival and proliferation signals, and to lead to profound cytoskeleton rearrangements. Binding of InlB to Met also activates downstream effectors such as class I PI 3-kinases and the mitogen-activated protein kinase (MAPK) pathway (Ireton et al. 1996, 1999). In cells lacking E-cadherin, the receptor for InlA, inhibition of PI 3-kinase has been shown to impair bacterial entry. Upon activation, Met becomes tyrosine phosphorylated and recruits adaptor molecules including Gab1, Shc, Cbl and CrkII, which will interact with the regulatory subunit of class I PI 3-kinase allowing activation of the catalytic subunit close to its substrate PtdIns(4,5)P_2 thus producing PtdIns(3,4,5)P_3 (Basar et al. 2005; Sun et al. 2005; Dokainish et al. 2007). The transient increase in this lipid second messenger will then, on one hand, activate the Akt survival pathway and, on the other hand, elicit important remodeling of the plasma membrane and actin dynamics required for bacterial internalization (Fig. 12.1). The regulation of actin dynamics downstream of PI 3-kinase activation is dependent on the small Rho GTPases, Rac and Cdc42 and their effectors (Hall 2005). Even though the direct target of PtdIns(3,4,5)P_3 is still unknown, it is likely that production of this lipid will lead to the recruitment of guanine nucleotide exchange factors (GEFs) activators of GTPases. The Rac/WAVE and Cdc42/N-WASP pathways will then activate the Arp2/3 actin-nucleation complex that localizes actin polymerization, which contributes to cytoskeleton and membrane remodeling required for engulfment of bacteria (Bershadsky 2004). Disassembly of F-actin under the regulation of cofilin will then allow proper internalization of the phagosome (Bierne et al. 2001). Study of *L. monocytogenes* entry into non-phagocytic cells demonstrated that they use components of the endocytic machinery such as clathrin-coated vesicles or lipid rafts that are in other cases used for internalization of small particles or solutes (Veiga and Cossart 2005). Lipid rafts are microdomains enriched in cholesterol and sphingolipid that contribute to membrane plasticity and organization of signaling platforms upon activation. In InlB-induced Met activation, they are not required to activate PI 3-kinase but appear to be important for Rac activation (Duncan et al. 2002; Rosenberg et al. 2000; Seveau et al. 2007). In the epithelial cell line Hep-2, InlB has also been shown to activate PLCγ1 and the production of InsP_3 leading to calcium mobilization (Bierne et al. 2000). The functional consequences of such an activation are still unknown, but inhibition of this pathway has no effect on bacterial entry.

---

virulent encapsulated *N. meningitis* with an unknown receptor (CD46?) and proteins such as ICAM-I/CD44. They also bind to the ErbB2 receptor, proliferate and form a colony, inducing clustering and activation of the receptors that recruit Src and PI3-kinase, thereby forming the so-called cortical plates. The PtdIns(3,4,5)P_3 generated induces activation of Cdc42 and Rac, which mediates actin polymerization. (2) Together with cortactin, the local actin dynamics leads to the formation of membrane protrusions that protect bacteria from the shear stress of the blood stream and participate in the uptake. **e** The fibronectin binding domains (FnBPs) of *Staphylococcus aureus* and *Streptococcus pyogenes* bind to fibronectin from the ECM, leading to integrin engagement and subsequent recruitment of Src/Fak, followed by activation of Cdc42 and Rac and actin polymerization. The involvement of PI 3-kinase in the complexes is still a matter of debate. **f** *Streptococcus pneumoniae* adhesin PspC binds to the pIgR receptor, recruiting PI 3-kinase and Cdc42 to the entry site. A second mean of entry involves binding of *S. pneumoniae* to vitronectin that then activates its receptor, the αvβ3 integrin, to complete action on the cytoskeleton *via* ILK, PI 3-kinase and Akt

Recruitment of PI 4-kinase IIα and IIβ has been described at the entry site of *L. monocytogenes* and on the phagosome (Pizarro-Cerdà et al. 2007). The precise mechanism of recruitment of these kinases is unknown but their product, PtdIns4P, which binds the clathrin adaptor AP-1, seems to be involved in *L. monocytogenes* engulfment. Indeed, PI 4-kinase IIα and IIβ knockdown (as well as AP-1 knockdown), but not type III PI 4-kinase knockdown, inhibits bacteria entry without significantly affecting InlB-mediated PtdIns(3,4,5)P_3 production (Pizarro-Cerdà et al. 2007).

In this context, clathrin might act either as a scaffold to control localization and/or activation of protein and lipid kinases or to concentrate Met molecules to reach a critical threshold in signal intensity important for bacterial entry. However, the kinetics and intensity of InlB and HGF-induced signaling are different. This discrepancy can be explained by the fact that InlB also binds other molecules on the cell surface that act as cofactors for Met activation. It has been shown that gC1qR, a receptor for the complement component C1q, could interact with InlB through the C-terminal dipeptide GW (GW module) of the protein and regulate bacteria uptake (Banarjee et al. 2004; Braun et al. 2000; Marino et al. 2002). Interestingly, the GW module binds glycosaminoglycans (GAGs) on the host cell surface, and GAGs are involved in the interaction of HGF with Met. Recent work demonstrated that the GW module of InlB could not only induce binding to cellular GAGs and gC1qR, but was also capable of linking bacteria-free InlB to the bacterium surface, leading to the hypothesis that displaced InlB could be concentrated at sites of bacterial contact together with Met and the LRR domain to amplify cellular response and help bacterial entry (Marino et al. 1999; Cossart and Sansonetti 2004).

L. monocytogenes surface also contains InlA, which utilizes the extracellular region of E-cadherin as a receptor (Mengaud et al. 1996). E-cadherin is a transmembrane cell-cell adhesion protein involved in the formation of adherent junctions in epithelial cells. Crystallography demonstrated that the LRR domain was responsible for InlA binding to E-cadherin as it was for InlB binding to Met (Schubert et al. 2002). E-cadherin is linked to the cytoskeleton through the catenin complex, α- and β-catenin, the latter being important for *L. monocytogenes* invasion (Lecuit et al. 2000). The action of this complex is mostly on the actin dynamics regulated *via* Arp2/3 complex and Rho GTPases. The host proteins ARHGAP10 and myosin VIIa have been localized to *L. monocytogenes* entry sites and demonstrated to be required for InlA-mediated bacterial uptake (Sousa et al. 2003, 2005). In Caco-2 cells, which is dependent on InlA for invasion, PI 3-kinase inhibition also affects bacterial entry. The mechanism by which this lipid kinase is activated is unknown, but it has been shown in other models that E-cadherin can directly activate a class I PI 3-kinase (De Santis et al. 2009).

Yersinia species are Gram-negative enteropathogens and three of them (*Y. pestis*, *Y. pseudotuberculosis*, *Y. enterocolitica*) are human pathogens causing different forms of diseases including pneumomic, septicemic and bubonic plague for *Y. pestis* and gastroenteritis episodes of various severity for the two other species.

The outer-membrane protein of the enteropathogen *Y. pseudotuberculosis* invasin binds to the β1 chain of integrins, thereby inducing integrin clustering and activation of downstream signaling *via* oligomerization (Isberg and Leong 1990). Integrin-engagement leads to tyrosine phosphorylations that are important for *Yersinia* uptake (Fig. 12.1). The focal adhesion kinase (FAK) gets recruited at the C-terminal tail of the β1 subunit together with Src kinase, PI 3-kinase and Rac (Alrutz et al. 2001; Alrutz and Isberg 1998). Although not clearly demonstrated, one can postulate that PI 3-kinase recruitment will allow generation of PtdIns(3,4,5)P_3 required for GEFs relocation and subsequent GTPases activation. Again, efficient invasin-mediated uptake will require the Rho-GTPases and Arp2/3 pathways to regulate actin dynamics. Interestingly, another GTPase is important in this process, Arf6, which regulates cellular trafficking. Here, it might play a role in activating the type I PtdIns4P 5-kinase α, which leads to local accumulation of PtdIns(4,5)P_2 to promote membrane and cytoskeleton rearrangements and direct closure of the phagocytic cup (Wong and Isberg 2003; Brown et al. 2001). As a consequence, a membrane targeted PtdIns(4,5)P_2 5-phosphatase decreases bacterial uptake. Besides being the *in vivo* substrate of PLC and class I PI 3-kinases, PtdIns(4,5)P_2 is a signaling molecule on its own interacting with and modulating ion channels and several actin cytoskeleton regulatory proteins including the WASP/Arp2/3 complex. Whether one of these PtdIns(4,5)P_2 functions is preferentially exploited by *Yersinia* remains to be established.

Uropathogenic Escherichia coli (UPEC) are responsible for the majority of urinary tract infections. Type 1 pilus adhesin FimH present on the surface of the bacterium is responsible for adhesion and invasion of bladder epithelial cells (Martinez et al. 2000). A receptor identified for FimH is the uroplakin Ia protein that is a component of lipid rafts (Min et al. 2002). Phosphorylation of FAK and its interaction with a class IA PI 3-kinase occurs in a uroplakin Ia- and FimH-dependent manner and are required for UPECs uptake. Within lipid rafts, uroplakin Ia interacts with caveolin-1 and Rac1, the later getting activated probably as a results of PtdIns(3,4,5)P_3 production (Duncan et al. 2004). Then, dynamics of the host cell cytoskeleton is engaged *via* recruitment of molecules such as vinculin, tensin, talin and α-actinin to promote entry of UPECs together with endocytosis of the uroplakin Ia. PI 3-kinase inhibition blocks the formation of the cytoskeleton stabilizing multiprotein complex involving α-actinin and vinculin and, in turn, the entry of UPEC into bladder cells (Fig. 12.1). However, binding of UPECs to lipid rafts has also been shown to involve another protein that could act as a receptor, CD48, which is a glycosylphosphatidylinositol (GPI)-anchored protein, suggesting again implication of lipid microdomains as important membrane components of bacterial entry (Malaviya et al. 1999; De Santis et al. 2009).

Neisseria meningitis is a Gram-positive bacterium responsible for severe sepsis and cerebrospinal meningitis because of its propensity to interact with endothelial cells, especially in brain capillaries. Systemic bacteria are subjected to drag forces generated by the blood stream, therefore, they evolved sophisticated and specific mechanisms to bind to endothelial target cells and generate membrane protrusion that will allow them to proliferate and colonize host cells (Lemichez et al. 2010). Type IV pili initiate the interaction of capsulated *N. meningitis* with endothelial cells by a receptor which has not yet been identified, even though some reports mention the cell surface receptor CD46 as a candidate (Kallstrom et al. 1997). Bacteria proliferate to form colonies at their site of attachment on the host cell. Then, retraction of the pili drives intimate association of the bacterium with the cell surface to promote cellular events leading to the formation of membrane protrusions that protect bacteria from the blood flow and initiate internalization of the pathogen (Merz et al. 1999). The cortical plaques formed bellow the colonies of *N. meningitis* are under the control of the levels of PtdIns(4,5)P_2 (Fig. 12.1). Indeed, bacteria induce the clustering of transmembrane proteins also found in lipid rafts such as CD44 (the hyaluronan receptor), intercellular-adhesion molecule-1 (ICAM1), vascular-cell adhesion protein 1 (VCAM1) or E-selectin, which C-terminus partakes in the recruitment of ezrin and moesin that link the membrane to the actin cytoskeleton (Yonemura et al. 1998, 1999; Matsui et al. 1999). Interestingly, ErbB2, a tyrosine kinase of the EGF receptor family, also gets activated through a ligand-independent homodimerization mechanism. As a result, Src, the Ras-MAPK cascade and class IA PI 3-kinase are activated (Hoffman et al. 2001). Then, actin polymerization and cortactin phosphorylation occur in response to activation of Rac and Cdc42, as previously described for other pathogens. The membrane projections generated by the interconnection of all the signaling modules switched on by *N. meningitis* resemble epithelial microvilli structures and they will provoke internalization of the pathogen within endothelial vacuoles (Lemichez et al. 2010).

The *Staphylococcus aureus* and *Streptococcus pyogenes* Gram-positive bacteria responsible for a range of streptococcal infections from minor skin infection to life-threatening diseases such as pneumonia, meningitis, endocarditis, toxic shock syndrome or sepsis. They produce cell-wall anchored proteins (MSCRAMMs or microbial surface components recognizing adhesive matrix molecules) that target the extracellular matrix (ECM) to bind target host cells. They utilize repeats in the C-terminal region of MSCRAMMs as a extended binding surface for fibronectin (Schwarz-Linek et al. 2004). These fibronectin-binding proteins (FnBPs, FnBPA for *S. aureus* and FnBP SfbI/F1 for *S. pyogenes*) mediate adhesion and uptake into host cells. After binding to fibronectin, the pathogens interact with the cell surface through receptors for the ECM component, mostly the integrin $\alpha_5\beta_1$ (Fowler et al. 2000; Sinha et al. 1999). The integrin $\alpha_v\beta_3$ has also been described as a target in the absence of β_1 to invade endothelial and epithelial cells (Ozeri et al. 1998). Upon engagement by fibronectin, integrins cluster and recruit FAK. The entry site will be organized through a signaling cascade which resembles the classical signal initiated upon integrin engagement (Fig. 12.1), with generation of phosphorylation events involving docking of Src kinase, paxillin and other proteins that regulate the cytoskeleton dynamics (Ozeri et al. 2001). Among those, the GTPases Rac and Cdc42 are activated. However PI 3-kinases are major regulators of Rho GTPases, their involvement in this case is still elusive. In the case of M3 serotypes *S. pyogenes* that lack the SfbI invasin, bacterial uptake is accompanied by local F-actin accumulation and formation of membrane protrusions at the entry site. Src kinases and Rac1 but not PI 3-kinases are then essential for internalization (Nerlich et al. 2009).

Streptococcus pneumoniae, which causes lobar pneumonia and invasive diseases, has evolved two complementary strategies to invade respiratory epithelial and endothelial cells. Through the adhesin PspC (pneumococcal surface protein C), it binds to the ectodomain of the human polymeric immunoglobulin receptor (pIgR) and invades the host cell. Experiments have demonstrated that a strong dynamic of both actin and tubulin are required for entry. PspC-pIgR-dependent ingestion of the pathogen depends on the activation of the class I PI3-kinase/Akt module, and also requires Cdc42 (Agarwal and Hammerschmidt 2009). Interestingly, *S. pneumoniae* can also bind ECM proteins. Indeed, binding to the adhesive glycoprotein vitronectin provides a second mean of colonization for the pathogen (Fig. 12.1). The $\alpha v\beta 3$ integrin was identified as the cellular receptor responsible for vitronectin-mediated uptake of pneumococci. Again, internalization requires cooperation of the actin cytoskeleton, the integrin-linked kinase (ILK), PI 3-kinase and Akt (Bergmann et al. 2009).

12.3 Phosphoinositide Metabolizing Enzyme Delivered to Host Cells by Intracellular Pathogens

A second strategy developed by several bacteria to enter and survive in host cells is through the delivery of phosphoinositide metabolizing enzymes. Interestingly, examples of such enzymes mainly concern phosphoinositide phosphatases, which may have some advantages compared to kinases, including the fact that they display non-energy demanding activity. So far, the best documented bacteria-injected phosphoinositide phosphatase are *Shigella flexneri* IpgD and *Salmonella enterica* SigD/SopB. A third phosphatase, named VPA0450, from *Vibrio parahaemolyticus* has been recently described (Broberg et al. 2010). These three enzymes have several points in common, they belong to bacteria affecting the gastrointestinal system, they are injected by a type III (T3SS) secretion system, and they target PtdIns(4,5)P_2. Among phosphoinositides, PtdIns(4,5)P_2 appears as a target of choice. It is rapidly and easily found in the first steps of the infection process as it is mainly present in the plasma membrane, and is involved in several cell functions, particularly membrane/cytoskeleton remodeling. Other bacteria like *Mycobacterium tuberculosis* and *Legionella micdadei* also deliver atypical phosphatases into their cellular host. These acid phosphatases show nevertheless a specificity towards phosphoinositides. In addition to the above mentioned bacterial effectors, the cellular pathogen *Plasmodium falciparum*, a protozoan, uses the PI 3-kinase PfPI3K as a phosphoinositide metabolizing enzyme effector.

The enterobacteria *Shigella flexneri* is the causative agent of bacillary dysentery, which accounts for 60% of the cases of shigellosis in the developing world. Using a T3SS and their dedicated injected effector proteins, bacteria achieve key steps of their infectious process, such as entry into epithelial cells and cell to cell spreading, apoptotic killing of macrophages, as well as regulation of host innate and adaptive immune responses. Once inside the cell, bacteria escape from the phagosome and multiply in the cytoplasm. Among the many effectors secreted by the pathogen is the phosphoinositide phosphatase IpgD. IpgD is 59% similar to SigD/SopB from *S. enterica*, and as the latter, shows sequence homology with 4- and 5-phosphoinositide phosphatases (Marcus et al. 2001) and harbors the CX_5R catalytic motif of dual-specific and protein-tyrosine phosphatase family also found in 3-phosphatases like PTEN or

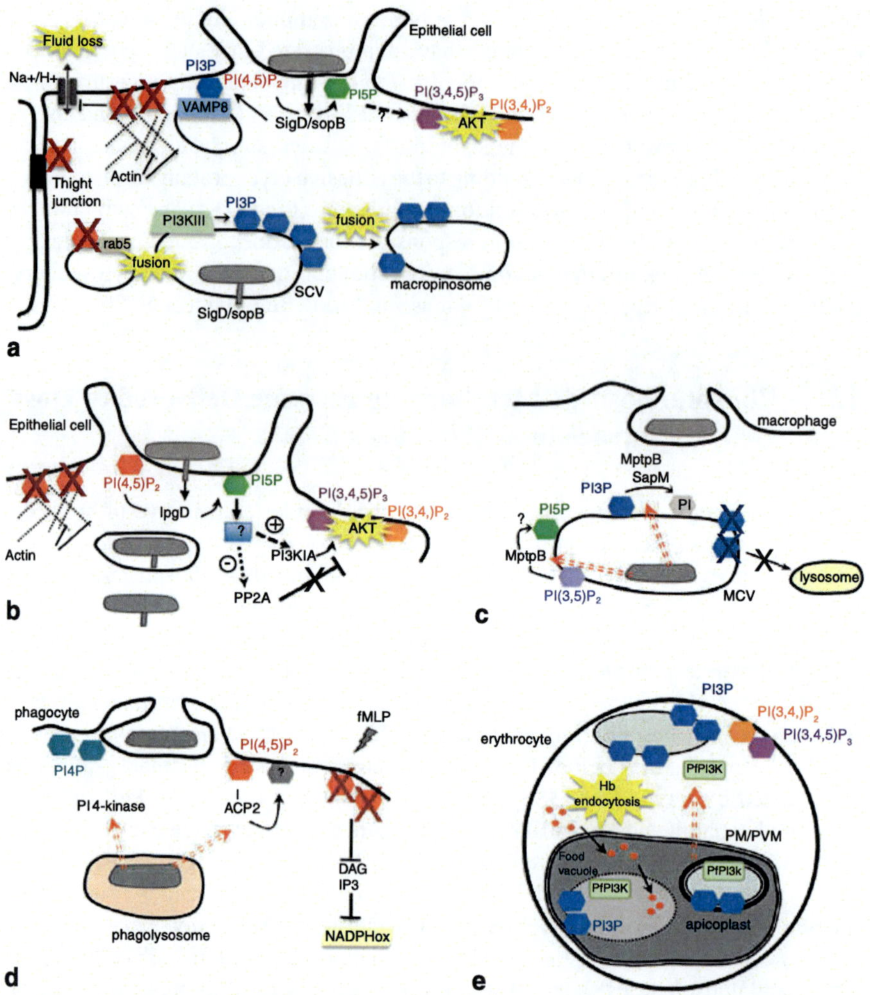

Fig. 12.2 Subversion of host cell phosphoinositide metabolism by cellular pathogens-delivered phosphoinositide metabolizing enzymes. **a** *Salmonella enterica* injects into the host cell the PtdIns(4,5)P_2 4-phosphatase SigD/SopB and generates PtdIns5P at the plasma membrane, that could be responsible for the increase in PtdIns(3,4)P_2 and PtdIns(3,4,5)P_3 and the activation of the Akt survival pathway. The phosphatase SigD/SopB also induces the generation of a pool of PtdIns3P (PI3P) at the ruffle formation sites, that recruits VAMP8-containing vesicles to facilitate uptake of the pathogen. The depletion of PtdIns(4,5)P_2 by SigD/SopB phosphatase activity (1) destabilize actin cytoskeleton-plasma membrane interactions to facilitate bacterial uptake (2) alters host cell tight junctions integrity and inhibits Na+/H+ exchange activity thereby contributing to fluid loss. Once in the host cell, the pathogen replicates in a salmonella-containing vacuole (SCV) and increases PtdIns3P to favor fusion with macropinosomes and escape the lysosomal degradative pathway. The SCV PtdIns3P pool is generated by a class III/vps34 PI 3-kinase that is recruited to the SCV through Rab5 containing vacuoles. The depletion of PtdIns(4,5)P_2 by SigD/SopB phosphatase activity favors Rab5 recruitment to the SCV through a yet unknown mechanism. **b** *Shigella flexneri* injects the PtdIns(4,5)P_2 4-phosphatase IpgD and generates PtdIns5P (PI5P). PtdIns5P activates the class IA PI 3-kinase (PI3KIA), increasing PtdIns(3,4)P_2 (PI(3,4)P_2) and PtdIns(3,4,5)P_3 (PI(3,4,5)P_3), leading to the activation of the Akt survival pathway, that is sustained

myotubularins. *In vitro*, IpgD shows a rather broad specificity, hydrolyzing multiple phosphoinositides and inositol phosphates (Marcus et al. 2001). However, when expressed in epithelial cells, either through infection or ectopic expression, it has been shown by biochemical analysis that IpgD efficiently hydrolyzes PtdIns(4,5)P_2 to generate PtdIns5P (Niebuhr et al. 2002), thereby changing the basal levels of this two phosphoinositides with strong consequences for the host cell. A remarkable fact is the accumulation of large amounts of the newly discovered (Rameh et al. 1997) and poorly characterized monophosphoinositide PtdIns5P, which is normally present at low basal levels ($< 15\%$ of PtdIns4P). It also suggests that PtdIns5P could have a function at the plasma membrane in addition to its role in the nucleus in the activation of the p53 apoptotic pathway through binding of the PHD domain of ING2 (Gozani et al. 2003; Jones et al. 2006). PtdIns5P produced by IpgD at the plasma membrane was shown to activate a class IA PI 3-kinase, increasing the levels of PtdIns(3,4)P_2 and PtdIns(3,4,5)P_3, thereby activating Akt (Pendaries et al. 2006) (Fig. 12.2). In addition, IpgD-produced PtdIns5P also induces PP2A phosphatase inhibition, thereby sustaining Akt activation in IpgD transfected cells (Ramel et al. 2009). So, through the injection of its effector IpgD, *S. flexneri* divert the antiapoptotic/survival pathway of the host cell to favour its replication and multiplication (Pendaries et al. 2006). The molecular mechanisms initiated by PtdIns5P increase are under investigation and appear to involve growth factor receptor signaling and modification of the trafficking to sustain survival signals (Ramel et al. 2011).

In addition, the morphological modifications observed in the host cell, like the reorganization of the actin cytoskeleton and the reduction of the membrane/cytoskeleton tension forces, have been linked to the concomitant decrease in PtdIns(4,5)P_2. Transfection of IpgD lead to the reduction of the tethering forces that anchor the cytoskeleton to the plasma membrane to facilitate ruffles and filopodia formation (Niebuhr et al. 2002) (Fig. 12.2). IpgD also potentiates the activation of the

---

by the inhibitory effect of PtdIns5P on the protein phosphatase PP2A. The concomitant depletion of PtdIns(4,5)P_2 (PI(4,5)P_2) induces a reorganization of the actin cytoskeleton and diminution of tension forces between the plasma membrane and the cytoskeleton to facilitate bacterial entry. **c** *Mycobacterium tuberculosis* replicates in a mycobacterium-containing vacuoles (MCV). By injecting the two PtdIns3P 3-phosphatases MptpB and SapM, the pathogen keeps the MCV PI3P levels low and inhibits the maturation of the vacuoles to lysosomes, thereby escaping destruction. In addition, MptpB has also another substrate PtdIns(3,5)P_2. **d** *Legionella micdadei* replicates into a phagolysosome and injects the PtdIns(4,5)P_2 phosphatase ACP2 in the host cytoplasm. Depletion of PtdIns(4,5)P_2 leads to a diminution of the second messengers diacylglycerol and inositol phosphate IP3 and inhibits the microbicidal response of the host cell. In addition, *L. micdadei* injects a kinase that increases plasma membrane PtdIns4P (PI4P) levels. **e** *Plasmodium falciparum*, at one of its development stages, resides in the erythrocyte and secretes the PfPI3K PI 3-kinase in the erythrocyte cytosol. The kinase generates large amounts of PtdIns3P, probably on vesicular compartments and lower amounts of PtdIns(3,4)P_2 and PtdIns(3,4,5)P_3 at the plasma membrane. By acting on host PtdIns3P, *P. falciparum* could control the trafficking pathways that lead to the endocytosis of hemoglobin to the parasite food vacuole. PfPI3K is also present in the food vacuole and in the apicoplast and controls PtdIns3P levels in these compartments small G proteins Rac and Cdc42 involved in cytoskeleton reorganization. However, a role for PtdIns5P in the cytoskeletal modifications should not be excluded as this phosphoinositide was shown to remodel actin stress fibers when injected in CHO cells (Sbrissa et al. 2004; Ramel et al. 2011).

Finally, it is noteworthy that upon *S. flexneri* infection the levels of PtdIns3P and PtdIns(3,5)P_2, two lipids involved in the control of vesicular trafficking, increase (Niebuhr et al. 2002), suggesting that the bacteria may subvert the enzymes involved in the control of these lipids (such as class III PI 3-kinase, PIKfyve or myotubularins) possibly to modulate the trafficking of the *S. flexneri* containing vacuole.

The enterobacteria *Salmonella enterica* causes illnesses in human including food-borne gastroenteritis and typhoid fever. The pathogen possesses two T3SS through which effector proteins are injected into the host cell. Unlike *S. flexneri*, once intracellular, the bacteria replicate in ≪Salmonella-containing vacuoles≫ (SCVs) that communicate with the host trafficking pathway. Among *S. enterica* effectors, SopB/SigD has been identified as a phosphoinositide phosphatase and shares homology with 4- and 5- phosphoinositide phosphatases catalytic domain. The phosphatase shows activity towards phosphoinositides and inositol phosphate substrates (Norris et al. 1998). *In vitro*, SopB/SigD hydrolyzes phosphoinositides, on the 4 and/or the 5-positions, but some controversy remains concerning its substrate selectivity (Norris et al. 1998; Marcus et al. 2001; Hernandez et al. 2004). In vivo, PtdIns(4,5)P_2 seems to be an important target of SopB/SigD as shown by the dissociation of the PtdIns(4,5)P_2 probe PLC-PH-GFP in cells infected with *S. enterica* or upon ectopic expression of SopB/SigD (Terebiznik et al. 2002). These data suggest an important decrease in the plasma membrane PtdIns(4,5)P_2 pool. It was later confirmed biochemically that the phosphatase hydrolyzes PtdIns(4,5)P_2 and can generate PtdIns5P *in vivo* (Mason et al. 2007), suggesting some similarities with IpgD. The local decrease in PtdIns(4,5)P_2 was shown to destabilize the cytoskeleton-plasma membrane interactions (Terebiznik et al. 2002) in order to facilitate the phagosomal vacuoles sealing and bacterial uptake. However, the disappearance of actin stress fibers observed upon SigD/SopB ectopic expression is independent of the phosphoinositide phosphatase activity (Aleman et al. 2005). The depletion of PtdIns(4,5)P_2 by SopB/SigD alters host cell tight junctions integrity and inhibits the Na^+/H^+ exchange activity, suggesting that the phosphatase could contribute to the fluid loss observed in *S. enterica*-induced diarrhea (Mason et al. 2007), a function also attributed to inositol 1,4,5,6 tetrakisphosphate ($InsP_4$), the product of the hydrolysis of inositol 1,3,4,5,6 pentakisphosphate ($InsP_5$) by SopB/SigD (Norris et al. 1998; Feng et al. 2001) (Fig. 12.2). The role of PtdIns5P in *S. enterica*-mediated infection has not been described yet. By homology with IpgD-produced PtdIns5P, one can propose that this lipid may contribute to activate the PI 3-kinase/Akt survival pathway which has been shown to be upregulated in *S. enterica* infected epithelial cells and upon SopB/SigD ectopic expression (Steele-Mortimer et al. 2000; Marcus et al. 2001; Knodler et al. 2005). Surprisingly, PtdIns(3,4)P_2 and PtdIns(3,4,5)P_3 production at plasma membrane ruffles of *S. enterica*-infected cells, is relatively insensitive to PI 3-kinase inhibitors as monitored by imaging using fluorescent probes (Mallo et al. 2008). These data suggest the implication of a wortmannin-insensitive PI 3-kinase and a potential inhibitory role of PtdIns5P on PtdIns(3,4,5)P_3 phosphatases (Carricaburu et al. 2003; Pendaries et al. 2006). A production of PtdIns3P at the plasma membrane ruffles and the nascent *S. enterica* containing vacuoles was also described (Pattni et al. 2001) (Fig. 12.2) and was shown to be dependent of SigD/SopB phosphatase activity (Hernandez et al. 2004). The PtdIns3P plasma membrane pool binds to the SNARE protein VAMP8, involved in homotypic fusion and exocytosis, and recruits VAMP8-containing vesicles to the site of ruffle formation to facilitate uptake of the pathogen by adding more membrane (Dai et al. 2007). By which metabolic pathway this pool of PtdIns3P is produced is still unknown, but it does not seem to arise from the hydrolyzes of PtdIns(3,4,5)P_3, PtdIns(3,5)P_2 or PtdIns(3,4)P_2 by SigD/SopB as initially proposed (Hernandez et al. 2004). This pool of PtdIns3P appears insensitive to PI 3-kinase inhibitors, as the invasion process is (Dai et al. 2007). High levels of PtdIns3P were also detected in the SCVs (Pattni et al. 2001), in a SigD/SopB phosphatase activity-dependent manner (Hernandez et al. 2004). However, the sensitivity of this particular PtdIns3P pool to PI 3-kinases inhibitors remains controversial (Pattni et al. 2001; Hernandez et al. 2004; Scott et al. 2002). Mallo et al. proposed that the SigD/SopB-dependent PtdIns3P pool on *S. enterica* containing vacuoles is generated by the recruitment of the class III PI 3-kinase Vps34 to the SCVs through the fusion of Rab5-positive vesicles (Mallo et al. 2008) (Fig. 12.2). In fact, the SigD/SopB phosphatase does not produce PtdIns3P directly but rather hydrolyzes PtdIns(4,5)$P_{2,}$ which depletion favours Rab5 recruitment to the SCVs, through a yet unknown mechanism (Mallo et al. 2008). In conclusion, the salmonella pathogen has found many ways to manipulate various phosphoinositides at different locations and at different times of infection and replication. More studies combining imaging and biochemical approaches should help to clarify the spatiotemporal aspects of the picture.

The marine proteobacteria *Vibrio paraheamolyticus* is the cause of seafood-borne disease leading to gastroenteritis in humans. Like *S. flexneri* and *S. enterica*, the pathogen injects effector proteins through two TSS3, rapidly causing host cell rounding and lysis. One of these effector proteins, VPA0450, was linked to host cell blebbing, a phenomenon already observed with the *S. flexneri* effector IpgD (Niebuhr et al. 2002), and cell lysis (Broberg et al. 2010). VPA0450 has homology in its catalytic site with inositol polyphosphate 5-phosphatases of the IPP5Cs family that dephosphorylates the 5 position of the inositol ring. Accordingly, the phosphatase was recently shown to hydrolyzes PtdIns(4,5)P_2 and PtdIns(3,4,5)P_3 but not PtdIns(3,4)P_2*in vitro* (Broberg et al. 2010). Ectopic expression of VPA0450 in Hela cells induced the cytosolic relocation of the PtdIns(4,5)P_2 probe GFP-PH(PLCδ1), suggesting that *in vivo* PtdIns(4,5)P_2 is indeed a true substrate. The authors proposed that VPA0450, through its capacity to hydrolyze PtdIns(4,5)$P_{2,}$ was causing membrane/cytoskeleton reorganization and cell blebbing at the early stage of infection. A VPA0450-H356A mutant in the putative catalytic site was unable to induce such morphological changes. Thus, in a similar way to IpgD, VPA0450 affects the host plasma membrane PtdIns(4,5)P_2 level, but based on its homology with 5-phosphatases it is expected to generate PtdIns4P instead of PtdIns5P. Further characterization of the function of this phosphatase should bring interesting complementary information on the mechanism of *V. paraheamolyticus* infection.

Mycobacterium tuberculosis is a pathogen affecting the respiratory system by infecting the lung and is the causative agent of tuberculosis. Mycobacteria enter macrophages and thanks to an unusual lipid-rich cell wall is not recognized and replicate in mycobacterium-containing vacuoles (MCV). The pathogen has developed two strategies that target lipids to escape lysosomes by maintaining low levels of PtdIns3P. The first strategy is related to the secretion of the lipid analogues lipoarabinomannan and phosphatidylinositol mannoside that are toxins affecting the trafficking pathway (see Subchapter 3) (Chua et al. 2004). The second strategy involves the delivery of two phosphoinositide phosphatases SapM and MptpB (Fig. 12.2). Initially, a PtdIns3P phosphatase activity was detected in culture supernatant filtrates from virulent *M. tuberculosis* infected macrophages and identified as SapM (Vergne et al. 2005), an acid phosphatase previously described but only partially characterized (Raynaud et al. 1998; Saleh and Belisle 2000). SapM phosphatase activity was studied towards monophosphoinositides and showed a specificity for PtdIns3P *in vitro* (Vergne et al. 2005). SapM is also present in the mycobacterium phagosomes where it inhibits phagosome maturation by reducing PtdIns3P levels (Vergne et al. 2005). Thus, levels of PtdIns3P are double controlled: first by lipoarabinomannan which indirectly regulates class III PI 3-kinase (see below) and second by secretion of a PtdIns3P lipid phosphatase. In addition to SapM, *M. tuberculosis* encodes another phosphatase that hydrolyzes PtdIns3P. This 30 kDa secreted phosphatase, called MptpB, is involved in maintaining the mycobacterial infection (Singh et al. 2003) and was first described as a tyrosine phosphatase (Koul et al. 2000). However, it was later evidenced to be a triple-specificity phosphatase dephosphorylating phosphotyrosine, phosphoserine/threonine and phosphoinositides (Beresford et al. 2007). MptpB bears the active motif signature CX_5R of dual-specific and protein-tyrosine phosphatase family. *In vitro*, MptpB binds PtdIns3P, PtdIns4P, PtdIns5P and PtdIns(3,5)P_2 with equal potency but preferentially hydrolyses PtdIns3P and PtdIns(3,5)P_2 (Beresford et al. 2007), indicating that bacteria have multiplied strategies to maintain low levels of PtdIns3P to unable phagosome maturation. However, the *in vivo* relevant substrates need to be fully determined to uncover the physiological function of the MptpB phosphatase in the scheme of infection. It is interesting to note that orthologues of MptpB are found in other mycobacteria and fungi defining a new family of atypical lipid phosphatases characterized by a HCXXGKDR active site (Beresford et al. 2010).

Legionella micdadei is one of the 50 species of the *Legionella* protobacteria family. These Gram-negative bacteria, at various degrees, can cause human pulmonary infections, *Legionella pneumophila* causing Legionnaires' disease and *L. micdadei* causing flu-like symptoms and possibly pneumonia, including Pittsburgh pneumonia. The pathogen infects host phagocytes (monocytes, macrophages, neutrophils) and multiplies in the phagolysosomes. *L. micdadei* was originally reported to contain an acid phosphatase blocking the superoxide anion production by neutrophils (Saha et al. 1985). Further studies showed that the 68 kDa phosphatase, named ACP2, was interfering with the phosphoinositide metabolism. Indeed, it was demonstrated that ACP2 dephophorylates PtdIns(4,5)P_2 thereby indirectly reducing the production of the second messengers diacylglycerol and inositol phosphate $InsP_3$ produced by phospholipase C (Saha et al. 1988). Moreover, ACP2 also directly hydrolyzes $InsP_3$. These results suggested that the pathogen could interfere with the neutrophil microbicidal responses (Fig. 12.2). Acid phosphatases were also described in *Francisella tularensis, Coxiella burnetii, and Leishmania donovani* and involved in the resistance of pathogens to their host cells. In the case of *L. dovani,* the phosphatase was also shown to dephosphorylate inositol phosphates (Das et al. 1986). In addition to the modification of host cell phosphoinositide metabolism by phosphatases, *L. micdadei* might also interfere by injecting a phosphoinositide kinase (Saha et al. 1989). A c-AMP-independent kinase was purified to homogeneity from *L. micdadei* and shows activity towards both phosphoinositides and proteins. Although the proteins substrates were not characterized, it seems that tubulin was a good substrate *in vitro*. The lipid kinase activity has a specificity for PtdIns and when the enzyme was added on intact neutrophils, it increased the concentration of PtdInsP in the neutrophil membrane by up to 87%. Interestingly, the bacterial lipid kinase behaves differently from the eukaryotic lipid kinases in terms of divalent cations requirement and detergents sensitivity. Although the authors suggested that the kinase is a PtdIns 4-kinase increasing PtdIns4P in neutrophil plasma membrane, the product formed by the kinase has not been identified formally. Finally, since these earlier reports, no additional studies were published on *L. micdadei* phosphoinositide phosphatase and kinase, leaving open the question of their biological functions in the pathogen scheme of invasion.

Plasmodium falciparum is a protozoan parasite transmitted by the female Anopheles mosquito and is one of the plasmodium species that causes malaria in humans. During one of its developmental stages, the parasite endocytoses large quantities of erythrocyte cytosol and digests hemoglobin in a vesicular compartment also called the food vacuole. The original observation concluding to the production of large amounts of PtdInsP and PtdInsP_2 in plasmodium-infected erythrocytes (Elabbadi et al. 1994) was later on confirmed by phosphoinositides analysis showing a large increase in PtdIns3P, that represented one third of total monophosphoinositides of infected red blood cells (Tawk et al. 2010). An ortholog of mammalian class III PI 3-kinase (Vps34) was found in a search against the *P. falciparum* genome database, encoded by the gene PFE0765w, and was named PfPI3K (Tawk et al. 2010; Vaid et al. 2010). Indeed, the kinase has an *in vitro* activity towards PtdIns, PtdIns4P and PtdIns(4,5)P_2 thereby generating PtdIns3P, PtdIns(3,4)P_2 and PtdIns(3,4,5)P_3, respectively (Vaid et al. 2010). Consistent with these observations, besides high amounts of PtdIns3P, low quantities of PtdIns(3,4)P_2 and PtdIns(3,4,5)P_3 were also detected in infected erythrocytes (Tawk et al. 2010). PfPI3K is present in the food vacuole and in vesicular compartments at the parasite membrane but is also exported to the host red blood cell (Fig. 12.2). The PFE0765w gene does not harbor an host target sequence (HTS) with a PEXEL motif, known to target parasite-secreted proteins to their host (Marti et al. 2004). However, the export was found to be brefeldin A sensitive, suggesting the involvement of a secretory pathway (Vaid et al. 2010). Altogether, these results suggest that the increase in PtdIns3P observed in infected erythrocyte is the consequence of the export of the parasite PfPI3K. The parasite, by acting on host PtdIns3P could control trafficking pathway of hemoglobin to the food vacuole, an important step in the parasite development.

12.4 Recruitment and/or Modulation of Host Cell Phosphoinositide Metabolizing Enzymes by Pathogens

Besides delivering bacterial encoded effectors that possess phosphoinositide metabolizing activities, some pathogens like *enteropathogenic Escherichia coli* (EPEC) have the capacity to mimic host receptor signaling motifs.

The diarrheal pathogens EPEC colonize the intestinal tracts of their hosts and attach to the surface of enterocytes. EPEC translocate effectors into the host cell through a T3SS to induce a localized loss of microvilli and form actin-rich protrusions called pedestals, on which the bacteria reside. EPEC multiply in the extracellular space and pedestal is though to block bacterial internalization. A recent study (Smith et al. 2009) shows that EPEC recruits the host cell PtdIns(3,4,5)P_3 5-phosphatase SHIP2 by inserting its intimin receptor (Tir) into the host cell plasma membrane. Upon interactions with intimin present at the bacterial surface, EPEC Tir clusterizes and increases host cell tyrosine kinases activities, leading to phosphorylation of the tyrosine residues 474 and 454 in the C-terminal part of the receptor. The phosphorylation of tyrosine 474 is important for the formation of a signaling complex leading to N-WASP/Arp2/3-mediated actin remodelling through recruitment of proteins such as Nck (Gruenheid et al. 2001). Recently, phosphorylation of tyrosine 454 has been shown to drive the recruitment of PI 3-kinase leading to production of PtdIns(3,4,5)P_3 beneath EPEC attachment sites (Sason et al. 2009). Interestingly, the regions encompassing tyrosines 483 and 511 in EPEC Tir shares sequence similarities with immunotyrosine-based activation motifs (ITIM). In immune cells, ITIM motifs are known to recruit several phosphatases, including SHIP1, to downregulate activating signals (Daeron et al. 2008). The clustering of EPEC Tir and its tyrosines 483 and 511 appears essential to mimic ITIM-containing receptors by recruiting SHIP2. However, unlike most ITIM-motifs, the phosphorylation of the two tyrosine residues is dispensable to interact with SHIP2. The role of SHIP2 in this model is not only to downregulate PtdIns(3,4,5)P_3-mediated effects. Indeed, the phosphatase has two complementary functions; it acts as a scaffold protein for cytoskeletal regulators such as Shc, and a 5-phosphatase producing PtdIns(3,4)P_2 at EPEC pedestal. Using a fluorescent probe (TAPP1-PH), it was shown that this lipid accumulates throughout the pedestal where it is capable to anchor the actin branching protein lamellipodin through its PH domain. The role of SHIP2 as a scaffolding protein is also critical for the regulation of the morphology of actin pedestals. Indeed, SHIP2 recruits Shc to these structures by a mechanism requiring the phosphorylation of Tir at Y474. Shc induces a signaling cascade necessary for actin pedestal rearrangement. Mutation of tyrosines 483 and 511 into phenylalanine in the ITIM-like domain of

Tir impairs SHIP2 recruitment and appropriate actin rearrangement in pedestal. Although these processes deserve further mechanistic analysis to better understand the tight compartmentalization and coordination of actin dynamics, they provide new insights into the sequence of molecular events occurring during EPEC association with the cell membrane and pedestal formation. However, the role of PI 3-kinase in pedestal formation and/or host cell survival remains incompletely understood. PI 3-kinase inhibitors as well as mutation on tyrosine 454 of Tir have minor effects on actin recruitment to sites of bacterial attachment (Sason et al. 2009; Campellone and Leong 2005). Moreover, PI 3-kinase recruitment and function are likely different in EPEC-infected professional phagocytes. In macrophages, PI 3-kinase inhibition by the pathogen appears important to prevent bacterial phagocytosis (Celli et al. 2001).

Other pathogens such as *Mycobacterium tuberculosis* adopt dual strategies involving phosphoinositide metabolizing enzymes and phosphoinositide analogues delivery to stop bacterial endocytic pathway. *M. tuberculosis*, the etiological agent of the severe pulmonary disease tuberculosis, replicates in vacuole characterized by early phagosomal stage features, with incomplete luminal acidification and weak activity of lysosomal hydrolases (Russell et al. 2002). The inhibition of the bactericidal endocytic pathway occurs through removal of PtdIns3P from the phagosome *via* secretion by the pathogen of the phosphatases SapM and MptpB, (see Subchapter 2), and the lipid lipoarabinomannan, a glycosylated PtdIns which indirectly inhibits class III PI 3-kinase Vps34 (Vergne et al. 2003; Hmam et al. 2004). Lipoarabinomannan has been shown to decrease cytosolic calcium rise in infected cells leading to reduction of the calmodulin-kinase II-dependent activation of class III PI 3-kinase Vps34. In turn, PtdIns3P production is reduced together with EEA1 recruitment to phagosomes, leading to impairment of the maturation and delivery of vacuolar proton ATPase and acid hydrolases to *Mycobacterium*-containing vacuoles (Vergne et al. 2003; Fratti et al. 2001, 2003). Another mycobacterial phosphoinositide used by the pathogen is phosphatidylinositol mannoside. This lipid specifically promotes fusion between early endosomes and phagosome, thus allowing communication and nutrient supply despite the trafficking inhibition due to PtdIns3P depletion (Vergne et al. 2004).

Another example of bacteria recruiting a phosphoinositide metabolizing enzyme involved in the regulation of the trafficking pathway in order to prevent acquisition of the lysosomal characteristics of bacteria-containing vacuoles is provided by *Legionella pneumophila*. It replicates within a specific *Legionella*-containing vacuole and uses a type IV secretion system and a number of effectors proteins to form this vacuole. Among the effectors translocated by this bacteria into host cells, the LpnE protein interacts with the N-terminal domain of the PtdIns(4,5)P_2 5-phosphatase OCRL1 (Weber et al. 2009a). OCRL1, a phosphoinositide phosphatase deficient in the hereditary oculocerebrorenal syndrome of Lowe, plays an important role in the control of the endocytic pathway and in the maintenance of the trans-Golgi network (McCrea and De Camilli 2009). LpnE plays a critical role in regulating host cell entry and vacuolar trafficking and its interaction with OCRL1 at the *Legionella*-containing vacuoles in macrophages suggests that this phosphatase may contribute to prevent acquisition of lysosomal characteristics by the vacuole. Interestingly, the bacterium *Chlamydiae*, which replicates in a specific vacuole, has also been shown to recruit OCRL1 and other enzymes producing PtdIns4P such as PtdIns 4-kinase type IIα to chlamydial inclusion by an unknown mechanism (Moorhead et al. 2010). These results suggest a role for the product of OCRL1, PtdIns4P, in providing a replication-competent vacuolar environment.

12.5 Pathogen Effectors Interacting with Host Cell Phosphoinositides

As is clear from this overview, many cellular pathogens can manipulate the levels of phosphoinositides. However, considering the central place of these lipids in the field of signaling and trafficking *via* protein interactions, it is not surprising that pathogens also secrete effector proteins capable to directly interact with phosphoinositides through specific domains. These domains are not systematically identifiable through similarity search in eukaryotic proteins data bases. For instance, several effectors injected in the host cell by *Legionella pneumophila* contain phosphoinositide binding domains which have no similarity with the eukaryotic ones. Substrate of lcm/Dot transporter C (SidC) and SdcA localise to the *Legionella*-containing vacuole and selectively interact with PtdIns4P *in vitro*. The binding site has been mapped and is termed P4C domain (Ragaz et al. 2008). This prokaryotic PtdIns4P probe, which appears easy to produce and is stable, may be a useful tool to image this lipid in eukaryotic cells, however it remains to be clearly demonstrated that it is indeed a highly selective probe for PtdIns4P in an eukaryotic environment. Pull-down experiments using phosphoinositide-coated beads led to the identification of several other PtdIns4P binding domains in *L. pneumophila* effectors (Machner and Isberg 2006; Murata et al. 2006). Particularly, the lcm/Dot translocated Rab1 guanine exchange factor SidM/DrrA was identified as a prominent PtdIns4P binding effector protein. The binding domain was identified in the C-terminal part of the protein and was called P4M (PtdIns4P-binding of SidM/DrrA). This domain can compete with SidC for binding to PtdIns4P on *Legionella*-containing vacuole (Brombacher et al. 2009). LidA, a protein promoting the guanine exchange factor activity of SidM, was also identified as a phosphoinositide binding protein, preferentially interacting with PtdIns3P and PtdIns4P. Furthermore, LpnE, the OCRL1-interacting protein, specifically binds PtdIns3P. Thus, several *L. pneumophila* effector proteins use their phosphoinositide binding capacities to anchor to Legionella-containing vacuole and modulate trafficking mechanisms.

Remarkably, the enteric protozoan parasite *Entamoeba histolytica*, a major cause of colitis and liver abscess, possesses twelve FYVE domain-containing proteins (EhFP1-12). Although the capacity of these FYVE domain containing proteins to interact with PtdIns3P has not been systematically analyzed, the expansion of a set of proteins harboring PtdIns3P interacting domains in *E. histolytica* may be considered with respect to the especially large complement of PtdIns3P 3-phosphatase myotubularin homologues found in this parasite (Kerk and Moorhead 2010). These observations suggest a role for PtdIns3P-mediated signaling in the regulation of *E.*

histolytica life style and host infection. Recently, one of these FYVE domain containing proteins, EhFP4, has been shown to be recruited to the phagocytic cup and to tunnel-like tubular structures linking the plasma membranes to the phagosome (Nakada-Tsukui et al. 2009). This protein interacts with Rho/Rac GTPases. Localization of EhFP4 to the phagocytic cup involves its binding to PtdIns4P via a C-terminal region, while its FYVE domain appears to modulate the binding selectivity of the protein to PtdIns4P (Nakada-Tsukui et al. 2009). Overall these data strongly suggest that several *E. histolytica* effectors are targeted to the *Legionella*-containing vacuole *via* phosphoinositide interaction to control its maturation.

An important issue is the biochemical and functional characterization of the different phosphoinositide binding effectors injected into host cells by the various pathogens as well as the use and development of cell biology approaches to monitor the spatiotemporal dynamics of this system.

More surprising yet, a very recent study (Kale et al. 2010) reports observations suggesting the existence of a novel mechanism allowing a series of effector proteins from plant and animal pathogens to enter into host cells, independently of the pathogen, *via* specific interaction with PtdIns3P present in the outer leaflet of the plasma membrane. Phosphoinositides are known to be mainly located in the inner leaflet of the plasma membrane. However, using specific biosensors the authors showed that PtdIns3P is relatively abundant in the outer surface of plant cell plasma membranes and in the outer leaflet of some mammalian cells. Effectors of fungal plant pathogens and oomycete causing many destructive plant diseases contain a RXLR motif that is required to bind to "external" PtdIns3P and to enter into host cells *via* a lipid raft-mediated endocytosis. Blocking the interaction of these effectors with PtdIns3P impairs their entry into both plant and human cells, suggesting potential therapeutic intervention. The challenge is now to demonstrate whether or not this new mechanism also concerns oomycetes and fungi that infect humans and other animals.

Finally, although strategies developed by viruses are not described in this chapter, it is interesting to note that recent data show the critical recruitment of type III PtdIns 4-kinase via specific viral proteins and its requirement to form specialized PtdIns4P-rich organelles essential for RNA viruses replication (Hsu et al. 2010; Trotard et al. 2009; Borawski et al. 2009).

12.6 Conclusions

The seven polyphosphoinositides and the set of phosphoinositide metabolizing enzymes allowing rapid and reversible interconversion between them provide a powerful and highly dynamic system to coordinate actin cytoskeleton rearrangements, cell polarity, membrane reorganization and vesicle trafficking as well as signaling pathways determining cell fate. Cellular pathogens have taken advantage of this sophisticated system to manipulate host cells and to establish the complex pathogen-host cell relationship that govern their survival and replication. According to their lifestyle, cellular pathogens manipulate the phosphoinositide metabolism in many different ways. They can produce and use their own phosphoinositide metabolizing enzyme or effector, or even lipids factors, that directly or indirectly recruit and/or modulate the activity of host cell enzymes, and they can also express phosphoinositide binding effectors.

It will be important in the future to combine different approaches including biochemistry, imaging and biophysics to get a step further in the comprehension of the mechanisms related to phosphoinositide changes in the different steps of cellular pathogens infection. These combined approaches may for instance help to analyze the impact of phosphoinositides changes on cell polarity, particularly on the establishment of epithelial cell shape since some pathogens may locally modify an apical membrane into a membrane with basolateral characteristics to facilitate colonization of the mucosal barrier (Gassama and Payrastre 2009).

Clearly, deciphering the different strategies elaborated by cellular pathogens to hijack the phosphoinositide metabolism and analyzing the functional consequences should provide clues to our understanding of fundamental cell mechanisms as these fascinating and versatile lipids link and coordinate membrane and cytoskeleton dynamics to cell responses.

References

Agarwal V, Hammerschmidt S (2009) Cdc42 and the phosphatidylinositol 3-kinase-Akt pathway are essential for PspC-mediated internalization of pneumococci by respiratory epithelial cells. J Biol Chem 284:19427–19436

Aleman A, Rodriguez-Escudero I, Mallo GV, Cid VJ, Molina M, Rotger R (2005) The amino-terminal non-catalytic region of Salmonella typhimurium SigD affects actin organization in yeast and mammalian cells. Cell Microbiol 7:1432–1446

Alrutz MA, Isberg RR (1998) Involvement of focal adhesion kinase in invasin-mediated uptake. Proc Natl Acad Sci U S A 95:13658–13663

Alrutz MA, Srivastava A, Wong KW, D'Souza-Schorey C, Tang M, Ch'ng LE, Snapper SB, Isberg RR (2001) Efficient uptake of Yersinia pseudotuberculosis via integrin receptors involves a Rac1-Arp2/3 pathway that bypasses N-WASP function. Mol Microbiol 42:689–703

Banarjee M, Copp J, Vuga D, Marino M, Chapman T, van der Geer P, Ghosh P (2004) GW domains of the Listeria monocytogenes invasion protein InlB are required for potenitation of Met activation. Mol Microbiol 52:257–271

Basar T, Shen Y, Ireton K (2005) Redundant role for Met docking site tyrosines and Gab1 PH domain in InlB-mediated entry of Listeria monocytogenes. Infect Immun 73:2061–2074

Beresford N, Patel S, Armstrong J, Szoor B, Fordham-Skelton AP, Tabernero L (2007) MptpB, a virulence factor from Mycobacterium tuberculosis, exhibits triple-specificity phosphatase activity. Biochem J 406:13–18

Beresford N, Saville C, Bennett HJ, Roberts IS, Tabernero L (2010) A new family of phosphoinositide phosphatases in microorganisms: identification and biochemical analysis. BMC Genomics 11:457–469

Bergmann S, Lang A, Rohde M, Agarwal V, Rennermeier C, Grashoff C, Preissner KT, Hammerschmidt S (2009) Integrin-linked kinase is required for vitronectin-mediated internalization of Streptococcus pneumoniae by host cells. J Cell Sci 122:256–267

Bershadsky A (2004) Magic touch: how does cell-cell adhesion trigger actin assembly? Trends Cell Biol 14:589–593

Bierne H, Dramsi S, Gratacap MP, Randriamampita C, Carpenter G, Payrastre B, Cossart P (2000) The invasion protein InlB from L. monocytogenes activates PLC-g1 downstream from PI 3-kinase. Cell Microbiol 2:465–476

Bierne H, Gouin E, Roux P, Caroni P, Yin HL, Cossart P (2001) A role for cofilin and LIM kinase in Listeria-induced phagocytosis. J Cell Biol 155:101–112

Borawski J, Troke P, Puyang X, Gibaja V, Zhao S, Mickanin C, Leighton-Davies J, Wilson CJ, Myer V, Cornellataracido I, Baryza J, Tallarico J, Joberty G, Bantscheff M, Schirle M, Bouwmeester T, Mathy JE, Lin K, Compton T, Labow M, Wiedmann B, Gaither LA (2009) Class III phosphatidylinositol 4-kinase alpha and beta are novel host factor regulators of hepatitis C virus replication. J Virol 83:10058–10074

Braun L, Ghebrehiwet B, Cossart P (2000) gC1q-R/32, a C1q-binding protein is a receptor for the InlB invasion protein of Listeria monocytogenes. EMBO J 19:1458–1466

Broberg CA, Zhang L, Gonzalez H, Laskowski-Arce MA, Orth K (2010) A Vibrio effector protein is an inositol phosphatase and disrupts host cell membrane integrity. Science 329:1660–1662

Brombacher E, Urwyler S, Ragaz C, Weber SS, Kami K, Overduin M, Hilbi H (2009) Rab1 guanine nucleotide exchange factor SidM is a major PtdIns(4)P-binding effector protein of Legionella pneumophila. J Biol Chem 284:4846–4856

Brown FD, Rozelle AL, Yin HL, Balla T, Donaldson JG (2001) Phosphatidylinositol 4, 5-bisphosphate and Arf6-regulated membrane traffic. J Cell Biol 154:1007–1017

Brumel JH, Grinstein S (2003) Role of lipid-mediated signal transduction in bacterial internalization. Cell Microbiol 5:287–297

Campellone KG, Leong JM (2005) Nck-independent actin assembly is mediated by two phosphorylated tyrosines within nteropathogenic Escherichia coli Tir. Mol Microbiol 56:416–432

Carlton JG, Cullen PJ (2005) Coincidence detection in phosphoinositide signalling. Trends Cell Biol 15:540–547

Carricaburu V, Lamia KA, Lo E, Favereaux L, Payrastre B, Cantley LC, Rameh LE (2003) The phosphatidylinositol (PI)-5-phosphate 4-kinase type II enzyme controls insulin signaling by regulating PI-3,4,5-trisphosphate degradation. Proc Natl Acad Sci U S A 100:9867–9872

Celli J, Olivier M, Finlay BB (2001) Enteropathogenic Escherichia coli mediates antiphagocytosis through the inhibition of PI 3-kinase-dependent pathways. EMBO J 20:1245–1258

Chua J, Vergne I, Master S, Deretic V (2004) A tale of two lipids: Mycobacterium tuberculosis phagosome maturation arrest. Curr Opin Microbiol 7:71–77

Cossart P, Sansonetti P (2004) Bacterial invasion: the paradigms of enterinvasive pathogens. Science 304:242–248

Daeron M, Jaeger S, Du Pasquier L, Vivier E (2008) Immunoreceptor tyrosine based inhibition motifs: a quest in the past and future. Immunol Rev 224:11–43

Dai S, Zhang Y, Weimbs T, Yaffe MB, Zhou D (2007) Bacteria-generated PtdIns(3)P recruits VAMP8 to facilitate phagocytosis. Traffic 8:1365–1374

Das S, Saha AK, Remaley AT, Glew RH, Dowling JN, Kajiyoshi M, Gottlieb M (1986) Hydrolysis of phosphoproteins and inositol phosphates by cell surface phosphatase of Leishmania donovani. Mol Biochem Parasitol 20:143–153

De Santis G, Miotti S, Mazzi M, Canevari S, Tomassetti A (2009) E-cadherin directly contributes to PI3K/AKT activation by engaging the PI3K-p85 regulatory subunit to adherens junctions of ovarian carcinoma cells. Oncogene 28:1206–1217

DeVinney R, Steele-Mortimer O, Finlay BB (2000) Phosphatases and kinases delivered to the host cell by bacterial pathogens. Trends Microbiol 8:29–33

Di Paolo G, De Camilli P (2006) Phosphoinositides in cell regulation and membrane dynamics. Nature 443:651–657

Dokainish H, Gavicherla B, Shen Y, Ireton K (2007) The carboxyl-terminal SH3 domain of the mammalian adaptor CrkII promotes internalization of Listeria monocytogenes through activation of host phosphoinositide 3-kinase. Cell Microbiol 9:2497–2516

Duncan MJ, Shin JS, Abraham SN (2002) Microbial entry through caveolae: variations on a theme. Cell Microbiol 4:783–791

Duncan MJ, Li G, Shin JS, Carson JL, Abraham SN (2004) Bacterial penetration of bladder epithelium through lipid rafts. J Biol Chem 279:18944–18951

Elabbadi N, Ancelin ML, Vial HJ (1994) Characterization of phosphatidylinositol synthase and evidence of a polyphosphoinositide cycle in Plasmodium-infected erythrocytes. Mol Biochem Parasitol 63:179–192

Feng Y, Wente SR, Majerus PW (2001) Overexpression of the inositol phosphatase SopB in human 293 cells stimulates cellular chloride influx and inhibits nuclear mRNA export. Proc Natl Acad Sci U S A 98:75–79

Fowler T, Wann ER, Joh D, Johansson SA, Foster TJ, Höök M (2000) Cellular invasion by Staphylococcus aureus involves a fibronectin bridge between the bacterial fibronectin-binding MSCRAMMs and host cell beta 1 integrins. Eur J Cell Biol 79:672–679

Fratti RA, Backer JM, Gruenberg J, Corvera S, Deretic V (2001) Role of phosphatidylinositol 3-kinase and Rab5 effectors in phagosomal biogenesis and mycobacterial phagosome maturation arrest. J Cell Biol 154:631–644

Fratti RA, Chua J, Vergne I, Deretic V (2003) Mycobacterium tuberculosis glycosylated phosphatidylinositol causes phagosome maturation arrest. Proc Natl Acad Sci U S A 100:5437–5442

Gassama-Diagne A, Payrastre B (2009) Phosphoinositides signalling pathways: promising role as builders of epithelial cell polarity. In Rev Cell Mol Biol 273:313–343

Gozani O, Karuman P, Jones DR, Ivanov D, Cha J, Lugovskoy AA, Baird CL, Zhu H, Field SJ, Lessnick SL, Villasenor J, Mehrotra B, Chen J, Rao VR, Brugge JS, Ferguson CG, Payrastre B, Myszka DG, Cantley LC, Wagner G, Divecha N, Prestwich GD, Yuan J (2003) The PHD finger of the chromatin-associated protein ING2 functions as a nuclear phosphoinositide receptor. Cell 114:99–111

Gruenheid S, DeVinney R, Bladt F, Goosney D, Gelkop S, Gish GD, Pawson T, Finlay BB (2001) Enteropathogenic E. coli Tir binds Nck to initiate actin pedestal formation in host cells. Nat Cell Biol 3:856–859

Hall A (2005) Rho GTPases and the control of cell behaviour. Biochem Soc Trans 33:891–895

Hamon M, Bierne H, Cossart P (2006) Listeria monocytogenes: a multifaceted model. Nat Rev Microbiol 4:423–434

Hernandez LD, Hueffer K, Wenk MR, Galan JE (2004) Salmonella modulates vesicular traffic by altering phosphoinositide metabolism. Science 304:1805–1807

Hilbi H (2006) Modulation of phosphoinositide metabolism by pathogenic bacteria. Cell Microbiol 8:1697–1706

Hmam Z, Sendide K, Talal A, Garcia R, Dobos K, Reiner NE (2004) Quantitative analysis of phagolysosome fusion in intact cells: inhibition by mycobacterial lipoarabinomannan and rescue by an 1alpha,25-dihydroxyvitamin D3-phosphoinositide 3-kinase pathway. J Cell Sci 117:2131–2140

Hoffman I, Eugene E, Nassif X, Couraud PO, Bourdoulous S (2001) Activation of ErbB2 receptor tyrosine kinase supports invasion of endothelial cells by Neisseria meningitis. J Cell Biol 155:133–143

Hsu, N-Y, Ilnytska O, Belov G, Santiana M, Chen, Y-H, Takvorian PM, Pau C, Van der Schaar H, Kaushik-Basu N, Balla T, Cameron CE, Ehrenfeld E, Van Kuppeveld, FJM, Altan-Bonnet N (2010) Viral reorganization of the secretory pathway generates distinct organelles for RNA replication. Cell 141:799–811

Ireton K, Payrastre B, Chap H, Ogawa W, Sakaue H, Kasuga M, Cossart P (1996) A role for phosphoinositide 3-kinase in bacterial invasion. Science 274:780–782

Ireton K, Payrastre B, Cossart P (1999) The Listeria monocytogenes protein InlB is an agonist of mammalian Phosphoinositide 3-kinase. J Biol Chem 274:17025–17032

Isberg RR, Leong JM (1990) Multiple β1 chain integrins are recptors for invasin, a protein that promotes bacterial penetration into mammalian cells. Cell 60:861–871

Jones DR, Bultsma Y, Keune WJ, Halstead JR, Elouarrat D, Mohammed S, Heck AJ, D'Santos CS, Divecha N (2006) Nuclear PtdIns5P as a transducer of stress signaling: an in vivo role for PIP4K beta. Mol Cell 23:685–695

Kale SD, Gu B, Capelluto DGS, Dou D, Feldman E, Rumore A, Arredondo FD, Hanlon R, Fudal I, Rouxel T, Lawrence CB, Shan W, Tyler BM (2010) External lipid PI3P mediates entry of eukaryotic pathogen effectors into plant and animal host cells. Cell 142:284–295

Kallstrom H, LiszwskiI MK, Atkinson JP, Jonsson AB (1997) Membrane cofactor protein (MCP or CD46) is a cellular pilus receptor for pathogenic Neisseria. Mol Microbiol 25:639–647

Kang PB, Azad AK, Torrelles JB, Kaufman TM, Beharka A, Tibesar E, DesJardin LE, Schlesinger LS (2005) The human macrophage mannose receptor directs mycobacterium tuberculosis lipoarabinomannan-mediated phagosome biogenesis. J Exp Med 202:987–999

Kerk D, Moorhead GBG (2010) A phylogenetic survey of myotubularin genes of eukaryotes: distribution, protein structure, evolution, and gene expression. BMC Evol Biol 10:196–212

Knodler LA, Finlay BB, Steele-Mortimer O (2005) The Salmonella effector protein SopB protects epithelial cells from apoptosis by sustained activation of Akt. J Biol Chem 280:9058–9064

Koul A, Choidas A, Treder M, Tyagi AK, Drlica K, Singh Y, Ullrich A (2000) Cloning and characterization of secretory tyrosine phosphatases of Mycobacterium tuberculosis. J Bacteriol 182:5425–5432

Lecuit M, Hurme R, Pizarro-Cerdà J, Ohayon H, Geiger B, Cossart P (2000) A role for alpha- and beta-catenins in bacterial uptake. Proc Natl Acad Sci U S A 97:10008–10013

Lemichez E, Lecuit M, Nassif X, Bourdoulous S (2010) Breaking the wall: targetting of the endothelium by pathogen bacteria. Nat Rev Microbiol 8:93–104

Lemmon MA (2003) Phosphoinositide recognition domains. Traffic 4:201–213

Machner MP, Isberg RR (2006) Targeting of host Rab GTPase function by the intravacuolar pathogen Legionella pneumophila. Dev Cell 11:47–56

Malaviya R, Gao Z, Thankavel K, Merwe PA van der, Abraham SN (1999) The mast cell tumor necrosis factor alpha response to FimH-expressing Escherichia coli is mediated by the glycosylphosphatidylinositol-anchored molecule CD48. Proc Natl Acad Sci U S A 96:8110–8115

Mallo GV, Espina M, Smith AC, Terebiznik MR, Aleman A, Finlay BB, Rameh LE, Grinstein S, Brumell JH (2008) SopB promotes phosphatidylinositol 3-phosphate formation on Salmonella vacuoles by recruiting Rab5 and Vps34. J Cell Biol 182:741–752

Marcus SL, Wenk MR, Steele-Mortimer O, Finlay BB (2001) A synaptojanin-homologous region of Salmonella typhimurium SigD is essential for inositol phosphatase activity and Akt activation. FEBS Lett 494:201–207

Marino M, Braun L, Cossart P, Ghosh P (1999) Structure of the InlB leucine-rich repeats, a domain that triggers host cell invasion by the bacterial pathogen L. monocytogenes. Mol Cell 4:1063–1072

Marino M, Banarjee M, Jonquieres R, Cossart P, Ghosh P (2002) GW domains of the Listeria monocytogenes invasion protein InlB are SH3-like and mediate binding to host ligand. EMBO J 21:5623–5634

Marti M, Good RT, Rug M, Knuepfer E, Cowman AF (2004) Targeting malaria virulence and remodeling proteins to the host erythrocyte. Science 306:1930–1933

Martinez JJ, Mulvey MA, Schilling JD, Pinkner JS, Hultgren SJ (2000) Type 1 pilus-mediated bacterial invasion of bladder epithelial cell. EMBO J 19:2803–2812

Mason D, Mallo GV, Terebiznik MR, Payrastre B, Finlay BB, Brumell JH, Rameh L, Grinstein S (2007) Alteration of epithelial structure and function associated with PtdIns(4,5)P_2 degradation by a bacterial phosphatise. J Gen Physiol 129:267–283

Matsui T, Yonamura S, Tsukita S (1999) Activated of ERM proteins in vivo by Rho involves phosphatidyl-inositol 4-phosphate 5-kinase and not ROCK kinases. Curr Biol 9:1259–1262

McCrea HJ, De Camilli P (2009) Mutations in phosphoinositide metabolizing enzymes and human disease. Physiology 24:8–16

Mengaud J, Ohayon H, Gounon P, Mège RM, Cossart P (1996) E-cadherin is the receptor for internalin, a surface protein required for entry of Listeria monocytogenes into epithelial cells. Cell 84:923–932

Merz AJ, Enns CA, So M (1999) Type IV pili of pathogenic Neisseria elicit cortical plaque formation in epithelial cells. Mol Microbiol 32:1316–1332

Min G, Stolz M, Zhou G, Liang F, Sebbel P, Stoffler D, Glockshuber R, Sun T-T, Aebi U, Kong X-P (2002) Localization of uroplakin Ia, the urothelial receptor for bacterial adhesin FimH on the six inner domains of the 16 nm urothelial plaque particle. J Mol Biol 317:697–706

Moorhead AM, Jung JY, Smirnov A, Kaufer S, Scidmore MA (2010) Multiple host proteins that function in phosphatidylinositol-4-phosphate metabolism are recruited to the chlamydial inclusion. Infect Immun 78:1990–2007

Murata T, Delprato A, Ingmundson A, Toomre DK, Lambright DG, Roy CR (2006) The Legionella pneumophila effector protein DrrA is a Rab1 guanine nucleotide-exchange factor. Nat Cell Biol 8:971–977

Nakada-Tsukui K, Okada H, Mitra BN, Nozaki T (2009) Phosphatidylinositol-phosphates mediate cytoskeletal reorganisation during phagocytosis via a unique modular protein consisting of RhoGEF/DH and FYVE domains in the parasitic protozoon entamoeba histolytica. Cell Microbiol 11:1471–1491

Nerlich A, Rohde M, Talay SR, Genth H, Just I, Chhatwal GS (2009) Invasion of endothelial cells by tissue-invasive M3 type group A streptococci requires Src kinase and activation of Rac1 by a phosphatidylinoditol 3-kinase independent mechanism. J Biol Chem 284:20319–20328

Niebuhr K, Giuriato S, Pedron T, Philpott DJ, Gaits F, Sable J, Sheetz MP, Parsot C, Sansonetti PJ, Payrastre B (2002) Conversion of PtdIns(4,5)P(2) into PtdIns(5)P by the S.flexneri effector IpgD reorganizes host cell morphology. EMBO J 21:5069–5078

Niemann HH, Jäger V, Buttler PJ, Heuvel J van den, Schmidt S, Ferraris D, Gherardi E, Heinz DW (2007) Structure of the human receptor tyrosine kinase met in complex with the Listeria invasion protein InlB. Cell 130:235–246

Norris FA, Wilson MP, Wallis TS, Galyov EE, Majerus PW (1998) SopB, a protein required for virulence of Salmonella dublin is an inositol phosphate phosphatase. Proc Natl Acad Sci U S A 95:14057–14059

Ozeri V, Rosenshine I, Mosher DF, Fässler R, Hanski E (1998) Roles of integrins and fibronectin in the entry of streptococcus pyogenes into cells via protein F1. Mol Microbiol 30:625–637

Ozeri V, Rosenshine I, Ben-Ze'ev A, Bokoch GM, Jou TS, Hanski E (2001) De novo formation of focal complex-like structures in host cells by invading Streptococci. Mol Microbiol 41:561–573

Pattni K, Jepson M, Stenmark H, Banting G (2001) A PtdIns(3)P-specific probe cycles on and off host cell membranes during Salmonella invasion of mammalian cells. Curr Biol 11:1636–1642

Payrastre B, Missy K, Giuriato S, Bodin S, Plantavid M, Gratacap MP (2001) Phosphoinositides key players in cell signalling in time and space. Cell Signal 13:377–387

Pendaries C, Tronchère H, Plantavid M, Payrastre B (2003) Phosphoinositide signaling disorders in human diseases. FEBS Lett 546:25–31

Pendaries C, Tronchere H, Arbibe L, Mounier J, Gozani O, Cantley L, Fry MJ, Gaits-Iacovoni F, Sansonetti PJ, Payrastre B (2006) PtdIns5P activates the host cell PI3-kinase/Akt pathway during Shigella flexneri infection. EMBO J 25:1024–1014

Pizarro-Cerdà J, Cossart P (2004) Subvertion of phosphoinositide metabolism by intracellular bacterial pathogens. Nat Cell Biol 6:1026–1033

Pizarro-Cerdà J, Payrastre B, Wang YJ, Veiga E, Yin HL, Cossart P (2007) Type II phosphatidylinositol 4-kinases promote Listeria monocytogenes entry into target cells. Cell Microbiol 9:2381–2390

Ragaz C, Pietsch H, Urwyler S, Tiaden A, Weber SS, Hilbi H (2008) The Legionella pneumophila phosphatidylinositol-4 phosphate-binding type IV substrate SidC recruits endoplasmic reticulum vesicles to a replication-permissive vacuole. Cell Microbiol 10:2416–2433

Rameh LE, Tolias KF, Duckworth BC, Cantley LC (1997) A new pathway for synthesis of phosphatidylinositol-4,5-bisphosphate. Nature 390:192–196

Ramel D, Lagarrigue F, Dupuis-Coronas S, Chicanne G, Leslie N, Gaits-Iacovoni F, Payrastre B, Tronchere H (2009) PtdIns5P protects Akt from dephosphorylation through PP2A inhibition. Biochem Biophys Res Commun 387:127–131

Ramel D, Lagarrigue F, Pons V, Mounier J, Dupuis-Coronas S, Chicanne G, Sansonetti PJ, Gaits-Iacovoni F, Tronchère H, Payrastre B (2011) Shigella flexneri Infection generates the lipid PI5P to alter endocytosis and prevent termination of EGFR signaling. Sci Signal 4:ra61

Raynaud C, Etienne G, Peyron P, Laneelle MA, Daffe M (1998) Extracellular enzyme activities potentially involved in the pathogenicity of Mycobacterium tuberculosis. Microbiology 144:577–587

Rosenberg CM, Brumell JH, Finlay BB (2000) Microbial pathogenesis: lipid rafts as pathogen portals. Curr Biol 10:R823–R825

Russell DG, Mwandumba HC, Rhoades EE (2002) Mycobacterium and the coat of many lipids. J Cell Biol 158:421–426

Saha AK, Dowling JN, LaMarco KL, Das S, Remaley AT, Olomu N, Pope MT, Glew RH (1985) Properties of an acid phosphatase from Legionella micdadei which blocks superoxide anion production by human neutrophils. Arch Biochem Biophys 243:150–160

Saha AK, Dowling JN, Pasculle AW, Glew RH (1988) Legionella micdadei phosphatase catalyzes the hydrolysis of phosphatidylinositol 4,5-bisphosphate in human neutrophils. Arch Biochem Biophys 265:94–104

Saha AK, Dowling JN, Mukhopadhyay NK, Glew RH (1989) Legionella micdadei protein kinase catalyzes phosphorylation of tubulin and phosphatidylinositol. J Bacteriol 171:5103–5110

Saleh MT, Belisle JT (2000) Secretion of an acid phosphatase (SapM) by mycobacterium tuberculosis that is similar to eukaryotic acid phosphatases. J Bacteriol 182:6850–6853

Sasaki T, Takasuga S, Sasaki J, Kofuji S, Eguchi S, Yamazaki M, Suzuki A (2009) Mammalian phosphoinositide kinases and phosphatases. Prog Lipid Res 48:307–343

Sason H, Milgrom M, Weiss AM, Melamed-Book N, Balla T, Grinstein S, Backert S, Rosenshine I, Aroeti B (2009) Enteropathogenic Escherichia coli subverts phosphatidylinositol 4,5-bisphosphate and phosphatidylinositol 3,4,5-trisphosphate upon epithelial cell infection. Mol Biol Cell 20:544–555

Sbrissa D, Ikonomov OC, Strakova J, Shisheva A (2004) Role for a novel signaling intermediate, phosphatidylinositol 5-phosphate in insulin-regulated F-actin stress fiber breakdown and GLUT4 translocation. Endocrinology 145:4853–4865

Schubert W-D, Urbanke C, Ziehm T, Beier V, Machner MP, Domann E, Wehland J, Chakraborty T, Heinz DW (2002) Structure of internalin, a major invasion protein of Listeria monocytogenes in complex with its human receptor E-cadherin. Cell 111:825–836

Schwarz-Linek U, Höök M, Potts JR (2004) The molecular basis of fibronectin-mediated bacterial adherence to host cells. Mol Microbiol 52:631–641

Scott CC, Cuellar-Mata P, Matsuo T, Davidson HW, Grinstein S (2002) Role of 3-phosphoinositides in the maturation of Salmonella-containing vacuoles within host cells. J Biol Chem 277:12770–12776

Seveau S, Tham TN, Payrastre B, Hoppe AD, Swanson JA, Cossart P (2007) A FRET analysis to unravel the role of cholesterol in Rac1 and PI 3-kinase activation in the InlB/Met signalling pathway. Cell Microbiol 9:790–803

Shen Y, Naujokas K, Park M, Ireton K (2000) InlB-dependent internalization of Listeria is mediated by the Met receptor tyrosine kinase. Cell 103:501–510

Sinha B, Francois PP, Nüsse O, Foti M, Hartford OM, Vaudaux P, Foster TJ, Lew DP, Herrmann M, Krause KH (1999) Fibronectin-binding protein acts as Staphylococcus aureus invasin via fibronectin bridging to integrin alpha5beta1. Cell Microbiol 1:101–117

Singh R, Rao V, Shakila H, Gupta R, Khera A, Dhar, N. Singh A, Koul A, Singh Y, Naseema M, Narayanan PR, Paramasivan CN, Ramanathan VD, Tyagi AK (2003) Disruption of mptpB impairs the ability of Mycobacterium tuberculosis to survive in guinea pigs. Mol Microbiol 50:751–762

Smith K, Humphreys D, Hume PJ, Koronakis V (2009) Enteropathogenic Escherichia coli recruits the cellular inositol phosphatase SHIP2 to regulate actin-pedestal formation. Cell Host Microbe 7:13–24

Sousa S, Cabanes D, E-Amraoui A, Petit C, Lecuit M, Cossart P (2003) Unconventional myosin VIIa and vezatin, two proteins crucial for Listeria entry into epithelial cells. J Cell Sci 117:2121–2130

Sousa S, Cabanes D, Archambaud C, Colland F, Lemichez E, Popoff M, Boisson-Dupuis S, Gouin E, Lecuit M, Legrain P, Cossart P (2005) ARHGAP10 is necessary for alpha-catenin recruitment at adherens junctions and for Listeria invasion. Nat Cell Biol 7:954–960

Steele-Mortimer O, Knodler LA, Marcus SL, Scheid MP, Goh B, Pfeifer CG, Duronio V, Finlay BB (2000) Activation of Akt/protein kinase B in epithelial cells by the Salmonella typhimurium effector sigD. J Biol Chem 275:37718–37724

Sun H, Shen Y, Dokainish H, Holgado-Madruga M, Wong A, Ireton K (2005) Host adaptor proteins Gab1 and CrkII promote InlB-dependent entry of Listeria monocytogenes. Cell Microbiol 7:443–457

Tawk L, Chicanne G, Dubremetz JF, Richard V, Payrastre B, Vial HJ, Roy C, Wengelnik K (2010) Phosphatidylinositol 3-phosphate an essential lipid in Plasmodium, localizes to the food vacuole membrane and the apicoplast. Eukaryot Cell 9:1519–1530

Terebiznik MR, Vieira OV, Marcus SL, Slade A, Yip CM, Trimble WS, Meyer T, Finlay BB, Grinstein S (2002) Elimination of host cell PtdIns(4,5)P(2) by bacterial SigD promotes membrane fission during invasion by Salmonella. Nat Cell Biol 4:766–773

Trotard M, Lepère-Douard C, Régeard M, Piquet-Pellorce C, Lavillette D, Cosset FL, Gripon P, Le Seyec J (2009) Kinases required in hepatitis C virus entry and replication highlighted by small interference RNA screening. FASEB J 23:3780–3789

Vaid A, Ranjan R, Smythe WA, Hoppe HC, Sharma P (2010) PfPI3K, a phosphatidylinositol-3 kinase from Plasmodium falciparum is exported to the host erythrocyte and is involved in hemoglobin trafficking. Blood 115:2500–2507

Van den Bout I, Divecha N (2009) PIP5K-driven PtdIns(4,5)P_2 synthesis: regulation and cellular functions. J Cell Sci 122:3837–3850

Veiga E, Cossart P (2005) Listeria hijacks the clathrin-dependent endocytic machinery to invade mammalian cells. Nat Cell Biol 7:894–900

Vergne I, Chua J, Deretic V (2003) Tuberculosis toxin blocking phagosome maturation inhibits a novel Ca^{2+}/calmodulin-PI3K hVps34 cascade. J Exp Med 198:653–659

Vergne I, Fratti RA, Hill PJ, Chua J, Belisle J, Deretic V (2004) Mycobacterium tuberculosis phagosome maturation arrest: mycobacterial phosphatidylinositol analog phosphatidylinositol mannoside stimulates early endosomal fusion. Mol Cell Biol 15:751–760

Vergne I, Chua J, Lee HH, Lucas M, Belisle J, Deretic V (2005) Mechanism of phagolysosome biogenesis block by viable mycobacterium tuberculosis. Proc Natl Acad Sci U S A 102:4033–4038

Weber SS, Ragaz C, Hilbi H (2009a) The inositol polyphosphate 5-phosphatase OCRL1 restricts intracellular growth of Legionella, localizes to the replicative vacuole and binds to the bacterial effector LpnE. Cell Microbiol 11:442–460

Weber SS, Ragaz C, Hilbi H (2009b) Pathogen trafficking pathways and host phosphoinositide metabolism. Mol Microbiol 71:1341–1352

Wong KW, Isberg RR (2003) Arf6 and phosphoinositol-4-phosphate-5-kinase activites permit bypass of the Rac1 requirement for b1 integrin-mediated bacterial uptake. J Exp Med 198:603–614

Yonemura S, Hirao M, Doi Y, Takahashi N, Kondo T, Tsukita K, Tsukita S (1998) Ezrin/radixin/moesin (ERM) proteins bind to a positively chatged amino acid cluster in the juxta-membrane cytoplasmic domain of CD44, CD43, and ICAM-2. J Cell Biol 140:885–895

Yonemura S, Tsukita S, Tsukita S (1999) Direct involvement of ezrin/radixin/moesin (ERM)-binding membrane proteins in the organization of microvilli in collaboration with activated ERM proteins. J Cell Biol 145:1497–1509

Chapter 13
Defining Signal Transduction by Inositol Phosphates

Stephen B. Shears, Sindura B. Ganapathi, Nikhil A. Gokhale, Tobias M. H. Schenk, Huanchen Wang, Jeremy D. Weaver, Angelika Zaremba and Yixing Zhou

Abstract $Ins(1,4,5)P_3$ is a classical intracellular messenger: stimulus-dependent changes in its levels elicits biological effects through its release of intracellular Ca^{2+} stores. The $Ins(1,4,5)P_3$ response is "switched off" by its metabolism to a range of additional inositol phosphates. These metabolites have themselves come to be collectively described as a signaling "family". The validity of that latter definition is critically examined in this review. That is, we assess the strength of the hypothesis that $Ins(1,4,5)P_3$ metabolites are themselves "classical" signals. Put another way, what is the evidence that the biological function of a particular inositol phosphate depends upon stimulus dependent changes in its levels? In this assessment, examples of an inositol phosphate acting as a cofactor (i.e. its function is not stimulus-dependent) do not satisfy our signaling criteria. We conclude that $Ins(3,4,5,6)P_4$ is, to date, the only $Ins(1,4,5)P_3$ metabolite that has been validated to act as a second messenger.

Keywords Adenosine deaminase · AKT · β-cells · Calcium · cAMP · CaMKII · Chloride channel · ClC3 · Compartmentalization · DNA repair · Endosomes · ERK · Frizzled receptor · $GAP1^{IP4BP}$ · mRNA export · $Ins(1,4,5)P_3$ · $Ins(1,4,5)P_3$ receptor · $Ins(1,3,4)P_3$ · $Ins(1,3,4,5)P_4$ · $Ins(1,3,4,5)P_4$ receptor · $Ins(1,4,5,6)P_4$ · $Ins(3,4,5,6)P_4$ · $Ins(1,3,4,5,6)P_5$ · $InsP_6$ · Insulin · IPMK · IPK2 · IP5K · ITP · ITPK1 · ITPKB · Lymphocytes · Ku · Neutrophils · Protein phosphatase · $PtdIns(4,5)P_2$ · $PtdIns(3,4,5)P_3$ · PH domain · PTEN · RASA3 · Transcription · Wnt ligand

13.1 Introduction

The receptor-dependent, $Ins(1,4,5)P_3$-mediated mobilization of intracellular Ca^{2+} stores (Streb et al. 1983) is now a textbook signal transduction pathway (see Taylor's 2011 chapter). When this signaling activity was discovered, it was certainly a

S. B. Shears (✉) · S. B. Ganapathi · N. A. Gokhale · T. M. H. Schenk · H. Wang · J. D. Weaver · A. Zaremba · Y. Zhou
Inositol Signaling Section, Laboratory of Signal Transduction, NIEHS, NIH, DHHS, Research Triangle Park, NC 27709, USA
e-mail: shears@niehs.nih.gov

T. Balla et al. (eds.), *Phosphoinositides II: The Diverse Biological Functions*, Subcellular Biochemistry 59, DOI 10.1007/978-94-007-3015-1_13,

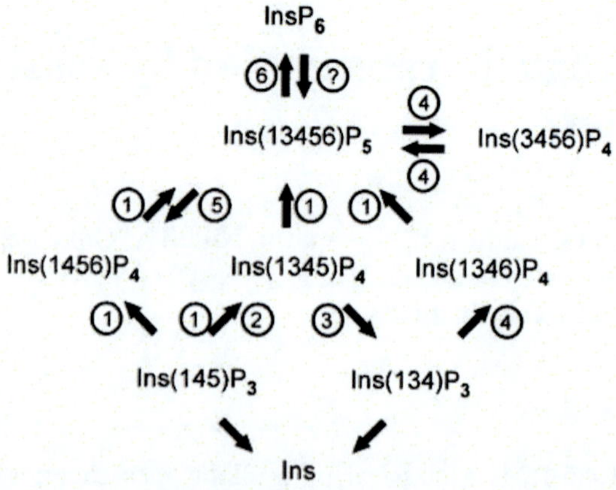

Fig. 13.1 Inositol phosphate metabolism. The figure shows the pathway of Ins(1,4,5)P_3 metabolism. The numbers in circles indicate the different enzymes that are involved: 1, IPK2/IPMK; 2, Ins(1,4,5)P_3 3-kinases; 3, Ins(1,4,5)P_3/ Ins(1,3,4,5)P_4 5-phosphatase; 4, ITPK1; 5, PTEN; 6, IP5K. There is a candidate for the question mark—MIPP—but it is uncertain how that enzyme can access its substrate (see text for details). The inositol pyrophosphates are not shown in this figure (they are the subject of a separate chapter (Saiardi 2011)). The enzymes that dephosphorylate Ins(1,4,5)P_3 and Ins(1,3,4)P_3 to inositol are not shown, as this review is only concerned with metabolites that have received attention as being cellular signals. Note that the positional specificity of IPK2/IPMK shows phylogenetic variation. In yeasts, Ins(1,4,5)P_3 is phosphorylated primarily to Ins(1,4,5,6)P_4, in mammals the product is predominantly Ins(1,3,4,5)P_4, and the enzyme in flies produces roughly equal quantities of both $InsP_4$ isomers

paradigm-altering concept. That novel idea was also highly contagious. After it emerged that Ins(1,4,5)P_3 is further phosphorylated to isomers of $InsP_4$, $InsP_5$ and $InsP_6$ (Fig. 13.1), we (Shears 1989) and others (Irvine and Schell 2001; York et al. 2001) proposed that some of these "higher" inositol phosphates might also have important signaling roles. This notion of "orphan signals" (Menniti et al. 1990) certainly captured the imagination (see Ismailov et al. 1996; York et al. 2001). Furthermore, the dramatic digital analogy that paints the inositol ring as a "six-bit signaling scaffold" (York et al. 2001; York 2006) implies that there are signaling roles for many members of the inositol phosphate family. However, the thesis of this review is that, except for one notable exception, we are still a long way from confirming that these "orphan" inositol phosphates truly function as classic second messengers.

In anticipation that our colleagues might raise their eyebrows in reaction to that last sentence, we will quickly provide some clarification. The second messenger concept (Robison et al. 1968) owes its existence to the discovery of cyclic AMP (Rall and Sutherland 1958; Sutherland and Rall 1958), a diffusible molecule that, in response to an extracellular stimulus, is generated at (or released from) a particular subcellular site. In this example, a change in the concentration of cAMP leads to the modification of a protein kinase activity. If a soluble inositol phosphate is to act as an intracellular signal it must also exhibit a stimulus-dependent change in its concentration. Additionally, the "information" encoded by the inositol phosphate should be converted—transduced—from one chemical form to another. These are significant criteria in the current context.

Take $InsP_6$ for example. This polyphosphate is an essential co-factor for adenosine deaminase, an mRNA editing enzyme; without $InsP_6$ at its core, the protein does not fold correctly and is devoid of catalytic activity (Macbeth et al. 2005). This is unarguably an important function, but there is no evidence that $InsP_6$ serves as a stimulus-dependent regulator of this enzyme. In fact, in add-back experiments, a maximal effect of $InsP_6$ upon deaminase activity was attained at a polyphosphate concentration of 1 μM (Macbeth et al. 2005), which is 10–50-fold less than the concentration that is always present in cells (Pittet et al. 1989; Oliver et al. 1992; Barker et al. 2004; Bunce et al. 1993). That is, the deaminase should always be fully $InsP_6$-activated by default. A similar argument can be made of the requirement that *S. cerevisiae* has for $InsP_6$, in order that mRNA can be exported out of the nucleus (Alcázar-Román et al. 2006; York et al. 1999). Here, $InsP_6$ acts by stimulating the ATPase activity of Dbp5, which is envisaged to be a molecular ratchet that pulls mRNA out of the nucleus (Yu et al. 2004). This function is fulfilled by only 0.1 μM $InsP_6$, a small fraction of the total cellular concentration. Again, the role of $InsP_6$ in this process is almost certainly as a cofactor rather than as the "regulator" or "signal" that it was originally proposed to be (York et al. 1999). Once more, let's acknowledge that these are important discoveries. They are just not examples of cell signaling activities.

Thus, in the current review it is our intention to assess the evidence that a biological response can be attributed to a stimulus-dependent alteration in the levels of a particular inositol phosphate. While $Ins(1,4,5)P_3$ clearly fulfils that criterion, it will not be discussed here as it is the dedicated subject of another chapter (Taylor 2011). The inositol "pyrophosphates" are also reserved for the attention of a separate chapter (Saiardi 2011). Finally, the emphasis in this review is on the inositol phosphates themselves and not the enzymes that synthesize them.

13.2 $Ins(1,3,4,5)P_4$

Receptor dependent Ca^{2+} mobilization arises not just by Ca^{2+} release from intracellular stores, but also through Ca^{2+} entry across the plasma membrane; the fact that the two processes are tightly linked lies at the heart of the "capacitative calcium entry" hypothesis (Putney 1986), which was subsequently refined and repackaged as "store-operated calcium entry" (Hoth and Penner 1992; Parekh and Penner 1997). In the immediate aftermath of $Ins(1,4,5)P_3$ being discovered to release Ca^{2+} from intracellular stores (Streb et al. 1983), a 3-kinase was discovered that phosphorylated $Ins(1,4,5)P_3$ (Irvine et al. 1986). At that time, it was not known how Ca^{2+} release was coupled to Ca^{2+} entry. Thus, $Ins(1,3,4,5)P_4$ became a prime suspect for signaling Ca^{2+} entry (Irvine 1986). The initial evidence seemed highly incriminating: stimulus-dependent increases in cellular levels of $Ins(1,4,5)P_3$ are followed shortly afterwards by several-fold increases in levels of $Ins(1,3,4,5)P_4$ (Batty et al. 1985).

That is a valuable credential if Ins(1,3,4,5)P_4 is to be a cellular signal. Furthermore, Irvine's group (Irvine and Moor 1986) reported that micro-injection of Ins(1,3,4,5)P_4 into sea urchin eggs raised the fertilization envelope in a manner that was dependent upon extracellular [Ca^{2+}]. To explain how Ins(1,3,4,5)P_4 might mediate a physiological activity that was dependent upon the Ca^{2+} outside the cell, it was proposed that this inositol phosphate somehow stimulates Ca^{2+} entry (Irvine and Moor 1986).

However, as (Schell 2010) noted in a recent review, we are still searching to identify an Ins(1,3,4,5)P_4-based signaling cascade, now more than 20 years after Irvine's initial experiments. Not in the least because Ins(1,3,4,5)P_4 has been a discordant beast, producing conflicting data in the hands of different groups. For example, the early results obtained from the sea urchin eggs proved impossible to reproduce in subsequent studies (Irvine and Moor 1987; Crossley et al. 1988). Irvine and colleagues were unable to account for this difficulty (Irvine and Moor 1987), and so they turned to other model systems in which they could test their hypothesis (Changya et al. 1989; Morris et al. 1987; Loomis-Husselbee et al. 1996). For example, they used Ca^{2+}-activated K^+ channels in mouse lacrimal acinar cells as a readout of the degree of Ca^{2+} mobilization, and they reported that Ins(1,3,4,5)P_4 augmented the Ins(1,4,5)P_3-induced Ca^{2+} response (Morris et al. 1987). Initially, a stimulation of Ca^{2+} entry by Ins(1,3,4,5)P_4 was put forward as the explanation (Morris et al. 1987), although subsequently (Changya et al. 1989) it was suggested to be more likely that Ins(1,3,4,5)P_4 somehow augmented intracellular Ca^{2+} mobilization. That idea was also consistent with a separate series of experiments performed with permeabilized L-1210 cells (Loomis-Husselbee et al. 1996). In contrast, Bird et al. (Bird et al. 1991; Bird and Putney 1996), who also worked with mouse lacrimal cells, reported that Ins(1,4,5)P_3 by itself maximally activates Ca^{2+}-activated K^+ channels; Ins(1,3,4,5)P_4 was not required. After some debate of this topic in the literature (see (Putney 1992; Irvine 1992)), Irvine and colleagues (Smith et al. 2000) later described how the two group's use of slightly different cell preparations might largely explain the contrary data. Putney's group (Bird et al. 1991) had studied cells that had been in primary culture for up to 24 h. In contrast, Irvine's group had used freshly isolated cells. So, which laboratory used the most appropriate cell preparation? Irvine and colleagues (Smith et al. 2000) accepted that the continuous morphology of the endoplasmic reticulum that is observed in cultured cells reflects a more physiologically-relevant model; this organelle is somewhat fragmented in freshly isolated lacrimal cells. Thus, Irvine et al. (Smith et al. 2000) accept that their model might be less biologically relevant. Nevertheless, they (Smith et al. 2000) argued, the effects of Ins(1,3,4,5)P_4 that they observed had to be an exaggeration of a physiological event, rather than an artifact. Perhaps, they suggested, Ins(1,3,4,5)P_4 acted by influencing the continuity of endoplasmic reticulum. However, later studies (Brough et al. 2005) did not support that hypothesis.

In their later studies Irvine and colleagues (Changya et al. 1989; Smith et al. 2000) somewhat de-emphasized a possible role for Ins(1,3,4,5)P_4 in regulating Ca^{2+} entry into cells. Others (Hermosura et al. 2000) have also shown that store-operated Ca^{2+} entry is not activated by Ins(1,3,4,5)P_4. Yet, there continue to be occasional electrophysiological demonstrations of plasma membrane Ca^{2+} currents being stimulated by Ins(1,3,4,5)P_4, although such data have not been placed in a physiologically-adequate context. In one example of this genre (Luckhoff and Clapham 1992), 30 μM Ins(1,3,4,5)P_4 was reported to enhance Ca^{2+}-activated Ca^{2+} current in excised inside-out patches. However, such high levels of Ins(1,3,4,5)P_4 are not biologically relevant. In most cases, cellular Ins(1,3,4,5)P_4 levels after receptor activation would not be expected to exceed 3–4 μM (Guse et al. 1993; Barker et al. 1992; Huang et al. 2007), and initial work with optical sensors do not reveal any compartmentalization of Ins(1,3,4,5)P_4 synthesis (Sakaguchi et al. 2010). Clapham's group (Luckhoff and Clapham 1992) were able to get lower concentrations of Ins(1,3,4,5)P_4 to activate a plasma membrane current when Mn^{2+} was used as a Ca^{2+} surrogate, but neither the ion selectivity of this current, nor its biophysical characteristics, match those of store-operated calcium entry (Parekh and Penner 1997).

During their work with lacrimal cells, Irvine and colleagues (Changya et al. 1989; Smith et al. 2000) went to great lengths to prove that their effects of Ins(1,3,4,5)P_4 upon Ca^{2+} mobilization could not be explained by inhibition of Ins(1,4,5)P_3 metabolism (Changya et al. 1989; Smith et al. 2000). It is therefore somewhat ironic that Penner and colleagues (Hermosura et al. 2000) concluded that in certain circumstances inhibition of Ins(1,4,5)P_3 metabolism by Ins(1,3,4,5)P_4 was indeed a genuine mechanism by which Ca^{2+} mobilization can be enhanced. More precisely, it was argued that Ins(1,3,4,5)P_4 set "a discriminatory time window for co-incidence detection that enables selective facilitation of Ca^{2+} influx by appropriately timed low-level receptor stimulation". However, to achieve these effects *in vitro*, 5–20 μM concentrations of Ins(1,3,4,5)P_4 were required (Hermosura et al. 2000), which, as mentioned above, are well above those that prevail inside cells.

Ins(1,3,4,5)P_4 has also been reported to *inhibit* Ca^{2+} signaling. For example, *in vitro* Ins(1,3,4,5)P_4 can inhibit Ins(1,4,5)P_3 from binding to its receptor (Hermosura et al. 2000); Putney's group (Bird and Putney 1996) had discovered this phenomenon some years earlier, but they had expressed concern that the levels of Ins(1,3,4,5)P_4 that were required were too high to be physiologically relevant. In yet another twist in this tale, Ins(1,3,4,5)P_4 was reported to promote Ca^{2+} re-uptake into the endoplasmic reticulum of permeabilized T51B liver cells (Hill et al. 1988; Boynton et al. 1990). That particular observation has never been reproduced in another cell type, and neither has a mechanistic rationale been developed.

Genetic perturbation of Ins(1,3,4,5)P_4 production has also failed to yield a consistent picture of the polyphosphate's putative role in Ca^{2+} mobilization. For example, Ca^{2+} signals in thymocytes were unaffected when Ins(1,3,4,5)P_4 synthesis was compromised upon knock-out of the type B Ins(1,4,5)P_3 3-kinase gene (*ITPKB*) (Pouillon et al. 2003). Yet, in B-lymphocytes from *Itpkb*$^{-/-}$ mice it was reported (Marechal et al. 2007) that there is a reduction in receptor-mediated Ca^{2+} signaling. Adding further to the confusion, Miller and colleagues (2007, 2009) observed the opposite effect; B-lymphocytes from their knock-out mice showed enhanced Ca^{2+} mobilization. The latter phenotype does have a quite facile explanation. Ca^{2+} mobilization might well be expected to be enhanced when the half-life of Ins(1,4,5)P_3 is prolonged following loss of a significant route of Ins(1,4,5)P_3 metabolism, i.e., an Ins(1,4,5)P_3 kinase. Moreover, cells that have more robust Ins(1,4,5)P_3 signals might be expected to have more depleted Ca^{2+} stores, and hence higher rates of store-dependent Ca^{2+} entry (Jia et al. 2008). However, Miller et al. (2007, 2009) interpreted their data quite differently. They argued that *Itpkb*$^{-/-}$ lymphocytes have lost an inhibitor of store-operated Ca^{2+} entry—Ins(1,3,4,5)P_4—and that, they concluded, is why Ca^{2+} mobilization is enhanced in those cells.

To more directly pursue their hypothesis, Miller et al. (2007, 2009) used a cell-permeant analogue of Ins(1,3,4,5)P_4. The addition of this analogue attenuated both receptor-dependent (anti-IgM) and receptor-independent (thapsigargin-mediated) increases in cytosolic [Ca^{2+}] (Miller et al. 2007, 2009). In many of their experiments, Miller et al. reported that cellular Ca^{2+} levels were reduced *immediately* upon the addition of cell-permeant Ins(1,3,4,5)P_4. The speed of those responses is, perhaps, unexpected, in view of a report (Li et al. 1997) that a delay of at least a minute should be expected, which, apparently, cannot be eliminated by increasing the concentration of the cell permeant analogue. One of the reasons for this lag is the time it takes for the analogue to diffuse across the membrane. Additionally, each phosphate group is "protected" by two butyryloxymethyl groups (Li et al. 1997), all eight of which must be removed by intracellular esterases before the Ins(1,3,4,5)P_4 can be liberated. These technical considerations raise a concern that the immediate effect of the cell permeant Ins(1,3,4,5)P_4 might be non-physiological. Yet, it does seems that the response is specific: Miller et al. (2007) added cell-permeant Ins(1,4,5,6)P_4 in control experiments. That had no effect upon the cell's Ca^{2+} responses. Nevertheless, a more direct demonstration of Ins(1,3,4,5)P_4 inhibiting a physiologically-relevant pathway for Ca^{2+} entry, for example in an electrophysiological assay, would be helpful.

Is it possible to rationalize all of these different reported effects of Ins(1,3,4,5)P_4 upon Ca^{2+} mobilization? And what are we to make of experimental observations where even the order of addition of Ins(1,4,5)P_3 and Ins(1,3,4,5)P_4 determines whether or not an effect upon Ca^{2+} fluxes is observed (Loomis-Husselbee et al. 1996)? One group (Hermosura et al. 2000) has argued that the somewhat confused and inconsistent literature reflects the actions of Ins(1,3,4,5)P_4 being complex and multifaceted. Perhaps, they say, the precise molecular actions of Ins(1,3,4,5)P_4 are cell-specific, varying with the strength of receptor activation, or subcellular localization of Ins(1,4,5)P_3-metabolic enzymes, or heterogeneity of intracellular Ca^{2+} stores, or differences in Ins(1,4,5)P_3 receptor subtypes, or other regulatory factors. However, we put it to the jury that the case for Ins(1,3,4,5)P_4 being a cellular signal for Ca^{2+} mobilization is unproven. Judgement should be reserved until a specific signaling activity can be reproducibly demonstrated using physiologically-relevant concentrations of Ins(1,3,4,5)P_4 and, moreover, an Ins(1,3,4,5)P_4-sensitive signaling entity with a defined role in Ca^{2+} signaling would need to be identified.

Of course, Ins(1,3,4,5)P_4 could have other signaling roles that do not involve Ca^{2+}. Indeed, *Itpkb*$^{-/-}$ mice are immunologically compromised (Pouillon et al. 2003). For example, thymocytes in *Itpkb*$^{-/-}$ mice synthesize almost no Ins(1,3,4,5)P_4 and their developmental program fails (Pouillon et al. 2003). *Itpkb*$^{-/-}$ mice also exhibit defective B-lymphocyte development (Miller et al. 2007; Marechal et al. 2007) and neutrophil migration is compromised (Jia et al. 2007). We will briefly discuss three recent developments that suggest there may be Ins(1,3,4,5)P_4 "receptors" that can help explain the nature of these *Itpkb*$^{-/-}$ phenotypes.

One intriguing protein to which Ins(1,3,4,5)P_4 binds tightly and specifically is the Ras-Gap that was originally named GAP1^{IP4BP} (Cullen et al. 1995). An exhaustive study in which the levels of GAP1^{IP4BP} were genetically manipulated (Walker et al. 2002) led to the conclusion that this protein did not exert any influence upon Ca^{2+} mobilization. Nevertheless, there has recently been renewed interest in GAP1^{IP4BP}, which has been re-christened as RASA3. The Ins(1,3,4,5)P_4-binding region of RASA3 is now known to be a PH domain, which also binds to PtdIns(4,5)P_2 (Cozier et al. 2000). Could competition between Ins(1,3,4,5)P_4 and PtdIns(4,5)P_2 have some signaling significance? In CHO cells, receptor-dependent PLC activation (and hence Ins(1,3,4,5)P_4 accumulation) did not affect the membrane-association of GAP1^{IP4BP}/RASA3 (Cozier et al. 2000). However, in a subsequent study with COS cells (Marechal et al. 2007), in which ITPKB was over-expressed, RASA3 was dislodged from the plasma membrane upon PLC activation. Furthermore, 30 min pre-incubation of cells with cell-permeant Ins(1,3,4,5)P_4 also caused RASA3 to translocate from the plasma membrane to the cytosol (Marechal et al. 2007). This may represent a new signaling function for Ins(1,3,4,5)P_4 in B-lymphocytes (Marechal et al. 2007). It was proposed that Ins(1,3,4,5)P_4 regulates the intracellular location and hence the activity of a RASA3-ERK signaling pathway that controls pro-apoptotic BIM gene expression (Marechal et al. 2007). As noted elsewhere (Sauer and Cooke 2010), the further development of this hypothesis would be helped by a demonstration that Ins(1,3,4,5)P_4 acts in this manner in lymphocytes.

Ins(1,3,4,5)P_4 has also been reported to be an inhibitory signal for neutrophil function, by competing with PtdIns(3,4,5)P_3 for binding to the PH domain of AKT (Jia et al. 2007). A cell-permeant Ins(1,3,4,5)P_4 analogue was shown to diminish receptor-dependent recruitment of AKT-PH domain to the plasma membrane (Jia et al. 2007). Moreover, neutrophils from *Itpkb*$^{-/-}$ mice exhibit up-regulated AKT activity because, it was argued, more of that kinase can translocate to the plasma membrane in the absence of Ins(1,3,4,5)P_4 (Jia et al. 2007).

Another proposed "receptor" for Ins(1,3,4,5)P_4 is the interleukin-2 tyrosine kinase ITK, which phosphorylates and activates PLC-γ. For ITK to regulate PLC-γ, the kinase must first translocate to the plasma membrane, courtesy of the affinity for PtdIns(3,4,5)P_3 of the protein's pleckstrin homology domain. Interestingly, this receptor-dependent translocation process is inhibited in thymocytes prepared from *Itpkb*$^{-/-}$ mice (Huang et al. 2007). In that latter study it was further reported that, *in vitro*, Ins(1,3,4,5)P_4 promotes the association of the PH-domain of ITK with an immobilized, short acyl-chain analogue of PtdIns(3,4,5)P_3. An attractive feature of this phenomenon is that it was observed when using concentrations of Ins(1,3,4,5)P_4 (1 μM) that are physiologically relevant. However, it is not clear exactly how Ins(1,3,4,5)P_4 has this effect. In fact, as discussed above, it would normally be expected that Ins(1,3,4,5)P_4 would *compete* with PtdIns(3,4,5)P_3 for the same ligand-binding domain. To resolve this apparent paradox, Huang et al. (2007) have proposed that, if ITK were to oligomerize, then the binding of Ins(1,3,4,5)P_4 to one subunit might allosterically enhance the affinity of another subunit for PtdIns(3,4,5)P_3. However, that idea is not consistent with FRET analysis that has revealed cytoplasmic ITK to be monomeric *in vivo* (Qi et al. 2006; Qi and August 2009). Oligomerization of ITK only occurs *after* the kinase has already translocated to the plasma membrane (Qi et al. 2006). The folded state of monomeric ITK (Qi and August 2009) suggests that it alters its conformation upon its transfer to the plasma membrane. Thus, the alternative "induced-fit" hypothesis that Huang et al. (2007) have proposed is arguably a more likely explanation for their *in vitro* data. Here, initial binding of Ins(1,3,4,5)P_4 is suggested to alter the conformation of ITK so that it gains an increased ligand affinity, but particularly for PtdIns(3,4,5)P_3. The viability of the latter proposal is critically dependent upon accurate determinations of the relative affinities of PtdIns(3,4,5)P_3 and Ins(1,3,4,5)P_4. It would therefore be useful to determine these binding parameters using an *in vitro* technique that is closer to a physiological context than is the use of a soluble PtdIns(3,4,5)P_3 analogue immobilized to beads. The preferred method (Narayan and Lemmon 2006) is to incorporate "natural" PtdIns(3,4,5)P_3 into phospholipid vesicles, and then determine the affinities of the lipid (and the competing headgroup) by using surface plasmon resonance. Additionally, rather than characterizing just the PH domain fragment of ITK (Huang et al. 2007), it would be more physiological to use full-length protein.

In all three of the examples described above—RASA3, AKT and ITK—it will be important to address the concern (Irvine et al. 2006) that an *Itpkb*$^{-/-}$ phenotype could be an unpredictable consequence of a loss of non-catalytic (scaffolding?) activities of the type B Ins(1,4,5)P_3 3-kinase, rather than a reduction in Ins(1,3,4,5)P_4 levels. This question could be pursued by studying if the phenotype of the *Itpkb*$^{-/-}$ cells can be rescued by expression of ITPKB, and not by a kinase-dead Itpkb mutant. That being said, it also needs to be established that the biological effects that have been attributed to Ins(1,3,4,5)P_4 are not actually performed by one or more of its metabolites (Fig. 13.1). For example, the elimination from cells of Ins(1,4,5)P_3 3-kinase activity can lead to a reduction in levels of Ins(1,3,4,5,6)P_5 and InsP_6 (Leyman et al. 2007). Ins(1,3,4)P_3 is another important metabolite of Ins(1,3,4,5)P_4. Loss of the latter in *Itpkb*$^{-/-}$ cells (Pouillon et al. 2003) will uncouple the link between PLC activity and the Ins(3,4,5,6)P_4-signaling cascade (see below). That is, cells that have reduced ITPK activity will also be encumbered by an inability to synthesize the Ins(3,4,5,6)P_4 signal. That could modify cell function in a number of ways (see below). To take one pertinent example, the ClC3 Cl^- channel that Ins(3,4,5,6)P_4 regulates (Mitchell et al. 2008) plays an important role in neutrophil migration (Volk et al. 2008) which, as mentioned above, is defective in *Itpkb*$^{-/-}$ cells.

13.3 Ins(1,4,5,6)P_4

In yeast (Odom et al. 2000; Saiardi et al. 2000) and in flies (Seeds et al. 2004), Ins(1,4,5,6)P_4 is formed from Ins(1,4,5)P_3 by the kinase activity of Ipk2. In a 2000 study (Odom et al. 2000) evidence was presented that this synthesis of Ins(1,4,5,6)P_4

was necessary for the function of an ArgR-Mcm1 transcriptional complex that regulated the expression of ornithine transaminase. The authors of that study assayed transcriptional control in wild-type and ipk2Δ cells using "growth on ornithine as a sole nitrogen source". However, this interpretation of the data has been disputed by others (Dubois et al. 2000), who reported that the slower growth of ipk2Δ cells was a general phenotype rather than being specific to the nutrient source. This is an important point; if slowed cell growth were to be a general phenotype, then it would no longer provide a specific readout of the expression of ornithine transaminase. Unfortunately, it has never been resolved why these two groups came to different conclusions. Yet, despite the controversy, the idea that the catalytic activity of Ipk2 might regulate transcription was taken up in subsequent studies: O'Shea and colleagues found that this kinase activity regulated Pho5 transcription (Steger et al. 2003). Additional genetic experiments indicated it was chromatin remodeling that was regulated by either Ins(1,4,5,6)P_4 and/or Ins(1,3,4,5,6)P_5 (another product of Ipk2 activity). These authors reported an increased accessibility of a *Cla* I restriction site in the *Pho5* promoter in nuclei that were isolated from cells shifted to *Pho5* inducing conditions (Steger et al. 2003). This increased *Cla* I accessibility was impaired in the *ipk2Δ* cells (Steger et al. 2003).

But are these phenotypes the *direct* consequence of altering the expression of the kinase activity of Ipk2? There are far-reaching effects upon many mRNA species in yeast strains in which the catalytic activity of Ipk2 is compromised (El Alami et al. 2003), and so it is important to separate primary regulatory events from secondary consequences. Additionally, metabolic homeostasis utilizes regulatory processes that control the expression of genes encoding metabolic enzymes. Thus, there are many links between gene regulation and metabolic status (McKnight 2003). It might be considered inevitable that control over phosphate supply to yeast, and the regulation of *Pho5* expression, must be intertwined with regulation of inositol phosphate synthesis, which of course is a phosphate-consuming process. The big question, therefore, is whether this apparent effect of Ipk2 upon chromatin remodeling reflects a global metabolic control process, or instead is this really a more specific utilization of inositol phosphate turnover to control gene expression? This query could be resolved if we had a molecular justification for the intriguing genetic effects that were described by O'Shea and colleagues (Steger et al. 2003).

A molecular mechanism by which Ipk2 might control *Pho5* expression has been put forward; it was proposed that Ins(1,4,5,6)P_4 and Ins(1,3,4,5,6)P_5 directly stimulate "nucleosome sliding" (Shen et al. 2003; Steger et al. 2003). Nucleosomes are the basic repetitive unit of chromatin: histone octomers around which are wrapped about 150 bp of DNA (Becker and Hörz 2002). Regulatory elements can be exposed when nucleosomes are nudged along the DNA helix by ATP-consuming, nucleosome remodeling factors. Wu and colleagues (Shen et al. 2003) studied nucleosome movement along a *Drosophila hsp70* DNA fragment driven by the yeast SWI/SNF chromatin remodeling complex. It was reported that 500 μM of either Ins(1,4,5,6)P_4 or Ins(1,3,4,5,6)P_5 stimulated this nucleosome sliding (Shen et al. 2003). Unfortunately, as discussed elsewhere (Shears 2004), Wu and colleagues (Shen et al. 2003) studied the effects of inositol phosphates at concentrations that are 500–1700 times higher than estimated cellular levels. In such circumstances, it is difficult for us to accept that those effects are physiologically relevant.

It is even more difficult to imagine Ins(1,4,5,6)P_4 being a signal that controls transcription in higher organisms. The mammalian homologue of Ipk2—often called IPMK—phosphorylates Ins(1,4,5)P_3 to Ins(1,3,4,5)P_4 (Fig. 13.1) rather than to Ins(1,4,5,6)P_4 (Saiardi et al. 2001; Nalaskowski et al. 2002). So mammalian cells do not use IPK2/IPMK to regulate levels of Ins(1,4,5,6)P_4. It is possible for Ins(1,4,5,6)P_4 to be synthesized by dephosphorylation of Ins(1,3,4,5,6)P_5, probably by the cytoplasmic and nuclear pools of PTEN that, in these particular locations, cannot access the alternative and better-known substrate, PtdIns(3,4,5)P_3 (Caffrey et al. 2001; Otto et al. 2007). In any case, there is no evidence that cellular levels of Ins(1,4,5,6)P_4 are receptor regulated (Menniti et al. 1990). (It has been published that Ins(1,4,5,6)P_4 levels are elevated in src-transformed fibroblasts (Mattingly et al. 1991), but that result has not been studied further).

There is an environmental pathogen that can perturb Ins(1,4,5,6)P_4 metabolism: Some years ago it was noted that the invasion of epithelial cells by *Salmonella* strongly activated the dephosphorylation of Ins(1,3,4,5,6)P_5 to Ins(1,4,5,6)P_4 (Eckmann et al. 1997). The evidence indicated that Ins(1,4,5,6)P_4 might augment the secretion of salt and fluid that accompanies *Salmonella* infection (Eckmann et al. 1997). Whether or not this phenomenon might have physiological rather than just pathological relevance has not been established. Subsequently, it was demonstrated that one of the proteins that is required for the pathogen's virulence, SopB, was responsible for dephosphorylating Ins(1,3,4,5,6)P_5 (Norris et al. 1998). A later study (Zhou et al. 2001) established that the main product was Ins(1,4,5,6)P_4. However, there was no evidence that virulence itself depends upon Ins(1,3,4,5,6)P_5 dephosphorylation (Zhou et al. 2001). Instead, cell invasion by *Salmonella* appeared to require inositol lipid dephosphorylation by SopB (Hernandez et al. 2004). Indeed, it now seems possible that Ins(1,3,4,5,6)P_5 is little more than an off-target substrate for SopB. In any case, the groups that work with SopB now focus on its role in metabolizing inositol lipids rather than the inositol phosphates (Hernandez et al. 2004). Taking all these data into account, we conclude that Ins(1,4,5,6)P_4 is not qualified to be described as a cellular signal.

13.4 Ins(1,3,4,5,6)P_5 and InsP_6

Most nucleated cells synthesize 15–50 μM of both Ins(1,3,4,5,6)P_5 and InsP_6 (Pittet et al. 1989; Oliver et al. 1992; Barker et al. 2004; Bunce et al. 1993). Undoubtably, the initial proposals that Ins(1,3,4,5,6)P_5 and InsP_6 might be cellular signals (Heslop et al. 1985; Vallejo et al. 1987; Michell et al. 1988) were strongly influenced by the manner in which these polyphosphates were first discovered in animal cells, that is, as a consequence of studying metabolism of Ins(1,4,5)P_3. However, intracellular signals typically are expected to exhibit significant stimulus-dependent changes in their concentrations. In contrast, cellular levels of Ins(1,3,4,5,6)P_5 and InsP_6 do not respond acutely to most extracellular stimuli, and even when they do, 25–35%

changes in their concentrations seem to be an upper limit (Larsson et al. 1997; Pittet et al. 1989).

One dramatic exception emerged in a study (Gao and Wang 2007) of certain signaling events that lie downstream of the so-called Frizzled receptors. Activation of Frizzleds by the Wnt ligands regulates many aspects of embryonic development and adult tissue homeostasis. Wnt ligands can activate PLC and stimulate inositol phosphate accumulation (Slusarski et al. 1997), but Ins(1,3,4,5,6)P_5 would normally be expected to be well-insulated from that response (Menniti et al. 1990). It was therefore unexpected when Gao and Wang (Gao and Wang 2007) demonstrated that Ins(1,3,4,5,6)P_5 levels increased up to 2.5-fold in the few minutes following activation of the (over-expressed) rat Fz1 receptor by Wnt3a. *In vitro* data indicated that the biological consequence of this increase in Ins(1,3,4,5,6)P_5 was activation of casein kinase II (CK2) and inhibition of GSK3β. The maximally effective concentration of Ins(1,3,4,5,6)P_5 in each case was approximately 50 μM, which is approximately what is normally estimated to be present inside mammalian cells (Oliver et al. 1992; Pittet et al. 1989). Both of those effects of Ins(1,3,4,5,6)P_5, if they occurred *in vivo*, would be expected to stabilize β-catenin, enhancing its transcriptional response to Wnt signaling (Gao and Wang 2007).

Let's first discuss the proposed regulation of CK2. It has been known for some years that Ins(1,3,4,5,6)P_5 can activate CK2 *in vitro*, although it was originally reported by Solyakov et al. (2004) that $InsP_6$ is more efficacious. However, neither Ins(1,3,4,5,6)P_5 nor $InsP_6$ affects the activity of purified, native CK2 (Solyakov et al. 2004; Gao and Wang 2007). Instead, it was reported that the inositol phosphates act by reversing the effect of an uncharacterized, heat-stable inhibitor of CK2 that is present in cell lysates (Solyakov et al. 2004). The lack of insight into either the nature of the inhibitor, or the mechanism of action of the polyphosphates, has prevented this hypothesis from developing further. Moreover, it is a popular viewpoint in the CK2 field that this kinase is normally constitutively active and therefore has no requirement to be stimulated (Ruzzene and Pinna 2010). Even if that prevailing opinion were to be incorrect, it is hard to see how a stimulus-dependent increase in Ins(1,3,4,5,6)P_5 levels would have any effect upon CK2 that should already be constitutively activated by endogenous $InsP_6$.

The inhibitory effect of Ins(1,3,4,5,6)P_5 upon GSK3β is more encouraging because of its specificity: neither Ins(1,4,5)P_3, Ins(1,3,4,5)P_4 nor $InsP_6$ had any effect (Gao and Wang 2007). The treatment of intact cells with inhibitors of Ins(1,3,4,5,6)P_5 synthesis also prevented Wnt3a from inhibiting GSK3β (Gao and Wang 2007). Since Ins(1,3,4,5,6)P_5 levels do not typically change in response to short-term receptor activation, it is possible that the response that Gao and Wang (Gao and Wang 2007) observed is specific to signaling through Frizzled receptors. As for possible mechanisms, Ins(1,3,4,5,6)P_5 did not inhibit purified GSK3β, so an intermediary seems to be required (Gao and Wang 2007). Further work on this topic would seem to be appropriate.

Aside from that isolated response of Ins(1,3,4,5,6)P_5 to activation of the Fz1 receptor, its steady-state levels—and also those of $InsP_6$—do not respond acutely to receptor activation (see above). Thus, when there are reports that Ins(1,3,4,5,6)P_5 and

$InsP_6$ have biological activity, it has been difficult to place these data in a signaling context. One illustrative example is a report that both Ins(1,3,4,5,6)P_5 and $InsP_6$ inhibit protein phosphatases (Larsson et al. 1997). How can this be of *regulatory* significance if the levels of these polyphosphates do not change acutely? (In this particular case, one might also ask how any signaling specificity could result from two inositol polyphosphates both being broad spectrum inhibitors of PP1, PP2A and PP5). Similarly, it is also difficult to place in a signaling context a report that Ins(1,3,4,5,6)P_5 and $InsP_6$ inhibit L-type Ca^{2+} channels (Quignard et al. 2003); a similar criticism can be made of proposals that $InsP_6$ is a "regulatory factor" in mRNA export and gene translation (Monserrate and York 2010; York et al. 1999). To be fair, we do note that others interpret these data rather differently. For example, it has been proposed that one role for $InsP_6$ is to "set" (Berggren and Barker 2008) the basal state of a number of signaling entities. In particular, there are a number of studies that argue $InsP_6$ establishes the default activities of various beta-cell signaling complexes (Berggren and Barker 2008). Barker and colleagues (2004) have also proposed that this putative global effector role for $InsP_6$ may be of regulatory significance as cells transit through the cell cycle, during which time they estimate that the level of $InsP_6$ may fluctuate by as much as threefold. Nevertheless, those cell-cycle dependent metabolic changes have not been tied to a specific signaling event.

Of course, the situation would be different if, as has been suggested (Larsson et al. 1997; Barker et al. 2002; Otto et al. 2007), the high total cellular levels of Ins(1,3,4,5,6)P_5 and $InsP_6$ mask stimulus-dependent alterations in smaller, discrete "signaling" pools of these compounds. That is, significant changes in "local" concentrations of Ins(1,3,4,5,6)P_5 and $InsP_6$ could be missed during the analysis of inositol phosphates in entire cell populations. Not so long ago, such a concept would have been labeled as heretical: how could a small and apparently freely-diffusible molecule not be uniformly distributed throughout the cell? However, it is now recognized that a concentration gradient of cAMP across a cell can be maintained by the spatial separation of the adenylyl cyclases from cAMP phosphodiesterases (Zaccolo et al. 2006). Is there any evidence for spatial heterogeneity of enzymes of inositol phosphate metabolism? Indeed there are. Arguably the most dramatic example is the receptor-dependent relocalization of the Ins(1,4,5)P_3 3-kinase in hippocampal neurones (Schell and Irvine 2006). In stimulated cells, the kinase moves away from the post-synaptic region of the neuronal spines and into the dendritic shaft. This translocation undoubtably influences Ins(1,4,5)P_3-dependent Ca^{2+} mobilization (Schell and Irvine 2006). But what about the enzymes that metabolize higher inositol phosphates? IPK2/IPMK (Nalaskowski et al. 2002; Odom et al. 2000) and IP5K (York et al. 1999; Brehm et al. 2007) are both concentrated in the nucleus. Moreover, the only known mammalian $InsP_6$ phosphatase—MIPP—is restricted to the lumen of the endoplasmic reticulum (Craxton et al. 1997; Ali et al. 1993). So it is also of interest that plants at least can utilize an ABC-transporter like protein to move $InsP_6$ across membranes (Nagy et al. 2009; Shi et al. 2007). It would be a significant breakthrough in this field if a mammalian homologue could be identified that transported $InsP_6$ across the endoplasmic reticulum so that it could be metabolized by MIPP. However, embryonic fibroblasts made from *Mipp*$^{-/-}$ mice showed only

30% higher levels of $InsP_6$ than wild-type animals; more discouragingly, the animals exhibited no obvious phenotype (Chi et al. 2000). Is it possible that mammals express another $InsP_6$ phosphatase that we've all missed? Certainly we are missing something: we (Yang et al. 2008) have reported that a 20–25% decrease in cell volume following hyperosmotic stress is accompanied by a proportionate decrease in the *amount* of cellular $InsP_6$, so that its *concentration* is not altered. This observation indicates that, when the cell deems it necessary, the metabolism of $InsP_6$ can be quite rapid. We really ought to find out how, and why.

Some time ago, Michell's group (Stuart et al. 1994) also considered this question of whether or not some inositol phosphates might be present inside cellular organelles. They ascertained that 80–90% of the cells's inositol phosphates, including Ins(1,3,4,5,6)P_5 and $InsP_6$, were immediately released into the surrounding medium when the plasma membrane was permeabilized. That observation argues strongly against inositol phosphates being inside membrane-delimited cellular organelles. It can also be argued that no more than about half of the $InsP_6$ pool in intact cells can be "hidden" from the cytoplasm, since the other half is readily accessible to soluble kinases that synthesize the inositol pyrophosphates (Menniti et al. 1993).

On the other hand, $InsP_6$ can bind to membranes, at least *in vitro* (Cooke et al. 1991). $InsP_6$ is also a structural component of certain cellular proteins (Macbeth et al. 2005); that particular pool of $InsP_6$ would not be expected to be freely exchangeable with the bulk phase. There are other cellular proteins that can bind $InsP_6$, which could also reduce its free concentration in the cytosol (Barker et al. 2002). The punctate intracellular distribution of endogenous IP5K, particularly in nucleoli and so-called stress granules (Brehm et al. 2007), also suggests that, to a degree at least, the synthesis of $InsP_6$ might be compartmentalized. Uncertainty over compartmentalization is unlikely to be resolved until appropriate sensors of the intracellular location of these polyphosphates can be developed. It is our opinion that the status of Ins(1,3,4,5,6)P_5 and $InsP_6$ as cellular signals depends upon this question being answered.

There is no doubt that Ins(1,3,4,5,6)P_5 and $InsP_6$ still fulfill important biological functions that do not depend upon dynamic changes in their concentrations. For example, both Ins(1,3,4,5,6)P_5 and $InsP_6$ are precursors for the inositol pyrophosphates, which are currently the recipients of considerable interest from the signaling community (Saiardi 2011). Another possible function for Ins(1,3,4,5,6)P_5 and $InsP_6$ (*in vitro* at least) is to compete with inositol lipids for binding to certain proteins (Komander et al. 2004; Kavran et al. 1998). It has been speculated that this phenomenon increases the signal-to-noise ratio for PtdIns(3,4,5)P_3-dependent functions (Irvine and Schell 2001; Komander et al. 2004). The idea is that binding of soluble inositol phosphates to a protein target helps keep it away from membranes until PtdIns 3-kinase activity is elevated by an appropriate stimulus (this concept is arguably analogous to the proposal (Berggren and Barker 2008) that Ins(1,3,4,5,6)P_5 and $InsP_6$ "set" the basal activities of certain signaling entities). Likewise, the binding of Ins(1,3,4,5,6)P_5 to PTEN may inhibit that enzyme's low protein phosphatase activity and possibly contribute to PTEN's cytoplasmic and nuclear localization in the absence of PtdIns 3-kinase signaling (Caffrey et al. 2001).

Also, as noted in the introduction, $InsP_6$ is an essential cofactor for adenosine deaminase (Macbeth et al. 2005), and, in yeast at least, $InsP_6$ stimulates mRNA export from the nucleus (Alcázar-Román et al. 2006; York et al. 1999). Additionally, by enhancing the interaction of Ku with other proteins, $InsP_6$ stimulates DNA repair through non-homologous end-joining (Cheung et al. 2008). However, since all of these functions for $InsP_6$ can be satisfied by just a small percentage of total cellular $InsP_6$ levels, it is our contention that in these cases this inositol polyphosphate more likely functions as a cofactor rather than as a dynamic "regulator".

In view of all of these activities of Ins(1,3,4,5,6)P_5 and $InsP_6$, it is not surprising that, in mammals, embryonic lethality results from the knock-out of IPK2/IPMK (Frederick et al. 2005) or IP5K (Verbsky et al. 2005a). The knock-down of Ip5K in zebrafish embryos is also phenotypically dramatic: there is disturbance of asymmetric Ca^{2+} signaling that is important for embryonic patterning (Sarmah et al. 2005). However, those genetic experiments in themselves do not speak to any specific *signaling* role of Ins(1,3,4,5,6)P_5 or $InsP_6$. It is possible that these phenotypes are, in part, consequences of the loss of non-catalytic activities of inositolphosphate kinases (Odom et al. 2000) and/or the absence of more highly phosphorylated metabolites, such as the inositol pyrophosphates, which also function in development (Sarmah and Wente 2010).

13.5 Ins(3,4,5,6)P_4

Cellular levels of Ins(3,4,5,6)P_4 are around 1 μM in resting cells, and they increase to the 5–10 μM range whenever PLC is activated (Ho and Shears 2002). Ins(3,4,5,6)P_4 is a concentration-dependent inhibitor of a CaMKII-activated Cl^- conductance that is located in the plasma membrane (Xie et al. 1996, 1998; Ho et al. 2001; Mitchell et al. 2008). At least in mammalian cells, the inhibition of Cl^- channel conductance by Ins(3,4,5,6)P_4 is an exquisitely specific regulatory process; it is not imitated by any of the many other inositol phosphates that exist inside cells (Ho and Shears 2002; Ho et al. 2000; Xie et al. 1996). In other words, it has been demonstrated that Ins(3,4,5,6)P_4 is a receptor-regulated signal, its biological target is known, and Ins(3,4,5,6)P_4 acts specifically. This inositol phosphate is undoubtedly an intracellular signal.

Ins(3,4,5,6)P_4 can only be formed in animal cells by receptor-dependent dephosphorylation of Ins(1,3,4,5,6)P_5 by an enzyme that is—unfortunately—known as ITPK1 (for Inositol Trisphosphate Kinase). This baptism by the Human Genome Nomenclature Committee seems to have been prompted by the fact that the protein was initially characterized as a 6-kinase activity that phosphorylates Ins(1,3,4)P_3 to Ins(1,3,4,6)P_4 (Shears et al. 1987; Balla et al. 1987). It is only recently that it has been determined that the trisphosphate kinase activity reflects a more complex, ADP-dependent phosphotransferase activity (Chamberlain et al. 2007; Ho et al. 2002). This is a unique phenomenon in the inositol phosphate field, that explains the molecular mechanism by which Ins(3,4,5,6)P_4 levels are coupled to receptor-regulated PLC activity (Fig. 13.2). In its ADP-bound form, ITPK1 dephosphorylates

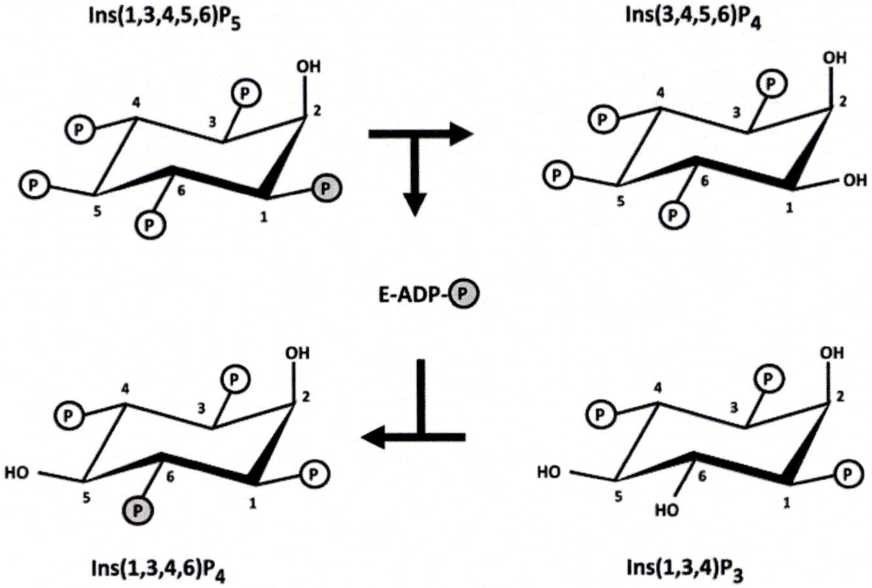

Fig. 13.2 The phosphotransferase activity of ITPK1. The figure shows the phosphotransferase activity that is catalyzed by ITPK1. The phosphate group that is transferred between from Ins(1,3,4,5,6)P_5 to Ins(1,3,4)P_3 is highlighted in by the *grey circle*

Ins(1,3,4,5,6)P_5 to Ins(3,4,5,6)P_4. The Ins(3,4,5,6)P_4 is released to the bulk phase in exchange for Ins(1,3,4)P_3, but the nucleotide—now ATP—remains bound. The tenacity of this binding of adenine nucleotide has been verified by crystallographic data showing that less than 10% of the nucleotide is solvent exposed (Miller et al. 2005; Chamberlain et al. 2007). Thus, the inorganic phosphate that is removed from Ins(1,3,4,5,6)P_5 is not released. Instead, it is passed on to the newly-bound Ins(1,3,4)P_3, thereby phosphorylating it to Ins(1,3,4,6)P_4, which the active-site then exchanges for a new molecule of Ins(1,3,4,5,6)P_5, and the entire phosphotransferase cycle is repeated (Fig. 13.2). Importantly, the rate at which Ins(1,3,4,5,6)P_5 is dephosphorylated to Ins(3,4,5,6)P_4 is stimulated as the rate-limiting concentration of phosphate acceptor—Ins(1,3,4)P_3—is increased (Ho et al. 2002). In turn, the cellular levels of Ins(1,3,4)P_3—a metabolite of Ins(1,4,5)P_3 (Fig. 13.1)—mirrors both the intensity and the duration of receptor-activated PLC activity (Batty et al. 1998; Batty and Downes 1994). In other words, the degree of PLC activity sets Ins(1,3,4)P_3 levels, which controls Ins(3,4,5,6)P_4 synthesis. This is the molecular basis for the integration of inositol phosphate signaling pathways via human ITPK1.

The Ins(1,3,4)P_3 6-kinase activity of ITPK1 also plays a metabolic role (Fig. 13.1) in maintaining the size of the cell's Ins(1,3,4,5,6)P_5 pool (Shears et al. 1987; Balla et al. 1987; Verbsky et al. 2005b). It is unclear if it is either this metabolic function, or the signaling activities of ITPK1, which explains why mice which are hypomorphic for the *Itpk1* allele are susceptible to neural tube defects (Wilson et al. 2009).

A complete knock-down of ITPK1 expression is apparently lethal (Verbsky et al. 2005b).

The best characterized biological end-point for Ins(3,4,5,6)P_4 action upon Cl^- transport is to regulate epithelial salt and fluid secretion (Vajanaphanich et al. 1994; Carew et al. 2000). However, the recent identification of ClC3 as the target of Ins(3,4,5,6)P_4, at least in mammalian cells (Mitchell et al. 2008), has greatly expanded the biological repertoire of this inositol phosphate. For example, ClC3 is responsible for the Ins(3,4,5,6)P_4-regulated Cl^- conductance in hippocampal neurones (Mitchell et al. 2008), which is thought to contribute to the overall regulation of the synaptic efficacy in generating action potentials (Wang et al. 2006). Long-term changes in synaptic efficacy comprise a cellular basis for information storage and memory formation (Bliss and Collingridge 1993). Thus, Ins(3,4,5,6)P_4 is a molecule that has the potential to affect neuronal development. It therefore seems pertinent that Ins(3,4,5,6)P_4 has also previously been suggested to have the characteristics of a "memory molecule", because its relatively slow rate of metabolism permit its physiological effects to long outlast the duration of the stimulus that initially prompts intracellular Ins(3,4,5,6)P_4 to accumulate (Ho and Shears 2002).

ClC3 that is in the plasma membrane may have other roles, such as tumor cell migration (Mao et al. 2008; Cuddapah and Sontheimer 2010) and the regulation of apoptosis (Claud et al. 2008). We can therefore anticipate that Ins(3,4,5,6)P_4 might also regulate these processes. It should be noted, however, that some of the workers in this field (Jentsch 2008; Jentsch et al. 2002) propose that ClC3 is not a plasmalemmal Cl^- channel *per se*, but instead a regulator of other Cl^- channels. That argument, if correct, does not devalue the role of Ins(3,4,5,6)P_4 in regulating ClC3 function. For example, in one cell type in which our own data are consistent with the ClC3 regulating other Cl^- channels, we have shown that ClC3 still mediates the effect of Ins(3,4,5,6)P_4 upon plasma membrane Cl^- fluxes (unpublished data).

While there is disagreement as to whether or not ClC3 is an independent plasma membrane Cl^- channel, it is well recognized that ClC3 also resides in intracellular vesicles such as insulin granules (Barg et al. 2001) and the early endosomal compartment (Zhao et al. 2007; Gentzsch et al. 2003; Stobrawa et al. 2001; Hara-Chikuma et al. 2005; Mitchell et al. 2008). Here, ClC3 contributes to endosomal acidification (Jentsch 2008), although there is uncertainty concerning the exact mechanism. Nevertheless, when a cell-permeant analogue of Ins(3,4,5,6)P_4 was added to cells so as to inhibit ClC3, the pH of certain vesicular sub-compartments became more alkaline (Renström et al. 2002; Mitchell et al. 2008). What is the biological significance of this regulation of intra-vesicular pH? With regard to insulin granules, it has been proposed that their intraluminal acidification is a priming process, without which they become less competent to fuse with the plasma membrane and release their cargo (Barg et al. 2001). In support of this idea, we have shown that alkalinization of insulin granules by Ins(3,4,5,6)P_4 has the effect of reducing insulin secretion from pancreatic β-cells (Renström et al. 2002). In many other cell types, the acidification of endosomes and secretory vesicles serves other important functions, including modulation of certain ligand-protein interactions during endocytosis, enzyme targeting, and H^+-coupled uptake of small molecules (such as neurotransmitters)

(Nishi and Forgac 2002; von and Sorkin 2007). It appears that we have only scratched the surface of our understanding of the biological importance of Ins(3,4,5,6)P_4.

It is unfortunate that we do not yet understand the mechanism by which Ins(3,4,5,6)P_4 prevents CaMKII from activating ClC3. Data obtained to date indicate that Ins(3,4,5,6)P_4 does not inhibit CaMKII activity *per se* (Xie et al. 1996; Ho et al. 2001; Ho and Shears 2002). Furthermore, in single channel analysis of CaMKII-activated Cl^- channels, Ins(3,4,5,6)P_4 was not inhibitory, so it is unlikely to act as a direct channel blocker (Ho et al. 2001). Presumably an intermediary protein mediates the action of Ins(3,4,5,6)P_4. However, our efforts to identify an Ins(3,4,5,6)P_4 "receptor" have so far been disappointingly fruitless (unpublished data). Our work on this important problem is ongoing.

13.6 Conclusions

The metabolic intermediates that accumulate during the dephosphorylation of Ins(1,4,5)P_3 and Ins(1,3,4)P_3 to inositol are not generally considered to be signaling molecules. The same *could* be true of at least some of the intermediates in the pathways of phosphorylation of Ins(1,4,5)P_3 to InsP_6. The concept that inositol is a combinatorial signaling scaffold (York 2006) is intellectually appealing, but it is still not obligatory that all of these inositol phosphates be cellular signals; only at least one of the end products might act in a signaling pathway.

We have argued here that there is no strong evidence that Ins(1,4,5,6)P_4 is a cellular signal. As for Ins(1,3,4,5)P_4, we have highlighted the confusing and often conflicting observations in the literature concerning proposed actions of this inositol phosphate. Under such circumstances, it is difficult to formulate a general signaling role. Ins(1,3,4,5,6)P_5 and InsP_6—particularly the latter—clearly have important roles as cofactors, but our conclusion is that we need more concrete evidence before we can claim that these molecules are truly cellular signals. As we have discussed, it may be that further information of cellular compartmentalization may be the savior of these molecule's signaling credentials. So, other than Ins(1,4,5)P_3, that leaves, in our opinion, Ins(3,4,5,6)P_4 as the only validated "classical" cellular signal from within this group of molecules.

References

Alcázar-Román AR, Tran EJ, Guo S, Wente SR (2006) Inositol hexakisphosphate and Gle1 activate the DEAD-box protein Dbp5 for nuclear mRNA export. Nat Cell Biol 8:645–647

Ali N, Craxton A, Shears SB (1993) Hepatic Ins(1,3,4,5)P_4 3-phosphatase is compartmentalized inside endoplasmic reticulum. J Biol Chem 268:6161–6167

Balla T, Guillemette G, Baukal AJ, Catt KJ (1987) Formation of inositol 1,3,4,6-tetrakisphosphate during angiotensin II action in bovine adrenal glomerulosa cells. Biochem Biophys Res Commun 146:199–205

Barg S, Huang P, Eliasson L, Nelson DJ, Obermüller S, Rorsman P, Thévenod F, Renström E (2001) Priming of insulin granules for exocytosis by granular chloride uptake and acidification. J Cell Sci 114:2145–2154

Barker CJ, Wong NS, Maccallum SM, Hunt PA, Michell RH, Kirk CJ (1992) The interrelationships of the inositol phosphates formed in WRK-1 stimulated rat mammary tumour cells. Biochem J 286:469–474

Barker CJ, Leibiger IB, Leibiger B, Berggren PO (2002) Phosphorylated inositol compounds in beta -cell stimulus-response coupling. Am J Physiol Endocrinol Metab 283:E1113–E1122

Barker CJ, Wright J, Hughes PJ, Kirk CJ, Michell RH (2004) Complex changes in cellular inositol phosphate complement accompany transit through the cell cycle. Biochem J 380:465–473

Batty IH, Downes CP (1994) The inhibition of phosphoinositide synthesis and muscarinic-receptor-mediated phospholipase C activity by Li+ as secondary selective consequences of inositol depletion in 1321N1 cells. Biochem J 297:529–537

Batty IH, Nahorski SR, Irvine RF (1985) Rapid formation of inositol 1,3,4-tetrakisphosphate following muscarinic receptor stimulation of rat cerebral cortical slices. Biochem J 232:211–215

Batty IH, Currie RA, Downes CP (1998) Evidence for a model of integrated inositol phospholipid pools implies an essential role for lipid transport in the maintenance of receptor-mediated phospholipase C activity in 1321N1 cells. Biochem J 330:1069–1077

Becker PB, Hörz W (2002) ATP-dependent nucleosome remodeling. Annu Rev Biochem 71:247–273

Berggren PO, Barker CJ (2008) A key role for phosphorylated inositol compounds in pancreatic beta-cell stimulus-secretion coupling. Adv Enzyme Regul 48:276–294

Bird GStJ, Putney JW Jr (1996) Effect of inositol 1,3,4,5-tetrakisphosphate on inositol trisphosphate-activated Ca^{2+} signaling in mouse lacrimal acinar cells. J Biol Chem 271:6766–6770

Bird GStJ, Rossier MF, Hughes AR, Shears SB, Armstrong DL, Putney JW Jr (1991) Activation of calcium entry into acinar cells by a non-phosphorylatable inositol trisphosphate. Nature 352:162–165

Bliss TV, Collingridge GL (1993) A synaptic model of memory: long-term potentiation in the hippocampus. Nature 361:31–39

Boynton AL, Dean NM, Hill TD (1990) Inositol 1,3,4,5-tetrakisphosphate and regulation of intracellular calcium. Biochem Pharmacol 40:1933–1939

Brehm MA, Schenk TM, Zhou X, Fanick W, Lin H, Windhorst S, Nalaskowski MM, Kobras M, Shears SB, Mayr GW (2007) Intracellular localization of human inositol 1,3,4,5,6-pentakisphosphate 2-kinase. Biochem J 408:335–345

Brough D, Bhatti F, Irvine RF (2005) Mobility of proteins associated with the plasma membrane by interaction with inositol lipids. J Cell Sci 118:3019–3025

Bunce CM, French PJ, Allen P, Mountford JC, Moor B, Greaves MF, Michell RH, Brown G (1993) Comparison of the levels of inositol metabolites in transformed haemopoietic cells and their normal counterparts. Biochem J 289:667–673

Caffrey JJ, Darden T, Wenk MR, Shears SB (2001) Expanding Coincident Signaling by PTEN through its Inositol 1,3,4,5,6-Pentakisphosphate 3-phosphatase Activity. FEBS Lett 499:6–10

Carew MA, Yang X, Schultz C, Shears SB (2000) Ins(3,4,5,6)P_4 inhibits an apical calcium-activated chloride conductance in polarized monolayers of a cystic fibrosis cell-line. J Biol Chem 275:26906–26913

Chamberlain PP, Qian X, Stiles AR, Cho J, Jones DH, Lesley SA, Grabau EA, Shears SB, Spraggon G (2007) Integration of inositol phosphate signaling pathways via human ITPK1. J Biol Chem 282:28117–28125

Changya L, Gallacher DV, Irvine RF, Potter BVL, Petersen OH (1989) Inositol 1,3,4,5-trisphosphate is essential for sustained activation of the Ca^{2+}-dependent K+ current in single internally perfused lacrimal cells. J Membr Biol 109:85–93

Cheung JC, Salerno B, Hanakahi LA (2008) Evidence for an inositol hexakisphosphate-dependent role for Ku in mammalian nonhomologous end joining that is independent of its role in the DNA-dependent protein kinase. Nucleic Acids Res 36:5713–5726

Chi H, Yang X, Kingsley PD, O'Keefe RJ, Puzas JE, Rosier RN, Shears SB, Reynolds PR (2000) Targeted deletion of Minpp1 provides new insight into the activity of multiple inositol polyphosphate phosphatase in vivo. Mol Cell Biol 20:6496–6507

Claud EC, Lu J, Wang XQ, Abe M, Petrof EO, Sun J, Nelson DJ, Marks JD, Jilling T (2008) Platelet-activating Factor induced chloride channel activation is associated with intracellular acidosis and apoptosis of intestinal epithelial cells. Am J Physiol Gastrointest Liver Physiol 294, G1191-G1200

Cooke F, Poyner DR, Hawkins P, Erlebach C, Hanley MR (1991) Inositol hexakisphosphate-membranes interactions: the role of metal ions. Biochem Soc Trans 19:152s

Cozier GE, Lockyer PJ, Reynolds JS, Kupzig S, Bottomley JR, Millard TH, Banting G, Cullen PJ (2000) GAP1IP4BP contains a novel group I pleckstrin homology domain that directs constitutive plasma membrane association. J Biol Chem 275:28261–28628

Craxton A, Caffrey JJ, Burkhart W, Safrany ST, Shears SB (1997) Cloning and expression of rat hepatic multiple inositol polyphosphate phosphatase. Biochem J 328:75–81

Crossley I, Swann K, Chambers E, Whitaker M (1988) Activation of sea urchin eggs by inositol phosphates is independent of external calcium. Biochem J 252:257–262

Cuddapah VA, Sontheimer H (2010) Molecular interaction and functional regulation of ClC-3 by Ca^{2+}/calmodulin-dependent protein kinase II (CaMKII) in human malignant glioma. J Biol Chem 285:11188–11196

Cullen PJ, Hsuan JJ, Truong O, Letcher AJ, Jackson TR, Dawson AP, Irvine RF (1995) Identification of a specific Ins(1,3,4,5)P_4-binding protein as a member of the GAP1 family. Nature 376:527–530

Dubois E, Dewaste V, Erneux C, Messenguy F (2000) Inositol polyphosphate kinase activity of Arg82/ArgRIII is not required for the regulation of the arginine metabolism in yeast. FEBS Lett 486:300–304

Eckmann L, Rudolf MT, Ptasznik A, Schultz C, Jiang T, Wolfson N, Tsien R, Fierer J, Shears SB, Kagnoff MF, Traynor-Kaplan A (1997) D-myo-inositol 1,4,5,6-tetrakisphosphate produced in human intestinal epithelial cells in response to Salmonella invasion inhibits phosphoinositide 3-kinase signaling pathways. Proc Natl Acad Sci U S A 94:14456–14460

El Alami M, Messenguy F, Scherens B, Dubois E (2003) Arg82p is a bifunctional protein whose inositol polyphosphate kinase activity is essential for nitrogen and PHO gene expression but not for Mcm1p chaperoning in yeast. Mol Microbiol 49:457–468

Frederick JP, Mattiske D, Wofford JA, Megosh LC, Drake LY, Chiou ST, Hogan BL, York JD (2005) An essential role for an inositol polyphosphate multikinase, Ipk2, in mouse embryogenesis and second messenger production. Proc Natl Acad Sci U S A 102:8454–8459

Gao Y, Wang HY (2007) Inositol Pentakisphosphate Mediates Wnt/beta-Catenin Signaling. J Biol Chem 282:26490–26502

Gentzsch M, Cui L, Mengos A, Chang XB, Chen JH, Riordan JR (2003) The PDZ-binding chloride channel ClC-3B localizes to the Golgi and associates with cystic fibrosis transmembrane conductance regulator-interacting PDZ proteins. J Biol Chem 278:6440–6449

Guse AH, Greiner E, Emmrich F, Brand K (1993) Mass changes of inositol 1,3,4,5,6-pentakisphosphate and inositol hexakisphosphate during cell cycle progression in rat thymocytes. J Biol Chem 268:7129–7133

Hara-Chikuma M, Yang B, Sonawane ND, Sasaki S, Uchida S, Verkman AS (2005) ClC-3 chloride channels facilitate endosomal acidification and chloride accumulation. J Biol Chem 280:1241–1247

Hermosura MC, Takeuchi H, Fleig A, Riley AQM, Potter BVL, Hirata M, Penner R (2000) InsP4 facilitates store-operated calcium influx by inhibition of InsP3 5-phosphatase. Nature 408:735–740

Hernandez LD, Hueffer K, Wenk MR, Galan JE (2004) Salmonella modulates vesicular traffic by altering phosphoinositide metabolism. Science 304:1805–1807

Heslop JP, Irvine RF, Tashjian AH, Berridge MJ (1985) Inositol tetrakis- and pentakisphosphates in GH4 cells. J Exp Biol 119:395–401

Hill TD, Dean NM, Boynton AL (1988) Inositol 1,3,4,5-tetrakisphosphate induces Ca^{2+} sequestration in rat liver cells. Science 242:1176–1178

Ho MWY, Shears SB (2002) Regulation of calcium-activated chloride channels by inositol 3,4,5,6-tetrakisphosphate. In: Fuller CM (ed) Current topics in membranes, 53. Academic Press, London, pp 345–363

Ho MWY, Carew MA, Yang X, Shears SB (2000) Regulation of chloride channel conductance by $Ins(3,4,5,6)P_4$; a phosphoinositide-initiated signalling pathway that acts downstream of $Ins(1,4,5)P_3$. In: Cockroft S (ed) Frontiers in molecular biology: biology of phosphoinositides. Oxford University Press, Oxford, pp 298–319

Ho MWY, Kaetzel MA, Armstrong DL, Shears SB (2001) Regulation of a human chloride channel: a paradigm for integrating input from calcium, CaMKII and $Ins(3,4,5,6)P_4$. J Biol Chem 276:18673–18680

Ho MWY, Yang X, Carew MA, Zhang T, Hua L, Kwon Y-U, Chung S-K, Adelt S, Vogel G, Riley AM, Potter BVL, Shears SB (2002) Regulation of $Ins(3456)P_4$ signaling by a reversible kinase/phosphatase. Curr Biol 12:477–482

Hoth M, Penner R (1992) Depletion of intracellular calcium stores activates a calcium current in mast cells. Nature 355:353–356

Huang YH, Grasis JA, Miller AT, Xu R, Soonthornvacharin S, Andreotti AH, Tsoukas CD, Cooke MP, Sauer K (2007) Positive regulation of Itk PH domain function by soluble IP4. Science 316:886–889

Irvine RF (1986) Calcium transients: mobilization of intracellular Ca^{2+}. Br Med Bull 42:369–374

Irvine RF (1992) Is inositol tetrakisphosphate the second messenger that controls Ca^{2+} entry into cells? In: Putney JW Jr (ed) Advances in second messenger and phosphoprotein research. Raven Press, New York, pp 161–185

Irvine RF, Moor RM (1986) Micro-injection of inositol 1,3,4,5-tetrakisphosphate activates sea urchin eggs by a mechanism dependent on external Ca^{2+}. Biochem J 240:917–920

Irvine RF, Moor RM (1987) Inositol(1,3,4,5)tetrakisphosphate-induced activation of sea urchin eggs requires the presence of inositol trisphosphate. Biochem Biophys Res Commun 146:284–290

Irvine RF, Schell M (2001) Back in the water: the return of the inositol phosphates. Nat Rev Mol Cell Biol 2:327–338

Irvine RF, Letcher AJ, Heslop JP, Berridge MJ (1986) The inositol tris/tetrakisphosphate pathway—demonstration of $Ins(1,4,5)P_3$ 3-kinase activity in animal tissues. Nature 320:631–634

Irvine RF, Lloyd-Burton SM, Yu JC, Letcher AJ, Schell MJ (2006) The regulation and function of inositol 1,4,5-trisphosphate 3-kinases. Adv Enzyme Regul 46:314–323

Ismailov II, Fuller CM, Berdiev BK, Shlyonsky VG, Benos DJ, Barrett KE (1996) A biologic function for an "orphan" messenger: D-myo-Inositol 3,4,5,6-tetrakisphosphate selectively blocks epithelial calcium-activated chloride current. Proc Nat Acad Sci U S A 93:10505–10509

Jentsch TJ (2008) CLC chloride channels and transporters: from genes to protein structure, pathology and physiology. Crit Rev Biochem Mol Biol 43:3–36

Jentsch TJ, Stein V, Weinreich F, Zdebik AA (2002) Molecular structure and physiological function of chloride channels. Physiol Rev 82:503–568

Jia Y, Subramanian KK, Erneux C, Pouillon V, Hattori H, Jo H, You J, Zhu D, Schurmans S, Luo HR (2007) Inositol 1,3,4,5-tetrakisphosphate negatively regulates phosphatidylinositol-3,4,5-trisphosphate signaling in neutrophils. Immunity 27:453–467

Jia Y, Loison F, Hattori H, Li Y, Erneux C, Park SY, Gao C, Chai L, Silberstein LE, Schurmans S, Luo HR (2008) Inositol trisphosphate 3-kinase B (InsP3KB) as a physiological modulator of myelopoiesis. Proc Natl Acad Sci U S A 105:4739–4744

Kavran JM, Klein DE, Lee A, Falasca M, Isakoff SJ, Skolnik EY, Lemmon MA (1998) Specificity and Promiscuity in phosphoinositide binding by pleckstrin homology domains. J Biol Chem 273:30497–30508

Komander D, Fairservice A, Deak M, Kular GS, Prescott AR, Peter DC, Safrany ST, Alessi DR, Aalten DM van (2004) Structural insights into the regulation of PDK1 by phosphoinositides and inositol phosphates. EMBO J 23:3918–3928

Larsson O, Barker CJ, Sj-oholm A, Carlqvist H, Michell RH, Bertorello A, Nilsson T, Honkanen RE, Mayr GW, Zwiller J, Berggren PO (1997) Inhibition of phosphatases and increased Ca^{2+} channel activity by inositol hexakisphosphate. Science 278:471–474

Leyman A, Pouillon V, Bostan A, Schurmans S, Erneux C, Pesesse X (2007) The absence of expression of the three isoenzymes of the inositol 1,4,5-trisphosphate 3-kinase does not prevent the formation of inositol pentakisphosphate and hexakisphosphate in mouse embryonic fibroblasts. Cell Signal 19:1497–1504

Li W, Schultz C, Llopis J, Tsien RY (1997) Membrane-permeant esters of inositol polyphosphates, chemical synthesis and biological applications. Tetrahedron 53:12017–12040

Loomis-Husselbee JW, Cullen PJ, Dreikausen UE, Irvine RF, Dawson AP (1996) Synergistic effects of inositol 1,3,4,5-tetrakisphosphate on inositol 2,4,5-trisphosphate-stimulated Ca^{2+} release do not involve direct interaction of inositol 1,3,4,5-tetrakisphosphate with inositol trisphosphate binding sites. Biochem J 314:811–816

Luckhoff A, Clapham DE (1992) Inositol 1,3,4,5-tetrakisphosphate activates an endothelial Ca^{2+} -permeable channel. Nature 355:356–358

Macbeth MR, Schubert HL, Vandemark AP, Lingam AT, Hill CP, Bass BL (2005) Inositol hexakisphosphate is bound in the ADAR2 core and required for RNA editing. Science 309:1534–1539

Mao J, Chen L, Xu B, Wang L, Li H, Guo J, Li W, Nie S, Jacob TJ, Wang L (2008) Suppression of ClC-3 channel expression reduces migration of nasopharyngeal carcinoma cells. Biochem Pharmacol 75:1706–1716

Marechal Y, Pesesse X, Jia Y, Pouillon V, Perez-Morga D, Daniel J, Izui S, Cullen PJ, Leo O, Luo HR, Erneux C, Schurmans S (2007) Inositol 1,3,4,5-tetrakisphosphate controls proapoptotic Bim gene expression and survival in B cells. Proc Natl Acad Sci U S A 104:13978–13983

Mattingly RR, Stephens LR, Irvine RF, Garrison JC (1991) Effects of transformation with the v-src oncogene on inositol phosphate metabolism in rat-1 fibroblasts: D-myo-inositol 1,4,5,6-tetrakisphosphate is increased in v-src transformed rat-1 fibroblasts and can be synthesised from d-myo-inositol 1,3,4-trisphosphate in cytosolic extracts. J Biol Chem 266:15144–15153

McKnight S (2003) Gene switching by metabolic enzymes—how did you get on the invitation list? Cell 174:150–152

Menniti FS, Oliver KG, Nogimori K, Obie JF, Shears SB, Putney JW Jr (1990) Origins of myo-inositol tetrakisphosphates in agonist-stimulated rat pancreatoma cells. Stimulation by bombesin of myo-inositol 1,3,4,5,6-pentakisphosphate breakdown to myo-inositol 3,4,5,6-tetrakisphosphate. J Biol Chem 265:11167–11176

Menniti FS, Miller RN, Putney JW Jr, Shears SB (1993) Turnover of inositol polyphosphate pyrophosphates in pancreatoma cells. J Biol Chem 268:3850–3856

Michell RH, King CE, Piper CJ, Stephens LR, Bunce CM, Guy GR, Brown G (1988) Inositol lipids and phosphates in erythrocytes and HL60 cells. J Gen Physiol 43:345–355

Miller GJ, Wilson MP, Majerus PW, Hurley JH (2005) Specificity determinants in inositol polyphosphate synthesis: crystal structure of inositol 1,3,4-trisphosphate 5/6-kinase. Mol Cell 18:201–212

Miller AT, Sandberg M, Huang YH, Young M, Sutton S, Sauer K, Cooke MP (2007) Production of Ins(1,3,4,5)P_4 mediated by the kinase Itpkb inhibits store-operated calcium channels and regulates B cell selection and activation. Nat Immunol 8:514–521

Miller AT, Beisner DR, Liu D, Cooke MP (2009) Inositol 1,4,5-trisphosphate 3-kinase B is a negative regulator of BCR signaling that controls B cell selection and tolerance induction. J Immunol 182:4696–4704

Mitchell J, Wang X, Zhang G, Gentzsch M, Nelson DJ, Shears SB (2008) An Expanded Biological Repertoire for Ins(3,4,5,6)P(4) through its Modulation of ClC-3 Function. Curr Biol 18:1600–1605

Monserrate JP, York JD (2010) Inositol phosphate synthesis and the nuclear processes they affect. Curr Opin Cell Biol 22:365–373

Morris AP, Gallacher DV, Irvine RF, Petersen OH (1987) Synergism of inositol trisphosphate and tetrakisphosphate in activating Ca^{2+}-dependent K+ channels. Nature 330:653–655

Nagy R, Grob H, Weder B, Green P, Klein M, Frelet A, Schjoerring JK, Brearley CA, Martinoia E (2009) The Arabidopsis ATP-binding cassette protein ATMRP5/ATABCC5 is a high-affinity inositol hexakisphosphate transporter involved in guard cell signaling and phytate storage. J Biol Chem 384:33614–33622

Nalaskowski MM, Deschermeier C, Fanick W, Mayr GW (2002) The human homologue of yeast ArgRIII protein is an inositol phosphate multikinase with predominantly nuclear localization. Biochem J 366:549–556

Narayan K, Lemmon MA (2006) Determining selectivity of phosphoinositide-binding domains. Methods 39:122–133

Nishi T, Forgac M (2002) The vacuolar (H^+)-ATPases—nature's most versatile proton pumps. Nat Rev Mol Cell Biol 3:94–103

Norris FA, Wilson MP, Wallis TS, Galyov EE, Majerus PW (1998) SopB, a protein required for virulence of Salmonella dublin, is an inositol phosphate phosphatase. Proc Natl Acad Sci U S A 95:14057–14059

Odom AR, Stahlberg A, Wente SR, York JD (2000) A role for nuclear inositol 1,4,5-trisphosphate kinase in transcriptional control. Science 287:2026–2029

Oliver KG, Putney JW Jr, Obie JF, Shears SB (1992) The interconversion of inositol 1,3,4,5,6-pentakisphosphate and inositol tetrakisphosphates in AR4–2J cells. J Biol Chem 267:21528–21534

Otto JC, Kelly P, Chiou ST, York JD (2007) Alterations in an inositol phosphate code through synergistic activation of a G protein and inositol phosphate kinases. Proc Natl Acad Sci U S A 104:15653–15658

Parekh AB, Penner R (1997) Store depletion and calcium influx. Physiol Rev 77:901–930

Pittet D, Schlegel W, Lew DP, Monod A, Mayr GW (1989) Mass changes in inositol tetrakis- and pentakisphosphate isomers induced by chemotactic peptide stimulation in HL-60 cells. J Biol Chem 264:18489–18493

Pouillon V, Hascakova-Bartova R, Pajak B, Adam E, Bex F, Dewaste V, Van Lint C, Leo O, Erneux C, Schurmans S (2003) Inositol 1,3,4,5-tetrakisphophate is essential for T lymphocyte development. Nat Immunol 4:1136–1143

Putney JW Jr (1986) A model for receptor-regulated calcium entry. Cell Calcium 7:1–12

Putney JW Jr (1992) Inositol phosphates and calcium entry. In: Putney JW (ed) Advances in second messenger and phosphoprotein research. Raven Press, New York, pp 143–160

Qi Q, August A (2009) The Tec family kinase Itk exists as a folded monomer in vivo. J Biol Chem 284:29882–29892

Qi Q, Sahu N, August A (2006) Tec kinase Itk forms membrane clusters specifically in the vicinity of recruiting receptors. J Biol Chem 281:38529–38534

Quignard JF, Rakotoarisoa L, Mironneau J, Mironneau C (2003) Stimulation of L-type Ca^{2+} channels by inositol pentakis- and hexakisphosphates in rat vascular smooth muscle cells. J Physiol 549:729–737

Rall TW, Sutherland EW (1958) Formation of a cyclic adenine ribonucleotide by tissue particles. J Biol Chem 232:1065–1076

Renström E, Ivarsson R, Shears SB (2002) Ins(3,4,5,6)P_4 inhibits insulin granule acidification and fusogenic potential. J Biol Chem 277:26717–26720

Robison GA, Butcher RW, Sutherland EW (1968) Cyclic AMP. Annu Rev Biochem 37:149–174

Ruzzene M, Pinna LA (2010) Addiction to protein kinase CK2: a common denominator of diverse cancer cells? Biochim Biophys Acta 1804:499–504

Saiardi A (2011) Phosphoinositides and inositol phosphates

Saiardi A, Caffrey JJ, Snyder SH, Shears SB (2000) Inositol polyphosphate multikinase (ArgRIII) determines nuclear mRNA export in Saccharomyces cerevisiae. FEBS Lett 468:28–32

Saiardi A, Nagata E, Luo HR, Sawa A, Luo X, Snowman AM, Snyder SH (2001) Mammalian inositol polyphosphate multikinase synthesizes inositol 1,4,5-trisphosphate and an inositol pyrophosphate. Proc Natl Acad Sci U S A 98:2306–2311

Sakaguchi R, Tainaka K, Shimada N, Nakano S, Inoue M, Kiyonaka S, Mori Y, Morii T (2010) An in vivo fluorescent sensor reveals intracellular ins(1,3,4,5)P_4 dynamics in single cells. Angew Chem Int Ed Engl 49:2150–2153

Sarmah B, Wente SR (2010) Inositol hexakisphosphate kinase-2 acts as an effector of the vertebrate Hedgehog pathway. Proc Natl Acad Sci U S A 107:19921–19926

Sarmah B, Latimer AJ, Appel B, Wente SR (2005) Inositol polyphosphates regulate zebrafish left-right asymmetry. Dev Cell 9:133–145

Sauer K, Cooke MP (2010) Regulation of immune cell development through soluble inositol-1,3,4,5-tetrakisphosphate. Nat Rev Immunol 10:257–271

Schell MJ (2010) Inositol trisphosphate 3-kinases: focus on immune and neuronal signaling. Cell Mol Life Sci 67:1755–1778

Schell MJ, Irvine RF (2006) Calcium-triggered exit of F-actin and IP(3) 3-kinase A from dendritic spines is rapid and reversible. Eur J Neurosci 24:2491–2503

Seeds AM, Sandquist JC, Spana EP, York JD (2004) A molecular basis for inositol polyphosphate synthesis in Drosophila melanogaster. J Biol Chem 279:47222–47232

Shears SB (1989) Metabolism of the inositol phosphates produced by receptor activation. Biochem J 260:313–324

Shears SB (2004) How versatile are inositol phosphate kinases? Biochem J 377:265–280

Shears SB, Parry JB, Tang EKY, Irvine RF, Michell RH, Kirk CJ (1987) Metabolism of D-myo-inositol 1,3,4,5-tetrakisphosphate by rat liver, including the synthesis of a novel isomer of myo-inositol tetrakisphosphate. Biochem J 246:139–147

Shen X, Xiao H, Ranallo R, Wu W-H, Wu C (2003) Modulation of ATP-dependent chromatin-remodeling complexes by inositol polyphosphates. Science 299:112–114

Shi J, Wang H, Schellin K, Li B, Faller M, Stoop JM, Meeley RB, Ertl DS, Ranch JP, Glassman K (2007) Embryo-specific silencing of a transporter reduces phytic acid content of maize and soybean seeds. Nat Biotechnol 25:930–937

Slusarski DC, Corces VG, Moon RT (1997) Interaction of Wnt and a Frizzled homologue triggers G-protein-linked phosphatidylinositol signalling. Nature 390:410–413

Smith PM, Harmer AR, Letcher AJ, Irvine RF (2000) The effect of inositol 1,3,4,5-tetrakisphosphate on inositol trisphosphate-induced Ca^{2+} mobilization in freshly isolated and cultured mouse lacrimal cells. Biochem J 347:77–82

Solyakov L, Cain K, Tracey BM, Jukes R, Riley AM, Potter BV, Tobin AB (2004) Regulation of casein kinase-2 (CK2) activity by inositol phosphates. J Biol Chem 279:43403–43410

Steger DJ, Haswell ES, Miller AL, Wente SR, O'Shea EK (2003) Regulation of chromatin remodeling by inositol polyphosphates. Science 299:114–116

Stobrawa SM, Breiderhoff T, Takamori S, Engel D, Schweizer M, Zdebik AA, Bösl MR, Ruether K, Jahn H, Draguhn A, Jahn R, Jentsch TJ (2001) Disruption of ClC-3, a chloride channel expressed on synaptic vesicles, leads to a loss of hippocampus. Neuron 29:185–196

Streb H, Irvine RF, Berridge MJ, Schulz I (1983) Release of Ca^{2+} from a nonmitochondrial store in pancreatic cells by inositol-1,4,5-trisphosphate. Nature 306:67–68

Stuart JA, Anderson KL, French PJ, Kirk CJ, Michell RH (1994) The intracellular distribution of inositol polyphosphates in HL60 promyeloid cells. Biochem J 303:517–525

Sutherland EW, Rall TW (1958) Fractionation and characterization of a cyclic adenine ribonucleotide formed by tissue particles. J Biol Chem 232:1077–1091

Taylor C (2011) Phosphoinositides and inositol phosphates

Vajanaphanich M, Schultz C, Rudolf MT, Wasserman M, Enyedi P, Craxton A, Shears SB, Tsien RY, Barrett KE, Traynor-Kaplan AE (1994) Long-term uncoupling of chloride secretion from intracellular calcium levels by Ins(3,4,5,6)P_4. Nature 371:711–714

Vallejo M, Jackson T, Lightman S, Hanley MR (1987) Occurrence and extracellular actions of inositol pentakis- and hexakisphosphate in mammalian brain. Nature 330:656–658

Verbsky J, Lavine K, Majerus PW (2005a) Disruption of the mouse inositol 1,3,4,5,6-pentakisphosphate 2-kinase gene, associated lethality, and tissue distribution of 2-kinase expression. Proc Natl Acad Sci U S A 102:8448–8453

Verbsky JW, Chang SC, Wilson MP, Mochizuki Y, Majerus PW (2005b) The pathway for the production of inositol hexakisphosphate in human cells. J Biol Chem 280:1911–1920

Volk AP, Heise CK, Hougen JL, Artman CM, Volk KA, Wessels D, Soll DR, Nauseef WM, Lamb FS, Moreland JG (2008) ClC-3 and IClswell are required for normal neutrophil chemotaxis and shape change. J Biol Chem 283:34315–34326

von Zastrow M, Sorkin A (2007) Signaling on the endocytic pathway. Curr Opin Cell Biol 19:436–445

Walker SA, Kupzig S, Lockyer PJ, Bilu S, Zharhary D, Cullen PJ (2002) Analysing the role of the putative inositol 1,3,4,5-tetrakisphosphate receptor GAP1IP4BP in intracellular Ca^{2+} homeostasis. J Biol Chem 277:48779–48785

Wang XQ, Deriy LV, Foss S, Huang P, Lamb FS, Kaetzel MA, Bindokas V, Marks JD, Nelson DJ (2006) CLC-3 channels modulate excitatory synaptic transmission in hippocampal neurons. Neuron 52:321–333

Wilson MP, Hugge C, Bielinska M, Nicholas P, Majerus PW, Wilson DB (2009) Neural tube defects in mice with reduced levels of inositol 1,3,4-trisphosphate 5/6-kinase. Proc Natl Acad Sci USA 106:9831–9835

Xie W, Kaetzel MA, Bruzik KS, Dedman JR, Shears SB, Nelson DJ (1996) Inositol 3,4,5,6-tetrakisphosphate inhibits the calmodulin-dependent protein kinase II-activated chloride conductance inT84 colonic epithelial cells. J Biol Chem 271:14092–14097

Xie W, Solomons KRH, Freeman S, Kaetzel MA, Bruzik KS, Nelson DJ, Shears SB (1998) Regulation of Ca^{2+}-dependent Cl- conductance in T84 cells: cross-talk between Ins(3,4,5,6)P_4 and protein phosphatases. J Physiol (Lond) 510:661–673

Yang L, Reece JM, Cho J, Bortner CD, Shears SB (2008) The nucleolus exhibits an osmotically regulated gatekeeping activity that controls the spatial dynamics and functions of nucleolin. J Biol Chem 283:11823–11831

York JD (2006) Regulation of nuclear processes by inositol polyphosphates. Biochim Biophys Acta 1761:552–559

York JD, Odom AR, Murphy R, Ives EB, Wente SR (1999) A phospholipase C-dependent inositol polyphosphate kinase pathway required for efficient messenger RNA export. Science 285:96–100

York JD, Guo S, Odom AR, Spiegelberg BD, Stolz LE (2001) An expanded view of inositol signaling. Adv Enzyme Regul 41:57–71

Yu X, Duan KL, Shang CF, Yu HG, Zhou Z (2004) Calcium influx through hyperpolarization-activated cation channels (I(h) channels) contributes to activity-evoked neuronal secretion. Proc Natl Acad Sci U S A 101:1051–1056

Zaccolo M, Di BG, Lissandron V, Mancuso L, Terrin A, Zamparo I (2006) Restricted diffusion of a freely diffusible second messenger: mechanisms underlying compartmentalized cAMP signalling. Biochem Soc Trans 34:495–497

Zhao Z, Li X, Hao J, Winston JH, Weinman SA (2007) The ClC-3 chloride transport protein traffics through the plasma membrane via interaction of an N-terminal dileucine cluster with clathrin. J Biol Chem 282:29022–29031

Zhou D, Chen L-M, Hernandez L, Shears SB, Galán JE (2001) A Salmonella inositol polyphosphatase acts in conjunction with other bacterial effectors to promote host-cell actin cytoskeleton rearrangements and bacterial internalization. Mol Microbiol 39:248–259

Chapter 14
Cell Signalling by Inositol Pyrophosphates

Adolfo Saiardi

Abstract Inositol serves as a module for the generation of a high level of molecular diversity through the combinatorial attachment and removal of phosphate groups. The array of potential inositol-containing molecules is further expanded by the generation of diphospho inositol polyphosphates, commonly referred as inositol pyrophosphates. All eukaryotic cells possess inositol pyrophosphates containing one or more diphospho- moieties. The metabolism of this class of molecules is highly dynamic, and the enzymes responsible for their metabolism are evolutionary conserved. This new, exciting class of molecules are uniquely chracterized by a high energetic diphospho- bound that is able to participate in phosphotrasfer reactions thereby generating pyrophosphorylation of protein. However, allosteric mechanisms of action have been also proposed. In the past decade several disparate nuclear and cytoplasmic functions have been attributed to inositol pyrophosphates, ranging from intracellular trafficking to telomere length control and from regulating apoptotic process to stimulating insulin secretion. The extraordinary range of cellular function controlled by inositol pyrophosphate underline their great importance.

Keywords Kinase · Phosphatase · IP_6 · IP_7 · IP_8 · Kcs1 · Vip1 · IP6K · PP-IP5K · DDP1 · DNA recombination · Stress responses · Telomere · Ribosome · Trafficking · Insulin · Phosphate homeostasis · Cell death · Development · Chemotaxis

14.1 Introduction

When inositol is phosphorylated, it contributes to the regulation of a surprising high number of physiological functions, either as cytosolic inositol polyphosphates (Irvine and Schell 2001; York 2006) or as lipid bound phosphoinositides (Di Paolo and De Camilli 2006; Gonzales and Anderson 2006). Inositol as a signalling module can be

A. Saiardi (✉)
MRC-LMCB, Department of Cell and Developmental Biology,
University College London, Gower Street, WC1E 6BT, London, UK
e-mail: dmcbado@ucl.ac.uk

T. Balla et al. (eds.), *Phosphoinositides II: The Diverse Biological Functions*,
Subcellular Biochemistry 59, DOI 10.1007/978-94-007-3015-1_14,

modified, not only by adding phosphates to each carbon group, but also by adding a β-phosphates to a pre-phosphorylated position of the six carbon inositol ring generating diphospho-inositol polyphosphates also called inositol pyrophosphate (Bennett et al. 2006). Inositol polyphosphates are usually analysed using strong anion exchange chromatography (SAX-HPLC), where salt gradients can elute the different poly-phosphorylated forms of inositol in function of their polarity/charge: in other words, the inositol polyphosphates elute in order of the number of phosphate groups attached to the inositol ring (Azevedo and Saiardi 2006). Although, an earlier literature report (Oliver et al. 1992) indicated the existence of inositol molecules more polar than inositol hexakisphosphate (IP_6; phytic acid), the discovery of inositol pyrophosphates is attributed to two independent laboratories that both published in 1993. One group indentified inositol pyrophosphate in the amoeba *Dictyostelium discoideum*, which contain very high levels of this class of molecule (Stephens et al. 1993). Contemporaneously, a second group identified inositol pyrophosphates in a mammalian pancreatoma cell line by using NaF as a clever metabolic trap to elevate inositol pyrophosphate levels (Menniti et al. 1993). Subsequent studies confirmed the widespread evolutionary conservation of the pyrophosphate-containing inositols. The slime mould *D. discoideum* represents the extreme of higher inositol polyphosphate metabolism with the pyrophosphate species reaching sub-molar concentrations (Laussmann et al. 2000). Inositol pyrophosphate complex metabolism is present in this amoeba, whereas the inositol trisphosphate $I(1,4,5)P_3$ (IP_3) role in controlling calcium release (Irvine 2003) from intracellular stores is apparently absent (Traynor et al. 2000). Therefore inositol pyrophosphate signalling evolutionarily precedes the IP_3 control of Ca^{++} release (Bennett et al. 2006) and likely also precedes many of the well-characterised lipid phosphoinositides signalling pathways. In eukaryotes, the enzymes responsible for the synthesis of inositol pyrophosphates have been identified in all organisms so far studied, even in the early-diverged eukaryote, the metamonada *Giardia lamblia* (Letcher et al. 2008). The ubiquitous presence of the synthetic enzymes indicates that as early as prokaryotic–eukaryotic specialization, this class of molecule was playing some important cellular function.

In yeast and mammalian cell, the inositol pyrophosphate species are present in sub-micro-molar concentrations, representing 2–8% of total IP_6 levels. Remarkably, however, in mammalian cells up to 50% of IP_6 cycles through the more phosphorylated inositol pyrophosphate every hour (Glennon and Shears 1993; Menniti et al. 1993), suggesting that a considerable investment of cellular energy is committed to keeping the inositol pyrophosphates level constant. The widespread distribution and their high turnover lead to the suggestion that inositol pyrophosphates are performing very basic and primitive functions (Bennett et al. 2006).

It has been suggested that inositol pyrophosphates work with interacting or binding partners (Lee et al. 2008; Luo et al. 2003), in similar fashion to cytosolic inositol polyphosphates such IP_3 (Mikoshiba et al. 1993) or lipid phosphoinositides (Lemmon 2008). Considering the growing number of inositol pyrophosphates isomers now identified, it is unlikely that there are specific binding partners for all of them, and it is more likely that some of them are metabolic intermediates, as proposed for other inositol polyphosphates (Shears 2004; Saiardi and Cockcroft 2008). The main, differentiating characteristic of inositol pyrophosphates is the "high energy" charge of the pyrophosphates moiety. This "high energy" peculiarity has lead many groups to suggest an energetic role similar to ATP (Stephens et al. 1993; Voglmaier et al. 1996). However, the sub-micromolar concentration of this class of molecule is unlikely to compete with or replace ATP. Nevertheless, the utilization of the pyrophosphates energy might occur in a different fashion, and the proposed phosphotransfer mechanism from inositol pyrophosphates to protein (Bhandari et al. 2007; Saiardi et al. 2004) or other cellular acceptor should to be seriously considered and studied more in detail.

In this review I will summarise our knowledge to date of this exciting research field, briefly describing their structure and the enzymes involved in the metabolism of inositol pyrophosphates, as well as and the multifaceted cellular functions attributed to these molecules. Although this synopsis will primarily focuses on inositol pyrophosphates, I also invite the interested reader to refer to the specific chapter on IP_3 and calcium signalling, inositol polyphosphates and their metabolising enzymes by C. Taylor (Chap. 1), S.B. Shears (Chap. 13) and to read the following inositol pyrophosphate reviews (Azevedo et al. 2010; Burton et al. 2009; Shears 2009; Barker et al. 2009).

14.2 Nomenclature

Inositol pyrophosphate is the common name used to describe an inositol molecule containing a diphosphate group: esters of diphosphoric acid ($H_4P_2O_7$), also known as pyrophosphoric acid, are call pyrophosphates. Although the biosynthesis of inositol pyrophosphate does not directly involve diphosphoric acid (instead it involves the transfer of the γ phosphate of ATP to a pre-existing phosphate on the inositol ring), the generation of a "pyrophosphate" moiety determines the common name. The chemically correct name suggested by the IUPAC (International Union of Pure and Applied Chemistry) commission on biochemical nomenclature to describe this class of molecule is diphosphoinositol polyphosphates (IUPAC-IUB 1977), as indicated in Table 14.1. The diphosphoinositol polyphosphates name was used originally by the Stephens group (Stephens et al. 1993). In contrast, the Shears group used the name "inositol polyphosphates pyrophosphate" to describe this class of molecule (Menniti et al. 1993) and this has since been abbreviated to inositol pyrophosphates; this simpler characterizing name has been favoured and widely used in the current literature. Because the current synopsis is aimed at a non-specialized audience, we will use inositol pyrophosphate throughout this chapter (see Table 14.1). Further clarification is needed for the abbreviations used to define specific inositol pyrophosphate molecules. For simplicity, I will use "I" as the abbreviation for inositol, instead of the "Ins" as indicated by IUPAC commission on chemical nomeclature; thus we will abbreviate the fully phosphorylated ring of inositol hexakisphosphate (Phytic acid) as IP_6.

Inositol pyrophosphates are a growing and complex family of molecules and in cells may coexist different species containing the same number of phosphates but

Table 14.1 Nomenclature

Name used	Other name	IUPAC name	Mammalian isomer	Full name
IP_6	$InsP_6$			Inositol hexakisphosphate, Phytic acid
IP_7	$InsP_7$	$PP\text{-}InsP_5$	$5PP\text{-}IP_5$	Diphosphoinositol pentakisphosphate
IP_8	$InsP_8$	$(PP)_2\text{-}InsP_4$	$1/3,5PP\text{-}IP_5$	Bis-diphosphoinositol tetrakisphosphate
$PPP\text{-}IP_5$				Triphosphoinositol pentakisphosphate
IP_5		$Ins(1,3,4,5,6)P_5$		Inositol pentakisphosphate
$PP\text{-}IP_4$		$PP\text{-}InsP_4$	$1/3PP\text{-}IP_4$, $5PP\text{-}IP_4$	Diphosphoinositol tetrakisphosphate
$(PP)_2\text{-}IP_3$		$(PP)_2\text{-}InsP_3$		Bis-diphosphoinositol trisphosphate
$PP\text{-}IP_3$				Diphosphoinositol trisphosphate
$(PP)_2\text{-}IP_2$				Bis-diphosphoinositol bikisphosphate

The left column list the inositol pyrophosphate nomenclature used in this chapter. To facilitate the reader I simplified the suffix for inositol from 'Ins' to 'I'. I refer to the most common of the six possible IP_5 isomers as reported in the table. Although, diphospho- represents the scientific correct prefix to indicate a pyrophosphate moiety; the suffix 'pyro' better translated the 'high energy' characteristic of these molecules and the majority of authors have used pyrophosphate to indicate this class of molecules

having diverse molecular organizations. Furthermore, two or more isomeric forms of the same inositol pyrophosphate can exist, creating even further complexity (see below paragraph on the molecular structures). Because it has become common to refer to the inositol polyphosphates as IPx were the 'x' represent the number of phosphates attached to the inositol ring, this rule will be employed to refer to the inositol pyrophosphates generated from the fully phosphorylate ring of IP_6. Thus the terminology IP_7 will be employed to refer to diphosphoinositol pentakisphosphate, which has one pyrophosphate moiety attached to the fully phosphorylate inositol ring (Fig. 14.1). IP_8 will be used to refer to bis-diphosphoinositol tetrakisphosphate molecule possessing two pyrophosphate moieties (Fig. 14.1) and so on. However, when required specific isomeric forms of IP_7 and IP_8 will be indicated by the position labelling of the pyrophosphate moiety; thus $5PP\text{-}IP_5$ will be used to indicate the isomer five of IP_7 and $5,6(PP)_2\text{-}IP_4$ to indicate the IP_8 isomer possessing the two pyrophosphate moieties attached to the carbon 5 and 6.

Conversely, to refer to the inositol pyrophosphates generated from inositol pentakisphosphate (IP_5), the pyrophosphates moiety will be specified; thus $PP\text{-}IP_4$ will be used to refer to the diphosphoinositol tetrakisphosphate molecule possessing six phosphate groups of which two are in a pyrophosphate moiety (Fig. 14.1). Table 14.1 summarizes the common terminology used in this chapter reporting also the correct IUPAC nomenclature for this class of molecules.

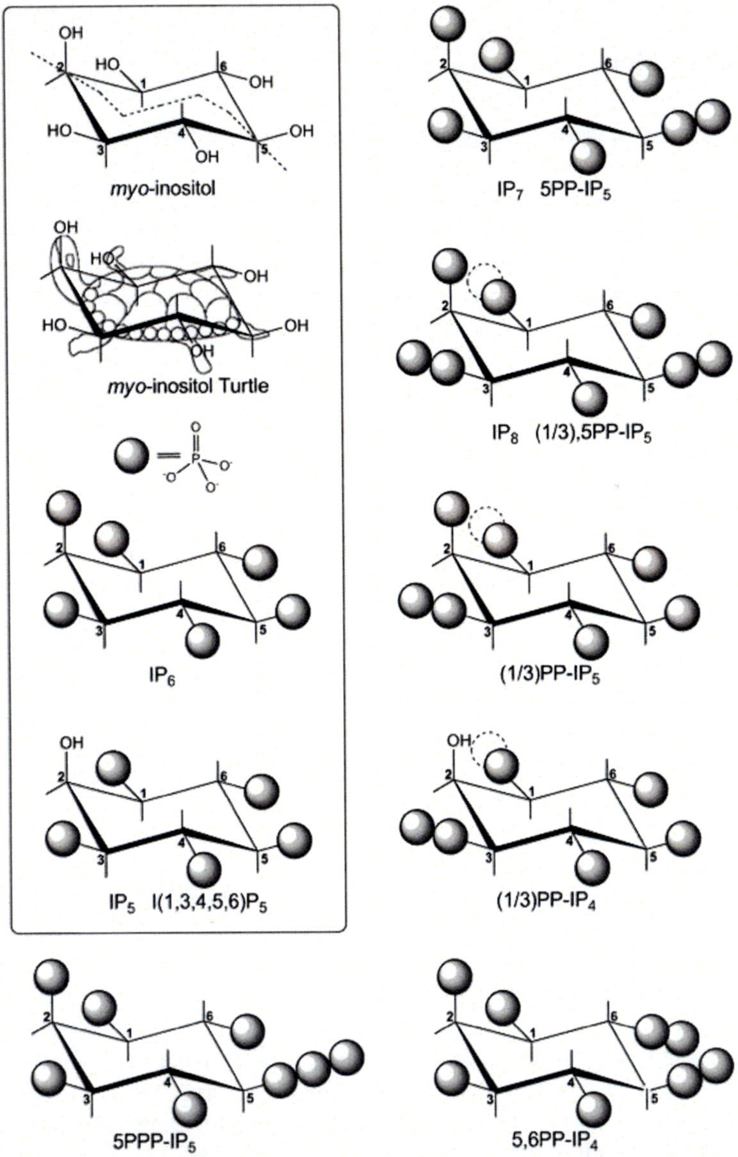

Fig. 14.1 Structures of Inositol Pyrophosphates. In the *box* are depicted the structured of *myo*-inositol and the Agranof's turtle that help to recognise the hydroxyl on position 2 (*turtle head*) perpendicular to the ring plane. The ax of symmetry on the inositol ring is represented by the *dashed line*, this symmetry make position 1/3 and 4/5 enantiomeric. In the box are also depicted the structure of inositol hexakisphosphates IP_6 and of the more common inositol pentakisphosphate isomer present in eukaryotic cells $I(1,3,4,5,6)P_5$. These high phosphorylated inositol polyphosphate are the main precursor of the inositol pyrophosphates. In mammalian cell IP_6 is the precursor of the IP_7 isomer 5PP-IP_4 (Albert et al. 1997; Draskovic et al. 2008) and of the IP_8 isomer (1/3),5$(PP)_2$-IP_4 (Lin et al. 2009), whereas IP_5 is the precursor of (1/3)PP-IP_4 (Draskovic et al. 2008), *dashed circle* indicate the alternative enantiomeric position. The structure of IP_7 isomer generated by Vip1/PP-IP5K is shown (1/3)PP-IP_5 (Lin et al. 2009), *dashed circle* indicate the alternative enantiomeric position. The structure of the IP_8 isomer present in *D. discoideum* 5,6$(PP)_2$-IP4 is shown at *bottom right* (Laussmann et al. 1998). The IP6Ks enzymes are also able to synthesize a tri-phosphate form 5PPP-IP_5 of 'IP_8' depicted at *bottom left* (Draskovic et al. 2008)

14.3 Enzymology

Two distinct classes of evolutionarily conserved enzyme are able to synthesize inositol pyrophosphates: the inositol hexakisphosphates kinases (IP6Ks) and the enzymes related to the yeast Vip1 protein. The scope of this paragraph is to briefly summarise the main characteristic of these kinases and of the diphosphoinositol-polyphosphate phosphohydrolases DDP1, an inositol pyrophosphates phosphatase. I invite the interested reader to read the following recent review for more information on these classes of enzymes (Azevedo et al. 2010; Barker et al. 2009). The IP6Ks, along with inositol polyphosphate multikinase (IPMK) and IP_3 3-kinases (IP_3-3K), define a single highly related gene family (pfam IPK, number PF03770) (Bennett et al. 2006). IP6Ks possess extraordinary catalytic flexibility: utilizing ATP as a phosphate donor, this class of enzymes is able to phosphorylate IP_6 to IP_7 and inositol pentakisphosphate (I(1,3,4,5,6)P_5) to PP-IP_4 (Saiardi et al. 1999, 2000a). Furthermore, this class of enzyme is able to use these products as substrates to generate more complex molecules containing two or more additional pyrophosphate moieties or even three-phosphate species (Draskovic et al. 2008; Saiardi et al. 2000a, 2001b). Mammalian species possess three IP6K isoforms referred to as IP6K1, 2, 3. However, there is a single IP6K gene called Kcs1 in the yeast *Saccharomyces cerevisiae*, which besides synthesizing IP7 was shown to be able to generate inositol pyrophosphate using either IP_3 or IP_4 as substrate (Seeds et al. 2005). It is worth noting that, at least in vitro, mammalian and yeast IPMK have also been shown to generate inositol pyrophosphate using IP_5 as substrate (Saiardi et al. 2001a; Zhang et al. 2001). A second class of enzyme that generates inositol pyrophosphates was identified in yeast, namely the protein Vip1 (Mulugu et al. 2007). The Vip1 protein possesses a dual-domain structure with an N-terminus adenosine triphosphate (ATP)-grasp domain where the kinase activity resides, and a C-terminus acidic-phosphatase domain (Mulugu et al. 2007) of not yet clarified function. This second class of enzyme is less promiscuous, in fact Vip1 is only able to generate inositol pyrophosphate from IP_6 and not from IP_5. However, they can metabolize the inositol pyrophosphates that are produced by the IP6Ks when IP_5 is used as substrate (Losito et al. 2009). Initially Vip1 type of enzyme were described as specific IP_6-kinases, additional studies has shown that Vip1 is not only able to convert IP_6 to IP_7 but also able to subsequently phosphorylate IP_7 to IP_8 (Choi et al. 2007; Fridy et al. 2007). In addition, kinetic studies carried out on the mammalian Vip1a isoform indicate that this protein uses IP_7 (5PP-IP_5 isomer generated by IP6Ks) as its primary substrate; therefore, the mammalian enzymes are also called PP-IP_5-kinases (PP-IP5K) in order to describe their physiological activity as confirmed *in vivo* by the analysis of *vip1*Δ mutant that accumulate the un-metabolized substrate IP_7 (Padmanabhan et al. 2009; Onnebo and Saiardi 2009). Little is known about the Vip1/PP-IP5K acid phosphatise domain. However, recent evidence indicates that the phosphatase domain of *Schizosaccharomyces pombe* Asp1 protein (Vip1 homologous) negatively regulates the N-terminal kinase domain in controlling the dimorphic switch (Pohlmann and Fleig 2010). Mammalian organisms possess two PP-IP_5Ka, b proteins, while yeast possess only one Vip1 enzyme.

Inositol pyrophosphates are hydrolyzed by the diphosphoinositol-polyphosphate phosphohydrolases (DIPPs) (Safrany et al. 1998). These phosphatases are also promiscuous enzymes that are able to degrade inositol pyrophosphate and nucleotide analogues such diadenosine hexaphosphate (Ap6A) (Caffrey et al. 2000; Fisher et al. 2002). DIPPs belong to the Nudix hydrolase family possessing the MutT protein motif (Bessman et al. 1996). This class of enzyme is well characterised in bacteria, which do not posses inositol pyrophosphates, and thus the main function is to protect prokaryotic cells from DNA damage degrading a wide range of nucleotide analogues. Four mammalian genes have been characterised encoding very active enzymes of about 20 kDa in size (Caffrey et al. 2000; Hidaka et al. 2002; Leslie et al. 2002; Safrany et al. 1998). In eukaryotes, DIPP's dual substrate characteristic might link inositol polyphosphate to nucleotide metabolism. The deletion strains of the single DIPP yeast protein, call Ddp1 in *S. cerevisiae* and Aps1 in *S. pombe*, possesses an increase level of inositol pyrophosphates (Ingram et al. 2003), but unchanged level of Ap6A, suggesting that inositol pyrophosphates are the primary substrate, at least in yeast. Interestingly, the *kcs1*Δ*ddp1*Δ double mutants possess unexpectedly large amounts of IP_7 (York et al. 2005) and this allowed the identification of Vip1 as a novel IP_6-kinases (Mulugu et al. 2007). This unexpected *kcs1*Δ*ddp1*Δ feature can be explained by the fact that DIPPs prefer to hydrolyse the diphosphate added by PP-IP_5Ks (Shears et al. 1995; Shears 2009), thus masking the rate of synthesis of one of the two different IP_7 isomers leading to IP_8 production. Further work is needed to fully appreciate the metabolic flux and the exact physiological role played by the DIPPs phoshohydrolase.

14.4 Structure

The six carbons inositol ring might acquire several different configurations. However in eukaryotic cells it is almost exclusively present the *myo*-inositol arrangement (Fig. 14.1). The *myo*-inositol structure is often depicted by the Agranof's turtle (Agranoff et al. 1958) possessing the position 2 hydroxyl group perpendicular to the carbons plane represented by the head of the turtle. In *myo*-inositol the other five hydroxyl groups are parallel to the ring plane and are represented by the arms, legs and tail of the turtle (Agranoff et al. 1958). The *myo*-inositol has an axis of symmetry through the 2- and the 5-position which makes the 1/3 and 4/6 position enantiomers, therefore indistinguishable using biophysical techniques.

Nuclear Magnetic Resonance (NMR) is the primary analytical method to determine the atomic organization in molecular structures. Unfortunately, NMR studies require large quantities of material, which are difficult to obtain from most cells. Consequently, structure determination of extracted inositol pyrophosphates has been possible only in amoeba such *D. discoideum* that possess large quantities (up to 0.1 mM) of inositol pyrophosphates that are generated in a developmentally regulated fashion (Laussmann et al. 2000). NMR spectroscopy analysis of inositol pyrophosphates in *D. discoideum* revealed a single IP_8 isomer possessing diphospho- moieties on carbon 5 and 6 5,6$(PP)_2$-IP_4 (Fig. 14.1) and two IP_7 isomers, namely 5PP-IP_5 and 6PP-IP_5 (Albert et al. 1997; Laussmann et al. 1997, 1998). In a second dictyosteloid amoeba, *Polysphondylium pallidum* beside possessing the inositol pyrophosphate discovered in *D. discoideum* it also posses (1/3),5$(PP)_2$-IP_4 isomer of IP_8 (Laussmann et al. 1998, 2000). Inositol pyrophosphate derived from *neo*-inositol have also been described in the *Entamoeba histolytica* that possess 2PP-IP_5 isomer of *neo*-IP_7 and 2,5(PP)2-IP_4 isomer of *neo*-IP_8 (Martin et al. 2000).

The micromolar to sub-micromolar concentration of inositol pyrophosphates in yeast and mammalian cells have not allowed direct NMR studies of purified inositol pyrophosphates from these organisms. However, the cloning of the kinases responsible for inositol pyrophosphate synthesis has allowed generating sufficient quantities to determine their structures. The IP6Ks enzymes, when IP_6 is used as substrate, generated the pyrophosphorylation of the 5-position sinthesizing the isomer 5PP-IP_5 of IP_7 (Draskovic et al. 2008). Interestingly this class of enzymes is also able to produce a triphosphate derivative 5PPP-IP_5 (Fig. 14.1) (Draskovic et al. 2008). When IP_5 is used as substrate, the IP6Ks prefer to pyrophosphorylate the 1/3 position and the 5 position, generating 1/3PP-IP_4 and 5PP-IP_4 (Draskovic et al. 2008) (1 and 3 are enantiomeric carbons). The yeast Vip1 and mammalian PP-IP5K enzymes are 1/3-kinases and so are responsible for the synthesis of the (1/3)PP-IP_5 isomer of IP_7 (Lin et al. 2009). Consequently, the 1 or 3 positions of the inositol ring can be pyrophosphorylated by IP6Ks when IP_5 is used as substrate and by PP-IP5K when IP_6 is used as substrate. Therefore, at least in the mammalian cells that possess similar levels of IP_5 and IP_6, there is the possibility of a higher degree of inositol pyrophosphate promiscuity with the presence of molecules of similar structure (1/3)PP-IP_5 and (1/3)PP-IP_4 generated by different enzymes. Chromatographic co-migration with standard has been used successfully to determine that the (1/3),5$(PP)_2$-IP_4 isomer of IP_8 is present in mammalian cells (Lin et al. 2009). However, the not radioactive metal-dye detection system (MDD-HPLC) (Mayr 1988) used in this study is not sensitive to fully appreciate inositol pyrophosphate complex metabolism (Shears 2009). The recent developed method to analyze inositol polyphosphates using polyacrylamide gel electrophoresis (PAGE) (Losito et al. 2009) permitted to easy identify previously uncharacterized inositol pyrophosphate. This novel technology as allowed a more accurate analysis of IP6Ks and PP-IP5K enzymatic reaction allowing, the identification of inositol pyrophosphates possessing nine or more phosphates groups, demonstrating that the enzymes responsible for inositol pyrophosphate synthesis are more versatile than previously described (Draskovic et al. 2008; Losito et al. 2009). Furthermore, the comparison of HPLC methodology with the new PAGE tecnology shows that the acidic conditions used in traditional chromatographic analyses lead to substantial degradation, and thus to a serious under-representation, of inositol pyrophosphates cellular species (Losito et al. 2009). Therefore, it is well possible that more inositol pyrophosphates molecular species will be discovered *in vivo* in future.

14.5 Regulation of Inositol Pyrophosphates Levels

An essential feature of well-established signalling molecules is their ability to respond rapidly and dynamically to a specific stimulus. In the seventeen years from their discovery, several circumstances that result in changed levels of inosiol pyrophosphate have been identified. These events can't be classified as classical extracellular stimuli in the fashion of receptor stimulation, but the 'circumstances' controlling inositol pyrophosphate intracellular concentrations might gives us important insight into the cellular roles of these molecules.

As previously mentioned the identification of inositol pyrophosphates in mammalian cells was made possible through the use of fluoride treatment that dramatically modulates the level of this class of molecules (Glennon and Shears 1993; Menniti et al. 1993). These experiments also revealed that up to 50% of the IP_6 pool is converted every hour to IP_7 (Glennon and Shears 1993; Menniti et al. 1993). Because IP_7 levele are only represent 2–8% of IP_6 concentration, IP_7 itself is turning over many times each hour. In primary hepatocytses it was calculated that the pool of IP_7 turns over ten times in a 40 min time period. Although these fluctuation of IP_7/IP_8 intracellular concentration introduced the inositol pyrophosphates as novel signalling molecules with 'molecular switch' activity, their physiological significance remains unknown. Fluoride inhibits the activity of the DIPPs phosphohydrolase that specifically cleaves the β-phosphate of the pyrophosphate moiety of IP_7 and IP_8 (Safrany et al. 1998). However, fluoride is a commonly used phosphatases inhibitor and so has many effects on cell signalling (Bollen and Stalmans 1988); consequently, the fluoride effect on inositol pyrophosphate metabolism is unclear. Nevertheless these data indicate that in mammalian cells inositol pyrophosphates are not static molecules: they turnover very rapidly and therefore the cells invest a considerable amount of energy to keep their steady state levels constant.

The most dramatic modulation of the levels of inositol pyrophosphates was observed in *D. discoideum*. Although the levels of inositol pyrophosphates show similar ratios to the precursor IP_6 levels in the vegetative state, compared to mammalian cells, the concentrations of both IP_7 and IP_8 increase considerably during starvation-induced aggregation (Laussmann et al. 2000). When starvation is sensed in *D. discoideum* cyclic-AMP (cAMP) is the chemotactic molecule secreted in order to induce aggregation (Van Haastert 1995). It was demonstrated in a later study that cAMP alone is sufficient to induce a rapid and considerable elevation in the levels of IP_7 and IP_8 when added to the extracellular space (Luo et al. 2003). Interestingly, it has also been observed that inositol pyrophosphate concentrations are regulated by changes in cAMP levels in mammalian cells although with opposite result. In the DTT_1MF-2 smooth muscle cell line, raising cAMP levels, either through the use of phosphodiesterase inhibitors or via the activation of β_2-adrenergic receptors, was able to reduce IP_8 levels, providing the first and only evidence between receptor activation, a second messenger, and inositol *pyro*phosphates turnover (Safrany and Shears 1998). Unfortunately, this early observation has not been studied further. It was soon realised, however, that there is an apparent independence of inositol *pyro*phosphates turnover from the most well characterised response of the inositol polyphosphates, the activation of phospholipase C (PLC) (Safrany and Shears 1998). These early studies are complemented by the more recent over expression analysis of a constitutively active Galpha(q)QL that hyper-stimulated PLC activity but does not induce an direct increase in inositol pyrophosphate levels (Otto et al. 2007).

A interesting correlation between inositol pyrophosphates levels and the cell cycle has been observed in mammalian cells, where two-fold higher levels were observed in G1 compared to S phase (Barker et al. 2004). This important observation need to be further expanded to understand whether inositol pyrophosphate dynamics during the cell cycle are a response to the metabolic requirement of particular phase of the cell cycle, or whether inositol pyrophosphate modulation represents a specific signal controlling cell cycle progression.

Several studies suggest that 'environmental stressors' such as heat shock or osmotic stress affects inositol pyrophosphate levels in mammalian cells (Choi et al. 2005; Pesesse et al. 2004). After induction of osmotic challenge by sorbitol there was a rapid accumulation of IP_8. Similarly (Pesesse et al. 2004), it was demonstrated that thermal stress in mammalian cells causes an increase in IP_8 (Choi et al. 2005). Although the MAP kinase pathway was initially suggested to be involved through the use of pharmacological inhibitors, it has been found that it was the off-target effect of the inhibitors on cellular energetic status that was responsible for the observed effects on IP_8 levels (Choi et al. 2008). The induction of bioenergetics stress decreases the rate of IP_8 synthesis in order to preserve energy and this is dominant over the increased level normally observed upon induction of osmotic stress (Choi et al. 2008). It is noteworthy that mitochondrial energy dynamics affect inositol pyrophosphate production; an early report identified a inositol polyphosphates peak eluting on SAX-HPLC after IP_6, thus an inositol pyrophosphate, to rapidly decrease after Antimycin A treatment (Oliver et al. 1992). This drug binds to mitochondria cytochrome C reductase blocking ATP production (Slater 1973). Likely it is the relatively high Km for ATP (1.4 mM) of the IP6Ks enzymes that makes the synthesis of inositol pyrophosphate sensitive to fluctuation of cellular ATP (Saiardi et al. 1999; Voglmaier et al. 1996).

It is quite surprising that in yeast, where challenges such as heat and osmotic stress are a constant environmental pressure, the levels of IP_8 do not respond to either stress (Pesesse et al. 2004). In contrast, yeast inositol pyrophosphates are modulated by oxidative stress as their levels decrease rapidly upon H_2O_2 treatment (Onnebo and Saiardi 2009). The effect of H_2O_2 was shown to be directly on the enzyme responsible for their synthesis, namely, the IP6Ks, as concentrations of H_2O_2 as low as 0.1 mM dramatically reduced their enzymatic activities *in vitro* (Onnebo and Saiardi 2009). The IP6Ks might work as sensors of the cellular redox status of the cells, since mutations in the conserved cysteine residues of mouse IP6K1 changed the enzymatic properties (Onnebo and Saiardi 2009). Another example of the involvement of the cellular redox status in regulating the levels of inositol pyrophosphates was established with the cloning and characterization of DIPP from *Drosophila melanogaster* (Winward et al. 2010). It has been suggested that the half-life of this enzyme and consequently the levels of IP_7 might be regulated by a redox-sensitive manner (Winward et al. 2010).

Also of particular relevance to yeast is the observation that intracellular inositol pyrophosphates levels may be controlled by levels of phosphate in the extracellular environment. It was recently reported that a decrease in extracellular inorganic phosphate concentration led to a dramatic rise in the intracellular level of inositol pyrophosphates in particular IP_7 (Mulugu et al. 2007). This result is counter-intuitive because a reduction of phosphate in the extracellular medium leads to a rapid reduction in the level of cellular ATP (Boer et al. 2010; Martinez et al. 1998), which is required for inositol pyrophosphates biosynthesis. More work is needed to clarify this matter because mounting evidence suggests a positive linear correlation (not the inverse) between inositol pyrophosphate and ATP cellular levels (see above) (Choi et al. 2008; Oliver et al. 1992).

14.6 Mechanisms of Action

Biological systems use two basic mechanisms to transduce signals events. Firstly, signals can be transmitted by the covalent modifications, such as phosphorylation, acetylation or ubiquination of proteins or by methylation of DNA. The second predominant mechanism used in signal transduction is through the binding of usually small molecules to specific protein targets, such as cAMP to protein kinase A or lipid hormone to specific receptors. The well-documented mechanism of action for signal transduction by the inositol polyphosphates family is via binding to a particular receptor, such as the classic binding of IP_3 to the IP_3-receptor (Irvine 2003; Mikoshiba et al. 1993). Binding partners for many other inositol phosphates have also been identified, especially for the lipid phosphatidylinositols such as the binding to pleckstrin homology (PH), phagocyte oxidase homology (PX) or FYVE (for Fab1, YOTB, Vac1 and EEA1) protein domains (Balla 2005; Lemmon 2008). It seems logical, therefore that inositol pyrophosphates may also signal through allosteric interactions with proteins, and indeed binding partners of IP_7 have been identified. However the peculiar presence of a highly energetic pyrophosphate bond has also suggested (Stephens et al. 1993) the possibility of covalently modifying proteins as a mechanism of action for this class of molecule.

14.6.1 Allosteric Interaction

IP_7 binds with high affinity to several proteins *in vitro*, including proteins that are important for controlling vesicular trafficking, such as AP180 (Ye et al. 1995), the Golgi coatomer (Fleischer et al. 1994) and the clathrin-assembly adaptors AP2 (Shears et al. 1995). However, these early *in vitro* binding experiments were performed in the absence of bivalent cations, which are physiologically important in coordinating the negative charge of the phosphorylated inositol ring (Torres et al. 2005). Furthermore, all the proteins that bind to IP_7 also bind to its precursor IP_6, albeit to a modest extent. The substantial lower cellular levels of inositol pyrophosphates species would necessitate a strict specificity of binding for IP_7 over IP_6, before any physiological relevance could be attributed to such binding. In addition, a dramatic difference in binding affinity for IP_7 over IP_6 may not be possible due to their relatively similar structures (Fig. 14.1), although it is conceptually more likely if more complex molecules such IP_8 were the functional binding partner. In *D. discoideum*, both IP_7 and IP_8 levels increase dramatically during aggregation (Laussmann et al. 2000) and a more modest selective binding affinity would be required to give the inositol pyrophosphates specificity for a binding partner during the aggregation stage of the life cycle. Indeed, the relevance of binding as a mechanism of action has been demonstrated for IP_7 competition with PI(3,4,5)P_3 binding to the PH-domain containing CRAC protein (Cytosolic Regulator of Adenylyl Cyclase) (Luo et al. 2001). The *in vivo* relevance was supported by the demonstration that the *D. discoideum* IP6K deletion mutant possessed an increased sensitivity to cAMP-induced aggregation (Luo et al. 2001) function controlled by the CRAC protein (see below). Although IP_7 was found to bind to various other PH-domain containing proteins, including mammalian Akt *in vitro* (Luo et al. 2001), the significance of this binding is less clear in mammalian cells where IP_6 is by far more abundant that IP_7. Furthermore, others were unable to reproduce the binding of IP_7 to Akt/Pkb (Downes et al. 2005) and was also demonstrated that IP_7 does not bind to the PH-domain of phosphoinositide-dependent protein kinase 1 (PDK1) (Komander et al. 2004).

Allosteric interaction has been also proposed as mechanism for controlling the Pho85/Pho80/Pho81 cyclin dependent kinase/cyclin/cyclin-dependent kinase inhibitor complex of the yeast *S. cerevisiae* (Lee et al. 2007, 2008). The binding of the IP_7 isomer (1/3)PP-IP_5 (generated by Vip1) to the Pho85/Pho80/Pho81 complex inhibit its action; as consequence the transcription factor Pho4 is no longer hyperphosphorylated by Pho85 and could enter the nucleus to activate several gene of the PHO pathway (Lee et al. 2007). Specifically, (1/3)PP-IP_5 enhances the inhibitory effect of the cyclin dependent kinase inhibitor Pho81, but is not clear if it is by interaction with both Pho81 and Pho85/Pho80 or by stabilising the trimeric complex (Lee et al. 2008). The inhibitory effect of (1/3)PP-IP_5 seems quite specific with an IC_{50} of 55 μM and different IP_7 isomers such as 5PP-IP_5 or IP_6 are far less efficient in inhibiting Pho85/Pho80/Pho81 (Lee et al. 2007). The levels of the (1/3)PP-IP_5 isomer of IP_7 increases during phosphate starvation (see chapter on modulation of inositol pyrophosphate level); however, as wisely observed by Dr. Shears, the elevation of IP_7 observed is unlikely to account for the 55 μM level necessary to inhibit the Pho85/Pho80/Pho81 complex (Shears 2009). Further studies are require to establish if the inhibitory regulation of Pho85/Pho80 by Pho81 enhanced by inositol pyrophosphates is a generic mechanism that can be expanded to mammalian cyclin-depended kinase/cyclin complexes.

14.6.2 Phosphotransfer

The presence of a high-energy pyrophosphate moiety has led to the suggestion that inositol pyrophosphates participate in a phosphotrasfer reaction (Stephens et al. 1993).

It was later demonstrated that the β-phosphate of the *pyro*phosphate group is donated to proteins in a reaction that do not require a protein kinase (Saiardi et al. 2004); IP_7, as well as other inositol pyrophosphates, specifically recognize and transfer the β-phosohate to target proteins in the presence of bivalent cations. This is an intrinsic ability of inositol pyrophosphate as demonstrated by computational studies that attribute a high phosphorylation potential to IP_7 due to the sterically- and electron-packed environment of the pyrophosphate moiety (Hand and Honek 2007). IP_7-mediated protein phosphorylation is a eukaryotic-specific process because, despite being kinase-independent, it was demonstrated that a priming event through a canonical kinase-ATP phosphorylation was required (Bhandari et al. 2007). The protein phosphorylation mediated by IP_7 is actually a pyrophosphorylation event as the phosphate donated by IP_7 is added to an existing phosphorylated serine residue (Bhandari et al. 2007). All the substrates of IP_7-mediated protein pyrophosphorylation that have been identified so far contain stretches of serine residues surrounded by acidic amino acids. The acidic residues may play a role in coordinating magnesium ions that are absolutely required for the reaction (Bhandari et al. 2007; Saiardi et al. 2004). Although these two studies were essential in the identification and *in vitro* characterization of this new mechanism of post-translation modification, they did not present direct evidence that this mechanism operates *in vivo* nor did they illuminate its function. More recently however, further work gave more insight into the functional role of protein pyrophosphorylation. The β-subunit of the AP-3 multi-protein complex (AP3B1) which is involved in regulating intracellular trafficking (Dell'Angelica 2009), was identified as target of IP_7 pyrophosphorylation: this modification of AP3B1 *in vivo* affected its mobility when subsequently analyzed by gel electrophoresis (Azevedo et al. 2009). Moreover, the interaction between AP3B1 and a newly identified binding partner, Kif3A a motor protein of the kinesin superfamily (Hirokawa et al. 2009), was shown to be negatively modulated by the pyrophosphorylation status of AP3B1 (Azevedo et al. 2009). The AP-3 complex (Dong et al. 2005) and Kif3A (Azevedo et al. 2009) participate in the intracellular trafficking of GAG, a major structural protein of the HIV-1 virus. Evidence of a function of IP_7-mediated pyrophosphorylation *in vivo* arose when it was demonstrated that an increase in the intracellular levels of IP_7, which decreases the interaction between AP3B1 and Kif3A, diminishes the release of HIV-1 virus like particles (Azevedo et al. 2009). Although similar in some aspects, protein phosphorylation and protein pyrophosphorylation have some distinct biochemical characteristics. For example, there is evidence that the pyrophosphorylated proteins may have unique properties: IP_7 pyrophosphorylated peptides are more acid labile and more resistant to phosphatases than are peptides phosphorylated by ATP-kinases (Bhandari et al. 2007). If this also applies to cellular proteins, pyrophosphorylation would allow this modification protein to be protected from general phosphatase (Burton et al. 2009).

It must be stressed that the two proposed mechanisms for signalling through IP_7, either by binding or by pyrophosphorylation of proteins, are not mutually exclusive and there is now good evidence for both. To overcome the problem of binding specificity described above, can be hypothesize a precise intracellular spatial control of the local levels of inositol pyrophosphates. A specific localization of IP6K and/or

PP-IP5K activity could, for example, induce a localized increase in the IP_7/IP_6 ratio and therefore a specificity of IP_7 binding could be possible within such a discreet region *in vivo*. In addition, it is plausible that the phosphohydrolase DIPPs would also be present in the vicinity to prevent diffusion of the IP_7 and maintain its localized increased ratio. In the case of protein pyrophosphorylation, the presumably low intracellular concentration of inositol pyrophosphates becomes less important, and the high turnover of pyrophosphates might be a direct consequence of protein pyrophosphorylation cycles.

14.7 Functions

In the previous sections I described basic aspects of inositol pyrophosphate biology and gave few indications of the functions controlled by inositol pyrophosphates. I will now illustrate the high heterogeneity of the physiological functions that are thought to be controlled by inositol pyrophosphates, thus underling the likely fundamental importance of this class of molecule.

In the past ten years, the yeast *S. cerevisiae* has been the preferred experimental organism used by many groups to identify cellular process affected by altered inositol polyphosphates metabolism. Before discussing the many inositol pyrophosphate functions, I will summarise few basic phenotypes resulting from the deletion of the yeast inositol pyrophosphate synthesizing enzymes *kcs1*Δ and *vip1*Δ. Deletion of the yeast IP6K enzyme, *kcs1*Δ results in cells with undetectable levels of inositol pyrophosphates, which possess a slow growth rate (Dubois et al. 2002; Saiardi et al. 2000a). The *kcs1*Δ cells are larger than Wild Type (WT) and are also hypersensitive to salt stress (Dubois et al. 2002). When *kcs1*Δ was transformed with a plasmid encoding wild type Kcs1 or mammalian IP6K1, the IP_7 and IP_8 levels were restored and defects rescued, whereas transformation with a catalytically inactive form of Kcs1 did not rescue the mutant phenotypes (Dubois et al. 2002; Saiardi et al. 2002). Interestingly, mutants with a deletion of the inositol polyphosphates multikinase *arg82*Δ (also known as IPMK or Ipk2) (Odom et al. 2000; Saiardi et al. 1999, 2000b), which don't metabolise IP_3 to IP_5, often exhibit phenotypes, similar to those observed in *kcs1*Δ cells. Conversely, mutants in the IP_5-2 kinases *ipk1*Δ that do not convert IP_5 to IP_6, and as consequence do not possess IP_7 and IP_8, have normal growth rates (York et al. 1999; Saiardi et al. 2002; Onnebo and Saiardi 2009). The similarity between *arg82*Δ and *kcs1*Δ cells and the behavioural difference of *ipk1*Δ have led to attribute many functions to PP-IP_4 generated by Kcs1 using IP_5 as substrate (Auesukaree et al. 2005; York et al. 2005). Inositol pentakisphosphates is virtually undetectable in WT yeast but accumulates in *ipk1*Δ cells (York et al. 1999; Saiardi et al. 2002). However, it is also possible that the different inositol pyrophosphates can functionally replace each other; thus WT an *ipk1*Δ possess similar growth rate even if they possess different inositol pyrophosphates species. The possibility that different inositol pyrophosphates might possess overlapping function is particularly relevant for the protein pyrophosphorylation mechanism of action because it was demonstrated that different inositol pyrophosphates can pyrophosphorylate the same protein (Bhandari et al. 2007). To support this concept, there are experiments showing that Kcs1 overexpression in *arg82*Δ partially overcame the metabolic block and rescued some phenotypic defects (Dubois et al. 2002). Although the IP_7 level were not rescued in these experiments, there is a production of some unusual inositol pyrophosphate species (Dubois et al. 2002; Seeds et al. 2005), supporting the concept that the different inositol pyrophosphates species may play overlapping functions.

Fewer functions have been attributed to pyrophosphates produced by the Vip1 enzyme, but this is certainly due to the more recent identification of this enzyme and the shorter time investigators have spent studying its physiological roles. Nevertheless, although in *S. cerevisiae vip1*Δ strain does not present major phenotypic defects, the Vip1 homologous Asp1 enzyme in *S. pombe* is important for the control of cell morphology by regulating the actin related protein (Arp) complexes (Mulugu et al. 2007). The *asp1*Δ strain shows temperature sensitive growth, cell wall defects and cell shape abnormalities; each of these phenotypes are rescued by plasmids carrying the Asp1 kinase but not the kinase-dead construct, proving that is the synthesis of the (1/3)PP-IP5 isomer of IP_7, generated by Asp1 that is important for the control of these functions (Mulugu et al. 2007).

For simplicity, I will review the different physiological aspects by subdividing them into categories, discussing all the evidence supporting a specific function even if the data originates from different model organism. Several of the functions attributed to inositol pyrophosphates described in the following paragraphs were characterized before the recent identification of Vip1/PP-IP5K enzymes. Consequently, these studies mainly address the function of inositol pyrophosphates generated by IP6Ks (Kcs1). However, IP6Ks and PP-IP5K enzymes operate sequentially to generate IP_8, thus it is likely that some of the cellular roles I will describe below are dependent on the action of both kinases.

14.7.1 DNA Recombination

The Kcs1 gene in yeast was initially identified as encoding a novel leucine zipper protein in a screen for mutations in genes that were able to suppress the observed increase in homologous recombination rate caused by mutations in the Protein Kinase C (Pkc1) (Huang and Symington 1995). The pkc1-4 allele had an increased rate of recombination, which was completely reversed by a second mutation in a gene designated Kinase C suppressor 1 (thus the name Kcs1). It was later shown that the inositol pyrophosphates products of Kcs1 are responsible for the DNA hyper-recombination in Pkc1 mutants allele (Luo et al. 2002). However, inositol pyrophosphates control DNA recombination only by reverting to the high rate of recombination in the pkc1-4 background. Yeast *kcs1*Δ possess homologous recombination rate similar to WT yeast.

14.7.2 Stress Responses

As discussed earlier osmotic stress and temperature shift were shown in mammalian cells to increase the cellular concentration of IP_8 (Choi et al. 2005, 2008; Pesesse et al. 2004). Initially this effect was linked to the activation of MEK signal transduction cascade but further work indicated that alteration of cellular energy homeostasis is responsible for the change in IP_8 levels (Choi et al. 2008). It is puzzling that osmotic and temperature shift don't affected the level of IP8 in yeast an organism constantly subjected to these environmental stress. In yeast, however, inositol pyrophosphates are important to control oxidative stress response. In fact both *kcs1*Δ and *vip1*Δ possess a higher threshold resistance to the lethal effects of hydrogen peroxide (H_2O_2) due to a sustain activation of Rad53 pathway (Onnebo and Saiardi 2009), a component of the yeast DNA-damage response (Branzei and Foiani 2006). The privileged activation of Rad53 in response to DNA damage in yeast deficient in inositol pyrophosphates results in the activation of DNA-damage response mechanism that resulted in a decreased in mutation rate. It has been demonstrated that low doses of H_2O_2 directly reduce the levels of inositol pyrophosphates *in vivo* which was linked with the inhibition of Kcs1 activity *in vitro* (Onnebo and Saiardi 2009). However, yeast mutants that do not possess inositol pyrophosphates are more sensitive to DNA damage caused by phleomycin, indicating that inositol pyrophosphates are not involved in directly sensing DNA damage (Onnebo and Saiardi 2009).

14.7.3 Telomere Control

Telomeres consist of long, repetitive, protective DNA sequences at the end of chromosomes that prevent DNA degradation (Monaghan 2010; Rhodes et al. 2002). The analysis of *kcs1*Δ yeast that do not possess detectable levels of inositol pyrophosphates revealed longer telomeres (Saiardi et al. 2005; York et al. 2005). One study analyzed the telomere length of different mutants such *arg82*Δ and *ipk1*Δ and the authors concluded that PP-IP_4 generated from IP_5 by Kcs1 specifically regulates telomere length (York et al. 2005). However, *S. cerevisiae* possesses extremely low levels if any of IP_5 (York et al. 1999). Therefore the production of PP-IP_4 is negligible in WT cells. Alternatively, as suggested by another study, telomere length is regulated by relative levels of inositol pyrophosphates, the absence of inositol pyrophosphates in *kcs1*Δ cells and their very high levels in the *ipk1*Δ strain (Saiardi et al. 2005). In WT yeast IP_7 represent 2–8% of its precursor IP_6; however in *ipk1*Δ strain PP-IP_4 represent 20–30% of its precursor IP_5 (Saiardi et al. 2002; York et al. 1999). Both studies demonstrated that the inositol pyrophosphate effect on telomere length requires the presence of Tel1 (Mallory and Petes 2000; Morrow et al. 1995), the yeast homologue of the protein ATM that is mutated in the human disease ataxia telangiectasia, as telomere length is decreased in the *tel1*Δ strain. If Kcs1 acts by generating inositol pyrophosphates to antagonise Tel1 activity, yeast with elevated levels of inositol pyrophoshates should have shorter telomeres. Indeed, *ipk1*Δ yeast, which have perturbed inositol phosphate metabolism but has high levels of inositol pyrophosphates, shows reduced telomere length. In yeast, two phosphatidylinositol 3-kinase-related kinase (PIKK) members, Mec1 and Tel1, are important for telomere maintenance (Craven et al. 2002; Mallory and Petes 2000). Interestingly, the *kcs1*Δ yeast are resistant to the generic inhibitors of PIKK kinases, caffeine and wortmannin (Rahal et al. 2008), indicating that inositol *pyro*phosphates may be responsible for mediating the lethal actions of these drugs by blocking PIKK activity (Saiardi et al. 2005).

The role of inositol pyrophosphate in controlling telomere length has been confirmed by a third study that suggested a possible signalling link between inositol pyrophosphates and lipid metabolism (Ponnusamy et al. 2008). Loss of very long-chain fatty acid (VLCFA) by deletion of fatty acid elongase 3 (*elo3*Δ) resulted in reduced telomere length. Interestingly, telomere shortening in *elo3*Δ cells was prevented in *elo3*Δ*kcs1*Δ double mutant strain (Ponnusamy et al. 2008). The exact mechanism of this functional interaction is unknown.

14.7.4 Ribosome Biogenesis

The main protein targets of IP_7-mediated pyrophosphorylatin identified are nucleolar proteins such the yeast Nsr1 and Srp40 and the mammalian Tcof1 (Treacle) and Nopp140 (Bhandari et al. 2007; Saiardi et al. 2004). These nucleolar proteins are important to control rRNA processing and in ribosomal biogenesis (Verheggen et al. 2001; Lo et al. 2006). However, there is yet no link between the function of these proteins and their pyrophosphorylation status. Nevertheless, A recent genetic interaction study revealed a role for Kcs1 and consequently of inositol pyrophosphates, in ribosomal biogenesis. The nucleolar Rrs1 protein is required for 25S rRNA processing and for the assembly of the yeast 60S ribosomal subunit. Analysis of polysomal profile of rrs1-1 allele at low temperature revealed an accumulation of the 40S subunits with a decrease in 80S monosomes; interestingly these defects are restored if the rrs1-1 allele is present in *kcs1*Δ background (Horigome et al. 2009). Although, this evidence supports a role for inositol pyrophosphate in controlling ribosome biogenesis, further work is necessary to demonstrate that pyrophosphorylation of specific nucleolar proteins affects ribosome biogenesis/functions.

14.7.5 Trafficking

Perhaps the first cellular role linked to inositol pyrophosphate was vesicular trafficking because soon after the discovery of this class of molecules several proteins involved in trafficking were identified as IP_7 binding partners (Ali et al. 1995; Fleischer et al. 1994; Shears et al. 1995). However these *in vitro* binding studies were performed using inappropriate experimental conditions as discussed earlier. Nevertheless, it was this early literature that lead, after the cloning of the IP6K, to investigate if *kcs1*Δ yeast was impaired in vesicular trafficking. These studies demonstrated altered vacuolar morphology with several smaller fragmented vacuoles (Saiardi et al. 2000a), as well as accumulation of membranous vesicular structures derived from the plasma membrane (Saiardi et al. 2002). Using antibodies against proteins associated with endosomal membranes, immunogold localization confirmed that these membranes were aberrant endosomal intermediates (Saiardi et al. 2002) suggesting that the endocytic pathway is affected. Indeed, ligand-dependent internalisation of the mating pheromone receptor Ste3 is slower than in WT and the processing is abnormal (Saiardi et al. 2002).

Work in mammalian models reinforced the connection between inositol pyrophosphate and vesicular trafficking because the beta subunit of the AP-3 complex AP3B1 is pyrophospholrylated by IP_7 (see chapter on mechanism of action) (Azevedo et al. 2009). The AP-3 complexhas been implicated in yeast in the regulation of the cargo-selective transport from the Golgi to the vacuole (Stepp et al. 1997) and in mammalian cells in the biogenesis and function of lysosome-related organelles, virus release, and to synaptic vesicle formation (Dell'Angelica 2009; Dong et al. 2005; Odorizzi et al. 1998). The function of the AP-3 complex is affected by the pyrophosphorylation status of its beta subunit AP3B1, through the modulation of the interaction between AP3B1 and Kif3A, a motor protein of the kinesin superfamily important for transport of organelles and cargo along microtubules (Hirokawa 2000). It was demonstrated that the regulation of the interaction between AP3B1 and Kif3A controls virus like particles (VLPs) release into the extracellular medium, most likely affecting its intracellular transport to the plasma membrane where it assembles and is released from. VLPs were generated by transfecting GAG, a major structural protein of the HIV-1 virus. HIV-1 Gag alone contains all of the determinants required to produce non-infectious VLP in the absence of other viral proteins (Gheysen et al. 1989). Increasing IP_7 levels by over-expression of the IP6K or decreasing IP_7 level by using ip6k1$^{-/-}$ knock-out mouse embryonic fibroblasts caused decrease or increase in HIV-1 GAG VLP release, respectively (Azevedo et al. 2009).

Another mammalian work that linked inositol pyrophosphates and vesicular trafficking derived from a yeast two hybrid screening in which GRAB was identified as an IP6K1 binding partner (Luo et al. 2001). GRAB is a physiological guanine nucleotide exchange factor (GEF) for Rab3A a small GTP-binding protein that is involved in synaptical vesicle exocytosis (Geppert et al. 1994). The regulation of GRAB function by the IP6K1 seems to be kinase independent; in fact, both the catalytically active or inactive form of IP6K is able to block the binding of GRAB to Rab3A (Luo et al. 2001). GRAB's ability to promote GDP release from Rab3A is independent from IP_7 production. Thus the IP6Ks enzyme might possess functional roles that are independent of their kinase activity. IP_7 was also shown to affect the trafficking of insulin containing granules but I will discuss this important function in the next section.

14.7.6 Insulin Signalling

An elegant electrophysiological study demonstrated that increasing IP_7 in pancreatic β-cells, by overexpressing the IP6Ks enzymes, stimulated exocytosis of insulin-containing granules from the readily releasable pool (Illies et al. 2007). This effect seems IP_7 specific because direct application with a patch pipette of several IP_7 isomers induces secretion with a half maximal concentration of 1 μM value that is in the estimated concentration range for IP_7 in mammalian cells (Illies et al. 2007). Overexpession of any of the three isoforms resulted in similar increases in exocytosis. However, only RNAi silencing of IP6K1, but not IP6K2, inhibited exocytosis, which indicated that IP6K1 is the important endogenous kinase. These data are supported by a mouse knock-out model, the ip6k1$^{-/-}$ mice are smaller than wild-type despite normal food intake and they have significantly lower levels of circulating blood insulin (Bhandari et al. 2008). Furthermore, male knock-out mice are also sterile with major defects observed in spermatogenesis. In contrast, ip6k2$^{-/-}$ mice appear to have regular growth rate, no defect in spermatogenesis and normal insulin blood levels (Morrison et al. 2009).

The first characterization of ip6k1$^{-/-}$ mice was carried out using six weeks old animals and although the authors reported lower insulin blood levels they didn't observed any change in glucose tolerance curve (Bhandari et al. 2008). However, analysis of ten months old animal revealed significant changes of glucose tolerance, thereby demonstrating that older mice acquire insulin hypersensitivity (Chakraborty et al. 2010). Interestingly, ip6k1$^{-/-}$ mice when exposed to a high fat diet become resistant to body weight gain due to reduced fat accumulation. Furthermore, several serum parameters normally associated with a high fat diet, including triglycerides and cholesterol levels, are lower in ip6k1$^{-/-}$ mice when compared to WT littermates. The authors also demonstrated that IP6K1 is important to control glucose homeostasis. When ip6k1$^{-/-}$ mice were fed with high fat diet, enhanced insulin-induced reduction of blood glucose was observed (Chakraborty et al. 2010). The physiological changes of global glucose homeostasis observed in ip6k1$^{-/-}$ mice have been associated with inositol pyrophosphate-dependent inhibition of AKT signalling. IP_7 inhibits *in vitro* PDK1-dependent phosphorylation of AKT with an astonishing IC_{50} of about 20 nM. Inhibition of AKT in turn, decreased mTOR signalling and increased GSK3b pathway (Chakraborty et al. 2010).

Two further observations support the link between insulin signalling and inositol pyrophosphates. Firstly a human genetic study identified the disruption of the IP6K1 gene in a family with type 2 diabetes (Kamimura et al. 2004). Secondly, the use of TNP (N(2)-(m-(trifluoromethy)lbenzyl) N(6)-(p-nitrobenzyl)purine), an IP6Ks pharmacological inhibitor, on the insulin secreting cell line Min6 resulted in a decrease in insulin release (Padmanabhan et al. 2009).

14.7.7 Phosphate Homeostasis

Inositol pyrophosphates are molecules with the unique property of possessing more phosphates that carbons atoms; IP_7 is often described as a molecule with 7 phosphate groups attached to the 6 carbon ring of inositol. Therefore, it may not be surprising that there is evidence in the literature that inositol pyrophosphates control cellular phosphate homeostasis.

One of the mammalian IP_6-kinase enzymes, specifically IP6K2 (Saiardi et al. 1999; Schell et al. 1999), was initially identified as PiUS (Phosphate inorganic Uptake Stimulator), a protein named after its role as a stimulator of inorganic phosphate uptake after transfection into a *Xenopus* oocyte (Norbis et al. 1997). It is unfortunate that this early work in the mammalian system has not been followed up by further work because the regulation of phosphate up-take from the intracellular medium appears to be an evolutionarily conserved feature. In fact, *kcs1*Δ yeast conversely exhibits a considerably reduced uptake of inorganic phosphate (Saiardi et al. 2004).

As discussed earlier (see section on the mechanism of action), the selective binding of (1/3)PP-IP_5 isoform of IP_7 generated by Vip1, to the Pho80/Pho85/Pho81 cyclin dependent kinase/cyclin/cyclin-dependent kinase inhibitor complex augments the inhibitory activity of Pho81 (Lee et al. 2007). The Pho80/Pho85 inhibition, in low phosphate conditions, results in a decrease in Pho4 phosphorylation, this transcription factor then can translocate from the cytoplasm into the nucleus (O'Neill et al. 1996) activating the PHO pathway. However, a recent genome-wide mapping of Pho4 and Rpo21 binding (subunit of RNA polymerase II) using a tiling array revealed that Pho4 is present in the nucleus even under high phosphate conditions to control transcription (Nishizawa et al. 2008). Interestingly this work revealed that Pho4 promote the transcription of antisense and intragenic RNAs in Kcs1 locus but not affect the transcription of Vip1. The Pho4 regulated Kcs1 antisense transcript might be responsible for the formation of the intragenic shorter Kcs1 mRNA (Nishizawa et al. 2008). The truncated Kcs1 enzyme generated in this condition might possess different enzymatic properties. These studies indicated that both the yeast inositol pyrophosphates synthesizing enzymes Vip1 and Kcs1 are important in regulating or are regulated by phosphate dependent transcription. However, the PHO transcriptional regulon is a *S. cerevisiae* specific paradigm and it does not appear to be conserved in mammals or even in the yeast *S. pombe* (Liu and Kipreos 2000; Santos et al. 1995). Consequently, this transcriptional control is unlikely to account for the link observed between basic phosphate homeostasis and inositol pyrophosphates in mammalian cells.

Beside the transcriptional control just described, there is probably a metabolic link between inositol pyrophosphates and cellular phosphates homeostasis. A yeast screen of deletion mutants based on the secreted acid phosphatase activity of Pho5 revealed that the enzymatic pathway responsible for IP_7 synthesis negatively regulates Pho5 synthesis or secretion (Auesukaree et al. 2005). Furthermore yeast not possessing inositol pyrophosphate, such as the *arg82*Δ and *kcs1*Δ strains possesses low levels of inorganic polyphosphates (PolyP) the main intracellular phosphate storage molecule. PolyP is a polymer containing from tens to hundreds of phosphate residues linked by the same phosphoanhydride, "high-energy" bonds, similar to the bond found in IP_7 or ATP (Rao et al. 2009). Recently the yeast PolyP polymerase has been identified in the vacuolar membrane transporter chaperone (VTC) complex (Hothorn et al. 2009). Interestingly, pyrophosphates (Pi-Pi) dramatically accelerate the polymerase reaction (Hothorn et al. 2009); it would therefore be interesting to determine if the pyrophosphates moiety of IP_7 can stimulate PolyP vacuolar synthesis in a similar fashion.

14.7.8 Chemotaxis

Disruption of the IP6K in *D. Discoideum* (Dictybase gene name *I6KA*) by homologous recombination abolished IP_7 and IP_8 levels that could be restored to wild-type levels upon complementation (Luo et al. 2003). Mutant cell growth was normal as was the shape and size of the *D. Discoideum* fruiting body. However starvation-induced aggregation occurred more rapidly in the IP6K null mutant than in the WT amoeba. Moreover, mutant cells posses an increased sensitivity to chemo-attractant, responding more rapidly to low concentrations of cAMP gradients than WT cells. The authors suggests that this phenotype could be a result of the ability of IP_7 to compete with $PI(3,4,5)P_3$ for the PH domain binding of CRAC (Luo et al. 2003), as described above. The absence of IP_7 in the null mutant would free the CRAC-PH domain to bind to $PI(3,4,5)P_3$ and its easier translocation to the membrane would facilitate the response to the chemo-attractant allowing the amoeba to aggregate faster. However recent work challenges the role $PI(3,4,5)P_3$ in regulating *D. Discoideum* chemotaxis because the deletion of all the six PI3Ks enzymes responsible for $PI(3,4,5)P_3$ biosynthesis resulted in amoeba with normal chemotaxis behaviour (Hoeller and Kay 2007). Therefore, the effect of IP_7 on chemotaxis cannot be simply explained by the regulation of CRAC function.

14.7.9 Regulation of Cell Death Mechanisms

Several studies have linked inositol pyrophosphates to mammalian cell death. IP6K2 was first identified as a positive regulator of apoptosis through a technical knock-down screening approach (Morrison et al. 2001, 2002). In ovarian carcinoma cells, IP6K2 deletions conferred protection against interferon IFN-α induced cell death and over-expression of full length IP6K2 enhanced the degree of apoptosis induced by both IFN-α and γ-irradiation. The nuclear localization of IP6K2 and the activation of the Apo2L/TRAIL (Apo2L/tumor-necrosis-factor-related apoptosis-inducing ligand, TNFSF10) pathways are required for IFN-α and γ-irradiation induced apoptosis (Morrison et al. 2005). However, the cytotoxicity of IP6K2 is associated with its translocation from the nucleous to mitochondria (Nagata et al. 2005), while no alterations are observed in the intracellular localisation of IP6K1 or IP6K3 (Nagata et al. 2005). These findings were extended by showing that the apoptotic function of IP6K2 is not restricted to a specific cell line or apoptotic pathway. Transfection with IP6K2 increased cell death in HEK293, HeLa, PC12, Jurkat T and HL60 cells exposed to different cell stressors such as hydrogen peroxide, staurosporine, etoposide and hypoxic-ischemia (Nagata et al. 2005). The cytotoxicity induced by the different cell stressors was associated with a rise in IP_6-kinase activity. When over-expressed all three isoforms of IP6Ks induced cell death in untreated cells but only RNAi down regulation of IP6K2 reduce the rate of cell death suggesting a primary role for IP6K2 in the regulation of cell death mechanism (Nagata et al. 2005). Thus, increasing inositol pyrophosphate cellular level sensitizes mammalian cells to stressors and facilitates the activation of cell death mechanism. Whereas decreasing inositol pyrophosphate by reducing the activity of IP6K, specifically IP6K2, result in a blockage of cell death mechanism.

The specific role of IP6K2 in controlling cell death mechanisms is likely to rely on the ability of this protein to interact with the molecular chaperone Heat Shock Protein 90 (HSP90) (Wandinger et al. 2008; Pearl et al. 2008), IP6K1 does not interact with HSP90 (Chakraborty et al. 2008). IP6K2 bind to HSP90 C-terminus using a unique 12-aa sequence comprise between amino acid 132 and 143. The HSP90 binding to IP6K2 inhibits IP6K2 catalytic activity. Drugs that bind to the C-terminus of HSP90, such as cisplatin and novobiocin, abolished HSP90-IP6K2 interaction, and resulted in the activation of IP6K2 which lead to cell death. This study proposed that the prosurvival actions of HSP90 (Pearl et al. 2008), reflect at least in part, the inhibition of IP6K2 catalytically activity (Chakraborty et al. 2008). Another important IP6K2 interacting partner has been identified with transcription factor and tumour suppressor p53 (Koldobskiy et al. 2010) that regulates expression of target genes, thereby promoting apoptotic cell death or cell cycle arrest (Michalak et al. 2005). The IP6K2 N-terminus amino-acid 1–67 specifically interact with p53. Cells expressing the dominant negative IP6K2(1–67) construct showed increased expression of the cyclin-dependent kinase inhibitor p21 in response to the chemiotherapy drug 5-fluorouracil. The authors demonstrated that p53-dependent apoptosis is modulated by IP6K2 by regulating the expression of the pro-arrest proteins p21. Although, IP6K2 regulates p53 by direct binding, the author indicated that IP6K2 enzymatic activity is important to influence p53 signalling (Koldobskiy et al. 2010).

Deregulation of cell death mechanism often leads to the development of uncontrolled growth and cancer. Despite, ip6k2$^{-/-}$ mice knock-out displaying normal growth and fertility, exposure to the carcinogen 4-nitroquinoline 1-oxide (4-NQO), resulted in an increase in the incidence of invasive squamous cell carcinoma formation in the oral cavity and esophagus (Morrison et al. 2009). The predisposition to the carcinogen of ip6k2$^{-/-}$ mice further supports the suggestion of a specific role of IP6K2 in regulating cell death.

Another observation linking inositol pyrophosphates to cancer derived from a screening aim to identify β-catenin transforming activity of target genes. Using subtractive hybridization technique IP6K2 was identified as one of the two genes unregulated after β-catenin is transfected in chicken embryo fibroblasts (Aoki et al. 2002). Oncogenic transformation by β-catenin is believed to result from the translocation of this protein from the cytoplasm to the nucleus where it combines with lymphoid-enhancer binding factor (LEF-1) to induce transcriptional changes (Aoki et al. 2002; Arce et al. 2006).

A very recent report suggested that inositol pyrophosphates enhance cell death by increasing autophagy (Nagata et al. 2010). The autophagic pathway represents an important pro-survival mechanism that helps to recycle cell metabolic components. However, it has been demonstrated that excessive autophagosome formation are associated with autophagic cell death (Tsujimoto and Shimizu 2005). An increased inositol pyrophosphates signalling induces the formation of autophagosomes through a decrease of mTOR phosphorylation (Nagata et al. 2010) the principal regulator of the autophagy process (Diaz-Troya et al. 2008).

14.7.10 Development

The generation of mice knock-out for ipmk$^{-/-}$ and ipk1$^{-/-}$, the enzymes responsible for IP_6 synthesis, resulted in embryonic lethality with defects in neuronal tube closure (Verbsky et al. 2005; Frederick et al. 2005), suggesting a developmental role for the higher phosphorylated forms of inositol. However, analysis of ip6k1$^{-/-}$ and ip6k2$^{-/-}$ mutant mice revealed the absence of embryonic defects and a mendelian distribution of the deleted allele in the litter (Bhandari et al. 2008; Morrison et al. 2009). Despite these results, the presence of three IP6Ks and two PP-IP5Ks in the mammalian genome suggests that functional redundancy of these enzymes might have precluded the easy identification of a developmental role for inositol pyrophosphates in the mammalian system.

However, a very recent report in zebrafish, demonstrated a pivotal role for inositol pyrophosphates in control early developmental processes in fish. Wente and coworker reported that down regulation of inositol pyrophosphates signalling using morpholino of fish IP6K2 gene resulted in embryonic craniofacial skeleton defects possibly caused by an alteration of Neuronal Crest Cell (NCC) migration (Sarmah and Wente 2010). The embryonic defects may be due to an alterations in the Hedgehog (Hh) signalling pathway, a molecular cue that induces developmental regulated expression of several tissue-specific markers (Riobo and Manning 2007). Moreover it has been demonstrated that IP6K2 regulate Hh signalling not only in zebrafish embryo but also in mammalian cells (Sarmah and Wente 2010).

14.8 Concluding Remarks

The extraordinary diversity of cell biology processes regulated by inositol pyrophosphates emphasizes their fundamental role in cell signalling. It also opens many questions (Burton et al. 2009; Shears 2009). The inositol pyrophosphates research field is still young and not fully investigated; I am convinced that many important discoveries are forthcoming because several main questions still need to be answered.

First, is an allosteric mechanism of action responsible for inositol pyrophosphate functions? Several different protein effectors are necessary to justify the disparate functions linked to inositol pyrophosphates. Obviously, all these proteins should possess a common module able to specifically recognise inositol pyrophosphate. Furthermore, it is likely that more than one protein domain is necessary to selectively recognise the different inositol pyrophosphates isomers associated to specific role. The specificity and selectivity of these hypothetic inositol pyrophosphate binding "modules" (receptors) should facilitate their identification.

Second, is protein pyrophosphorylation functionally relevant? This mechanism of action breaks the 'dogma' that inositol polyphosphates work by allosteric interaction like IP_3 and its receptor or PIP_3 with PH domains, and perhaps for this reason protein pyrophosphorylation has encountered a number of criticisms. IP_7-driven protein pyrophosphorylation does not compete nor is complementary to the well-established ATP-kinase dependent mechanism of protein phosphorylation. Protein pyrophosphorylation might have a unique function in signal transduction. However, no direct evidences of protein pyrophosphorylation have been produced *in vivo* so far. The demonstration that protein pyrophosphorylation occurs in eukaryotic cells is one of the most exciting and challenging goal of future research.

Is there a common function for inositol pyrophosphates? IP_3 controls many disparate cellular functions by regulating intracellular Ca^{++} signalling. However, ultimately the only purpose of IP_3 is to open IP_3-receptor Ca^{++} channels. Similarly, inositol pyrophosphates might control a single and very basic cell biology aspect that subsequently regulates the many functions described in this chapter. Recently, Dr. Shears proposed to name inositol pyrophosphates 'metabolic messenger' (Shears 2009). Similarly, I previously hypothesized that protein pyrophosphorylation may be a sensor of cellular well-being that link inositol pyrophosphate to metabolism (Bennett et al. 2006). However, direct proves that inositol pyrophosphates are metabolic messengers and that protein pyrophosphorylation links metabolism with signalling are still missing.

Finally, the most challenging aspect of inositol pyrophosphate research is the development of novel technical tools that will allow the study of this class of molecules. The cloning of the enzymes responsible for inositol pyrophosphates metabolism certainly has hugely facilitated our research. However, too often, there is the tendency to work on the 'Molecular Biology' aspects of these enzymes forgetting that the real objective should be to understand the function of the inositol pyrophosphate molecule itself.

Acknowledgements I thank Drs. A.W. Mudge, A. Riccio, C. Azevedo for critical reading the manuscript and the members of the laboratory for discussions and suggestions. The work is funded by the Medical Research Council support of the Cell Biology Unit.

References

Agranoff BW, Bradley RM, Brady RO (1958) The enzymatic synthesis of inositol phosphatide. J Biol Chem 233:1077–1083

Albert C, Safrany ST, Bembenek ME, Reddy KM, Reddy K, Falck J, Brocker M, Shears SB, Mayr GW (1997) Biological variability in the structures of diphosphoinositol polyphosphates in Dictyostelium discoideum and mammalian cells. Biochem J 327(Pt 2):553–560

Ali N, Duden R, Bembenek ME, Shears SB (1995) The interaction of coatomer with inositol polyphosphates is conserved in Saccharomyces cerevisiae. Biochem J 310(Pt 1):279–284

Aoki M, Sobek V, Maslyar DJ, Hecht A, Vogt PK (2002) Oncogenic transformation by beta-catenin: deletion analysis and characterization of selected target genes. Oncogene 21:6983–6991

Arce L, Yokoyama NN, Waterman ML (2006) Diversity of LEF/TCF action in development and disease. Oncogene 25:7492–7504

Auesukaree C, Tochio H, Shirakawa M, Kaneko Y, Harashima S (2005) Plc1p, Arg82p, and Kcs1p, enzymes involved in inositol pyrophosphate synthesis, are essential for phosphate regulation and polyphosphate accumulation in Saccharomyces cerevisiae. J Biol Chem 280:25127–25133

Azevedo C, Saiardi A (2006) Extraction and analysis of soluble inositol polyphosphates from yeast. Nat Protoc 1:2416–2422

Azevedo C, Burton A, Ruiz-Mateos E, Marsh M, Saiardi A (2009) Inositol pyrophosphate mediated pyrophosphorylation of AP3B1 regulates HIV-1 Gag release. Proc Natl Acad Sci U S A 106:21161–21166

Azevedo C, Szijgyarto Z, Saiardi A (2010) The signaling role of inositol hexakisphosphate kinases (IP(6)Ks). Adv Enzyme Regul 51:74–82

Balla T (2005) Inositol-lipid binding motifs: signal integrators through protein–lipid and protein–protein interactions. J Cell Sci 118:2093–2104

Barker CJ, Wright J, Hughes PJ, Kirk CJ, Michell RH (2004) Complex changes in cellular inositol phosphate complement accompany transit through the cell cycle. Biochem J 380:465–473

Barker CJ, Illies C, Gaboardi GC, Berggren PO (2009) Inositol pyrophosphates: structure, enzymology and function. Cell Mol Life Sci 66:3851–3871

Bennett M, Onnebo SM, Azevedo C, Saiardi A (2006) Inositol pyrophosphates: metabolism and signaling. Cell Mol Life Sci 63:552–564

Bessman MJ, Frick DN, O'handley SF (1996) The MutT proteins or "Nudix" hydrolases, a family of versatile, widely distributed, "housecleaning" enzymes. J Biol Chem 271:25059–25062

Bhandari R, Saiardi A, Ahmadibeni Y, Snowman AM, Resnick AC, Kristiansen TZ, Molina H, Pandey A, Werner JK Jr, Juluri KR, Xu Y, Prestwich GD, Parang K, Snyder SH (2007) Protein pyrophosphorylation by inositol pyrophosphates is a posttranslational event. Proc Natl Acad Sci U S A 104:15305–15310

Bhandari R, Juluri KR, Resnick AC, Snyder SH (2008) Gene deletion of inositol hexakisphosphate kinase 1 reveals inositol pyrophosphate regulation of insulin secretion, growth, and spermiogenesis. Proc Natl Acad Sci U S A 105:2349–2353

Boer VM, Crutchfield CA, Bradley PH, Botstein D, Rabinowitz JD (2010) Growth-limiting intracellular metabolites in yeast growing under diverse nutrient limitations. Mol Biol Cell 21:198–211

Bollen M, Stalmans W (1988) Fluorine compounds inhibit the conversion of active type-1 protein phosphatases into the ATPMg-dependent form. Biochem J 255:327–333

Branzei D, Foiani M (2006) The Rad53 signal transduction pathway: replication fork stabilization, DNA repair, and adaptation. Exp Cell Res 312:2654–2659

Burton A, Hu X, Saiardi A (2009) Are inositol pyrophosphates signalling molecules? J Cell Physiol 220:8–15

Caffrey JJ, Safrany ST, Yang X, Shears SB (2000) Discovery of molecular and catalytic diversity among human diphosphoinositol-polyphosphate phosphohydrolases. An expanding Nudt family. J Biol Chem 275:12730–12736

Chakraborty A, Koldobskiy MA, Sixt KM, Juluri KR, Mustafa AK, Snowman AM, van Rossum DB, Patterson RL, Snyder SH (2008) HSP90 regulates cell survival via inositol hexakisphosphate kinase-2. Proc Natl Acad Sci U S A 105:1134–1139

Chakraborty A, Koldobskiy MA, Bello NT, Maxwell M, Potter JJ, Juluri KR, Maag D, Kim S, Huang AS, Dailey MJ, Saleh M, Snowman AM, Moran TH, Mezey E, Snyder SH (2010) Inositol pyrophosphates inhibit Akt signaling, regulate insulin sensitivity and weight gain. Cell 143:897–910

Choi K, Mollapour E, Shears SB (2005) Signal transduction during environmental stress: InsP(8) operates within highly restricted contexts. Cell Signal 17:1533–1541

Choi JH, Williams J, Cho J, Falck JR, Shears SB (2007) Purification, sequencing, and molecular identification of a mammalian PP-InsP5 kinase that is activated when cells are exposed to hyperosmotic stress. J Biol Chem 282:30763–30775

Choi K, Mollapour E, Choi JH, Shears SB (2008) Cellular energetic status supervises the synthesis of bis-diphosphoinositol tetrakisphosphate independently of AMP-activated protein kinase. Mol Pharmacol 74:527–536

Craven RJ, Greenwell PW, Dominska M, Petes TD (2002) Regulation of genome stability by TEL1 and MEC1, yeast homologs of the mammalian ATM and ATR genes. Genetics 161:493–507

Dell'Angelica EC (2009) AP-3-dependent trafficking and disease: the first decade. Curr Opin Cell Biol 21:552–559

Di Paolo G, De Camilli P (2006) Phosphoinositides in cell regulation and membrane dynamics. Nature 443:651–657

Diaz-Troya S, Perez-Perez ME, Florencio FJ, Crespo JL (2008) The role of TOR in autophagy regulation from yeast to plants and mammals. Autophagy 4:851–865

Dong X, Li H, Derdowski A, Ding L, Burnett A, Chen X, Peters TR, Dermody TS, Woodruff E, Wang JJ, Spearman P (2005) AP-3 directs the intracellular trafficking of HIV-1 Gag and plays a key role in particle assembly. Cell 120:663–674

Downes CP, Gray A, Fairservice A, Safrany ST, Batty IH, Fleming I (2005) The regulation of membrane to cytosol partitioning of signalling proteins by phosphoinositides and their soluble headgroups. Biochem Soc Trans 33:1303–1307

Draskovic P, Saiardi A, Bhandari R, Burton A, Ilc G, Kovacevic M, Snyder SH, Podobnik M (2008) Inositol hexakisphosphate kinase products contain diphosphate and triphosphate groups. Chem Biol 15:274–286

Dubois E, Scherens B, Vierendeels F, Ho MM, Messenguy F, Shears SB (2002) In Saccharomyces cerevisiae, the inositol polyphosphate kinase activity of Kcs1p is required for resistance to salt stress, cell wall integrity, and vacuolar morphogenesis. J Biol Chem 277:23755–23763

Fisher Di, Safrany ST, Strike P, Mclennan AG, Cartwright JL (2002) Nudix hydrolases that degrade dinucleoside and diphosphoinositol polyphosphates also have 5-phosphoribosyl 1-pyrophosphate (PRPP) pyrophosphatase activity that generates the glycolytic activator ribose 1,5-bisphosphate. J Biol Chem 277:47313–47317

Fleischer B, Xie J, Mayrleitner M, Shears SB, Palmer DJ, Fleischer S (1994) Golgi coatomer binds, and forms K(+)-selective channels gated by, inositol polyphosphates. J Biol Chem 269:17826–17832

Frederick JP, Mattiske D, Wofford JA, Megosh LC, Drake LY, Chiou ST, Hogan BL, York JD (2005) An essential role for an inositol polyphosphate multikinase, Ipk2, in mouse embryogenesis and second messenger production. Proc Natl Acad Sci U S A 102:8454–8459

Fridy PC, Otto JC, Dollins DE, York JD (2007) Cloning and characterization of two human VIP1-like inositol hexakisphosphate and diphosphoinositol pentakisphosphate kinases. J Biol Chem 282:30754–30762

Geppert M, Bolshakov VY, Siegelbaum SA, Takei K, De Camilli P, Hammer RE, Sudhof TC (1994) The role of Rab3A in neurotransmitter release. Nature 369:493–497

Gheysen D, Jacobs E, De Foresta F, Thiriart C, Francotte M, Thines D, De Wilde M (1989) Assembly and release of HIV-1 precursor Pr55gag virus-like particles from recombinant baculovirus-infected insect cells. Cell 59:103–112

Glennon MC, Shears SB (1993) Turnover of inositol pentakisphosphates, inositol hexakisphosphate and diphosphoinositol polyphosphates in primary cultured hepatocytes. Biochem J 293 (Pt 2):583–590

Gonzales ML, Anderson RA (2006) Nuclear phosphoinositide kinases and inositol phospholipids. J Cell Biochem 97:252–260

Hand CE, Honek JF (2007) Phosphate transfer from inositol pyrophosphates InsP5PP and InsP4(PP)2: a semi-empirical investigation. Bioorg Med Chem Lett 17:183–188

Hidaka K, Caffrey JJ, Hua L, Zhang T, Falck JR, Nickel GC, Carrel L, Barnes LD, Shears SB (2002) An adjacent pair of human NUDT genes on chromosome X are preferentially expressed in testis and encode two new isoforms of diphosphoinositol polyphosphate phosphohydrolase. J Biol Chem 277:32730–32738

Hirokawa N (2000) Stirring up development with the heterotrimeric kinesin KIF3. Traffic 1:29–34

Hirokawa N, Noda Y, Tanaka Y, Niwa S (2009) Kinesin superfamily motor proteins and intracellular transport. Nat Rev Mol Cell Biol 10:682–696

Hoeller O, Kay RR (2007) Chemotaxis in the absence of PIP3 gradients. Curr Biol 17:813–817

Horigome C, Ikeda R, Okada T, Takenami K, Mizuta K (2009) Genetic interaction between ribosome biogenesis and inositol polyphosphate metabolism in Saccharomyces cerevisiae. Biosci Biotechnol Biochem 73:443–446

Hothorn M, Neumann H, Lenherr ED, Wehner M, Rybin V, Hassa PO, Uttenweiler A, Reinhardt M, Schmidt A, Seiler J, Ladurner AG, Herrmann C, Scheffzek K, Mayer A (2009) Catalytic core of a membrane-associated eukaryotic polyphosphate polymerase. Science 324:513–516

Huang KN, Symington LS (1995) Suppressors of a Saccharomyces cerevisiae pkc1 mutation identify alleles of the phosphatase gene PTC1 and of a novel gene encoding a putative basic leucine zipper protein. Genetics 141:1275–1285

Illies C, Gromada J, Fiume R, Leibiger B, Yu J, Juhl K, Yang SN, Barma DK, Falck JR, Saiardi A, Barker CJ, Berggren PO (2007) Requirement of inositol pyrophosphates for full exocytotic capacity in pancreatic beta cells. Science 318:1299–1302

Ingram SW, Safrany ST, Barnes LD (2003) Disruption and overexpression of the Schizosaccharomyces pombe aps1 gene, and effects on growth rate, morphology and intracellular diadenosine 5′,5′′′-P1,P5-pentaphosphate and diphosphoinositol polyphosphate concentrations. Biochem J 369:519–528

Irvine RF (2003) 20 years of Ins(1,4,5)P3, and 40 years before. Nat Rev Mol Cell Biol 4:586–590

Irvine RF, Schell MJ (2001) Back in the water: the return of the inositol phosphates. Nat Rev Mol Cell Biol 2:327–338

IUPAC-IUB (1977) Commission on biochemical nomenclature. Nomenclature of phosphorus-containing compounds of biochemical importance (Recommendations 1996). Proc Natl Acad Sci U S A 74:2222–2230

Kamimura J, Wakui K, Kadowaki H, Watanabe Y, Miyake K, Harada N, Sakamoto M, Kinoshita A, Yoshiura K, Ohta T, Kishino T, Ishikawa M, Kasuga M, Fukushima Y, Niikawa N, Matsumoto N (2004) The IHPK1 gene is disrupted at the 3p21.31 breakpoint of t(3;9) in a family with type 2 diabetes mellitus. J Hum Genet 49:360–365

Koldobskiy MA, Chakraborty A, Werner JK Jr, Snowman AM, Juluri KR, Vandiver MS, Kim S, Heletz S, Snyder SH (2010) p53-mediated apoptosis requires inositol hexakisphosphate kinase-2. Proc Natl Acad Sci U S A 107(49):20947–20951

Komander D, Fairservice A, Deak M, Kular GS, Prescott AR, Peter Downes C, Safrany ST, Alessi DR, van Aalten DM (2004) Structural insights into the regulation of PDK1 by phosphoinositides and inositol phosphates. Embo J 23:3918–3928

Laussmann T, Reddy KM, Reddy KK, Falck JR, Vogel G (1997) Diphospho-myo-inositol phosphates from Dictyostelium identified as D-6-diphospho-myo-inositol pentakisphosphate and D-5,6-bisdiphospho-myo-inositol tetrakisphosphate. Biochem J 322(Pt 1):31–33

Laussmann T, Hansen A, Reddy KM, Reddy KK, Falck JR, Vogel G (1998) Diphospho-myo-inositol phosphates in dictyostelium and polysphondylium: identification of a new bisdiphospho-myo-inositol tetrakisphosphate. FEBS Lett 426:145–150

Laussmann T, Pikzack C, Thiel U, Mayr GW, Vogel G (2000) Diphospho-myo-inositol phosphates during the life cycle of dictyostelium and polysphondylium. Eur J Biochem 267:2447–2451

Lee YS, Mulugu S, York JD, O'Shea EK (2007) Regulation of a cyclin-CDK-CDK inhibitor complex by inositol pyrophosphates. Science 316:109–112

Lee YS, Huang K, Quiocho FA, O'Shea EK (2008) Molecular basis of cyclin-CDK-CKI regulation by reversible binding of an inositol pyrophosphate. Nat Chem Biol 4:25–32

Lemmon MA (2008) Membrane recognition by phospholipid-binding domains. Nat Rev Mol Cell Biol 9:99–111

Leslie NR, Mclennan AG, Safrany ST (2002) Cloning and characterisation of hAps1 and hAps2, human diadenosine polyphosphate-metabolising Nudix hydrolases. BMC Biochem 3:20

Letcher AJ, Schell MJ, Irvine RF (2008) Do mammals make all their own inositol hexakisphosphate? Biochem J 416:263–270

Lin H, Fridy PC, Ribeiro AA, Choi JH, Barma DK, Vogel G, Falck JR, Shears SB, York JD, Mayr GW (2009) Structural analysis and detection of biological inositol pyrophosphates reveal that the family of VIP/diphosphoinositol pentakisphosphate kinases are 1/3-kinases. J Biol Chem 284:1863–1872

Liu J, Kipreos ET (2000) Evolution of cyclin-dependent kinases (CDKs) and CDK-activating kinases (CAKs): differential conservation of CAKs in yeast and metazoa. Mol Biol Evol 17:1061–1074

Lo SJ, Lee CC, Lai HJ (2006) The nucleolus: reviewing oldies to have new understandings. Cell Res 16:530–538

Losito O, Szijgyarto Z, Resnick AC, Saiardi A (2009) Inositol pyrophosphates and their unique metabolic complexity: analysis by gel electrophoresis. PLoS One 4:e5580

Luo HR, Saiardi A, Nagata E, Ye K, Yu H, Jung TS, Luo X, Jain S, Sawa A, Snyder SH (2001) GRAB: a physiologic guanine nucleotide exchange factor for Rab3A, which interacts with inositol hexakisphosphate kinase. Neuron 31:439–451

Luo HR, Saiardi A, Yu H, Nagata E, Ye K, Snyder SH (2002) Inositol pyrophosphates are required for DNA hyperrecombination in protein kinase c1 mutant yeast. Biochemistry 41:2509–2515

Luo HR, Huang YE, Chen JC, Saiardi A, Iijima M, Ye K, Huang Y, Nagata E, Devreotes P, Snyder SH (2003) Inositol pyrophosphates mediate chemotaxis in dictyostelium via pleckstrin homology domain-PtdIns(3,4,5)P3 interactions. Cell 114:559–572

Mallory JC, Petes TD (2000) Protein kinase activity of Tel1p and Mec1p, two Saccharomyces cerevisiae proteins related to the human ATM protein kinase. Proc Natl Acad Sci U S A 97:13749–13754

Martin JB, Laussmann T, Bakker-Grunwald T, Vogel G, Klein G (2000) Neo-inositol polyphosphates in the amoeba Entamoeba histolytica. J Biol Chem 275:10134–10140

Martinez P, Zvyagilskaya R, Allard P, Persson BL (1998) Physiological regulation of the derepressible phosphate transporter in Saccharomyces cerevisiae. J Bacteriol 180:2253–2256

Mayr GW (1988) A novel metal-dye detection system permits picomolar-range h.p.l.c. analysis of inositol polyphosphates from non-radioactively labelled cell or tissue specimens. Biochem J, 254:585–591

Menniti FS, Miller RN, Putney JW Jr, Shears SB (1993) Turnover of inositol polyphosphate pyrophosphates in pancreatoma cells. J Biol Chem 268:3850–3856

Michalak E, Villunger A, Erlacher M, Strasser A (2005) Death squads enlisted by the tumour suppressor p53. Biochem Biophys Res Commun 331:786–798

Mikoshiba K, Furuichi T, Miyawaki A, Yoshikawa S, Nakade S, Michikawa T, Nakagawa T, Okano H, Kume S, Muto A et al (1993) Structure and function of inositol 1,4,5-trisphosphate receptor. Ann N Y Acad Sci 707:178–197

Monaghan P (2010) Telomeres and life histories: the long and the short of it. Ann N Y Acad Sci 1206:130–142

Morrison BH, Bauer JA, Kalvakolanu DV, Lindner DJ (2001) Inositol hexakisphosphate kinase 2 mediates growth suppressive and apoptotic effects of interferon-beta in ovarian carcinoma cells. J Biol Chem 276:24965–24970

Morrison BH, Bauer JA, Hu J, Grane RW, Ozdemir AM, Chawla-Sarkar M, Gong B, Almasan A, Kalvakolanu DV, Lindner DJ (2002) Inositol hexakisphosphate kinase 2 sensitizes ovarian carcinoma cells to multiple cancer therapeutics. Oncogene 21:1882–1889

Morrison BH, Tang Z, Jacobs BS, Bauer JA, Lindner DJ (2005) Apo2L/TRAIL induction and nuclear translocation of inositol hexakisphosphate kinase 2 during IFN-beta-induced apoptosis in ovarian carcinoma. Biochem J 385:595–603

Morrison BH, Haney R, Lamarre E, Drazba J, Prestwich GD, Lindner DJ (2009) Gene deletion of inositol hexakisphosphate kinase 2 predisposes to aerodigestive tract carcinoma. Oncogene 28:2383–2392

Morrow DM, Tagle DA, Shiloh Y, Collins FS, Hieter P (1995) TEL1, an S. cerevisiae homolog of the human gene mutated in ataxia telangiectasia, is functionally related to the yeast checkpoint gene MEC1. Cell 82:831–840

Mulugu S, Bai W, Fridy PC, Bastidas RJ, Otto JC, Dollins DE, Haystead TA, Ribeiro AA, York JD (2007) A conserved family of enzymes that phosphorylate inositol hexakisphosphate. Science 316:106–109

Nagata E, Luo HR, Saiardi A, Bae BI, Suzuki N, Snyder SH (2005) Inositol hexakisphosphate kinase-2, a physiologic mediator of cell death. J Biol Chem 280:1634–1640

Nagata E, Saiardi A, Tsukamoto H, Satoh T, Itoh Y, Itoh J, Shibata M, Takizawa S, Takagi S (2010) Inositol hexakisphosphate kinases promote autophagy. Int J Biochem Cell Biol 42:2065–2071

Nishizawa M, Komai T, Katou Y, Shirahige K, Ito T, Toh EA (2008) Nutrient-regulated antisense and intragenic RNAs modulate a signal transduction pathway in yeast. PLoS Biol 6:2817–2830

Norbis F, Boll M, Stange G, Markovich D, Verrey F, Biber J, Murer H (1997) Identification of a cDNA/protein leading to an increased Pi-uptake in Xenopus laevis oocytes. J Membr Biol 156:19–24

O'Neill EM, Kaffman A, Jolly ER, O'Shea EK (1996) Regulation of PHO4 nuclear localization by the PHO80-PHO85 cyclin-CDK complex. Science 271:209–212

Odom AR, Stahlberg A, Wente SR, York JD (2000) A role for nuclear inositol 1,4,5-trisphosphate kinase in transcriptional control. Science 287:2026–2029

Odorizzi G, Cowles CR, Emr SD (1998) The AP-3 complex: a coat of many colours. Trends Cell Biol 8:282–288

Oliver KG, Putney JW Jr, Obie JF, Shears SB (1992) The interconversion of inositol 1,3,4,5,6-pentakisphosphate and inositol tetrakisphosphates in AR4–2J cells. J Biol Chem 267:21528–21534

Onnebo SM, Saiardi A (2009) Inositol pyrophosphates modulate hydrogen peroxide signalling. Biochem J 423:109–118

Otto JC, Kelly P, Chiou ST, York JD (2007) Alterations in an inositol phosphate code through synergistic activation of a G protein and inositol phosphate kinases. Proc Natl Acad Sci U S A 104:15653–15658

Padmanabhan U, Dollins DE, Fridy PC, York JD, Downes CP (2009) Characterization of a selective inhibitor of inositol hexakisphosphate kinases: use in defining biological roles and metabolic relationships of inositol pyrophosphates. J Biol Chem 284:10571–10582

Pearl LH, Prodromou C, Workman P (2008) The Hsp90 molecular chaperone: an open and shut case for treatment. Biochem J 410:439–453

Pesesse X, Choi K, Zhang T, Shears SB (2004) Signaling by higher inositol polyphosphates. Synthesis of bisdiphosphoinositol tetrakisphosphate ("InsP8") is selectively activated by hyperosmotic stress. J Biol Chem 279:43378–43381

Pohlmann J, Fleig U (2010) Asp1, a conserved 1/3 inositol polyphosphate kinase, regulates the dimorphic switch in Schizosaccharomyces pombe. Mol Cell Biol 30:4535–4547

Ponnusamy S, Alderson NL, Hama H, Bielawski J, Jiang JC, Bhandari R, Snyder SH, Jazwinski SM, Ogretmen B (2008) Regulation of telomere length by fatty acid elongase 3 in yeast. Involvement of inositol phosphate metabolism and Ku70/80 function. J Biol Chem 283:27514–27524

Rahal EA, Henricksen LA, Li Y, Turchi JJ, Pawelczak KS, Dixon K (2008) ATM mediates repression of DNA end-degradation in an ATP-dependent manner. DNA Repair (Amst) 7:464–475

Rao NN, Gomez-Garcia MR, Kornberg A (2009) Inorganic polyphosphate: essential for growth and survival. Annu Rev Biochem 78:605–647

Rhodes D, Fairall L, Simonsson T, Court R, Chapman L (2002) Telomere architecture. EMBO Rep 3:1139–1145

Riobo NA, Manning DR (2007) Pathways of signal transduction employed by vertebrate Hedgehogs. Biochem J 403:369–379

Safrany ST, Shears SB (1998) Turnover of bis-diphosphoinositol tetrakisphosphate in a smooth muscle cell line is regulated by beta2-adrenergic receptors through a cAMP-mediated, A-kinase-independent mechanism. Embo J 17:1710–1716

Safrany ST, Caffrey JJ, Yang X, Bembenek ME, Moyer MB, Burkhart WA, Shears SB (1998) A novel context for the 'MutT' module, a guardian of cell integrity, in a diphosphoinositol polyphosphate phosphohydrolase. Embo J 17:6599–6607

Saiardi A, Cockcroft S (2008) Human ITPK1: a reversible inositol phosphate kinase/phosphatase that links receptor-dependent phospholipase C to Ca2+-activated chloride channels. Sci Signal 1:pe5

Saiardi A, Erdjument-Bromage H, Snowman AM, Tempst P, Snyder SH (1999) Synthesis of diphosphoinositol pentakisphosphate by a newly identified family of higher inositol polyphosphate kinases. Curr Biol 9:1323–1326

Saiardi A, Caffrey JJ, Snyder SH, Shears SB (2000a) The inositol hexakisphosphate kinase family. Catalytic flexibility and function in yeast vacuole biogenesis. J Biol Chem 275:24686–24692

Saiardi A, Caffrey JJ, Snyder SH, Shears SB (2000b) Inositol polyphosphate multikinase (ArgRIII) determines nuclear mRNA export in Saccharomyces cerevisiae. FEBS Lett 468:28–32

Saiardi A, Nagata E, Luo HR, Sawa A, Luo X, Snowman AM, Snyder SH (2001a) Mammalian inositol polyphosphate multikinase synthesizes inositol 1,4,5-trisphosphate and an inositol pyrophosphate. Proc Natl Acad Sci U S A 98:2306–2311

Saiardi A, Nagata E, Luo HR, Snowman AM, Snyder SH (2001b) Identification and characterization of a novel inositol hexakisphosphate kinase. J Biol Chem 276:39179–39185

Saiardi A, Sciambi C, Mccaffery JM, Wendland B, Snyder SH (2002) Inositol pyrophosphates regulate endocytic trafficking. Proc Natl Acad Sci U S A 99:14206–14211

Saiardi A, Bhandari R, Resnick AC, Snowman AM, Snyder SH (2004) Phosphorylation of proteins by inositol pyrophosphates. Science 306:2101–2105

Saiardi A, Resnick AC, Snowman AM, Wendland B, Snyder SH (2005) Inositol pyrophosphates regulate cell death and telomere length through phosphoinositide 3-kinase-related protein kinases. Proc Natl Acad Sci U S A 102:1911–1914

Santos RC, Waters NC, Creasy CL, Bergman LW (1995) Structure-function relationships of the yeast cyclin-dependent kinase Pho85. Mol Cell Biol 15:5482–5491

Sarmah B, Wente SR (2010) Inositol hexakisphosphate kinase-2 acts as an effector of the vertebrate Hedgehog pathway. Proc Natl Acad Sci U S A 107(46):19921–19926

Schell MJ, Letcher AJ, Brearley CA, Biber J, Murer H, Irvine RF (1999) PiUS (Pi uptake stimulator) is an inositol hexakisphosphate kinase. FEBS Lett 461:169–172

Seeds AM, Bastidas RJ, York JD (2005) Molecular definition of a novel inositol polyphosphate metabolic pathway initiated by inositol 1,4,5-trisphosphate 3-kinase activity in Saccharomyces cerevisiae. J Biol Chem 280:27654–27661

Shears SB (2004) How versatile are inositol phosphate kinases? Biochem J 377:265–280

Shears SB (2009) Diphosphoinositol polyphosphates: metabolic messengers? Mol Pharmacol 76:236–252

Shears SB, Ali N, Craxton A, Bembenek ME (1995) Synthesis and metabolism of bis-diphosphoinositol tetrakisphosphate in vitro and in vivo. J Biol Chem 270:10489–10497

Slater EC (1973) The mechanism of action of the respiratory inhibitor, antimycin. Biochim Biophys Acta 301:129–154

Stephens L, Radenberg T, Thiel U, Vogel G, Khoo KH, Dell A, Jackson TR, Hawkins PT, Mayr GW (1993) The detection, purification, structural characterization, and metabolism of diphosphoinositol pentakisphosphate(s) and bisdiphosphoinositol tetrakisphosphate(s). J Biol Chem 268:4009–4015

Stepp JD, Huang K, Lemmon SK (1997) The yeast adaptor protein complex, AP-3, is essential for the efficient delivery of alkaline phosphatase by the alternate pathway to the vacuole. J Cell Biol 139:1761–1774

Torres J, Dominguez S, Cerda MF, Obal G, Mederos A, Irvine RF, Diaz A, Kremer C (2005) Solution behaviour of myo-inositol hexakisphosphate in the presence of multivalent cations. Prediction of a neutral pentamagnesium species under cytosolic/nuclear conditions. J Inorg Biochem 99:828–840

Traynor D, Milne JL, Insall RH, Kay RR (2000) Ca(2+) signalling is not required for chemotaxis in Dictyostelium. Embo J 19:4846–4854

Tsujimoto Y, Shimizu S (2005) Another way to die: autophagic programmed cell death. Cell Death Differ 12(Suppl 2):1528–1534

Van Haastert PJ (1995) Transduction of the chemotactic cAMP signal across the plasma membrane of Dictyostelium cells. Experientia 51:1144–1154

Verbsky J, Lavine K, Majerus PW (2005) Disruption of the mouse inositol 1,3,4,5,6-pentakisphosphate 2-kinase gene, associated lethality, and tissue distribution of 2-kinase expression. Proc Natl Acad Sci U S A 102:8448–8453

Verheggen C, Mouaikel J, Thiry M, Blanchard JM, Tollervey D, Bordonne R, Lafontaine DL, Bertrand E (2001) Box C/D small nucleolar RNA trafficking involves small nucleolar RNP proteins, nucleolar factors and a novel nuclear domain. Embo J 20:5480–5490

Voglmaier SM, Bembenek ME, Kaplin AI, Dorman G, Olszewski JD, Prestwich GD, Snyder SH (1996) Purified inositol hexakisphosphate kinase is an ATP synthase: diphosphoinositol pentakisphosphate as a high-energy phosphate donor. Proc Natl Acad Sci U S A 93:4305–4310

Wandinger SK, Richter K, Buchner J (2008) The Hsp90 chaperone machinery. J Biol Chem 283:18473–18477

Winward L, Whitfield WG, Mclennan AG, Safrany ST (2010) Oxidation of the diphosphoinositol polyphosphate phosphohydrolase-like Nudix hydrolase Aps from Drosophila melanogaster induces thermolability – a possible regulatory switch? Int J Biochem Cell Biol 42:1174–1181

Ye W, Ali N, Bembenek ME, Shears SB, Lafer EM (1995) Inhibition of clathrin assembly by high affinity binding of specific inositol polyphosphates to the synapse-specific clathrin assembly protein AP-3. J Biol Chem 270:1564–1568

York JD (2006) Regulation of nuclear processes by inositol polyphosphates. Biochim Biophys Acta 1761:552–559

York JD, Odom AR, Murphy R, Ives EB, Wente SR (1999) A phospholipase C-dependent inositol polyphosphate kinase pathway required for efficient messenger RNA export. Science 285:96–100

York SJ, Armbruster BN, Greenwell P, Petes TD, York JD (2005) Inositol diphosphate signaling regulates telomere length. J Biol Chem 280:4264–4269

Zhang T, Caffrey JJ, Shears SB (2001) The transcriptional regulator, Arg82, is a hybrid kinase with both monophosphoinositol and diphosphoinositol polyphosphate synthase activity. FEBS Lett 494:208–212

Glossary

A glossary of terms is provided for readers who are not experts of the inositol lipid field.

***myo*-Inositol:** This is one of nine possible stereoisomers of inositol which is a cyclohexanehexol. *Myo*-inositol is the most commonly occurring stereoisomer in nature, therefore, the IUPAC-approved abbreviation "Ins" refers to *myo*-inositol (Nomenclature Committee of the International Union of Biochemistry, 1989, http://www.chem.qmul.ac.uk/iupac/cyclitol/myo.html). The conformation of *myo*-inositol is the so-called "chair" conformation with five equatorial and one axial hydroxyl groups. This conformation, and the numbering of the hydroxyls have been best visualized by Agranoff (1978) who compared the ring to a turtle and the hydroxyls to the appendages. Here, the numbering starts with the right front flipper going counterclockwise. The head of the turtle then corresponds to the axial hydroxyl at the 2nd position. It is notable that *myo*-inositol has an axis of symmetry going through the 2nd and 5th carbons. The numbering used refers to the D-enantiomers but it is important to remember that because of this symmetry D-Ins1*P* is the same as L-Ins3*P* and, therefore, isomers (such as Ins1*P* and Ins3*P*) that are enantiomeric twins cannot be separated with conventional HPLC methods.

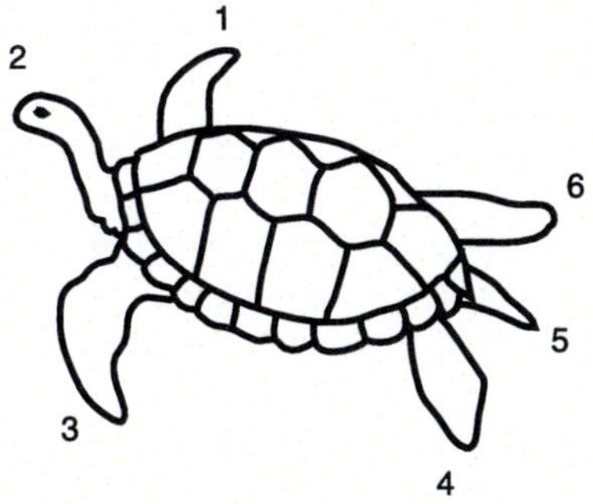

Agranoff's turtle

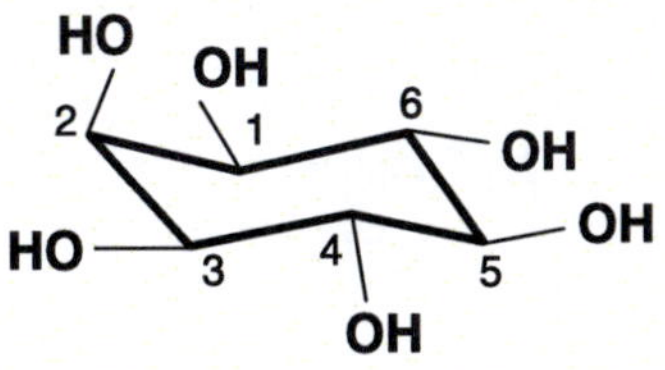

myo-inositol

T. Balla et al. (eds.), *Phosphoinositides II: The Diverse Biological Functions*, Subcellular Biochemistry 59, DOI 10.1007/978-94-007-3015-1,

Phosphatidylinositol: This is the base molecule for all phosphoinositides. The recommended abbreviation is PtdIns (http://www.chem.qmul.ac.uk/iupac/misc/phos2t8.html#t4) but the early literature often uses "PI" as the abbreviation. This short form is still in use in the context of kinases that phosphorylate PtdIns or its phosphorylated derivatives, such as in PI 3-kinases or PI 4-kinases. PtdIns consists of a diacylglycerol backbone in which the 1- and 2-positions of the glycerol are most often esterified with a stearoyl- and arachidonyl- fatty acid chains, respectively, and the *myo*-inositol ring is linked to the 3rd- position of the glycerol via a phosphodiester bond formed with the 1st hydroxyl of inositol. PtdIns can be phosphorylated in all but the 2nd and 6th positions of the inositol ring, giving rise to the seven known phosphoinositides.

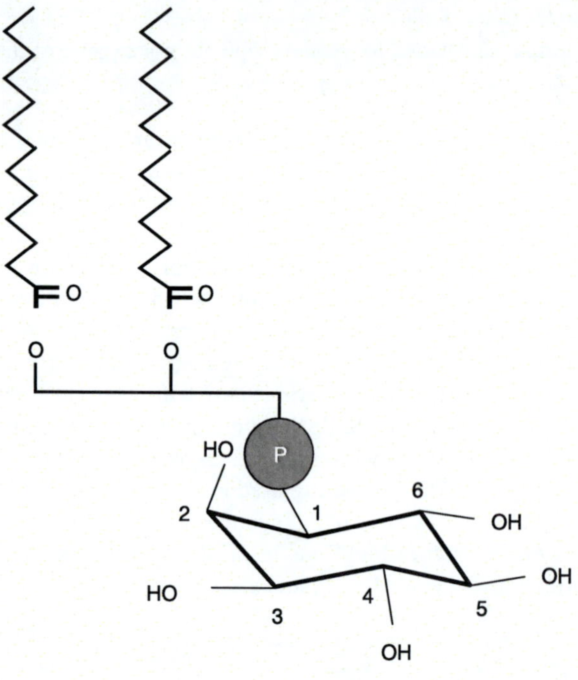

PtdIns

Polyphosphoinositides: This refers to any of the further phosphorylated PtdIns regardless of the number and positions of the phosphate groups. Sometimes they are abbreviated as PPIs but mostly in the 80's literature but this is still the recommended abbreviation (see in Michell et al. 2005).

Phosphoinositides: This is a term often used to designate collectively PtdIns and all of its phosphorylated derivatives regardless of their isomerism. "PI" is the abbreviation used lately for phosphoinositides but it often causes confusion so it should be avoided. There is no consensus abbreviation for this term that includes both PtdIns and the PPIs. The individual forms of PPIs are abbreviated specifying the positions phosphorylated on the inositol ring. For example, phosphatidylinositol 4,5-bisphosphate [PtdIns(4,5)P_2] is a double phosphorylated PtdIns with positions 4- and 5- phosphorylated.

It is worth pointing out here a few rules about the terminology: the numbers on the inositol ring are not "primed", since there is no other ring in the structure that would require the use of "prime" to discriminate the rings (as opposed to multiring structures such as the nucleosides). The number of phosphates are indicated by the prefixes "bis-" (latin for twice) "tris-" (greek for three times) to indicate that these molecules contain the indicated number of phosphates but all placed individually at various positions. (contrast this with ATP which is a "triphosphate" with the phosphates linked to one another). There are no higher numbers than trisphosphates in phosphoinositides but there are in the soluble inositol phosphates, for which the numbers continue as "tetrakis-", "pentakis-" and "hexakis–" 'kis' being a greek prefix for –times. In the abbreviations it is recommended to italicise the "*P*" if it designates a phosphomonoester (see Michell et al. (2005) in the list below for more details on nomenclature recommendations).

Old literature used terms such as "DPI" and "TPI" for diphosphoinositide and triphosphoinositide, respectively. These correspond to mono- and bis-phosphorylated PtdIns from a time when the exact configurations of the phosphates were not known.

Phosphoinositide kinases: Phosphoinositide kinases add a phosphate to a specific position onto the inositol ring of phosphoinositides using ATP. The kinases are named after the position they phosphorylate and hence we distinguish 3-, 4- and 5-kinases (no primes!!). There is an inherent inconsistency about the abbreviations used to designate these enzymes. For example, the term "PI 3-kinase" is used to refer to any of the 3-kinases, regardless of the substrates they phosphorylate. Since there are PI 3-kinases that can only phosphorylate PtdIns (and not further phosphorylated forms) (the Class III PI 3-kinases) they are also named PtdIns 3-kinases. However, the Class I PI 3-kinases that phosphorylate PtdIns(4,5)P_2 are rarely called PtdIns(4,5)P_2 3-kinases and in most articles unspecified "PI 3-kinase" refers to the Class I enzymes. In contrast, PI 4-kinases can only phosphorylate PtdIns (and not further phosphorylated forms) in which case it would be more correct to call them PtdIns 4-kinases. However, because of historical reasons, these inconsistencies are tolerated even if they defy logic based on current knowledge. The list of the various forms and classes of PI kinases are summarized in the respective chapters.

Phosphoinositide phosphatases: Phosphoinositide phosphatases remove one or more phosphates from PPIs. They can be specific to the position of the phosphate they remove and the substrate they can use. Some will dephosphorylate only PPIs while others can also use the water-soluble inositol phosphates as substrates. Phosphatases are usually named after the position of phosphate they attack such as 5-phosphatases or 3-phosphatases. Some PI phosphatases are not position specific, such as the monophosphatases (see Chapters 7 and 8 in Volume I for more details).

Phospholipase C: These enzymes (PLCs) hydrolyze PtdIns (or PPIs) by cleaving the phosphodiester group such that they leave diacylglycerol behind and release the inositol headgroup, which carries the phosphate still attached at the 1-position (or other phosphates if the substrate is any of the PPIs). To discriminate from other PLCs that use other phospholipids as substrate (such as PC-PLC), PLCs that hydrolyze phosphoinositides are called PI-PLCs. This, however, also causes some confusion, since mammalian PI-PLCs are believed to hydrolyze primarily PtdIns(4,5)P_2 *in vivo* (although they can also hydrolyze PtdIns and PtdIns4*P* *in vitro*). However, there are bacterial PLC enzymes that will use either PtdIns or phosphatidylinositol glycan (GPI) linkages but cannot hydrolyze polyphosphoinositides. The literature that deals with the bacterial enzymes uses the term PI-PLC to emphasize that the bacterial enzymes are specific for PtdIns or GPI. So the term "PI-PLC" means two different enzyme groups depending on whether used in mammalian or prokaryotic studies. However, in most cases PLC without any designation refers to the mammalian phosphoinositide-specific PLCs.

Inositol 1,4,5-trisphosphate: Ins(1,4,5)P_3 is the water soluble molecule liberated after PLC-mediated hydrolysis of PtdIns(4,5)P_2. This molecule has a receptor located in the ER membrane that also is a Ca^{2+} channel and which is gated by Ins(1,4,5)P_3 binding. Ins(1,4,5)P_3 is a bona fide second messenger liberated upon stimulation of cell surface receptors coupled to PLC activation.

PH domain: PH domains (for pleckstrin homology domains) are protein modules of roughly 150 amino acid length that were first recognized in pleckstrin (Tyers et al. 1988). These were the first protein modules that were shown to bind PPIs. Many PH domains can recognize and bind phosphoinositides with variable specificities earning these domains the reputation of being PPI binding modules. Although several PH domains can, indeed, recognize PIs with high affinity and specificity, many PH domains show promiscuous PPI recognition and many do not bind PIs at all. Moreover, PH domains also recognize proteins and often bind proteins and lipids simultaneously (Lemmon 2004).

FYVE domain: This was the second protein module identified with specific PPI recognition, namely to recognize PtdIns3*P* (Burd and Emr 1998). Its name originated from the four molecules (three from baker's yeast) in which this module was first described (Fab1, YOTB, Vac1 and EEA1). FYVE domains use two Zn^{2+} ions to stabilize their structure and they are also called FYVE zinc fingers. They show structural similarities to the C1 domains that recognize diacylglycerol (Misra and Hurley 1999; Kutateladze et al. 1999).

PX domain: Phox-homology domains were also recognized as capable of binding PtdIns3*P*. They were initially found in sorting nexins (Ponting 1996) and NADPH oxidase subunits (Bravo et al. 2001; Ellson et al. 2001; Kanai et al. 2001), but they are present in a large variety of signaling molecules. PX domains can also bind other phospholipids, such as PtdOH and PtdIns(3,4)P_2, and they also interact with proteins (Vollert and Uetz 2004).

Phosphoinositide binding protein domains: In addition to the above defined protein modules, several other modular protein domains have been identified as phosphoinositide effectors (Lemmon 2008). Because of their increasing number they will not be listed here but can be found in the individual Chapters.

References

Agranoff BW (1978) Trends Biochem Sci 3(12):N283–N285
Bravo J, Karathanassis D, Pacold CM, Pacold ME, Ellson CD, Anderson KE, Butler PJG, Lavenir I, Perisisc O, Hawkins PT, Stephens L, Williams RL (2001) Mol Cell 8:829–839
Burd CG, Emr SD (1998) Mol Cell 2:157–162
Ellson CD, Gobert-Gosse S, Anderson KE, Davidson K, Erdjument-Bromage H, Tempst P, Thuring JW, Cooper MA, Lim ZY, Holmes AB, Gaffney PRJ, Coadwell J, Chilvers ER, Hawkins PT, Stephens LR (2001) Nat Cell Biol 3:679–682
Kanai F, Liu H, Field SJ, Akbary H, Matsuo T, Brown GE, Cantley LC, Yaffe MB (2001) Nat Cell Biol 3:675–678
Kutateladze TG, Ogburn KD, Watson WT, deBeer T, Emr SD, Burd CG, Overduin M (1999) Mol Cell 3:805–811
Lemmon MA (2004) Biochem Soc Trans 32:707–711
Lemmon MA (2008) Nat Rev Mol Cell Biol 9(2):99–111
Michell RH, Heath VL, Lemmon MA, Dove SK (2005) Trends Biochem Sci 31(1):52–63
Misra S, Hurley JH (1999) Cell 97:657–666
Ponting, CP (1996) Prot Sci 5:2353–2357
Tyers M, Rachubinski RA, Stewart MI, Varrichio AM, Shorr RG, Haslam RJ, Harley CB (1988) Nature 333(6172):470–473
Vollert CS, Uetz P (2004) Mol Cell Proteomics 3(11):1053–1064

Index

T. Balla et al. (eds.), *Phosphoinositides II: The Diverse Biological Functions*, Subcellular Biochemistry 59, DOI 10.1007/978-94-007-3015-1,

C

D